Lecture Notes in Bioinformatics 2812

Edited by S. Istrail, P. Pevzner, and M. Waterman

Subseries of Lecture Notes in Computer Science

Springer
Berlin
Heidelberg
New York
Hong Kong
London
Milan
Paris
Tokyo

Gary Benson Roderic Page (Eds.)

Algorithms in Bioinformatics

Third International Workshop, WABI 2003
Budapest, Hungary, September 15-20, 2003
Proceedings

Springer

Series Editors

Sorin Istrail, Celera Genomics, Applied Biosystems, Rockville, MD, USA
Pavel Pevzner, University of California, San Diego, CA, USA
Michael Waterman, University of Southern California, Los Angeles, CA, USA

Volume Editors

Gary Benson
The Mount Sinai School of Medicine
Department of Biomathematical Sciences
Box 1023, One Gustave L. Levy Place, New York, NY 10029-6574, USA
E-mail: benson@camelot.mssm.edu

Roderic Page
University of Glasgow, Institute of Biomedical and Life Sciences
Division of Environmental and Evolutionary Biology
Glasgow G12 8QQ, Scotland
E-mail: r.page@bio.gla.ac.uk

Cataloging-in-Publication Data applied for

A catalog record for this book is available from the Library of Congress.

Bibliographic information published by Die Deutsche Bibliothek.
Die Deutsche Bibliothek lists this publication in the Deutsche Nationalbibliografie; detailed bibliographic data is available in the Internet at <http://dnb.ddb.de>.

CR Subject Classification (1998): F.1, F.2.2, E.1, G.1, G.2, G.3, J.3

ISSN 0302-9743
ISBN 3-540-20076-2 Springer-Verlag Berlin Heidelberg New York

Springer-Verlag Berlin Heidelberg New York
a member of BertelsmannSpringer Science+Business Media GmbH

http://www.springer.de

Typesetting: Camera-ready by author, data conversion by Boller Mediendesign
Printed on acid-free paper SPIN: 10949993 06/3142 5 4 3 2 1 0

Preface

We are pleased to present the proceedings of the *Third Workshop on Algorithms in Bioinformatics (WABI 2003)*, which took place on September 15–20, 2003 in Budapest, Hungary. The WABI workshop was part of the four-conference meeting, ALGO 2003, which was locally organized by Dr. János Csirik, Head of the Department of Computer Science, József Attila University, Budapest. See `http://www.conferences.hu/ALGO2003/algo_2003.htm` for more details.

WABI focuses on discrete algorithms that address important problems in molecular biology, genomics, and genetics, that are founded on sound models, that are computationally efficient, that have been implemented and tested in simulations and on real datasets, and that provide new biological results. The workshop goals are to present recent research and identify and explore directions for future research.

We received 78 submissions in response to the call for papers and 36 were accepted. We would like to sincerely thank the authors of all submitted papers and the conference participants. We especially thank a terrific program committee for their diligent and thorough work in reviewing and selecting the papers. We were fortunate to have on the program committee the following distinguished group of researchers:

Amihood Amir (Bar Ilan University, Israel)
Alberto Apostolico (Purdue University)
Pierre Baldi (University of California, Irvine)
Gary Benson (Mount Sinai School of Medicine, New York; Co-chair)
Benny Chor (Tel Aviv University)
Nadia El-Mabrouk (University of Montreal)
Olivier Gascuel (LIRMM-CNRS, Montpellier)
Raffaele Giancarlo (Università di Palermo)
David Gilbert (University of Glasgow)
Jan Gorodkin (The Royal Veterinary and Agricultural University, Denmark)
Roderic Guigó (Institut Municipal d'Investigacions Mèdiques, Barcelona)
Dan Gusfield (University of California, Davis)
Jotun Hein (University of Oxford)
Daniel Huson (Tübingen University)
Simon Kasif (Boston University)
Gregory Kucherov (INRIA-Lorraine/LORIA)
Gad Landau (University of Haifa)
Thierry Lecroq (Université de Rouen)
Bernard M.E. Moret (University of New Mexico, Albuquerque)
Vincent Moulton (Uppsala University, Sweden)
Roderic Page (University of Glasgow; Co-chair)
Sophie Schbath (INRA, Jouy-en-Josas)

Charles Semple (University of Canterbury, New Zealand)
Jens Stoye (Universität Bielefeld)
Fengzhu Sun (University of Southern California)
Alfonso Valencia (Centro Nacional de Biotecnología, Madrid)
Jacques Van Helden (Université Libre de Bruxelles)
Louxin Zhang (National University of Singapore)

The program committee's work was greatly assisted by the helpful reviews provided by Federico Abascal, Ali Al-Shahib, Rumen Andonov, Ora Arbell, Sebastian Böcker, David Bryant, Peter Calabrese, Robert Castelo, Kwok Pui Choi, Miklós Csürös, Minghua Deng, Tobias Dezulian, Nadav Efraty, P.L. Erdos, Revital Eres, Eleazer Eskin, Jose Maria Fernandez, Pierre Flener, Jakob Fredslund, Dan Gieger, Robert Giegerich, Vladimir Grebinskiy, Pawel Herzyk, Mark Hoebeke, Robert Hoffmann, Katharina Huber, Michael Kaufmann, Carmel Kent, Jens Lagergren, Yinglei Lai, Xiaoman Li, Gerton Lunter, Rune Lyngsoe, Laurent Mouchard, Pierre Nicolas, Laurent Noe, Sebastian Oehm, Christian N.S. Pedersen, Johann Pelfrene, Shalom Rackovsky, Mathieu Raffinot, Kim Roland Rasmussen, Christian Rausch, Knut Reinert, Olivier Sand, Klaus-Bernd Schürmann, Steven Skiena, Dina Sokol, Y.S. Song, W. Szpankowski, Helene Touzet, Michael Tress, Juris Viksna, Lusheng Wang, Zohar Yakhini, Kui Zhang, and Michal Ziv-Ukelson.

We also thank the WABI steering committee, Olivier Gascuel, Raffaele Giancarlo, Roderic Guigó, Dan Gusfield, and Bernard Moret, for inviting us to co-chair the conference and for their help in carrying out that task.

We are particularly indebted to Kevin Kelliher of the Mount Sinai School of Medicine, New York, for his conscientious administration of the CyberChair software used to manage the review process, and Robert Castelo at Universitat Pompeu Fabra, Barcelona, for generously sharing the scripts he developed to generate the final copy of last year's WABI. Production of the proceedings was greatly assisted by Richard Koch's wonderful TeXShop software, and large quantities of Peet's coffee.

Thanks again to all who participated to make WABI 2003 a success. It has been, for us, a challenging and rewarding experience.

July 2003 Gary Benson and Roderic Page

Table of Contents

Comparative Genomics

Database Searching

Gene Finding and Expression

Genome Mapping

Pattern and Motif Discovery

Phylogenetic Analysis

Polymorphism

Protein Structure

Sequence Alignment

String Algorithms

A Local Chaining Algorithm and Its Applications in Comparative Genomics

Mohamed Ibrahim Abouelhoda[1] and Enno Ohlebusch[2]

[1] Faculty of Technology, University of Bielefeld, P.O. Box 10 01 31, 33501 Bielefeld, Germany
mibrahim@techfak.uni-bielefeld.de
[2] Faculty of Computer Science, University of Ulm, 89069 Ulm, Germany
eo@informatik.uni-ulm.de

Abstract. Given fragments from multiple genomes, we will show how to find an optimal local chain of colinear non-overlapping fragments in sub-quadratic time, using methods from computational geometry. A variant of the algorithm finds all significant local chains of colinear non-overlapping fragments. The local chaining algorithm can be used in a variety of problems in comparative genomics: The identification of regions of similarity (candidate regions of conserved synteny), the detection of genome rearrangements such as transpositions and inversions, and exon prediction.

1 Introduction

Given the continuing improvements in high-throughput genomic sequencing and the ever-expanding sequence databases, new advances in software tools for post-sequencing functional analysis are being demanded by the biological scientific community. Whole genome comparisons have been heralded as the next logical step toward solving genomic puzzles, such as determining coding regions, discovering regulatory signals, and deducing the mechanisms and history of genome evolution. However, before any such detailed analysis can be addressed, methods are required for comparing such large sequences. If the organisms under consideration are closely related (that is, if no or only a few genome rearrangements have occurred) or one compares regions of conserved synteny, then global alignments can, for example, be used for the prediction of genes and regulatory elements. This is because coding regions are relatively well preserved, while non-coding regions tend to show varying degree of conservation. Non-coding regions that do show conservation are thought important for regulating gene expression, maintaining the structural organization of the genome and possibly have other, yet unknown functions. Several comparative sequence approaches using alignments have recently been used to analyze corresponding coding and non-coding regions from different species, although mainly between human and mouse. These approaches are based on software-tools for aligning DNA-sequences [4,12,20,22,23,29]; see [9] for a review. To cope with the shear volume of data, most of the software-tools use an anchor-based method that is composed of three

G. Benson and R. Page (Eds.): WABI 2003, LNBI 2812, pp. 1–16, 2003.

phases: (1) computation of fragments (regions in the genomes that are similar), (2) computation of an optimal chain of colinear non-overlapping fragments: these are the anchors that form the basis of the alignment, (3) alignment of the regions between the anchors.

For diverged genomic sequences, however, a global alignment strategy is likely predestined to failure for having to align non-syntenic and unrelated regions in an end-to-end colinear approach. In this case, either local alignments are the strategy of choice or one must first identify syntenic regions, which then can be individually aligned (syntenic regions are regions in two or more genomes in which orthologous genes occur in the same order). However, both alternatives are faced with obstacles. As we shall see in the next section, current local alignment programs suffer from a huge running time. The problem of automatically finding syntenic regions requires a priori knowledge of all genes and their locations in the genomes—a piece of information that is often not available. (It is beyond the scope of this paper to discuss the computational difficulties of gene prediction and the accurate determination of orthologous genes.)

In this paper, we address the problem of automatically finding regions of similarity in large genomic DNA sequences. Our solution is based on a modification of the anchor-based global alignment method. As in that method, we first efficiently compute fragments. These fragments will often be exact matches (maximal unique matches as in MUMmer [11,12], maximal multiple exact matches as in MGA [22], or exact k-mers as in GLASS [4]), but one may also allow substitutions (yielding gap-free fragments as in DIALIGN [23]) or even insertions and deletions (as the BLASTZ-hits [28] that are used in PipMaker [29]). Each of the fragments has a weight that can, for example, be the length of the fragment (in case of gap-free fragments) or its statistical significance. In the second phase, however, instead of computing a global optimal chain of colinear non-overlapping fragments, we compute significant local chains. We call a local chain significant if its score (the sum of the weights of the fragments in the chain, where gaps between the fragments are penalized) exceeds a user-defined threshold. Under stringent thresholds, significant local chains of colinear non-overlapping fragments represent candidate regions of conserved synteny. If one aligns these individually, one gets good local alignments.

We see several advantages of our method:

1. It can handle multiple genomes.
2. It can handle any kind of fragments.
3. In contrast to most other methods used in comparative genomics, it is not heuristic: It correctly solves a clearly defined problem.

We would like to point out that the automatic identification of regions of similarity is a first step toward an automatic detection of genome rearrangements. In the applications section, we will demonstrate how our method can be used to detect genome rearrangements such as transpositions (where a section of the genome is excised and inserted at a new position in the genome, without changing orientation) and inversions (where a section of the genome is excised, reversed in orientation, and re-inserted). The chaining algorithms presented here are implemented in the module *Gcomp* of the program MGA [22].

2 Related Work

Early methods for locating regions of similarity between two genomes are based on semi-automatic techniques using programs like BLAST [3] or FASTA [25]. These techniques divide the two genomes under consideration into windows (say of 1000 bp each), which are then compared in an all-against-all fashion. Such a technique was used, for example, to identify the transpositions (for bacteria also called translocations) between the genomes of the bacteria *M.genitalium* and *M.pneumonia* [17]. However, this technique is too time consuming to cope with large genomes.

To fully automate this technique, the program WABA (Wobble Aware Bulk Aligner) has been developed [20]. The key feature of WABA is that wobble bases are treated differently from other bases. The third base in a codon is called *wobble base* because mutations in this base are often silent in the sense that they do not change the corresponding amino acid (due to the redundancy of the genetic code). WABA has been developed specifically for separately aligning 229 different sequences from *C. briggsae* (8 megabases total length, 34,722 bp average length) against 97 megabases of the *C. elegans* genome. A computation of these alignments takes about 12 days on a Pentium III 450 MHz computer. WABA treats the bigger genome as "target" and the smaller genome as "query". The query genome is divided into 2000 bp blocks and two consecutive blocks overlap at 1000 bp. Then a three-phase procedure is applied. In the first phase, homologies between each block and the target sequence are searched for. The hits of every block against the target sequence is generated. In WABA, the hits are 8-mers that are not required to match exactly but may contain a mismatch every three bases (to take wobble bases into account). These hits are further grouped into clumps that are scored by the square of the number of hits occurring in it. In the second phase, clumps with a score above some threshold are aligned in an extended window of 5000 bp using a pairwise seven-state Hidden Markov Model. In the third phase, overlapping alignments are joined.

A different idea for a faster identification of such regions of similarity was introduced in the program ASSIRC (Accelerated Search for SImilarity Regions in Chromosomes) [31]. ASSIRC also has three phases. In the first phase, seeds (pairs of identical k-mers) are identified using standard hashing techniques. In the second phase, every seed is extended separately in each genome using a random walk procedure. The random walk procedure can be visualized in a two dimensional Euclidean space as a series of displacement vectors (the direction and amplitude of a displacement vector depends on the character in the extended region). The extension stops when the two series of displacement vectors diverge "too much". In the third phase, the resulting regions are aligned using standard dynamic programming algorithms.

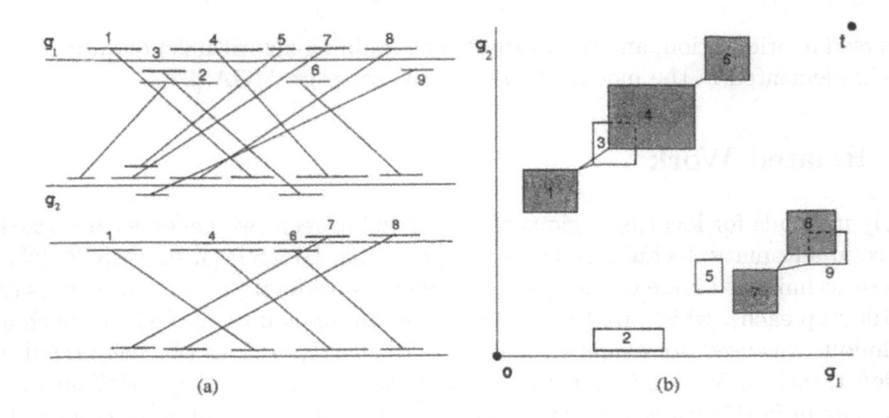

Fig. 1. Computation of an optimal local chain of colinear non-overlapping fragments. The optimal local chain is composed of the fragments 1, 4, and 6.

Recently, in MUMmer II [12] a new module called NUCmer has been implemented to produce alignments of incomplete genomes (draft genomes consisting of several contigs). NUCmer clusters MUMs (fragments) together if they are separated in both genomes by no more than a user-specified distance. This clustering technique can be viewed as a primitive local chaining procedure.

The program CHAOS [8] was developed to provide the DIALIGN [23] alignment program with an anchor-based global alignment method. CHAOS uses a heuristic approach to identify local sequence similarities from fragments. From these similarities, a set of anchors is chosen for a DIALIGN alignment. The heuristics uses k-mers allowing for a user-specified number of mismatches (the default parameter is $k = 7$ with no mismatch allowed) as seeds (fragments) and processes them in ascending order w.r.t. their diagonal number. For every seed, all the preceding seeds that have a diagonal number within a user-specified gap criterion and are located within a user-specified distance are considered and the one (if there is one) that terminates a chain of maximum score is connected to the current seed.

The problem of all the above-mentioned approaches is that

1. they cannot handle multiple genomes,
2. they cannot handle arbitrary fragments,
3. they employ ad hoc methods as heuristics.

3 Basic Concepts and Definitions

For any point $p \in \mathbb{R}^k$, let $p.x_1, p.x_2, \ldots, p.x_k$ denote its coordinates. If $k = 2$, the coordinates of p will also be written as $p.x$ and $p.y$. A hyper-rectangle (called hyperrectangular domain in [26]) is the Cartesian product of intervals on distinct coordinate axes. A hyper-rectangle $[l_1 \ldots h_1] \times [l_2 \ldots h_2] \times \ldots \times [l_k \ldots h_k]$ (with $l_i < h_i$ for all $1 \leq i \leq k$) will also be denoted by $R(p, q)$, where $p = (l_1, \ldots, l_k)$ and

$q = (h_1, \ldots, h_k)$ are its two extreme corner points. In the problem we consider, all points are given in advance (off-line). Therefore, it is possible to map the points into $\mathbb{N}^k$, called the *rank space*; see, e.g., [10]. Every point $(x_1, x_2, \ldots, x_k)$ is mapped to point $(r_1, r_2, \ldots, r_k)$, where r_i, $1 \leq i \leq k$, is the index (or rank) of point p in a list that is sorted in ascending order w.r.t. dimension x_i. This transformation takes $O(kn \log n)$ time and $O(n)$ space because one has to sort the points k times. Thus, we can assume that the points are already transformed to the rank space.

For $1 \leq i \leq k$, S_i denotes a string of length $|S_i|$. In our application, S_i is the DNA sequence of a genome. $S_i[l_i \ldots h_i]$ is the substring of S_i starting at position l_i and ending at position h_i. An *exact fragment* (or *multiple exact match*) f consists of two k-tuples *beg*(f)$= (l_1, l_2, \ldots, l_k)$ and *end*(f)$= (h_1, h_2, \ldots, h_k)$ such that $S_1[l_1 \ldots h_1] = S_2[l_2 \ldots h_2] = \ldots = S_k[l_k \ldots h_k]$, i.e., the substrings are identical. It is maximal, if the substrings cannot be simultaneously extended to the left and to the right in every S_i. If mismatches are allowed in the substrings, then we speak of a *gap-free fragment*. If one further allows insertions and deletions (so that the substrings may be of unequal length), we will use the general term *fragment*. As mentioned in the introduction, many algorithms have been developed to efficiently compute all kinds of fragments and the algorithms presented here work for arbitrary fragments.

A fragment f of k genomes can be represented by a hyper-rectangle in $\mathbb{R}^k$ with the two extreme corner points *beg*(f) and *end*(f), where each coordinate of the points is a non-negative integer. In the following, the words *number of genomes* and *dimension* will thus be used synonymously. With every fragment f, we associate a positive weight $f.weight \in \mathbb{R}$. This weight can, for example, be the length of the fragment (in case of gap-free fragments) or its statistical significance. Fig. 1 shows fragments of two genomes.

In what follows, we will often identify the point *beg*(f) or *end*(f) with the fragment f. For example, if we speak about the weight of a point *beg*(f) or *end*(f), we mean the weight of the fragment f. At some places, we will consider the points $\mathsf{0} = (0, \ldots, 0)$ (the origin) and $\mathsf{t} = (|S_1|, \ldots, |S_k|)$ (the terminus) as fragments with weight 0. For these fragments, we define $beg(\mathsf{0}) = \bot$, $end(\mathsf{0}) = \mathsf{0}$, $\mathsf{0}.score = 0$, $beg(\mathsf{t}) = \mathsf{t}$, $end(\mathsf{t}) = \bot$, and $\mathsf{t}.score = 0$.

Definition 1. *We define the relation* $\ll$ *on the set of fragments by* $f \ll f'$ *if and only if* $end(f).x_i < beg(f').x_i$ *for all* $1 \leq i \leq k$. *If* $f \ll f'$, *then we say that* f precedes f'. *We further define* $\mathsf{0} \ll f \ll \mathsf{t}$ *for every fragment* f *with* $f \neq \mathsf{0}$ *and* $f \neq \mathsf{t}$.

Definition 2. *A* chain C *of colinear non-overlapping fragments (or chain for short) is a sequence of fragments* $f_1, f_2, \ldots, f_\ell$ *such that* $f_i \ll f_{i+1}$ *for all* $1 \leq i < \ell$. *The score of* C *is* $score(C) = \sum_{i=1}^{\ell-1}(f_i.weight - g(f_{i+1}, f_i))$, *where* $g(f_{i+1}, f_i)$ *is the cost of connecting fragment* f_i *to* f_{i+1} *in the chain. We will call this cost* gap cost.

The gap cost that will be used in this paper is defined below, but we stress that our approach can also deal with other gap costs (for other gap costs, see [1]). For two points $p, q \in \mathbb{R}^k$, their distance in the L_1 metric is defined by

$$d_1(p,q) = \sum_{i=1}^{k} |p.x_i - q.x_i|$$

For two fragments $f \ll f'$ we define $g_1(f', f) = d_1(beg(f'), end(f))$. That is, the cost for a gap between two fragments is the distance between the end and start point of the two fragments in the L_1 metric.

Definition 3. *Let n fragments, a weight function, and a gap cost function be given.*

- *The* global chaining problem *is to determine a chain of maximum score starting at the origin* 0 *and ending at terminus* t. *Such a chain will be called* optimal global chain.
- *The* local chaining problem *is to determine a chain of maximum score* ≥ 0. *Such a chain will be called* optimal local chain.

Note that if the gap cost function is the constant function 0, then an optimal local chain must also be an optimal global chain, and vice versa.

4 The Local Chaining Algorithm

In this section, we present the core algorithm of this paper, while the efficient computation of the scores of all fragments will be addressed later.

Definition 4. *We define*

$$f'.score = \max\{score(C) : C \text{ is a chain ending with } f'\}$$

A chain C ending with f' and satisfying $f'.score = score(C)$ will be called optimal chain ending with f', *or simply* optimal chain *if f' can be inferred from the context.*

Lemma 5. *The following equality holds:*

$$f'.score = f'.weight + \max\{0, f.score - g_1(f', f) : f \ll f'\} \tag{1}$$

Proof. Let $C' = f_1, f_2, \ldots, f_\ell, f'$ be an optimal chain that is ending with f', i.e., $score(C') = f'.score$. Because the chain that solely consists of fragment f' has score $f'.weight \geq 0$, we must have $score(C') \geq f'.weight$. If $score(C') = f'.weight$, then $f.score - g_1(f', f) \leq 0$ for every fragment f that precedes f', because otherwise it would follow $score(C') > f'.weight$. Hence equality (1) holds in this case. So suppose $score(C') > f'.weight$. Clearly, $score(C') = f'.weight + score(C) - g_1(f', f_\ell)$, where $C = f_1, f_2, \ldots, f_\ell$. It is not difficult to see that C must be an optimal chain that is ending with f_ℓ because otherwise C' would not be optimal. Therefore, $score(C') = f'.weight + f_\ell.score - g_1(f', f_\ell)$. If there were a fragment f that precedes f' such that $f.score - g_1(f', f) > f_\ell.score - g_1(f', f_\ell)$, then it would follow that C' is not optimal. We conclude that equality (1) holds.

With the help of Lemma 5, we obtain an algorithm that solves the local chaining problem.

Algorithm 6 *Finding an optimal local chain*

for every fragment f'
  determine a fragment f *such that* $f.score - g_1(f', f)$ *is maximal among all*
    fragments that precede f'
  $max := \max\{0, f.score - g_1(f', f)\}$
  if $max > 0$ **then** $f'.prec := f$ **else** $f'.prec := NULL$
  $f'.score := f'.weight + max$
determine a fragment $\tilde{f}$ *such that* $\tilde{f}.score$ *is maximal among all fragments*
report an optimal local chain by following the pointers $\tilde{f}.prec$ *until*
  a fragment f *with* $f.prec = NULL$ *is reached*

It is not difficult to modify the preceding algorithm, so that it can report all chains whose score exceeds some threshold T (in Algorithm 6, instead of determining a fragment $\tilde{f}$ of maximum score, one determines all fragments whose score exceeds T). Such chains will be called *significant local chains.*

The scores of n fragments can be computed by a simple dynamic programming algorithm in $O(n^2)$ time. Because the computation of a maximum of all the scores and the subsequent computation of an optimal local chain can be done in linear time, the overall running time is quadratic in the number n of fragments. This is a severe drawback if n is large. To align amino acid sequences (proteins), Zhang et al. [32] introduced a recurrence similar to (1) and presented a geometric-based algorithm to solve it. It constructs an optimal chain using space division based on *kd*-trees, a data structure known from computational geometry. However, a rigorous analysis of the running time of the algorithm is difficult because the construction of the chain is embedded in the *kd*-tree structure. Another chaining algorithm, devised by Myers and Miller [24], falls into the category of *sparse dynamic programming* [15]. Their algorithm is based on the line-sweep paradigm, and uses orthogonal range search supported by *range trees* instead of *kd*-trees. It is the only chaining algorithm for $k > 2$ sequences that runs in sub-quadratic time $O(n \log^k n)$, "but the result is a time bound higher by a logarithmic factor than what one would expect" [14]. In particular, for $k = 2$ sequences it is one log-factor slower than previous chaining algorithms [15], which require only $O(n \log n)$ time. In [1], we have presented a global chaining algorithm that takes $O(n \log^{k-2} n \log\log n)$ time and $O(n \log^{k-2} n)$ space (for $k = 2$ genomes, the algorithm takes $O(n \log n)$ time and $O(n)$ space). This, surprisingly, did not only reduce the time and space complexities of Myers and Miller's algorithm by a log-factor but actually improved the time complexity by a factor $\frac{\log^2 n}{\log \log n}$. In essence, this improvement was achieved by (1) incorporating gap costs into the weight of fragments, so that it is enough to determine a maximum function value over a semi-dynamic set (instead of a dynamic set) and (2) a combination of fractional cascading with the efficient priority queues of [19,30], which yields a more efficient search than on ordinary range trees.

In the following, we present an algorithm that efficiently computes the scores of all fragments. Our algorithm can employ the *kd*-tree, the range tree, or any other data structure supporting orthogonal range searches. When the *kd*-tree is used, the algorithm takes $O((k-1)n^{2-\frac{1}{k-1}})$ time and $O(n)$ space in the worst case. Using the range tree, our algorithm takes $O(n \log^{k-2} n \log \log n)$ time and $O(n \log^{k-2} n)$ space. For $k = 2$ genomes, both implementations of the algorithm take $O(n \log n)$ time and $O(n)$ space in the worst case.

4.1 Chaining Based on Range Maximum Queries

Because our algorithm is based on orthogonal range search for maximum, we have to recall the following notion. Given a set of points in $\mathbb{R}^k$ with associated score, a *range maximum query* (RMQ) asks for a point of maximum score that lies in a hyper-rectangle $R(p,q)$. In the following, RMQ will also denote a procedure that takes two points p and q as input and returns a point of maximum score in the hyper-rectangle $R(p,q)$.

We will use the line-sweep paradigm. Suppose that the start and end points of the fragments are sorted w.r.t. their x_1 coordinate. Then, processing the points in ascending order of their x_1 coordinate simulates a line (plane or hyper-plane in higher dimensions) that sweeps the points w.r.t. their x_1 coordinate. If a point has already been scanned by the sweeping line, it is said to be *active*; otherwise it is said to be *inactive*. During the sweeping process, the x_1 coordinates of the active points are smaller than the x_1 coordinate of the currently scanned point s. If s is the start point of fragment f', then an optimal chain ending at f' can be found by a RMQ over the set of active end points of fragments (this will be proven in Lemma 9). Since $p.x_1 < s.x_1$ for every active end point p, the RMQ need not take the first coordinate into account. In other words, the RMQ is confined to the range $R(0, (s.x_2, \ldots, s.x_k))$, so that the dimension of the problem is reduced by one.

But the question is how to integrate the gap costs into our RMQ based approach. If we would explicitly compute $g_1(f', f)$ for every pair of fragments with $f \ll f'$, then this would yield a quadratic time algorithm. Thus, it is necessary to express the gap costs implicitly in terms of weight information attached to the points. We achieve this by using the *geometric cost* of a fragment f, which we define in terms of the terminus point t as $gc(f) = d_1(\mathsf{t}, end(f))$.

Lemma 7. *Let f, $\tilde{f}$, and f' be fragments such that $f \ll f'$ and $\tilde{f} \ll f'$. Then we have $\tilde{f}.score - g_1(f', \tilde{f}) > f.score - g_1(f', f)$ if and only if the inequality $\tilde{f}.score - gc(\tilde{f}) > f.score - gc(f)$ holds.*

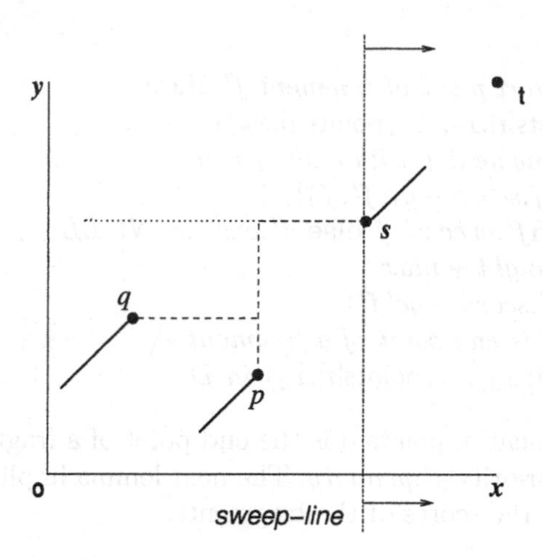

Fig. 2. Points p and q are active end points of the fragments f and $\tilde{f}$. The start point s of fragment f' is currently scanned by the sweeping line and t is the terminus point.

Proof.

$$
\begin{aligned}
& \tilde{f}.score - g_1(f', \tilde{f}) > f.score - g_1(f', f) \\
\Leftrightarrow\ & \tilde{f}.score - \textstyle\sum_{i=1}^{k} (beg(f').x_i - end(\tilde{f}).x_i) > f.score \\
& \qquad\qquad\qquad\qquad - \textstyle\sum_{i=1}^{k} (beg(f').x_i - end(f).x_i) \\
\Leftrightarrow\ & \tilde{f}.score - \textstyle\sum_{i=1}^{k} (\mathrm{t}.x_i - end(\tilde{f}).x_i) > f.score - \textstyle\sum_{i=1}^{k} (\mathrm{t}.x_i - end(f).x_i) \\
\Leftrightarrow\ & \tilde{f}.score - gc(\tilde{f}) > f.score - gc(f)
\end{aligned}
$$

The second equivalence follows from adding $\sum_{i=1}^{k} beg(f').x_i$ to and subtracting $\sum_{i=1}^{k} \mathrm{t}.x_i$ from both sides of the inequality. Fig. 2 illustrates the lemma for $k = 2$.

Note that the value $gc(f)$ is known in advance for every fragment f, because t is fixed. Hence, we define $f.priority = f.score - gc(f)$ for every fragment f, and the RMQ will return the point of maximum priority.

To manipulate the point set during the sweeping process, we need a semi-dynamic data structure D that stores the end points of the fragments and efficiently supports the following two operations: (1) activation of a point and (2) RMQ over the set of active points. The following algorithm is based on such a data structure.

Algorithm 8 *Computation of the scores of n fragments*

Sort all start and end points of the n fragments in ascending order w.r.t. their x_1 *coordinate and store them in the array* points;
Store all end points of the fragments (ignoring their x_1 *coordinate) as inactive in the* $(k-1)$*-dimensional data structure D.*

for $i := 1$ **to** $2n$
  if $\mathsf{points}[i]$ *is the start point of fragment* f' **then**
    $q := \mathit{RMQ}(0, (\mathsf{points}[i].x_2, \ldots, \mathsf{points}[i].x_k))$
    determine the fragment f *with* $end(f) = q$
    $max := \max\{0, f.score - g_1(f', f)\}$
    if $max > 0$ **then** $f'.prec := f$ **else** $f'.prec := NULL$
    $f'.score := f'.weight + max$
    $f'.priority := f'.score - gc(f')$
  else * $\mathsf{points}[i]$ *is end point of a fragment* *\
    activate $(\mathsf{points}[i].x_2, \ldots, \mathsf{points}[i].x_k)$ *in* D

In the **else**-statement, if $\mathsf{points}[i]$ is the end point of a fragment f', then it is activated in D with priority $f'.priority$. The next lemma implies that Algorithm 8 correctly computes the scores of the fragments.

Lemma 9. *If* $\mathit{RMQ}(0, beg(f'))$ *returns the end point of fragment* $\tilde{f}$*, then we have* $\tilde{f}.score - g_1(f', \tilde{f}) = \max\{f.score - g_1(f', f) : f \ll f'\}$.

Proof. $\mathtt{RMQ}(0, beg(f'))$ returns the end point of fragment $\tilde{f}$ such that $\tilde{f}.priority = \max\{f.priority : f \ll f'\}$. Since $f.priority = f.score - gc(f)$ for every fragment f, it is an immediate consequence of Lemma 7 that $\tilde{f}.score - g_1(f', \tilde{f}) = \max\{f.score - g_1(f', f) : f \ll f'\}$.

The complexity of Algorithm 8 depends of course on how the data structure D is implemented. In [1], an implementation of D using the range tree is presented that supports RMQ with activation for n points in time $O(n \log^{d-1} n \log\log n)$ and space $O(n \log^{d-1} n)$, where $d > 1$ is the dimension. Because in our chaining problem $d = k - 1$, Algorithms 8 and 6 both take $O(n \log^{k-2} n \log\log n)$ time and $O(n \log^{k-2} n)$ space.

If D is implemented as a *kd*-tree that supports RMQ with activation, the algorithms require, for n points and $d > 1$, $O(dn^{2-\frac{1}{d}})$ time and $O(n)$ space in the worst case. The worst case time analysis of the range search using the *kd*-tree is presented in [21]. Hence, the algorithms take $O((k-1)n^{2-\frac{1}{k-1}})$ time and $O(n)$ space. For small k, a collection of programming tricks can speed up the running time in practice; for more details we refer the reader to [5].

In case $k = 2$, D is simply a priority queue over the rank space of all points. As explained in [1], the priority queue of [19,30] can support RMQ with activation for n points in $O(n \log\log n)$ time and $O(n)$ space. But the transformation to the rank space and the sorting procedure in Algorithm 8 require $O(n \log n)$ time, and thus dominate the overall time complexity. That is, for $k = 2$ Algorithm 8 takes $O(n \log n)$ time and $O(n)$ space.

4.2 Processing the Output

As already mentioned, the local chaining algorithm can easily be modified to report all local chains whose score exceeds some threshold T. In this case, however, an additional problem arises: Several chains can share one or more fragments, so that the output can be quite complex. To avoid this, we first cluster the local chains. A cluster consists of all local chains that start with the same fragment. Instead of reporting all the local chains in a cluster, we report only a chain of maximum score as a representative of that cluster. Clustering and reporting the representative chains requires linear time and space. In Fig. 1, for example, the chains 1,4,6 and 7,8 are reported.

5 Applications

5.1 Finding Translocation Events

In a comparative study [17] of the genomes of *M.genitalium* (580,074 bp) and *M.pneumonia* (816,394 bp), it was shown that most of the former genome is contained in the latter but some large segments are not conserved in order, i.e., there are translocation events (transpositions). These segments were first identified by all gene-to-gene comparisons using BLAST and FASTA after excluding non-coding regions. This comparison was repeated again in [11] using three methods applied directly to the DNA sequences of the two genomes without determining ORFs. In the first method, the two genomes were divided into windows of 1000 bp. Then all these windows were compared to each other using FASTA. In the second method, MUMs of fixed length (25 bp) were plotted to visually locate the translocated segments. But in both methods the translocations were difficult to identify. In the third method, MUMs of different lengths were gradually generated (starting with the longest MUMs) until the translocation pattern could be visually identified. Based on this visualization, these segments were manually separated and aligned independently using the program MUMmer.

We ran our program to automatically locate such translocations. Fig. 3 (left) shows the fragments of the two genomes. These fragments are maximal exact matches of minimum length of 12 bp. It is obvious that the plot is too noisy to clearly observe the translocated segments. The right part of the figure shows the local chains in a 2D plot. It is clear that the local chaining procedure has filtered out the noise and identified the translocated segments. To demonstrate the benefit of generating local chains rather than a global one, we also generated the global chain between the two genomes and, as expected, the global chain yielded misleading results in the translocated regions (data not shown). We ran the program with the following parameters: the fragments were weighted by 8 times their lengths and threshold $T = 50$. The chaining procedure was fast: it took about 8 seconds to chain 97176 fragments. The generation of the fragments took less than half a minute. For all applications, we used a Pentium III 933 MHz CPU and the fragments were generated as described in [2].

5.2 Detecting Inversions

It was observed in [13] that inversions in bacterial genomes are centered around the origin and/or the terminus of replication; see also [18]. Consequently, if

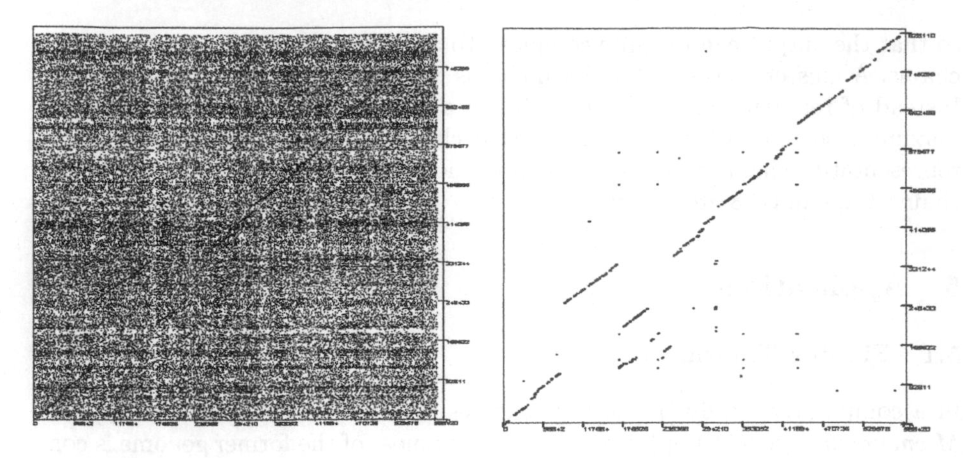

Fig. 3. Left: the fragments of the two genomes *M.genitalium* (x-axis) and *M.pneumonia* (y-axis). Right: The local chains showing the translocated segments.

homologous regions are plotted in two dimensions, then it is easy to observe that they are shaping an X-pattern centered around the origin and/or the terminus of replication of the two genomes. In [13], these patterns were clearly observed in a two dimensional dot plot of homologous proteins of two bacteria. Every dot in this plot corresponded to two proteins whose similarity is above some specified threshold (the program BLAST was used for this task). By contrast, it is not easy to observe the pattern when plotting DNA fragments. This is because the plot is very noisy, i.e., the fragments of small length appear to be uniformly distributed in it; see Fig. 4 (left).

Therefore, we used our algorithm to test the above-mentioned biological hypothesis directly on the DNA sequences using short fragments. We first generated two sets of fragments. The first, which we call the *forward fragment set*, is the set of maximal exact matches obtained from S_1 and S_2, where S_1 and S_2 are the sequences of the first and second genome, respectively. The second, which we call the *inverted fragment set*, is the set of maximal exact matches obtained from S_1 and S_2^{-1}, where S_2^{-1} is the reverse complement of S_2. Fig. 4 (left) shows the forward and inverted fragments (of at least 15 bp) of the two genomes *E.coliK12* (4,639,221 bp) and *V.cholerae* chromosome I (2,961,149 bp). It is obvious that no pattern can by visually identified. We ran our local chaining procedure over each set independently. Fig. 4 (right) shows the local chains of the two sets of fragments. It is easy to observe the X-pattern and to identify that it is centered around the termini of replication of the bacterial genomes. The termini of replication of *E.coli* and *V.cholerae* are nearly at positions 1.6 Mbp [6] and 1.5 Mbp [16], respectively. The program used fragments whose weights were 10 times their lengths and threshold $T = 50$. It took about 4 seconds to chain 50978 fragments and less than a minute to generate the fragments.

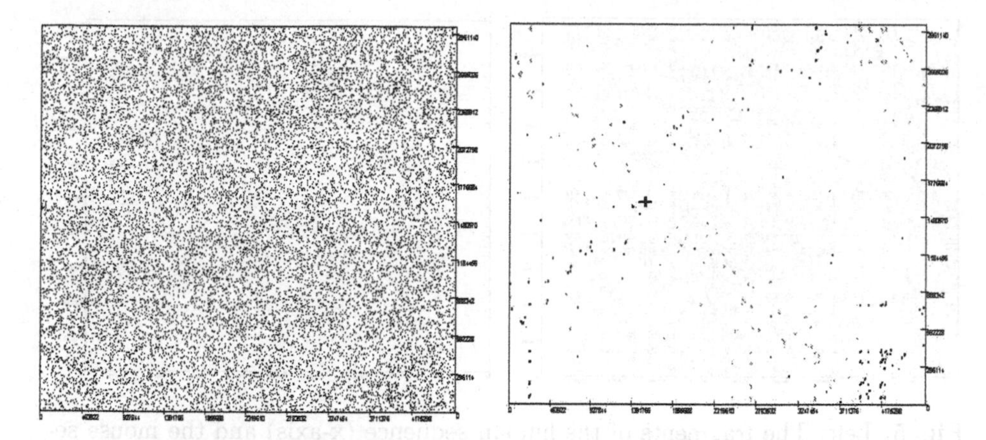

Fig. 4. Left: The fragments of the *E.coli* (x-axis) and the *V.cholerae* (y-axis). Right: The local chains. The forward regions are drawn in black and the reversed regions are colored (see the original postscript). The cross lies nearly on the termini of replication of the two genomes. Note that the bacterial genomes are circular.

5.3 Finding Anchors and Masking Out Repetitions

Most of the tools that are capable of performing large scale alignments are restricted to DNA sequences of closely related organisms. For eukaryotic genomes, the global alignments provided by these tools usually suffer from two problems: (1) the inability to deal with repetitive elements and (2) the failure of recognizing short regions of significant similarity (exons). The first problem is usually tackled, in a preprocessing step, by masking out the known types of repeats using programs like RepeatMasker; this is done, for example, in the programs PipMaker [29] and AVID [7]. Unfortunately, this step is very time consuming. This is especially true of genomes whose repeat content is very large, e.g., the human genome. But no other (built-in) procedure is available to handle such repetitions during the construction of the global alignment.

The second problem often occurs in the comparison of eukaryotic genomes, where exons are small regions that are easily missed by end-to-end global alignment tools. For example, when the authors of [27] compared two sequences of mouse (AF139987) and human (AF045555), it was found that the global chaining procedure of PipMaker misses the 10th exon of the gene RFC2 (nucleotides 104514 ... 104583 in the mouse sequence and nucleotides 80855 ... 80924 in the human genome) and connects another region to the chain instead. A recent approach to tackle this problem is to start by first identifying regions of significant local similarity. From these regions a set of *anchors* is selected to guide the global alignment; see, for example, the programs GLASS [4], CHAOS [8], OWEN [27], and AVID [7]. However, the methods by which the tools identify the anchors are still ad-hoc and differ significantly from tool to tool.

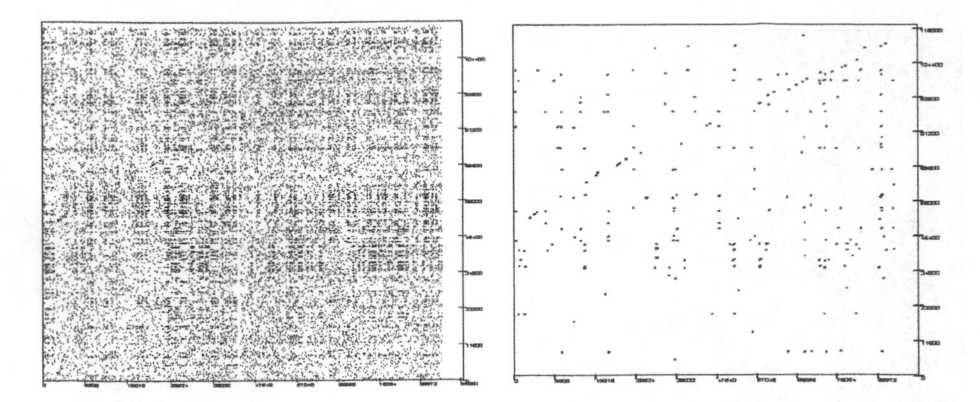

Fig. 5. Left: The fragments of the human sequence (x-axis) and the mouse sequence (y-axis). Right: The local chains.

To efficiently cope with the two problems, algorithmic support is required to improve the sensitivity of such tools without affecting their running time. In our opinion, this support must be based on a solid theoretical basis and we believe that our local chaining procedure meets the requirements. To demonstrate this, we ran the local chaining procedure to compare the aforementioned sequences of mouse (AF139987) and human (AF045555). We generated fragments as maximal exact matches of minimum length of 10 bp. The left side of Fig. 5 shows a 2D plot of the fragments and the right side shows the output of the local chaining procedure. One can see a grid-like structure and the chain of exons (all the exons of RFC2 are identified). The grid like structure is a result of repetitions: vertical dots show repetitions in the mouse genome while horizontal dots correspond to repetitions in the human genome. To sum up, (1) the local chains can work as *potential anchors* and (2) the repetitions are nothing but local chains that can geometrically be identified and filtered out. Our program used the threshold $T = 50$ and every fragment was weighted by 10 times its length. It ran only few seconds including fragments generation (3 seconds for chaining 50651 fragments).

References

1. M. I. Abouelhoda and E. Ohlebusch. Multiple genome alignment: Chaining algorithms revisited. In *Proceedings of the 14th Annual Symposium on Combinatorial Pattern Matching*, LNCS 2676, pages 1–16. Springer-Verlag, 2003.
2. M. I. Abouelhoda, S. Kurtz, and E. Ohlebusch. The enhanced suffix array and its applications to genome analysis. In *Proceedings of the Second Workshop on Algorithms in Bioinformatics*, LNCS 2452, pages 449–463. Springer-Verlag, 2002.
3. S. F. Altschul, W. Gish, W. Miller, E. W. Myers, and D. J. Lipman. A basic local alignment search tool. *J. Mol. Biol.*, 215:403–410, 1990.

4. S. Batzoglou, L. Pachter, J. P. Mesirov, B. Berger, and E. S. Lander. Human and mouse gene structure: Comparative analysis and application to exon prediction. *Genome Research*, 10:950–958, 2001.
5. J. L. Bently. K-d trees for semidynamic point sets. In *6th Annual ACM Symposium on Computational Geometry*, pages 187–197. ACM, 1990.
6. F. R. Blattner et al. The complete genome sequence of Escherichia coli K-12. *Science*, 277(5331):1453–1474, 1997.
7. N. Bray, I. Dubchak, and L. Pachter. AVID: A global alignment program. *Genome Research*, 13:97–102, 2003.
8. M. Brudno and B. Morgenstern. Fast and sensitive alignment of large genomic sequences. In *Proceedings of the IEEE Computer Society Bioinformatics Conference*, pages 138–150. IEEE, 2002.
9. P. Chain, S. Kurtz, E. Ohlebusch, and T. Slezak. An applications-focused review of comparative genomics tools: Capabilities, limitations and future challenges. *Briefings in Bioinformatics*, 4(2), 2003.
10. B. Chazelle. A functional approach to data structures and its use in multidimensional searching. *SIAM Journal on Computing*, 17(3):427–462, 1988.
11. A. L. Delcher, S. Kasif, R. D. Fleischmann, J. Peterson, O. White, and S. L. Salzberg. Alignment of whole genomes. *Nucleic Acids Res.*, 27:2369–2376, 1999.
12. A. L. Delcher, A. Phillippy, J. Carlton, and S. L. Salzberg. Fast algorithms for large-scale genome alignment and comparison. *Nucleic Acids Res.*, 30(11):2478–2483, 2002.
13. J. A. Eisen, J. F. Heidelberg, O. White, and S. L. Salzberg. Evidence for symmetric chromosomal inversions around the replication origin in bacteria. *Genome Biology*, 1(6):1–9, 2000.
14. D. Eppstein. http://www.ics.uci.edu/ eppstein/pubs/p-sparsedp.html.
15. D. Eppstein, R. Giancarlo, Z. Galil, and G. F. Italiano. Sparse dynamic programming. I:Linear cost functions; II:Convex and concave cost functions. *Journal of the ACM*, 39:519–567, 1992.
16. J. F. Heidelberg et al. DNA sequence of both chromosomes of the cholera pathogen Vibrio cholerae. *Nature*, 406:477–483, 2000.
17. R. Himmelreich, H. Plagens, H. Hilbert, B. Reiner, and R. Herrmann. Comparative analysis of the genomes of the bacteria Mycoplasma pneumoniae and Mycoplasma genitalium. *Nucleic Acids Res.*, 25:701–712, 1997.
18. D. Hughes. Evaluating genome dynamics: The constraints on rearrangements within bacterial genomes. *Genome Biology*, 1(6):reviews 0006.1–0006.8, 2000.
19. D. B. Johnson. A priority queue in which initialization and queue operations take $O(\log \log D)$ time. *Math. Sys. Theory*, 15:295–309, 1982.
20. W. J. Kent and A. M. Zahler. Conservation, regulation, synteny, and introns in a large-scale C.briggsae-C.elegans genomic alignment. *Genome Research*, 10:1115–1125, 2000.
21. D. T. Lee and C. K. Wong. Worst-case analysis for region and partial region searches in multidimensional binary search trees and balanced quad trees. *Acta Informatica*, 9:23–29, 1977.
22. M. Höhl, S. Kurtz, and E. Ohlebusch. Efficient multiple genome alignment. In *Proceedings of the 10th International Conference on Intelligent Systems for Molecular Biology*, pages 312–320. Bioinformatics, 18(Supplement 1), 2002.
23. B. Morgenstern. A space-efficient algorithm for aligning large genomic sequences. *Bioinformatics*, 16:948–949, 2000.

24. E. W. Myers and W. Miller. Chaining multiple-alignment fragments in subquadratic time. *Proceedings of the 6th ACM-SIAM Symposium on Discrete Algorithms*, pages 38–47, 1995.
25. W. R. Pearson and D. J. Lipman. Improved tools for biological sequence comparison. *Proc. Natl. Acad. Sci. USA*, 85:2444–2448, 1988.
26. F. P. Preparata and M. I. Shamos. *Computational geometry: An introduction.* Springer-Verlag, New York, 1985.
27. M. A. Roytberg, A. Y. Ogurtsov, S. A. Shabalina, and A. S. Kondrashov. A hierarchical approach to aligning collinear regions of genomes. *Bioinformatics*, 18:1673–1680, 2002.
28. S. Schwartz, J. K. Kent, A. Smit, Z. Zhang, R. Baertsch, R. Hardison, D. Haussler, and W. Miller. Human-mouse alignments with BLASTZ. *Genome Research*, 13:103–107, 2003.
29. S. Schwartz, Z. Zhang, K. A. Frazer, A. Smit, C. Riemer, J. Bouck, R. Gibbs, R. Hardison, and W. Miller. PipMaker—A web server for aligning two genomic DNA sequences. *Genome Research*, 10(4):577–586, 2000.
30. P. van Emde Boas. Preserving order in a forest in less than logarithmic time and linear space. *Information Processing Letters*, 6(3):80–82, 1977.
31. P. Vincens, L. Buffat, C. Andre, J. P. Chevrolat, J. F. Boisvieux, and S. Hazout. A strategy for finding regions of similarity in complete genome sequences. *Bioinformatics*, 14:715–725, 1998.
32. Z. Zhang, B. Raghavachari, R. C. Hardison, and W. Miller. Chaining multiple-alignment blocks. *J. Computational Biology* , 1:51–64, 1994.

Common Intervals of Two Sequences

Gilles Didier

Laboratoire Genome et Informatique - CNRS UMR 8116
Tour Evry2 - 523, place des Terrasses de l'Agora
91034 EVRY CEDEX
didier@genopole.cnrs.fr

Abstract. Looking for the subsets of genes appearing consecutively in two or more genomes is an useful approach to identify clusters of genes functionally associated. A possible formalization of this problem is to modelize the order in which the genes appear in all the considered genomes as permutations of their order in the first genome and find k-tuples of contiguous subsets of these permutations consisting of the same elements: the common intervals. A drawback of this approach is that it doesn't allow to take into account paralog genes and genomic internal duplications (each element occurs only once in a permutation). To do it we need to modelize the order of genes by sequences which are not necessary permutations.
In this work, we study some properties of common intervals between two general sequences. We bound the maximum number of common intervals between two sequences of length n by n^2 and present an $O(n^2 \log(n))$ time complexity algorithm to enumerate their whole set of common intervals. This complexity does not depend on the size of the alphabets of the sequences.

1 Introduction

The comparison of the sequences of genes of two or more organisms, in particular finding locally conserved subsets of genes in several genomes, can be used to predict protein function or just to determine cluster of genes functionally associated [3,6,7]. A way to find such clusters is to represent these sequences of genes by permutations and to find the consecutive elements conserved in these permutations, which are called common intervals [5]. A limitation of this type of modelization is that an element of a permutation can occur only once within it. Consequently paralog genes or internal duplication in a genome cannot be modelized by permutation. This yields to define and study a notion of common intervals of general sequences.

But firstly, let us recall that the question of finding all common intervals of two permutations of n elements was studied and solved by algorithms having $O(n^2)$ or $O(n + K)$ time complexity, where K is the effective number of common intervals [8]. This number K is less or equal than $\binom{n}{2}$ in the case of two permutations. So in the worst case, any algorithm solving this question will

G. Benson and R. Page (Eds.): WABI 2003, LNBI 2812, pp. 17–24, 2003.

run in $O(n^2)$ time. Enumerating the common intervals of a given number (≥ 2) of permutations was solved in [4], and some more theoretical applications and developments of this type of algorithms in [2].

We are interested here in a similar problem over two general sequences. The main difference with permutations is that elements can occur more than a single time. Because of this, techniques used in the permutations case cannot be directly transposed. We first give a bound for the number of common intervals between two sequences and an algorithm to enumerate them. The algorithmic complexity is $O(n^2 \log(n))$ in the worst case, where n is the length of the sequences. In the case of two permutations, this algorithm runs in $O(n^2)$ time.

A recent algorithm, developed by Amir *et al* in [1], solves the same question in $O(n|\Sigma| \log(n) \log(|\Sigma|))$, where $|\Sigma|$ is the size of the alphabet of the sequences, by using an efficient coding of the (sub-)alphabets (called *fingerprints*) of their subsequences. Depending on the size of the alphabet relatively to the length of the sequences, their algorithm or the present one is the fastest.

2 Definitions and Notations

Let s be a finite sequence over a certain set of elements called symbols, we note $|s|$ its length and index it from 1 to $|s|$: $s = s_1 s_2 \dots s_{|s|}$. The alphabet of s, noted $\mathcal{A}_s$, is the set of symbols appearing in this sequence: $\mathcal{A}_s = \{s_l | 1 \leq l \leq |s|\}$. To simplify the statements of definitions and proofs, we add a special symbol δ, never occurring in the considered sequences, at their beginnings and their ends: *i.e.* we fix $s_0 = s_{|s|+1} = \delta$ for each sequence s. Nevertheless, in the following, 0 and $|s|+1$ won't be considered as positions of s and $\delta \notin \mathcal{A}_s$.

For a pair (i, j) of integers with $1 \leq i \leq j \leq |s|$ we note $[i, j]$ the set $\{i, i+1, \dots, j\}$. The corresponding *subsequence* of s, noted $s_{[i,j]}$ is the sequence $s_i s_{i+1} \dots s_j$. A subsequence $s_{[i,j]}$ is said *alphabetically deliminated* if $s_{i-1} \notin \mathcal{A}_{s_{[i,j]}}$ and $s_{j+1} \notin \mathcal{A}_{s_{[i,j]}}$.

Let be two sequences s and s', a pair $([i, j], [i', j'])$ is a *common interval* of s and s' if it satisfies:

1. $s_{[i,j]}$ and $s'_{[i',j']}$ are alphabetically deliminated ;
2. $\mathcal{A}_{s_{[i,j]}} = \mathcal{A}_{s'_{[i',j']}}$.

We note $\mathcal{C}(s, s')$ the set of common intervals of s and s'. Considering only alphabetically deliminated subsequences involves a sort of maximality property over intervals. More precisely, if $([i, j], [i', j'])$ and $([k, l], [i', j'])$ (resp. and $([i, j], [k', l'])$) belong to $\mathcal{C}(s, s')$ then the intervals $[k, l]$ and $[i, j]$ (resp. $[k', l']$ and $[i', j']$) are disjoint or equal.

For the following sequences:

$$s = \texttt{a b c b d a d b c a}$$
$$s' = \texttt{b a c a d b a b d c}$$

The set of common intervals is:

$\mathcal{C}(s, s') = \{([1,1],[2,2]),([1,1],[4,4]),([1,1],[7,7]),([1,2],[1,2]),([1,2],[6,8]),([1,4],[1,4]),([1,10],[1,10]),\dots\}$

3 Number of Common Intervals

By definition, the alphabetically deliminated subsequences starting at a position i of s cannot end after the first occurrence of s_{i-1} following i. For each position i of s, let us denote by $\mathbf{o}^i$ the sequence of the set of symbols occurring between i (included) and the first occurrence of s_{i-1} following i (excluded), ordered according to their first occurrences. The main interest of this sequence is that the alphabets of alphabetically deliminated subsequences starting at i are the sets $\{\mathbf{o}^i_1, \ldots, \mathbf{o}^i_m\}$ for all integer $m \in \{1, \ldots, |\mathbf{o}^i|\}$.

We note $\mathbf{p}_i$ the application which associates to each $m \in \{1, \ldots, |\mathbf{o}^i| - 1\}$, the position preceding the first occurrence of $\mathbf{o}^i_{k+1}$ following i and to $|\mathbf{o}^i|$ the position preceding the first occurrence of s_{i-1} following i.

Remark 1 *The set of alphabetically deliminated subsequences of s starting at i is $\{[i, \mathbf{p}_i(m)] \,|m \in \{1, \ldots, |\mathbf{o}^i|\}\}$ and we have $\mathcal{A}_{s_{[i,\mathbf{p}_i(m)]}} = \{o^i_1, \ldots, o^i_m\}$ for all integer $m \in \{1, \ldots, |o^i|\}$.*

We associate to each symbol $x \in \mathcal{A}_{s'}$ the integer called *i-rank*, noted $\mathbf{r}_i(x)$ and defined as the position of x in $\mathbf{o}^i$ if x occurs in it and as $+\infty$ if not. We fix always $\mathbf{r}_i(\delta)$ to $+\infty$. The restriction of the application $\mathbf{r}_i$ to the set of symbols with finite images is one to one.

$\mathbf{o}^1$ = a b c d
$\mathbf{o}^2$ = b c d
$\mathbf{o}^3$ = c
$\mathbf{o}^4$ = b d a
...

	1	2	3	4
$\mathbf{p}_1$	1	2	4	10
$\mathbf{p}_2$	2	4	5	–
$\mathbf{p}_3$	3	–	–	–
$\mathbf{p}_4$	4	5	8	–
...	...	...	...	...

	a	b	c	d
$\mathbf{r}_1$	1	2	3	4
$\mathbf{r}_2$	$+\infty$	1	2	3
$\mathbf{r}_3$	$+\infty$	$+\infty$	1	$+\infty$
$\mathbf{r}_4$	3	1	$+\infty$	2
...	...	...	...	...

Fig. 1. Construction of $\mathbf{o}^i$, $\mathbf{p}_i$ and $\mathbf{r}_i$ of the sequence s in the preceding example.

An interval $[a, b]$ is said to be *i-complete* if $\{\mathbf{r}_i(s'_l) | l \in [a, b]\} = \{1, \ldots, m\}$, where $m = \max\{\mathbf{r}_i(s'_l) | l \in [a, b]\}$. If $[a, b]$ is i-complete then $\mathcal{A}_{s'_{[a,b]}} = \mathcal{A}_{s_{[i,\mathbf{p}_i(m)]}}$. So an interval i-complete and alphabetically deliminated of s' is common with an interval of s starting at i.

For each position k of s', we note $\mathbf{I}_i(k)$ the interval $[a, b]$ containing k such that the subsequence $s'_{[a,b]}$ of s' verifies:

1. $\mathbf{r}_i(s'_l) \leq \mathbf{r}_i(s'_k)$ for all $l \in [a, b]$;
2. $\mathbf{r}_i(s'_{a-1}) > \mathbf{r}_i(s'_k)$ and $\mathbf{r}_i(s'_{b+1}) > \mathbf{r}_i(s'_k)$.

The collection of intervals $\mathbf{I}_i$ is hierarchically-nested.

Lemma 1 *If $\mathbf{I}_i(k)$ and $\mathbf{I}_i(k')$ are two intervals of $\mathbf{I}_i$, then $\mathbf{I}_i(k) \cap \mathbf{I}_i(k') \in \{\emptyset, \mathbf{I}_i(k), \mathbf{I}_i(k')\}$. More precisely, if $\mathbf{I}_i(k) \cap \mathbf{I}_i(k') \neq \emptyset$ we have:*

1. $\mathbf{I}_i(k) \subset \mathbf{I}_i(k')$ *if and only if* $\mathbf{r}_i(s'_k) \leq \mathbf{r}_i(s'_{k'})$;
2. $\mathbf{I}_i(k) \supset \mathbf{I}_i(k')$ *if and only if* $\mathbf{r}_i(s'_k) \geq \mathbf{r}_i(s'_{k'})$;
3. $\mathbf{I}_i(k) = \mathbf{I}_i(k')$ *if and only if* $\mathbf{r}_i(s'_k) = \mathbf{r}_i(s'_{k'})$.

Proof. We just prove the assertion 1, the other ones being direct consequences. By definition, the maximum i-rank over $\mathbf{I}_i(k)$ (resp.$\mathbf{I}_i(k')$) is reached in position k (resp. k').If $\mathbf{I}_i(k) \supset \mathbf{I}_i(k')$ then $\mathbf{r}_i(s'_k) \geq \mathbf{r}_i(s'_{k'})$.

Let $\mathbf{I}_i(k) = [a,b]$ and $\mathbf{I}_i(k') = [a',b']$ be two intervals of $\mathbf{I}_i$ such that $\mathbf{I}_i(k) \cap \mathbf{I}_i(k') \neq \emptyset$. Assume that $\max\{s'_l | l \in \mathbf{I}_i(k)\} = \mathbf{r}_i(s'_k) \leq \mathbf{r}_i(s'_{k'}) = \max\{s'_l | l \in \mathbf{I}_i(k')\}$. As $\mathbf{I}_i(k) \cap \mathbf{I}_i(k') \neq \emptyset$, we have either $a \in [a',b']$ or $b \in [a',b']$. If $a \in [a',b']$ then b cannot be strictly greater than b', in the other case, $s'_{b'}$, by definition greater than $\max\{s'_l | l \in \mathbf{I}_i(k')$ would belong to $[a,b]$, which would be in contradiction with the hypothesis. In the same way, if $b \in [a',b']$ then a cannot be strictly smaller than a' ; and the assertion 1 is proved.

The collection $\mathbf{I}_i$ can be represented as a tree of which nodes are associated to an interval $\mathbf{I}_1(k)$, or eventually to several intervals if they are confounded, and where the set of all the leafs having a given node $\mathbf{I}_1(k)$ as ancestor corresponds to the set of positions of s' bounded by it (figure 2).

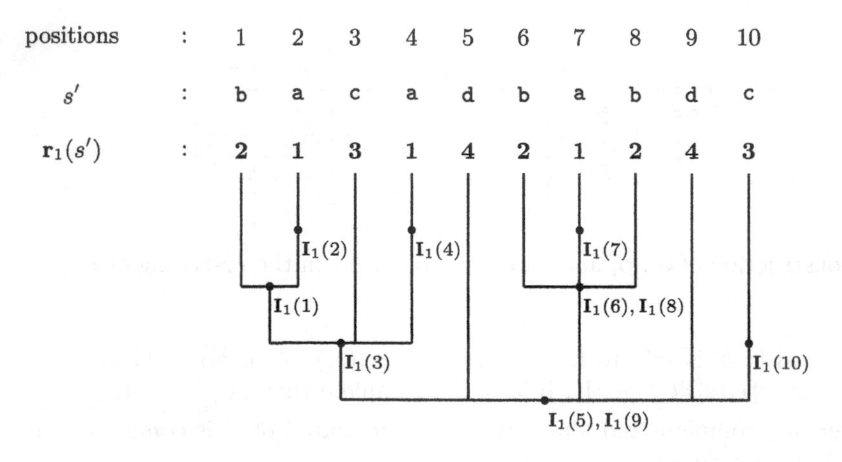

Fig. 2. Tree representation of the hierarchy of the $\mathbf{I}_1(k)$ of the example.

Lemma 2 *Let s and s' be two sequences. If $([i,j],[a,b])$ is a common interval of s and s' then there is a position k of s' such that $[a,b] = \mathbf{I}_i(k)$*

Proof. If $([i,j],[a,b]) \in \mathcal{C}(s,s')$ then the subsequences $s_{[i,j]}$ and $s'_{[a,b]}$ are alphabetically deliminated and verify $\mathcal{A}_{s_{[i,j]}} = \mathcal{A}_{s'_{[a,b]}} = \mathcal{A}$. By definition of $\mathbf{r}_i$, there is no symbol in $\mathcal{A}$ of infinite i-rank and each integer between 1 and $m = \max\{\mathbf{r}_i(x)|x \in \mathcal{A}\}$ appears as an i-rank of a symbol of $\mathcal{A}$: *i.e.* $\{\mathbf{r}_i(x)|x \in \mathcal{A}\} = \{1,\ldots,m\}$ and $[a,b]$ is i-complete. As $s'_{[a,b]}$ is alphabetically deliminated, the symbols s'_{a-1} and s'_{b+1} are not in $\mathcal{A}$. Moreover, each rank in $\{1,\ldots,m\}$ has an unique antecedent in $\mathcal{A}_{s'}$ and we have necessarily $\mathbf{r}_i(s'_{a-1})$ and $\mathbf{r}_i(s'_{b+1})$ strictly greater than m, equal to $\max\{\mathbf{r}_i(s'_l)|l \in [a,b]\}$. Let $k \in [a,b]$ be such that $\mathbf{r}_i(s'_k) = m$, it follows from the preceding and the definition of $\mathbf{I}_i(k)$ that $[a,b] = \mathbf{I}_i(k)$.

As for each position i of s, there are at most $|s'|$ subsequences $\mathbf{I}_i(k)$, we obtain the following bound.

Theorem 1 *Let s and s' be two sequences. The cardinal of $\mathcal{C}(s,s')$ is smaller than $|s| \times |s'|$.*

This bound is certainly not optimal. But as in the particular case of two n-permutations the optimal bound is $\binom{n}{2}$, the optimal bound for two sequences of length n grows as $O(n^2)$.

4 Algorithm

Let k be a position of s' with a finite i-rank $\mathbf{r}_i(s'_k)$, the *left-neighbour* (resp. the *right-neighbour*) of k is the greatest position smaller (resp. the smallest position greater) than k where a symbol of i-rank $\mathbf{r}_i(s'_k)+1$ occurs if it exists (if there is not such symbol then k won't have a right- and/or a left-neighbour).

If k and k' are positions of s', the *i-distance* between k and k' is the maximum i-rank of symbols occurring in the subsequence $s'_{[k,k']}$ if $k \leq k'$ or in $s'_{[k',k]}$ if $k \geq k'$. The smallest distance of a position k of finite i-rank and a position of i-rank $\mathbf{r}_i(s'_k)+1$ is reached either between k and its left-neighbour or between k and its right-neighbour.

Let j be a position of finite i-rank and L (resp. R) the i-distance between p and its left-neighbour (resp. its right-neighbour) if it exists and $+\infty$ if not. If R and L are both infinite, j don't have a *successor*. In the other case, the *successor* of j is:

- its left-neighbour if $L \leq R$;
- its right-neighbour if $L > R$.

An *i-path* of order m is a sequence p of m positions of s' verifying:

1. $\mathbf{r}_i(s'_{p_l}) = l$ for all $l \in \{1,\ldots m\}$;
2. for all $l \in \{1,\ldots m-1\}$, p_{l+1} is the successor of p_l.

Lemma 3 *If p and q are two i-paths such that $p_j = q_j$ for an integer $j \in \{1,\ldots,\min(|p|,|q|)\}$ then $p_l = q_l$ for all $l \in \{j,\ldots,\min(|p|,|q|)\}$.*

Proof. By definition an i-path can be extend in only one way which depends only on its last element.

Theorem 2 *An interval $\mathbf{I}_i(k)$ is i-complete if and only if it contains an i-path of order $\mathbf{r}_i(s'_k)$.*

Proof. By definition, if $\mathbf{I}_i(k)$ contains an i-path of order $\mathbf{r}_i(s'_k) = \max\{\mathbf{r}_i(s'_l)|l \in \mathbf{I}_i(k)\}$, then it is complete.

Reciprocally let us assume $\mathbf{I}_i(k)$ to be complete. We will proceed by induction over the length of an i-path of $\mathbf{I}_i(k)$. In particular, there is at least a position $p_1 \in \mathbf{I}_i(k)$ such that $\mathbf{r}_i(p_1)$ and an i-path of order 1 belong to $\mathbf{I}_i(k)$. Assume that the i-path of order $l < \mathbf{r}_i(s'_k)$ is included in $\mathbf{I}_i(k)$. As $\mathbf{I}_i(k)$ is complete, there is at least a position $p \in \mathbf{I}_i(k)$ with a i-rank equal to $l+1$. As the i-distance of p_l and any position not in $\mathbf{I}_i(k)$ is greater than $\mathbf{r}_i(s'_k)$ (by definition of $\mathbf{I}_i(k)$). If p is the position of $\mathbf{I}_i(k)$ of i-rank $l+1$ having the smallest i-distance d from p_l which is the left- or right-neighbour of p_l, then p is the successor of p_l and we have an i-path of order $l+1$.

The lemma 2 and the theorem 2 ensure that $([i,j], [a,b])$ is a common interval of s and s' if and only if there exists a position k of s such that $[a,b] = \mathbf{I}_i(k)$ and $\mathbf{I}_i(k)$ contains an i-path p of order $\mathbf{r}_i(s'_k)$. Without loss of generality we can choose $k = p_{|p|}$ ($\mathbf{r}_i(p_{|p|}) = \mathbf{r}_i(k)$, $\mathbf{r}_i(p_{|p|}) \in \mathbf{I}_i(k)$ and lemma 1).

The main idea of our algorithm is, for each position i of s, to test if the elements of each i-path p of s' are included in $\mathbf{I}_i(p_{|p|})$. From the preceding, if this inclusion is granted then we have a common interval of the form $([i, \mathbf{p}_i(\mathbf{r}_i(s'_{p_{|p|}}))], \mathbf{I}_i(p_{|p|}))$ (see remark 1) and all the common intervals will be find by this way.

To perform this we need first to compute, fixing a position i of s, all the intervals $\mathbf{I}_i$. This can be done in a linear time and additional space using a stack S (see the part **Compute the table of intervals $\mathbf{I}_i$ and the table *LeftDistance*** in the algorithm).

Another step is the computing of the successor of each position of s'. To do it, we need to determine the left- and right-neighbours of each position of s', which can be classically done in a linear time. The main difficulty is to compute the i-distances between each position and its left- and right-neighbours. In the algorithm these distances are stored in the tables *LeftDistance* and *RightDistance*. The table *LeftDistance* is compute in the same time and using the same stack than the intervals. The table *RightDistance* is compute in the same way. It need for each position to perform a binary search in the stack. So the complexity of this part of the algorithm is in the worst case $O(|s'|\log(|s'|))$. The table of successors is easy to compute in a linear time using the tables *LeftDistance* and *RightDistance*.

The last step is straightforward. Starting by all the positions of i-rank 1 of s', we extend iteratively all the i-paths starting from these and test the preceding inclusion at each iteration. Following the lemma 3, we don't need to consider each position more than once in an i-path (*PositionParsed* is a boolean table maintained to avoid such useless iterations). Complexity of this part becomes linear with the length of s' in time and space.

Algorithm 1 Compute the set of common intervals between two sequences

```
for all position i of s do
  Compute the rank table r_i, iMax the max of finite i-ranks and the corresponding positions p_i
  Compute the chain tables LeftNeighbour and RightNeighbour
  Initialize the tables LeftDistance and RightDistance to +∞
  Initialize the stack S
  [Compute the table of intervals I_i and the table LeftDistance]
  for k = 0 to |s'| + 1 do
    if r_i(s'_k) < iMax and LeftNeighbour(k) is not EMPTY then
      t ← the smallest position in the stack ≥ LeftNeighbour(k) (computed by binary search)
      LeftDistance[k] ← r_i(s'_t)
    end if
    while r_i(s'_top(S)) < r_i(s'_k) do
      I_i(top(S)).end ← k − 1
      pop(S)
    end while
    if r_i(s'_top(S)) ≠ r_i(s'_k) then
      I_i(k).start ← top(S) + 1
    else
      I_i(k).start ← I_i(top(S)).start
    end if
    push(k, S)
  end for
  [Compute the table RightDistance]
  for k = |s'| + 1 to 0 do
    if r_i(s'_k) < iMax and RightNeighbour(k) is not EMPTY then
      t ← the greatest position in the stack ≤ RightNeighbour(k) (computed by binary search)
      RightDistance[k] ← r_i(s'_t)
    end if
    while r_i(s'_top(S)) < r_i(s'_k) do
      pop(S)
    end while
    push(k, S)
  end for
  [Compute the table Successor]
  for all position k of s' do
    if r_i(s'_k) < iMax and min(LeftDistance[k], RightDistance[k]) < +∞ then
      if LeftDistance[k] ≤ RightDistance[k] then
        Successor(k) ← LeftNeighbour(k)
      else
        Successor(k) ← RightNeighbour(k)
      end if
    else
      Successor(k) ← EMPTY
    end if
  end for
  [Parse the i-paths and output the common intervals]
  Initialize the table PositionParsed to FALSE
  for all position j of i-rank 1 of s' do
    BoundL ← j ; BoundR ← j
    repeat
      BoundL ← min(j, BoundL) ; BoundR ← max(j, BoundR)
      if [BoundL, BoundR] ⊂ I_i(j) then
        output ([i, p_i(r_i(s'_k))], I_i(j))        [a common interval]
      end if
      PositionParsed(j) ← TRUE ; j ← Successor(j)
    until j is EMPTY or PositionParsed(j) is TRUE
  end for
end for
```

For each position i of s, the greater complexity to compute common intervals with left bound i in their s-component is $O(|s'| \log(|s'|))$. If the length of the sequences s and s' are $O(n)$, the general complexity of the algorithm is $O(n^2 \log(n))$.

Acknowledgements

We thank Mathieu Raffinot for introducing the problem, references and discussions and Marie-Odile Delorme for helpful comments and suggestions. We also thank the referees for comments and indicating the reference [1].

References

1. A. Amir, A. Apostolico, G. Landau and G. Satta, Efficient Text Fingerprinting Via Parikh Mapping, *Journal of Discrete Algorithms*, to appear.
2. A. Bergeron, S. Heber, J. Stoye, Common intervals and sorting by reversals: a marriage of necessity, *ECCB 2002*, 54-63.
3. T. Dandekar, B. Snel, M. Huynen and P. Bork, Conservation of gene order: A fingerprint of proteins that physically interact, *Trends Biochem. Sci.* **23** (1998), 324-328.
4. S. Heber, J. Stoye, Finding All Common Intervals of k Permutations, *CPM 2001*, 207-218.
5. S. Heber, J. Stoye, Algorithms for Finding Gene Clusters, *WABI 2001*, 252-263.
6. A. R. Mushegian and E. V. Koonin, Gene order is not conserved in bacterial evolution, *Trends Genet.* **12** (1996), 289-290.
7. J. Tamames, M. Gonzales-Moreno, J.Mingorance, A. Valencia, Conserved clusters of functionally related genes in two bacterial genomes, *J. Mol. Evol.* **44** (1996), 66-73.
8. T. Uno and M. Yagira, Fast algorithms to enumerate all common intervals of two permutations, *Algorithmica*, **26**(2) (2000), 290-309.

A Systematic Statistical Analysis of Ion Trap Tandem Mass Spectra in View of Peptide Scoring

Jacques Colinge, Alexandre Masselot, and Jérôme Magnin

GenePro Inc., Pré de la Fontaine 2, CH-1219 Meyrin, Switzerland
Jacques.Colinge@geneprot.com

Abstract. Tandem mass spectrometry has become central in proteomics projects. In particular, it is of prime importance to design sensitive and selective score functions to reliably identify peptides in databases. By using a huge collection of 140 000+ peptide MS/MS spectra, we systematically study the importance of many characteristics of a match (peptide sequence/spectrum) to include in a score function. Besides classical match characteristics, we investigate the value of new characteristics such as amino acid dependence and consecutive fragment matches. We finally select a combination of promising characteristics and show that the corresponding score function achieves very low false positive rates while being very sensitive, thereby enabling highly automated peptide identification in large proteomics projects. We compare our results to widely used protein identification systems and show a significant reduction in false positives.

1 Introduction

Tandem mass spectrometry (MS/MS) combined with database searching has become central in proteomics projects. Such projects aim at discovering all or part of the proteins present in a certain biological tissue, e.g. tears or plasma. Before MS/MS can be applied, the complexity of the initial sample is reduced by protein separation techniques like 2D-page or liquid chromatography (LC). The proteins of the resulting simpler samples are digested by an enzyme that cleaves the proteins at specific locations. Trypsin is frequently used for this purpose. MS/MS analysis is performed on the digestion products, which are named peptides. Alternatively, early digestion can be applied and peptide separation techniques used. In both cases, the peptides are positively ionized and fragmented individually [18] and, finally, their masses as well as the masses of their fragments are measured. Such masses constitute a data set, the experimental MS/MS spectrum, that is specific to each peptide. The MS/MS spectra can be used to identify the peptides by searching into a database of peptide sequences. By extension, this procedure allows to identify the proteins [9,16].

MS/MS database searching usually involves the comparison of the experimental MS/MS spectrum with theoretical MS/MS spectra, computed from the

G. Benson and R. Page (Eds.): WABI 2003, LNBI 2812, pp. 25–38, 2003.

peptide sequences found in the database. A (peptide) score function or scoring scheme is used to rate the matching between theoretical and experimental spectra. The database peptide with the highest score is usually considered as the correct identification, provided the score is high enough and/or significant enough.

Clearly, the availability of sensitive and selective score functions is essential to implement reliable and automatic MS/MS protein identification systems. In [3] we proposed a generic approach (OLAV) to design such score functions. This approach is based on standard signal detection techniques [22]. In this paper we apply it to LC electrospray ionization ion trap (LC-ESI-IT) mass spectra and we study the relative interest of various quantities we can compute when we compare theoretical and experimental spectra. We finally select a combination of such quantities and establish the performance of the corresponding score function by performing large-scale computations. For reference purposes, we give results obtained with Mascot [20], a widely used commercial protein identification program (available from Matrix Sciences Ltd), and we report the performance we obtain on a generally available data set [13] for which Sequest [6,28] (available from ThermoFinnigan) results have been published [13,12].

Currently available protein identification systems can be classified into three categories: heuristic systems, systems based on a mathematical model and hybrid systems. In the heuristic category there are well known commercial programs: Mascot, Sequest and SONAR MS/MS [7]. Sequest and SONAR correlate theoretical and experimental spectra directly, without involving a model. Mascot includes a limited model [19] as well as several heuristics intended to capture some properties related to signal intensity and consecutive fragment matches. Model-based systems use stochastic models to assess the reliability of matches. In this category we find: MassLynx (available from Micromass Limited [25]), SCOPE [2], ProbId [29] and SHERENGA [4]. SCOPE considers fragment matches as independent events and estimates a likelihood by assuming a Gaussian distribution of mass errors. MassLynx uses a Markov chain to estimate the correct match likelihood and to model consecutive fragment matches. ProbId uses Bayesian techniques to estimate the probability a match is correct. It integrates several elementary observations like peak intensities and simultaneous detection of fragments in several series. SHERENGA estimates a likelihood ratio by considering every fragment match as an independent Bernoulli random variable. [8] improves over SHERENGA by considering signal intensity and neutral losses. The hybrid category generally uses multivariate analysis techniques to filter the results returned by heuristic systems [1,12,14,17,23].

The knowledge of which are the essential quantities to include in a score function is certainly beneficial to most of the approaches above.

According to the relative performance of the various score functions we tested, the most important quantity to include in a score function is the probability to detect each ion type. The next quantity is the intensity of detected fragment: intense fragment must match with probabilities depending on the ion type. Then, different extra quantities improve performance: probability to detect a fragment depending on its amino acid composition, probability to observe consecutive fragment matches. By combining these quantities in a naive Bayesian classifier, we design a score function that has a false positive rate as low as 3% is the true positive rate is fixed at 95%. On data set [13], the false positive rate is inferior to 0.5%.

It is difficult to compare peptide score functions without testing them on the same data set. As a matter of fact, MS/MS data are noisy and of variable precision. Hence, the absolute performance of a given score function may change depending on data set quality. According to our experience, the relative advantage of a score function compared to another one is generally stable from data sets to data sets. To allow readers to compare our results with their own experience, we report them by using an available data set or with a standard algorithm tested on the same set. We observe a strong advantage in favor of the approach we propose.

2 Mass Spectrometry Concepts

2.1 ESI Ion Trap Instruments

Current peptide ionization methods that are common in proteomics include electrospray ionization (ESI) and matrix assisted laser desorption ionization (MALDI) [10]. Several technologies are also available for selecting and fragmenting the accelerated peptides, one of which is quadrupole ion trap (IT) [11,26]. IT represents a significant and growing portion of the mass spectrometers used in proteomics. The approach presented in [3] is not specific to ESI-IT mass spectra.

ESI produces positively charged ions, whose charge states are mainly two or three. An IT instrument breaks peptides by low-energy collision-induced dissociation (CID) [26]. The fragmentation process yields several ion types (a, b, y) [18], depending on the exact cleavage location. The proportion of each ion type produced changes with the MS/MS technology. Additionally, certain amino acids can loose one water (H_2O) or ammonia (NH_3) molecule. Consequently, fragment ion masses can be shifted by -18 Da and/or -17 Da.

Every mass spectrometry instrument produces a signal that can be assumed to be continuous. Peak detection software is used for extracting peptide or fragment masses from this signal. The list of extracted masses is named a peak list. When we refer to a spectrum, we always refer to the corresponding peak list, which is the primary data for identification.

2.2 Matching Theoretical and Experimental Spectra

We do not describe here how to compute theoretical spectra [26]. It is sufficient to know that, given an amino acid sequence, there exit precise rules to compute the mass of every possible fragment of each ion type. The theoretical spectrum consists of the masses of fragments for a selected set S of ion types. S is instrument technology dependent. S also depends on the peptide charge state.

We name the comparison of an experimental spectrum with a theoretical spectrum a *match*. A match can be either correct or random. From the match we compute several quantities that are then used by the score function. These quantities are modeled as random variables. It is convenient to represent them by a random vector E.

The score function is intended to distinguish between correct and random matches. This problem can be viewed as an hypothesis testing problem. In [3] we propose to build score functions as log-likelihood ratios as this approach yields optimal decision rules [22], provided E probability distributions are known in the correct (H_1) and random (H_0) cases. In practice, we have to approximate these two distributions. Nevertheless, we believe that log-likelihood ratios are very effective for peptide scoring, which is confirmed by the high performance we achieve, see Section 4. In [3] we give other arguments to justify this point of view.

3 Statistical Modeling

3.1 Data Set

We analyzed by proteomics two pools of 2.5 liters of plasma. One control pool and one diseased pool (coronary artery disease), each containing roughly 50 selected patients. Multidimensional liquid chromatography was applied, yielding roughly 13 000 fractions per pool, which were digested by trypsin and analyzed by mass spectrometry (LC-ESI-IT) using 40 Bruker Esquire 3000 instruments.

The set of ion trap mass spectra we use is made of 146 808 correct matches, 33 000 of which have been manually validated. The other matches have been automatically validated by a procedure, which, in addition to fixed thresholds, includes biological knowledge and statistics about the peptides that were validated manually. There are 3329 singly charged peptides (436 distinct), 82 415 doubly charged peptides (3039 distinct) and 61 064 triply charged peptides (2920 distinct).

Every performance reported in this paper is obtained by randomly selecting independent training and test sets, which sizes are 3000 and 5000 matches respectively (1000/2329 for charge 1). This procedure is repeated 5 times and the results averaged. We also checked that both model parameters and performance barely change from set to set.

In order to validate one important hypothesis at Section 3.6, we use another set of 1874 doubly charged peptides (73 distinct) and 4107 triply charged peptides (90 distinct). The spectra were acquired on 4 Bruker Esquire 3000+ instruments. We refer to this data set as data set B. Data set [13] has been generated by a ThermoFinnigan LCQ ion trap instrument.

3.2 Theoretical Spectrum and Neutral Losses

As we mentioned in Section 2.1, certain amino acids may lose water or ammonia (a so-called neutral loss). By considering the chemical structure of amino

Table 1. Neutral loss statistics. Relative abundance in Cys_CAM, Asn, Gln, Arg, Ser and Thr between b-17 and b, b-18 and b, etc. Other amino acids are not enriched significantly. Mean and standard deviation are computed from all amino acid enrichments. *Cys_CAM.

Ions	CysC*	Asn	Gln	Arg	Mean	std dev	Ions	Ser	Thr	Mean	std dev
Singly charged peptides											
a-17	1.2	1.7	1.3	0.7	1.00	0.25	a-18	1.2	1.4	0.98	0.24
b-17	1.2	2.1	1.5	1.9	1.08	0.37	b-18	1.2	1.3	0.94	0.19
y-17	1.0	1.9	1.2	2.3	1.13	0.42	y-18	1.2	1.0	1.04	0.14
Doubly charged peptides											
a-17	1.0	1.9	0.9	0.4	1.02	0.28	a-18	1.2	1.2	0.99	0.21
b-17	1.2	1.6	1.2	0.9	1.00	0.18	b-18	1.2	1.2	0.93	0.19
y-17	1.2	1.5	1.5	0.9	1.04	0.19	y-18	1.1	1.2	1.00	0.13
Triply charged peptides											
b-17	1.0	1.4	0.9	0.9	1.00	0.20	b-18	1.4	1.1	0.95	0.20
y-17	0.8	1.5	1.3	0.8	0.99	0.23	y-18	1.1	1.6	0.94	0.28
b^{++}-17	0.9	1.0	1.0	1.0	0.99	0.05	b^{++}-18	1.1	1.1	0.98	0.06
y^{++}-17	1.0	1.0	1.0	0.8	0.99	0.11	y^{++}-18	1.0	1.2	0.99	0.13

acids [26], we observe that Arg (R), Asn (N) and Gln (Q) may loose ammonia, and Ser (S) and Thr (T) may loose water. In order to break disulfur bonds, Cys (C) are modified. A common modification is S-carboxamidomethyl cysteines (Cys_CAM, +57 Da) whose chemical structure [26] suggests a potential loss of ammonia. In the data sets we use (except [13]), Cys are modified as Cys_CAMs.

To be able to compute realistic theoretical spectra, we check which of the amino acids above loose water or ammonia significantly. This point has been already considered by [27] for doubly charged peptides and based on a much smaller data set. We follow a similar approach, i.e. we assume that every amino acid may loose water or ammonia and we compute the amino acid composition of matched ions a, b, y and a,b,y-17,18. Finally, the amino acid compositions are compared to find significant enrichments. The results are shown in Table 1 and we decide to exclude Cys_CAM. Asn is kept although it is only significant for singly charged peptides. We checked that including it provides a small benefit for singly charged peptides without penalizing higher charge states (data not shown).

3.3 Comparing Peptide Score Functions

To supplement the peptide scores with p-values, we have introduced in [3] a method for generating random matches and thus random scores. The general principle is the following: given a match, we generate a fixed number of random peptide sequences, with appropriate masses and possible PTMs, by using a Markov chain. The random peptides are matched with the original spectrum to estimate the random score distribution. Now, to compare the relative performance of various score functions, we compute, for each correct match, the ratio of the correct match score with the best of 10 000 random match scores.

3.4 A Basic Reference Score Function

We define a first score function L_1, which we will refer to as the minimal score function. Every potentially improved score function will be compared to this one. L_1 is a slight extension (charge state dependence) of a score function introduced in [4] and it can be derived as follows. We assume that fragment matches (tolerance given) constitute independent events and the probability of these events depends on the ion type $\theta \in S$ and the peptide charge state z. We denote this probability $p_\theta(z)$. Now, let $s = a_1 \cdots a_n$ be a peptide sequence and a_i its amino acids. The probability of a correct match between s and an experimental spectrum is estimated by taking the product of $p_\theta(z)$ for every matched fragment and of $1 - p_\theta(z)$ for every unmatched fragment. The null-model is identical with random fragment match probabilities $r_\theta(z)$. We find

$$L_1 = \log\left(\prod_{i=1}^{n}\left[\prod_{\theta\in M(s,i)}\frac{p_\theta(z)}{r_\theta(z)}\prod_{\theta\in S(s,i)-M(s,i)}\frac{1-p_\theta(z)}{1-r_\theta(z)}\right]\right).$$

$S(s,i) \subset S$ is the set of ion types ending at amino acid a_i, $M(s,i) \subset S(s,i)$ is the set of ion types matching experimental fragment mass. $S(s,i)$ may be a proper subset of S because certain ions are not always possible depending on the fragment last amino acid (neutral loss). $p_\theta(z), \theta \in S$, are learnt from a set of correct matches. The probabilities of random fragment matches $r_\theta(z)$ are learnt from random peptides.

We use relative entropy in bit $H_\theta(z) = p_\theta(z)\log_2(p_\theta(z)/r_\theta(z))$ to measure the importance of each ion type. For $z = 1, 2, 3$, we empirically determined a threshold that is a constant divided by the average peptide length given z. The performance of L_1 and the list of ion types selected are robust with respect to threshold variations. In decreasing order of $H_\theta(z)$, we use: ($z = 1$) y, b-18, b, y-17, b-17, ($z = 2$) y, b, b-18, b-17, y-18, y-17 and ($z = 3$) y, b^{++}, b, b^{++}-18, y^{++}, b^{++}-17, y^{++}-18, y-18, y^{++}-17, b-18. The performance results are shown in Table 2.

3.5 Consecutive Fragment Matches

The score function L_1 is based on a very strong simplifying assumption: fragment matches are independent events. In theory (very simplified), if an amino acid in the peptide has been protonated, then successive fragments should also contain this protonation site and hence be detected. In reality, the protonation sites are not the same for every copy of a peptide and a probabilistic approach should be followed.

A natural improvement of L_1 would be a model able to "reward" consecutive matches, while tolerating occasional gaps. This can be achieved by using a

Table 2. Performance comparison. Percentage of correct match scores that are equal (ratio=1.0), 20% superior (1.2) or 40% superior (1.4) to the best of 10 000 random match scores, which are generated for each correct match. As the test sets we use comprise numerous good matches, which are treated easily, we also report performance on matches having a L_1 score between 0 and 10. Such lines are marked by an asterisk*. Lines marked with ^{B} refer to the complementary data set B.

	Charge 1			Charge 2			Charge 3		
Function	1.0	1.2	1.4	1.0	1.2	1.4	1.0	1.2	1.4
L_1	75.0	56.2	42.1	97.4	94.7	90.4	96.4	94.4	91.9
L_{consec}	73.3	54.8	40.8	97.8	95.0	91.1	97.8	96.2	93.9
L_{intens}	79.5	57.5	40.9	97.8	95.3	91.2	97.7	95.8	93.5
$L_{1,\text{class}}$	73.2	57.0	46.5	97.6	95.2	91.9	96.4	94.7	91.8
L_{iClass}	76.6	57.5	42.7	97.8	94.9	90.8	97.5	96.0	93.1
L_1^*	66.5	47.7	36.7	83.9	79.3	75.0	82.7	77.8	74.0
L_{consec}^*	73.8	55.6	42.5	85.5	79.3	75.7	89.0	84.8	82.4
L_{intens}^*	71.1	48.5	35.2	83.9	78.9	76.3	87.0	83.7	79.4
$L_{1,\text{class}}^*$	65.5	50.5	41.9	87.8	84.2	81.2	84.2	80.7	76.6
L_{iClass}^*	67.7	48.5	36.8	84.9	79.6	75.7	86.3	83.0	77.6
L_1^B				98.4	96.5	93.7	98.7	94.7	85.9
L_{intens}^B				99.9	99.8	99.4	99.9	99.9	99.3
L_1^{B*}				88.8	83.1	79.8	95.4	85.1	82.8
L_{intens}^{B*}				99.2	98.1	96.5	98.9	98.9	97.7

Markov chain (MC) or a hidden Markov model (HMM) [5]. In this perspective, as observed in [3], it may be advantageous to unify several ion types in one generalized ion type to better capture the consecutive fragment match pattern. For instance, one may want to consider ion types b, b-17, b-18, b^{++} as one general ion type B.

We consider one MC and two HMMs, see Figure 1, and we denote by L_2 the log-likelihood ratio of models for consecutive matches. Given a choice of model (MC or HMM), we define a new score function $L_{\text{consec}} = L_1 L_2$, $L_2 = \prod_{\theta \in S'} L_{2,\theta}$, where S', the set of generalized ion types, and $L_{2,\theta}$, the corresponding log-likelihood ratios.

For each peptide charge state, we test every combination of the models hmmA, hmmJ and mcA both (Fig. 1) for the alternative (H_1) and the null hypotheses (H_0), with orders $n = 2, 3, 4$, order(H_0 model) $\leq$ order(H_1 model). That is 84 L_2 models in total. We empirically set S' to ($z = 1$) Y={y}, ($z = 2$) B={b, b-18, b-17}, Y={y, y-18, y-17} and ($z = 3$) B={b, b^{++}, b^{++}-18, b^{++}-17}, Y={y, y^{++}, y^{++}-18}. The parameters are learnt by expectation maximization (Baum-Welch Algorithm, [5]).

We found no unique combination that would dominate the other ones. Several combinations perform well and, as a general tendency, we have observed that HMMs have a slight advantage for the H_1-model, whereas MCs are sufficient for the H_0-model. The performance of the various combinations is simi-

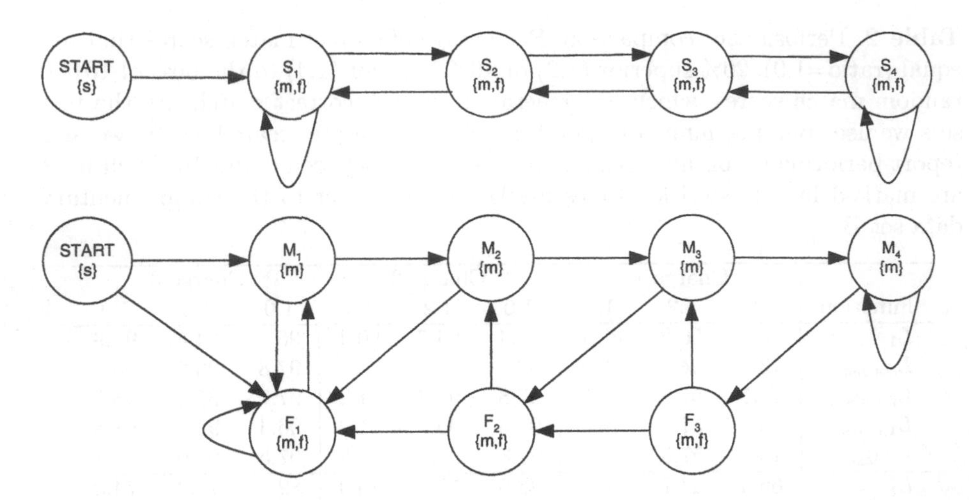

Fig. 1. Consecutive fragment matches models. Two models aimed at capturing successive fragment match patterns: hmmJ (top) and hmmA (bottom). Circles represent states and the letters between curly brackets are the emitted symbols. The structures shown correspond to what we name order 4. The symbol 'm' emitted by a state M_i represents a correct fragment match, while the symbol 'm' emitted by a state F_i represents a random fragment match (this is even possible in a correct peptide match). 'f' represents a theoretical fragment mass not matched in the experimental data.The model mcA is identical to hmmA except for the states F_i, which only emit the symbol 'f'. The structure of these models is designed to allow for higher match probability after a first match has been found. It is also designed for accepting a few missing matches in a longer match sequence.

lar, therefore indicating the intrinsic importance of considering consecutive "no matter" the exact method. The performance of the best combinations ($z = 1$: hmmA(3)/mcA(2), $z = 2$: hmmA(4)/mcA(3), $z = 3$: hmmA(4)/hmmJ(2)) is shown in Table 2. In every case, the transition and emission probabilities nicely fit the model structures. For hmmJ(2), Y ions and charge 2, we find the transitions START to S_1 (probability 1), S_1 to S_1 (0.59), S_1 to S_2 (0.41), S_2 to S_2 (0.88), S_2 to S_1 (0.12) and emissions S_1 ('m' with probability 0.05, 'f' 0.95), S_2 ('m' 0.97, 'f' 0.03).

3.6 Signal Intensity

It is well known among the mass spectrometry community that different ion types have different typical signal intensity. In the case of tryptic peptides, C-terminal ion types (x, y, z) naturally produce more intense peaks. This is due to the basic tryptic cleavage sites (Lys, Arg), which facilitate protonation. [27] and [8] even report fragment relative length intensity dependence.

Here we use a simple model that orders the experimental peaks by intensity and then split them into 5 bins. We obtain $L_{\mathbf{intens}} = L_1 L_3$, $L_3 = \prod_{\theta \in S''} L_{3,\theta}$, S'', a set of ion types, and $L_{3,\theta}$, the corresponding log-likelihood ratios. By selecting ion types for their significance (relative entropy), we set S'' to ($z = 1$) b, b-17, b-18, y, y-17, y-18, ($z = 2$) b, y and ($z = 3$) b-17, b^{++}, y, y-17, y^{++}. The performance of $L_{\mathbf{intens}}$ is shown in Table 2.

Although signal intensity improves performance, we expected a more spectacular change. By further investigating, we found a direct explanation for this disappointing result. Bruker peak detection software allows for exporting the n most intense peaks above noise level into the peak list. At the time we generated our main data set, n was set to 100. Ion types like b-18, b-18, y-17, y-18, b^{++}, y^{++} are generally important for scoring, although their signal is much less intense than b or y signal [27]. Now, given that longer peptides statistically have more protonation sites, it is clear that singly charged peptides are shorter. Therefore, $n = 100$ is sufficient to cover most of the fragment masses. On the other hand, it turns out that it is not sufficient to include enough fragment masses in the model $L_{\mathbf{intens}}$ for $z = 2, 3$. We used the complementary data set B, which was generated with $n = 200$. We denote by $L^B_{\mathbf{intens}}$ the model $L_{\mathbf{intens}}$ trained and tested on this set. The results shown in Table 2 nicely confirm the explanation. Since the spectra in data set B were acquired on a different (better) instrument and the samples were made of purified proteins (not a biological sample), we repeat the performance of L_1 (renamed L^B_1) for reference purpose.

In theory, L_3 includes L_1 and one could expect that L_3 performance is similar to $L_{\mathbf{intens}}$. In practice, L_3 performance is very inferior to $L_{\mathbf{intens}}$ (less than 66% for a ratio of 1 in Table 2), even on data set B. The reason is that the pattern captured by L_1 is more or less always available, whereas the intensity pattern is more variable.

3.7 Amino Acid Dependence

Depending on their amino acid sequence, fragments may be more or less easily detected. The actual dependence involves several phenomena (dissociation, ionization). A model making use of the whole fragment sequence would contain too many parameters. It is commonly accepted that the last amino acid of a fragment (cleavage site) plays a significant role in the above mentioned phenomena. We limit the number of model parameters by only considering the last amino acid of a fragment and by grouping them in classes.

Basic score revisited. We designed an improved version of L_1, which we name $L_{1,\mathrm{class}}$, that uses parameters $p_\theta(z, c)$, $r_\theta(z, c)$, c a set of amino acids, and whose performance is shown in Table 2. We empirically determined the following amino acid classes (amino acids with similar probabilities): (N-tern ions) 'AFHILMVWY', 'CDEGNQST', 'KPR' and (C-term ions) 'HP', 'ACFIMDEGLNQSTVWY', 'KR'.

Consecutive fragment matches revisited. It is possible to extend hmmA, hmmJ and mcA by replacing their states by one state per amino acid class to separate amino acids that inhibit fragmentation from amino acids that favor fragmentation. As we prefer to stay with simple and robust models, we have not implemented the amino acid dependent versions of hmmA, hmmJ and mcA.

Signal intensity revisited. The model we introduced in Section 3.6 can be extended as we did for $L_{1,\text{class}}$, thus obtaining a model $L_{\text{iClass}} = L_1 L_{3,\text{class}}$. The performance of the latter is reported in Table 2. L_{iClass} performs much worse than L_{intens}, what we explain by the fact that, although, the last amino acid plays a major role in the fragment dissociation phenomena, it is no strong relation with signal intensity. Signal intensity is more a consequence of the ion type and the entire peptide sequence.

4 An Efficient Score

L_1 is significantly improved by considering signal intensity. Consecutive fragment matches as well as the amino acid dependent version of L_1 also improve the performance. Accordingly, we tested 4 combinations, which are $C_1 = L_{\text{intens}}$, $C_2 = L_{1,\text{class}} L_2$, $C_3 = L_{\text{intens}} L_3$ and $C_4 = L_{1,\text{class}} L_2 L_3$, see Figure 2.

The receiver operating characteristics (ROC) curves of Figure 2 (top) are obtained by setting a threshold on match p-values. The correct match p-values are computed by searching the peptides of our data set against a database of 15 000 human proteins with variable Cys_CAM and oxidation (Met, His, Trp) modifications. The random match p-values are computed by searching against a database of 15 000 random proteins with the same variable modifications and by taking the best match. The random protein sequences were obtained by training an order 3 MC on the human protein database. From Figure 2 we observe that C_4 is the best combination. C_4 performance on data set and database [13] is shown in Figure 2 (bottom).

5 Discussion

In this work we designed score functions by assuming the independence of many statistical quantities. This choice allowed us to rapidly design efficient score functions as naive Bayesian classifier. This can be seen as preparative work to identify key contributors to successful peptide score functions in order to design future models, comprising fewer simplifying assumptions. We found that ion type probabilities, signal intensity and consecutive fragment matches are essential to a peptide score function. We omitted one aspect of a match that may be pertinent: the simultaneous detection of several ion types [29] (neutral loss, complementary fragments). Peptide elution times may also provide additional information to select correct matches [21]. New approaches based on the mobile proton model are under development, see for instance [24], and it should be possible to combine them with what we presented here.

The general applicability of modern stochastic score functions has not been addressed here. In particular, two central questions might limit their value: the

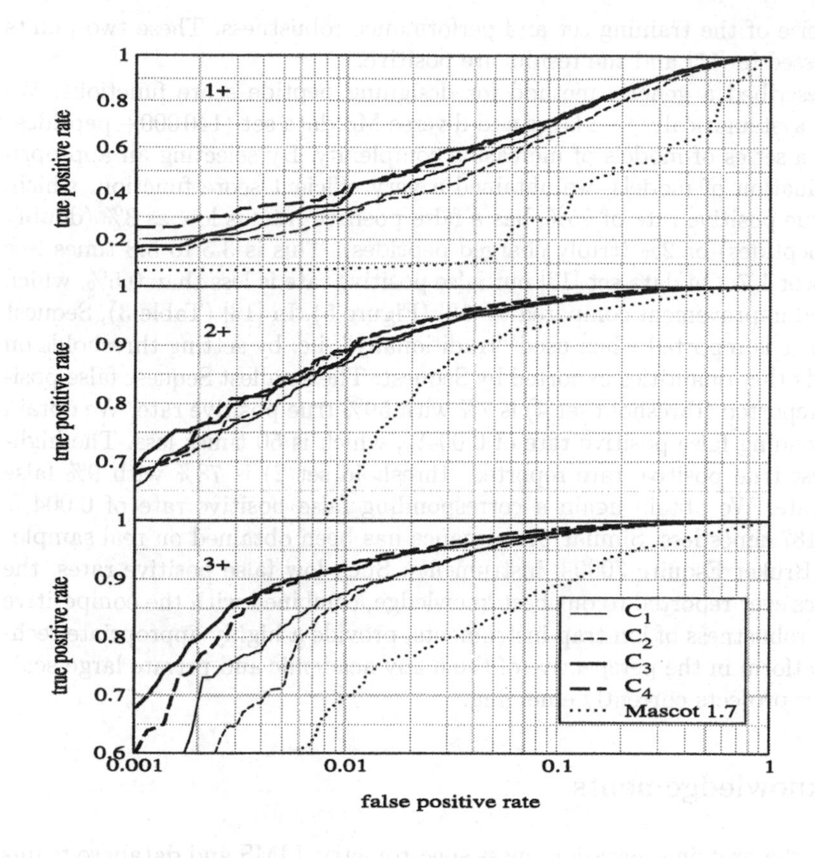

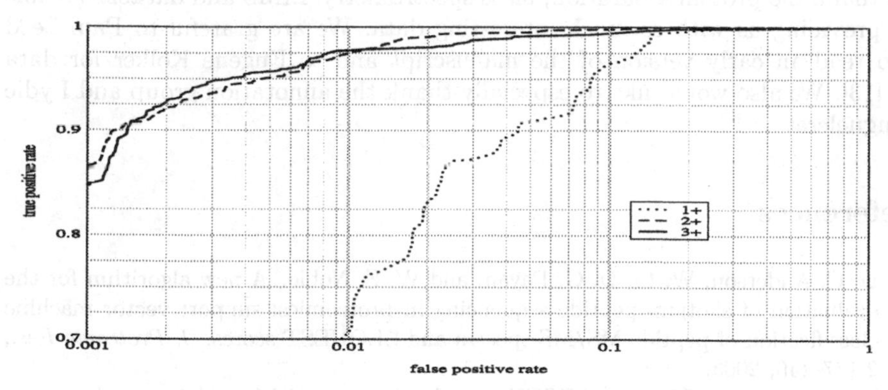

Fig. 2. ROC curves. **Top.** Score functions $C_{1,2,3,4}$ on Bruker Esquire 3000 data. $C_4 = L_{1,\text{class}}L_2L_3$ is the best score function at every charge state. At charge states 2 and 3, if we fix a true positive rate of 95%, the improvement is 3.5-fold (charge 2) or 5.8-fold (charge 3) over Mascot. At charge state 1, if we accept a false positive rate of 5%, 4 times more peptides are identified. **Bottom.** ThermoFinnigan LCQ data [13]. The improvement is 14-fold compared to [12].

minimal size of the training set and performance robustness. These two points are addressed in [15] and the results are positive.

We described a generic method for designing peptide score functions. We applied it systematically to a large and diverse MS data set (140 000+ peptides) to design a series of models of increasing complexity. By selecting an appropriate combination of models, we obtained a very efficient score function, which, given a true positive rate of 95%, has a false positive rate as low as 3% (doubly charged peptides) or 2% (triply charged peptides). This is 3.5 to 5.8 times less than Mascot 1.7. On data set [13] our false positive rate is less than 0.5%, which is a 14-fold improvement compared to [12] (Figure 5). In [13] (Table 3), Sequest performance is reported when used "traditionally", i.e. by setting thresholds on Xcorr and other quantities exported by Sequest. The smallest Sequest false positive rate reported (threshold set 4) is 2% with 59% true positive rate. We obtain a corresponding false positive rate of 0.004%, which is 50 times less. The highest Sequest true positive rate reported (threshold set 2) is 78% with 9% false positive rate. We obtain again a corresponding false positive rate of 0.004%, which is 187 times less. Similar performance has been obtained on real samples by using Bruker Esquire 3000+ instruments. Such low false positive rates, the lowest ones ever reported to our best knowledge, combined with the competitive price and robustness of ion trap instruments, provide a highly appropriate technology platform in the perspective of the many academic and private large-scale proteomics projects currently emerging.

6 Acknowledgements

We thank the protein separation, mass spectrometry, LIMS and database teams for providing us with so much interesting data. We are grateful to Paul Seed who read an early version of the manuscript and to Eugene Kolker for data set [13]. We also would like to especially thank the annotation group and Lydie Bougueleret.

References

1. D. C. Anderson, W. Li, D. G. Payan, and W. S. Noble. A new algorithm for the evaluation of shotgun peptide sequencing in proteomics: support vector machine classification of peptide MS/MS spectra and SEQUEST scores. *J. Proteome Res.*, 2:137–146, 2003.
2. V. Bafna and N. Edwards. SCOPE: a probabilistic model for scoring tandem mass spectra against a peptide database. *Bioinformatics*, 17:S13–S21, 2001.
3. J. Colinge, A. Masselot, M. Giron, T. Dessingy, and J. Magnin. OLAV: Towards high-throughput MS/MS data identification. *Proteomics*, to appear in August, 2003.
4. V. Dancik, T. A. Addona, K. R. Clauser, J. E. Vath, and P. A. Pevzner. De novo peptide sequencing via tandem mass spectrometry: a graph-theoretical approach. *J. Comp. Biol.*, 6:327–342, 1999.

5. R. Durbin et al. *Biological sequence analysis.* Cambridge University Press, Cambridge, 1998.
6. J. K. Eng, A. J. McCormack, and J. R. III Yates. An approach to correlate tandem mass spectral data of peptides with amino acid sequences in a protein database. *J. Am. Soc. Mass Spectrom.*, 5:976–989, 1994.
7. H. L. Field, D. Fenyö, and R. C. Beavis. RADARS, a bioinformatics solution that automates proteome mass spectral analysis, optimises protein identifications, and archives data in a relational database. *Proteomics*, 2:36–47, 2002.
8. M. Havilio, Y. Haddad, and Z. Smilansky. Intensity-based statistical scorer for tandem mass spectrometry. *Anal. Chem.*, 75:435–444, 2003.
9. W. J. Henzel et al. Identifying protein from two-dimensional gels by molecular mass searching of peptide fragments in protein sequence databases. *Proc. Natl. Acad. Sci. USA*, 90:5011–5015, 1993.
10. P. James. *Mass Spectrometry.* Proteome Research. Springer, Berlin, 2000.
11. R. S. Johnson et al. Collision-induced fragmentation of $(m + h)^+$ ions of peptides. Side chain specific sequence ions. *Intl. J. Mass Spectrom. and Ion Processes*, 86:137–154, 1988.
12. A. Keller, A. I. Nesvizhskii, E. Kolker, and R. Aebersold. Empirical statistical model to estimate the accuracy of peptide identification made by MS/MS and database search. *Anal. Chem.*, 74:5383–5392, 2002.
13. A. Keller, S. Purvine, A. I. Nesvizhskii, S. Stolyar, D. R. Goodlett, and E. Kolker. Experimental protein mixture for validating tandem mass spectral analysis. *OMICS*, 6:207–212, 2002.
14. D. C. Liebler, B. T. Hansen, S. W. Davey, L. Tiscareno, and D. E. Mason. Peptide sequence motif analysis of tandem MS data with the SALSA algorithm. *Anal. Chem.*, 74:203–210, 2002.
15. A. Masselot, J. Magnin, M. Giron, T. Dessingy, D. Ferrer, and J. Colinge. OLAV: General applicability of model-based MS/MS peptide score functions. In *Proc. 51st Am. Soc. Mass Spectrom.*, Montreal, 2003.
16. A. L. McCormack et al. Direct analysis and identification of proteins in mixture by LC/MS/MS and database searching at the low-femtomole level. *Anal. Chem.*, 69:767–776, 1997.
17. R. E. Moore, M. K. Young, and T. D. Lee. Qscore: An algorithm for evaluating sequest database search results. *J. Am. Soc. Mass Spectrom.*, 13:378–386, 2002.
18. I. A. Papayannopoulos. The interpretation of collision-induced dissociation mass spectra of peptides. *Mass Spectrometry Review*, 14:49–73, 1995.
19. D. J. Papin, P. Hojrup, and A. J. Bleasby. Rapid identification of proteins by peptide-mass fingerprinting. *Curr. Biol.*, 3:327–332, 1993.
20. D. N. Perkins, D. J. Pappin, D. M. Creasy, and J. S. Cottrell. Probability-based protein identification by searching sequence databases using mass spectrometry data. *Electrophoresis*, 20:3551–3567, 1999.
21. K. Petritis, L. J. Kangas, P. L. Fergusson, G. A. Anderson, L. Paša-Tolić, M. S. Lipton, K. J. Auberry, E. F. Strittmatter, Y. Shen, R. Zhao, and R. D. Smith. Use of artificial neural networks for the accurate prediction of peptide liquid chromatography elution times in proteome analysis. *Anal. Chem.*, 75:1039–1048, 2003.
22. H. V. Poor. *An Introduction to Signal Detection and Estimation.* Springer, New York, 1994.
23. R. G. Sadygov, J. Eng, E. Durr, A. Saraf, H. McDonald, M. J. MacCoss, and J. Yates. Code development to improve the efficiency of automated MS/MS spectra interpretation. *J. Proteome Res.*, 1:211–215, 2002.

24. F. Schütz, E. A. Kapp, J. E. Eddes, R. J. Simpson, T. P. Speed, and T. P. Speed. Deriving statistical models for predicting fragment ion intensities. In *Proc. 51st Am. Soc. Mass Spectrom.*, Montreal, 2003.
25. J. K. Skilling. Improved methods of identifying peptides and protein by mass spectrometry. European Patent Application EP 1,047,107,A2., 1999.
26. A. P. Snyder. *Interpreting Protein Mass Spectra.* Oxford University Press, Washington DC, 2000.
27. D. L. Tabb, L. L. Smith, L. A. Breci, V. H. Wysocki, D. Lin, and J. Yates. Statistical characterization of ion trap tandem mass spectra from doubly charged tryptic peptides. *Anal. Chem.*, 75:1155–1163, 2003.
28. J. Yates, J. K., and Eng. Identification of nucleotides, amino acids, or carbohydrates by mass spectrometry. United States Patent 6,017,693, 1994.
29. N. Zhang, R. Aebersold, and B. Schwikowski. ProbId: A probabilistic algorithm to identify peptides through sequence database searching using tandem mass spectral data. *Proteomics*, 2:1406–1412, 2002.

Vector Seeds: An Extension to Spaced Seeds Allows Substantial Improvements in Sensitivity and Specificity

Broňa Brejová, Daniel G. Brown, and Tomáš Vinař

School of Computer Science, University of Waterloo, Waterloo ON N2L 3G1 Canada
{bbrejova, browndg, tvinar}@math.uwaterloo.ca

Abstract. We present improved techniques for finding homologous regions in DNA and protein sequences. Our approach focuses on the core region of a local pairwise alignment; we suggest new ways to characterize these regions that allow marked improvements in both specificity and sensitivity over existing techniques for sequence alignment. For any such characterization, which we call a *vector seed*, we give an efficient algorithm that estimates the specificity and sensitivity of that seed under reasonable probabilistic models of sequence. We also characterize the probability of a match when an alignment is required to have multiple hits before it is detected. Our extensions fit well with existing approaches to sequence alignment, while still offering substantial improvement in runtime and sensitivity, particularly for the important problem of identifying matches between homologous coding DNA sequences.

1 Introduction

We study techniques for faster and more sensitive pairwise local alignment. Recent advances [10, 8, 5] have demonstrated modifications of the basic approach introduced in BLAST [1] that lead to significant improvements in both sensitivity and running time of local alignment. Here, we present a framework unifying and further extending these approaches, leading to even better performance.

The traditional approach for fast local alignment problem is represented by the BLASTN [1] program. BLASTN first identifies all pairs of short exact matches between the two sequences (hits). Every hit is then extended to a longer alignment, and alignments with high scores are reported, while those with low scores are discarded. High scoring alignments that do not contain a hit cannot be found by this approach.

Sensitivity can be increased by decreasing the required length of the hit; however, this also increases the number of spurious hits (decreasing *specificity*) and thus also increases the running time. Thus, there is a tradeoff between sensitivity and specificity induced by particular definition of a hit.

Recently, several researchers have reported alterations to the hit definition that improve sensitivity without decreasing specificity. Kent [8] in his program BLAT allows a *fixed number of mismatches* in the region that makes up a hit. For example, we may require at least 11 matches in a region of length 12.

G. Benson and R. Page (Eds.): WABI 2003, LNBI 2812, pp. 39–54, 2003.

PatternHunter [10] uses so called *spaced seeds*, which allow arbitrary numbers of mismatches in fixed positions of the hit. For example, a region of length 9 of an alignment is a hit to the PatternHunter seed 110110001 if there is a match on the first, second, fourth, fifth and ninth positions of the region. The other positions (the zeros in the seed) are not relevant.

Ma et al. [10] also introduce the idea of *optimizing the seed*, i.e. choosing the seed with given specificity which has the highest sensitivity under a given probabilistic model of alignment (region with 70% similarity of length 64). In follow-up work, more realistic probabilistic models of alignments (Markov chains, hidden Markov models) have been shown to yield optimal seeds with better performance on real data [4, 5]. (For our purpose, the performance of a seed is its sensitivity and specificity.) These methods also allow to create seeds tailored for particular application such as finding homologous protein coding regions.

Protein alignments are scored by substitution matrices such as BLOSUM62 [6] that define different scores for different matching and mismatching amino acid pairs. Therefore techniques considering only matches and mismatches for finding hits do not work very well. BLASTP [1] defines a hit as several consecutive positions with total score exceeding a given *threshold*.

These techniques are often supplemented by requiring *two non-overlapping hits* that are evenly spaced in both sequences. This method also increases sensitivity compared to a single stronger hit [2].

Here, we generalize these approaches into a new model of *vector seeds* (Section 2). Our model is general, allowing us to produce seeds for nucleotide alignments that incorporate positions that are required to match, positions that are free to vary, sets of positions that allow for a maximum number of variation, and more. Moreover, vector seeds allow easy transfer of techniques developed specifically for match/mismatch models into models based on scoring matrices. For example, we can apply spaced seed ideas to protein homology search.

We also provide an algorithm to predict the sensitivity of a vector seed, given a probabilistic model of alignments (Section 3). Our algorithm extends the original algorithm used to compute the sensitivity of PatternHunter's spaced seeds [7] to the case of vector seeds. We have previously extended this algorithm to more realistic probabilistic models of alignments, such as HMMs [4], and the sensitivity of our new seed models can also be computed for HMMs. This algorithm can be used to find the seed with the best predicted sensitivity for a given family of vector seeds. We further extend the algorithm to allow it to predict the sensitivity of two-hit alignment methods.

Our extensions universally allow greater sensitivity and specificity over existing pairwise alignment methods. For coding DNA alignments, we greatly improve over the performance of BLAT or PatternHunter seeds specifically chosen for this problem [4], allowing false positive rates several times smaller than previously existed, or offering large advantages in specificity with comparable sensitivity (Section 4). Our methods offer substantially improved performance over BLAT or PatternHunter, with minimal additional required changes.

2 Alignments and Vector Seeds

Vector seeds are a new way of defining a *hit*, the conserved part of an alignment that triggers alignment extension in the homology search program. A good definition of hit allows efficient identification of all hits in a sequence database and leads to high sensitivity and specificity. In this section we introduce vector seeds as well as probabilistic models for predicting their performance.

Vector seeds. To define a hit in the vector seed model, we represent ungapped pairwise local alignments as a sequence of real numbers, each corresponding to a position in the alignment. Alignments may potentially contain gaps. However, the hit must be located inside a single ungapped region of an alignment. Here, we model only individual ungapped fragments of such alignments.

In the simplest case, we represent pairwise alignments as binary sequences. Zero represents a mismatch and one a match. For protein alignments, we represent the alignment between sequences $Y = y_1 y_2 \dots y_n$ and $Z = z_1 z_2 \dots z_n$ by the sequence of positional scores, $(s_{y_1,z_1}, s_{y_2,z_2}, \dots, s_{y_n,z_n})$, where $S = (s_{i,j})$ is the scoring matrix. We call such sequence of positional scores an *alignment sequence*. Now we are ready to formally define a vector seed.

Definition 1. A vector seed *is an ordered pair* $Q = (v, T)$, *where* v *is the seed vector* $(v_1, v_2, \dots, v_\ell)$ *of real numbers and* T *is the seed threshold value.*

An alignment sequence $X = (x_1, x_2, \dots, x_n)$ hits the seed Q at position p *if* $\sum_{i=1}^{\ell}(v_i \cdot x_{p+i-1}) \geq T$. *That is, the dot product of the seed vector and the alignment sequence of length* ℓ *beginning at position* p *is at least the threshold* T. *The number of nonzero positions in the vector* v *is the* support *of the seed.*

Vector seeds generalize the spaced seeds of PatternHunter, the mismatching seeds of BLAT, and the minimum word score seeds used by BLASTP. For example, the BLAT seed that requires seven matching positions in nine consecutive positions in a nucleotide alignment is the vector seed $((1,1,1,1,1,1,1,1,1), 7)$. The PatternHunter seed 110110001 can be represented as the vector seed $((1,1,0,1,1,0,0,0,1), 5)$. The BLASTP rule that a hit is three consecutive positions having total score at least 13 corresponds to the vector seed $((1,1,1), 13)$.

However, vector seeds can also encode more complicated concepts. For example, if the alignment sequence is binary, the vector seed $((1,2,0,1,2,0,1,2), 8)$ requires matches in all positions with seed vector value of two, but allows one mismatch in the three positions with value one. The positions with value zero are not relevant to a hit. Or, if the alignment sequence is over the values $\{0, 1, 10\}$, then the seed $((10,10,1,10), 301)$ matches either the alignment vector $(10,10,1,10)$ or the vector $(10,10,10,10)$, but no others.

Of course, more complicated vector seeds than these rather simple examples could be developed; the framework is general enough to allow vector seeds matching any half-space of R^ℓ. However, for simplicity, we will focus on a few families of vector seeds: for nucleotides, we will consider seed vectors with only zeros, ones and twos, where the total number of allowed mismatches is at most one or two, while for amino acids, we will consider short binary seed vectors.

Vector seeds are not universally expressive. For example, the seed $((1,1,0,1,1,0,1,1,0,1,1),8)$ corresponds to requiring matches in the first two positions of each of four consecutive codons. There is no way in the vector seed model to encode the requirement that three of the four codons are matched this way; the seed $((1,1,0,1,1,0,1,1,0,1,1),6)$ also allows one mismatch each in two codons.

Identifying hits in a sequence database. Assume we are given two sequences (or sequence databases) and want to find all hits between them. If hits are required to be exact matches of length k, the common approach is to create a hash table of all k-mers in one of the sequences and then search for each k-mer of the other sequence in the table. If hits are not exact matches (such as in BLAT or BLASTP), we can take each k-mer in the second sequence, generate a list of k-mers that would produce a hit and search for these k-mers in the hash table.

This approach extends to the vector seed scenario. Notice that we need to hash only characters on positions corresponding to non-zero elements in the vector seed. Hence, we seek vector seeds with small support that allow for a small number of hash table entries to be examined for each position in a query sequence. Otherwise, our results would not be of practical use!

In particular, if most examined hash table entries are expected to be empty, the time to find the false positives will dominate the time required to verify whether seed hits are false positives; this is undesirable. For example, all alignments of at least 50% identity and length at least 100 contain a hit to the vector seed with vector $(1,1,\ldots,1)$ of length 100 and with threshold 50. This seed will also be useless, as we would have to find matches to each 100 base pair sequence, and searching for such sequences could not be done efficiently.

Seed probabilities. So far, we have described how to use a single vector seed to generate a set of hits. However, given a family of seeds, we desire the seed that will perform best. To allow such optimization, we represent properties of alignments by a probabilistic model and search for a seed maximizing probability of at least one hit in an alignment sampled from the model. However, to control runtime, one must also control the false positive rate. Given two probabilistic models, one modeling alignments of unrelated sequences, and one for true alignments, we seek seeds with high sensitivity to true alignments, and low false positives.

Probabilistic models of true alignments. We model gap-free local alignments with probabilistic processes that generate sequences of real numbers. We investigate three models for local alignments.

The simplest is the model introduced by PatternHunter for nucleotide alignments, with alignments of length 100 and each position having probability 0.3 of being zero (mismatch) and 0.7 of being one (match), independently.

The other model we use for nucleotide alignments is a three-periodic model of alignments in protein coding regions, where each triplet is emitted as a unit, chosen from a probability distribution over $\{0,1\}^3$. Each triplet is independent of the others in this model. Such models can be used to effectively model the conservation in coding alignments, which are of key importance [9]. Recently [4], we have shown that the optimal spaced seed for coding alignments is quite different from the one that optimizes PatternHunter's model, and showed that this simple codon model is moderately effective at representing these alignments. To estimate seed sensitivity, we represent this probabilistic model as an 8-state hidden Markov model emitting individual binary characters.

For protein sequences, we represent the alignment as a sequence of BLOSUM62 scores ranging from −4 to 11, and we use a positionally independent model similar to PatternHunter's. An amino acid scoring matrix implies a probability distribution on pairs of residues being aligned in true alignments. In particular, an entry $s_{i,j}$ in a scoring matrix implies that the probability of residue i aligning with residue j in related sequences is approximately $b^{s_{i,j}}$ times the probability of them aligning by chance, for some base b [2], and we use this observation to compute a distribution on positional scores implied by the matrix.

Alignments are typically made up of more than one ungapped fragment, separated by gaps, and in our experiments, we treat the number and lengths of these fragments as random variables. Assume we know the probability $p_Q(\ell)$ that an ungapped alignment of a fixed length ℓ, generated by a probabilistic model M, has a match to a given vector seed Q. These probabilities can be used to compute the probability that a seed matches an alignment sequence whose length is a random variable L, by simply computing $P_Q(L) = \sum_\ell p_Q(\ell) \Pr[L = \ell]$. We have noted that alignments between homologous proteins usually consist of one long ungapped fragment and some number of shorter fragments. If the lengths of the long and short fragments are from known distributions L and S, and the number of short fragments is a random variable F, we can compute the probability that an alignment has any matches:

$$1 - (1 - P_Q(L)) \left(\sum_{n=0}^{\infty} (1 - P_Q(S))^n \Pr[F = n] \right).$$

Background model. To control the false positive rate, we need to be able to compute the probability p that a hit of a given seed occurs purely by chance at a given pair of unrelated sequence positions. The expected number of spurious hits is then pnm where n and m are the lengths of the two sequences.

For both nucleotide alignment models, our background probability distribution is a simple noise model, with zero emitted with probability 0.75, and one with probability 0.25. For protein alignments, we use the probabilities of the background distribution built into the BLOSUM62 matrix to determine the probability that residues are aligned by chance, and then use that to compute the score distribution.

Given this background probability, we can either use the algorithm given in the next section to compute the probability of a random match between two sequences, assuming that the sequence length is ℓ, or, if the seed is particularly simple, we can just compute it directly. For example, in the background model for nucleotide alignments, the probability of a 0/1 seed with support s and threshold T having a match at a random place is just

$$\sum_{k=T}^{s} \binom{s}{k} 0.25^k 0.75^{s-k}.$$

3 Computing Sensitivity for Vector Seeds

Here, we show how to compute the sensitivity of a vector seed to detect alignments that come from a position-independent alignment model. Our method is analogous to the original Keich et al. [7] algorithm, except that the alphabet has changed and need not be binary, and that the definition of a hit is the more complicated dot product property. In recent work [4], we show how to extend the original Keich et al. algorithm to the case where the alignment sequence is generated by a hidden Markov model. The extension to HMMs for vector seeds is straightforward, and we omit it for brevity.

Suppose that the probabilistic model generates an alignment sequence $X = (x_1, x_2, \ldots, x_n)$, and that the value in each position is chosen from a small finite set D of real numbers. For each $d \in D$, $\Pr[x_i = d] = p_d$, and all positions are independent. We seek the probability that sequence X generated by this process has a hit of a given vector seed $Q = (v, T)$, where $|v| = \ell$. Let D^* be the set of all sequences of numbers from D whose length is at most ℓ. In the analysis of our algorithm's runtime, we assume that we can represent vectors from D^* as integers and manipulate them in constant time. This assumption is reasonable in our experimental setting.

We compute the probability by dynamic programming, where the subproblem is the probability $P_Q(k, Z)$ that a sequence of length k, which begins with a given sequence Z from D^*, hits the seed Q. We are looking for $P_Q(n, \lambda)$, the probability of a hit when we make no conditions on the string, and the string is of length n. (We use λ to denote the empty sequence.)

We first identify sequences which are *guaranteed hits* and *possible hits*. Let F be a set of all sequences Z from D^* for which all extensions Z' to sequences from D^ℓ have $Z' \cdot v \geq T$. (That is, all extensions are a hit of the seed.) Let M be a set of all sequences Z from D^* for which there *exists* an extension Z' from D^l with $Z' \cdot v \geq T$. (Members of Z *can* be extended to seed hits.)

With these definitions, $P_Q(k, Z)$ can be computed as follows:

1. If $k < \ell$, then $P_Q(k, Z) = 0$.
2. Otherwise, if $Z \in F$, then $P_Q(k, Z) = 1$.
3. Otherwise, if $Z \notin M$, then $P_Q(k, Z) = P_Q(k-1, Y)$, where $Z = z_1 \ldots z_m$ and $Y = z_2 \ldots z_m$.
4. Otherwise, $P_Q(k, Z) = \sum_{d \in D} p_d P_Q(k, Z \cdot d)$.

The computation can be rearranged so that the third case is never reached, by always shifting forward enough positions when d is added to the end of Z in the fourth case. We move to the longest suffix of $Z \cdot d$ that is in M, and skip each instance of the third case. For each entry in M, and each value in D, this skip value can easily be computed in $O(l)$ time.

Sets F and M can be found in $O(|M|)$ time. We initially compute the best and worst possible score for each suffix of the seed Q. Initially, F is empty and M contains the empty string. In each iteration we choose one string X from M and using the pre-computed scores we identify which elements d from D permit or require an extension of Xd to a sequence of length ℓ to hit the seed.

Thus, for an arbitrary vector seed Q, the algorithm computes $P_Q(n, \lambda)$ in $O(|M|(\ell+|D|n))$ time, if the entries of the dynamic programming table are stored in a data structure with $O(1)$ access. Note that the running time is dependent on the size of M; however, seeds with many matching strings are less useful in practice. Also, we need only keep the table P_Q for ℓ values of k, so the memory requirement is $O(|M|\ell)$.

Some algorithm extensions. We note four other fairly straightforward extensions. The first is to seed pairs, where two seeds, Q_1 and Q_2, must both have a hit in the sequence. We first compute the P_{Q_1} and P_{Q_2} matrices, and then, when we reach a prefix Z that is in the set F of guaranteed hits for either seed, we keep the prefix Z, and switch to requiring a hit to the other seed. The overall runtime is at most twice the runtime to compute P_{Q_1} and P_{Q_2}.

We can also compute the probabilities when a hit of either of the two seeds is required, or expand the set of matching strings in a variety of ways, by simply changing the sets M and F. For example, one can easily examine a single seed and the BLAST-style vector seed $(1^k, k)$, by simply adding the sequence 1^k to F and all of its $k-1$ prefixes to M.

We can also consider multi-hit models. Here, we require that the alignment contains at least p hits at least ℓ positions apart. This can be incorporated by keeping matrices $P_{Q,a}$, where $P_{Q,a}(k, Z)$ is the probability of at least a hits in a sequence of length k starting with Z. The recurrence is the same as before, except for the following modifications. First, $P_{Q,1}(k, X) = P_Q(k, X)$. Second, if $Z \in F$ and $a > 1$, then $P_{Q,a}(k, Z) = P_{Q,a-1}(k - \ell, \lambda)$. This is because if a sequence of length k starts with a hit, we need $a - 1$ hits in the rest of the sequence. We investigate the theoretical properties of two-hit BLAST and its variations, for both protein and nucleotide models. We also study two-hit vector seeds in general, comparing their sensitivity and specificity to one-hit models.

Finally, the probabilities p_d also need not be the same for all positions, as long as positions are independent. One can instead incorporate a position-specific probability distribution on D. This is equivalent to computing the sensitivity of a seed to a position-specific score matrix, if we assume that the true positives are generated with the probabilities implied by the score matrix. Of course, such scoring matrices are simply special cases of HMMs, which our algorithms can also expand to cover, using the techniques in our recent paper [4], but the algorithm is especially straightforward for these profiles.

4 Experiments

We performed four experiments to verify the usefulness of vector seeds. Our first two experiments investigate their predicted performance in the simple Pattern-

Hunter model of alignments. In one case, we compute the best single seeds, and in the other case, we study pairs of seeds that join together well.

In our other experiments, we used models trained for DNA coding regions and for proteins, and we computed both theoretical performance of seeds and actual performance on real sequences.

In each experiment, we computed the probability of one hit and of two hits for the seeds we considered. We also computed the false positive probabilities for these seeds (assuming that two-hit models were satisfied when two matches occurred within 100 positions of each other). In experiments on real data, we also computed how many alignments from our set can be detected by each seed.

4.1 Predicted Performance in the PatternHunter Model

One hit required. First, we studied all vector seeds (v, T) with vector entries zero or one, support s satisfying $8 \leq s \leq 15$, threshold T satisfying $s-2 \leq T \leq s$, and whose length is at most $\min\{s+4, 17\}$. We have evaluated sensitivity of these seeds in a simple PatternHunter model.

Our results are summarized in Figure 1 and Table 1. Seeds with both permitted mismatches and the structure of spaced seeds have a large advantage over either alone. For example, the no-mismatch seed PH-10 has false negative rate 22.4% and false positive rate 9.54×10^{-7}. By contrast, the two-mismatch vector seed VS-13-15 has false negative rate 4.11% with false positive rate 9.23×10^{-7}.

This seed may not be practical, as there are 4^{15} possible hash table entries, but the more practical one mismatch seed VS-11-12 has false negative rate 4.9%, with twice the false positive rate (2.2×10^{-6}). This is to be compared to BLAT-11-12 with roughly the same specificity, but much lower sensitivity (almost three times as many false negatives).

Spaced seeds permitting errors are much more useful than unspaced seeds allowing errors. For example, the one-mismatch seed BLAT-11-12, with one hit, has false negative rate 14.8%, while the best vector seed, VS-11-12, has false negative rate 4.9%, three times lower. Both have the same false positive rate, and are equally simple to implement.

Two hits of the same seed required. The situation is even more dramatic if we may require two hits in the alignment. For example, the seed VS-9-11, allowing two mismatches, has unacceptably high false positive rate for one hit. If we require two hits, however, the false positive rate drops to 1.58×10^{-6}, comparable that for one hit to the VS-11-12 seed. Yet the false negative rate is an astonishing 0.1%. While there is some overhead involved in throwing out the many single hits that aren't extended, this can still be done extremely quickly.

If one seeks much better specificity, VS-11-12, with two hits, allows false positive rate 4.8×10^{-10}, over three orders of magnitude better than for one hit to PH-10, with comparable false negative rate (19.1%).

The vector seeds with support between ten and twelve may be appropriate for practice with two hit models. If an input sequence is large, say 20 Mb, the expected number of entries in each site of a hash table will be at least one for seeds of these supports, and the added work to identify double hits should be

Seed				One hit		Two hits	
Vector	T	Support	Name	False negative	False positive	False negative	False positive
1111111111	10	10	BLAST-10	42.0%	9.54×10^{-7}	77.9%	9.10×10^{-11}
11111001101011	10	10	PH-10	22.4%	9.54×10^{-7}	55.6%	9.10×10^{-11}
111111111111111	13	15	BLAT-13-15	11.3%	9.23×10^{-7}	34.4%	8.52×10^{-11}
11111101111011111	13	15	VS-13-15	4.11%	9.23×10^{-7}	18.9%	8.52×10^{-11}
1110111011010111 1	12	13	VS-12-13	10.4%	5.96×10^{-7}	34.5%	3.56×10^{-11}
111111111111	11	12	BLAT-11-12	14.8%	2.21×10^{-6}	41.9%	4.86×10^{-10}
1111110011010111	11	12	VS-11-12	4.89%	2.21×10^{-6}	19.1%	4.86×10^{-10}
101111011001111	9	11	VS-9-11	< 0.01%	1.26×10^{-4}	0.1%	1.58×10^{-6}

Table 1. Theoretical performance of seeds of different support and with different allowed number of mismatches. BLAST-X – unspaced seeds of support X; BLAT-X-Y – unspaced seeds with allowed mismatches; PH-X – optimized spaced seeds of support X; VS-X-Y – optimized vector seeds (allowing both spaced seeds and mismatches).

moderate for these seeds. Our tentative recommendation is two hits to the seed VS-11-12, with false negative rate 19.1% (better than one hit to the best seed of support 10), and false positive rate 4.86×10^{-10}.

Interestingly, we found that the sensitivity of a seed to one hit is an excellent predictor of its sensitivity to two hits. The sensitivity of a seed to two hits is quite consistently close to the cube of the sensitivity to one hit, with correlation coefficient $r^2 = .9982$. The fourteen seeds of support 11 allowing no errors with highest sensitivity to one hit are also the best for sensitivity to two hits; this pattern is consistent for other supports and seed lengths. This suggests that one need only consider sensitivity of seeds to one hit, perhaps computing sensitivity to two hits for appealing seeds.

Two seeds are better than one. One can use a pair of seeds instead of one, allowing matches to either seed. We can avoid twice completing alignments that hit both seeds, so runtime will roughly double for false positives and not change at all for true positives found with both seeds. We considered adding a different seed of support 13 with threshold 12 to the seed VS-12-13, which has false negative rate 10.4% by itself.

The results are shown in Table 2. The best pair halves the false negative rate while the worst augmentation (non-spaced seed with one mismatch allowed) only improves false negatives slightly, yet will still double runtime. Interestingly, one of the best seeds to augment the seed with is its mirror image. This seed has the same sensitivity by itself as VS-12-13.

Of course, there is no evidence that the best seed pair includes the best solo seed in it; however, it is a reasonable heuristic. (The best pair found here was also superior to 1000 random pairs of seeds.) It is also sensible in the context that one is unhappy with the results of a search merely using the first seed.

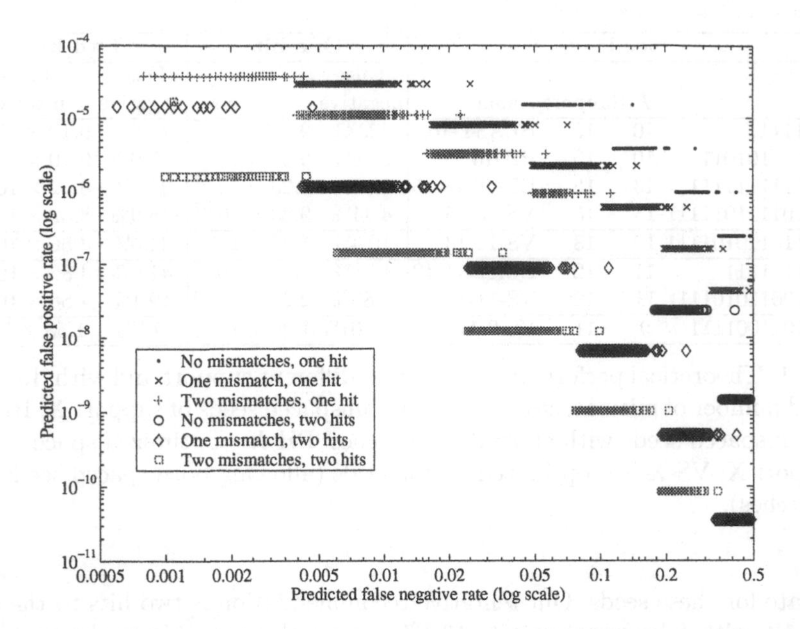

Fig. 1. False positive and false negative rates of many seeds with zero, one or two mismatches permitted, according to the simple Bernoulli model of alignments. Seeds in a horizontal row in the figure have the same support, threshold, and number of hits; the unspaced BLAT or BLAST seeds always have the lowest sensitivity. Two hits to the best one-mismatch seed of support 12 are found in 81% of alignments, comparable to two hits to the best PatternHunter seed of weight 9 or one hit to the best PatternHunter seed of weight 10, yet these have orders of magnitude more false positives.

Seed vector	False Negative	False Positive
No augmentation	10.4%	5.96×10^{-7}
1111111111111 (worst)	8.7%	1.18×10^{-6}
1110010111111111 (median)	6.1%	1.16×10^{-6}
11110101101110111 (mirror)	5.6%	1.18×10^{-6}
11101110110101111 (best)	5.4%	1.16×10^{-6}

Table 2. Theoretical performance of two-seed models when a seed of support 13 and threshold 12 is added to the seed VS-12-13 from Table 1.

4.2 DNA Seeds for Coding Alignments

We conducted further experiments on a data set consisting of alignments of homologous coding regions from human and fruit fly. The initial set contained 339 protein alignments. One protein alignment can yield several DNA alignments (or *fragments*) because the coding regions for each protein can be interrupted by non-coding introns. The final data set contained 972 un-

Seed Vector	T	Support	Name	Actual False Negative	Predicted False Negative	Predicted False Positive
1111111111	10	10	BLAST-10	56.7%	71.4%	9.54×10^{-7}
1101100011011011	10	10	CPH-10	14.0%	32.8%	9.54×10^{-7}
11011011000011011001011011	13	15	CVS-13-15	4.0%	19.1%	9.23×10^{-7}
11011012000012011001011011	15	15	CVS2-15-15	4.9%	21.1%	7.0×10^{-7}
11011011000011011000011011	13	14	CVS-13-14	13.1%	33.1%	1.6×10^{-7}
11011012000011011000011012	15	14	CVS2-15-14	13.8%	34.4%	1.4×10^{-7}
12012012000012012001012012	20	15	CVS2-20-15	9.3%	34.6	2.7×10^{-7}
11111111111	10	11	BLAT-10-11	16.2%	32.4%	7.9×10^{-6}
111111111	9	9	BLAST-9	46.3%	60.2%	3.82×10^{-6}
11001011000011011	9	9	CPH-9	8.14%	24.0%	3.82×10^{-6}
11011000011000011011011	11	12	CVS-11-12	5.2%	19.1%	2.2×10^{-6}
12011000011000012011011	13	12	CVS2-13-12	5.6%	20.5%	2.2×10^{-6}
12022012000012 (two hits)	12	8	CVS2-12-8	6.0%	29.3%	2.3×10^{-6}

Table 3. Theoretical and actual performance of various coding detection seeds. BLAST-X – unspaced seeds of support X; BLAT-X-Y – unspaced seeds with allowed mismatches; CPH-X – spaced seeds of support X optimized in coding-aware model; CVS-X-Y – vector seeds (allowing spaced seeds and mismatches) optimized in coding-aware model; CVS2-X-Y – same as CVS, except values from $\{0, 1, 2\}$ are allowed.

gapped fragments in the training set and 810 gapped fragments in the testing set, after discarding weak and short fragments. A detailed description of the data set can be found elsewhere [4] and the data set can be obtained at `http://www.bioinformatics.uwaterloo.ca/supplements/03seeds`. We model aligned coding regions by the three-periodic model described earlier. We used the training set to estimate the probability that a codon triplet has a given alignment pattern and the alignment length distribution.

First we have investigated 1372 binary vector seeds (v, T) with supports $s \in \{10, \ldots, 15\}$ and $T \geq s - 2$. The seeds we have investigated have codon structure: they can be divided into triplets, where each triplet is either $(0, 1, 0)$, $(1, 1, 0)$ or $(0, 0, 0)$. In real alignments, the second codon position is most likely to be conserved and the third often varies.

We compared the theoretical performance of the codon-aware spaced seeds versus their performance on our test set. Here, we also found quite striking advantages of vector seeds over seeds that are unspaced or that do not allow error. The model is a good predictor of the performance of a seed, though our seeds do better than predicted, as the model is not aware of highly conserved parts of alignments. Results for some interesting seeds are shown in Table 3.

We then chose the best seeds allowing mismatches, at each support/threshold combination (the seeds denoted by CVS-X-Y in Table 3), and explored the effect of fixing the middle positions of some of their codons, by setting the corresponding position in the seed vector to two and raising the threshold by one. Results for this experiment are shown in Table 3. It shows the performance of a collec-

Property	N	mean	std. dev.
Number of fragments	566	5.80	4.23
Length of longest fragment	566	59.6	20.4
Length of other fragments	2718	20.4	13.1

Table 4. Parameters characterizing the gamma distributions that approximate the properties of protein alignments from our data set.

tion of BLAST and BLAT seeds, optimized spaced seeds, and optimized vector seeds, chosen for sensitivity to this model.

This experiment highlights the expressive richness of vector seeds. The seed CVS2-15-15 in Table 3 matches fully 95% of alignments. This is comparable to the sensitivity of the BLAST-7 seed, whose false positive rate is ninety times higher. It is also preferred over the BLAT-10-11 seed for both sensitivity and specificity, and over the CPH-10 seed, which has comparable specificity.

Lastly, we note that the seeds we examined are also overwhelmingly preferred for two-hit study as well. While they are long enough that they reduce sensitivity below 65% in two-hit models still better than the BLAST-9 seed, they also admit tens of thousands of times fewer false positives. One can also use shorter seeds for use with two hits, which still allows for simpler hash table structures. The last seed in Table 3 is especially appealing; the support for this seed is just eight, making for very simple hash table structures, while the performance is comparable to one hit to the longer seed CVS2-13-12 in the table.

4.3 Seeds for Protein Alignments

We also studied the use of vector seeds for protein alignments. Our data set consisted of randomly chosen BLASTP 2.0.2 alignments [2] of pairs of human sequences, taken from 8654 human proteins in the SWISSPROT database [3], release 40.38. We chose 566 alignments with score between 50 and 75 bits using the BLOSUM62 scoring matrix, with a requirement that there are at most eight alignments from each collection of genes connected by BLASTP hits. (This restriction avoids choosing too many matches from one family of very common proteins; without it, the majority of alignments would be from only one family.)

To train our probabilistic model we estimated the distribution of the number of ungapped fragments, the length of the longest ungapped fragments, and the length of all other fragments as gamma distributions from this set of alignments. These parameters are summarized in Table 4. Other parameters of the probability distribution were obtained directly from BLOSUM62 matrix.

We investigated 237 seeds of vector length at most six with between three and five ones and the remaining positions all zero. We chose values of T between 11 and 18 for seeds with 3 ones, between 12 and 22 for seeds with 4 ones, and between 15 and 25 for seeds with 5 ones. There is an observational bias in our experiments since the test set consists of alignments found by two-hit BLASTP. This guarantees that the two-hit BLASTP seed will match all alignments!

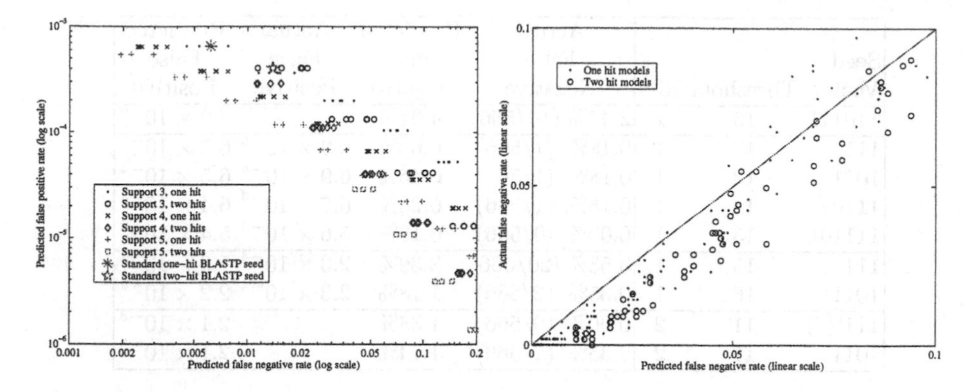

Fig. 2. Theoretical and actual performance of seeds for protein alignments. The left graph shows the predicted false positive rate and false negative rate for several seeds of different supports and threshold. The commonly used one-hit and two-hit BLASTP seeds are shown in red; both are dominated by vector seeds with the same support and by vector seeds with different support. The right graph compares the actual performance of seeds for protein alignments with their theoretical performance. The model does a good job of predicting the sensitivity of seeds, though for the seeds with highest sensitivity, the accuracy is worse. This is due to observation bias (alignments were found by BLASTP), and because the data set is of size 566.

The advantage of spaced seeds over unspaced seeds is not as dramatic as for nucleotide matches, as seen in Figure 2 and Table 5. This is because BLASTP seed matches for proteins are more independent than BLASTN hits are for nucleotides. For example, the probability of a pair of amino acids scoring at least +5 is only 0.22, in homologous protein sequence, while the probability of a 1 is 0.7 in the PatternHunter model. Thus, for proteins, immediately following a hit, there is a lower probability of another hit.

Also, a disadvantage of spaced seeds is that they offer fewer positions for a possible match; when ungapped fragments are as short as they are for our protein alignments (with fragments of length 20 being reasonably common), this reduction affects seed sensitivity.

We also performed an experiment to verify that the false positive rates predicted were close to those found in reality. We chose 100 proteins at random and built a hash table for each one-hit seed in Table 5. We then counted the number of seed hits when we chose 100 other random proteins. For all seeds, the false positive rate was within 10% of what was predicted by the background model.

Still, there is some advantage to the use of spaced seeds. The spaced seed $((1,1,1,0,1),14)$ offers a theoretical advantage of 61% fewer false positives over the standard BLAST seed,$((1,1,1),13)$, with slightly lower false negative rate; for that matter, the obvious spaced seed of support 3, $((1,0,1,1),13)$, is also preferred over the standard BLAST seed, with 27% fewer missed alignments (though the difference is small).

Seed Vector	Threshold	Hits	Actual False Negative	Predicted False Negative	Actual False Positive	Predicted False Positive
110111	15	2	2.47% (14/566)	4.04%		2.9×10^{-5}
111	13	1	0.18% (1/566)	0.62%	6.9×10^{-4}	6.5×10^{-4}
1011	13	1	0.18% (1/566)	0.45%	6.9×10^{-4}	6.5×10^{-4}
11101	14	1	0.18% (1/566)	0.24%	6.7×10^{-4}	6.4×10^{-4}
111101	15	1	0.00% (0/566)	0.19%	5.6×10^{-4}	5.4×10^{-4}
111	15	1	3.53% (20/566)	3.39%	2.0×10^{-4}	1.9×10^{-4}
10111	16	1	0.35% (2/566)	1.18%	2.3×10^{-4}	2.2×10^{-4}
111 (*)	11	2	0.00% (0/566)	1.38%		2.1×10^{-3}
1011	11	2	0.35% (2/566)	1.14%		2.1×10^{-3}

Table 5. Theoretical and actual performance of various protein seeds. The unspaced seeds are commonly used BLAST seeds, which are potentially improved upon by the spaced seeds listed after them. We also show the false positive rates for the one hit models in comparisons between 200 unrelated proteins.

The advantage of two-hit models is not as great as for nucleotides, either. One notable discovery is that seeds depending on more than three positions have greater sensitivity than those depending on just three. Given that there are only 160,000 amino acid 4-mers, for large protein databases, most hash table entries would be populated if one used seeds depending on four positions.

If one is willing to tolerate an error rate comparable to the theoretical performance of the one-hit BLAST seed of threshold 15 (approximately 3.4% false negatives), the one-hit seed $((1,0,1,1,1),16)$, offers comparable false positives, with one third the false negative rate (1.2%). Or, if one is willing to use seeds of support 5, the two-hit seed $((1,1,0,1,1,1),15)$ offers 4.0% false positives and seven times fewer false positives in theory.

A data set not derived from BLASTP. To try to avoid the observation bias we noted before, we aligned sequences that are not reported as aligned by BLASTP, but which are connected by a sequence of alignments that *are* reported by BLAST. If BLAST defines a graph on the set of proteins, these are nonadjacent nodes in the same connected component. We aligned all such pairs from connected components of size at most 30, and discovered 396 alignments in our target score range, again using the BLOSUM62 scoring matrix.

However, all match the one-hit BLASTP seed with threshold 13, and all but three match the two-hit BLASTP seed with threshold 11. Presumably, all were incorrectly thrown out during one of BLASTP's many filtering steps.

In this new data set, we do see some advantage to spaced seeds, especially among less sensitive seeds. For example, the spaced vector seed $((1,1,0,1),16)$ matches 381 (96.2%) of these alignments, while the unspaced seed $((1,1,1),16)$ matches 367 (92.6%) of them. Similarly, the spaced support 4 seed $((1,1,0,1,1),$ $18)$ matches 382 alignments (96.4%), while the unspaced seed $((1,1,1,1),18)$ matches 373 (94.2%). We found similar results for two hit models.

Predictions versus reality. Finally, the simple model of sequences allows good predictions of seed sensitivity. Figure 2 shows the predicted and actual sensitivity of the seeds we studied. The model generally does a good job at predicting the sensitivity of the seeds, except for the most sensitive seeds. This is expected, since our test set includes only 566 alignments. Our set is also subject to observation bias: the alignments we tested were found with BLASTP, which guarantees that they have a hit to the two-hit seed $((1,1,1),11)$!

5 Conclusion

We have presented an extension to previous models for hits that are used in large-scale local alignments. Our vector seeds offer a much wider vocabulary for seed matches than previously studied seeds. For example, they allow certain positions to be more important than others, they allow a fixed number of mismatches in some positions and an arbitrary number in others, and more.

Our extensions to spaced seeds or seeds with constrained mismatches allow substantially improved pairwise local alignment, with vastly improved sensitivity. Especially with the coding sequence nucleotide alignments (possibly one of the most important large-scale local alignment problems), alignment programs using our seeds can reduce their false negative rates by over half, with no change to false positives over BLAT or PatternHunter.

We have also shown an algorithm that allows us to predict the sensitivity and specificity of a vector seed on probabilistic models of true and random alignments. This allows us to choose an optimal seed for a given alignment task. Our algorithm is an extension to the original Keich *et al.* algorithm for predicting seed sensitivity in simple models. Our method is practical as long as the number of alignment sequences matching a given seed is moderate. Extensions to our algorithm allow one to predict the sensitivity of a seed in multihit models, or when using multiple seeds.

We show that spaced seeds can be helpful for proteins as well. The improvements are not as dramatic as for nucleotides, mostly because the seeds themselves are so short, yet they are still useful, and if one is willing to allow a moderate false positive rate, spaced seeds are strongly preferred over unspaced seeds. We also show the contexts under which two-hit models are preferred over one-hit models for proteins.

Our results offer substantial improvement over the current state of the art, with minimal change required in coding.

Future work. Finally, we discuss a few extensions which we have only begun to consider. In PSI-BLAST [2], alignment phases after the first are based on position-specific scoring matrices, which model the probabilities expected in sequence matching a profile. Much as with the standard BLASTP model of aligning sequences that comes from the scoring matrix, here as well, one may desire a seed that has a higher probability of matching the sequence than the usual $(1^k, T)$ consecutive seed. It is impractical to compute the probabilities for thousands of seeds for each profile, yet it is quite reasonable to compute the match rates for each of a small set of diverse good seeds, and use the best of these seeds in that round. (One may also compute predicted false positive rates, using the standard background probabilities with the new scoring matrices.)

We are also interested in whether the representation of vector seed hits as half-spaces in a lattice can help in optimizing them. Given a certain length, support and threshold, is it possible to find the best seed, even for a simple position-independent model, without using essentially exhaustive search? While the exhaustive search is possible for small seed families, for the vector seed models, this becomes absurd as the families of possible seeds grows to the millions. Clearly, one could use heuristic methods. However, sometimes, there is a large difference between the best seed and seeds at the 99th percentile, so one would want a very good heuristic.

Finally, we demonstrated that the vector seeds are not universally expressive. For example, the is no way in the vector seed model to require that three of the four codons are match in the first two positions each. Is it possible to extend vector seeds so that they are more expressive?

Acknowledgments. This work has been supported by a grant from the Human Science Frontier Program, grant from the Natural Sciences and Engineering Research Council of Canada, and by an Ontario Graduate Scholarship. We would like to thank anonymous referees for many constructive comments on the paper.

References

[1] S. F. Altschul, W. Gish, W. Miller, E. W. Myers, and D. J. Lipman. Basic local alignment search tool. *Journal of Molecular Biology*, 215(3):403–410, 1990.

[2] S. F. Altschul, T. L. Madden, A. A. Schaffer, J. Zhang, Z. Zhang, W. Miller, and D. J. Lipman. Gapped BLAST and PSI-BLAST: a new generation of protein database search programs. *Nucleic Acids Research*, 25(17):3389–3392, 1997.

[3] A. Bairoch and R. Apweiler. The SWISS-PROT protein sequence database and its supplement TrEMBL in 2000. *Nucleic Acids Research*, 28(1):45–48, 2000.

[4] B. Brejová, D. Brown, and T. Vinař. Optimal spaced seeds for hidden Markov models, with application to homologous coding regions. In *Proceedings of the 14th Annual Symposium on Combinatorial Pattern Matching (CPM)*, 2003. To appear.

[5] J. Buhler, U. Keich, and Y. Sun. Designing seeds for similarity search in genomic dna. In *Proceedings of the 7th Annual International Conference on Computational Biology (RECOMB)*, pages 67–75, 2003.

[6] S. Henikoff and J. G. Henikoff. Amino acid substitution matrices from protein blocks. *Proceedings of the National Academy of Sciences of the United States of America*, 89(22):10915–10919, 1992.

[7] U. Keich, M. Li, B. Ma, and J. Tromp. On spaced seeds. Unpublished.

[8] W. J. Kent. BLAT–the BLAST-like alignment tool. *Genome Research*, 12(4):656–664, 2002.

[9] I. Korf, P. Flicek, D. Duan, and M. R. Brent. Integrating genomic homology into gene structure prediction. *Bioinformatics*, 17 Suppl 1:S140–8, 2001.

[10] B. Ma, J. Tromp, and M. Li. PatternHunter: faster and more sensitive homology search. *Bioinformatics*, 18(3):440–445, March 2002.

A Stochastic Approach to Count RNA Molecules Using DNA Sequencing Methods

Boris Hollas and Rainer Schuler

Abteilung Theoretische Informatik, Universität Ulm, D-89069 Ulm, Germany
{schuler,hollas}@informatik.uni-ulm.de

Abstract. Estimating the number of congeneric mRNA molecules in a preparation is an essential task in DNA technology and biochemistry. However, DNA sequencing methods can only distinguish molecules with different sequences or of different lengths. Recently, it was shown that it is possible to combine RNA molecules with short DNA molecules (tags) such that, with high probability, any two RNA molecules are combined with tags having different sequences. For this technique, we propose a statistical estimator and a confidence interval to determine the number of mRNA molecules in a preparation.
In a second approach, the mRNA molecules are lengthened by attaching a random number of linker oligonucleotides. The original number of mRNA molecules is then determined by the number of different lengths obtained from the experiment. We also give estimator and confidence interval for this method.
Both methods can be implemented using recent as well as established methods from DNA technology. The methods can also be applied to a larger number of molecules without the need to exhaust the complete preparation. The computation of estimators and confidence intervals can be accomplished by dynamic programming.

1 Introduction

Using DNA sequencing methods it is possible to determine the number of different nucleic acid molecules in a preparation. However, it remains difficult to estimate the number of molecules (or the concentration) of a single mRNA species. Methods currently used (e.g., measuring uv-absorption, gel electrophoresis with DNA size markers, northern blotting, linear PCR, or chip hybridization) are not sufficiently reliable or may require more RNA than available. If the number of molecules is increased by PCR, the estimation error roughly doubles in each amplification round. In this case, a quantitative analysis of the probe is even more difficult.

Recently, it was shown that it is possible to sequence many DNA molecules with different sequences simultaneously [3,2,10]. Short oligonucleotides (DNA molecules) are attached to the DNA molecules. If the number of different oligonucleotide sequences is large enough (100 times larger than the number of DNA molecules), almost all oligonucleotides that are combined with a DNA molecule

G. Benson and R. Page (Eds.): WABI 2003, LNBI 2812, pp. 55–62, 2003.

will contain a unique sequence. The oligonucleotide sequences which are selected by this process are a random subset of the set of all oligonucleotide sequences. Furthermore, the sequences can be chosen such that the combined molecules can be separated by the oligonucleotide sequences (by hybridization to reverse complementary sequences which are bound to the surface of micro beads). This fact is used in [3], for further DNA sequencing.

As discussed in [7], this technique can be used to determine the number of molecules of a single mRNA species. Molecules with the same sequence of base pairs are combined with oligonucleotides with many different (tag-) sequences. The number of different tag sequences coupled to the mRNA molecules depends on the number of mRNA molecules in the preparation. However, in some cases several mRNA molecules might be coupled with oligonucleotides having the same sequence.

A different technique to determine the number of molecules is proposed in [7]. In this approach molecules of different length are prepared from the RNA molecules. This can be achieved, e.g., by ligation of multimeric linkers which can be used to assemble DNA of different length and sequences. Short linker oligonucleotides are attached to a molecule until ligation is terminated by adding a stop linker. The lenght of a molecule is random and depends on the ratio of stop linkers. Again, in some cases several molecules will have the same length.

In this paper we consider the relationship between the observed number of different tag sequences or molecule lengths and the number of molecules in the preparation.

Uniform Model. In the first approach described above, the combination of molecules with oligonucleotides containing different tag sequences is random. Moreover, each tag sequence is chosen with a certain probability, depending on the length and the concentration of oligonucleotides with that sequence. If all oligonucleotides containing the tag sequences are of the same length and each tag sequence is present in the same quantity then each tag sequence is chosen with the same probability. The maximal number of tag sequences can be quite large, e.g., 10^6 tag sequences can be distinguished by a DNA sequencing method (MPSS) [3]. We describe the outcome of the experiment with the uniform model.

In the *uniform model*, s denotes the number of molecules (we wish to estimate) and T_i denotes the tag sequence of molecule i, $1 \leq i \leq s$. We assume that the tag sequences are independently and uniformly (equiprobably) selected. Furthermore, we assume that all random variables T_i are identically distributed, i.e., that the ratio of the tag sequences does not change.

Geometric Model. In the second approach, DNA molecules are lengthened at random. Linker oligonucleotides and stop linkers are mixed with a certain ratio and added to the DNA molecules. Linker oligonucleotides are attached to the end of the molecule until the lengthening is terminated by adding a stop linker. The probability that the assembly is terminated depends on the concentration of the terminating linkers. Thus, we obtain a random number of different lengths. The maximal number of different lengths that can be distinguished, e.g., by gel electrophoresis, is approximately 1000. For the mathematical modelling, we neglect several experimental factors, e.g., that the addition of stop linkers is length dependent and that the maximum length of the molecules is bounded.

For the *geometric model*, let again s denote the number of molecules. The random variables $L_1, \ldots, L_s$ denote the number of linker oligonucleotides that have been attached to the molecules. We assume that molecules are independently lengthened with termination probability $1-p$, thus we assume that $L_1, \ldots, L_s$ are independent and geometrically distributed. Hence, $L_1, \ldots, L_s$ are geometrically distributed. We denote the maximal number of different lengths that can be distinguished experimentally by $k_{\max}$.

For both models, we derive a maximum-likelihood-estimator and an approximate confidence interval for the unknown number of molecules.

2 Uniform Distribution

Let $T_1, \ldots, T_s$ be random variables that are independent and uniformly distributed on the set of tags $\{1, \ldots, t\}$, that is, all tags are equiprobable, and let

$$N_s = |\{T_1, \ldots, T_s\}|$$

be the (random) number of different tag sequences obtained from an experiment on s sequences. Accordingly, denote by

$$\mathcal{P}_{s,n} = \{(t_1, \ldots, t_s) \mid 1 \leq t_1, \ldots, t_s \leq t \wedge |\{t_1, \ldots, t_s\}| = n\}$$

the set of tags assuming exactly n different values. Then

$$\begin{aligned} P(N_s = n) &= \sum_{(t_1,\ldots,t_s)\in\mathcal{P}_{s,n}} P(T_1 = t_1, \ldots, T_s = t_s) \\ &= (\frac{1}{t})^s S_{s,n} \binom{t}{n} n! \end{aligned}$$

whereby $S_{s,n}$ is the number of ways to partition the set $\{t_1, \ldots, t_s\}$ into n nonempty subsets (*Stirling numbers of the second kind* [6]).

Note that $\{T_1 = T_2 = 5,\ T_3 = 1\}$ and $\{T_1 = 1,\ T_2 = T_3 = 5\}$ are different events, each having the same probability.

To calculate $P(N_s = n)$ we consider the two cases $T_s \in \{T_1, \ldots, T_{s-1}\}$ and $T_s \notin \{T_1, \ldots, T_{s-1}\}$. Since $S_{s,n} = S_{s-1,n-1} + nS_{s-1,n}$ and $\binom{t}{n} = \binom{t}{n-1}\frac{t+1-n}{n}$, we get

$$\begin{aligned} P(N_s = n) &= (\frac{1}{t})^{s-1} S_{s-1,n-1} \binom{t}{n-1} (n-1)! \frac{t+1-n}{t} \\ &\quad + (\frac{1}{t})^{s-1} S_{s-1,n} \binom{t}{n} n! \frac{n}{t} \\ &= P(N_{s-1} = n-1)\frac{t+1-n}{t} + P(N_{s-1} = n)\frac{n}{t} \end{aligned}$$

For the edges ($s = 1$ or $n = 1$) holds

$$P(N_1 = n) = \begin{cases} 1 & n = 1 \\ 0 & n > 1 \end{cases} \quad \text{and} \quad P(N_s = 1) = (\frac{1}{t})^{s-1}$$

Though this recursion can readily be computed, it is much more efficient with regards to time and space to build up an array containing the values $P(N_s = n)$ (dynamic programming [4]). Note that, since necessarily $s \geq n$, only half of the array elements need to be computed. Then, a maximum-likelihood-estimator

$$\hat{s}(n) = \arg \max_{1 \leq s \leq s_{max}} P(N_s = n)$$

for the unknown number of sequences s, given n different tags obtained, can be computed. The upper limit s_{max} must be sufficiently large to ensure that the maximum is really attained.

Apart from the most likely number of sequences $\hat{s}$, we are also interested in the reliability of this result. A $(1 - \alpha)$ confidence interval for s must satisfy

$$\begin{aligned} 1 - \alpha &\leq P(\hat{s} + l \leq s \leq \hat{s} + r) \\ &= P(s - r \leq \hat{s} \leq s - l) \end{aligned}$$

Remember that $\hat{s} = \hat{s}(N_s)$. Thus, we get

$$P(\hat{s}^{-1}(s - r) \leq N_s \leq \hat{s}^{-1}(s - l)) \tag{1}$$

However, N_s depends on the unknown parameter s. Therefore, we construct an approximate confidence interval by replacing s with $\hat{s} = \hat{s}(n)$ in (1) and compute a, b such that

$$P(a \leq N_{\hat{s}} \leq b) \geq 1 - \alpha$$

Then, $\hat{s}(a)$ and $\hat{s}(b)$ are the bounds for this confidence interval.

Figure 1 shows the most probable number of sequences $\hat{s}$ (thick line) together with the bounds for the approximate confidence interval (thin lines). The confidence interval was constructed according to the density of $N_{\hat{s}}$. Thus the confidence interval has minimum length. The chart on the top shows the results for 500 tag sequences, the chart on the bottom for 1000 tag sequences. n denotes the number of different tags that result from the experiment.

3 Geometric Distribution

Let $L_1, \ldots, L_s$ be independent and geometrically distributed random variables, that is $P(L = l) = p^l q$, $l \geq 0$, for $p \in (0, 1)$ and termination probability $q = 1 - p$. Let $N_s = \{L_1, \ldots, L_s\}$ denote the number of different values (numbers) obtained from an experiment on s sequences. Let $p_{s,n,k} := P(N_s = n \wedge \forall i \in \{1, 2, \ldots, s\}\, L_i \leq k)$. For k holds $1 \leq k \leq k_{\max}$, whereby $k_{\max}$ is the maximal number of different linker length detectable. With current technology $k_{\max}$ is approximately 1000.

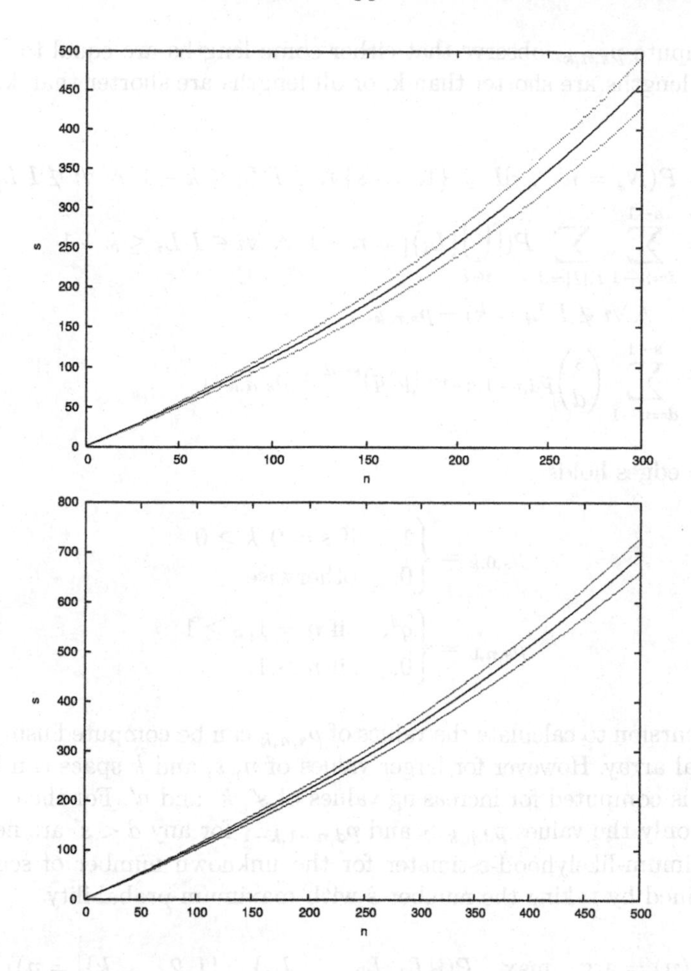

Fig. 1. Uniformly distributed lengths.

For the number $|\{L_1, L_2, \ldots, L_s\} \cap \{1, 2, \ldots, k_{\max}\}|$ of different values obtained from the experiments we get

$$P(|\{L_1, L_2, \ldots, L_s\} \cap \{1, 2, \ldots, k_{\max}\}| = n)$$

$$= \sum_{d=0}^{s} P(\exists I \subseteq \{1, \ldots, s\}, |I| = d \ \wedge$$

$$\forall i \notin I \ L_i > k_{\max} \ \wedge \ \forall i \in I \ L_i \leq k_{\max} \ \wedge \ |\bigcup_{i \in I} \{L_i\}| = n)$$

$$= \sum_{d=0}^{s} \binom{s}{d} P(L_1 > k_{\max})^{s-d} \cdot p_{d,n,k_{\max}}$$

since $P(L_1 \leq k_{\max}) = \sum_{i=0}^{k_{\max}} p^i q = q \frac{1-p^{k_{\max}+1}}{1-p} = 1 - p^{k_{\max}+1}$, this equals

$$= \sum_{d=0}^{s} \binom{s}{d} p^{(k_{\max}+1)(s-d)} \cdot p_{d,n,k_{\max}}$$

To compute $p_{s,n,k}$, observe that either some lengths are equal to k and the remaining lengths are shorter than k, or all lengths are shorter than k.

$$\begin{aligned} p_{s,n,k} &= P(N_s = n \wedge \exists I \subseteq \{1,\ldots,s\} \forall i \in I\; L_i \leq k-1 \wedge \forall i \notin I\; L_i = k) \\ &= \sum_{d=n-1}^{s-1} \sum_{I,|I|=d} P(|\bigcup_{i \in I}\{L_i\}| = n-1 \wedge \forall i \in I\; L_i \leq k-1 \\ &\qquad \wedge \forall i \notin I\; L_i = k) + p_{s,n,k-1} \\ &= \sum_{d=n-1}^{s-1} \binom{s}{d} p_{d,n-1,k-1} \cdot (p^k q)^{s-d} + p_{s,n,k-1} \end{aligned}$$

For the edges holds

$$p_{s,0,k} = \begin{cases} 1, & \text{if } s = 0, k \geq 0 \\ 0, & \text{otherwise.} \end{cases}$$

$$p_{s,n,0} = \begin{cases} q^s, & \text{if } n = 1, s \geq 1 \\ 0, & \text{if } n > 1. \end{cases}$$

The recursion to calculate the values of $p_{s,n,k}$ can be computed using a three-dimensional array. However for larger values of n, s, and k space can be saved. The array is computed for increasing values of s', k' and n'. For the calculation of $p_{s',n',k'}$ only the values $p_{d,n,k-1}$ and $p_{d,n-1,k-1}$ for any $d < s'$ are needed.

A maximum-likelyhood-estimater for the unknown number of sequences s can be defined by taking the number $\hat{s}$ with maximum probability.

$$\hat{s}(n) = \arg \max_{1 \leq s \leq s_{\max}} P(|\{L_1, L_2, \ldots, L_s\} \cap \{1, 2, \ldots, k\}| = n))$$

The upper limit s_{max} must be sufficiently large to ensure that the maximum is really attained.

Using the approach from section 2 we construct an approximate confidence interval by replacing s with $\hat{s} = \hat{s}(n)$ and compute an interval a, b such that

$$P(a \leq |\{L_1, L_2, \ldots, L_{\hat{s}}\} \cap \{1, 2, \ldots, k\}| \leq b) \geq 1 - \alpha$$

Then, $\hat{s}(a)$ and $\hat{s}(b)$ are the bounds for this confidence interval.

Figure 2 shows the most probable number of sequences $\hat{s}$ (thick line) together with the bounds for the approximate confidence interval (thin lines). The chart on the top shows the results for termination probability $1 - p = 1/50$ and maximum length $k_{\max} = 800$, the chart on the bottom for $1 - p = 1/80$ and $k_{\max} = 1100$. n denotes the number of different tags that result from the experiment. For these numbers the probability that all lengths are less than $k_{\max}$ is at least 0.99998 and 0.9998, respectively.

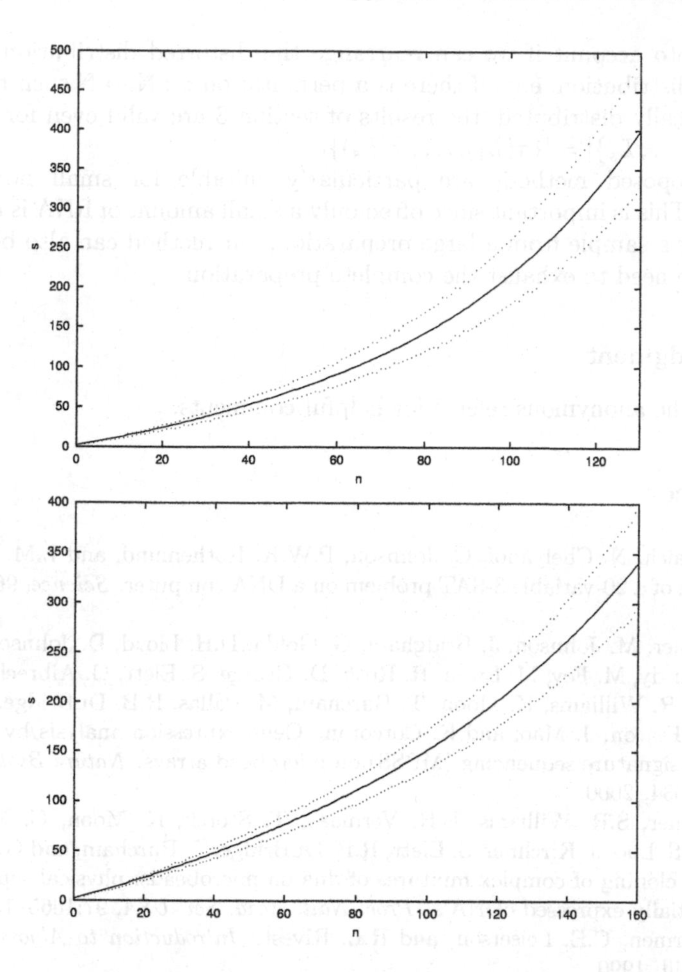

Fig. 2. Geometrically distributed lengths.

4 Discussion

The technology to obtain the data is already established. Oligonucleotides which contain the tag sequences can be synthesized (by repeatedly adding 12 different sequences of length 4) [3]. Detecting the number of different tag sequences using the MPSS method from [3] is only feasible in laboratories with sophisticated equipment. However, smaller number of tag sequences can be separated using DNA chips.

Similarly, the technique to assemble DNA sequences of different length is well established in DNA sequence analysis [9]. We note however, that the exact mixing ratio of terminating linkers has to be obtained experimentally. Furthermore, the assembly of multimeric linkers, i.e., the addition of linker oligonucleotides and stop linkers is dependent on the length of the molecules. That is, the lengths of the molecules are not truly geometrically distributed. This can be taken into account if we can rearrange the distorted distribution P_L to a geometric distribution: e.g., if there is a permutation $\pi : \mathbb{N} \to \mathbb{N}$ such that $\pi(L)$ is geometrically distributed, the results of section 3 are valid even for P_L since $N_s = |\{L_1, \ldots, L_s\}| = |\{\pi(L_1), \ldots, \pi(L_s)\}|$.

The proposed methods are particularly suitable for small numbers of molecules. This is important since often only a small amount of RNA is available. By drawing a sample from a large preparation, our method can also be applied without the need to exhaust the complete preparation.

Acknowledgment

We thank the anonymous referee for helpful comments.

References

1. R.S. Braich, N. Chelyapof, C. Johnson, P.W.K. Rothemund, and L.M. Adleman. Solution of a 20-variable 3-SAT problem on a DNA computer. *Science*, 96:478–479, 2002.
2. S. Brenner, M. Johnson, J. Bridgham, G. Golda, D.H. Lloyd, D. Johnson, S. Luo, S. McCurdy, M. Foy, M. Ewan, R. Roth, D. George, S. Eletr, G. Albrecht, E. Vermaas, S.R. Williams, K. Moon, T. Burcham, M. Pallas, R.B. DuBridge, J. Kirchner, K. Fearon, J. Mao, and K. Corcoran. Gene expression analysis by massively parallel signature sequencing (MPSS) on microbead arrays. *Nature Biotechnology*, 18:630–634, 2000.
3. S. Brenner, S.R. Williams, E.H. Vermaas, T. Storck, K. Moon, C. McCollum, J. Mao, S. Luo, J. Kirchner, S. Eletr, R.B. DuBridge, T. Burcham, and G. Albrecht. In vitro cloning of complex mixtures of dna on microbeads: physical separation of differentially expressed cDNA's. *Proc. Natl. Acad. Sci. USA*, 97:1665–1670, 2000.
4. T.H Cormen, C.E. Leiserson, and R.L. Rivest. *Introduction to Algorithms.* Mc Graw Hill, 1990.
5. W. Feller. *An Introduction to Probability Theory and Its Applications*, volume II. Wiley, 1971.
6. R. Graham, D. Knuth, and O. Patashnik. *Concrete Mathematics.* Addison-Wesley, 1998.
7. H.Hug and R. Schuler. Measurement of the number of molecules of a single mRNA species in a complex mRNA preparation. *Journal of Theoretical Biology*, 221(4): 615 - 624, 2003.
8. William H. Press, Brian P. Flannery, Saul A. Teukolsky, and William T. Vetterling. *Numerical Recipes in C.* Cambridge University Pree, 2nd edition, 1999.
9. F. Sanger, S. Nicklen, and A.R. Coulson. DNA sequencing with chain-terminating inhibitors. *Proc. Natl. Acad. Sci. USA*, 74:5463–5467, 1977.
10. S. Tyagi. Taking consensus of mRNA populations with microbeads. *S. Nat. Biotechnol*, 18:597–598, 2000.
11. S. Wilks. *Mathematical Statistics.* Wiley, 2nd edition, 1963.

A Method to Detect Gene Structure and Alternative Splice Sites by Agreeing ESTs to a Genomic Sequence

Paola Bonizzoni[1], Graziano Pesole[2], and Raffaella Rizzi[1]

[1] Dipartimento di Informatica Sistemistica e Comunicazione
Università degli Studi di Milano - Bicocca
Via Bicocca degli Arcimboldi 8,
20126 Milano - Italy
bonizzoni @disco.unimib.it,raffaella.rizzi@unimib.it

[2] Dipartimento di Scienze Biomolecolari e Biotecnologie,
Università degli Studi di Milano
Via Celoria 26,
20135 Milano - Italy
graziano.pesole@unimi.it

Abstract. In this paper we propose a new approach to the problem of predicting constitutive and alternative splicing sites by defining it as an optimization problem (MEFC). Then, we develop an algorithm to detect splicing sites based on the idea of using a combined analysis of a set of ESTs alignments to a genomic sequence instead of considering single EST alignments. In this way we require that all ESTs alignments must agree, i.e. are *compatible* to a plausible exon-intron structure of the genomic sequence. Indeed, we show that a progressive and independent alignment of ESTs may produce unsupported splicing forms. Our method has been implemented and experimental results show that it predicts alternative splicings with high accuracy and in a small amount of time. More precisely, compared to published splicing data, the method confirms validated data while in many cases it provides novel splicing forms supported by several ESTs alignments.

1 Introduction

Eukaryotic genes consists of coding and non coding regions of DNA, called exons and introns, respectively. During transcription, each intron is spliced out of pre mRNA, while the informative portions of the gene, the exons, are joined together into a mRNA sequence. This process is known as splicing, while the junction between exons and introns on the genomic sequence are called *splicing sites*. Alternative splicing is recognized as an important mechanism used to expand protein diversity in organisms, since it can generate from a single gene multiple transcripts encoding different proteins [3]. Indeed, different combinations of donor and acceptor splice sites can be used producing several transcript

G. Benson and R. Page (Eds.): WABI 2003, LNBI 2812, pp. 63–77, 2003.

isoforms from the same gene. Recent large scale genomic studies show that alternative splicing occurs in 30-60% of human genes and it seems that this percentage is even probably higher [3]: this discover suggests that alternative splicing may play an important role in producing the functional complexity of human genome. This hypothesis is supported by recent results that show that alternative splicing forms are specific to tissues thus regulating the functional characteristics of proteins [11]. Other studies show that changes in the alternative splicing common forms are related to human diseases [7]. For these reasons there is a growing interest in discovering alternative splicing forms in human genes, as well as in collecting splicing data. Online databases to access alternative splicing information are now available [2] to biologists and are quite important in the study of alternative splicing. In fact, little is known about the functional significance of splicing forms mainly when they are tissue-specific [2]. Motivated by the growing interest of biologists in understanding alternative splicing role in functional regulation, computer science researchers are developing tools to predict alternative splicing forms [9,5] or to support the analysis of such a mechanism by providing representation of splicing forms [4]. A certain number of programs to produce alternative splicing have been proposed: they predict the exon-intron structure of the genomic sequence by a progressive alignment of ESTs to a genomic sequence and enumerate the alternative splicing forms derived by the different introns determined by each single EST alignment. The execution time and the reliability of the solution produced differentiate these programs. The main goal of a software tool for predicting alternative splicings is to obtain a correct alignment to a genomic sequence of a large set of ESTs data in a relative small amount of time. Actually, several matters may complicate this goal: let us discuss some of them below. An expressed sequence tag EST should be identical to the genome sequence from which it derives, but the alignment of an EST to the original sequence is not univocally determined because of these facts. First of all, EST sequences do not exactly match to the genomic sequence because of sequencing errors (deletions, insertions and even mutations) and the fragmentary nature of ESTs. By quoting [9], "furthermore, there are often small repeated sequences near exon boundaries, so that the exact exon boundaries cannot be determined unambiguously from alignment information alone". The detection of the exact location of exon boundaries is complicated by other elements: the presence of quite large intron regions, the biological phenomenon of duplications or the occurrence of paralogous genes. Indeed, all these elements may increase the number of possible alignments of ESTs portions to the genomic sequence. Some work has been done to speed up the process of aligning a single EST sequence to a genomic sequence [5]. But, as observed in [10], few work, if none, has been done on the combined analysis of all ESTs alignments to reconstruct the exon-intron structure of the genomic sequence. In fact, software tools for gene prediction are usually based on a progressive and independent alignment of each EST for the gene. But this approach has some main drawbacks w.r.t. the above mentioned issues. An independent alignment of ESTs may not consider all possible exon-intron combinations due to different ESTs alignments, thus producing an incorrect alignment even when the predicted exons sequences may be correct, but their location is not. Example 1 of section 2 illustrates this situation, by showing that ESTs alignments in a cluster for a putative gene cannot be performed independently. On the other hand, tools reporting all possible exon-intron combinations (i.e. all possible alignments) produce a redundancy in the information (sometimes exponential) and require a high execution time to discover compatible alignments [4].

In this paper, we face the above problem by proposing a new approach to predict the exon-intron structure of a genomic sequence, based on the idea of using a combined analysis of all ESTs alignments while producing the alignment of a single new EST to the genomic sequence. The combined analysis aims to produce only alignments that make all ESTs agree or *compatible* to a common plausible exon-intron structure. In this sense, our method is different from the one suggested in [10], where a combined analysis of ESTs alignments is done after all ESTs alignments have been produced. Our approach is based on a new formalization of the problem of detecting splicing sites as an optimization problem, (MEFC), since an optimization criteria may takes into account the specific critical matter illustrated in example 1. We prove that the MEFC problem is NP-hard in general [6]. Then, we provide an efficient algorithm to solve the MEFC problem when the number of occurrences of repeated sequences is bounded. Based on this algorithm, the software tool ASPic (Alternative, Splicing, PredICtion) has been designed and implemented. It must be pointed out that ASPic software is mainly designed to process an entire set or cluster of ESTs, since the agreement of multiple ESTs alignments w.r.t. splicing forms is fundamental to validate them.

Our method calculate locations of splicing sites with high sensitivity and accuracy, while retaining computational efficiency close to the one of recent software tools such as [5]. Indeed, the method could require more computational time only for particular instances.

2 Predicting Alternative Splicing from ESTs: Preliminaries

Our approach to alternative splicing prediction is based on the idea of discovering the exon-intron structure of a genomic sequence by finding alignments for a set of ESTs that agree to each other or are compatible w.r.t. such a structure.

In this section we formalize these notions and define the problem of predicting the exon-intron structure from a set of ESTs (cluster) and a genomic sequence.

Let us first consider the problem in absence of errors, that is in the case ESTs sequences do not contain errors. In the following we use G to denote a genomic sequence, that is a sequence over alphabet $\Sigma = \{a, c, g, t\} \cup \{n\}$, with n a symbol used to denote a missing data. Given a genomic sequence G, an *exon-factorization of* G of length n, denoted G_E is a sequence $< f_1, \cdots, f_n >$ of n substrings f_i of G, called *exons*, such that $G = I_1 f_1 I_2 f_2 I_3 \cdots I_n f_n I_{n+1}$. Each substring I_j, for $1 \leq j \leq n+1$ is called *intron*.

Definition 1 (EST factorization). *Let G be a genomic sequence. A sequence S over Σ, with $S = s_1 s_2 \cdots s_k$, is an EST sequence* agreeing with *G iff there is an exon-factorization $G_E =< f_1, \cdots, f_n >$ and a sequence of k indices $1 \leq i_1 < i_2 < i_3 < \cdots < i_k \leq n$ such that:*

(1) $s_t = f_{i_t}$, for each $1 < t < k$, and

(2) s_1 is a suffix of f_{i_1} and s_k is a prefix of f_{i_k}.

The sequence $exon(S) =< s_1, s_2, \cdots, s_k >$ is called a factorization of S compatible to *G_E.*

Given the above definition 1, we say that each factor s_t *aligns to* f_{i_t} and hence $< f_{i_1}, f_{i_2}, \cdots, f_{i_k} >$ is an alignment of EST S against G. Moreover, factors s_1 and s_k are called *external factors*, while $s_2, \cdots, s_{k-1}$ are *internal* factors. Whenever, errors may be present in ESTs sequences, assuming that the alignment of factor s_t against f_{i_t} has errors measured by the edit distance $edit(s_t, f_{i_t})$, bounded by a value *ebound*, then in the above definition 1 statements (1) and (2) must be replaced by the following ones:

1. $edit(s_t, f_{i_t}) \leq ebound$ for each $1 < t < k$, and
2. $edit(s_1, f'_{i_1}) \leq ebound$, and $edit(s_k, f'_{i_k}) \leq ebound$, where f'_{i_1} is a suffix of f_{i_1} and f'_{i_k} is a prefix of f_{i_k}.

The exon-factorization of a genomic sequence G could not be unique, due to different possible intron regions determined by a pattern of alternative splicing that potentially may occur in nature. This pattern, illustrated in figure 1, is known as *competing 3' or 5'* and consists in the fact that exons have multiple 3' or 5' splicing sites that are alternatively used when ESTs are derived from the genomic sequence. Formally, a competing 5' (or 3') site occurs for an exon f_i of G whenever there exists an exon f_j such that is a prefix (or suffix) of f_i (see figure 1). Clearly, competing $3'$ and $5'$ could occur together for the same exon f_j, in which case there exists an exon f_i that is a proper factor (substring) of f_j.

We model this situation by extending the previous notion of exon-factorization of a genomic sequence to include such patterns of alternative splicing, or *splicing variants.*

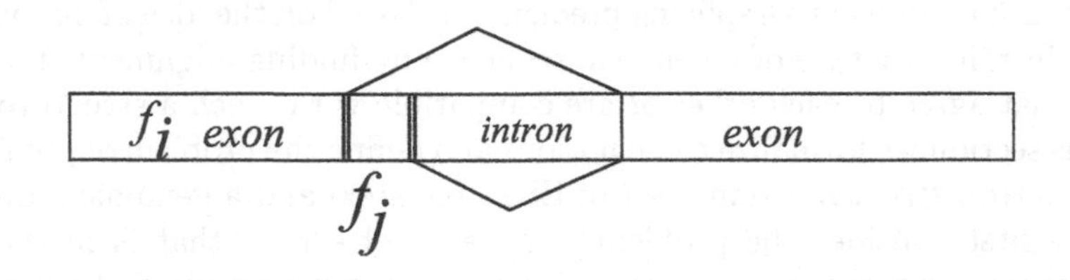

Fig. 1. Exons f_j and f_i derive from competing 5' sites, that is two intron regions are possible, one producing exon f_j and a larger region producing exon f_i

Definition 2 (splicing variants). *Let* $G_E =< f_1, \cdots, f_n >$ *be an exon-factorization of a genomic sequence* G. *A* splicing variant *of* G_E *is an exon-factorization* $G'_E =< f'_1, \cdots, f'_n >$ *such that for each* i, $1 \leq i \leq n$, f'_i *is a substring (prefix, suffix or factor) of* f_i.

Then, we say that a set E of ESTs factorizations is *compatible* with G, if there exists an exon-factorization G_E of G such that each sequence in E is compatible to G_E or to some splicing variant of G_E. We say also that E is compatible to G by G_E.

Problem I: ESTs factorizations compatible to G (EFC)
Input: G a genomic sequence, $\mathcal{S} = \{S_1, S_2, \cdots, S_m\}$, a set of ESTs sequences.
Output: an exon-factorization G_E of G and a set E of $ESTs$ factorizations for S such that E is compatible with G by G_E, whenever such G_E exists.

Since, the above problem could not admit a solution, clearly, we may be interested in a maximum subset of sequences for which problem EFC has a solution.

Problem IA: Maximum set of ESTs factorizations compatible to G (SEFC)
Input: G a genomic sequence, $\mathcal{S} = \{S_1, S_2, \cdots, S_m\}$, a set of ESTs sequences.
Output: an exon-factorization G_E of G and a set E of $ESTs$ factorizations for a maximum size subset $S' \subseteq S$ such that E is compatible with G by G_E.

The following optimization problem derives by the need of finding a correct ESTs' alignment whenever portions of an EST sequence can be aligned in more than one place along the genomic sequence because of sequence repeats. Indeed, it is reasonable to assume that a correct exon-intron structure for a genomic sequence is the one that allows to align all ESTs sequences by forbidding that common EST's regions may be aligned to distinct places on the genomic sequence. Clearly, this requirement can be satisfied by minimizing the total number of exons of an exon-factorization of a genomic sequence. This situation is illustrated by example 1.

Problem II: Minimum Exon ESTs factorizations compatible to G (MEFC)
Input: G genomic sequence, $\mathcal{S} = \{S_1, S_2, \cdots, S_m\}$, set of ESTs sequences agreeing with G.
Output: a minimum length exon-factorization G_E of G and a set E of $ESTs$ factorizations that is compatible with G.

The example 1 illustrates a situation that can occur frequently on real data, since many ESTs admit factorizations with external factors that are as long as 10 to 30 bp, thus such factors may have multiple possible alignments (mainly when ESTs contain errors) to the genomic sequence.

Example 1. Assume that S_1, S_2 and S_3 are sequences in S agreeing with a genomic sequence, such that $S_1 = A_1D_1C$, $S_2 = ABD_1$ and $S_3 = A_2DC$, where $A = A'A_1 = A''A_2$, $D = D'D_1$ and $C = C'C_1$. Moreover, let $G = I_1A_1I_2AI_3A_2I_4BI_5D_1I_6C_1I_7DI_8CI_9$.

Then, the solution of the MEFC problem is given by the exon-factorization $G_E = < A, B, D, C >$, since there are factorizations of each S_i that are compatible to G_E or to splicing variants of G_E, as for example $< A_1, D_1, C >$ is compatible to $< A_1, B, D_1, C >$ which is a splicing variant of G_E. Note that an approach to the exon-intron prediction problem that aligns each s_i independently to G, may result in an exon-factorization with more exons than the MEFC solution, for example $< A_1, A, B, D_1, D, C >$ is another possible exon-factorization for S.

We prove that the problem MEFC is NP-hard (see appendix). Thus, we propose an heuristic to solve MEFC that is implemented in our ASPic algorithm.

3 The Algorithm

In this section we describe the general ideas of the heuristic for solving the MEFC problem. Many details will be omitted because of limits of space. The general structure of the ASPic method is an iterative processing of each EST in the instance $\mathcal{S} = \{S_1, \cdots, S_n\}$, where at each iterative step i a set E_i of factorizations for $\{S_1, \cdots, S_i\}$ compatible with a partial exon-factorization of G is computed. The generic step of the iteration consists in finding the next factor s_j of a partial EST factorization $< s_1, \cdots, s_{j-1} >$ and its alignment, or corresponding exon, along the genomic sequence. The new computed exon is compared with all previously computed exons, since each EST factorization must be compatible to the same final exon-factorization. Since, the MEFC problems requires to minimize the total number of exons of the genomic sequence, a set E of ESTs factorizations is compatible to an exon-factorization in which the total number of computed exons is bounded w.r.t. a specific upper bound, MAXEXONS. Whenever, the computed exon gives a factorization not compatible, the current factorization or previous ESTs factorizations are recomputed in order to get a compatible set E_i. Such a process is done by backtracking, that is allowing the iteration to restart from a previous computed EST factor, and by trying alternative alignments to the genomic sequence for finding an EST factor.

Example 2. Consider the set S of ESTs and the genomic sequence G of example 1 and assume that MAXEXONS is 4. Assume that $< A_1, A, D_1, C >$ is the exon factorization computed at step i and the partial factorization $< A >$ for S_2 is computed. Now, since the next factor B of S_2 gives a total number of 5 exons, then backtracking must be used to recompute a factorization of S_1 that produces the exon-factorization $< A, B, D_1, C >$.

Before giving more details on the general algorithm, let us illustrate the general ideas of the basic procedure used to align a single EST.

3.1 ESTs Alignment: The Basic Algorithm

Let us first introduce some notation used to describe the algorithm.

Given $G = g_1 \cdots g_M$ the genomic sequence and $S = a_1 \cdots a_N$ an EST agreeing with G, then $G[i,j]$ denotes the substring $g_i \cdots g_j$ of G, while $S[i,j]$ denotes the substring $a_i \cdots a_j$ of S.

Given $x = S[i,j]$, by $\mathbf{ALIGN[x,G]}$ we denote the set of all substrings of G to which x can be aligned within an error specified by *ebound*: then each element in $\mathbf{ALIGN[x,G]}$ is called *a genomic-copy of x*. In other words $\mathbf{ALIGN[x,G]}$ represents all possible alignments of string x to the genomic sequence. Given a substring y of S (or G), let us denote by $Left_S(y)$ and $Right_S(y)$ (or $Left_G(y)$ and $Right_G(y)$, respectively), the initial and final position on S (or G) of y.

The Initialization

First of all, the algorithm starts with an initial pre-processing of the genomic sequence G that consists in building an hash-table, denoted $HASH(G)$, containing the occurrences in G of each substring of G of a specified length (usually up to 10b): each distinct substring of fixed length is called *a component.* Given, a suffix of G, $G[i,M]$, by $HASH(G,i)$ we denote the portion of the hash-table related to components that occur in $G[i,M]$. The algorithm (see the procedure EST-ALIGNMENT) uses a component to find a position on the genomic sequence where to align an EST factor. Indeed, since a component is of bounded length and usually its occurrences on the genomic are quite few (usually below 2), a substring of an EST factor that is a component is also a substring (i.e. an error free alignment) of the genomic sequence. Consequently, components allow to find the position of EST factor alignments.

The algorithm locates the intron regions by validating the splice sites using the $gt-ag$ rule by which each intron region starts with a gt and ends with an ag, since as pointed out in [1], such a rule is confirmed by a huge percentage of splicing sites in mammalian. Thus a second hash-table for all gt and ag occurrences on the genomic sequence, called $HASH(pattern)$, is initially computed and stored. The most recent ASPic implementation of the algorithm also consider other "non canonical" splice sites, whenever the use of the rule $gt-ag$ does not allow to locate correct exons.

Given an EST sequence S, the factorization $< s_1, \cdots, s_n >$ of S is obtained by first computing each internal factor s_j one after the other (if no backtracking is required) as described by step 1.

Basic step 1: computing an internal factor s_j of an EST factorization

Let $z \in \mathbf{ALIGN[s_{j-1}, G]}$ be the genomic-copy (exon) of factor s_{j-1} and let $R = Right_G(z)$. Moreover, let $S' = S[Right_S(s_{j-1}), N]$ be the suffix of S where factor s_j must be placed. Clearly, $Left_S(s_j) = Right_S(s_{j-1})$.

Step 1.1: finding a component $comp_j$ on S'.

This step consists in finding a first occurrence of a component $comp_j$ in the hash-table $HASH(G,R)$ such that $comp_j$ is a substring of S', see Figure 2. In other words, we look for a substring of S that follows the computed factor s_{j-1} and has its leftmost occurrence in the suffix of the genomic sequence that follows the genomic-copy of s_{j-1}. This step is realized by the call to the procedure *Find-component* over instance S' and $HASH(G,R)$: it computes a pair of values, (genomic-index,$comp_j$), where genomic-index is the left position on G of the leftmost occurrence in $HASH(G,R)$ of $comp_j$, $G[l,r]$.

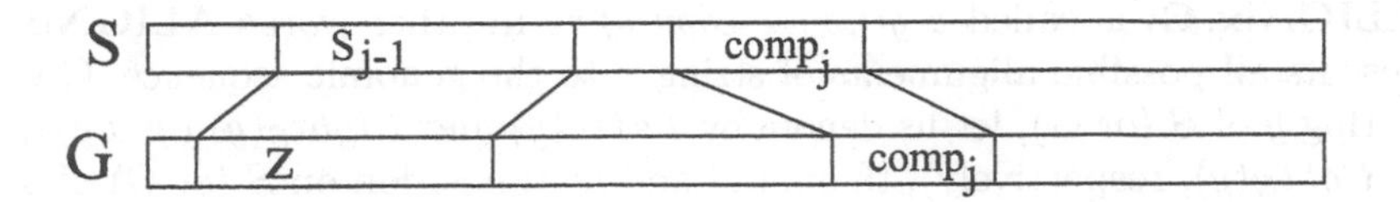

Fig. 2. Alignment of the component $comp_j$ over the genomic sequence G and the sequence S

Step 1.2: finding intron boundaries for factor s_j.
Given, $G[l,r] = comp_j$, a left intron splice site is determined by computing a left extension **ag**x of $G[l,r]$ on G, and a prefix x' of S' before $comp_j$ occurrence such that $xG[l,r] \in$ **ALIGN[x'comp$_j$, G]**. Similarly, a right intron splice site is determined by computing a right extension y**gt** of $G[l,r]$ on G and a right extension y' of $comp_j$ on S', such that $xG[l,r]y \in$ **ALIGN[x'comp$_j$y', G]**. This step is realized by the procedure *Find-exon* over instance G and $comp_j$ (see procedure EST-ALIGNMENT): it produces $s_j = x'comp_jy'$ and $xcomp_jy$, which is the alignment of s_j on G (computed-exon) (see Figure 3).
Whenever no intron boundaries are found using $comp_j$, then Step 1.1 is repeated, by considering a new component occurrence in $HASH(G,R+1)$, i.e. in a successive genomic position.
Note that each alignment is found using an efficient dynamic programming algorithm computing the edit-distance in linear time and space (see [8] for a discussion on similar approaches).

Step 1.3: computing the external factors s_1 and s_n of S.
Since, external factors are usually prefixes and suffixes of exons of G_E, the alignment of external factors is computed in a step successive to the computation of all internal factors of the set E of factorizations compatible to G_E. Indeed, an external factor may be aligned to a previously computed exons of G_E or it can produce a new factor. In this last case, the exact location of the exon must be carefully computed, because most ESTs errors are mainly present at the EST ends.

Basic step 2: adding s_j to a partial EST factorization $S' =< s_1, \cdots, s_{j-1} >$ compatible to an exon-factorization G_E
This step of the algorithm consists in verifying the following two properties:
(1) - adding s_j does not increase the number of exons over the MAXEXONS upper bound,
(2) - $< s_1, \cdots, s_{j-1}, s_j >$ is still compatible to an exon-factorization G_E.
The second step consists in comparing the genomic-copy of the new factor s_j

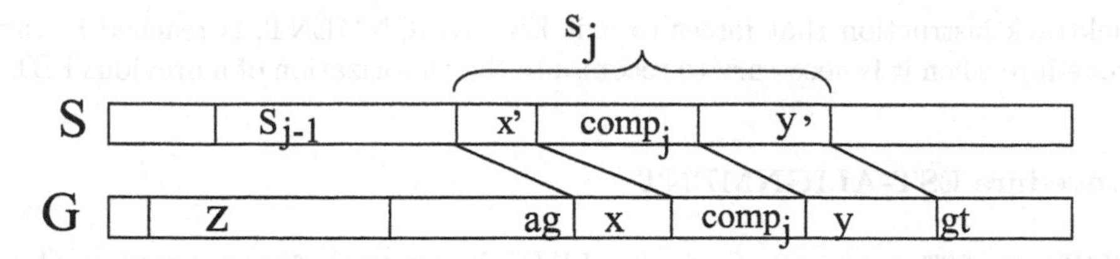

Fig. 3. Representation of the factor $s_j = x'comp_jy'$ and its alignment to the genomic-copy $xcomp_jy$ on G

with all previously computed exons. First of all, such comparison allows to count correctly the number of exons, as required in step (1). Secondarily, it is used to check the alignment of a factor near the intron boundaries. Indeed, the comparison of s_j with a previous computed exon x that overlaps s_j allows to verify whether the intron boundaries of s_j can be extended to those of x, or viceversa. Otherwise, s_j may induce a true splicing variant, because of the competing $3'$ or $5'$ pattern.

Basic step 3: computing an EST factorization for S_{i+1} compatible with a set E_i of ESTs factorizations for $\{S_1, \cdots, S_i\}$.

This step is illustrated below by the pseudo-code of the procedure EST-ALIGNMENT. Such a procedure is over an instance consisting of an EST S, an index $INDEX$ for an initial position of S, index k for an initial position of G and j denoting factor s_j. Then, it computes a factorization $< s_j, \cdots, s_n >$ of a suffix $S[INDEX, N]$ of S by aligning it to the genomic sequence $G[k, M]$.

The factorization $< s_2, \cdots, s_{n-1} >$ of all internal factors of an EST S is obtained by a call to the procedure EST-ALIGNMENT, having in input MIN-LENGTH, the minimum length of the first external factor of the factorization of S, the initial position of the genomic sequence, the index 2 of factor s_2.

The factorization of S is realized by the *repeat until* iteration. By the basic step 1 described above, a new factor s_j is computed. Hence the basic step 2 is applied to test whether to add the genomic-copy of s_j, i.e. the new computed-exon, to the exon-factorization G_E. This step is realized by the third *if*. Indeed, whenever the new computed exon gives an EST factorization compatible with the actual exon-factorization, then a successive factor is computed. Otherwise, the factor s_j is recomputed and is located in a new genomic position, represented by the increment of index k.

The fail to place a factor s_j is tested by *if* (1), in which case the previous factor s_{j-1} is recomputed. Recomputing the factor s_{j-1} means to consider a different EST factorization (i.e. genomic alignment). This step is represented in the procedure by the call to *Update-factorization*(est-suff, k), required to consider a new est suffix and genomic index to repeat the procedure iteration. The backtrack instruction that forces to exit EST-ALIGNMENT, is reached in the procedure when it is necessary to recompute the factorization of a previous EST.

Procedure EST-ALIGNMENT

```
INPUT: an EST sequence S, j, k, INDEX,an exon-factorization G_E,

  i := INDEX; /* initial position on S to look up for the first exon*/
  est-suff:= S[i,N];   /* suffix of S to look up for factor s_j*/
   repeat
  begin
    (genomic-index, comp_j):= Find-component(est-suff,HASH(G,k));
     if (1) (genomic-index != NIL) then
    begin
      (s_j, computed-exon):= Find-exon(G,comp_j);
      if (2) (computed-exon != NIL) then
        if (3) (Compatible(computed-exon, G_E)) then
          begin
          G'_E:= Update-exons(G_E, computed-exon); /* new exon is
added to G_E*/
          k := genomic-index; /* new genomic suffix */
          j := j + 1; /* new factor index*/
          est-suffix: = S[Right_S(s_j),N]; /* est-suff after factor s_j*/
          end
    end
    else
      begin
      j := j - 1;
      Update-factorization(est-suff,k);
       if(j=1) then /* it is not possible to place the first factor s_2 */
       backtrack();   /*backtrack to align previous ESTs */
       end
    k:= k+1; /* successive alignments start from positions in G[k+1,M] */
  end
  until EST-factorized
```

Main procedure: computing the set E of ESTs factorizations for S compatible with G_E.

The main procedure consists in the following iterative step. Initially, the exon-factorization of G, G_E is an empty sequence. We pose i to be the index of the first EST S_1 in S, that is $i = 1$ and $INDEX$ equal to a parameter $MINLENGTH$, which is the minimum length of the first external factor of an EST. Then, the procedure EST-ALIGNMENT is iterated for $i = 1$ to $|S|$ in order to compute the internal factors $< s_2, \cdots, s_n >$ for each EST S_i.

But, whenever a generic iteration of the procedure EST-ALIGNMENT applied to the EST S_i executes the **backtrack**() instruction, then the procedure EST-ALIGNMENT is applied to previous EST S_{i-1} in order to look for a new factorization of S_{i-1}, obtained by starting from the last factor and eventually going back to the first internal factor of S_{i-1}, if this is possible. Clearly, there may be successive applications of the backtrack instruction, by which it is possible to go back to an EST of index $l < i$, and to provide a new factorization of S_l that allows to obtain a final set E_i compatible to G_E.

4 Experimental Results

The algorithm described in the paper has been implemented in C and we have called it ASPIc (Alternative Splicing PredICtion).

Table 1. Splice sites detected by ASAP and ASPic: part I

Genomic	*Asap*	*Aspic*	*Novel*	*Genomic*	*Asap*	*Aspic*	*Novel*
ABCB10	11	11	0	ACADM	14	13	3
ACTN2	20	19	5+	ADAM15	12	12	3
ADAMTS4	7	7	1	ADIR	8	8	2
ADORA1	7	7	2	ADORA3	1	1	1
ADPRT	23	23	2	AGL	37	37	1+
AGRN	34	31	2+	AGT	3	3	1
AHCYL1	18	18	1	AIM1L	1	1	0
AKR7A2	5	5	2	ALDH9A1	10	10	2
ALPL	13	13	0+	AMPD1	12	12	0
AMY2A	6	5	4+	AMY2B	8	8	1
ANGPTL1	4	4	1	ANGPTL3	5	5	0
ANXA9	12	12	1+	AP4B1	11	11	3
APCS	0	0	1	ARHC	12	12	1
ARHGEF2	21	21	1+	ARHGEF11	0	0	1
ARHGEF16	9	9	3	ARNT	19	17	4
ARPC5	2	1	1	ARTN	6	4	2
ATP1B1	6	6	3	ATP2B4	21	21	2
AUTL1	10	10	2+	C1orf10	1	1	0
C1orf26	16	16	0	C1QB	4	4	0
Cab45	5	4	0	CAPZA1	10	10	1+
CASQ1	9	9	1	CHC1	20	18	3+
CTRC	7	7	1	DKFZp434D177	18	17	0
DMRTA2	0	0	0	DPH2L2	9	9	1
EPHA2	15	15	3+	EYA3	14	14	0
FBXO2	8	7	1	FCGR3B	3	3	2

Table 2. Splice sites detected by ASAP and ASPic: part II

Genomic	*Asap*	*Aspic*	*Novel*	*Genomic*	*Asap*	*Aspic*	*Novel*
FLJ10349	3	3	1	FLJ10407	21	21	0
FLJ10709	15	15	1	FLJ20321	11	11	0
FLJ23323	7	7	1	FUCA1	7	7	1+
GAC1	4	3	0	GBP2	11	11	3+
GMEB1	8	8	0+	HES2	2	2	1
HNRPR	15	15	2	LGALS8	11	10	1
LYPLA2	13	13	1	MASP2	10	10	1
MGC10334	0	0	1	MGC4796	10	10	0
MOV10	28	28	1	NPPB	1	1	0
PAFAH2	11	10	1+	PALMD	6	6	1+
PEX10	5	5	2	PINK1	6	6	1
PRDM16	8	6	0	PTPRU	19	19	1+
SDC3	3	1	1	SDHB	8	8	1
SERPINC1	7	7	0	SFPQ	8	8	2+
SNFT	1	1	0	TARDBP	14	11	0
VAMP3	5	4	0				

4.1 Comparing ASPIC Data with ASAP

We analyzed the accuracy of data produced by ASPIC by comparing such data with the ones provided by ASAP database. Indeed, ASAP is one of the largest human alternative splicing database available and it is easy accessible by the URL http://www.bionformatics.ucla.edu/ASAP. Currently, ASAP is based on the analysis of alternative splicing in human from detailed alignment of EST onto the genomic sequence and it provides precise exon-intron structure, validated by comparison with NCBI Assembly. Moreover, a large number of splicing sites detected on EST sequences are claimed to be novel. In order to compare ASPIC data with ASAP, we choose a list of more than 150 EST clusters of chromosome 1 from UniGene database, while the genomic sequences for the genes have been directly taken from ASAP database in almost all cases. ASAP's detected splicing sites for a given EST cluster and a genomic sequence are viewed by giving different detailed tables and mainly two kind of data: (1) for each pair of splice sites, the number of ESTs and ESTs identifiers detecting them by an alignment to the genomic sequence are reported, (2) for a specific EST, the factorization and the splice sites induced by each factor are reported. Then, ASPIC has been compared to ASAP by producing tables according to the above types of data. The tables 1, 2 report some of the results obtained for Chromosome 1. In the tables, Unigene ESTs clusters are identified by the gene symbol, (e.g. Hs.1710 is identified by ABCB10). For each EST cluster, the column named *Asap* reports the total number of splicing sites detected by ASAP, while column named *Aspic* gives the number of such splices detected also by ASPic software. Finally, the last column *novel* reports the number of splices (corresponding in many cases to the presence of novel exons) not reported in ASAP database, but which are confirmed by a high number of ESTs alignments produced by ASPic software. The number of ESTs for each cluster examined by ASPic is reported in the Execution Time tables 3 and 4. Novel splice sites are usually confirmed by around 20 ESTs. But, some splice sites are only confirmed by one to four ESTs; in such cases significant alignments have been observed upstream and downstream 5' and 3' splice sites. The comparison between locations of splicing sites in ASAP and ASPic is accurate in almost all cases, where the genomic sequence processed by ASPic is the same used to locate ASAP splicing sites. By the symbol + we mark the tables entries related to cases where we have observed that the coordinates of the genomic sequence used by ASAP to locate introns are different from the ones used by ASPic. Detailed results of the experiments and ASPic software are available at the webpage: http://bioinformatics.bio.disco.unimib.it/~bonizzoni/asp.html. We plan future experiments aimed to point out the use of the optimization criteria when the genomic sequences contain highly repeated sequences.

Table 3. Execution times over ESTs clusters from Chromosome 1

Genomic	*Time cluster (sec)*	*Number of ESTs*	*Time EST (sec)*	*Genomic length*	*Genomic*	*Time cluster*	*Number of ESTs*	*Time EST*	*Genomic length*
ABCB10	31	125	0.248	46104	ACADM	29	309	0.094	42800
ACTN2	149	244	0.611	93820	ADAM15	31	370	0.084	13480
ADAMTS4	23	48	0.479	13360	ADIR	70	241	0.290	17905
ADORA1	81	99	0.818	80879	ADORA3	8	38	0.211	8209
ADPRT	344	826	0.416	51360	AGL	114	184	0.620	77598
AGRN	138	532	0.259	36460	AGT	116	573	0.202	15680
AHCYL1	110	687	0.160	41689	AIM1L	0	25	0.000	7360
AKR7A2	22	320	0.069	12068	ALDH9A1	65	468	0.139	40648
ALPL	19	109	0.174	80157	AMPD1	3	52	0.058	26480
AMY2A	560	405	1.383	287011	AMY2B	3	147	0.020	12080
ANGPTL1	11	94	0.117	24160	ANGPTL3	6	88	0.068	11440
ANXA9	9	84	0.107	17740	AP4B1	13	180	0.072	14080
APCS	20	169	0.118	5014	ARHC	58	814	0.071	8080
ARHGEF11	0	814	0.007	113600	ARHGEF16	7	81	0.086	18240
ARHGEF2	51	334	0.153	25640	ARNT	87	263	0.331	68160
ARPC5	72	696	0.103	13600	ARTN	6	43	0.140	7920
ATP1B1	79	718	0.110	30000	ATP2B4	29	490	0.059	61600
AUTL1	24	197	0.122	188038	C1orf10	9	40	0.225	9120
C1orf26	40	88	0.455	134960	C1QB	18	259	0.069	12289
Cab45	59	977	0.060	11058	CAPZA1	86	649	0.133	33500
CASQ1	5	110	0.045	15360	CHC1	39	230	0.170	43100
CTRC	3	89	0.034	12211	DKFZ(...)	39	140	0.279	48640

4.2 Execution Time

The algorithm may require exponential time when processing a genomic sequence that contains an unbounded number of long sequence repeats and backtracking is used many times because of the large number of possible alignments. But, because of the use of canonical patterns for splice sites (i.e. gt-ag rule) and since in practice there is a limited number of possible alignments compatible with all sequences, the algorithm turns out to be very efficient as shown below. The performance of ASPIC have been evaluated by installing the software on a single processor PENTIUM IV, with a clock rate of 1600 MHZ, 256 MB of main memory, and running Linux.

Table 4. Execution times over ESTs clusters from Chromosome 1

Genomic	*Time cluster (sec)*	*Number of ESTs*	*Time EST (sec)*	*Genomic length*	*Genomic*	*Time cluster*	*Number of ESTs*	*Time EST*	*Genomic length*
DMRTA2	3	3	1.000	6960	DPH2L2	60	288	0.208	7440
EPHA2	158	219	0.721	116480	EYA3	23	58	0.397	65228
FBXO2	10	147	0.068	10320	FCGR3B	69	217	0.318	11920
FLJ10349	50	224	0.223	14916	FLJ10407	243	298	0.815	76960
FLJ10709	48	222	0.216	26560	FLJ20321	37	38	0.974	22080
FLJ23323	22	135	0.163	27680	FUCA1	35	254	0.138	55538
GAC1	161	67	2.403	72133	GBP2	3189	249	12.807	435338
GMEB1	54	149	0.362	44238	HES2	2	8	0.250	16080
HNRPR	158	595	0.266	38640	LGALS8	64	381	0.168	28944
LYPLA2	152	413	0.368	8480	MASP2	8	40	0.200	24720
MGC10334	18	111	0.162	5360	MGC4796	228	423	0.539	25711
MOV10	99	345	0.287	29908	NPPB	1	25	0.040	5520
PAFAH2	50	174	0.287	33820	PALMD	43	182	0.236	52482
PEX10	52	313	0.166	11840	PINK1	66	412	0.160	22063
PRDM16	50	20	2.500	30880	PTPRU	102	143	0.713	82613
SDC3	274	300	0.913	13280	SDHB	68	369	0.184	39400
SERPINC1	10	129	0.078	17440	SFPQ	152	916	0.166	13958
SNFT	29	37	0.784	17440	TARDBP	236	658	0.359	15440
VAMP3	105	445	0.236	14240					

Tables 3, 4 report the execution time for instances given by some ESTs clusters and genomic sequences of chromosome 1.

Acknowledgments We thank Gianluca Della Vedova for his helpful suggestions on the preliminary design of ASPic software and Gabriele Ravanelli for providing a PERL library to visualize ASPic data. This work is supported by FIRB project "Bioinformatica: Genomica e Proteomica".

References

1. Burset, M., Seledtsov, I.A., Solovyev, V.V.: SpliceDB: database of canonical and non-canonical mammalian splice sites. Nucl. Acids. Res. **29** (2001) 255 – 259
2. Christopher, L., Levan, A., Barmak, M., Yi, X.: ASAP:the Alternative Splicing Annotation Project. Nucleic Acids Research **31:1** (2003) 101 – 105
3. Graveley, B.: Alternative Splicing: increasing diversity in the proteomic world. Trends Genet. **17:4** (2001) 100 – 107

4. Heber, S., Alekseyev, M., Sze, S., Tang, H., Pevzner, P.: Splicing graphs and EST Assembly Problem. Bioinformatics **1:1** (2002) 1 – 8
5. Ogasawara, J., Morishita, S.: Fast and Sensitive Algorithm for Aligning ESTs to Human Genome. bookFirst IEEE Computer Society Bioinformatics Conference (2002) 43 – 53
6. Papadimitriou, C.H.: Computational Complexity. Addison-Wesley (1994)
7. Qiang, X., Modrek, B., Lee, L.: Genome-wide detection of tissue-specific alternative splicing in the human transcriptome. Nucleic Acids Research **30:17** (2002) 3754 – 3766
8. Setubal, J., Meidanis, J.: Introduction to Computational Molecular Biology. PWS Publishing Company (1997)
9. Wheelan, S., Church, D.J., Ostell J.M.: Spidey: A tool for mRNA-to-Genomic Alignments. Genome Research **1:1** (2001) 1 – 6
10. Wheeler, R.: A Method of Consolidating and Combining EST and mRNA. Alignments to a Genome to Enumerate Supported Splice Variants book Proc. WABI 2002, Lectures Notes in Computer Science **2452** (2002) 201 – 209
11. Xie, H., Zhu, W., Wasserman, A., Grebinskiy, V., Olson, A., Mintz, L.: Computational Analysis of Alternative Splicing Using EST Tissue Information. Genomics **80** (2002) 327 – 330

5 Appendix

Theorem 1. *The* **MEFC** *problem is MAXSNP-hard.*

Proof. Sketch of the proof. In the following we give a linear reduction from the NODE COVER problem on cubic graphs to MEFC. It follows that MEFC is MAXSNP-hard (NODE COVER on cubic graphs is MAXSNP-hard [6]), thus implying that the problem cannot be approximated within an arbitrary constant factor. Let $\mathcal{G} = (V, A)$ be a cubic graph with vertex set $V = \{v_1, v_2, \cdots, v_n\}$ and edge set $A = \{e_1, e_2, \cdots, e_m\}$. Then, an instance (S, G) of the MEFC problem is constructed from $\mathcal{G}$ as follows: to each edge $e_k = (v_i, v_j)$ is associated the EST sequence $s_{e_k} = v_i C e_k B v_j$. The genomic sequence G consists of $G_1 C G_V B G_2$, where: given $e_k = (v_i, v_j)$, define $x_{e_k} = v_i C e_k C v_j$, and $y_{e_k} = v_i B e_k B v_j$, then $G_1 = x_{e_1} x_{e_2} \cdots x_{e_m}$, $G_2 = y_{e_1} y_{e_2} \cdots y_{e_m}$, and $G_V = v_1 v_2 \cdots v_n$. Then the set $\mathcal{S}$ of EST consists of $\mathcal{S}=\{s_{e_k}, e_k \in A) \cup \{s = CB\}$. Let $N = \{v_{i1}, v_{i2}, \cdots, v_{iN}\}$ be a solution of the NODE COVER problem on graph $\mathcal{G}$ and let $L(A) = \{l_1, l_2, \cdots, l_{m1}\}$ be the set of indices of edges of $\mathcal{G}$ that have a right vertex in N, and $R(A) = \{r_1, r_2, \cdots, r_{m2}\}$ be the set of indices of edges of $\mathcal{G}$ that have a left vertex in N. Moreover, given $e_k = (v_i, v_j)$, define $x(k) = v_i C e_k$ and $y(k) = e_k B v_j$. Then, the following Claim holds:

$G_E = < x(l_1), \cdots, x(l_{m1}), C, v_{i1}, v_{i2}, \cdots, v_{iN}, B, y(r_1), \cdots, y(r_{m2}) >$ is an exon-factorization with $m + 2 + |N|$ factors for the instance $(G, \mathcal{S})$ of the MEFC-problem iff N is a node cover for graph $\mathcal{G}$.

Optimal DNA Signal Recognition Models with a Fixed Amount of Intrasignal Dependency

Broňa Brejová, Daniel G. Brown, and Tomáš Vinař

School of Computer Science, University of Waterloo Waterloo ON N2L 3G1 Canada
{bbrejova, browndg, tvinar}@math.uwaterloo.ca

Abstract. We study new probabilistic models for signals in DNA. Our models allow dependencies between multiple non-adjacent positions, in a generative model we call a higher-order tree. Computing the model of maximum likelihood is equivalent in our context to computing a minimum directed spanning hypergraph, a problem we show is NP-complete. We instead compute good models using simple greedy heuristics. In practice, the advantage of using our models over more standard models based on adjacent positions is modest. However, there is a notable improvement in the estimation of the probability that a given position is a signal, which is useful in the context of probabilistic gene finding. We also show that there is little improvement by incorporating multiple signals involved in gene structure into a composite signal model in our framework, though again this gives better estimation of the probability that a site is an acceptor site signal.

1 Introduction

Accurate detection of DNA signals is essential for the proper identification of important features found in DNA. Here, we study new probabilistic models for detecting such signals based on optimizing directed spanning hypertrees in hypergraphs that represent all possible dependencies of bounded cardinality among positions of the signal. We evaluate performance of our new models on human splice site recognition. Our new models offer modest improvement over existing techniques, most notably in improving the accuracy of the probabilistic model, rather than its usefulness as a classifier.

Our study is motivated by gene finding, where we desire signal detectors based on generative probabilistic models that use only small window around the functional site for their prediction. Altough it is often possible to increase signal detector prediction accuracy by considering properties of the wider sequence (such as the coding potential of the region upstream from donor splice site), such information is already considered in other parts of a gene finder.

Generative probabilistic models commonly used for signal detection include position weight matrices [20,21], maximum decomposition trees [5], and Chow-Liu trees [8,7]. Here, we extend these techniques to a wider class of models. Our new models, described in Section 2, allow for more complicated dependencies among the positions of a signal than were previously modeled. In particular, we

G. Benson and R. Page (Eds.): WABI 2003, LNBI 2812, pp. 78–94, 2003.

model dependencies within a signal as a directed hypertree whose nodes are the positions of the signal. The root of the hypertree has no dependencies on other positions, while all other positions are dependent on the positions represented by the tail of the hyperedge that is incident on them. When the hyperedges are just normal directed edges, this is equivalent to the Chow-Liu tree models studied in this context by Cai et al. [7], but allowing multiple dependencies at a position allows more richness in the set of possible signal models being considered.

Unfortunately, computing the optimal signal model for a training set from the set of hypertrees we study is NP-hard, as we show in Section 3. However, one can either use integer programming (which is practical as long as the number of dependencies a position has is small) or simple greedy heuristics in practice. For the problems we have considered, the difference between the optimal model and the one found by greedy heuristics was small.

We have tested our models by using human chromosome 22, which is extremely well annotated. Our experimental results are found in Section 4. In practice, the improvement of our hypertree-based models over more standard signal detection algorithms, such as second order position weight matrices (PWMs), is modest. In this sense, our results are similar to the results of Cai *et al.* [7], who show that tree dependency structures do not improve much over first order PWMs.

However, we are able to offer a notable improvement in the prediction of the probability that a given position is a signal. This us useful for gene finding, given that gene finders typically integrate this predicted probability into their other inference engines. If one can improve reliability of the scores returned by signal detectors, perhaps one can improve *de novo* gene finding as well.

In particular, when tested on a long testing sequence, if our model for donor sites predicts that a possible sequence is a donor with probability 6%, then with probability 6.7% it actually *is* such a signal. In contrast, when the more standard second order PWM predicts a donor signal with the same probability, it is actually a donor signal with probability 7.5%. This pattern consists across several different families of signals, and at many probability values.

Our models are general enough to be adapted to other probabilistic inference scenarios beyond gene finding. For example, they are appopriate for modeling dependencies among positions in protein domains, for purpose of quickly screeing a database of proteins for a match to a non-consecutively dependent motif. We are currently investigating their integration into a gene finder.

2 Models of Dependencies in Signals

A useful way to represent biological signals is by probabilistic generative models. For our purposes, signals are short windows of fixed length in the sequence located near the biologically significant boundary or functional site. For example, donor and acceptor splice signals are located at exon-intron and intron-exon boundaries respectively.

The simplest generative model for signals is the *position weight matrix* (PWM) [20,21]. A PWM gives the probability that each position in the signal is a particular character, and allows all positions to be independent. If $r(i, b)$ is the probability that the ith position in the signal is the base b, then the probability that the model generates a a given sequence S of length n is $\prod_{i=1}^{n} r(i, S_i)$.

The basic assumption of the PWM is that the probability of generating the character depends only on its position in the signal. Many researchers (*e.g.*, [6,22]) demonstrated that this assumption is false. In fact, there exist dependencies in signals between positions which are several bases apart in the sequence. Higher order PWMs [22] allow the incorporation of dependencies between adjacent positions in the sequence. For example, in a first order PWM, each position is dependent on its immediate predecessor in the sequence; to generate the signal, one starts with the first position; then each subsequent character is picked using a probability distribution based on its predecessor.

Here, we investigate an extension of the PWMs to allow for multiple dependencies among positions in the signal which are not adjacent, and also to allow for dependencies between pairs of related signals. Such signals, especially corresponding donor and acceptor sites in introns, may be mutually dependent, and we wish to capture this information.

In our model, each position in the signal is dependent on a fixed set of other positions and there are no cyclic dependencies. We can view such a model as a directed acyclic graph (DAG), where nodes represent positions in the signal, and edges represent dependencies between the positions. Each node is assigned a probability distribution of bases depending only on the bases on positions which are its immediate ancestors in the DAG (see examples in Figure 1). We call the underlying DAG the *signal model topology*, and the maximum in-degree of a node in such a graph is the model's *order*. Such models are also called *Bayesian networks* and are extensively used in machine learning [13].

Models with order zero are exactly PWMs. If the order is 1, then the underlying topology is a directed tree. Such model was used for classification of splice sites before by Cai et al. [7] and Agarwal and Bafna [1]. The special case where the topology is restricted to paths was used for identifying transcription factor binding sites by Ellrott et al. [12]. PWMs of k-th order can be expressed as k-th order signal models (see Figure 1 for an example of a second order PWM). These network models are actually quite general: for example, the maximum dependency decomposition (MDD) model used by Burge [6] to model donor splice sites can be expressed as a fourth-order model. However, this is not very practical; such a model would contain many parameters not used by the MDD model. In general, we call all order k models the *higher order trees of k-th order* (or HOT-k).

To generate a signal in one of these models, one iteratively finds positions whose antecedents have been fixed, and then chooses the sequence value at that position dependent on the predecessors. This can of course be done in $O(kn)$ time, if the signal is of length n and k is the order of the model.

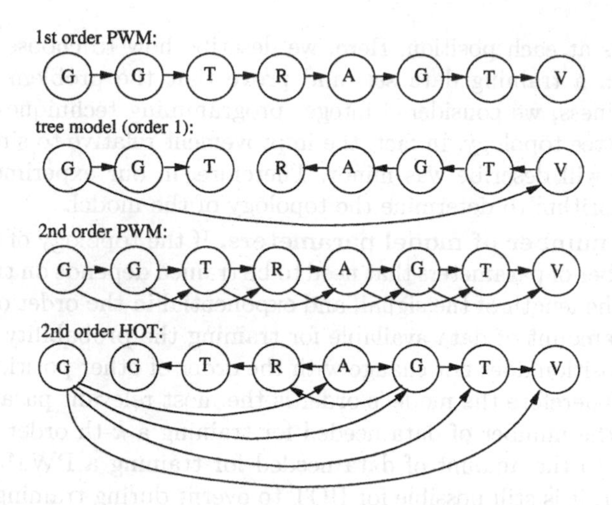

Fig. 1. Examples of different model topologies for donor signal positions (−1, +7).

Similarly, one can compute the probability that the model generates a given sequence S by multiplying the dependent probabilities. If the ith position in the sequence is dependent on positions $d_{i,1} \ldots d_{i,k_i}$, and we denote $r(i, b, x_1, \ldots x_{k_i}) = \Pr[S_i = b | d_{i,1} = x_1, d_{i,2} = x_2, \ldots, d_{i,k_i} = x_{k_i}]$, the probability of generating the sequence S by the model is

$$\Pr(S|+) = \prod_{i=1}^{n} r(i, s_i, s_{d_{i,1}}, s_{d_{i,2}}, \ldots, s_{d_{i,k_i}}).$$

When using such model as a classifier, one also wants to compute the probability that the seen sequence S constitutes the signal. This can be done using Bayes rule:

$$\Pr(+|S) = \frac{\Pr(S|+) \cdot \Pr(+)}{\Pr(S|+) \cdot \Pr(+) + \Pr(S|-) \cdot (1 - \Pr(+))}.$$

Here, $\Pr(+)$ is a frequency of the signal occuring in the sequence, and $\Pr(S|-)$ is the probability of the sequence S occuring in a background model (in our work, we use a fifth-order position-independent Markov chain as the background model).

3 Choosing the Best Model

Here, we investigate methods for estimation of parameters of HOT models to maximize likelihood of the training data set. Once the topology of the model is fixed, we only need to count the frequencies from the training data. One needs to pick a good choice for the model's topology itself, before computing the parameters at each position. Here, we describe how to choose the optimal topology, given a training data set, and prove that the problem is NP-hard. Given its hardness, we considered integer programming techniques to find the optimal hypertree topology. In fact, the improvement relative to simpler greedy algorithms we will describe was minor. Therefore, in our experiments, we use the greedy algorithm to determine the topology of the model.

Note on the number of model parameters. If the topology of the model is fixed, the number of parameters that need to be trained depends on two variables: it is linear in the length of the signal, and exponential in the order of the model. However, the amount of data available for training the probability distribution of any given position does not change with the order of other positions or length of the signal. Therefore the model's order is the most relevant parameter in the training, and the number of data needed for training a k-th order HOT model is comparable to the amount of data needed for training a PWM of the same order. However, it is still possible for HOT to overfit during training because we are searching for model over a wider family (namely, all possible topologies).

Connection to hypergraphs. To formalize the problem of finding the best topology, we formulate the problem in terms of hypergraphs. A hypergraph is a pair $\mathcal{H} = (V, \mathcal{E})$, where V is a set of vertices, and $\mathcal{E} = \{E_1, E_2, \ldots, E_m\}$ is a set of directed hyperedges. Each *directed hyperedge* $E = (T, h)$ has a tail T, which is a subset of V, and a head h, which is a single vertex.[1] Let the *order of a directed hyperedge* be the cardinality of its tail.

A *directed hypertree* is a hypergraph $\mathcal{H}$, where all nodes are the tail of at most one hyperdge, and the directed graph which can be obtained by replacing every hyperedge $(\{v_1, \ldots, v_k\}, v)$ with k edges (v_1, v), (v_2, v), $\ldots$, (v_k, v) is acyclic. A *spanning directed hypertree*, for us, is a directed hypertree where each vertex is a head of a hyperedge (note, that a hyperedge can have an empty tail). Spanning directed hypertrees are exactly equivalent to spanning outtrees in ordinary directed graphs.

Let us define a directed hypergraph analogous to the complete graph. The complete hypergraph is a hypergraph that contains all hyperedges (T, h) of order at most k.

There is an easy correspondence between the spanning directed hypertrees and HOT models: for a given vertex, all incoming edges in the HOT model can be represented as a single hyperedge. Such hyperedges form a spanning directed hypertree. Now we can make the following statement.

Theorem 1. *Let $\mathcal{H}$ be a complete hypergraph of order k on a set of vertices representing positions in the signal. Let weight of every hyperedge $(T, h) \in \mathcal{H}$ be $H(T \cup \{h\}) - H(T)$, where $H(P)$ is the entropy of the signal positions from set P in the training set of signals $S^{(1)}, \ldots, S^{(m)}$.*

---

[1] Sometimes, directed hyperedges are defined as a pair (T, H), where both tail T and head H are sets of vertices, rather than head being always a single vertex [14]. However, we will use the simpler definition here.

Then the directed acyclic graph M^* corresponding to the minimum spanning hypertree $\mathcal{M}^*$ of graph $\mathcal{H}$ yields a topology of the HOT model of order k with the maximum likelihood of the training set $S^{(1)}, \ldots, S^{(m)}$.

Proof. For any set of positions P, let $f_P(x_P)$ be the number of occurences of string x_P at positions in P in the training set $S^{(1)}, \ldots, S^{(m)}$. In this notation, the entropy $H(P)$ can be expressed as

$$H(P) = -\sum_{x_P} \frac{f_P(x_P)}{m} \cdot \log \frac{f_P(x_P)}{m}.$$

We have noted earlier that in the maximum likelihood model with fixed topology M, the string S has probability

$$\Pr(S \mid M) = \prod_{(T,h)\in\mathcal{E}(M)} \frac{f_{T\cup\{h\}}(S_{T\cup\{h\}})}{f_T(S_T)},$$

where $\mathcal{E}(M)$ is the set of hyperedges in the corresponding spanning directed hypertree. To maximize the likelihood of generating $S^{(1)}, \ldots, S^{(m)}$, we have to maximize over all possible model topologies M:

$$\Pr(S^{(1)}, \ldots, S^{(m)} \mid M) = \prod_{i=1}^{m} \Pr(S^{(i)} \mid M) = \prod_{i=1}^{m} \prod_{(T,h)\in\mathcal{E}(M)} \frac{f_{T\cup\{h\}}(S^{(i)}_{T\cup\{h\}})}{f_T(S^{(i)}_T)}$$

$$= \prod_{(T,h)\in\mathcal{E}(M)} \frac{\prod_{x_{T\cup\{h\}}} f_{T\cup\{h\}}(x_{T\cup\{h\}})^{f_{T\cup\{h\}}(x_{T\cup\{h\}})}}{\prod_{x_T} f_T(x_T)^{f_T(x_T)}}$$

We want to maximize $\Pr(S^{(1)}, \ldots, S^{(m)} \mid M)$ which is equivalent to minimizing $-(1/m)\log\Pr(S^{(1)}, \ldots, S^{(m)} \mid M)$:

$$-\frac{1}{m}\log\Pr(S^{(1)}, \ldots, S^{(m)} \mid M) =$$

$$\sum_{(T,h)\in\mathcal{E}(M)} \left[-\sum_{x_{T\cup\{h\}}} \frac{f_{T\cup\{h\}}(x_{T\cup\{h\}})}{m} \log f_{T\cup\{h\}}(x_{T\cup\{h\}}) \right] + \left[\sum_{x_T} \frac{f_T(x_T)}{m} \log f_T(x_T) \right]$$

$$= \sum_{(T,h)\in\mathcal{E}(M)} H(T,h) - H(T),$$

which is exactly what we wanted to prove. □

Solving the minimum spanning directed hypertree problem. Theorem 1 shows, how to reformulate the problem of finding the graphical model maximizing the likelihood of our training data to the problem of finding the minimum spanning directed hypertree problem.

To solve this problem, first consider its special case, where $k = 1$. In this case, we are looking for the minimum spanning directed tree of a complete graph.

Chow and Liu [8] showed, that the problem is equivalent to finding the minimum spanning undirected tree, where weight of the edge (u, v) is $H(u, v) - H(u) - H(v)$. This can be easily done by Prim's algorithm (see, e.g. [10]) in $O(n^2 \log n)$ time.

Unfortunately, in the general case, this method cannot be extended. The following theorem shows that the problem is hard, even for $k = 2$.

Theorem 2. *Finding the minimum spanning directed hypertree in the hypergraph is NP-hard, even if all the edges are of degree at most 2.*

Proof. We prove NP-hardness by reduction from a special case of the problem of minimum feedback arc set restricted to directed graphs with indegree at most 2. Minimum feedback arc set problem is to find a minimum edge set whose removal makes the graph acyclic. It was proven NP-complete by a reduction from vertex cover [17] and this reduction can be easily modified to produce graphs with indegree at most 2, thus proving the special case of minimum feedback arc is also NP-hard.

To prove that the minimum spanning directed hypertree is NP-hard we take a directed graph G with indegree at most 2 and create a hypergraph $\mathcal{H}$ on the same set of vertices. For every vertex v we find the set X of tails of edges incoming to v in G. By our assumption X has at most 2 elements. For each subset A of X we create a hyperedge (A, v) with cost $|X| - |A|$. To complete the graph, we add all other possible hyperedges with tails of size at most 2, but with a large cost C, so that they will not be chose.

As established earlier, each spanning hypertree of $\mathcal{H}$ corresponds to a directed acyclic graph M and clearly M is a subgraph of G, if it does not use one of the very high-cost edges. Conversely, if we remove a feedback arc set from G we get a directed acyclic graph that correspond to some spanning hypertree of $\mathcal{H}$. Moreover for every edge deleted from G the cost of the corresponding hypertree increases by one. Therefore graph G has a feedback arc set of size at most k if and only if $\mathcal{H}$ has a spanning hypertree of cost at most k. □

As a result of Theorem 2, we must either consider slow solution algorithms, or use heuristic methods to find the optimal hypertree topology. We present each of these ideas in what follows.

Exact algorithms for finding the optimal hypertree. We first identified optimal hypertrees by using an integer linear programming algorithm. Integer linear programming is guaranteed to give us the optimal hypertree, yet it uses potentially exponential runtime [19].

Our IP model of the problem includes two kinds of variables. One family of variables models the acyclicity of the spanning hypergraph, by requiring that we order the nodes in the hypergraph, and require the heads of hyperedges to be later in the ordering than their tails. The second family of variables model the hyperedges of the graph. We require that each chosen hyperedge be properly ordered, and that each node have only one incoming hyperedge, to again ensure the hypertree property.

In particular, we assign a decision variable $b_{i,j}$ to each pair of distinct positions, i and j. This variable is set to 1 exactly when i comes before j in the ordering of nodes, and 0 otherwise. There is a variable $a_{T,h}$ for each possible directed hyperedge $E = (T, h)$, where $|T| \leq k$, and where $h \notin T$. When $a_{T,h} = 1$, the hyperedge E is in the chosen spanning hypertree, while when it is 0, the hyperedge is not in the chosen hypertree.

We model the ordering constraint on the nodes by requiring that the order relationship is antisymmetric and that there are no 3-cycles in it. In particular, we require that $b_{i,j} + b_{j,i} = 1$ for all pairs i and j and that $b_{i,j} + b_{j,k} + b_{k,i} \leq 2$ for all triplets i, j and k of distinct nodes.

The requirement that chosen hyperedges are properly ordered is modeled by the constraint $a_{T,h} \leq b_{x,h}$ for all nodes x in T. This requires that all nodes in the head of the hyperedge are before the tail node, for chosen hyperedges.

Finally, we require that every node has an incoming edge (in the case of the tree's root, this will have no tail nodes). This is the constraint $\sum_{E:E=(T,h)} a_{T,h} = 1$ for all nodes h.

The cost of a chosen hypertree is $\sum_{E=(T,h)} a_{T,h} w_{T,h}$, where $w_{T,h}$ is the cost of including the hyperedge $E = (T, h)$ in the tree.

This gives the integer linear program:

$$\min \sum_{E=(T,h)} w_{T,h} a_{T,h}, \quad \text{subject to:}$$

$$\begin{aligned}
b_{i,j} + b_{j,i} &= 1, \text{ for all pairs } i \text{ and } j, \\
b_{i,j} + b_{j,k} + b_{k,i} &\leq 2, \text{ for all triplets } i, j \text{ and } k, \\
a_{T,h} &\leq b_{x,h}, \text{ for all hyper edges } E = (T, h) \text{ and nodes } x \text{ in } T, \\
\sum_{E:E=(T,h)} a_{T,h} &= 1, \text{ for all nodes } h, \\
a_{T,h} &\in \{0, 1\}, \text{ for all hyperedges } E = (T, h), \\
b_{i,j} &\in \{0, 1\}, \text{ for pairs of nodes } i \text{ and } j.
\end{aligned}$$

We used the integer programming solver CPLEX [15] to solve moderate-sized instances of these problems, where $k = 2$ (so all hyperedges have at most two head nodes). The runtime of the optimization procedure was only a few hours, and we were able to compute optimal hypertrees for many of the signals we discuss in Section 4. We note that computing these optimal signal models is something one does not have to do often, so spending a reasonable amount of computational effort on finding the right model is appropriate, assuming runtime is at all reasonable.

However, the improvement over the simple greedy algorithm we discuss in the next subsection was not large enough in our actual examples to justify its much larger runtime. In some applications, however, the added quality given by integer programming might be appropriate.

Simple heuristic. In practice, we have used a simple greedy heuristic to compute good HOT models. We start with a single node in the spanning hypertree $\mathcal{T}$. In each iteration, we add one more vertex v into the hypertree $\mathcal{T}$, where v is the head of the shortest hyperedge (T, h) such that $T \in \mathcal{T}$ and $h \notin \mathcal{T}$. This can be implemented in $O(kn^{k+1})$ time. Since the hypertrees generated in this way can differ depending on the starting vertex, we run the algorithm for every vertex as a starting point and choose the shortest resulting hypertree.

Note, that this is just a simple variation of Prim's algorithm for maximum spanning trees [10], however, unlike in the case of undirected spanning trees, in the case of directed spanning hypertrees this process does not guarantee the optimal result. This simple heuristic gave hypertrees with good performance often enough that we did not use the integer programming model in practice. It is certainly reasonable to consider more complicated heuristics for the problem.

Again, once we had the chosen spanning directed hypertree topology, we then computed the positional probabilities by using the empirical values from the training data set.

Note on undirected graphical models. Inference of the optimal topology of fixed order has been studied for undirected graphical models in the context of machine learning. Srebro and Karger showed that the problem is NP-hard and presented approximation algorithms for it [16]. Independently, Bach and Jordan [4] described the problem and used practical heuristic algorithms to solve it. They also applied their solution to a simple biological signal data set. Finally, Andersen and Fleischner stuied the problem of finding a minimum spanning undirected hypertree [3], who showed that this problem is also NP-hard. However, the result is not related, because of significant differences in properties of undirected and directed hypertrees.

4 Experiments

We tested the usefulness of higher order models for identifying splicing signals in human DNA. In general, our experiments show that there are some cases where including multiple dependencies and not requiring them to be adjacent can help with sensitivity of models, but in general, the improvement is quite slight; this is similar to the results of Cai *et al.* [7], who found this with tree models.

However, one useful finding is that higher order tree models can be better in predicting the probability that a given position actually *is* a signal. This probability can be used in other programs, such as probabilistic gene finders. Perhaps our most interesting result is that using a higher order tree model allows substantially better prediction of the probability that a position is a donor site than is available with other measures.

We also show that including branch site models does not aid much in detecting acceptor sites, but that this inclusion, again, helps with estimating the probability that sites are acceptor sites.

Data sets. We used two sets of annotated human sequences in our experiments. The first is the Genscan data set of Burge and Karlin [5]. This data set is

Name	Abbreviation	Length	Donor sites		Acceptor sites	
			True	False	True	False
Genscan (training)	B97T	2.9 MB	1,237	146,204	1,236	212,144
Genscan (testing)	B97	0.6 MB	339	30,319	339	45,512
Chromosome 22 (training)	C22T	19.0 MB	1,952	1,000,646	1,965	1,546,120
Chromosome 22 (testing)	C22	15.8 MB	1348	814,665	1366	1,229,695

Table 1. Characteristics of used data sets. Only canonical donor (GT) and acceptor (AG) sites were considered as true sites. We list any two-base sequence of GT or AG in the sequences as a false donor or false acceptor site, respectively.

divided into two parts: a training set (B97T) and a smaller testing set (B97). Our second data set is the Sanger Centre annotation of Human chromosome 22, release 3.1b [11]. We divided the chromosome, masked for repeats, into 73 pieces, each of roughly 500KB in length, and then randomly divided these into training (C22T) and testing (C22) sets. We present basic characteristics of the data sets in Table 1. These characteristics were used to compute $\Pr(+)$ in our experiments.

The C22/C22T data set contains significantly more false donor and acceptor sites per true site than does the Genscan set. This is because the B97/B97T set contains mostly genic sequence, with very short intergenic regions before and after the genes. On the other hand, the C22/C22T data set is an entire chromosome, where genes are a small fraction of all DNA. As we will see, this significantly affects the specificity of our models and prevents us from directly comparing the results from the two data sets.

To avoid overfitting, we only trained first-order models on the training set B97T, while we used both first-order and second-order models with C22T. Note, that the test set B97 is curated and contains only sequences with high-quality annotation. On the other hand, the test data set C22 is an annotation of a whole chromosome, and may contain several errors.

Accuracy measures. We used two measures of accuracy in our experiments. The first is the sensitivity and specificity of classification. Our models can be used as classifiers when we set a threshold on $\Pr(+|S)$. Sequences with score below the threshold are classified as not being signals, while sequences with score above the threshold are classified as signals.

Let us denote the number of true positives as TP, false positives FP, true negatives TN, and false negatives FN. We measure classification accuracy with the sensitivity $SN = TP/(TP + FN)$ and specificity $SP = TP/(TP + FP)$. If we lower the threshold, sensitivity increases and specificity decreases. Thus for a given model, we can plot a curve depicting the tradeoff, and compare different models.

Our interest in gene finding and generative probabilistic models has inspired another measure of accuracy. In these applications, signal models are not only used as classifiers, but the estimation of $\Pr(+|S)$, the probability that the eval-

	Sensitivity level				
Model	20%	35%	50%	70%	90%
PWM-0	50.0%	39.2%	31.7%	17.7%	7.9%
PWM-1	52.1%	44.2%	33.3%	24.3%	13.0%
tree	54.1%	47.4%	33.6%	26.0%	12.9%
MDD (*)	54.3%	-	36.0%	-	13.4%

Table 2. Specificity of models for donor splice sites on data set B97. We include the results for the maximum dependency decomposition model on the same set, from [6], for comparison.

uated sequence is a signal, is also used. Therefore, this estimate should be as close to the actual probability that the sequence is a signal as possible.

To evaluate accuracy of the predicted signal probability, we first divide the range of scores into buckets (in logarithmic scale). In each bucket, we compute the number of positive and negative examples among the sequences whose estimated probability causes them to be placed in the bucket. Then we compare the predicted signal probability corresponding to the bucket with the actual fraction of positive examples belonging to the bucket (actual signal probability). We quantify this measure using the standard Pearson correlation weighted by number of samples in each bucket.

In what follows, we abbreviate the various chosen signal topology models as follows: PWM-k is the chosen position weight matrix of kth order, TREE is the model obtained by optimizing the first order HOT model, and HOT-2 is the second order HOT model selected by our heuristics.

4.1 Donor Site Experiments

For our experiments with donor sites, we represented donor sites by a window of length 12, from the position −4 to the position +7 (consensus `VMAG|GTRAGTRN`). We developed first-order models, using the smaller B97T data set, and higher-order models, using the larger C22T set.

First order models. We trained the PWM-0 and first-order models (PWM-1, TREE) on the training set B97T and evaluated their performance on testing set B97. We found two non-adjacent dependencies in the TREE model: positions +3 and +5 both depend on the position −1, rather than on one of their direct neighbors.

Figure 2 shows the classification power of these models. The tree model improves on both models that do not consider non-adjacent dependencies (PWM-0, PWM-1); the improvement is more apparent in the regions around 40% and 70% sensitivity (see Table 2). We note that the improvement obtained at the 35% level of sensitivity by the TREE model is comparable to the improvement reported for the maximum dependency decomposition model at 50% level by [6].

Higher order models. To evaluate second order models, we used the data set C22T to train models of order at most 2 (PWM-0, PWM-1, TREE, PWM-2, and HOT-2). We then evaluated their prediction accuracy on the testing set C22. We

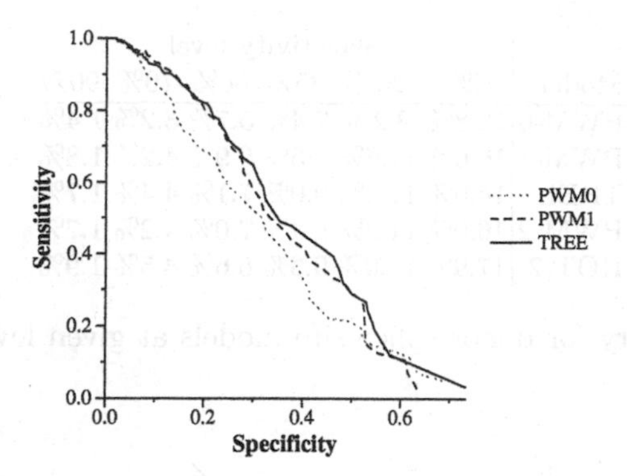

Fig. 2. Sensitivity vs. specificity for donor splice site on data set B97.

found several non-adjacent dependencies (*i.e.*, dependencies at distance greater than 2) in model HOT-2. Position -4 depended on $+7$, positions $-3 \ldots -1$ on $+3$, and positions $+4 \ldots +5$ on -1.

Table 3 shows the models' classification performance. The performance of all models except PWM-0 is roughly equivalent. This equivalence is caused mostly by the much higher presence of false positives in this data set. HOT-2 has a slightly higher specificity for low and high values of sensitivity, and a slightly lower specificity in the middle range.

We notice real improvement in our second evaluation measure: the comparison of the predicted signal probability with the real signal probability (Figure 3). The figure shows that in general, most models slightly underpredict the probability for low-scoring possible signals, while greatly overpredict the probability for high-scoring possible signals. This is the case for all tested models except the HOT-2 model, which consistently slightly underpredicts at all scores. The advantage of the HOT-2 model can also be seen in the correlation coefficient between the predicted and actual signal probability (see Figure 3): the HOT-2 model has a strongly higher correlation (0.955) in our measure than does the second best model (PWM-2, with correlation coefficient 0.911).

The HOT-2 model has approximately the same classification power as other models, while showing significant improvement in predicting accurately the signal probability. We thus suggest that it is a better model for incorporating into gene finding programs and other probabilistic frameworks than the other models.

We also compared the HOT-2 model to other models on a data set that had fewer false positives. We trained all models on the C22T data set (to avoid overfitting) and then tested them on the B97 data set. We show the results in Figure 4. They show that HOT-2 has highest specificity at high sensitivity settings (for sensitivities more than 50%), while at lower sensitivities (less than 50%), the TREE model has the best specificity. The advantage of the HOT-2 model at the high levels of sensitivity is small, however.

Model	Sensitivity level 5%	20%	35%	50%	70%	90%
PWM-0	12.8%	9.2%	7.4%	5.7%	3.2%	1.4%
PWM-1	16.0%	11.6%	9.5%	6.9%	4.2%	1.8%
TREE	14.0%	11.3%	9.0%	7.1%	4.4%	1.7%
PWM-2	16.0%	11.2%	9.4%	7.0%	4.2%	1.7%
HOT-2	17.9%	11.0%	9.3%	6.6%	4.5%	1.9%

Table 3. Specificity for donor splice site models at given levels of sensitivity, on data set C22.

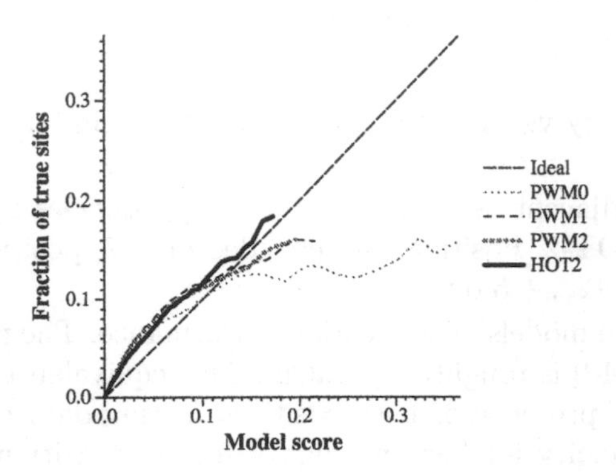

Fig. 3. Predicted signal probability versus actual signal probability for donor splice site signal on data set C22. For example, the bucket with average score 0.066 in the HOT2 model contains 258 samples, out of which 20 are true donor sites. This corresponds to the y-coordinate 0.078 = 20/258. The correlation coefficients between the predicted probability and the actual probability are 0.955 for HOT2, 0.911 for PWM2, 0.890 for PWM1 and 0.827 for PWM0.

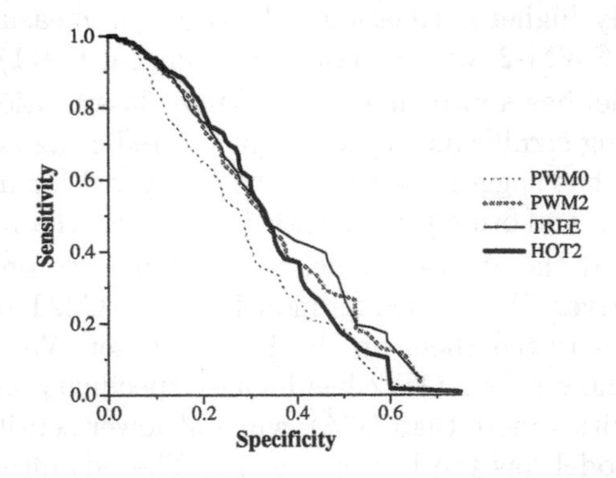

Fig. 4. Accuracy statistics for donor splice site, training on data set C22, testing on data set B97.

4.2 Acceptor Sites and Branch Sites

Next, we studied acceptor sites, using a window of size 13 from position −10 to position +2 (consensus `YYYYYYVCAG|GNN`). Burge [6] pointed out that the acceptor site does not exhibit strong non-adjacent dependencies in the signal. In this case, our experiments confirmed that the possibility of non-adjacent dependencies in the models TREE and HOT-2 does not gain any advantage, and the performance of all models except PWM-0 is roughly equivalent on both data sets.

We also tried to enhance prediction of acceptor sites by trying to locate branch sites in the sequences, upstream of the acceptor site. Since branch sites are not generally annotated, we used an iterative refining procedure described by Zhang [22] to train a model for recognizing branch sites. We started with a simple model based on the consensus sequence NNNYTVAYYYYYYYYY (in a window of size 16, from positions −6 to +9). In each iteration, we located the best-scoring sequence in the sequence window from 50 positions before each acceptor site to 10 positions before the acceptor site, and used this sequence to train the new model. We performed three iterations to train branch site models. Lastly, we paired the possible branch sites we found with their corresponding acceptor sites and built new models for both signals together that allow inter-signal dependencies.

To estimate the signal probability of each possible acceptor site A, we first located the most likely branch site B in the window from 10 to 50 bases upstream from the possible acceptor A and then estimated $\Pr(+|B, A)$ of both signals B and A together.

Several interesting dependencies appeared in the optimal HOT-2 model. There are dependencies of the acceptor positions −10, −6, −5, −3, and +2 on the branch site position −1, of the acceptor position −9 on the branch site position +9, of the acceptor position −7 on the branch site positions +6 and +9, and of the branch site position −2 on the acceptor site position −1.

However, considering these dependencies do not improve classification power significantly, according to our experiments. In general, using branch site models improved specificity very slightly (usually less than 1%). Interestingly, however, the correlation between predicted and actual probability of the signal is much improved in all models by using the branch site; we show these data in Table 4.

4.3 Dependencies among Three Signals

Previous studies have shown a relationship between the strengths of corresponding donor and acceptor sites [22,6,9]. Inspired by this, we investigated whether dependencies between positions of these signals can be found. We trained our models on a composite signal of donor, branch, and acceptor sites extracted from the same introns. The models were used to classify whether given pair of donor and acceptor come from the same intron. However, the dependencies between the signals were so weak that they do not facilitate such classification with any reasonable degree of accuracy. Table 5 shows that the dependencies between the

Model	Data set B97 without branch	Data set B97 with branch	Data set C22 without branch	Data set C22 with branch
PWM0	0.825	0.930	0.760	0.920
PWM1	0.882	0.906	0.920	0.961
TREE	0.872	0.912	0.918	0.959
PWM2	–	–	0.920	0.952
HOT-2	–	–	0.925	0.955

Table 4. Correlation coefficient between predicted signal probability and actual signal probability, for models of acceptor sites on the data sets B97 and C22 that do and do not include branch site enhancement.

Signal	Intra-signal adjacent	Intra-signal adjacent	Inter-signal
Donor	0.0597	0.0535	branch: 0.0066
			acceptor: 0.0094
Branch	0.1044	0.0303	acceptor: 0.0136
Acceptor	0.0835	0.0210	

Table 5. The strongest dependencies found among signals. Shown are the strongest dependencies between adjacent positions, non-adjacent intrasignal positions and intersignal positions. The values shown in the table are differential entropy $H(i)+H(j)-H(i,j)$; this is the value used in determining the best tree structure. Higher table entries correspond to stronger dependencies.

signals are much weaker than the dependencies between both non-adjacent and adjacent positions within the same signal.

5 Conclusions

We have presented a new model for biological signals found in DNA. Our new approach allows for positions in signals to be dependent on a small number of other positions, which need not be their direct predecessors. This model is an extension of both higher order position weight matrices, which are typically used for this problem, and of the Chow-Liu trees studied by Cai et al. [7] for this purpose.

Our models are based on directed hypertrees of dependence structure. For a given training data set, we have shown that it is NP-hard to compute the optimal dependency model from the family we consider, as long as positions are allowed to depend on multiple sites. However, in practice, we are able to compute good models, either by using integer programming or much simpler greedy heuristics.

In practice, we find that the addition of flexibility in characterizing dependencies in a model is modestly helpful, but not profoundly so. There is some advantage over the simply ordered model of position weight matrices, especially in the case of donor site prediction. We were also able to demonstrate that the signal probability estimates given by our new models are much more reliable predictors of actual probability of the sequence being a signal than the scores given by PWMs or tree models. This is useful in the context of probabilistic gene finding, where this estimation is one of many that are integrated into the gene finding process.

Future work. Several questions remain from our work. The first has to do with overtraining. Is it possible to estimate the size of data set needed to avoid overfitting these more complicated models? Clearly, they depend on more parameters than the corresponding higher order PWMs, since one must also infer the hypertree topology. On a related note, can overfitting be prevented by combining parameters from several orders, as is done in interpolated Markov chains [18]?

Second, in this work we chose a model with maximum likelihood. This resulted in models that were good predictors of a probability that a given site is a signal, but their discriminative power was not improved compared to other methods. Perhaps, the discriminative power could be improved by estimating parameters using different optimization criterion, as is suggested by Akutsu et al. [2] for PWMs.

Third, inter-signal dependencies between donor, branch, and acceptor sites do not help us to improve prediction. Another context in which intersignal dependencies might be discovered is in the core promoter, which is a region in DNA sequence, composed of several smaller signals located close to each other, that work together in transcription initiation. Our methods may help with recognition of core promoters.

Fourth, can we apply hypertree PWMs to other classification or prediction problems? Chow-Liu trees have been used successfully in numerous applications, which suggests that the more rich hypertree PWMs may, as well.

Lastly, we are currently investigating the effect of integrating these models together with probabilistic gene finders.

Acknowledgments. This work has been supported by a grant from the Human Science Frontier Program, a grant from the Natural Sciences and Engineering Research Council of Canada, and by an Ontario Graduate Scholarship.

References

1. P. Agarwal and V. Bafna. Detecting non-adjoining correlations within signals in DNA. In *Proceedings of the Second Annual International Conference on Research in Computational Molecular Biology (RECOMB 1998)*, pages 2–8. ACM Press, 1998.
2. Tatsuya Akutsu, Hideo Bannai, Satoru Miyano, and Sascha Ott. On the complexity of deriving position specific score matrices from examples. In A. Apostolico and M. Takeda, editors, *Combinatorial Pattern Matching, 13th Annual Symposium (CPM 2002)*, volume 2373 of *Lecture Notes in Computer Science*, pages 168–177. Springer, 2002.
3. L. D. Andersen and H. Fleischner. The NP-completeness of finding A-trails in Eulerian graphs and of finding spanning trees in hypergraphs. *Discrete Applied Mathematics*, 59:203–214, 1995.

4. F. R. Bach and M. I. Jordan. Thin junction trees. In Thomas G. Dietterich, Suzanna Becker, and Zoubin Ghahramani, editors, *Proceedings of NIPS 2001*, pages 569–576. MIT Press, 2001.
5. C. Burge and S. Karlin. Prediction of complete gene structures in human genomic DNA. *Journal of Molecular Biology*, 268:78–94, 1997.
6. C. B. Burge. Modeling dependencies in pre-mRNA splicing signals. In S. L Salzberg, D. B. Searls, and S. Kasif, editors, *Computational Methods in Molecular Biology*, pages 129–164. Elsevier, Amsterdam, 1998.
7. D. Cai, A. Delcher, B. Kao, and S. Kasif. Modeling splice sites with Bayes networks. *Bioinformatics*, 16(2):152–158, 2000.
8. C. K. Chow and C. N. Liu. Approximating discrete probability distributions with dependence trees. *IEEE Transactions on Information Theory*, IT-14(3):462–467, 1968.
9. F. Clark and T. A. Thanaraj. Categorization and characterization of transcript-confirmed constitutively and alternatively spliced introns and exons from human. *Human Molecular Genetics*, 11(4):451–454, 2002.
10. T. H. Cormen, C. E. Leiserson, R. L. Rivest, and C. Stein. *Introduction to Algorithms.* MIT Press, Cambridge, Mass., 2nd edition, 2001.
11. I. Dunham et al. The DNA sequence of human chromosome 22. *Nature*, 402:489–495, 1999.
12. K. Ellrott, C. Yang, F. M. Sladek, and T. Jiang. Identifying transcription factor binding sites through Markov chain optimization. In *Proceedings of the European Conference on Computational Biology (ECCB 2002)*, pages 100–109, 2002.
13. N. Friedman, D. Geiger, and M Goldszmidt. Bayesian network classifiers. *Machine Learning*, 29:131–163, 1997.
14. G. Gallo, G. Longo, S. Pallottino, and S. Nguyen. Directed hypergraphs and applications. *Discrete Applied Mathematics*, 42:177–201, 1993.
15. ILOG Inc. CPLEX optimizer, 2000. Computer software.
16. D. Karger and N. Srebro. Learning Markov networks: Maximum bounded tree-width graphs. In *Proceedings of the Twelfth Annual Symposium on Discrete Algorithms (SODA 2001)*, pages 392–401. SIAM, 2001.
17. R. M. Karp. Reducibility among combinatorial problems. In R. E. Miller and J. W. Thatcher, editors, *Complexity of Computer Computations*, pages 85–103, New York, 1972. Plenum Press.
18. S. L. Salzberg, A. L. Delcher, S. Kasif, and O. White. Microbial gene identification using interpolated Markov models. *Nucleic Acids Research*, 26(2):544–548, 1998.
19. A. Schrijver. *Theory of Linear and Integer Programming.* Wiley and sons, 1986.
20. R. Staden. Computer methods to aid the determination and analysis of DNA sequences. *Biochemical Society Transactions*, 12(6):1005–1008, 1984.
21. G. D. Stormo, T. D. Schneider, L. E. Gold, and A. Ehrenfeucht. Use of the 'Perceptron' algorithm to distinguish translational initiation sites in *E. coli. Nucleic Acids Research*, 10(9):2997–3011, 1982.
22. M. Q. Zhang. Statistical features of human exons and their flanking regions. *Human Molecular Genetics*, 7(5):919–932, 1998.

New Algorithm for the Simplified Partial Digest Problem*

J. Błażewicz[1,2] and M. Jaroszewski[1,2]

[1] Insitute of Computing Science,
Poznań University of Technology,
Piotrowo 3a, 60-965 Poznań, Poland
{Jacek.Blazewicz, Marcin.Jaroszewski}@cs.put.poznan.pl
[2] Institute of Bioorganic Chemistry,
Polish Academy of Sciences,
Noskowskiego 12, 61-704 Poznań, Poland

Abstract. In the paper, the problem of genome mapping is considered. In particular, the restriction site approach is used for this purpose. A new, efficient algorithm for solving the Simplified Partial Digest Problem is presented. The ideal data as well as data with measurement errors can be handled by this algorithm. An extensive computational experiment proved a clear superiority of the presented algorithm over other existing approaches. In addition, a thorough analysis of the Simplified Partial Digest Problem and a discussion of common experimental errors are given.

1 Introduction

Creation of a physical map is one of the basic steps of genome sequencing process. Such a map of a DNA strand contains the information about locations of short, specific subsequences called markers. There are many ways of physical map construction. One of them is based upon splitting a target strand into many shorter ones, called clones, that overlap each other. Next, each clone is subject to hybridization with a set of short DNA fragments, called probes, unique within the target DNA. The information upon which the original ordering of clones is reconstructed is the set of probes that bind to each clone. Methods and algorithms based on foregoing approach are presented among others in [1,13].

Another way to construct physical maps consists of a digestion of a DNA molecule with restriction enzymes. These enzymes cut DNA strands within short, specific patterns called restriction sites. After digestion, the lengths of obtained fragments are measured and serve as the basic information in the process of a reconstruction of the original ordering. In practice, several variants of this approach are used. Two of the best known are: *the double digest* and *the partial digest*.

* The research has been partially supported by a KBN grant.

G. Benson and R. Page (Eds.): WABI 2003, LNBI 2812, pp. 95–110, 2003.

In the former problem two restriction enzymes are used. A target DNA is amplified, perhaps using a PCR reaction, and the copies are divided into three sets. Molecules from the first set are digested by one enzyme, molecules from the second set are digested by the other enzyme and molecules from the third set are cut by both enzymes. All digestions are complete for the time span of each reaction is long enough to allow the enzyme to cut the target strand at each occurrence of the restriction site. As the result one obtains three collections of short DNA fragments that correspond to three digestion processes. The lengths of these fragments are measured during a gel electrophoresis process and recorded as three multisets. On the basis of this data locations of restriction sites in the target DNA are reconstructed. Unfortunately, from the combinatorial point of view this is a hard problem and, in addition, a number of equivalent solutions grows exponentially with the length of a strand being mapped [2,6,8,9,10,12,16].

A good alternative is, thus, *the partial digest method* where only one enzyme is used [11,14,15]. After amplification, a target DNA is divided into three sets. Molecules from each set are digested by the same enzyme, but the time span allowed for digestion differs among sets. The reaction times are chosen in such a way that in one of the sets most DNA strands were cut exactly once and in another set, exactly twice. For the third set, the reaction time span must be sufficient to let the enzyme cut all molecules in all ocurrences of the restriction site. As the result one gets three collections of restriction fragments. Again, the most important information are lengths of restriction fragments measured during a gel electrophoresis process. The restriction mapping problem based upon above-mentioned biochemical experiment is known as PDP (the Partial Digest Problem). Efficient backtracking algorithm for solving PDP that fills out the matrix of distances between restriction sites was designed by S. Skiena and co-workers [14]. While the algorithm is known to have an exponential complexity in the worst case, on average, however, it performs quite well. In addition, a modification of the above algorithm was proposed by S. Skiena and G. Sundaram [15] that yields very promising results in the presence of measurement errors. The computational complexity of PDP assuming error-free input data is an open question [13]. M. Cieliebak and co-workers proved the NP-hardness of PDP in the presence of unbounded errors [5].

In this paper, we propose an improved algorithm for solving *the simplified partial digest problem (SPDP)*, introduced in [3], where only two digestion processes are performed. In what follows, we will compare the new algorithm to the one discussed in [3] and to the backtracking algorithm designed by S. Skiena and G. Sundaram [15]. The comparison will include two cases. The first one deals with ideal data, involving no experimental errors, the second deals with data containing measurement errors. As an error model we assume the one discussed in [15] to unify the presentation and to enable comparison with the best approach to PDP known so far.

An organization of the paper is as follows. In Section 2, a description of biochemical experiment and a mathematical formulation of SPDP are given. In Section 3, the algorithm for ideal data case is presented along with a discussion of common experimental errors. Considering length measurement errors to be the most important ones, we adopt the model of errors similar to the one presented in [15]. In Section 4, computational results are given and the comparison with the results obtained in [3] and [15] is done.

2 The Problem and Its Basic Features

In the Simplified Partial Digest Problem (SPDP), similarly as in the Partial Digest Problem, only one restriction enzyme is used [3]. However, in case of SPDP only two digestions are performed. We will call them, respectively, a short digestion and a complete digestion. After amplification, the copies of a target strand are split into two sets. The goal of a short digestion is to have almost all molecules from one of the sets cut in at most one ocurrence of the restriction site. This is assured by properly chosen time span of the reaction. Molecules from the other set are cut in all ocurrences of the restriction site due to the long reaction time span (a complete digestion). Then, as in other methods, the lengths of restriction fragments obtained are measured during a gel electrophoresis process.

Let us now define the problem more formally. Let $\Gamma = \{\gamma_1, \gamma_2, \ldots, \gamma_{2n}\}$ be a multiset of fragment lengths (excluding the length of a whole DNA strand) that are obtained out of a short digestion and let $\Lambda = \{\lambda_1, \lambda_2, \ldots, \lambda_{n+1}\}$ be a multiset of fragment lengths obtained out of a complete digestion, where n denotes a number of restriction sites in the target DNA. Furthermore, let us sort the elements of multiset Γ in non-decreasing order. In this way we obtain list $A = \langle a_1, a_2, \ldots, a_{2n} \rangle$. It is easy to observe that in the ideal case (assuming no experimental errors) $a_i + a_{2n-i+1} = l$, $i = 1, \ldots, n$, where l denotes a length of the target DNA strand. The restriction fragments whose lengths are equal to, respectively, a_i and a_{2n-i+1} will be called *complementary* and a pair of complementary fragments will be denoted by $\{a_i, a_{2n-i+1}\}$, $i = 1, \ldots, n$. Obviously, each pair corresponds to exactly one restriction site in the target molecule. Unfortunately, the real ordering of complementary fragments within a pair is not known as the actual information gets lost during digestion processes. Thus, let $P_i = \langle a_i, a_{2n-i+1} \rangle$ and $P_{2n-i+1} = \langle a_{2n-i+1}, a_i \rangle$ denote permutations of pair $\{a_i, a_{2n-i+1}\}$, $i = 1, \ldots, n$. Next let us label restriction sites as $r_1, \ldots, r_n$ in such a way that condition: $s \leq t \Rightarrow a_s \leq a_t$, $s, t = 1, \ldots, n$ would hold for any restriction sites r_s and r_t. Additionally, let us arbitrarily label one of the ends of the target molecule as r_0, the other as r_{n+1} and let us call them, respectively, the left end and the right end of the map. In what follows, by the notion of *a labeled site* we will understand any restriction site or any one of the ends of the map. Let $\Sigma = \{a_1, \ldots, a_n, a_i - a_1, \ldots, a_i - a_{i-1}, l - a_n, l - a_n - a_1, \ldots, l - a_n - a_{n-1}\}$, $i = 2, \ldots, n$ denote a set of all distances between any two restriction sites r_s and r_t, $s = 1, \ldots, n-1$, $t = s+1, \ldots, n$ and between any restriction site r_s and the nearest end of the map (be it r_0 or r_{n+1}); only for r_n two distances: between r_0 and r_n and between r_n and r_{n+1}, respectively, are included. Some elements of Σ may be identical with regard to the value but still represent distances between different restriction sites and thus Σ is a set. Furthermore, let

$\Phi = \{\phi_1, \phi_2, \ldots, \phi_{2^n}\}$, $\phi_k = \langle c_{k1}, c_{k2}, \ldots, c_{k(n+1)} \rangle$, $c_{ki} \in \Sigma$, $i = 1, \ldots, n+1$, $k = 1, \ldots, 2^n$, denote a set of all *feasible orderings* of n restriction sites. A feasible ordering is defined as the result of *a composition* of n permutations, each permutation being obtained out of a different complementary pair. A composition of two permutations $P_i = \langle a_i, a_{2n-i+1} \rangle$ and $P_j = \langle a_j, a_{2n-j+1} \rangle$ is defined as triple $\langle a_i, a_j - a_i, a_{2n-j+1} \rangle$ if $a_i < a_j$ or triple $\langle a_j, a_i - a_j, a_{2n-i+1} \rangle$ whenever $a_j < a_i$, $i = 1, \ldots, 2n, j = 1, \ldots, 2n$, $i \neq j$. Using the foregoing definition, one can easily extend the notion of the composition to the case where i permutations are involved, $i = 3, \ldots, n$. In addition, one should note that values c_{ki} and $c_{k(i+1)}$, $i = 1, \ldots, n$, $k = 1, \ldots, 2^n$ are correlated. As the foregoing dependency is essential to the problem formulation, we will examine it a little bit further. Let us consider elements c_{ki} and $c_{k(i+1)}$ of ϕ_k, $i = 2, \ldots, n-1$, $k = 1, \ldots, 2^n$. The values of elements c_{ki} and $c_{k(i+1)}$ are determined by locations of restriction sites we will denote by r_s, r_t and r_t, r_u. Let $\{a_s, a_{2n-s+1}\}$, $\{a_t, a_{2n-t+1}\}$ and $\{a_u, a_{2n-u+1}\}$ denote complementary pairs that correspond, respectively, to restriction sites r_s, r_t and r_u. There are four different cases we will consider:

1. Restriction sites r_s, r_t and r_u are located in the left half of the map (the half that begins with r_0). In this case c_{ki} equals $a_t - a_s$ and $c_{k(i+1)}$ equals $a_u - a_t$.
2. Restriction sites r_s, r_t and r_u are located in the right half of the map (the half that ends with $r_{(n+1)}$). In this case we obtain $c_{ki} = a_s - a_t$ and $c_{k(i+1)} = a_t - a_u$.
3. Restriction sites r_s, r_t are located in the left half of the map, while restriction site r_u is located in the right half of the map. In this case $c_{ki} = a_t - a_s$ and $c_{k(i+1)} = l - a_u - a_t$.
4. Restriction site r_s is located in the left half of the map, while restriction sites r_t, r_u are located in the right half of the map. In the mentioned case c_{ki} equals $l - a_t - a_s$ and $c_{k(i+1)}$ equals $a_t - a_u$.

One can also construct similar rules for $i = 1$ as well as $i = n$.

Clearly, Φ contains all feasible as well as unacceptable solutions of SPDP. At the end, let us sort the elements of multiset Λ in non-decreasing order to obtain list $B = \langle b_1, b_2, \ldots, b_{n+1} \rangle$.

Using the above denotations, *the Simplified Partial Digest Problem* (the search version) can be formulated as follows:

Find element ϕ of Φ that satisfies the following criterion:

The list of elements of ϕ, sorted non-decreasingly, is identical to list B.

The above notions are illustrated in Fig. 1.

One can prove that a number of different feasible solutions of the Simplified Partial Digest Problem grows exponentially with a number of restriction sites for some instances.

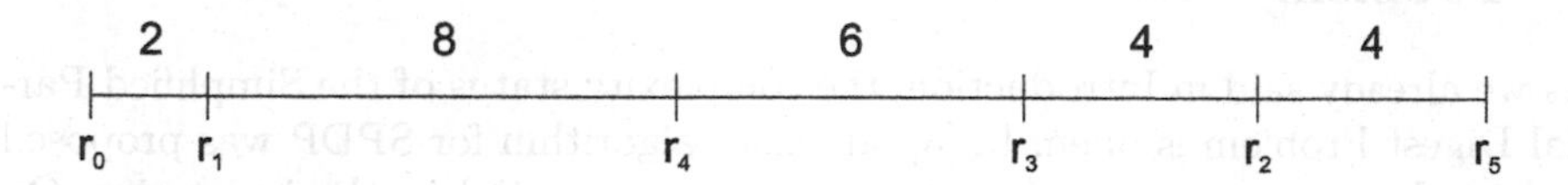

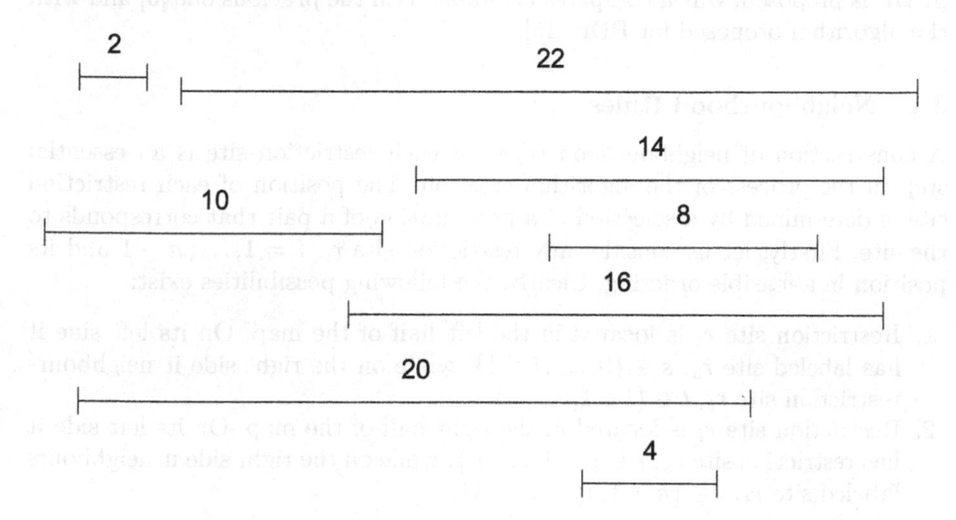

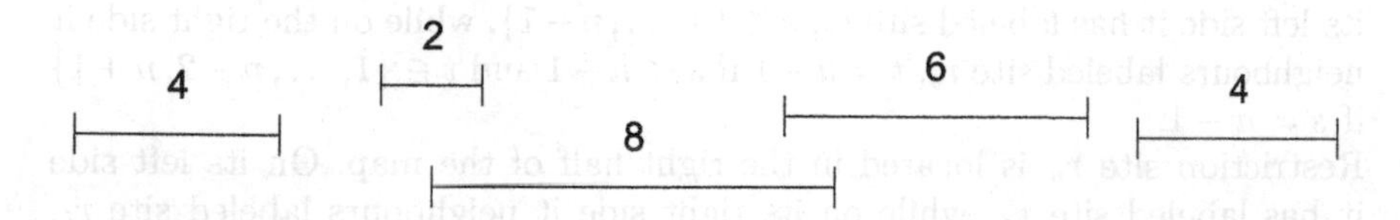

Fig. 1. An exemplary instance of the Simplified Partial Digest Problem with number of restriction sites n equal to 4 and length l of a target molecule equal to 24. $\Gamma = \{2, 22, 10, 8, 20, 4, 16, 14\}$, thus $A = \langle 2, 4, 8, 10, 14, 16, 20, 22\rangle$, while $\Lambda = \{2, 6, 4, 8, 4\}$ and hence $B = \langle 2, 4, 4, 6, 8\rangle$. There are four complementary pairs: $\{2, 22\}$, $\{4, 20\}$, $\{8, 16\}$, $\{10, 14\}$, each pair having exactly two permutations: $P_1 = \langle 2, 22\rangle$, $P_2 = \langle 4, 20\rangle$, $P_3 = \langle 8, 16\rangle$, $P_4 = \langle 10, 14\rangle$, $P_5 = \langle 14, 10\rangle$, $P_6 = \langle 16, 8\rangle$, $P_7 = \langle 20, 4\rangle$, $P_8 = \langle 22, 2\rangle$. Set of distances $\Sigma = \{2, 4, 8, 10, 14, 2, 6, 8, 12, 4, 6, 10, 2, 6\}$. Set Φ contains $2^4 = 16$ elements, with $\phi_1 = \langle 2, 2, 4, 2, 14\rangle$, $\phi_2 = \langle 2, 2, 4, 6, 10\rangle$, $\phi_3 = \langle 2, 2, 6, 6, 8\rangle$, $\phi_4 = \langle 2, 2, 10, 2, 8\rangle$, $\phi_5 = \langle 2, 6, 2, 10, 4\rangle$, $\phi_6 = \langle 2, 8, 6, 4, 4\rangle$, $\phi_7 = \langle 2, 6, 6, 6, 4\rangle$, $\phi_8 = \langle 2, 12, 2, 4, 4\rangle$ and with elements from ϕ_9 to ϕ_{16} inversely ordered with regard to elements, respectively, from ϕ_1 to ϕ_8. Only two solutions to the problem exist: ϕ_6 and ϕ_{14}, the former one being a composition of permutations: P_1, P_4, P_6, P_7 and the later one being a composition of permutations: P_8, P_5, P_3, and P_2

3 New Algorithm for the Simplified Partial Digest Problem

As we already said in Introduction, the complexity status of the Simplified Partial Digest Problem is open. In [3], an exact algorithm for SPDP was proposed and in the worst case its running time was exponential in the input size. On average, however, this time was much lower. In this paper, new algorithm for SPDP is proposed, which compares favorably with the previous one [3] and with the algorithm proposed for PDP [15].

3.1 Neighbourhood Rules

A construction of neighbourhood rules for each restriction site is an essential step in the process of the algorithm creation. The position of each restriction site is determined by a selection of a permutation of a pair that corresponds to the site. Firstly, let us consider any restriction site r_i, $i = 1, \ldots, n-1$ and its position in a feasible ordering. Clearly, the following possibilities exist:

1. Restriction site r_i is located in the left half of the map. On its left side it has labeled site r_s, $s \in \{0, \ldots, i-1\}$, while on the right side it neighbours restriction site r_t, $t \in \{i+1, \ldots, n\}$.
2. Restriction site r_i is located in the right half of the map. On its left side it has restriction site r_s, $s \in \{i+1, \ldots, n\}$, while on the right side it neighbours labeled site r_t, $t \in \{n+1, 1, \ldots, i-1\}$.

Now, let us analyze the situation for restriction site r_n:

1. Restriction site r_n is located in the left half or in the middle of the map. On its left side it has labeled site r_s, $s \in \{0, \ldots, n-1\}$, while on the right side it neighbours labeled site r_t, $t = n-1$ if $s < n-1$ and $t \in \{1, \ldots, n-2, n+1\}$ if $s = n-1$.
2. Restriction site r_n is located in the right half of the map. On its left side it has labeled site r_s, while on its right side it neighbours labeled site r_t, $t \in \{1, \ldots, n-1, n+1\}$, $s = n-1$ if $t \neq n-1$ and $s \in \{0, \ldots, n-2\}$ whenever $t = n-1$.

The neighbourhood rules are based upon analyses presented above. To make the rules as brief and consistent as possible, we will use Boolean variables. Let α_{ji} denote a Boolean variable that is equal to one whenever difference $a_j - a_i$ is present in a feasible ordering, and is equal to zero otherwise, $i = 1, \ldots, n-1$, $j = i+1, \ldots, n$. Furthermore, let α_{j0} and $\alpha_{(n+1)i}$ denote Boolean variables that are equal to one whenever, respectively, $a_j - a_0$ and $l - a_n - a_i$ are present in a feasible ordering, $i = 0, \ldots, n-1$, $j = 1, \ldots, n$, $a_0 = 0$. Otherwise, respective variables are equal to zero. Notice that equality: $\alpha_{j0} = 1$ may hold even if restriction site r_j is not a neighbour of r_0. In such a case r_j is located in the right half of the map and neighbours r_{n+1} on the right side. Adding the fact that each restriction site has exactly one neighbour on either side, we can formulate the neighbourhood rules as follows:

1. r_1: $\alpha_{10} \cap (\alpha_{21} \cup \alpha_{31} \cup \ldots \cup \alpha_{n1} \cup \alpha_{(n+1)1}) = 1$. Additionally, $(\alpha_{s1} \cap \alpha_{t1}) = 0$, $s, t = 2, \ldots, n+1$ and $s \neq t$.
2. r_2: $(\alpha_{20} \cup \alpha_{21}) \cap (\alpha_{32} \cup \alpha_{42} \cup \ldots \cup \alpha_{n2} \cup \alpha_{(n+1)2}) = 1$, i.e. $(\alpha_{20} \cup \alpha_{21}) = 1$ and $(\alpha_{32} \cup \alpha_{42} \cup \ldots \cup \alpha_{n2} \cup \alpha_{(n+1)2}) = 1$. Additionally, $(\alpha_{20} \cap \alpha_{21}) = 0$ and $(\alpha_{s2} \cap \alpha_{t2}) = 0$, $s, t = 3, \ldots, n+1$ and $s \neq t$.
3. ...
4. r_k, $k = 3, \ldots, n-2$: $(\alpha_{k0} \cup \alpha_{k1} \cup \alpha_{k2} \cup \ldots \cup \alpha_{k(k-1)}) \cap (\alpha_{(k+1)k} \cup \alpha_{(k+2)k} \cup \ldots \cup \alpha_{nk} \cup \alpha_{(n+1)k}) = 1$,i.e. $(\alpha_{k0} \cup \alpha_{k1} \cup \alpha_{k2} \cup \ldots \cup \alpha_{k(k-1)}) = 1$ and $(\alpha_{(k+1)k} \cup \alpha_{k+2)k} \cup \ldots \cup \alpha_{nk} \cup \alpha_{(n+1)k}) = 1$. Additionally, $(\alpha_{sk} \cap \alpha_{tk}) = 0$, $s, t = k+1, \ldots, n+1, s \neq t$ and $(\alpha_{ks} \cap \alpha_{kt}) = 0$, $s, t = 0, \ldots, k-1$ and $s \neq t$.
5. ...
6. r_{n-1}: $(\alpha_{(n-1)0} \cup \alpha_{(n-1)1} \cup \alpha_{(n-1)2} \cup \ldots \cup \alpha_{(n-1)(n-2)}) \cap (\alpha_{n(n-1)} \cup \alpha_{(n+1)(n-1)}) = 1$, i.e. $(\alpha_{(n-1)0} \cup \alpha_{(n-1)1} \cup \alpha_{(n-1)2} \cup \ldots \cup \alpha_{(n-1)(n-2)}) = 1$ and $(\alpha_{n(n-1)} \cup \alpha_{(n+1)(n-1)}) = 1$. Additionally, $(\alpha_{n(n-1)} \cap \alpha_{(n+1)(n-1)}) = 0$ and $\alpha_{(n-1)s} \cap \alpha_{(n-1)t}$, $s, t = 0, \ldots, n-2$ and $s \neq t$.
7. r_n: $(\alpha_{n0} \cup \alpha_{n1} \cup \alpha_{n2} \cup \ldots \cup \alpha_{n(n-2)} \cup \alpha_{n(n-1)}) \cap (\alpha_{(n+1)0} \cup \alpha_{(n+1)1} \cup \alpha_{(n+1)2} \cup \ldots \cup \alpha_{(n+1)(n-2)} \cup \alpha_{(n+1)(n-1)}) = 1$. Additionally, $(\alpha_{n(n-1)} \cup \alpha_{(n+1)(n-1)}) = 1$, $(\alpha_{ns} \cap \alpha_{nt}) = 0$, $(\alpha_{(n+1)s} \cap \alpha_{(n+1)t}) = 0$ and $(\alpha_{ns} \cap \alpha_{(n+1)s}) = 0$, $s, t = 0, \ldots, n-1$ and $s \neq t$.
8. The additional rule: $(\alpha_{20} \cup \ldots \cup \alpha_{(n+1)0}) = 1$ and $(\alpha_{s0} \cap \alpha_{t0}) = 0$, $s, t = 2, \ldots, n+1$ and $s \neq t$.

3.2 Construction of Distance Matrix M

For each restriction site r_j, $j = 1, \ldots, n$, the neighbourhood rules contain the product of two sums, each sum representing the set of feasible candidates for the left or the right neighbour of the restriction site. At most one element of each sum equals one for, as it has been pointed out above, each restriction site neighbours exactly one labeled site on the left and exactly one labeled site on the right. Moreover, at least one element of each sum equals one, as each restriction site clearly must have a neighbour on either side. For $j = 1, \ldots, n-1$ let $\Delta_j = \langle \delta_{j0}, \ldots, \delta_{j(j-1)}, 0, \ldots, 0 \rangle$ and $\Omega_j = \langle \omega_{(j+1)j}, \ldots, \omega_{(n+1)j}, 0, \ldots, 0 \rangle$, $\delta_{ji} = a_j - a_i$, $i = 0, \ldots, j-1$, $\omega_{(n+1)j} = l - a_n - a_j$ and $\omega_{ij} = a_i - a_j$, $i = j+1, \ldots, n$, denote the n dimensional vectors of differences that correspond, respectively, to the first and the second sum in the product given for restriction site r_j. Furthermore, let us extend the above denotations to restriction site r_n. Thus, let $\Delta_n = \langle \delta_{n0}, \ldots, \delta_{n(n-1)} \rangle$ and $\Omega_n = \langle \omega_{(n+1)0}, \ldots, \omega_{(n+1)(n-1)} \rangle$, $\delta_{ni} = a_n - a_i$, $i = 0, \ldots, n-1$, $\omega_{(n+1)i} = l - a_n - a_i$, $i = 0, \ldots, n-1$ denote the n dimensional vectors of differences that correspond to the first and the second sum in the product given for restriction site r_n. Moreover, let $\Omega_0 = \langle \omega_{20} \cup \ldots \cup \omega_{(n+1)0} \rangle$, $\omega_{(n+1)0} = l - a_n$ and $\omega_{i0} = a_i$, $i = 2, \ldots, n$ denote the n dimensional vector that corresponds to sum $\alpha_{20} \cup \ldots \cup \alpha_{(n+1)0}$. Square matrix M of distances between labeled sites, based on the neighbourhood rules, is constructed in the following way:

1. The i-th row of matrix M equals Δ_{i+1}, $i = 1, \ldots, n-1$.
2. The n-th row of matrix M equals Ω_n.

One can prove the following properties of matrix M:

1. j-th column of M equals Ω_{j-1}, $j = 1, \ldots, n$.
2. Construction of a feasible solution of SPDP consists in picking up exactly one element in each row and exactly one element in each column of the matrix, however, these elements may not be chosen arbitrarily.

The algorithm to be described in the next subsection is based on the above property 2.

3.3 The Algorithm

The proposed algorithm operates on matrix M of distances between restriction sites in order to eliminate all elements, but one, in each row and each column (see property 2 above), according to the rules that are specified below. In what follows by the notion of a main element we will understand such an element of M that forms any feasible solution of SPDP. One can prove that there is always at least one main element in each row and in each column of the matrix, however, no two main elements in any row or in any column form the same solution (see property 2 above). Due to the ambiguity that arises whenever it is impossible to establish the main element in a row or in a column (to establish means to distinguish the main element, that belongs to a solution constructed, from other non-zero elements in a row or in a column basing upon available knowledge), the algorithm performs random choices that may lead to non-feasible solutions to the Simplified Partial Digest Problem. Thus, the algorithm is recursive and enables backtracking to the stage where a false random choice has been made if a non-feasible solution has been obtained. Firstly, we will introduce some basic notions and definitions, then the stages of the algorithm will be presented.

Let p denote a level of recurrence. At the beginning $p = 1$. Let $f(p)$, $f(p) = 1, \ldots, n-1$ denote an index of the first row for which it is impossible to establish the main element at current level of recurrence p (one can prove that any unprocessed row at a current level of recurrence may be selected for that purpose). Furthermore, let $N_M(value)$ and $N_B(value)$ denote, respectively, a number of occurrences of elements of matrix M and a number of occurrences of elements of list B that are equal to $value$. Additionally, while considering any element m_{ij} of matrix M we will assume that $j = 1, \ldots, i+1$ for $i = 1, \ldots, n-1$ and $j = 1, \ldots, n$ for $i = n$. Let v denote an auxiliary variable used to indicate whether, at current level of recurrence p, a random choice of the main element (in $f(p)$-th row) is necessary.

Now let us present some definitions:

A disposable element - positive element m_{ij} for which $N_B(m_{ij}) = 0$. Thus, a disposable element does not have its counterpart on list B and, as list B contains only the lengths of fragments of the original ordering, a disposable element cannot form any feasible solution.

A solitary element of a column - positive element m_{ij} for which the following holds: $\neg(\exists k : m_{kj} > 0)$, where $k = j-1, \ldots, i-1, i+1, \ldots, n$ if $j \neq 1$ and $k = 1, \ldots, i-1, i+1, \ldots, n$ otherwise. Any solitary element of a column forms a feasible solution of SPDP (see property 2 of the matrix), unless a false random choice has been made before.

A solitary element of a row - positive element m_{ij} for which the following holds: $\neg(\exists k : m_{ik} > 0)$, where $k = 1, \ldots, j-1, j+1, \ldots, i+1$ if $i \neq n$ and $k = 1, \ldots, j-1, j+1, \ldots, n$ otherwise. Any solitary element of a row forms a feasible solution of SPDP (see property 2 of the matrix), unless a false random choice has been made before.

A unique element - positive element m_{ij} for which the following holds: $N_M(m_{ij}) = N_B(m_{ij})$. Any unique element forms a feasible solution of SPDP (see property 2 of the matrix), unless a false random choice has been made before.

A deletion - an operation that may proceed in two separate ways, depending on the kind of the element being deleted: a deletion of a disposable element consists in an assignment of 0 to the corresponding entry of the matrix, while a deletion of any other element of M or element of B consists in an assignment of $-p$ to the proper entry. These assignments allow for easy backtracking in the case a non-feasible solution of SPDP were constructed.

A consistency check - an operation performed on positive element m_{ij}, just deleted, which consists in comparison of both $N_M(m_{ij})$ and $N_B(m_{ij})$. The result of a check is negative whenever $N_M(m_{ij}) < N_B(m_{ij})$, since in such a case there are too few elements of the matrix that equal m_{ij} to construct any acceptable solution of SPDP.

A labeling - an operation performed on main element m_{ij} of a constructed solution. A labeling consists in an assignment of $-p$ to the proper entry of the matrix and in a deletion of an element of list B that equals m_{ij}. The deletion is performed in order to prevent a usage of elements of the matrix that equal m_{ij} too many times in the constructed solution.

Now, we may give a description of Algorithm 1 that finds a feasible solution of SPDP. On Fig. 2, a high-level description of the algorithm is given, while performed steps are described below in a greater detail.

Algorithm 1

1. Delete all disposable elements from matrix M.
2. Label m_{ij}. Delete all positive elements, but m_{ij}, in i-th row and j-th column of the matrix. After a deletion of each element perform the consistency check for this element. Mark i-th row and j-th column as processed. If the result of any check is negative, a feasible solution cannot be constructed anymore, thus, proceed to step 7.

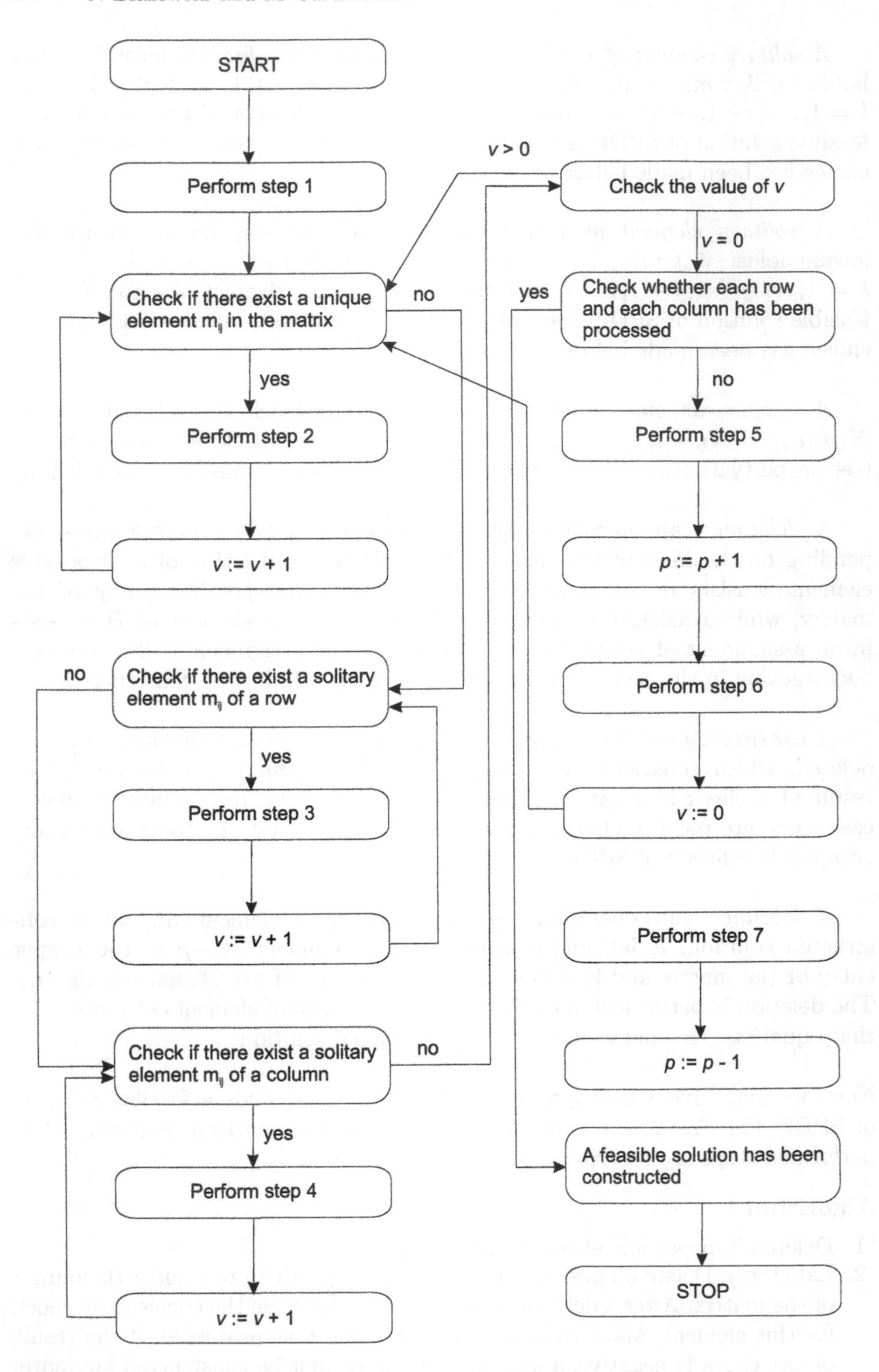

Fig. 2. A high-level description of the algorithm

3. Label m_{ij}. Delete all positive elements, but m_{ij}, in j-th column of the matrix. After a deletion of each element perform the consistency check for this element. Mark i-th row and j-th column as processed. If the result of any check is negative, a feasible solution cannot be constructed anymore, thus, proceed to step 7.
4. Label m_{ij}. Delete all positive elements, but m_{ij}, in i-th row of the matrix. After a deletion of each element perform the consistency check for this element. Mark i-th row and j-th column as processed. If the result of any check is negative, a feasible solution cannot be constructed anymore, thus, proceed to step 7.
5. Choose the first, not selected previously, positive element $m_{f(p)j}$, where $j = 1, \ldots, f(p)+1$, in $f(p)$-th row of the matrix. If no such element is found, a feasible solution cannot be constructed anymore, thus, proceed to step 7.
6. Label $m_{f(p)j}$. Delete all positive elements, but $m_{f(p)j}$, in $f(p)$-th row and j-th column of the matrix. After a deletion of each element perform the consistency check for this element. Mark $f(p)$-th row and j-th column as processed. If the result of any check is negative, a feasible solution cannot be constructed anymore, thus, proceed to step 7.
7. Backtrack to the stage of the last random choice undoing all the changes made during current level of recurrence p. They are fairly easy to track as each time an element of M or B is deleted, value $-p$ is assigned to the corresponding entry.

The above algorithm finds only the first feasible solution of the considered problem, but it can be modified to find all of them as it has been done for the needs of testing. Due to the ambiguities that may arise during construction of a solution, the algorithm has an exponential complexity in the worst case. However, its mean behavior is much better, what has been verified by the series of tests, results of which are shown in Section 4.

Algorithm 1 in its basic form could handle ideal data. However, it can be modified to handle data containing errors. This issue is briefly discussed below. The most common experimental errors are caused by imprecise measurement of the lengths of restriction fragments. According to [4], a 2%-5% relative measurement error is achievable. Thus, we have adopted the same model of measurement errors as given in [15], assuming a relative measurement error to range from 0.5% to 5% of the measured length.

Another type of errors is caused by missing restriction fragments. This can be due to approximately equal lengths of fragments, which cannot be then discriminated on a gel. On the other hand, this type of errors may be caused by the fact that certain sites are less likely to be cut than the others, so some fragments might not occur in sufficient quantity to be measured. However, in our SPDP approach such fragments are rather easily traced because there are only two digestion processes, respectively, with a quite short time span and with a very long one, which result in a relatively low number of restriction fragments as compared to the standard Partial Digest Problem. What is more, resulting fragments of a short digestion must form complementary pairs of the known total length. Adding the fact that the number of restriction sites is known, one can usually reconstruct missing fragments.

Summing up, we assume that measurement errors are the most important ones and for the model of such errors adopted from [15] we propose the algorithm that is very similar to the one for error-free input data. There are only two differences:

1. Instead of numbers, intervals are used that correspond to the lower and upper bound of the length of each fragment obtained out of the biochemical experiment (i. e. for measured length l and relative error r, an interval $\langle l_{min}, l_{max} \rangle$ such that $l(1-r) \leq l_{min} \leq l$ and $l \leq l_{max} \leq l(1+r)$ is assumed).
2. All operations performed by the algorithm for ideal data have been changed to fit the rules of interval arithmetic.

In the next section, extensive computational experiments are described, that characterize favorably the approach proposed.

4 Computational Results

In this section, we present the results of tests of the new algorithm in the case of error-free as well as noisy input data. The algorithm has been implemented in DJGPP C++. The tests were run on the PC with Pentium 670MHz processor and 128MB RAM under Windows 98 system. For the sake of comparison, the results obtained for the algorithm solving PDP, presented in [15], and those obtained for the first algorithm solving SPDP, presented in [3], have been added. Running times, numbers of equivalent solutions and relative error rates (in the case of noisy data) of all algorithms are compared. The number of equivalent solutions can be decreased by applying the grouping percentage factor [15], but it can also lead to the loss of feasible solutions. Since no grouping factor has been implemented for both algorithms solving SPDP, the results presented for PDP correspond to the grouping percentage factor equal to zero. In the tables below A_{PDP} stands for the algorithm presented in [15], A_{SPDP} stands for the algorithm presented in [3]. A_{NEW} denotes our just proposed approach, while *msec* stands for miliseconds.

4.1 Error-Free Input Data

At first, we have compared all three algorithms, i.e. solving PDP and two our SPDP approaches on randomly generated instances. Table 1 presents running times for tested algorithms.
The value of each entry in Table 1 represents an average of ten runs of the algorithm. We see that A_{SPDP} as well as A_{NEW} outperform the other algorithm.

The other series of tests for error-free input data have been performed for A_{NEW} only. The data have been obtained by cutting DNA chains taken from [7] in the sites recognizable by restriction enzymes AluI, HaeIII, HhaI and NlaIII. The lengths of restriction fragments created in such a way are indistinguishable

Table 1. Computational results for randomly generated instances (error-free input data)

	Running time		
n	A_{PDP}[sec]	A_{SPDP}[msec]	A_{NEW}[msec]
10	0.02 – 0.03	< 10	< 10
12	0.02 – 0.06	< 10	< 10
14	0.03 – 0.05	< 10	< 10
16	0.03 – 0.06	< 10	< 10
18	0.05 – 0.08	< 10	< 10
20	0.05 – 0.08	< 10	< 10

from data coming from an ideal biochemical experiment. Table 2 shows the results of computations for 11 instances based on real DNA chains, whose lengths range from 2, 7 kbp to 8, 9 kbp.

Table 2. Computational results of SPDP for instances based on real DNA chains (A_{NEW}, error-free input data)

Accession number in GenBank	Restriction enzyme	Number of restriction sites	Computation time [msec]
L13460F	HaeIII	5	< 10
K00470F	HhaI	8	< 10
K00470F	NlaIII	12	< 10
J00277F	NlaIII	19	< 10
D26561F	HaeIII	21	< 10
K00470F	HaeIII	21	< 10
L13460F	NlaIII	22	< 10
D26561F	NlaIII	33	< 10
D26561F	AluI	36	< 10
J00277F	HhaI	38	< 10
J00277F	HaeIII	99	< 10

4.2 Data with Measurement Errors

As a test instance for erroneous data, the restriction map of bacteriophage λ, see [15], was used, as well as randomly generated instances with even number of restriction sites n ranging from 10 to 20. In all cases length l of any restriction fragment was replaced by an interval $\langle l_{min}, l_{max} \rangle$ such that $l(1-r) \leq l_{min} \leq l$ and $l \leq l_{max} \leq l(1+r)$, where r denotes a relative error rate. Table 3 shows the results of computations for the restriction map of λ bacteriophage cut by enzyme HindIII. It resulted in seven restriction sites, the distances between adjacent sites being, respectively, $23130, 2027, 2322, 9416, 564, 125, 6557, 4361$ (cf. [15]). For each instance of the problem, numbers of equivalent solutions and running times are presented. Again, both algorithms for SPDP compare favorably with the other algorithm as far as running times are concerned, while having the same numbers of equivalent solutions.

Table 3. Computational results for erroneous instances based on bacteriophage λ (∞ means that no solution has been found in 5 minutes time span)

	Running time			Number of solutions		
Relative error r	A_{PDP}[sec]	A_{SPDP}[msec]	A_{NEW}[msec]	A_{PDP}	A_{SPDP}	A_{NEW}
1.0%	0.04	< 10	< 10	1	2	1
2.0%	0.53	< 10	< 10	1	2	2
3.0%	5.55	< 10	< 10	2	2	2
4.0%	24.17	< 10	< 10	2	2	2
5.0%	∞	< 10	< 10	–	2	2

Additionally, we have performed similar tests, for A_{NEW} only, on randomly generated instances with number of restriction sites n equal to 10 and relative measurement error rate r ranging from 1% to 5%. The results of the test are shown in Table 4.

Table 4. Computational results for randomly generated erroneous instances (n = 10, A_{NEW})

	Running time[msec]	Number of solutions	
Relative error r		minimal	maximal
1.0%	< 10	1	2
2.0%	< 10	1	2
3.0%	< 10	1	8
4.0%	< 10	3	10
5.0%	< 10	12	48

Next all algorithms have been compared on randomly generated erroneous instances. The results of the tests are reported in Table 5. Ten different problem instances were created for each value of n and r. The left bound of the given time interval denotes the best (shortest) and the right bound the worst (longest) running time.

Table 5. Computational results of SPDP for randomly generated instances (∞ means that no solution has been found in 5 minutes time span

	$r = 0,5$			$r = 1,0$		
n	A_{PDP}[sec]	A_{SPDP}[msec]	A_{NEW}[msec]	A_{PDP}[sec]	A_{SPDP}[msec]	A_{NEW}[msec]
10	$0 - 1,10$	< 10	< 10	$0 - 7,70$	< 10	< 10
12	$0 - 4,42$	< 10	< 10	$10,6 - 131$	< 10	< 10
14	$0 - 127$	< 10	< 10	∞	< 10	< 10
16	$75,9 - 94,9$	< 10	< 10	∞	< 10	< 10
18	∞	< 10	< 10	∞	< 20	< 10
20	∞	< 20	< 10	∞	< 20	< 60
	$r = 1,5$			$r = 2,0$		
n	A_{PDP}[sec]	A_{SPDP}[msec]	A_{NEW}[msec]	A_{PDP}[sec]	A_{SPDP}[msec]	A_{NEW}[msec]
10	$0 - 25,5$	< 10	< 10	$0 - 152$	< 10	< 10
12	∞	< 10	< 10	∞	< 20	< 10
14	∞	< 40	< 10	∞	< 400	< 50
16	∞	< 20	< 10	∞	< 100	< 50
18	∞	< 460	< 60	∞	< 5400	< 60
20	∞	< 12060	< 60	∞	< 83000	< 110

5 Conclusions

In the paper, the Simplified Partial Digest Problem, important for genome mapping, has been considered. The new algorithm for SPDP has been described. Again, computational experiments prove its clear superiority over the other approaches: PDP one [15] and the previous SPDP algorithm [3]. The advantage of the new algorithm is evident for both: error-free data and data with measurement errors.

6 Acknowledgement

The authors are greatly indepted to Michael Waterman for his valuable comments and a discussion of the approach.

References

1. Alizadeh, F., Karp, R.M., Weisser, D.K., Zweig, G.: Physical mapping of chromosomes using unique end-probes. Journal of Computational Biology **2** (1995) 159–184
2. Bellon, B.: Construction of restriction maps. Comput. Appl. Biol. Sci. **4** (1988) 111–115
3. Błażewicz, J., Formanowicz, P., Jaroszewski, M., Kasprzak, M., Markiewicz, W.T.: Construction of DNA restriction maps based on a simplified experiment. Bioinformatics **5** (2001) 398–404
4. Chang, W.I., Marr, T.: Personal communication to S. Skiena (1992)

5. Cieliebak, M., Eidenbenz, S., Penna, P.: Noisy Data Make the Partial Digest Problem NP-hard. Technical Report 381, ETH Zurich, Department of Computer Science (2002)
6. Dix, T.I., Kieronska, D.H.: Errors between sites in restriction site mapping. Comput. Appl. Biol. Sci. **4** (1988) 117–123
7. http://www.ncbi.nlm.nih.gov/GenBank/
8. Goldstein, L., Waterman, M.S.: Mapping DNA by stochastic relaxation. Adv. Appl. Math. **8** (1987) 194–207
9. Grigorjev, A.V., Mironov, A.A.: Mapping DNA by stochastic relaxation: a new approach to fragment sizes. Comput. Appl. Biol. Sci. **6** (1990) 107–111
10. Pevzner, P.A.: Physical mapping and alternating eulerian cycles in colored graphs. Algorithmica **13** (1995) 77–105
11. Rosenblatt, J., Seymour, P.: The structure of homeometric sets. SIAM J. Alg. Disc. Math. **3** (1982) 343–350
12. Schmitt, W., Waterman, M.S.: Multiple solutions of DNA restriction mapping problem. Adv. Appl. Math. **12** (1991) 412–427
13. Setubal, J., Meidanis, J.: Introduction to Computational Biology. PWS Publishing Company, Boston (1997)
14. Skiena, S.S., Smith, W.D., Lemke, P.: Reconstructing sets from interpoint distances. In: Proc. Sixth ACM Symp. Computational Geometry (1990) 332–339
15. Skiena, S.S., Sundaram, G.: A partial digest approach to restriction site mapping. Bulletin of Mathematical Biology **56(2)** (1994) 275–294
16. Waterman, M.S.: Introduction to Computational Biology. Maps, Sequences and Genomes. Chapman & Hall, London (1995)

Noisy Data Make the Partial Digest Problem NP-hard*

Mark Cieliebak[1], Stephan Eidenbenz[2**], and Paolo Penna[1]

[1] Institute of Theoretical Computer Science, ETH Zurich,
cieliebak@inf.ethz.ch, penna@inf.ethz.ch
[2] Los Alamos National Laboratory,
eidenben@lanl.gov

Abstract. The problem to find the coordinates of n points on a line such that the pairwise distances of the points form a given multi-set of $\binom{n}{2}$ distances is known as PARTIAL DIGEST problem, which occurs for instance in DNA physical mapping and de novo sequencing of proteins. Although PARTIAL DIGEST was – as a combinatorial problem – already proposed in the 1930's, its computational complexity is still unknown. In an effort to model real-life data, we introduce two optimization variations of PARTIAL DIGEST that model two different error types that occur in real-life data. First, we study the computational complexity of a minimization version of PARTIAL DIGEST in which only a subset of all pairwise distances is given and the rest are lacking due to experimental errors. We show that this variation is NP-hard to solve exactly. This result answers an open question posed by Pevzner (2000). We then study a maximization version of PARTIAL DIGEST where a superset of all pairwise distances is given, with some additional distances due to inaccurate measurements. We show that this maximization version is NP-hard to approximate to within a factor of $|D|^{\frac{1}{2}-\varepsilon}$ for any $\varepsilon > 0$, where $|D|$ is the number of input distances. This inapproximability result is tight up to low-order terms as we give a trivial approximation algorithm that achieves a matching approximation ratio.

1 Introduction

The PARTIAL DIGEST problem is one of the most intriguing problems in computational biology: on the one hand, it is a basic problem with relevant applications in DNA mapping and in protein sequencing; on the other hand, its computational complexity is a long–standing open problem. In the PARTIAL DIGEST problem we are given a multiset D of distances and are asked to find coordinates of points on a line such that D is exactly the multiset of all pairwise distances of these points. More formally, the PARTIAL DIGEST problem can be defined as follows.

---

* A preliminary version of this paper has been published as Technical Report 381, ETH Zurich, Department of Computer Science, October 2002.

** LA-UR-03:1157; work done while at ETH Zurich.

G. Benson and R. Page (Eds.): WABI 2003, LNBI 2812, pp. 111–123, 2003.

Definition 1 (Partial Digest). *Given an integer m and a multiset*[1] *of $k = \binom{m}{2}$ positive integers $D = \{d_1, \ldots, d_k\}$, is there a set of m integers $P = \{p_1, \ldots, p_m\}$ such that $\{|p_i - p_j| \mid 1 \leq i < j \leq m\} = D$?*

For example, if $D = \{2, 5, 7, 7, 9, 9, 14, 14, 16, 23\}$, then $P = \{0, 7, 9, 14, 23\}$ is one feasible solution (cf. Figure 1).

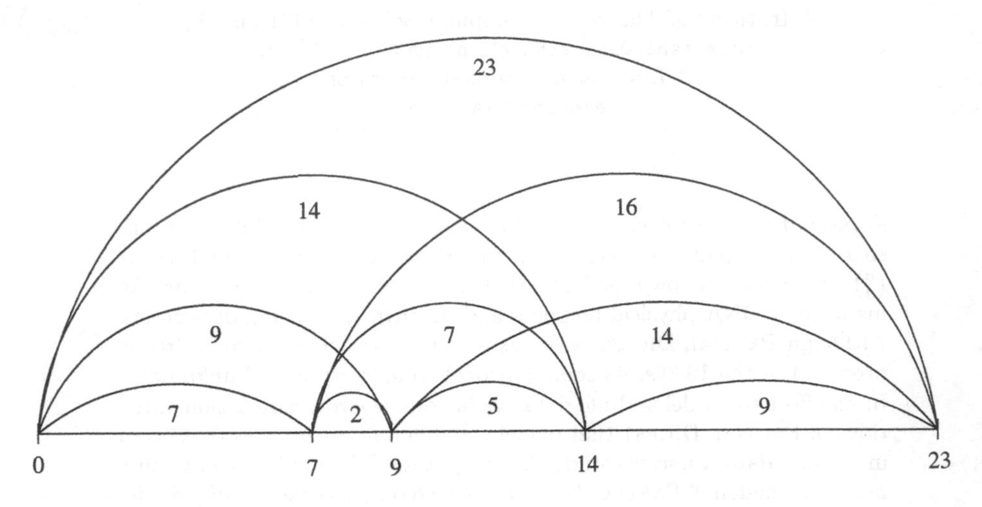

Fig. 1. Example for Partial Digest

Recently, the Partial Digest problem has received increasing attention due to its applications in computational biology, namely physical mapping of DNA and de novo sequencing of proteins (see below). However, in its pure combinatorial formulation the Partial Digest problem has been studied for a long time: It appears already in the 1930's in the sphere of X-ray crystallography (acc. to [23]); the problem is very closely related to the theory of homometric sets[2] [20]; and finally, it is also known as "turnpike problem", where we are given the pairwise distances of cities along a highway, and we want to find their ordering along the road [8].

We refer to the problem as Partial Digest due to it applications in the study of the structure of DNA molecules. Indeed, given a large DNA molecule, restriction enzymes can be used to generate a physical map of the molecule. A restriction enzyme cuts a DNA molecule at specific patterns, the restriction sites. For instance, the enzyme Eco RI cuts at occurrences of the pattern *GAATTC*. Under appropriate experimental conditions, e.g. by exposing the enzyme for different time periods or by using very small amounts of the enzyme, *all* fragments

[1] We will denote multisets like sets, since the fact of being a multiset is not crucial for our purposes.

[2] Two (noncongruent) sets of points are homometric if they generate the same multiset of pairwise distances.

between each two restriction sites are created. This process is called *partial digestion* (in contrast to full digestion, where the enzyme is applied long enough to cleave at all restriction sites). The lengths of the fragments, i.e., their number of nucleotides, are then measured by using gel electrophoresis. This leaves us with the multiset of distances between all restriction sites, and the objective is to reconstruct the original ordering of the fragments in the DNA molecule, which is the PARTIAL DIGEST problem.

The PARTIAL DIGEST problem occurs as well in de novo sequencing of proteins using tandem mass spectrometry: Given a probe with many copies of a single protein, we first use an enzyme like trypsin to digest the proteins. This leaves us with a set of protein fragments, called peptides. We separate the peptides by their mass, using tandem mass spectrometry. Then we break up these peptides into even smaller fragments using collision induced dissociation (CID). The mass/charge ratio of these fragments are measured using mass spectrometry again, resulting in a *tandem mass spectrum* of the peptide, which can be used to determine the amino acid sequence of the peptide (*de novo sequencing*). In the dissociation step, each single peptide can break up between any two amino acids in the peptide. If each single peptide breaks up exactly once (e.g., peptide $AEKGCWTR$ can break up into fragments $AEKG$ and $CWRT$, or into fragments AE and $KGCWTR$), then only fragments occur that are prefixes and suffixes of the peptide sequence. In this case there exist efficient algorithms for de novo sequencing [6, 17]. However, in real life experiments a single peptide does not only break up once, but it can break up several times, yielding internal fragments as well [3, 4, 14]. In the example, peptide $AEKGCWTR$ might break up into fragments AEK, GC and WTR. For this reason, we do not only obtain prefixes and suffixes in the spectrum, but all possible substrings of the peptide sequencs. Hence, the problem to find the appropriate sequence of amino acid for such a spectrum is equivalent to the PARTIAL DIGEST problem.

For the sake of simplicity, we will refer in this paper only to the setup of partial digestion experiments for DNA molecules. It is obvious that similar types of noise occur in tandem mass spectrometry data as well.

In reality, the partial digest experiment cannot be conducted under ideal conditions as outlined above, and thus errors occur in the data. In fact, there are four types of errors [9, 11, 13, 24, 26]:

Additional fragments An enzyme may erroneously cut in some cases at a site that is similar, but not exactly equivalent to a restriction site; thus, some distances will be added to the data even though they do not belong to it. Furthermore, fragments can be added through contamination with biological material, such as DNA from unrelated sources.

Missing fragments It can happen that a particular restriction site does not get cut in combination with all other restriction sites; then only one large fragment occurs in the data instead of the two (or even more) smaller fragments. Furthermore, fragments cannot be detected by gel electrophoresis if their amount is insufficient to be detected by common staining techniques.

Finally, small fragments may remain undetected at all since they run off at the end of the gel.

Fragment length Using gel electrophoresis, it is almost impossible to determine the exact length of a fragment. Typical error ranges are between 2% and 7% of the fragment length.

Multiplicity detection Determining the proper multiplicity of a distance from the brightness of its spot in the gel is a non–trivial problem.

In this paper, we define two optimization variations of PARTIAL DIGEST, where one variation models addition errors and the other models omission errors or missing fragments. Each variation allows only for one type of error to occur, and we will prove hardness results for both variations, implying that no polynomial-time algorithm can guarantee to find optimum or even nearly optimum solutions. For the third type of errors, it is known that the PARTIAL DIGEST problem becomes NP-hard if length measurements are erroneous [7], while we are not aware of any results on multiplicity errors. Intuitively, the problem of modeling "real-life" instances – in which *all* error types can occur – is even harder than having only one error type.

The MIN PARTIAL DIGEST SUPERSET problem models the situation of omissions, where we are given data in which some distances are missing, and we search for a set of points such that the number of omitted distances is minimum. It is formally defined as follows.

Definition 2 (MIN PARTIAL DIGEST SUPERSET). *Given a multiset of k positive integers $D = \{d_1, \ldots, d_k\}$, find the minimum m such that there is a set of m integers $P = \{p_1, \ldots, p_m\}$ with $D \subseteq \{|p_i - p_j| \mid 1 \leq i < j \leq m\}$.*

For example, if $D = \{2, 5, 7, 7, 9, 14, 23\}$, then the solution shown in Figure 1 would be a minimum solution for the MIN PARTIAL DIGEST SUPERSET instance D. On the other hand, if $D' = \{2, 7, 9, 9, 16\}$, then the points shown in the figure would cover all distances from D', but there exist solutions with less points that cover D', e.g. points $P' = \{0, 2, 9, 18\}$ (yielding distance multiset $\{2, 7, 9, 9, 16, 18\}$).

The MAX PARTIAL DIGEST SUBSET problem models the situation of additions, where we are given data in which some wrong distances were added, and we search for a set of points such that the number of added distances is minimum. A formal definition is as follows.

Definition 3 (MAX PARTIAL DIGEST SUBSET). *Given a multiset of k positive integers $D = \{d_1, \ldots, d_k\}$, find the maximum m such that there is a set of m integers $P = \{p_1, \ldots, p_m\}$ with $\{|p_i - p_j| \mid 1 \leq i < j \leq m\} \subseteq D$.*

Our two variations of the PARTIAL DIGEST problem allow the multiset of pairwise distances in a solution to be either a superset (i.e., to cover all given distances in D plus additional ones) or a subset (i.e., to contain only some of the distances in D) of the input multiset D. If a polynomial-time algorithm existed for either MIN PARTIAL DIGEST SUPERSET or MAX PARTIAL DIGEST SUBSET, we could use this algorithm to solve the original PARTIAL DIGEST problem as well: any YES instance of PARTIAL DIGEST is an instance of both problems above whose optimum is $\binom{m}{2}$; any NO instance of PARTIAL DIGEST is an instance of MAX PARTIAL DIGEST SUBSET (resp., MIN PARTIAL DIGEST SUPERSET) whose optimum is at most $\binom{m}{2} - 1$ (resp., at least $\binom{m}{2} + 1$). However, we show that such algorithms cannot exist, unless P = NP: We first show that computing the optimal solution for the MIN PARTIAL DIGEST SUPERSET problem is NP-hard, by proposing a reduction from the NP-complete problem EQUAL SUM SUBSETS. In a sense, our result provides an answer to the open problem 12.116 in the book by Pevzner [18], which asks for an algorithm to reconstruct a set of points, given a subset of their pairwise distances. We strengthen our hardness result by considering the t-PARTIAL DIGEST SUPERSET problem, where we restrict the cardinality of a solution to at most t, for some fixed parameter t; in this case, the problem remains NP-hard for *any* fixed $t = |D|^{\frac{1}{2}+\varepsilon}$ and any $\varepsilon > 0$. This result is tight in a sense, since any solution (even from the original PARTIAL DIGEST) must have at least cardinality $t = \Omega(|D|^{\frac{1}{2}})$. As for the MAX PARTIAL DIGEST SUBSET problem, we show that there is no polynomial–time algorithm for this problem that guarantees an approximation ratio[3] of $|D|^{\frac{1}{2}-\varepsilon}$ for any $\varepsilon > 0$, unless P = NP, by proposing a gap-preserving reduction from MAXIMUM CLIQUE. The problem MAXIMUM CLIQUE is very hard to approximate, and our reduction transfers the inapproximability of MAXIMUM CLIQUE to MAX PARTIAL DIGEST SUBSET. We also point to a trivial approximation algorithm that achieves a matching asymptotic approximation ratio. Thus, our result is tight up to low-order terms. Our inapproximability result means not only that can we not expect a polynomial-time algorithm that finds the optimum solution, but we cannot even expect a polynomial-time algorithm for MAX PARTIAL DIGEST SUBSET that finds solutions that are a factor $|D|^{\frac{1}{2}-\varepsilon}$ off the optimum.

Our hardness results show that a polynomial-time algorithm for the original PARTIAL DIGEST (if any) cannot be obtained by looking at the natural optimization problems we considered here. If any such algorithm exists, then it must exploit some combinatorial properties of PARTIAL DIGEST instances that do not hold for these optimization problems.

The exact computational complexity of PARTIAL DIGEST is a long–standing open problem: It can be solved in pseudopolynomial time[4] [15, 20]; there exists a backtracking algorithm (for exact or erroneous data) that has expected running time polynomial in the number of distances [23, 24], but exponential worst case running time [27]; it can be formalized by cut grammars, which have one additional symbol δ, the *cut*, that is neither a non–terminal nor a terminal symbol [21]; and finally, if the points are not on a line but in d-dimensional space, then the problem is NP-hard for some $d \geq 2$ [23]. However, for the original

[3] The approximation ratio of an algorithm $\mathcal{A}$ for any instance I is $\frac{OPT(I)}{\mathcal{A}(I)}$, where $\mathcal{A}(I)$ is the number of points in the solution of algorithm $\mathcal{A}$, and $OPT(I)$ is the number of points in an optimal solution.

[4] I.e., polynomial in the largest number of the input, but not necessarily polynomial in the bit length of the largest number.

PARTIAL DIGEST problem, neither a polynomial–time algorithm nor a proof of NP-completeness is known [5, 8, 17, 18, 19, 22].

In the biological setting of partial digestion, many experimental variations have been studied: Double digestion, where two different enzymes are used [22]; probed partial digestion, where probes (markers) are hybridized to partially digested DNA [1, 16]; simplified partial digest, where clones are cleaved in either one or in all restriction sites [5]; labeled partial digestion, where both ends of the DNA molecule are labeled before digestion [17]; and multiple complete digestion, where many different enzymes are used [10]. For a good survey on the PARTIAL DIGEST problem, see [23]; and for more recent discussions on the problem, see [18] and [22].

The paper is organized as follows: In Section 2 we present the hardness results of MIN PARTIAL DIGEST SUPERSET. In Section 3 we provide the (in-) approximability results on MAX PARTIAL DIGEST SUBSET. Finally, we conclude and present some open problems in Section 4.

2 NP-hardness of MIN PARTIAL DIGEST SUPERSET

In this section we show that MIN PARTIAL DIGEST SUPERSET is NP-hard by proposing a reduction from EQUAL SUM SUBSETS. We start with some notation.

A *multiset* with elements 1, 1, 3, 5, 5, and 8 is denoted by $\{1,1,3,5,5,8\}$. Subtracting an element from a multiset will remove it only once (if it is there), thus $\{1,1,3,5,5,8\} - \{1,4,5,5\} = \{1,3,8\}$. Given a set of integers $X = \{x_1,\ldots,x_n\}$, the *distance multiset* $\Delta(X)$ is defined as the multiset of all distances of X, i.e., $\Delta(X) := \{|x_i - x_j| \mid 1 \leq i < j \leq n\}$. We denote the sum of the elements of a set X of integers by sum(X), i.e., sum$(X) := \sum_{x \in X} x$. Finally, we say that a set of points P *covers* distance multiset D if $D \subseteq \Delta(P)$.

We first show that the minimum cardinality of a point set that covers all distances in a given multiset D cannot be to large: Let $D = \{d_1,\ldots,d_k\}$. If m is the minimal number such that a set P of cardinality m with $D \subseteq \Delta(P)$ exists, then $m \leq k+1$: We set $p_0 = 0, p_i = p_{i-1} + d_i$ for $1 \leq i \leq k$, and $P_{triv} = \{p_0,\ldots,p_k\}$, i.e., we simply put all distances from D in a chain "one after the other" (cf. Figure 2). In P_{triv}, each distance d_i induces a new point, and we use one additional starting point 0. Obviously, set P_{triv} covers D and has cardinality $k+1$.

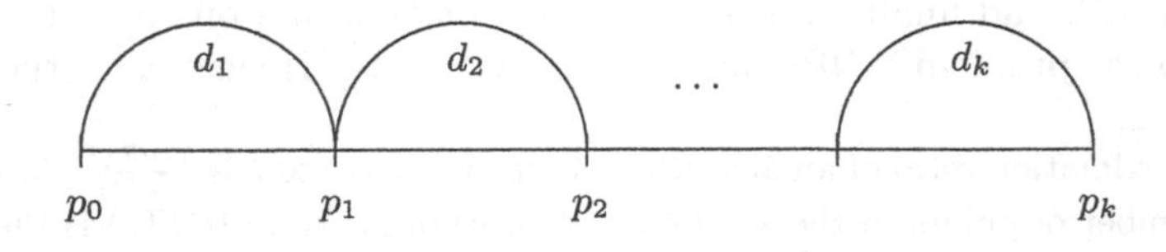

Fig. 2. Trivial solution for a distance multiset D.

Observe that PARTIAL DIGEST can be easily reduced to MIN PARTIAL DIGEST SUPERSET: Given an instance D of PARTIAL DIGEST of size $|D| = k$, there is a solution for D if and only if the minimal solution for the MIN PARTIAL DIGEST SUPERSET instance D has size $m = \frac{1}{2} + \sqrt{\frac{1}{4} + 2k}$ (in this case, $k = \binom{m}{2}$).

Theorem 1. MIN PARTIAL DIGEST SUPERSET *is* NP*-hard.*

Proof. We reduce EQUAL SUM SUBSETS to MIN PARTIAL DIGEST SUPERSET, where EQUAL SUM SUBSETS is the NP-complete problem [25] that is defined as follows: Given a set of n numbers $A = \{a_1, \ldots, a_n\}$, are there two disjoint nonempty subsets $X, Y \subseteq A$ such that $\text{sum}(X) = \text{sum}(Y)$?

Given an instance $A = \{a_1, \ldots, a_n\}$ of EQUAL SUM SUBSETS, we set $D = A$ (and $k = n$), and claim the following: There is a solution for the EQUAL SUM SUBSETS instance A if and only if a minimal solution for the MIN PARTIAL DIGEST SUPERSET instance D has at most n points.

"only if" part: Let X and Y be a solution for the EQUAL SUM SUBSETS instance. Assume w.l.o.g. that $X = \{a_1, \ldots, a_r\}$ and $Y = \{a_{r+1}, \ldots, a_s\}$ for some $1 \leq r < s \leq n$. We construct a set P that covers D and that has at most cardinality n. Similarly to the construction of P_{triv}, we line up the distances from D. In this case, *two* chains start at point 0: those distances from X and those from Y (cf. Figure 3); the remaining distances from $D - (X \cup Y)$ are at the end of the two chains.

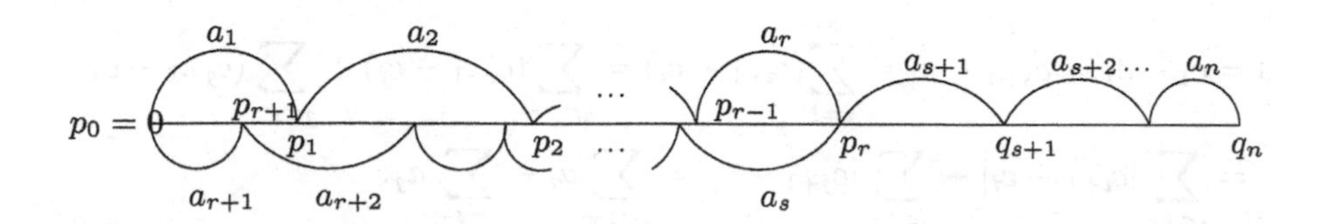

Fig. 3. Solution if there are two sets of equal sum.

Set $P = \{p_0, \ldots, p_{s-1}, q_{s+1}, \ldots, q_n\}$ is the corresponding set of points. Notice that there is no point "p_s" in set P, since the two chains corresponding to X and Y share two points, namely $p_0 = 0$ and their common endpoint p_r.

Obviously, P is a set of cardinality n. Moreover, by construction (cf. Figure 3), it holds that $D = \{a_1, \ldots, a_n\} \subseteq \Delta(P)$.

"if" part: Let $P = \{p_1, \ldots, p_m\}$ be an optimal solution for the MIN PARTIAL DIGEST SUPERSET instance with $m < n + 1$. Since P covers D, for each $a \in D$ there is a pair (p, q) of points $p, q \in P$ such that $a = |p - q|$. For each $a \in D$, we choose one such pair and say that it is *associated* with value a. We define a graph $G = (V, E)$ with $V = P$ and

$$E = \{(p, q) \mid (p, q) \text{ is associated with some } a \in D\},$$

i.e., G contains only those edges corresponding to some distance in D. Thus, $|V| = m$ and $|E| = |D| = n$. Since $m < n+1$, this graph contains a cycle. We show that such a cycle induces a solution of the EQUAL SUM SUBSETS instance.

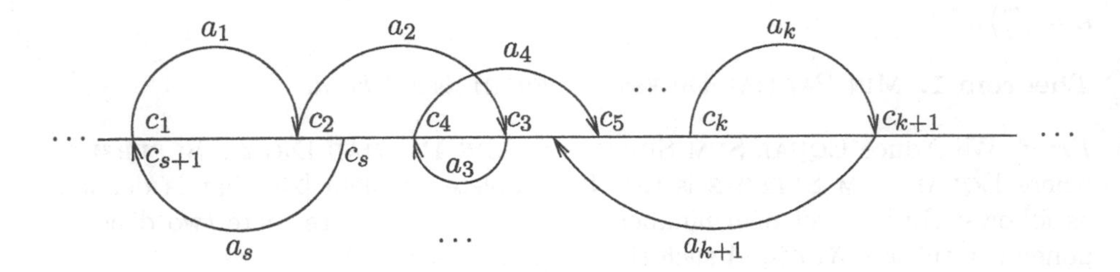

Fig. 4. A solution containing a cycle yields two subsets of equal sum: the overall lenght of right jumps equals to the overall length of left jumps.

Let $C = c_1, \ldots, c_s$ be a cycle in G (see Fig. 4). Then $|c_{i+1} - c_i| \in D$, for all $1 \le i \le s$ (with some abuse of notation we consider $c_{s+1} = c_1$). Assume w.l.o.g. that $|c_{i+1} - c_i|$ is associated with a_i, for $1 \le i \le s$. We define $I^+ = \{i \in \{1, \ldots, s\} \mid c_{i+1} > c_i\}$ and $I^- = \{j \in \{1, \ldots, s\} \mid c_{j+1} < c_j\}$, i.e., we partition the edges in the cycle into two sets, those that are oriented to the left (I^-) and those that are oriented to the right (I^+). This yields

$$0 = c_1 - c_1 = c_{s+1} - c_1 = \sum_{i=1}^{s}(c_{i+1} - c_i) = \sum_{i \in I^+}(c_{i+1} - c_i) + \sum_{j \in I^-}(c_{j+1} - c_j)$$
$$= \sum_{i \in I^+}|c_{i+1} - c_i| - \sum_{j \in I^-}|c_{j+1} - c_j| = \sum_{i \in I^+} a_i - \sum_{j \in I^-} a_j.$$

Sets $X := \{a_i \mid i \in I^+\}$ and $Y := \{a_j \mid j \in I^-\}$ yield equal sums, and thus a solution of the EQUAL SUM SUBSETS instance. □

In the previous proof, we distinguished whether a minimal solution uses at most n points, or $n+1$ points. It is even possible to "decrease" this boundary to some value t that is still sufficiently large. In fact, we can show that t-PARTIAL DIGEST SUPERSET is NP-hard for every $0 < \varepsilon < \frac{1}{2}$ if we set t to be at least $f(|D|) = |D|^{\frac{1}{2}+\varepsilon}$. Observe that for a distance multiset D, a minimal set of points covering D has cardinality at least $\frac{1}{2} + \sqrt{\frac{1}{4} + 2|D|} \approx |D|^{\frac{1}{2}}$. Moreover, the PARTIAL DIGEST problem is equivalent to t-PARTIAL DIGEST SUPERSET with $t = \frac{1}{2} + \sqrt{\frac{1}{4} + 2|D|} = O\left(|D|^{\frac{1}{2}}\right)$. The proof will be given in the full version of this paper. It is a reduction from EQUAL SUM SUBSETS where we "blow up" the instance of MIN PARTIAL DIGEST SUPERSET used in the proof above by adding an appropriate number of additional distances that do not interfere.

3 (In-) Approximability of Max Partial Digest Subset

In this section, we show that Max Partial Digest Subset is almost as hard to approximate as Maximum Clique, and we give a trivial approximation algorithm that achieves a matching approximation ratio.

We need to introduce some notation for large numbers first. The numbers are expressed in the number system of some base Z. We denote by $\langle a_1, \ldots, a_n \rangle$ the number $\sum_{1 \leq i \leq n} a_i Z^{n-i}$; we say that a_i is the i-th digit of this number. We will choose base Z large enough such that adding up numbers in our proof will not lead to carry-digits from one digit to the next. Therefore, we can add numbers digit by digit. The same holds for scalar products. For example, having base $Z = 27$ and numbers $\alpha = \langle 3, 5, 1 \rangle, \beta = \langle 2, 1, 0 \rangle$, then $\alpha + \beta = \langle 5, 6, 1 \rangle$ and $3 \cdot \alpha = \langle 9, 15, 3 \rangle$. We will allow different bases for each digit. We define the concatenation of two numbers by $\langle a_1, \ldots, a_n \rangle \circ \langle b_1, \ldots, b_m \rangle := \langle a_1, \ldots, a_n, b_1, \ldots, b_m \rangle$, i.e., $\alpha \circ \beta = \alpha Z^m + \beta$, where m is the number of digits in β. Let $\Delta_n(i) := \langle 0, \ldots, 0, 1, 0, \ldots, 0 \rangle$ be the number that has n digits, all 0's except for the i-th position where the digit is 1. Moreover, $\mathbb{1}_n := \langle 1, \ldots, 1 \rangle$ has n digits, all 1's, and $\mathbb{0}_n := \langle 0, \ldots 0 \rangle$ has n zeros.

We construct a gap-preserving reduction (as introduced in [2]) from Max Clique to Max Partial Digest Subset. Max Clique is the problem of finding a maximum complete subgraph from a given graph. It cannot be approximated by any polynomial-time algorithm with an approximation ratio of $n^{1-\varepsilon}$ for any $\varepsilon > 0$, where n is the number of vertices of the input graph, unless $\mathsf{P} = \mathsf{NP}$ [12]. Our reduction is gap-preserving, which means that the inapproximability of Max Clique is transfered to Max Partial Digest Subset.

Suppose we are given a graph $G = (V, E)$ with vertex set $V = \{v_1, \ldots, v_n\}$ and edge set $E \subseteq V \times V$. We construct an instance D of Max Partial Digest Subset by creating a number $d_{i,j} = \mathbb{0}_i \circ \mathbb{1}_{j-i} \circ \mathbb{0}_{n-j}$ with base $Z = n^2 + 1$ for each $(v_i, v_j) \in E, j > i$.

Let OPT be the size of the maximum clique in G (i.e., the number of vertices in the maximum clique), let OPT' be the maximum number of points that can be placed on a line such that all pairwise distances appear in D, let $k > 0$ be an integer, and let $\varepsilon > 0$. The following two lemmas show how the reduction works.

Lemma 1. $OPT \geq kn^{1-\varepsilon} \implies OPT' \geq kn^{1-\varepsilon}$

Proof. Assume we are given a clique in graph G of size $kn^{1-\varepsilon}$. We construct a solution for the corresponding Maximum Partial Digest instance D by positioning a point at position $v_i' = \mathbb{1}_i \circ \mathbb{0}_{n-i}$ for each vertex v_i in the clique. This yields a feasible solution for D, since – for $j > i$ – each distance $v_j' - v_i' = \mathbb{0}_i \circ \mathbb{1}_{j-i} \circ \mathbb{0}_{n-j} = d_{i,j}$ between two points v_j' and v_i' corresponds to an edge in G and is therefore encoded as distance $d_{i,j}$ in D. □

Lemma 2. $OPT < k \implies OPT' < k$

Proof. We prove the contraposition, i.e., $OPT' \geq k \implies OPT \geq k$. Suppose we are given a solution of the Max Partial Digest Subset instance consisting of k points $p_1 < \ldots < p_k$ on the line, where we assume w.l.o.g. that $p_1 = \mathbb{0}_n$. By definition distance $p_k - p_1$ must be contained in the distance set D and thus two indices $i_{\min}$ and $j_{\max}$ must exist with $d_{i_{\min}, j_{\max}} = p_k - p_1$. Each of the points $p_2, \ldots, p_{k-1}$ from the solution has the following properties:

1. It only has zeros and ones in its digits, as the distance to point p_1 would not be in D otherwise.
2. It only has zeros in the first $i_{\min}$ digits, as the distance to point p_k would not be in D otherwise.
3. It contains at most a single continuous block of ones in its digits, as the distance to point p_1 would not be in D otherwise.

The points $p_2, \ldots, p_{k-1}$ also have the property that they are either all of the form $\mathbb{0}_{i_{\min}} \circ \mathbb{1}_l \circ \mathbb{0}_{j_{\max}-l-i_{\min}} \circ \mathbb{0}_{n-j_{\max}}$ or all of the form $\mathbb{0}_{i_{\min}} \circ \mathbb{0}_l \circ \mathbb{1}_{j_{\max}-l-i_{\min}} \circ \mathbb{0}_{n-j_{\max}}$, where $i_{\min} \leq l \leq j_{\max}$. If both forms existed in a solution, i.e., at least one point of each form existed, then the distance between points of different form would not be in D, since at least one digit would not be 0 or 1.

We construct a vertex set V' that will turn out to be a clique by letting $v_{i_{\min}}$ and $v_{j_{\max}}$ be in this set V'. Additionally, for each $p_{k'}$ for $k' = 2, \ldots k-1$, where $p_{k'}$ is of the form $\mathbb{0}_{i_{\min}} \circ \mathbb{1}_{l'} \circ \mathbb{0}_{j_{\max}-l'-i_{\min}} \circ \mathbb{0}_{n-j_{\max}}$ or $\mathbb{0}_{i_{\min}} \circ \mathbb{0}_{l'} \circ \mathbb{1}_{j_{\max}-l'-i_{\min}} \circ \mathbb{0}_{n-j_{\max}}$, where $i_{\min} \leq l' \leq j_{\max}$, we let $v_{l'+i_{\min}}$ be in the vertex set V'.

In order to see that the vertex set V' is a clique, consider the difference $p_{k'} - p_{k''}$ of any two points with $k' > k''$, where $p_{k'}$ has led to the inclusion of vertex $v_{l'}$ into the set and $p_{k''}$ has led to the inclusion of vertex $v_{l''}$ into the clique. This difference is exactly $d_{l',l''}$ for both possible forms, and thus the edge $(v_{l'}, v_{l''})$ is in E. □

The promise problem of Max Clique, in which we are promised that the size of the maximum clique in a given graph G is either at least $kn^{1-\varepsilon}$, or less than k, and we are to decide which is true, is NP-hard to decide [12]. Lemmas 1 and 2 transform this promise problem of Max Clique into a promise problem of Max Partial Digest Subset, in which we are promised that in an optimum solution of D either at least $kn^{1-\varepsilon}$ or less than k points can be placed on a line. This promise problem of Max Partial Digest Subset is NP-hard to decide as well, since a polynomial-time algorithm for it could be used to decide the promise problem of Max Clique.[5] Thus, unless P = NP, Max Partial Digest Subset cannot be approximated with an approximation ratio of:

$$\frac{kn^{1-\varepsilon}}{k} = n^{1-\varepsilon} \geq |D|^{\frac{1}{2}-\varepsilon},$$

where $|D|$ is the number of distances in instance D, since we could decide the corresponding promise problem in polynomial time otherwise. We have shown the following:

[5] The concept of gap-preserving reductions is an alternative way to view a reduction. It provides a formal framework for preserving inapproximability ratios between two optimization problems. For details, see [2].

Theorem 2. MAX PARTIAL DIGEST SUBSET *cannot be approximated by any polynomial-time algorithm with an approximation ratio of* $|D|^{\frac{1}{2}-\varepsilon}$ *for any* $\varepsilon > 0$, *where* $|D|$ *is the number of input distances, unless* $\mathsf{P} = \mathsf{NP}$.

A trivial approximation algorithm for a MAX PARTIAL DIGEST SUBSET instance $D = \{d_1, \ldots, d_{|D|}\}$ that simply places two points at distance d_1 from each other achieves a matching approximation ratio of $O(|D|^{\frac{1}{2}})$.

4 Conclusion and Open Problems

We have shown that the optimization problems MIN PARTIAL DIGEST SUPERSET and MAX PARTIAL DIGEST SUBSET are NP-hard. Moreover, the maximization problem is not approximable within reasonable bounds, unless $\mathsf{P} = \mathsf{NP}$. This answers the problem 12.116 left open in [18], and gives rise to new open questions:

1. Since our optimization variations model different error types that (always) occur in real-life data, our hardness results suggest that real-life PARTIAL DIGEST problems are in fact instances of NP-hard problems. However, the backtracking algorithm from [23] seems to run in polynomial-time for real-life instances. How can this be explained? What relevant properties do real-life instances have that prevent them from becoming intractable?
2. What is the best approximation ratio for MIN PARTIAL DIGEST SUPERSET?
3. Using gel electrophoresis or mass spectrometry, it is very hard to determine the correct multiplicity of a distance. This yields the following variation of PARTIAL DIGEST: we are given a *set* of distances, and for each distance a multiplicity, and we ask for points on a line such that the multiplicities of the corresponding distance set do not differ "to much" from the given multiplicities. What is the computational complexity of this problem?
4. Is there a polynomial–time algorithm for the PARTIAL DIGEST problem if we restrict the input to be a *set* of distances (instead of a multiset), i.c., if we know in advance that each two distances are pairwise distinct?

Finally and obviously, the main open problem is still the computational complexity of PARTIAL DIGEST.

Acknowledgments We would like to thank Dirk Bongartz and Walter Unger for pointing us to the PARTIAL DIGEST problem, and Sacha Baginsky, Aris Pagourtzis and Peter Widmayer for their help in this work.

References

[1] F. Alizadeh, R. M. Karp, L. A. Newberg, and D. K. Weisser. Physical mapping of chromosomes: A combinatorial problem in molecular biology. In *Symposium on Discrete Algorithms*, pages 371–381, 1993.

[2] S. Arora and C. Lund. Hardness of approximations. In D. Hochbaum, editor, *Approximation Algorithms for* NP-*Hard Problems*, pages 399–446. PWS Publishing Company, 1996.
[3] V. Bafna and N. Edwards. On de novo interpretation of tandem mass spectra for peptide identification. In 7th *Annual International Conference on Computational Biology (RECOMB 03)*, pages 9–18, 2003.
[4] S. Baginsky. Personal communication, 2003.
[5] J. Błażewicz, P. Formanowicz, M. Kasprzak, M. Jaroszewski, and W. T. Markiewicz. Construction of DNA restriction maps based on a simplified experiment. *Bioinformatics*, 17(5):398–404, 2001.
[6] T. Chen, M. Kao, M. Tepel, J. Rush, and G. M. Church. A dynamic programming approacht to de novo peptide sequencing via tandem mass spectrometry. In 11th *SIAM-ACM Symposium on Discrete Algorithms (SODA)*, pages 389–398, 2000.
[7] M. Cieliebak and S. Eidenbenz. Measurement errors make the partial digest problem NP-hard, manuscript, to be published, 2003.
[8] T. Dakić. *On the turnpike problem.* PhD thesis, Simon Fraser University, 2000.
[9] T. I. Dix and D. H. Kieronska. Errors between sites in restriction site mapping. *Computer Applications in the Biosciences (CABIOS)*, 4(1):117–123, 1988.
[10] D. Fasulo. *Algorithms for DNA Restriction Mapping.* PhD thesis, University of Washington, 2000.
[11] J. Fütterer. Personal communication, 2002.
[12] J. Håstad. Clique is hard to approximate within $n^{1-\epsilon}$. In *Proc. of the Symposium on Foundations of Computer Science*, 1996.
[13] J. Inglehart and P. C. Nelson. On the limitations of automated restriction mapping. *Computer Applications in the Biosciences (CABIOS)*, 10(3):249–261, 1994.
[14] P. James. *Proteome Research: Mass Spectrometry.* Springer, 2001.
[15] P. Lemke and M. Werman. On the complexity of inverting the autocorrelation function of a finite integer sequence, and the problem of locating n points on a line, given the $\binom{n}{2}$ unlabelled distances between them. Preprint 453, Institute for Mathematics and its Application IMA, 1988.
[16] L. Newberg and D. Naor. A lower bound on the number of solutions to the probed partial digest problem. *Advances in Applied Mathematics (ADVAM)*, 14:172–183, 1993.
[17] G. Pandurangan and H. Ramesh. The restriction mapping problem revisited. *Journal of Computer and System Sciences (JCSS)*, to appear 2002. Special issue on Computational Biology.
[18] P. Pevzner. *Computational Molecular Biology.* MIT Press, 2000.
[19] P. A. Pevzner and M. S. Waterman. Open combinatorial problems in computational molecular biology. In *Proc. of the Third Israel Symposium on Theory of Computing and Systems ISTCS*, pages 158–173. IEEE Computer Society Press, 1995.
[20] J. Rosenblatt and P. Seymour. The structure of homometric sets. *SIAM Journal of Algorithms and Discrete Mathematics*, 3(3):343–350, 1982.
[21] D. B. Searls. Formal grammars for intermolecular structure. In *Proceedings of the International IEEE Symposium on Intelligence in Neural and Biological Systems*, 1995.
[22] J. Setubal and J. Meidanis. *Introduction to Computational Molecular Biology.* PWS Boston, 1997.
[23] S. S. Skiena, W. Smith, and P. Lemke. Reconstructing sets from interpoint distances. In *Sixth ACM Symposium on Computational Geometry*, pages 332–339, 1990.

[24] S. S. Skiena and G. Sundaram. A partial digest approach to restriction site mapping. *Bulletin of Mathematical Biology*, 56:275–294, 1994.
[25] G. J. Woeginger and Z. L. Yu. On the equal-subset-sum problem. *Information Processing Letters*, 42:299–302, 1992.
[26] L. W. Wright, J. B. Lichter, J. Reinitz, M. A. Shifman, K. K. Kidd, and P. L. Miller. Computer-assisted restriction mapping: an integrated approach to handling experimental uncertainty. *Computer Applications in the Biosciences (CABIOS)*, 10(4):435–442, 1994.
[27] Z. Zhang. An Exponential Example for a Partial Digest Mapping Algorithm. *Journal of Computational Biology*, 1(3):235–239, 1994.

Pattern Discovery Allowing Wild-Cards, Substitution Matrices, and Multiple Score Functions

Alban Mancheron and Irena Rusu

I.R.I.N., Université de Nantes, 2, Rue de la Houssinière, B.P. 92208,
44322 Nantes Cedex 3, France
{Mancheron, Rusu}@irin.univ-nantes.fr

Abstract. Pattern discovery has many applications in finding functionally or structurally important regions in biological sequences (binding sites, regulatory sites, protein signatures etc.). In this paper we present a new pattern discovery algorithm, which has the following features:

- it allows to find, in exactly the same manner and without any prior specification, patterns with fixed length gaps (i.e. sequences of one or several consecutive wild-cards) and contiguous patterns;

- it allows the use of any pairwise score function, thus offering multiple ways to define or to constrain the type of the searched patterns; in particular, one can use substitution matrices (PAM, BLOSUM) to compare amino acids, or exact matchings to compare nucleotides, or equivalency sets in both cases.

We describe the algorithm, compare it to other algorithms and give the results of the tests on discovering binding sites for DNA-binding proteins (ArgR, LexA, PurR, TyrR respectively) in *E. coli*, and promoter sites in a set of Dicot plants.

1 Introduction

In its simplest form, the pattern discovery problem can be stated as follows: given t sequences $S_1, S_2, \ldots, S_t$ on an arbitrary alphabet, find subsequences (or *patterns*) $o_1 \in S_1, o_2 \in S_2, \ldots, o_t \in S_t$ such that $o_1, o_2, \ldots, o_t$ are similar (according to some criterion). More complex formulations include a quorum constraint: given q, $1 \leq q \leq t$, at least q similar patterns must be found.

The need to discover similar patterns in biological sequences is straightforward (finding protein signatures, regulatory sites etc. [2]). Major difficulties are the NP-hardness of the problem, the diversity of the possible criteria of similarity, the possible types of patterns. All the variations in the formulation of the problem cannot be efficiently handled by the same, unique algorithm.

The consequence is that many pattern discovery algorithms have appeared during the last decade (Gibbs sampling, Consensus, Teiresias, Pratt, Meme, Motif, Smile, Winnower, Splash ...). See [2] for a survey. Most of these algorithms define a quite precise model (that is, type of the searched pattern, similarity

G. Benson and R. Page (Eds.): WABI 2003, LNBI 2812, pp. 124–138, 2003.

notion, score function), even if exceptions exist (e.g., Pratt allows the use of one among five score functions). Therefore, each such algorithm treats one aspect of the pattern discovery problem, although when they are considered together these algorithms form an interesting toolkit... providing one knows to use each of the tools. Indeed, an annoying problem occurs for a biologist while using most of these algorithms: (s)he needs to have an important prior knowledge about the pattern (s)he wants to discover, in order to fix the parameters. Some algorithms do not discover patterns with gaps (Winnower [10], Projection [3]), while others need to know if gaps exist and how many are they (Smile [9], Motif [13]); some algorithms ask for the length of the searched pattern or the number of allowed errors (Smile [9], Winnower [10], Projection [3], Gibbs sampling [7]); some algorithms find probabilistic patterns (Gibbs Sampling [7], Meme [1], Consensus [5]); some others often return a too large number of answers (Splash [4], Teiresias [11], Pratt [6], Smile [9]); some of them allow to use a degenerate alphabet and equivalency sets (Teiresias [11], Pratt [6]), and only one of them (Splash [4]) allows to choose a substitution matrix between amino acids or nucleic acids.

While the biologist often does not have the prior knowledge (s)he is asked for, (s)he usually has another knowledge related to the biological context: usually (s)he knows the biological significance of the common pattern (s)he is looking for (binding site, protein signature etc.); sometimes the given sequences have a particular composition (a character or family of characters appear much more often than some others); sometimes (s)he has an idea about the substitution matrix (s)he would like to use. These features *could* be taken into account while defining the score functions, but can we afford to propose a special algorithm for every special score function?

A way to avoid the two drawbacks cited before would be to propose an algorithm which:

i) needs a minimum prior knowledge on the type of the discovered patterns;

ii) allows a large flexibility with respect to the choice of the score function,

and, of course, is still able to quickly analyse the space of possible solutions in order to find the best ones. In other words, such an algorithm should be able, given a minimal description of the biological pattern, to infer a correct model (the type of the searched pattern, the similarity notion between patterns), to choose the correct score function among many possible score functions, and to (quickly) find the best patterns, that is, the ones which correspond to the searched biological patterns.

Unfortunately, when the facility of use (for a biologist) becomes a priority, and the suppleness of the algorithm (due to the need to define many types of patterns) becomes a necessity, the difficulty of the algorithmic problem grows exponentially: (1) how to deal with all the possible types of patterns, (2) what score function to use in what cases, (3) what searching procedure could be efficient in all these cases?

The algorithm we propose attempts to give a first (and, without a doubt, very partial) answer to these questions.

(1) Certainly, our algorithm is not able to deal with all the possible types of patterns; but it only imposes *one* constraint, which can be very strong or extremely weak depending on one integer parameter e. No condition on the length of the patterns, or on the number and/or the length of the gaps is needed; these features are deduced by the algorithm while analysing the input sequences and their similar regions.

(2) We use a score function f to evaluate the similarity degree between two patterns x_1, x_2 of the same length; f can be chosen arbitrarily, and can be applied either to the full-length patterns x_1, x_2, or only to the parts of x_1, x_2 corresponding to *significant* indices. The significant indices are defined using an integer parameter Q, and are intended to identify the parts of the patterns which are common to many (at least Q%) of the input sequences.

(3) The searching procedure is supposed to discover rather general patterns, but it needs to have a minimum piece of information about the similarity notion between the patterns. There are mainly two types of similarity: the *direct similarity* asks that $o_1, o_2, \ldots, o_t$ (on $S_1, S_2, \ldots, S_t$ respectively) be similar with respect to some global notion of similarity; the *indirect similarity* asks that a (possibly external) pattern s exists which is similar to each of $o_1, o_2, \ldots, o_t$ with respect to some pairwise notion of similarity. For efficiency reasons, we chose an alternative notion of similarity, which is the indirect similarity with an internal pattern, i.e. located on one of the input sequences (unknown initially). See Remark 2 for comments on this choice.

Remark 1. The parameters e, f, Q that we introduce are needed to obtain a model (pattern type, similarity notion, score function) and to define a searching procedure, but they are *not* to be asked to the user (some of them could possibly be estimated by the user, and the algorithm is then able to use them, but if the parameters are not given then the algorithm has to propose default values).

Remark 2. The indirect similarity with an internal pattern is less restrictive than, for instance, the direct similarity in algorithms like Pratt or Teiresias, where all the discovered patterns $o_1, o_2, \ldots, o_t$ have to have exactly the *same* character on certain positions i (every o_j is, in this case, an internal pattern to which all the others are similar). And, in general, our choice can be considered less restrictive than the direct similarity requiring the pairwise similarity of every pair (o_i, o_j), since we only ask for a similarity between *one* pattern and all the others. Moreover, the indirect similarity with a (possibly external) pattern can be simulated using our model by adding to the t input sequences a new sequence, artificially built, which contains all the candidates (but this approach should only be used when the number of candidates is small).

Finally, notice that our algorithm is *not* intended to be a generalization of any of the existing algorithms. It is based on a different viewpoint, in which the priority is the flexibility of the model, in the aim to be able to adapt the model to a particular biological problem.

In Section 2 we present the algorithm and the complexity aspects. The discussion of the experimental results may be found in Section 3. Further possible improvements are discussed in Section 4.

2 The Algorithm

2.1 Main Ideas

The algorithm is based on the following main ideas:

1. *Separate the similarity notion and the score function.*
That means that building similar patterns, and selecting only the interesting ones are two different things to do in two different steps. This is because we are going to define one notion of similarity, and many different score functions.

2. *Overlapping patterns with overlapping occurrences form a unique pattern with a unique occurrence.*
In order to speed up the treatment of the patterns, we ask the similarity notion to be defined such that if patterns x_1, x_2 overlap on p positions and have occurrences (that is, patterns similar with them) x'_1, x'_2 that overlap on the same p positions, then the union of x'_1, x'_2 is an occurrence of the union of x_1, x_2. This condition is similar to the ones used in alignment algorithms (BLAST, FASTA), where one needs that a pattern can be extended on the left and on the right, and still give a unique pattern (and not a set of overlapping patterns). This condition is not necessary, but is interesting from the viewpoint of the running time.

3. *Use the indirect similarity with an internal pattern (say o_1 on S_1).*
This choice allows us to define a unique search procedure for all types of patterns we consider.

4. *The non-interesting positions in the internal pattern o_1 (which is similar to $o_2, \ldots, o_t$) are considered as gaps.*
This is already the case in some well known algorithms: in Pratt and Teiresias a position is interesting if it contains the same character in all the similar patterns; in Gibbs sampling, a position is interesting if it contains a character whose frequency is statistically significant. In our algorithm, one can define what "interesting position" means in several different ways: occurrence of the same character in at least $Q\%$ of the patterns; or no other character but A or T in each pattern; or at least $R\%$ of A; and so on. This is a supplementary possibility offered by the algorithm, which can be refused by the user (then the searched patterns will be contiguous).

2.2 The Basic Notions

We consider t sequences $S_1, S_2, \ldots, S_t$ on an alphabet Σ. For simplicity, we will assume that all the sequences have the same length n. A subsequence x of a sequence S is called a *pattern*. Its *length* $|x|$ is its number of characters, and, for given indices (equivalently, positions) $a, b \in \{1, 2, \ldots, |x|\}$, $x[a..b]$ is the (sub)pattern of x defined by the indices i, $a \le i \le b$. When $a = b$, we simply write $x[a]$.

We assume that we have a collection of similarity sets covering Σ (either the collection of all singletons, as one often does for the DNA alphabet; or the two sets of purines and pyrimidines, as for the degenerated DNA alphabet; or equivalency sets based on chemical/structural features; or similarity sets obtained from a substitution matrix by imposing a minimal threshold). The similarity of the characters can either be unweighted (as is usually the case for the DNA alphabet) or weighted (when the similarity sets are obtained using a substitution matrix).

Definition 1. *(match, mismatch)* Two patterns x, y of the same length k are said to *match* in position $i \le k$ if $x[i]$ and $y[i]$ are in the same similarity set; otherwise x, y are said to *mismatch* in position i.

Definition 2. *(similarity)* Let $e > 0$ be an integer. Two patterns x, y are *similar with respect to* e if they have at most e consecutive mismatches.

From now on, we will consider that e is fixed and we will only say *similar* instead of *similar with respect to* e. Notice that the pairwise similarity notion above can be very constraining (when e is small), but can also be very weakly constraining (when e is large).

To define the indirect similarity between a set of patterns (one on each sequence), we need to identify one special sequence among the t given sequences. Say this sequence is S_1.

Definition 3. *(indirect similarity, common pattern, occurrences)* The patterns $o_1 \in S_1$, $o_2 \in S_2$, ..., $o_t \in S_t$ of the same length k are *indirectly similar* if, for each $j \in \{1, 2, \ldots, t\}$, o_1 and o_j are similar. In this case, o_1 is called a *common pattern* of $S_1, S_2, \ldots, S_t$, and $o_1, o_2, \ldots, o_t$ are its *occurrences.*

To define the interesting positions i of a common pattern (see 4. in subsection 2.1.), we use - in order to fix the ideas - one of the possible criteria: we ask that the character in position i in the common pattern exceed a certain threshold of matched characters in $o_1, o_2, \ldots, o_t$.

Definition 4. *(consensus, gap)* Let $0 \le Q \le 100$ be an integer and let $o_1 \in S_1$ of length k be a common pattern of $S_1, S_2, \ldots, S_j$ $(1 \le j \le t)$ with occurrences $o_1, o_2, \ldots, o_j$ respectively. Denote

$$A_{i,j} = |\{l \ , \ o_1[i] \text{ matches } o_l[i], \ 1 \le l \le j\}|, \ \text{ for every } 1 \le i \le k$$
$$R_{(Q,j)} = \{i \ , \ A_{i,j} \ge Qt/100, \ 1 \le i \le k\}.$$

Then the part of o_1 corresponding to the indices in $R_{(Q,j)}$, noted $o_1|R_{(Q,j)}$, is the *consensus with respect to* Q of $o_1, o_2, \ldots, o_j$; the contiguous parts not in the consensus are called *gaps*.

The use of the parameter Q allows to add (if one wishes) to the compulsory condition that $o_1, o_2, \ldots, o_j$ have to be similar to o_1, the supplementary condition that *only* the best conserved positions in o_1 are to be considered as interesting (the other positions are assimilated to *wild-card positions*, that is, they can be filled in with an arbitrary character). Different values of the pair (e, Q) allow to define a large range for the similar patterns and their consensus, going from a very weakly constrained notion of pairwise similarity and a very low significance of each character in the consensus, to a strongly constrained notion of similarity and a strong significance of each character in the consensus.

2.3 The Main Procedure

The main procedure needs a permutation π of the sequences. Without loss of generality we will describe it assuming that π is $1, 2, \ldots, t$. The sequences are successively processed to discover $o_1, o_2, \ldots, o_t$ in this order such that o_1 is a maximal common pattern of $S_1, S_2, \ldots, S_t$ with a high score $\mathcal{S}$ (defined below). Here "maximal" means with respect to extension to the left and to the right. The "high" score (as opposed to "best") is due to the use of a heuristic to prune the search tree.

The idea of the algorithm is to consider each pattern in S_1 as a possible solution, and to successively test its similarity with patterns in $S_2, \ldots, S_t$. At any step j $(2 \leq j \leq t)$, either one or more occurrences of the pattern are found (and then o_1 is kept), or only occurrences of some of its sub-patterns are found (and then o_1 is split in several new candidates). Each pattern is represented as an interval of indices on the sequence where it occurs.

Definition 5. *(offset, restriction)* Let $\mathcal{O}$ be a set of disjoint patterns on the sequence S_1 (corresponding to a set of candidates o_1), let $j \in \{1, 2, \ldots, t\}$ and let $\delta \in \{-(n-1), \ldots, n-1\}$ be an integer called an *offset*. Then the *restriction* of $\mathcal{O}$ with respect to S_j and δ is the set

$$\mathcal{O}|(j,\delta)=\{[a..b],\ [a..b] \text{ is a maximal subpattern of some pattern in } \mathcal{O} \text{ such that } 1 \leq a+\delta \leq b+\delta \leq n \text{ and } S_1[a..b] \text{ is similar to } S_j[a+\delta..b+\delta]\}.$$

Note that $\mathcal{O}|(j, \delta)$ is again a set of disjoint intervals (otherwise non disjoint intervals could be merged to make a larger interval). These are the maximal parts of the intervals in $\mathcal{O}$ which have occurrences on S_j with respect to the offset δ.

Each $[a..b] \in \mathcal{O}|(j, \delta)$ gives a (possibly new) candidate $o_1 = S_1[a..b]$ whose occurrence on S_j is $o_j = S_j[a + \delta..b + \delta]$.

Every such candidate (still called o_1) is evaluated using a score function, and only the most promising of them continue to be tested in the next steps. The score function is complex, since it has to take into account both the "local" similarities between o_1 and each occurrence o_j (i.e. the matches and their weights, the mismatches, the lengths of the gaps in o_1 etc.) and the "global" similarity between all the occurrences $o_1, o_2, \ldots, o_j$ (represented by the consensus with respect to Q and its gaps). Then, the evaluation of o_1 at step j has two components: first, the similarity between o_1 and o_j is evaluated using a *Q-score*; then, a global score of o_1 at step j is computed based on this Q-score and on the global score of o_1 at step $j-1$. The Q-score is defined below. The global score is implicit, and is calculated by the algorithm.

Let $0 \leq Q \leq 100$ and suppose we have a score function f defined on every pair of patterns of the same length. In order to evaluate the score of $o_1 \in S_1$ and $o_j \in S_j$ of length k assuming that $o_2 \in S_2, \ldots, o_{j-1} \in S_{j-1}$ have already been found, we define:

Definition 6. *(Q-significant index)* A *Q-significant index* of o_1, o_j is an index $i \in \{1, \ldots, k\}$ such that either (1) $i \in R_{(Q,j)}$ and o_1, o_j match in position i; or (2) $i \notin R_{(Q,j)}$ and o_1, o_j mismatch in position i. The set containing all the Q-significant indices of o_1, o_j is noted $I_{(Q,j)}$.

A Q-significant index is an index that we choose to take into account when computing the Q-score between o_1, o_j: in case (1), the character $o_1[i]$ is representative for the characters in the same position in $o_1, o_2, \ldots, o_j$ (since it matches in at least Q percent of the sequences), so we have to strengthen its significance by rewarding (i.e. giving a positive value to) the match between $o_1[i]$ and $o_j[i]$; in case (2), $o_1[i]$ is not representative for the characters in the same position in $o_1, o_2, \ldots, o_j$ and we have to strengthen its non-significance by penalizing (i.e. giving a negative value to) the new mismatch.

The reward or penalization on each Q-significant index is done by the score function f, which is arbitrary but should preferably take into account the matches (with their weights) and the mismatches. As an example, it could choose to strongly reward the matches while weakly penalizing the mismatches, or the converse, depending on the type of common pattern one would like to obtain (a long pattern possibly with many gaps, or a shorter pattern with a very few gaps).

To define the Q-score, note $x|I_{(Q,j)}$ the word obtained from x by concatenating its characters corresponding to the indices in $I_{(Q,j)}$.

Definition 7. *(Q-score)* The Q-score of o_1, o_j is done by

$$f_Q(o_1, o_j) = f(o_1|I_{(Q,j)}, o_j|I_{(Q,j)}).$$

The algorithm works as follows. At the beginning, the set $\mathcal{O}$ of candidates on S_1 contains only one element, the entire sequence S_1. This candidate is refined, by comparing it to S_2 and building the sets of new candidates $\mathcal{O}|(2, \delta_2)$, for every possible offset δ_2 (i.e $\delta_2 \in \{-(n-1), \ldots, n-1\}$). The most interesting (after evaluation) of these sets are further refined by comparing *each* of them to S_3 (with offsets $\delta_3 \in \{-(n-1), \ldots, n-1\}$) and so on. A tree is built in this way, whose root contains only one candidate (the entire sequence S_1), whose internal nodes at level L correspond to partial solutions (that is, patterns o_1 on S_1 which have occurrences on $S_2, \ldots, S_L$ and have a high value of the implicit score function computed by the algorithm) and whose leaves contain the solutions (that is, the patterns o_1 on S_1 which have occurrences on $S_2, \ldots, S_t$ and have a high value of the implicit score function computed by the algorithm).

To explain and to give the main procedure, assume that we have for each node u :

- its list of offsets $\delta(u) = \delta_2\delta_3 \ldots \delta_{L(u)}$, corresponding to the path between the root and u in the tree ($L(u)$ is the level of u); the offset δ_j signifies that if $o_1 = S_1[a..b]$ then $o_j = S_j[a + \delta_j..b + \delta_j]$ is its occurrence on S_j with offset δ_j;

- its set of disjoint patterns $\mathcal{O}(u)$ (the current candidates on S_1);

- its score $score(u)$;

- its parent $parent(u)$ and its (initially empty) children $child_p$, $-(n-1) \leq p \leq n-1$, each of them corresponding to an offset between S_1 and $S_{L(u)+1}$.

The procedure Choose_and_Refine below realizes both the branching of the tree, and its pruning. That is, it shows how to build *all* the children of a node u which is already in the tree, but it only keeps the most "promising" of these children. These two steps, branching and pruning, are realized simultaneously by the procedure; however, it should be noticed that, while the branching method is unique in our algorithm, the pruning method can be freely defined (according to its efficiency).

The branching of the tree is done one level at once. For each level L from 2 to t (see step 6), the list $List'$ contains the (already built) nodes at level $L-1$, and the list $List$ (initially empty) receives successively the most promising nodes v at level L, together with their parent u (a node from $List'$) and the corresponding offset δ between S_1 and S_L. Then, for each node u at level $L-1$ (that is, in $List'$), all its possible children v (that is, one for each offset δ) are generated and their set of candidates $\mathcal{O}(v)$ and scores $score(v)$ are calculated (lines 10 to 13).

Here comes the pruning of the tree: only the most "promising" of these nodes v will be kept in $List$ (using an heuristics), and further developed. The objective is to find the leaves z with a high implicit score, that is, the ones for which $score(z)$, computed successively in line 13, is high (and closed to the best score). In the present version of the procedure, we chose the simplest heuristic: we simply define an integer P and consider that a node v on level L is promising (lines 14 to 16) if it has one of the P best scores on level L. Therefore, the successively generated nodes v are successively ranged in $List$ by decreasing order of their scores, and only the first P nodes of $List$ are kept (line 16). These nodes are finally introduced in the tree (lines 19, 20). This simple heuristic allows to obtain, in practical situations, better results than many of the algorithms cited in Section 1 (see Section 3, where the predictive positive value and the sensibility of the results are also evaluated). Moreover, it has very low cost and provides a bound P on the number of nodes which are developed on each level of the tree. We think therefore that this heuristic represents a good compromise between the quality of the results and the running time. Further improvements, using for instance an A* approach, can and will be considered subsequently.

Finally (lines 23, 24), the sets of candidates in each leaf are considered as solutions and are ranged according to the decreasing order of their scores.

Procedure Choose_and_Refine

```
Input: a permutation π = {S1, S2, ..., St}; integers Q (0 ≤ Q ≤ 100) and P.
Output: a set Sol of solutions o1 on S1 (decreasingly ordered w.r.t. the
   final implicit score) and, for each of them, the sequence δ2δ3 ... δt.
 1. begin
 2.      generate the root r of the tree T;      /* the tree to be built */
 3.      Sol := ∅; L ← 1;                        /* L is the current level */
 4.      δ(r) ← λ; O(r) ← {[1..n]}; score(r) ← +∞; /*λ is the empty string */
 5.      List ← {(r, 0, λ)}; List' ← List;
 /* List is the decreasingly ordered list of the P most promising nodes */
 6.      while L < t do
 7.         List ← ∅; L ← L + 1;
 8.         for each (u, a, b) ∈ List' do
 9.            for δ ← −(n − 1) to n − 1 do
10.               generate a new node v;
11.               δ(v) ← δ(u).δ; O(v) ← O(u)|(L, δ);
12.               m ←       max        {fQ(o1, oL), o1 = S1[a..b], oL = SL[a + δ..b + δ]};
                      [a..b] ∈ O(v)
13.               score(v) ← min{score(u), m};
14.               if (score(v) > 0) and
                                 (|List| < P or score(v) > min_{w∈List} score(w)) then
15.                  insert (v, u, δ) into the ordered list List ;
16.                  if |List| = P + 1 then remove its last element;
17.            endfor
18.         endfor;
19.         for each (v, u, δ) in List do
20.            child_δ(u) ← v; parent(v) ← u;
21.         List' ← List;
22.      endwhile;
23.      Sol ← ∪_{(u,a,b)∈List} O(u);
24.      order the elements o1 in Sol by decreasing order of
            min{score(parent(u)), fQ(o1, ot)}
         (where o1 ∈ O(u) and ot is its occurrence on St);
25. end.
```

Remark 3. It can be noticed here that, if o_1 is found in a leaf u, only *one* occurrence of o_1 is identified on each sequence S_j. In case where at least one sequence S_j exists such that o_1 has more than one occurrence on S_j, two problems can appear: (1) the same pattern o_1 is (uselessly) analysed several times at level $j+1$ and the following ones; (2) o_1 counts for two or more solutions among the P (assumed distinct) best ones we are searching for. Problem (1) can be solved by identifying, in the nodes at level L, the common candidates; this is done by fusioning the list of candidates for all the nodes of a given level. In the same way we solve problem (2) and, in order to find *all* the occurrences of a solution o_1 on all sequences, we use a (more or less greedy) motif searching algorithm. The theoretical complexity (see Subsection 2.5) is not changed in this way.

2.4 The Cycles and the Complete Algorithm

We call a *cycle* any execution of the algorithm Choose_and_Refine for a given permutation π (P and Q are the same for all cycles). Obviously, a cycle approximates the best common patterns which are located on the reference sequence, $S_{\pi(1)}$. Different cycles can have different reference sequences, and can thus give different common patterns. To find the best, say, p common patterns, independently on the sequence they are located on, at least t cycles are necessary; but they are not sufficient and it is inconceivable to perform all the $t!$ possible cycles.

Instead, we use a refining procedure (another heuristic) which tries to deduce the p best permutations, that is, the permutations $\pi_1, \pi_2, \ldots, \pi_p$ which give the common patterns with p best scores. The main idea is to execute the procedure Choose_and_Refine on randomly generated permutations with reference sequences respectively $S_1, S_2, \ldots, S_t, S_1, S_2, \ldots$ and so on, until t^2 cycles are executed without modifying the p current best scores. Then the reference sequences $S_{i_1}, S_{i_2}, \ldots, S_{i_p}$ of the permutations which gave the best scores can be fixed to obtain the first sequence of the permutations $\pi_1, \pi_2, \ldots, \pi_p$ respectively. To obtain the second sequence S_{j_k} of the permutation π_k $(1 \leq k \leq p)$, the same principle is applied to randomly generated permutations whose first sequence is S_{i_k} and second sequence is $S_1, S_2, \ldots, S_{i_k-1}, S_{i_k+1}, \ldots S_t, S_1, S_2, \ldots$ and so on. When $(t-1)^2$ cycles have been executed without modifying the current best score, the second sequence of π_k is fixed and, in fact, it can be noticed that it is not worth fixing the other sequences in π_k (the score will be slightly modified, but not the common pattern). The best permutations $\pi_1, \pi_2, \ldots, \pi_p$ are then available.

The complete algorithm then consists in applying the refining procedure to approximate the p best permutations, and to return the sets *Sol* corresponding to these permutations. Section 3 discusses the performances of the complete algorithm on biological sequences and on randomly generated i.i.d. sequences with randomly implanted patterns.

2.5 The Complexity

Our simple heuristic for pruning the tree insures that the number of developed vertices is in $O(Pt)$, so that the steps 9 to 17 are executed $O(Pt)$ times. The most expensive of them are the steps 11 and 12. They can be performed in $O(n^2 min\{(\Sigma_{\alpha \in \Sigma} p_\alpha^2)e, 1\})$ using hash tables to compute the offsets and assuming that all S_j have i.i.d. letters with the probability of each character $\alpha \in \Sigma$ denoted p_α ([8], [15], [16]). Thus the running time of the procedure Choose_and_Refine is in $O(Ptn^2 min\{(\Sigma_{\alpha \in \Sigma} p_\alpha^2)e, 1\})$. When the letters are equiprobable, we obtain $O(Ptn^2 min\{e|\Sigma|^{-1}, 1\})$.

The number of cycles executed by our refining procedure is limited by pt^2N (where N is the total number of times the best scores change before the two first sequences of a permutation π_k are fixed). Practically, N is always in $O(t)$, so that the running time of the complete algorithm, when the p best results are required, can be (practically) considered in $O(pPt^4n^2 min\{e|\Sigma|^{-1}, 1\})$. The required memory space is in $O(tn^2)$.

Table 1. The 6 sets of input sequences

Set no.	Cardinality (t)	DNA of	Length(min, max)	Site type	Protein
1	9	*E. coli*	[53,806]	binding	ArgR
2	8	*E. coli*	[53,806]	binding, in tandem	ArgR
3	16	*E. coli*	[100,3841]	binding	LexA
4	18	*E. coli*	[299,5795]	binding	PurR
5	8	*E. coli*	[251,2594]	binding	TyrR
6	131	*Dicot plants*	[251, 251]	promoter	

3 Experimental Results

We present here the results of the simplest implementation of the algorithm, that is, using the nucleotide alphabet $\{A, C, G, T\}$ where each character is similar only to itself, and the similarity counts for 1 (while the non-similarity counts for 0). We implemented several pairwise score functions f, mainly by defining two components for each such function: a component f_1 which indicates how to reward the contiguous parts only containing matches, and a component f_2 which indicates how to penalize the contiguous parts only containing mismatches.

Example. For instance, if $f_1(a) = a\sqrt{a}$, $f_2(a) = a$, the value of f for $x = ACGGCTCTGGAT$ and $y = CCGGTACTGGCT$ is equal to $-f_2(1) + f_1(3) - f_2(2) + f_1(4) - f_2(1) + f_1(1) = 10.19$ since we have mismatches in positions (1), (5, 6), (11), and matches in positions (2, 3, 4), (7, 8, 9, 10), (12).

Biological sequences. To test our algorithm on DNA sequences, we used six input sets; five of them contain (experimentally confirmed) binding sites for DNA-binding proteins in *E. coli* and the last one contains (experimentally confirmed) promoter sites in Dicot plants (see Table 1). For each of these sets, the sequence logo [14] of the binding (respectively promoter) sites can be seen in Figure 1. Sets 1 to 5 were chosen among the 55 sets presented in [12] (whose binding sites can be found on the page *http://arep.med.harvard.edu/ecoli_matrices/*); the only criteria of choice were the number and length of the sequences in a set (we prefered these parameters to have important values, in order to insure a large search space for our algorithm). Since the sets of sequences were not available on the indicated site, each sequence was found on GenBank by applying BLASTN on each binding site, and this is one of the reasons why we limited our tests to 5 over 55 sets of sequences. In order to enrich the type of researched patterns, we added the 6th set, whose 131 sequences (corresponding to Dicot plants) can be downloaded on *http://www.softberry.com/berry.phtml*.

We firstly executed our algorithm on Set 1, with different values for e and Q, and with different pairwise score functions f (including the default score function in Pratt). We learned on this first example the best values for the parameters: $e = 5$, $Q = 70$, f formed by $f_1(a) = f_2(a) = a$. With these parameters and with $P = 1, p = 1$ (thus asking only for the best solution), we performed the executions on the 5 other input sets (20 executions on each set, in order to be

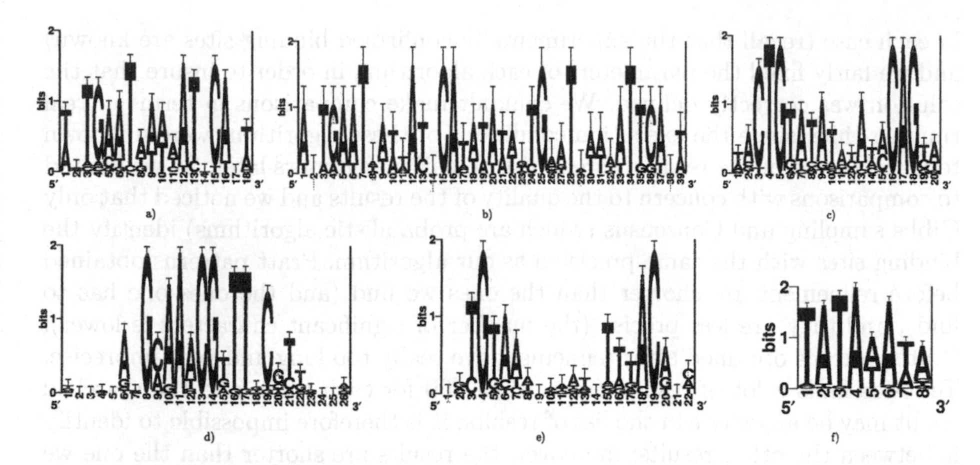

Fig. 1. The sequence logos for the binding sites of a) ArgR; b) ArgR (binding site in tandem); c) LexA; d) PurR; e) TyrR and for the promoter sites in f) a set of Dicot plants.

sure that different randomly generated permutations in the refining procedure do not yield different results); we found the consensi below (that means that *each* of the 20 executions on a given set found the same consensus). These results have to be compared with the logos, by noticing that the insignificant characters on the logos have few chances to be present in the consensus.

- For set 1, we found TGAATxAxxATxCAxxTxxxxTGxxT; the positions of the pattern on the sequences are very well found; the important length of the pattern (when compared to its logo) is explained by noticing that 8 of the 9 sequences contain the pattern in the logo twice (consecutively), while the 9th sequence contains only the beginning of the second occurrence.

- For set 2, TGAATxAxxATxCAxxTAxxxTGxxTxxxAATxCA was found; the length of the pattern is good, the locations are the good ones. The two distinct parts of the tandem are not visible here since the distance between them is simply considered as a gap in the consensus.

- For set 3, we found ACTGTATATxxAxxCAG; the length of the pattern is good, the positions on the sequences are the good ones.

- For set 4, we found GCxAxCGTTTTC; the algorithm finds very well the well conserved characters in the pattern and their positions on the sequences, but cannot find the beginning of the pattern because of the too weak similarity at this place.

- For set 5, we found TGTAAAxxAAxxTxTAC; the length of the pattern is good, the positions on the sequences are the good ones.

For these five sets, the results can be considered as being very good, when compared to the results obtained by the other pattern discovery algorithms. To make these comparisons, we restricted to the algorithms which are available via Internet (execution via a web page). We manually analysed the expected results in each case (recall that the experimentally confirmed binding sites are known) and we fairly fixed the parameters of each algorithm, in order to insure that the solution was correctly defined. We couldn't make comparisons in terms of real running time, since the execution conditions of these algorithms were unknown to us and some of the results were sent by email many hours later. So we limited to comparisons with concern to the quality of the results and we noticed that only Gibbs sampling and Consensus (which are probabilistic algorithms) identify the binding sites with the same precision as our algorithm. Pratt patterns obtained before refinement are shorter than the ones we find (and the ones one has to find), and they are less precise (the number of significant characters is lower); Pratt patterns obtained after refinement are really too long and very imprecise. Teiresias finds a lot of results (more than 200 for each test), and the searched result may be anywhere in the list of results; it is therefore impossible to identify it between the other results; moreover, the results are shorter than the one we find, and less precise. The too large number of results is also the drawback of Smile. Meme gives the expected result only in one case (Set 3).

- For set 6, our algorithm is the only one which identifies the pattern searched TATAAAT (good length, good significant characters), with the only problem that it finds much more locations (this is because the pattern is very frequent). The other algorithms find much longer patterns which sometimes include the searched pattern.

Conclusion. We deduce that on these examples the results of our algorithm are comparable with the best results of the best existing algorithms. The Q-significant letters are clearly found; the other ones are not indicated but they obviously could be computed by the algorithm. The identification of the biological patterns with our algorithm for $P = 1, p = 1$ in all these examples insures that the formal definition of the biological signals is correct; the *good* formal result (i.e. the one corresponding to the biological signal) is not lost in a long list of formal results (as is the case for algorithms like Teiresias and Smile). Obviously, if P, p have other values, the best result will be the same, and some other results will be added, by decreasing order of their score.

Randomly generated input sets. Our algorithm uses two heuristics (one in the procedure Choose_and_Refine, the other one in the refining procedure). To evaluate the performances of the algorithm (and thus of the heuristics), we evaluated the predictive positive value (PPV) and the sensibility (S) of its results, for $P = p = 1$ (thus looking only for the best pattern). We executed our algorithm on 1000 arbitrarily generated input sets (at least 10 i.i.d. sequences of length at least 100 in each set) where we implanted on each sequence several randomly mutated patterns (of given scores and lengths) at randomly chosen positions. The average PPV and S values we obtained for these executions are respectively 0.95 and 0.91, which means that, on average, 95% of the pattern found by the algorithm correspond to the implanted pattern with best score, and that 91 % of the implanted pattern with best score correspond to the pattern found by the algorithm. More precisely, that means that the found pattern is (almost) the implanted pattern with best score (it is on the same sequence, it recovers about 91 % of the indices which form the implanted pattern), but it is sometimes a little bit shifted (on the right or on the left), and one possible explanation (additionally to the one that our method is not exact) is that the insertion of the pattern with best score sometimes created in the sequence a neighboring pattern with an even better score.

4 Further Improvements

We did not insisted until now on the supplementary features of some other algorithms with respect to ours. We already have an algorithm which is competitive with the existing ones, but it could be improved by (1) introducing a quorum q corresponding to a smaller percentage of sequences which have to contain the pattern; (2) using a better heuristic for pruning the tree (for instance, an A* approach); (3) testing the algorithm on many different biological inputs, in order to (non-automatically) learn the best parameters for each type of biological pattern; (4) investigating under which conditions one can improve our algorithm to use an indirect similarity with an external pattern, by introducing a supplementary, artificially built, sequence containing all the candidates (their number should be small enough, possibly because a pre-treatment procedure has been applied). Some of these possible ameliorations are currently in study, the other ones will be considered subsequently.

References

1. T. L. Bailey, C. Elkan: Unsupervised learning of multiple motifs in biopolymers using expectation maximization. *Machine Learning*, 21 (1995), 51-80.
2. B. Brejová, C. DiMarco, T. Vinar, S. R. Hidalgo, G. Hoguin, C. Patten: Finding patterns in biological sequences. Tech. Rep. CS798g, University of Waterloo (2000).
3. J. Buhler, M. Tompa: Finding motifs using random projections. *Proceedings of RECOMB 2001*, ACM Press (2001) 69-76.
4. A. Califano: SPLASH : Structural pattern localization analysis by sequential histograms. *Bioinformatics* 16(4) (2000) 341-357.
5. G. Hertz, G. Stormo: Identifying DNA and protein patterns with statistically significant alignments of multiple sequences. *Bioinformatics* 15 (1999) 563-577.
6. I. Jonassen: Efficient discovery of conserved patterns using a pattern graph. *Computer Applications in the Biosciences* 13 (1997) 509-522.
7. C. Lawrence, S. Altschul, M. Boguski, J. Liu, A. Neuwald, J. Wootton: Detecting subtle sequence signal: a Gibbs sampling strategy for multiple alignment. *Science* 262 (1993) 208-214.
8. D.J. Lipman, W.R. Pearson: Rapid and sensitive protein similarity search. *Sciences* 227 (1985) 1435-1441.
9. L. Marsan, M.-F. Sagot: Extracting structured motifs using a suffix tree. *Proceedings of RECOMB 2000*, ACM Press (2000) 210-219.
10. P. A. Pevzner, S.-H. Sze: Combinatorial approaches to finding subtle signals in DNA sequences. *Proceedings of ISMB* (2000) 269-278.

11. I. Rigoutsos, A. Floratos: Combinatorial pattern discovery in biological sequences: The TEIRESIAS algorithm. *Bioinformatics* 14(1) (1998) 55-67.
12. K. Robison, A.M. McGuire, G.M. Church: A comprehensive library of DNA-binding site matrices for 55 proteins applied to the complete *Escherichia coli* K-12 genome. *J. Mol. Biol.* 284 (1998) 241-254.
13. H.O. Smith, T. M. Annau, S. Chandrasegaran: Finding sequence motifs groups of functionally related proteins. *Proc. Nat. Ac. Sci. USA* 87 (1990) 826-830.
14. T.D. Schneider, R.M. Stephens: Sequence logos: a new way to display consensus sequence. *Nucl. Acids Res* 18 (1990) 6097-6100.
15. M. Waterman: Introduction to computational biology: maps, sequences and genomes. Chapman & Hall (2000).
16. W. Wilbur, D. Lipman: Rapid similarity searches of nucleic acid and protein data banks. *Proceeding of National Academy of Science*, 80 (1983) 726-730.

A Combinatorial Approach to Automatic Discovery of Cluster-Patterns

Revital Eres[1], Gad M. Landau[1,2], and Laxmi Parida[3]

[1] Department of Computer Science, Haifa University, Haifa 31905, Israel
revitale@cslx.haifa.ac.il
[2] Department of Computer and Information Science, Polytechnic University, Six MetroTech Center, Brooklyn, NY 11201-3840, USA
landau@poly.edu
[3] Computational Biology Center, IBM TJ Watson Research Center, Yorktown Heights, New York 10598, USA
parida@us.ibm.com

Abstract. Functionally related genes often appear in each others neighborhood on the genome, however the order of the genes may not be the same. These groups or clusters of genes may have an ancient evolutionary origin or may signify some other critical phenomenon and may also aid in function prediction of genes. Such gene clusters also aid toward solving the problem of local alignment of genes. Similarly, clusters of protein domains, albeit appearing in different orders in the protein sequence, suggest common functionality in spite of being nonhomologous. In the paper we address the problem of automatically discovering clusters of entities be it genes or domains: we formalize the abstract problem as a discovery problem called the πpattern problem and give an algorithm that automatically discovers the clusters of patterns in multiple data sequences. We take a model-less approach and introduce a notation for *maximal* patterns that drastically reduces the number of valid cluster patterns, without any loss of information, We demonstrate the automatic pattern discovery tool on motifs on *E Coli* protein sequences.

Key Words: Design and analysis of algorithms, combinatorial algorithms on words, discovery, data mining, clusters, patterns, motifs.

1 Introduction

Genes that appear together consistently across genomes are believed to be functionally related: these genes in each others neighborhood often code for proteins that interact with one another suggesting a common functional association. However, the order of the genes in the chromosomes may not be the same. In other words, a group of genes appear in different permutations in the genomes [15,12,16]. For example in plants, the majority of snoRNA genes are organized in polycistrons and transcribed as polycistronic precursor snoRNAs [3]. Also, the olfactory receptor(OR)-gene superfamily is the largest in the mammalian genome. Several of the human OR genes appear in cluster with ten or

G. Benson and R. Page (Eds.): WABI 2003, LNBI 2812, pp. 139–150, 2003.

more members located on almost all human chromosomes and some chromosomes contain more than one cluster [5].

As the available number of complete genome sequences of organisms grows, it becomes a fertile ground for investigation along the direction of detecting gene clusters by comparative analysis of the genomes. A gene G is compared with its orthologs G' in the different organism genomes. Even phylogenetically close species are not immune from gene shuffling, such as in *Haemophilus influenzae* and *Escherichia Coli* [20,17]. Also, a multicistronic gene cluster sometimes results from horizontal transfer between species [10] and multiple genes in a bacterial operon fuse into a single gene encoding multi-domain protein in eukaryotic genomes [15].

If the functions of genes say G_1G_2 is known, the function of its corresponding ortholog clusters $G_2'G_1'$ may be predicted. Such positional correlation of genes as clusters and their corresponding orthologs have been used to predict functions of ABC transporters [19] and other membrane proteins [7].

The local alignment of nucleic or amino acid sequences, called the *multiple sequence alignment* problem, is based on similar subsequences; however the *local alignment of genomes* [13] is based on detecting locally conserved gene clusters. A measure of gene similarity is used to identify the gene orthologs. For example genes $G_1G_2G_3$ may be aligned with $G_3'G_1'G_2'$: such an alignment is never detected in subsequence alignments.

Domains are portions of the coding gene (or the translated amino acid sequences) that correspond to a functional sub-unit of the protein. Often, these are detectable by conserved nucleic acid sequences or amino acid sequences. The conservation helps in a relative easy detection by automatic motif discovery tools. However, the domains may appear in a different order in the distinct genes giving rise to distinct proteins. But, they are functionally related due to the common domains. Thus these represent functionally coupled genes such as forming operon structures for co-expression [18,4].

In the paper we address the problem of automatically discovering clusters of genes or domains. A similar problem is addressed in [11] that integrates data from different sources such as gene expression data and metabolic pathways and works on a single genome at a time. Yet another variation has been addressed as the problem of finding common intervals in multiple permutations [6]. In this paper, we formalize the abstract problem as a discovery problem called the πpattern problem and give an algorithm that automatically discovers the clusters of patterns (that appear in various permuted forms in the instances) in multiple data sequences. As there is not enough knowledge about forming an appropriate model to filter the meaningful from the apparently meaningless clusters, we take a model-less approach and introduce a notation for *maximal* patterns that drastically reduces the number of valid cluster patterns, without any loss of information, making it easier to study the results from an application viewpoint.

We demonstrate the automatic pattern discovery tool on motifs on *E Coli* protein sequences. It is interesting to observe that permutations involving as many as eight motifs are discovered. Although its biological significance is yet to be established, nevertheless it appears to be an interesting phenomenon.

Roadmap. In the next section we formalize the problem. In the following section we introduce our notion of maximality and its associated notation so that there is no loss of information. We next describe the algorithm and then give some experimental results, open problems and conclusions.

2 The πPattern Problem

We begin by giving some definitions.

Let $S = s_1 s_2 \ldots s_n$ be a string of length n, and $P = p_1 p_2 \ldots p_m$ a pattern, both over alphabet $\{1, ..., |\Sigma|\}$.

Definition 1. *($\Pi(s)$, $\Pi'(s)$) Given a string s on alphabet Σ,*

$$\Pi(s) = \{\alpha \in \Sigma \mid \alpha = s[i], \textit{ for some } 1 \leq i \leq |s|\} \textit{ and}$$

$$\Pi'(s) = \{\alpha(t) \mid \alpha \in \Pi(s), t \textit{ is the number of times that } \alpha \textit{ appears in } s\}$$

For example if $s = abcda$, $\Pi(s) = \{a, b, c, d\}$.

If $s = abbccdac$, $\Pi'(s) = \{a(2), b(2), c(3), d\}$. Note that d appears only once and we ignore the annotation altogether.

Definition 2. *(p-occurs) A pattern P p-occurs (permuted occurrence) in a string S at location i if: $\Pi'(P) = \Pi'(s_i \ldots s_{i+m-1})$.*

Definition 3. *(πpattern) Given an integer K, a Pattern P is a πpattern on S if:*

- *$|P| > 1$, we rule out the trivial single character patterns.*
- *P p-occurs at some $k' \geq K$ distinct locations on S. $\mathcal{L}_p = \{i_1, i_2, \ldots, i_{k'}\}$ is the location list of p.*

For example consider Π'(P) $=\{a(2), b(3), c\}$, and the string $S = aacbbbxxabcb$ ab. Clearly P p-occurs at positions 1 and 9.

The Problem of Permutation Pattern (πPattern) Discovery. Given a string S and $K < n$, find all πpatterns of S together with their location lists.

For example, if $S = abcdbacdabacb$, then $P = \{a, b, c\}$ is a 4-πpattern with location list $\mathcal{L}_p = \{1, 5, 10, 11\}$.

The total number of πpatterns is $O(n^2)$, but is this number actually attained? Consider the following example.

Example 1. Let $S = abcdefghijabdcefhgij$ and $k = 2$. The πpatterns below show that their number could be quadratic in the size of the input.

$$
\begin{array}{ll}
P_1 = \{a,b\}, & \mathcal{L}_{p_1} = \{1,11\} \\
P_2 = \{a,b,c,d\}, & \mathcal{L}_{p_2} = \{1,11\} \\
P_3 = \{a,b,c,d,e\}, & \mathcal{L}_{p_3} = \{1,11\} \\
P_4 = \{a,b,c,d,e,f\}, & \mathcal{L}_{p_4} = \{1,11\} \\
P_5 = \{a,b,c,d,e,f,g,h\}, & \mathcal{L}_{p_5} = \{1,11\} \\
P_6 = \{a,b,c,d,e,f,g,h,i\}, & \mathcal{L}_{p_6} = \{1,11\} \\
P_7 = \{a,b,c,d,e,f,g,h,i,j\}, & \mathcal{L}_{p_7} = \{1,11\} \\
P_8 = \{b,c,d\}, & \mathcal{L}_{p_8} = \{2,12\} \\
P_9 = \{b,c,d,e,f\}, & \mathcal{L}_{p_9} = \{2,12\} \\
P_{10} = \{b,c,d,e,f,g,h\}, & \mathcal{L}_{p_{10}} = \{2,12\} \\
P_{11} = \{b,c,d,e,f,g,h,i,j\}, & \mathcal{L}_{p_{11}} = \{2,12\} \\
P_{12} = \{c,d\}, & \mathcal{L}_{p_{12}} = \{3,13\} \\
P_{13} = \{c,d,e\}, & \mathcal{L}_{p_{13}} = \{3,13\} \\
P_{14} = \{c,d,e,f\}, & \mathcal{L}_{p_{14}} = \{3,13\} \\
P_{15} = \{c,d,e,f,g,h\}, & \mathcal{L}_{p_{15}} = \{3,13\} \\
P_{16} = \{c,d,e,f,g,h,i\}, & \mathcal{L}_{p_{16}} = \{3,13\} \\
P_{17} = \{c,d,e,f,g,h,i,j\}, & \mathcal{L}_{p_{17}} = \{3,13\} \\
P_{18} = \{e,f\}, & \mathcal{L}_{p_{18}} = \{5,15\} \\
P_{19} = \{e,f,g,h\}, & \mathcal{L}_{p_{19}} = \{5,15\} \\
P_{20} = \{e,f,g,h,i,j\}, & \mathcal{L}_{p_{20}} = \{5,15\} \\
P_{21} = \{f,g,h\}, & \mathcal{L}_{p_{21}} = \{6,16\} \\
P_{22} = \{f,g,h,i,j\}, & \mathcal{L}_{p_{22}} = \{6,16\} \\
P_{23} = \{g,h\}, & \mathcal{L}_{p_{23}} = \{7,17\} \\
P_{24} = \{g,h,i,j\}, & \mathcal{L}_{p_{24}} = \{7,17\} \\
P_{25} = \{i,j\}, & \mathcal{L}_{p_{25}} = \{9,19\}
\end{array}
$$

3 Maximal Patterns

We give a general definition of maximality which holds even for different kinds of substring patterns such as rigid, flexible, with or without wild cards [14].

In the following, assume that $\mathcal{P}$ is the set of all πpatterns on a given input string S.

Definition 4. $P_a \in \mathcal{P}$ *is* **non-maximal** *if there exists* $P_b \in \mathcal{P}$ *such that: (1) each p-occurrence of* P_a *on* S *is covered by a p-occurrence of* P_b *on* S*, (each occurrence of* P_a *is a substring in an occurrence of* P_b*) and, (2) each p-occurrence of* P_b *on* S *covers* $l \geq 1$*, p-occurrence(s) of* P_a *on* S*. A pattern* P_b *that is not non-maximal is* **maximal**.

Clearly, $\Pi'(P_a) \subset \Pi'(P_b)$. Although it seems counter-intuitive, but it is possible that $|\mathcal{L}_{p_a}| < |\mathcal{L}_{p_b}|$. Consider the input $S = abcdebca\ldots\ldots abcde$. $P_a = \{d,e\}$ p-occurs only two times but $P_b = \{a,b,c,d,e\}$ p-occurs three times and by the definition P_a is non-maximal with respect to P_b.

To illustrate the case of $l > 1$ in the definition, consider

$$S = abcdbac\ldots\ldots abcabcd\ldots\ldots abcdabc.$$

$P_a = \{a, b, c\}$ p-occurs two times in the first and third, and, four times in the second p-occurrence of $P_b = \{(a)2, (b)2, (c)2, d\}$. Also, by the definition, P_a is non-maximal with respect to P_b.

We further claim that such a non-maximal pattern P_a can be "deduced" from P_b and the p-occurrences of P_a on S can be estimated to be within the p-occurrences of P_b. This will be shown to be a consequence of Theorem 2 in the next section.

Theorem 1. *Let $\mathcal{M} = \{P_j \in \mathcal{P} |\ P_j$ is maximal$\}$. $\mathcal{M}$ is unique.*

This is straightforward to see. This result holds even when the patterns are substring patterns.

In Example 1, pattern P_7 is the only maximal πpattern in S.

3.1 Maximality Notation

Recall that in case of substring patterns, the maximal pattern very obviously indicates the non-maximal patterns as well. For example a maximal pattern of the form *abcd* implicates *ab*, *bc*, *cd*, *abc*, *bcd* as possible non-maximal patterns, unless they have occurrences not covered by *abcd*. Do maximal πpatterns have such an obvious form? In this section we introduce a special notation based on observations discussed below. We next demonstrate how this notation makes it possible to represent maximal πpatterns.

Theorem 2. *Let $Q \in \mathcal{P}$ and $\mathcal{Q} = \{Q' |\ Q'$ is non-maximal w.r.t $Q\ \}$. Then there exists a permutation, $\overline{Q}$, of $\Pi'(Q)$ such that for each element $Q' \in \mathcal{Q}$, a permutation of $\Pi'(Q')$ is a substring of $\overline{Q}$.*

Proof: Without loss of generality, let the ordering of the elements be as the one in the leftmost occurrence of Q on S as $\overline{Q}$. Clearly, there is a permutation of $\Pi'(Q')$ that is a substring of $\overline{Q}$, else Q' is not a non-maximal pattern by the definition. □

Corollary 1. *The ordering is not necessarily complete. Some elements may have no order with respect to some others.*

Consider $S = abcdef\ldots\ldots cadbfe\ldots\ldots abcdef$. Then $P_1 = \{a, b, c, d\}$, $P_2 = \{e, f\}$ and $P_3 = \{a, b, c, d, e, f\}$ are the πpatterns with three occurrences each on S. Then the intervals denoted by brackets can be represented as

$$({}_3({}_1a, b, c, d)_1, ({}_2e, f)_2)_3$$

where the elements within the brackets can be in any order. A pair of brackets $({}_i\ldots)_i$ corresponds to the πpattern P_i. An element is either a character from the alphabet or bracketed elements.

Corollary 2. *A representation that captures the order of the elements of Q along with the intervals that correspond to each Q′ encodes the entire set Q.*

This representation will appropriately annotate the ordering. The representation using brackets works except that there may intersecting intervals that could lead to clutter. When the intervals intersect, the brackets need to be annotated. For example, $(a(b,d)c)$ can have at least two distinct interpretations: (1) $({}_1a({}_2b,d)_2c)_1$, or, (2) $({}_1a({}_2b,d)_1c)_2$.

Consider the input string $S = abcd......dcba......abcd$. The πpatterns are $P_1 = ab$, $P_2 = bc$, $P_3 = cd$, $P_4 = abc$, $P_5 = bcd$, $P_6 = abcd$, each occurring three times. Using only annotated brackets will yield a clutterful representation as follows:

$$({}_6({}_1({}_4a({}_2({}_5b)_1({}_3c)_2)_4d)_3)_5)_6 \tag{1}$$

The annotation of the brackets is required to keep the pairing of the brackets unambiguous. It is clear that if two intervals intersect, then the intersection elements are immediate neighbors of the remaining elements. For example if $({}_1a({}_2b,c)_1d)_2$, then (b,c) must be immediate neighbors of (a) as well as (d). We introduce a symbol '-' to denote immediate neighbors, then the intervals never intersect. Further, they do not need to be annotated if they do not intersect. Thus the previous example can be simply given as a-(b,c)-d. The earlier clutterful representation of Equation 1 can be cleanly put as

$$a\text{-}b\text{-}c\text{-}d$$

Next, consider Example 1. Using the notation, there is only one maximal πpattern given by $M = a\text{-}b\text{-}(c,d)\text{-}e\text{-}f\text{-}(g,h)\text{-}i\text{-}j$ at locations 1 and 11 on S. Notice that $\Pi(P_7) = \Pi(M)$ and every other πpattern can be deduced from M.

4 The Algorithm

The input of the algorithm is a set of strings of total length n. In order to simplify the explanation we consider one string S of length n over an alphabet Σ.

The algorithm computes the maximal πpatterns in S. It has two stages: (1) Find all the πpatterns in S, and (2) Find the maximal πpatterns in S. In our implementation, in Stage 2 we use a straightforward computation using location lists of all the πpatterns in S obtained at Stage 1. The location lists of each pair of πpatterns are checked to find if one πpattern is covered by the other one. Assume that stage 1 outputs p πpatterns, and the maximum length of a location list is ℓ, stage 2 runs in $O(p^2\ell)$ time. From now on, only Stage 1 will be discussed.

We assume that the size of the longest pattern is L. Step ℓ $(2 \le \ell \le L)$ of Stage 1, finds πpatterns of length ℓ. The computation is based on an algorithm given by Amir et. al. [1].

The algorithm moves a window of size ℓ along string S, adding and deleting a letter in each iteration. This is similar to the algorithm for computing the sum of every consecutive ℓ elements of an array,

The algorithm maintains an array $NAME[1 \ldots |\Sigma|]$ where $NAME[q]$ keeps count of the number of appearances of letter q in the current window. Hence, the sum of the values of the elements of $NAME$ is ℓ. In each iteration the window shifts one letter to the right, and at most 2 variables of $NAME$ are changed one is increased by one (adding the rightmost letter) and one is decreased by one (deleting the leftmost letter of the previous window). Note that for a given window $s_a s_{a+1} \ldots s_{a+\ell-1}$ $NAME$ represents $\Pi'(s_a s_{a+1} \cdots s_{a+\ell-1})$. There is one difference between $NAME$ and Π', in Π' only the letters of Π are considered and in $NAME$ all letters of Σ are considered, but the values of letters that are not in Π are zero. At iteration j we define $NAME$ to represents the substring $s_j \ldots s_{j+\ell-1}$.

Observation: Substrings of S, of length ℓ, that are permutations of the same string are represented by the same $NAME$.

We have explained how the $NAME$s of all substrings of length ℓ of S are computed. However, we still have to find the $NAME$s that appear more than K times.

Each distinct $NAME$ is given a unique name - an integer in the range $0 \ldots n$. The names are given by using the *naming* technique [2,8], which is a modified version of the algorithm of Karp, Miller and Rosenberg [9].

4.1 The Naming Technique

Assume, for the sake of simplicity, that $|\Sigma|$ is a power of 2. (If $|\Sigma|$ is not a power of 2, $NAME$ can be extended to an appropriate size by concatenating to its end repeated -1. The size of the resulting array is no more than twice the size of the original array.)

A name is given to each subarray of size 2^i that starts on a position $j2^i + 1$ in the array, where $0 \leq i \leq \log|\Sigma|$ and $0 \leq j < |\Sigma|/2^i$. Names are given first to subarrays of size 1 then $2, 4, \ldots, |\Sigma|$, at the end a name is given to the entire array.

A subarray of size 2^i is a concatenation of 2 subarrays of size 2^{i-1}. The names of these 2 subarrays are used as the input for the computation of the name of the subarray of size 2^i. The process may be viewed as constructing a complete binary tree, which we will refer to as a *naming tree*. The leaves of the tree (level 0) are the elements of the initial array. Node x in level i is the parent of nodes $2x - 1$ and $2x$ in level $i - 1$.

Our naming strategy is as follows. A name is a pair of previous names. At level j of the naming, we compute the name of subarray $NAME_1 NAME_2$ of size 2^j, where $NAME_1$ and $NAME_2$ are consecutive subarrays of size 2^{j-1} each. We give as names the natural numbers in increasing order. Notice that every level only uses the names of the level below it, thus the names we use at every level are numbers from the set $\{1, ..., n\}$.

To give an array a name, we need only to know if the pair of names of the composing subarrays has appeared previously. If it did, then the array gets the name of this pair. Otherwise, it gets a new name. It is necessary, therefore, to show a quick way to dynamically access pairs of numbers from a bounded range universe. This is discussed in Section 4.2

Example 2. Let $\Sigma = \{a, b, c, d, e, f, g, h, i, j, k, \ell, m, n, o, p\}$, $|\Sigma| = 16$. Assume a substring $cbo\ell jikgik\ell j$ of S, the array $NAME$ that represents this substring is:

0	1	1	0	0	0	1	0	2	2	2	2	0	0	1	0

Below is the result of naming the above $NAME$.

<table>
<tr><td colspan="16">11</td></tr>
<tr><td colspan="8">9</td><td colspan="8">10</td></tr>
<tr><td colspan="4">6</td><td colspan="4">7</td><td colspan="4">8</td><td colspan="4">7</td></tr>
<tr><td colspan="2">2</td><td colspan="2">3</td><td colspan="2">4</td><td colspan="2">3</td><td colspan="2">5</td><td colspan="2">5</td><td colspan="2">4</td><td colspan="2">3</td></tr>
<tr><td>0</td><td>1</td><td>1</td><td>0</td><td>0</td><td>0</td><td>1</td><td>0</td><td>2</td><td>2</td><td>2</td><td>2</td><td>0</td><td>0</td><td>1</td><td>0</td></tr>
</table>

Suppose the window move adds the letter n, In the diagram below we indicate in boldface the names that changed as a result of the change to $NAME$.

<table>
<tr><td colspan="16">14</td></tr>
<tr><td colspan="8">9</td><td colspan="8">13</td></tr>
<tr><td colspan="4">6</td><td colspan="4">7</td><td colspan="4">8</td><td colspan="4">6</td></tr>
<tr><td colspan="2">2</td><td colspan="2">3</td><td colspan="2">4</td><td colspan="2">3</td><td colspan="2">5</td><td colspan="2">5</td><td colspan="2">2</td><td colspan="2">3</td></tr>
<tr><td>0</td><td>1</td><td>1</td><td>0</td><td>0</td><td>0</td><td>1</td><td>0</td><td>2</td><td>2</td><td>2</td><td>2</td><td>0</td><td>1</td><td>1</td><td>0</td></tr>
</table>

From example 2 one can see that a single change in $NAME$ causes at most $\log|\Sigma|$ names to change, since there is at most one name change in every level.

Time. We conclude that at every iteration, only $O(\log|\Sigma|)$ names need to be handled, since only two elements of array $NAME$ are changed.

We have seen that the name of the $NAME$ array can be maintained at a cost of $O(\log|\Sigma|)$ per iteration. What has to be found is whether the updated $NAME$ array gets a new name, or a name that appeared previously. Before we show an efficient implementation of this task, let us bound the maximum number of different names our algorithm needs to generate for a fixed window size ℓ.

Lemma 1. *[1] The maximum number of different names generated by our algorithm's naming of size ℓ window on a text of length n is $O(n\log|\Sigma|)$. The maximum number of names generated at a fixed level j in the naming tree is $O(n)$.*

4.2 The Pair Recognition Problem

We have seen earlier that it is necessary to show a quick way to dynamically access pairs of numbers from a bounded range universe. Formally, we would like a solution to the following problem:

Definition 5. The dynamic pair recognition problem *is the following:*
INPUT: A sequence of queries $\{(a_j, b_j)\}_{j=1}^{\infty}$*, where* $a_j, b_j \in \{1, ..., j\}$*.*
OUTPUT: Dynamically decide, for every query (a_j, b_j)*, whether there exist* $c,\ c < i$ *such that* $(a_j, b_j) = (a_c, b_c)$*.*

At any point j the pairs we are considering all have their first element no greater than j. Thus, accessing the first element can be done in constant time by direct access. This suggest "gathering" all pairs in trees rooted at their first element. However, if we make sure these trees are ordered by the second element and balanced, we can find elements by binary search in time that is logarithmic in the tree size.

Time: The above solution, for the pair recognition algorithm, requires, for solving each query (a_j, b_j), a search on a balanced search tree with all previous queries whose first pair element is a_j. In our case, since in every level there are at most $O(n)$ different numbers, the time for searching such a balanced tree is $O(\log |BAL[a]|) = O(\log(n))$.

The above technique gives names to many parts of $NAME$. We will keep a special attention to leaves, in the balanced search trees, that represent names of the entire array $NAME$. In Step ℓ all the patterns of one πpattern will reach the same leaf. We add to such a leaf a counter that finds if the number of occurrences is at least K, and a location list.

4.3 Time Complexity

Stage 1 of our algorithm runs L times. In a step ℓ we first initialize $NAME$ and the naming tree in $O(\ell + |\Sigma|)$ time and then compute $n - \ell$ iterations. Each iteration includes at most two changes in $NAME$, and the computation of $O(\log |\Sigma|)$ names. Computing a name takes $O(\log n)$ time. Hence the total running time of our algorithm is $O(Ln \log |\Sigma| \log n)$.

5 Open Problems

The answers to the following two questions would be interesting and remain open at this time: (1) It is known that the maximum number of πpatterns is $O(n^2)$, but what is the maximum number of maximal πpatterns? (2) The present algorithm works in two phases, it first detects all the πpatterns and then extracts the maximal πpatterns from this set. Is it possible to devise an algorithm that integrates the two steps to produce only the maximal πpatterns?

6 Experimental Results

We show some preliminary results on *E Coli* protein sequences. The input to our system is substring patterns detected on pruned set of *E Coli* sequences: in this pruned set, no pair of sequences is ninety percent or more similar in the sequences using standard sequence similarity measures. There are 8,394 protein sequences with a total of about 1,391,900 amino acids in the data set. The following parameters were used to obtain the substring patterns. (1) quorum: the patterns appear at least five times, (2) wild card density: the patterns have no more than two wild cards in a window of twelve bases. The number of such substring patterns is 207. The input sequences are now viewed as sequences of motifs/domains with a possibility of multiple occurrences at a location. Thus the alphabet size for this problem is 207. The πpattern discovery tool is run on this input file to yield the result. The input files are available from following site: `www.cs.nyu.edu/~parida/res/public/data/` as "ecobase.dat.gz" and "ecobase.mtfs.gz".

Table 1 shows the result of discovering πpatterns on this data where the alphabet is the motif/domain with parameters as described above. Figure 1 shows an example of a permuted πpattern of size 6.

Size of πPatterns	Total number of πPatterns	Number of Maximal πPatterns	Percentage of Maximal πPatterns
2	161	98	61%
3	129	53	41%
4	95	55	58%
5	43	17	40%
6	27	19	70%
7	15	11	67%
8	7	7	100%

Table 1. πpatterns on motifs/domains of the *E Coli* protein sequences.

7 Conclusions

Related genes often appear in each others neighborhood on the genome, however the order of the genes may not be the same. Such gene clusters also aid toward solving the problem of local alignment of genes. Similarly, clusters of protein domains, albeit appearing in different orders in the protein sequence, suggest common functionality in spite of being nonhomologous. In the paper we have addressed the problem of automatically discovering clusters as a discovery problem called the πpattern problem and give an algorithm that automatically discovers the clusters of patterns in multiple data sequences. We have taken a model-less approach and introduced a notation for maximal patterns that drastically reduces the number of valid cluster patterns. We conclude with two open problems and some preliminary results of the automatic pattern discovery tool on motifs on *E Coli* protein sequences.

Acknowledgments. We are grateful to the anonymous referees for their comments that has improved the paper. Also, one of the authors greatly benefitted from earlier discussions with Jens Stoye.

Example of a permuted maximal πpattern of size 6 in *E Coli* sequences:
(325) 35 {53 36} 81 136 {**72** 8} *35* {159 21} 36 109 {140 82 57}

(498) *35* {159 21} {53 36} 81 136 {**72** 8} *35* 187 {159 21} 36 109 {166 145 140 82 79 74 71 57}

6 4 [35 36 72 81 136 159] (325 1) (325 2) (498 0) (498 1)

35 *GETL..VGESGSGKS.T*
36 *VGESGSGKS.T*
72 **IADEPTT.LDV**
81 PHQLSGG..QRV
136 *LSGG.RQRV.IA*
159 *LVG.SGSGKS.T*

line 325

VLAVENLNIAFMQDQQKIAAVRNLSFSLQR*GETLAIVGESGSGKSVT*ALALMRLLEQAGGLV
QCDKMLLQRRSREVIELSEQNAAQMRHVRGADMAMIFQEPMTSLNPVFTVGEQIAESIRLHQ
NASREEAMVEAKRMLDQVRIPEAQTILSRYPHQLSGGMRQRV*MIA*MALSCRPAVL**IADE**
PTTALDVTIQAQILQLIKVLQKEMSMGVIFITHDMGVVAEIADRVLVMYQGEAVETGTVEQ
IFHAPQHPYTRALLAAVPQLGAMKGLDYPRRFPLISLEHPAKQAPPIEQKTVVDGEPVLRVR
NLVTRFPLRSGLLNRVTREVHAVEKVSFDLWP*GETLSLVGESGSGKSTT*GRALLRLVESQGG
EIIFNGQRIDTLSPGKLQALRRDIQFIFQDPYASLDPRQTIGDSIIEPLRVHGLLPGKDAAARVAW
LLERVGLLPEHAWRYPHEFSGGQRQRICIARALALNPKVIIADEAVSALDVSIRGQIINLLLDLQR
DFGIAYLFISHDMAVVERISHRVAVMYLGQIVEIGPRRAVFENPQHPYTRKLLAAVPVAEPSRQR
PQRVLLSDDLPSNIHLRGEEVAAVSLQCVGPGHYVAQPQSEYAFMRR

line 498

MTQTLLAIENLSVGFRHQQTVRTVVNDVSLQIEA*GETLALVGESGSGKSVT*ALSILRLLPSPP
VEYLSGDIRFHGESLLHASDQTLRGVRGNKIAMIFQEPMVSLNPLHTLEKQLYEVLSLHRGMRR
EAARGEILNCLDRVGIRQAAKRLTDYPHQLSGGERQRV*MIA*MALLTRPELL**IADEPTTAL**
DVSVQAQILQLLRELQGELNMGMLFITHNLSIVRKLAHRVAVMQNGRCVEQNYAATLFASPTH
PYTQKLLNSEPSGDPVPLPEPASTLLDVEQLQVAFPIRKGILKRIVDHNVVVKNISFTLRA*GETL*
*GLVGESGSGKSTT*GLALLRLINSQGSIIFDGQPLQNLNRRQLLPIRHRIQVVFQDPNSSLNPRLN
VLQIIEEGLRVHQPTLSAAQREQQVIAVMHEVGLDPETRHRYPAEFSGGQRQRIAIARALILKPSL
IILDEPTSSLDKTVQAQILTLLKSLQQKHQLAYLFISHDLHVVRALCHQVIILRQGEVVEQGPCAR
VFATPQQEYTRQLLALS

Fig. 1. An example to show how a pair of domains (motifs) numbered. The top two lines numbered 325 and 498 represent the input line numbers in the data. Each number in the row represents a domain (motif). Numbers in braces represents multiple occurrence of the domains: for example domains 53 and 36 occur at the same location. The next line shows that it is a pattern of size 6 (ie six domains in the permutation) occurring in four locations, twice in sequence 325 and twice in sequence 498. The next six lines show the mapping of the domain numbers to the actual domains. The permuted domains are displayed on the original sequences at the bottom. Using the maximality notation the pattern is represented as: (36-81-136-72), 35, 159.

The first author was partially supported by the Israel Science Foundation grant 282/01 and by the FIRST Foundation of the Israel Academy of Science and Humanities. The second author was partially supported by NSF grant CCR-0104307, by the Israel Science Foundation grant 282/01, by the FIRST Foundation of the Israel Academy of Science and Humanities, and by IBM Faculty Partnership Award.

References

1. A. Amir, A. Apostolico, G. M. Landau, and G. Satta. Efficient text fingerprinting via parikh mapping. *Journal of Discrete Algorithms*, 2003. to appear.
2. A. Apostolico, C. Iliopoulos, G. M. Landau, B. Schieber, and U. Vishkin. Parallel construction of a suffix tree with applications. *Algorithmica*, 3:347–365, 1988.
3. Brown, Clark, Leader, Simpson, and Lowe. *RNA*, 7:1817–1832, 2001.
4. T Dandekar, B Snel, M Huynen, and P Bork. *Trends Biochem. Sci.*, 23:324–328, 1998.
5. S Giglio, K W Broman, N Matsumoto, V Calvari, G Gimelli, T Neuman, H Obashi, L Voullaire, D Larizza, R Giorda, J L Weber, D H Ledbetter, and O Zuffardi. Olfactory receptor-gene clusters, genomic-inversion polymorphisms, and common chromosme rearrangements. *Am. J. Hum. Genet.*, 68(4):874–883, 2001.
6. Steffen Heber and Jens Stoye. Finding all common intervals of k permutations. In *Proc. of the Twelfth Symp. on Comp. Pattern Matching*, volume 2089 of *Lecture Notes in Computer Science*, pages 207–218. Springer-Verlag, 2001.
7. D. Kihara and M. Kanehisa. *Genome Res*, 10:731–743, 2000.
8. Z. M. Kedem, G. M. Landau, and K. V. Palem. Parallel suffix-prefix matching algorithm and application. *SIAM Journal of Computing*, 25(5):998–1023, 1996.
9. R. Karp, R. Miller, and A. Rosenberg. Rapid identification of repeated patterns in strngs, arrays and trees. In *Symposium on Theory of Computing*, volume 4, pages 125–136, 1972.
10. J. G. Lawrence and J. R. Roth. *Genetics*, 143:1843–1860, 1996.
11. Akhiro Nakaya, Susumo Goto, and Minoru Kanehisa. Extraction of corelated gene clusters by mulitple graph comparison. *Genome Informatics*, No 12:44–53, 2001.
12. R Overbeek, M Fonstein, M Dsouza, G D Pusch, and N Maltsev. The use of gene clusters to infer functional coupling. *Proc. Natl. Acad. Sci. USA*, 96(6):2896–2901, 1999.
13. H. Ogata, W. Fujibuchi, and S. Goto. *Nucleic Acids Res*, 28:4021–4028, 2000.
14. Laxmi Parida. Some results on flexible-pattern matching. In *Proc. of the Eleventh Symp. on Comp. Pattern Matching*, volume 1848 of *Lecture Notes in Computer Science*, pages 33–45. Springer-Verlag, 2000.
15. E M Marcott M Pellegrini, H L Ng, D W Rice, T O Yeates, and D Eisenberg. Detecting protein function and protein-protein interactions. *Science*, 285:751–753, 1999.
16. B. Snel, G Lehmann, P Bork, and M A Huynen. A web-server to retrieve and display repeatedly occurring neighbourhood of a gene. *Nucleic Acids Research*, 28(18):3443–3444, 2000.
17. J L Siefert, K A Martin, F Abdi, W R Widger, and G E Fox. *J. Mol. Evol.*, 45:467–472, 1997.
18. J Tamames, G Casari, C Ouzounis, and A Valencia. *J. Mol. Evol.*, 44:66–73, 1997.
19. K. Tomii and M. Kanehisa. *Genome Res*, 8:1048–1059, 1998.
20. H Watanbe, H Mori, T Itoh, and T Gojobori. *J. Mol. Evol.*, 44:S57–S64, 1997.

Dynamic Programming Algorithms for Two Statistical Problems in Computational Biology

Sven Rahmann[1,2]

[1] Department of Computational Molecular Biology
Max-Planck-Institute for Molecular Genetics
Ihnestraße 63–73, D-14195 Berlin, Germany
Sven.Rahmann@molgen.mpg.de
[2] Department of Mathematics and Computer Science
Freie Universität Berlin

Abstract. We present dynamic programming algorithms for two exact statistical tests that frequently arise in computational biology. The first test concerns the decision whether an observed sequence stems from a given profile (also known as position specific score matrix or position weight matrix), or from an assumed background distribution. We show that the common assumption that the log-odds score has a Gaussian distribution is false for many short profiles, such as transcription factor binding sites or splice sites. We present an efficient implementation of a non-parametric method (first mentioned by Staden) to compute the exact score distribution. The second test concerns the decision whether observed category counts stem from a specified Multinomial distribution. A branch-and-bound method for computing exact p-values for this test was presented by Bejerano at a recent RECOMB conference. Our contribution is a dynamic programming approach to compute the *entire distribution* of the test statistic, allowing not only the computation of exact p-values for all values of the test statistic simultaneously, but also of the power function of the test. As one of several applications, we introduce p-value based sequence logos, which provide a more meaningful visual description of probabilistic sequences than conventional sequence logos do.

1 Introduction

From a bird's eye perspective, many statistical methods in computational biology operate as follows: Given a notion of *signal* and a search space, find all (approximate) occurrences of the signal in the search space, and annotate each occurrence by its significance (p-value). Consider the following examples.

1. Position weight matrices (PWMs, also called profiles or position specific scoring matrices, PSSMs) are a widely used probabilistic description for signals in sequences, such as transcription factor binding sites [11] or splice sites [9].
2. (a) To measure tissue specificity of an alternatively spliced isoform of a particular gene, we compare the tissue distribution of the isoform transcripts

G. Benson and R. Page (Eds.): WABI 2003, LNBI 2812, pp. 151–164, 2003.

to the tissue distribution of all transcripts of the gene. (b) To determine whether a particular column of a multiple sequence alignment is informative in comparison to a certain background distribution, we again compare the observed nucleotide (or amino acid) distribution to the given background. In both cases, the signal we are looking for is a difference between the two distributions, as measured by an appropriate test statistic.

In each example, we test the *null hypothesis* H_0 that no signal is present against the *alternative* that the signal is present. The null hypothesis is specified by an appropriate probabilistic null model, and we write $\mathbb{P}_{H_0}(E)$ to denote the probability of event E under the null model. The decision to reject or accept the null hypothesis is based on an observation x and reached by examining a test statistic $s = S(x) \in \mathbb{R}$, which maps the observation to a real value (a "score"), where we assume that higher values of s correspond to stronger signal evidence. We decide that the signal occurs whenever s exceeds a certain threshold. Of course, this event can also happen when in fact the null hypothesis holds. This is called a *type-I error* or a *false positive.* We let X denote a random observation drawn according to the null model, so the score $S = S(X)$ becomes a random variable. The type-I error probability $\mathbb{P}_{H_0}(S \geq s)$ that S exceeds the observed score s, is called the p-value of s. A small p-value says that the observed score is improbable under the null model, and is interpreted as significant evidence of a signal occurrence. For the above examples, the situation is as follows.

1. The observation is a sequence of the same length as the PSSM. The null model is specified by a random i.i.d. model. The score $S(x)$ is a log-odds score comparing the likelihood of the profile and the null model for the observed data.
2. We observe a vector of tissue (nucleotide, amino acid) counts $x = (x_1, \ldots, x_k)$, where x_i denotes the number of transcripts from tissue i (nucleotides, amino acids i). The null model is specified by a background distribution of tissues (nucleotides, amino acids), and the score is a measure of deviation of the observed and null distribution, such as the χ^2-statistic or the generalized log-likelihood statistic (see below).

When we are dealing with large amounts of data, such as on a genomic scale, we need precise measures to separate signal from noise. In this sense, raw score values s are hard to interpret, and they should be converted to their p-value $\mathbb{P}_{H_0}(S \geq s)$.

Computing or estimating p-values is often a hard but unavoidable problem by itself, and the field of algorithms for statistics is steadily growing. After some preliminary remarks (Section 2), we recall attention to a simple algorithm for non-parametric computation of p-values for gapless profile scores (Example 1 above) in Section 3. While this method dates at least back to Staden [10], it has apparently been largely forgotten and replaced by a more convenient but dangerous Gaussian approximation (Section 3, Fig. 1). We point out that the method is more generally applicable than noted previously: Not only can we use it to compute p-values, but also the power function of the test, and the "cost" of using a "wrong" profile.

The main contribution of this paper is found in Section 4, where we describe a dynamic programming algorithm to compute the exact p-value and power function of a test for a specified multinomial distribution. Our method improves upon Bejerano's recent work at RECOMB'03 [1], since we obtain the whole score distribution instead of only one p-value at a single point. In addition to the manifold applications of this test mentioned in [1], we present a new method to visualize sequence profiles based on the exact p-values in Section 5. Section 6 concludes.

2 Preliminaries

In the following sections, we consider probability distributions on a finite subset of real numbers (score values). We first give a few details on the representation of probability distributions in the computer, and on the implementation of certain basic operations (addition in log-space; efficient computation of binomial probabilities). These remarks may be skipped on a first reading.

Score Rounding. We assume that all score values are rounded to a certain granularity ε (which could be as small as the machine precision), meaning that a score s is rounded to

$$\mathrm{rd}_\varepsilon(s) := \varepsilon \cdot \lfloor s/\varepsilon + 1/2 \rfloor.$$

The running time of our algorithms depends linearly on $1/\varepsilon$, i.e., for one additional digit of precision, the running time increases by a factor of 10. In practice, $\varepsilon = 0.01$ to $\varepsilon = 0.0001$ turn out to be sufficiently accurate.

Vector representation of probability distributions. Assume that a random score variable S takes minimum value $\underline{S}$ and maximum value $\overline{S}$ (both of which are integer multiples of ε); then the set of potential scores is $\Gamma = \{\underline{S} + \gamma\varepsilon \mid \gamma = 0, \dots, (\overline{S} - \underline{S})/\varepsilon\}$. The probability distribution function (pdf) of S, $f(\sigma) := \mathbb{P}(S = \sigma)$ with $\sigma \in \Gamma$ can then be represented as a vector

$$f = (f_0, f_1, \dots, f_{(\overline{S}-\underline{S})/\varepsilon}) \quad \text{with} \quad f_\gamma \equiv f(\underline{S} + \gamma\varepsilon).$$

While our algorithms use this vector representation, we shall nevertheless write $f(\sigma)$ instead of f_γ to refer to $\mathbb{P}(S = \sigma)$ when $\sigma = \underline{S} + \gamma\varepsilon$.

Probabilities in log-space. When we expect that some of the probabilities in a pdf become very small, it is preferable to represent them by their logarithms. In standard double precision, the smallest representable double precision number is about 10^{-323}, and if we need to represent smaller values, the whole computation can be done in log-space where $\ln(10^{-323}) \approx -743.747$. Multiplication of probabilities reduces to a simple addition in log-space, but a comment about addition is in order (cf. [1, 2]). To compute $\tilde{z} := \ln z$ from $z := x + y$ when only $\tilde{x} := \ln x$ and $\tilde{y} := \ln y$ are available, we have of course theoretically $\tilde{z} = \ln(\exp(\tilde{x}) + \exp(\tilde{y}))$, but we must allow for the fact that the exponentials are not representable by double precision. Assuming $x \geq y$, we have

$$\tilde{z} = \tilde{x} + \ln(1 + y/x) \quad \text{with} \quad 0 \leq y/x \leq 1.$$

If the ratio $r := y/x = \exp(\tilde{y} - \tilde{x})$ is not representable by a double precision number, we have $\tilde{z} = \tilde{x}$. If r is representable but very small, we have the approximation $\ln(1+r) \approx r$ and hence $\tilde{z} = \tilde{x} + r$. If r is moderately small (so that computing $1+r$ would still result in a loss of precision), it is better not to call the ln-function with argument $1+r$, but to compute the defining alternating series $\sum_{k=1}^{\infty} (-1)^k r^k/k$ directly; a few terms suffice since $r \ll 1$. As $r \nearrow 1$, we can use a convenient call to $\ln(1+r)$ or use the transformation $\rho := r/(2+r) \leq 1/3$ and use the power series $\log(1+r) = \log((1+\rho)/(1-\rho)) = 2 \cdot \sum_{k \text{ odd}} \rho^k/k$, which converges rapidly since the sign of the terms does not alternate.

Binomial probabilities. Let $b_{n,\rho}(m)$ be the probability that a Binomial random variable with parameters n and ρ takes the value m, i.e., $b_{n,\rho}(m) := \binom{n}{m} \cdot \rho^m \cdot (1-\rho)^{n-m}$. This probability is more conveniently computed in log-space where

$$\ln(b_{n,\rho}(m)) = g(n+1) - g(m+1) - g(n-m+1) + m \cdot \ln \rho + (n-m) \cdot \ln(1-\rho).$$

Here $g(x)$ is the logarithm of the Gamma function $\Gamma(x) = \int_0^{\infty} t^{x-1} e^{-t}\, dt$, for which $\Gamma(n+1) = n!$ for integers n. Efficient numerical methods to directly compute $g(x)$ are given in [6], so we may assume that Binomial coefficients can be computed in constant time.

3 Distribution of Profile Scores

We describe a simple dynamic programming algorithm to compute the distribution of profile scores under both background H_0 and signal model. Essentially, this method dates back to the late 1980s [10], but is included here because it does not seem to be widely used. In fact, it has become common practice to assume that the background distribution is a Gaussian, which is asymptotically true for long profiles under mild regularity conditions [4], but may lead to large relative errors in p-value estimation for short profiles. Since most profiles describing transcription factor binding sites [11] or splice sites [9] are quite short, this has become a severe practical problem.

Fig. 1 illustrates the shape of background and signal distribution for certain transcription factor binding sites from the TRANSFAC database [11].

Formally, a profile of length n over an alphabet Σ of size k can be specified in various ways as a $k \times n$ matrix. The *count matrix* C is the basis for creating a profile. The entry C_{ij} specifies how many times the i-th letter Σ_i was observed at position j. The true *profile matrix* P is obtained by normalizing the columns of C to sum to 1, i.e., $P_{ij} = C_{ij} / \sum_k C_{kj}$, and specifies a position specific probability distribution on the letters. We test whether a given Σ-word $W = (W_1, \ldots, W_n)$ of length n is generated by P, or by a background distribution π, which is

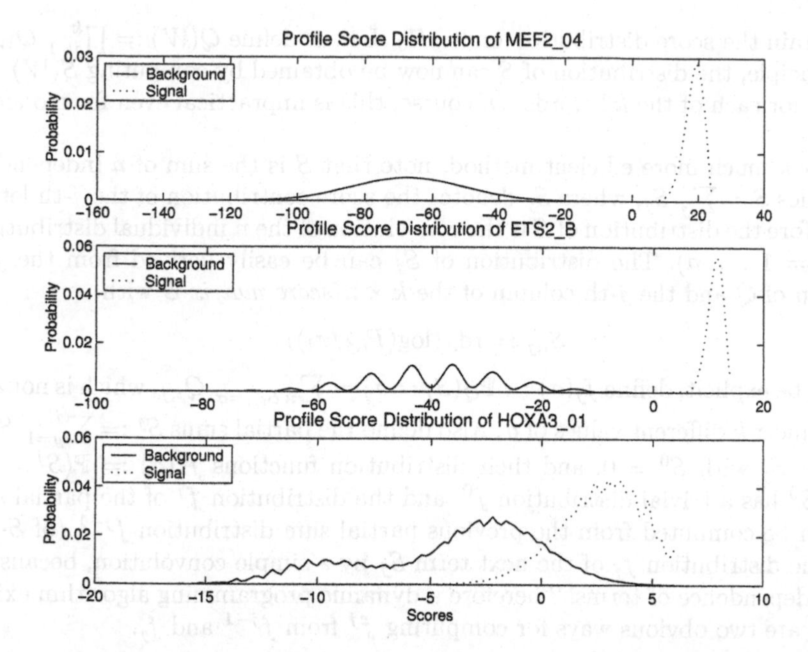

Fig. 1. Score distribution of certain transcription factor binding sites (MEF2_04, ETS2_B, and HOXA3_01) from the TRANSFAC database, both under the null hypothesis of a uniform background and the alternative of sequences generated according to the profile signal itself. From top to bottom, profile quality decreases, as seen by the increasing overlap of the distributions. In the second and third example, we also see that the background distribution does not follow a Gaussian distribution. Results based on such an assumption may therefore produce wrong significance or power estimates. Our non-parametric approach avoids distributional assumptions.

assumed to be the same at every position. Thus we write null hypothesis H_0 and alternative K as

$$H_0 : W \sim \pi, \qquad K : W \sim P.$$

By the Neyman-Pearson lemma, the uniformly most powerful test is obtained when we consider the log-likelihood ratio statistic

$$S(W) := \log \frac{\mathbb{P}_K(W)}{\mathbb{P}_{H_0}(W)} = \sum_{j=1}^{n} \log \left(P_{W_j,j} / \pi_{W_j} \right), \tag{1}$$

and reject H_0 when $S(W)$ becomes sufficiently large.

Assuming that W is a random word generated according to an arbitrary position specific distribution matrix Q, the score $S = S(W)$ becomes a random variable whose distribution we want to compute. For example, when $Q = P$ we obtain the distribution of S under K, and setting Q to the n-fold repetition of π, we obtain the score distribution under H_0. Let us define $Q(W) := \prod_{j=1}^{k} Q_{W_j,j}$. In principle, the distribution of S can now be obtained by computing $S(W)$ and $Q(W)$ for each of the k^n words. Of course, this is impractical even for moderate n.

For a much more efficient method, note that S is the sum of n independent variables $S = \sum_j S_j$, where S_j denotes the score contribution of the j-th letter. Therefore the distribution of S is the convolution of the n individual distributions S_j ($j = 1, \ldots, n$). The distribution of S_j can be easily derived from the j-th column of Q and the j-th column of the $k \times n$ *score matrix* $\mathcal{S}$ with

$$\mathcal{S}_{i,j} := \mathrm{rd}_\varepsilon \left(\log(P_{i,j}/\pi_i)\right).$$

To be explicit, define $f_j(\sigma) := \mathbb{P}_Q(S_j = \sigma) = \sum_{i:\, \mathcal{S}_{i,j}=\sigma} Q_{i,j}$, which is nonzero for at most k different values of σ. Also define the partial sums $S^j := \sum_{y=1}^{j} S_y = S^{j-1} + S_j$ with $S^0 = 0$, and their distribution functions $f^j(\sigma) := \mathbb{P}(S^j = \sigma)$. Now S^0 has a trivial distribution f^0, and the distribution f^j of the partial sum S^j can be computed from the previous partial sum distribution f^{j-1} (of S^{j-1}) and the distribution f_j of the next term S_j by a simple convolution, because of the independence of terms. Therefore a dynamic programming algorithm exists. There are two obvious ways for computing f^j from f^{j-1} and f_j.

1. Use the discrete fast Fourier transform (DFFT) to reduce the convolution to a simple multiplication in frequency space. This method is best if both arguments of the convolution are non-sparse distributions.
2. As noted above, f_j is nonzero for at most k different elements (e.g., for DNA profiles $k = 4$). So it is more economical to compute the convolution explicitly:

$$f^j(\sigma) := \sum_{i=1}^{k} f^{j-1}(\sigma - \mathcal{S}_{i,j}) \cdot Q_{i,j}.$$

The latter method is very practical for DNA profiles. Assume that in each column of the score matrix the difference of maximum and minimum score is bounded by a constant R. Then the computation runs in $O(kn^2R/\varepsilon)$ total time, since the final distribution vector f^n has $O(nR/\varepsilon)$ elements.

If we require that the entries of the background distribution are all nonzero, there are no positively infinite values in $\mathcal{S}$, but we may have $\mathcal{S}_{i,j} = -\infty$ whenever $P_{i,j} = 0$. This situation can be avoided by adding so-called pseudo-counts to the initial count matrix. Alternatively, we can check for $-\infty$ scores in each step of the computation and treat them separately. As mentioned in Section 2, all probability distributions f^j are represented as vectors, and if we expect to encounter very small probabilities, the whole computation can be carried out in log-space. To summarize, we have

Theorem 1. *Given an arbitrary $k \times n$ position specific score matrix of granularity ε with maximum score range R in each column, and an arbitrary $k \times n$ profile (probabilistic sequence description) Q, the exact score distribution over all k^n words of length n generated according to Q can be computed in $O(kn^2R/\varepsilon)$ time.*

Application areas. Since the position specific alphabet distribution Q that is used to compute the score distribution need not be the same as the background π, the method is not restricted to p-value ($\mathbb{P}_{H_0}(S \geq s)$) computation; a fact that seems to have gone unnoticed previously.

By choosing $Q = P$ (the profile itself), we obtain the test's power $\mathbb{P}_K(S \geq s)$ on the alternative K and hence the theoretical probability $\mathbb{P}_K(S < s)$ of a type-II error (false negative) when we accept the assumption that P is an accurate description of the signal. What's more, we can measure how well a profile separates signal from noise by comparing the score distributions under H_0 and K and finding an appropriate measure of their overlap. High overlap means low signal-to-noise separation and low quality. A possible visualization of the obtained information is shown in Fig. 1. A systematic study of all transcription factor binding site profiles of the TRANSFAC database is in preparation [7].

For the most powerful test, the score matrix S should contain the log-odds of the possible observations under H_0 and K. In general, the profile description of K is derived from empirical data and is only an estimate of the true profile (still assuming that the signal can be accurately described by a gapless profile). Assume we know that the true profile is in fact P, but the score matrix S' with $S'_{i,j} = \ln(P'_{i,j}/\pi_i)$ is based on an perturbation P' of P. By comparing the score distribution under K based on the optimal (true) score matrix S and on the estimated score matrix S', we can estimate how sensitively the test reacts to small errors in the profile.

4 An Exact Test for a Specified Multinomial Distribution

In this section we consider the test whether given observed counts $x = (x_1, \ldots, x_k)$ with $\sum_i x_i = n$ are derived from a multinomial distribution with parameters n and a probability vector $p = (p_1, \ldots, p_k)$. We write $\mathcal{M}(n, p)$ for this distribution. When $k = 2$, the parameter vector has the form $p = (p_1,\, 1-p_1)$, and we write $\mathcal{B}(n, p_1)$ for the resulting Binomial distribution. Our work is motivated by Bejerano's recent paper at RECOMB 2003 [1], which also mentions several applications of this test in biosequence analysis. Let us review the ingredients.

Null hypothesis H_0 and alternative K are given by

$$H_0 : X = (X_1, \ldots, X_k) \sim \mathcal{M}(n, p), \quad K : X \sim \mathcal{M}(n, q) \text{ for a } q \neq p.$$

The deviation between observed $q(x) = (q_1, \ldots, q_k) := (x_1/n, \ldots, x_k/n)$ and null distribution $p = (p_1, \ldots, p_k)$ is measured by the scaled generalized log-likelihood ratio statistic

$$S(x) := 2 \log \frac{\mathbb{P}_{\mathcal{M}(n,q(x))}(x)}{\mathbb{P}_{\mathcal{M}(n,p)}(x)} = 2n \sum_{i=1}^{k} q_i \log(q_i/p_i) \tag{2}$$

$$= \sum_{i=1}^{k} 2\, x_i \log(x_i/(np_i)), \tag{3}$$

where $\mathcal{M}(n, \theta)$ denotes the Multinomial distribution with parameters n and $\theta = (\theta_1, \ldots, \theta_k)$, so

$$\mathbb{P}_{\mathcal{M}(n,\theta)}(x) = \binom{n}{x_1, \ldots, x_k} \cdot \prod_{i=1}^{k} \theta_i^{x_i} = n! \cdot \prod_{i=1}^{k} \theta_i^{x_i} / x_i!.$$

Under H_0 and for $n \to \infty$, the distribution of S converges to a chi-square distribution with $k-1$ degrees of freedom. For small and moderate values of n, however, this approximation can be quite poor, depending on the degree of non-uniformity of p [1]. Therefore there is a need to compute the exact distribution of S in a non-asymptotic and non-parametric way.

We simply write $p(x)$ for $\mathbb{P}_{\mathcal{M}(n,p)}(x)$. The straightforward combinatorial way to compute the p-value belonging to an observation x^* with score $s = S(x^*)$ is to evaluate $\mathbb{P}_{H_0}(S \geq s) = \sum_x p(x) \cdot \mathbb{I}\{S(x) \geq s\}$, where the sum extends over all k-tuples $x = (x_1, \ldots, x_k)$ with nonnegative counts that sum to n. Therefore this sum consists of $\binom{n+k-1}{k-1} = O(n^k)$ terms.

Bejerano [1] uses a branch-and-bound algorithm to reduce the number of terms that must be evaluated: Given a partial assignment of counts to a prefix of x, such as $x = (7, 2, x_3, \ldots, x_k)$, we can easily obtain a lower bound L and an upper bound U of $S(x)$ for all vectors x with the specified initial counts. If $U < s$, then none of the vectors of this type contributes to the p-value, and we need not examine the subtypes. If $L \geq s$, then all vectors of this type contribute to the p-value, and we can add their probabilities in one step. Only if $L < s \leq U$, the subtypes of the current partial type must be evaluated. This approach seems to work well in practice when only one p-value or a few p-values for extreme values of s are required. However, we do not gain access to the whole distribution of S under H_0.

We present a simple dynamic programming algorithm to compute the score distribution of the test for $\mathcal{M}(n, p)$ under an *arbitrary* Multinomial distribution $\mathcal{M}(n, \theta)$. Taking $\theta = p$ allows us to compute p-values of the test; taking other vectors θ allows us to evaluate the power function of the test. Our approach is based on the following property of the Multinomial distribution.

Lemma 1. *Let the random vector $X = (X_1, \ldots, X_k)$ be distributed according to a Multinomial distribution with parameters n and $\theta = (\theta_1, \ldots, \theta_k)$.*

Then X_1 has a Binomial distribution $\mathcal{B}(n, \theta_1)$, and the conditional distribution of X_j given $(X_1, \ldots, X_{j-1})$ is $\mathcal{B}(N_j^*, \theta_j^*)$, where

$$N_j^* = n - \sum_{i=1}^{j-1} X_i \quad \text{and} \quad \theta_j^* = \theta_j / \left(1 - \sum_{i=1}^{j-1} \theta_i\right). \tag{4}$$

In particular, the conditional distribution of X_j given $(X_1, \ldots, X_{j-1})$ depends only on the sum $X_1 + \cdots + X_{j-1}$.

Proof. The shortest proof uses the fact that a Multinomial random variable is obtained from k independent Poisson random variables $X_1, \ldots, X_k$ by conditioning on a fixed sum n of their values. By additionally fixing $X_1, \ldots, X_{j-1}$, we see that the remaining vector $(X_j, \ldots, X_k)$ follows again a Multinomial law, and its one-dimensional marginals are the claimed Binomials. For a more detailed outline, see Feller's classic [3]. □

A consequence of Lemma 1 is that we can sample from a Multinomial distribution by successively sampling from different Binomial distributions. To exploit the lemma for our problem, we write

$$S = \sum_{i=1}^{k} S_i, \quad \text{where} \quad S_i = 2\, X_i \, \log(X_i/(np_i))$$

by Eq. (3). Note that the term S_i depends only on X_i, which takes values between 0 and n, so we pre-compute a $k \times (n+1)$ score matrix $\mathcal{S} = (\mathcal{S}_{i,y})$ of granularity ε by setting $\mathcal{S}_{i,y} := \mathrm{rd}_\varepsilon(2y \log(y/(np_i)))$ (the column corresponding to $y = 0$ consists entirely of zeros).

Let us define $X^j := \sum_{i=1}^{j} X_i$ and $S^j := \sum_{j=1}^{j} S_i$. We define the joint probabilities $f_m^j(\sigma) := \mathbb{P}_{\mathcal{M}(n,\theta)}(S^j = \sigma, X^j = m)$.

Lemma 2 (Recursion for f_m^j). *Let $b_{n,\rho}(m)$ be the probability that a Binomial random variable with parameters n and ρ takes the value m, i.e., $b_{n,\rho}(m) := \binom{n}{m} \cdot \rho^m \cdot (1-\rho)^{n-m}$, and let θ^* be defined as in Eq. (4). Let $\mathbb{I}\{c\} := 1$ if condition c is true and $\mathbb{I}\{c\} := 0$ otherwise. Then*

$$f_m^1(\sigma) = b_{n,\theta_1}(m) \cdot \mathbb{I}\{\mathcal{S}_{1,m} = \sigma\} \qquad (m = 0, \ldots, n), \qquad (5)$$

$$f_m^j(\sigma) = \sum_{y=0}^{m} b_{n-(m-y),\theta_j^*}(y) \cdot f_{m-y}^{j-1}(\sigma - \mathcal{S}_{j,y}) \qquad \begin{matrix}(j = 2, \ldots, k-1) \\ (m = 0, \ldots, n)\end{matrix}, \qquad (6)$$

$$f_m^k(\sigma) \equiv 0 \qquad (m = 0, \ldots, n-1), \qquad (7)$$

$$f_n^k(\sigma) = \sum_{y=0}^{n} f_{n-y}^{k-1}(\sigma - \mathcal{S}_{k,y}) \; = \; \mathbb{P}_{H_0}(S = \sigma). \qquad (8)$$

Proof. We write $\mathbb{P}$ shorthand for $\mathbb{P}_{\mathcal{M}(n,\theta)}$. For (5), we have by definition that $f_m^1(\sigma) = \mathbb{P}(S_1 = \sigma, X_1 = m) = \mathbb{I}\{\mathcal{S}_{1,m} = \sigma\} \cdot b_{n,\theta_1}$. For (6) we have

$$\begin{aligned} f_m^j(\sigma) &= \mathbb{P}(S^j = \sigma, X^j = m) \\ &= \sum_{y=0}^{m} \mathbb{P}(X_j = y,\; X^{j-1} = m - y,\; S^{j-1} = \sigma - \mathcal{S}_{j,y}) \\ &= \sum_{y=0}^{m} \left[\mathbb{P}(X_j = y \mid X^{j-1} = m - y,\; S^{j-1} = \sigma - \mathcal{S}_{j,y}) \right. \\ &\qquad \left. \cdot\, \mathbb{P}(X^{j-1} = m - y,\; S^{j-1} = \sigma - \mathcal{S}_{j,y}) \right], \end{aligned}$$

which is the claimed sum by Lemma 1. To prove (7) and (8), note that the last binomial parameter is $\theta_k^* = 1$, so we are certain to obtain all remaining counts (for a sum of n) from the last category k. □

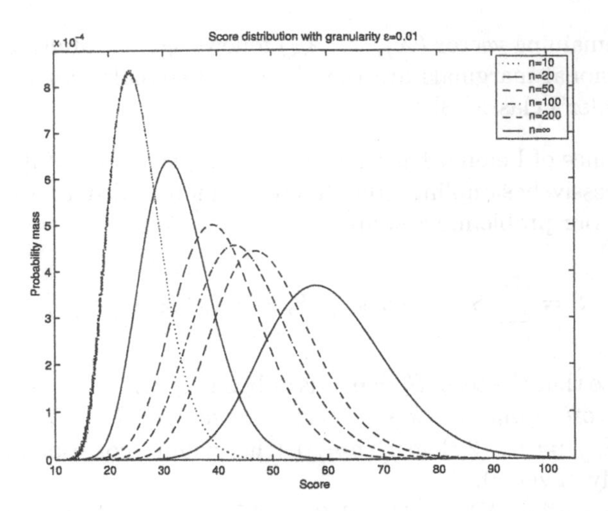

Fig. 2. Score distribution of the generalized log-likelihood ratio test (cf. Eq. (3)) for a certain parameter vector p with 61 components ranging from $2.08 \cdot 10^{-6}$ to 0.1136 for sample sizes of $n = 10, 20, 50, 100, 200$, and $n \to \infty$ (the chi-square distribution with 60 degrees of freedom). With granularity $\varepsilon = 0.01$, the computation took 2.4 sec, 11.6 sec, 2.1 min, 37.7 min, and 9.23 hrs.

Theorem 2. *Let $\mathcal{S} = (\mathcal{S}_{j,y})$ with $1 \leq j \leq k$ and $0 \leq y \leq n$ be an arbitrary $k \times (n+1)$ score table of granularity ε, and define the random score variable S by $S := \sum_{j=1}^{k} \mathcal{S}_{j,X_j}$, where $X = (X_1, \dots X_k) \sim \mathcal{M}(n, \theta)$.*

Assuming that all scores s in column y of $\mathcal{S}$ are bounded by $|s| \leq Ry$, the exact score distribution over all $\binom{n+k-1}{k-1}$ valid configurations of X can be computed in $O(kn^3 R/\varepsilon)$ time.

Proof. Note that $O(kn)$ distributions f_m^j are computed, and the computation of element $f_m^j(\sigma)$ requires $O(n)$ time assuming that the Binomial coefficients are available in constant time. Since the score contributed by X_j is bounded in absolute value by RX_j and $\sum_j X_j = n$, the range of S is contained in an interval of length $O(nR)$, and each distribution vector uses $O(nR/\varepsilon)$ elements. □

Note that the running time already includes the scaling factor of n in the test statistic (cf. Eq. (2)), which is necessary to ensure convergence to a non-trivial limiting distribution of S under H_0 for $n \to \infty$. Otherwise the running time would reduce to $O(kn^2 R/\varepsilon)$.

Sample application. To illustrate the necessity of an exact non-parametric method for this test, we examined the frequencies $p = (p_1, \dots, p_k)$ of the tissues from which the EST sequences of the GeneNest database [5] were originally obtained. This "background" tissue probability vector has $k = 61$ components, ranging from $2.08 \cdot 10^{-6}$ to 0.1136. Assume that a certain gene is represented in

Fig. 3. A conventional sequence logo of a translation initiation site profile taken from `http://genio.informatik.uni-stuttgart.de/GENIO/logo/`.

the database by n independently and randomly sampled ESTs. Then we may ask whether the tissue distribution of this gene's ESTs deviates significantly from the background distribution. From the vector $x = (x_1, \ldots, x_k)$ of tissue counts with $\sum_i x_i = n$ and compute the test statistic $S(x)$ by Eq. (3). When the counts x are derived from a $\mathcal{M}(n,p)$ distribution (null hypothesis), the score is distributed as shown in Fig. 2 (for $n = 10, 20, 50, 100, 200$, and in the limit $n \to \infty$). If for example less than 1% of the total probability mass lies to the right of the observed score $S(x)$, the observed distribution deviates significantly from the background distribution. In a similar way, we may attempt to detect tissue-specific splice variants of a gene. More applications of this test are given in [1] and in the following section.

5 Sequence Logos Based on p-values

Sequence logos such as the one in Fig. 3 are a well known way to visualize a profile or a multiple alignment [8]. The height of each stack of letters is given by the relative entropy between the observed distribution of letters at a particular profile position or alignment column and an assumed background distribution (on the nucleotides often the uniform distribution). Assume that the alphabet consists of k letters and the background frequency of the i-th letter is $p_i > 0$. Consider a position where the i-th letter is observed x_i times with $\sum_i x_i = n$. Let $q_i := x_i/n$. Then the stack height for this position is the relative entropy (information content) $H(q \,\|\, p) = \sum_{i=1}^{k} q_i \cdot \log_2(q_i/p_i)$ bits, where $0 \cdot \log_2(0/p_i) := 0$ by continuity. The height of the the i-th letter within the stack is simply fraction q_i of the total stack height.

The disadvantage of conventional sequence logos defined in this way is that we lose all information about the number n of observations that the profile position (alignment column) is based on, since we only compare probability distributions. Therefore we propose the following modification of sequence logos.

Definition 1 (p-value logos). *To determine the stack height of a position with observed count vector $x = (x_1, \ldots, x_k)$ with $\sum_i x_i = n$ with respect to a*

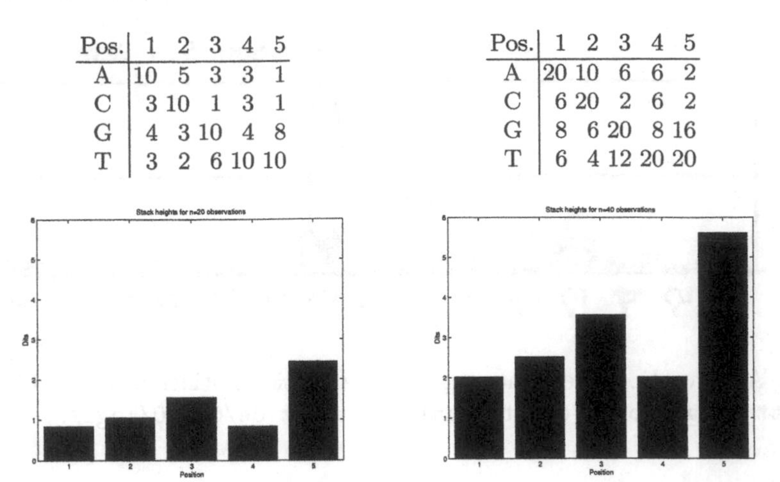

Pos.	1	2	3	4	5
A	10	5	3	3	1
C	3	10	1	3	1
G	4	3	10	4	8
T	3	2	6	10	10

Pos.	1	2	3	4	5
A	20	10	6	6	2
C	6	20	2	6	2
G	8	6	20	8	16
T	6	4	12	20	20

Fig. 4. Two profiles with the same position specific nucleotide distributions, but with different total numbers of observations. While the conventional sequence logos of both profiles would be identical, the p-value based logos allow direct visual comparison of the profiles (only stack heights are shown; here against the uniform background distribution). Significance (a p-value of less than 0.01) corresponds to more than 2 dits. In particular, only the distribution at position 5 deviates significantly from a uniform background distribution when based on 20 counts (left), but the deviation at positions 2 and 3 becomes significant when based on 40 counts.

given background distribution p, test whether $x \sim \mathcal{M}(n,p)$ using the method of Section 4: Compute the scaled relative entropy score $S(x)$ by Eq. (3), convert $S(x)$ to its p-value $\mathbb{P}_{\mathcal{M}(n,p)}(S \geq S(x))$, and use a stack height of

$$h(x) := -\log_{10}(\mathbb{P}_{\mathcal{M}(n,p)}[S \geq S(x)]) \text{ dits.}$$

The height of the i-th letter within the stack remains fraction x_i/n of $h(x)$, as for conventional sequence logos.

The advantage of this definition is that the stack height is directly related to the significance of the measured deviation. A p-value of 10^{-t} corresponds directly to a stack height of t dits or $3.322 \cdot t$ bits. This allows us to visually compare profiles or alignment columns based on different numbers of observations. For example, consider the two profiles in Fig. 4, based on $n = 20$ and $n = 40$ observations, respectively. Since the position specific probability distributions are identical, so are the classical sequence logos, not taking into account that 40 observations provide stronger evidence than 20 observations (always assuming independence of observations).

Hence, p-value based sequence logos allow direct visual comparison of profiles based on different numbers of observation counts. They will also be useful for visualizing how informative each column of a multiple alignment is when some sequences do not extend over the whole alignment and the total observation count varies within the alignment.

6 Discussion and Conclusion

We have described two simple dynamic programming algorithms for two frequently arising statistical testing problems in computational biology. While the method to compute profile scores is not new [10], its potential has apparently not been fully realized yet (see *Application Areas* in Section 3). The computed score distributions can be used to select optimal score thresholds during profile searches. When a long sequence is searched, each window of length n is scored against the profile. This leads to multiple testing issues which we have not discussed here.

The efficient algorithm to obtain the full exact test statistic distribution of the test for a Multinomial distribution constitutes an improvement over recent work [1], which only computes a single p-value with a branch-and-bound approach. Furthermore, our method allows to compare the power functions of different test statistics such as $S(x)$ from Eq. (3) and the chi-square statistic $S_{\chi^2}(x) = \sum_{i=1}^{k} (x_i - np_i)^2/(np_i)$, both of which converge to a chi-square distribution with $k-1$ degrees of freedom for $n \to \infty$ under H_0. When n becomes large, however, and only a single p-value or a few p-values are required, Bejerano's method is still an alternative, as our method is slowed by the $O(n^3)$ time bound.

A simple trick can be used to reduce the running time to $O(n^2)$: While the score range scales with n, we know that for $n \to \infty$ the score distribution converges in fact to a chi-square distribution which does not depend on n. Therefore, we know that the probability of large scores becomes zero, and we can neglect these high score ranges. Furthermore, sorting the null distribution p into decreasing order speeds up the computation in practice, because the initial score ranges will be small.

We have introduced a new p-value based way of visualizing probabilistically described sequences that allows better comparison of two profile or alignment positions based on different numbers of observations.

The methods described here are based on rounding the test statistic to a certain granularity ε; the running time depends linearly on $1/\varepsilon$. We point out that the *scores*, not the probabilities are rounded. On the rounded set of scores, the probabilities are as exact as possible.

Prototypical MATLAB source code for our methods is freely available and can be obtained from `http://logos.molgen.mpg.de`. We are working on interactive WWW forms that generate plots such as those in Figs. 1 and 2 on request, and another web application to generate p-value based sequence logos.

Acknowledgments. I thank Tobias Müller for very helpful discussions, Hannes Luz for motivation, Thomas Meinel and Christine Steinhoff for help with graphics, and Martin Vingron for motivation and support. The comments of the anonymous referees were also very helpful.

References

[1] G. Bejerano. Efficient exact p-value computation and applications to biosequence analysis. In *RECOMB 2003 Proceedings*, pages 38–47. ACM Press, April 2003.

[2] R. Durbin, S. Eddy, A. Krogh, and G. Mitchison. *Biological Sequence Analysis.* Cambridge University Press, Cambridge, 1998.

[3] W. Feller. *An Introduction to Probability Theory and Its Applications*, volume 1. John Wiley and Sons, second edition, 1971.

[4] L. Goldstein and M. Waterman. Approximations to profile score distributions. *Journal of Computational Biology*, 1(1):93–104, 1994.

[5] S. A. Haas, T. Beissbarth, E. Rivals, A. Krause, and M. Vingron. GeneNest: automated generation and visualization of gene indices. *Trends Genet.*, 16(11):521–523, 2000.

[6] W. H. Press, B. P. Flannery, S. A. Teukolsky, and W. T. Vetterling. *Numerical Recipes in C.* Cambridge University Press, second edition, 1993.

[7] S. Rahmann, T. Müller, and M. Vingron. On the power and quality of profiles with applications to transcription factor binding site detection. Unpublished Manuscript, 2003.

[8] T. D. Schneider and R. M. Stephens. Sequence logos: A new way to display consensus sequences. *Nucl. Acids Res.*, 18:6097–6100, 1990.

[9] R. Staden. Computer methods to locate signals in nucleic acid sequences. *Nucleic Acids Research*, 12:505–519, 1984.

[10] R. Staden. Methods for calculating the probabilities of finding patterns in sequences. *CABIOS*, 5:89–96, 1989.

[11] E. Wingender, X. Chen, R. Hehl, H. Karas, I. Liebich, V. Matys, T. Meinhardt, M. Prüss, I. Reuter, and F. Schacherer. TRANSFAC: an integrated system for gene expression regulation. *Nucleic Acids Res.*, 28:316–319, 2000.

Consensus Networks: A Method for Visualising Incompatibilities in Collections of Trees

Barbara Holland[1] and Vincent Moulton[2]

[1] Allan Wilson Centre for Molecular Ecology and Evolution, Massey University, New Zealand.
B.R.Holland@massey.ac.nz

[2] The Linnaeus Centre for Bioinformatics, Uppsala University, Box 598, 751 24 Uppsala, Sweden.
vincent.moulton@lcb.uu.se

Abstract. We present a method for summarising collections of phylogenetic trees that extends the notion of consensus trees. Each branch in a phylogenetic tree corresponds to a bipartition or split of the set of taxa labelling its leaves. Given a collection of phylogenetic trees, each labelled by the same set of taxa, all those splits that appear in more than a predefined threshold proportion of the trees are displayed using a median network. The complexity of this network is bounded as a function of the threshold proportion. We demonstrate the method for a collection of 5000 trees resulting from a Monte Carlo Markov Chain analysis of 37 mammal mitochondrial genomes, and also for a collection of 80 equally parsimonious trees resulting from a heuristic search on 53 human mitochondrial sequences.

1 Introduction

A central task in evolutionary biology is the construction of phylogenetic trees and, accordingly, many methods have been developed for performing this task. Quite often these methods produce a collection of trees rather than a point estimate of an optimal tree, since such a tree with no measure of reliability may not be particularly helpful. Examples of methods producing collections of trees include Monte Carlo Markov Chain (MCMC) methods [15], [13], and bootstrapping [9]. Heuristic or exact searches [22] can also produce collections of trees if the optimal solution is not unique.

Large collections of trees can be difficult to interpret and draw conclusions from. Thus, when faced with such a collection, it is common practice to construct a consensus tree, i.e., a tree that attempts to reconcile the information contained within all of the trees. Many ways have been devised for constructing consensus trees (see [6] for a comprehensive, recent overview). However, they all suffer from a common limitation: By summarizing all of the given trees by a single output tree, information about conflicting hypotheses is necessarily lost in the final representation.

G. Benson and R. Page (Eds.): WABI 2003, LNBI 2812, pp. 165–176, 2003.

Motivated by this problem we have developed a new approach to visualizing collections of trees that naturally generalizes consensus trees. This approach is based on the construction of phylogenetic networks, networks that are regularly used by biologists to visualize and analyze complex phylogenetic data sets. In particular, we will focus on the use of median networks [3] to visualize collections of trees as we now describe.

2 Methods

First we summarize some necessary concepts.

2.1 Background

Suppose that X is a finite set of taxa. A *split* $A|B$ of X is a bipartition of X, i.e., a partition of X into two non-empty sets or parts A and B with $A \cup B = X$ and $A \cap B = \emptyset$. We call a collection of splits a *split system* for short. A *phylogenetic tree* (on X) is a tree with leaves labelled by X. Each edge of a phylogenetic tree naturally gives rise to a split, since its removal results in two trees, each one being labelled by the elements in one part of a split. We say that a phylogenetic tree *displays* a split if there is an edge in the tree that gives rise to the split. A split system is called *compatible* if there is a phylogenetic tree that displays every split in the system. If this is the case then there is a unique such tree for which the edges are in one-to-one correspondence with the splits in the given system (see e.g., [21, pg 44]). We say that a split system is *incompatible* if it does not contain any subset of cardinality two that is compatible. Note that a split system which is not compatible, need not be incompatible.

It is possible to represent split systems on X by various networks [2], [3], [16]. In particular, a canonical *median network* [4] can be associated to any split system on X. These networks were originally designed for the analysis of mitochondrial data [4] and have also been used to analyze chloroplast data [12]. In a median network, certain vertices are labelled by the elements of X and, in a way similar to phylogenetic trees, splits are represented by classes of parallel edges. Figure 1 illustrates a simple median network on 5 taxa. The median network associated with a split system has several attractive properties. For example, it is a tree if and only if the split system is compatible (in which case it is the unique tree corresponding to the split system), and it is a hypercube if and only if the split system is incompatible [5]. In fact, for a general split system a median network lies somewhere between the extremes of being a tree and a hypercube since each incompatible subsystem of splits with cardinality k corresponds to a k-cube in the network. Moreover, the median network associated with a split system is straight-forward to generate using an algorithm first introduced in [4], which has been implemented in the freely available program Spectronet [11].

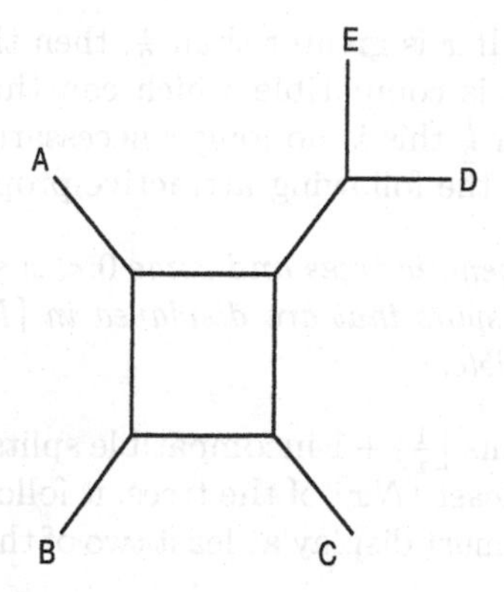

Fig. 1. The median network associated with the split system: $AB|CDE$, $ABC|DE$, $BC|ADE$, $A|BCDE$, $B|ACDE$, $C|ABDE$, $D|ABCE$, $ABCD|E$. The two horizontal parallel edges correspond to the split $AB|CDE$, and the two vertical parallel edges correspond to the split $BC|ADE$.

Since for visual purposes the complexity of the median network associated with a split system is directly related to the degree of incompatibility of the split system (since high dimensional hypercubes are rather difficult to visualize), it is useful to quantify this incompatibility as follows. For k a positive integer, we say that a split system is *k-compatible* if it contains no incompatible subsystem of $k+1$ splits. The concept of k-compatibility was introduced and studied in [8]. Clearly a k-compatible split system is compatible if and only if $k = 1$, in which case its associated median network is a tree, but, as k increases, the associated median network can become progressively more complex. Note that if X has cardinality n, then a (1-)compatible split system on X contains at most $2n-3$ splits, a 2-compatible split system on the same set contains at most $4n-10$ splits and, for general k, it will contain at most $n(1 + k\log_2(n))$ splits, cf. [8]. Hence for low values of n and k the number of splits in a k-compatible split system on X will not be excessively large, again making the associated median network easier to visualize.

2.2 Consensus Networks

Given a collection of phylogenetic trees, two common methods for computing a consensus tree are the *strict consensus* method, which outputs the tree displaying only those splits that are displayed by all of the input trees, and the *majority-rule consensus* method, which outputs the tree displaying only those splits that are displayed in more than half of the input trees. These two methods can be viewed as being members of a one-parameter family of consensus methods in which a split system S_x is generated that contains precisely those splits that are displayed by more than proportion x of the trees (for strict consensus $x = 1$, and for majority-rule $x = \frac{1}{2}$). If x is greater than $\frac{1}{2}$, then the consensus method results in a split system that is compatible which can thus be displayed by a tree. However, if x is less than $\frac{1}{2}$ this is no longer necessarily the case, although the split system S_x does have the following attractive property.

Theorem 1. *Given N phylogenetic trees and some $0 < x \leq 1$, let S_x denote the split system containing those splits that are displayed in $\lceil Nx \rceil$ or more of these trees. Then S_x is $\lfloor \frac{1}{x} \rfloor$-compatible.*

Proof: Suppose that S_x contains $\lfloor \frac{1}{x} \rfloor + 1$ incompatible splits. Then, since each of these splits is displayed by at least $\lceil Nx \rceil$ of the trees, it follows by the Pigeonhole Principle that one of the trees must display at least two of the incompatible splits. But this is impossible. □

For obvious reasons, we will call the median network associated with S_x a *consensus network*. In order to visualize the contribution that each split makes to the collection of trees in question, we usually weight the edges in this network corresponding to a given split according to the frequency with which it occurs in the trees. This last result indicates a way in which to control the visual complexity of the consensus network associated with S_x. For instance, if we only accept splits that appear in more than $\frac{1}{4}$ of the input trees, then $S_{\frac{1}{4}}$ will be 4-compatible, so that the associated median network is guaranteed to contain cubes only of dimension 3 or less. Note that in the case where $x = 0$, i.e. the split system S contains all splits from all N trees, S is N-compatible.

2.3 Greedy Consensus Networks

We now turn to the practical matter of how to select a split system to be represented by a consensus network. One possibility is to simply select the parameter x described in the previous section by trial and error, and this seems to work reasonably well in practice. A more attractive approach might be to try and select, for fixed k, a maximal k-compatible subset of splits in the split system consisting of all splits displayed by a given collection of trees. However, this is computationally hard even in case $k = 1$ (see e.g., [6]). Even so there are various heuristic approaches possible extending those used to construct consensus trees. We now describe one of these methods.

Consensus trees can be constructed using a *greedy* approach, which can be easily extended to construct networks. We begin by recalling the strategy for constructing a *greedy consensus tree* (cf. [6]). Given a collection of trees, list all splits displayed by at least one of the trees in order of frequency, so that those splits displayed by the largest number of trees come first (with ties broken arbitrarily). A compatible split system is then built up by starting at the beginning of the list and adding in splits one at a time that are compatible with all of the splits in the current split system, ignoring splits that are incompatible with any of the splits in the current system. The tree displaying the resulting compatible split system is the greedy consensus tree.

We construct a *k-greedy consensus network* for a fixed positive integer k in a similar manner, including splits in order of frequency provided they do not lead to a subset of $k+1$ incompatible splits. As with greedy consensus trees, this approach will also suffer from the fact that if two distinct splits occur with equal frequency, they will be chosen in arbitrary order which can lead to different results (see [6] for more details). In practice we found it useful to stop trying to add further splits after the first split inducing a subset of $k+1$ incompatible splits was obtained; this prevented the main features of the network being obscured by many edges of relatively small weight (results not shown).

2.4 Implementation

Code has been developed to read a list of trees in Newick format (bracket notation) and produce the corresponding weighted split system in nexus format. (Python script available from b.r.holland@massey.ac.nz). This nexus file can then be read by the program Spectronet [11] which displays the associated consensus network.

3 Results

We present two representative examples to illustrate the method.

3.1 MCMC Analysis

Our first example comes from a Monte Carlo Markov Chain (MCMC) analysis [15], [13] of 37 mammal mitochondrial genomes [19]. We used the software MrBayes [14] under a general time-reversible model with gamma distributed rates across sites to generate a chain of 1,000,000 trees; of these every hundredth tree was recorded. We discarded the first half of these trees to provide for a burn in period, leaving 5000 trees in our collection.

Figures 2a-2d show the consensus networks corresponding to the split systems S_x for $x = 1, 0.5, 0.25$ and 0.1. In an MCMC analysis the proportion of times an edge appears in a tree in the chain is interpreted as its posterior probability of being in the true tree, hence the length of the edges in the network are proportional to their posterior probability. Note that all the external edges have posterior probability 1, as they necessarily appear in all of the trees in the collection.

The marsupials (opossum, possum, wallaroo, bandicoot) and the platypus form an outgroup to the placental mammals. We can see in Figures 2a-2d that while the more recent divergences are well resolved, the order of the deeper divergences and the position of the root of the placentals is unresolved. Using the complete data set, the outgroup breaks the rodents into two groups, this is thought to be a long branch attraction artefact, and indeed, when the outgroup taxa are removed the rodents form a single group [19], [20]. Although the strict consensus tree (Figure 2a) and majority-rule consensus tree (Figure 2b) give

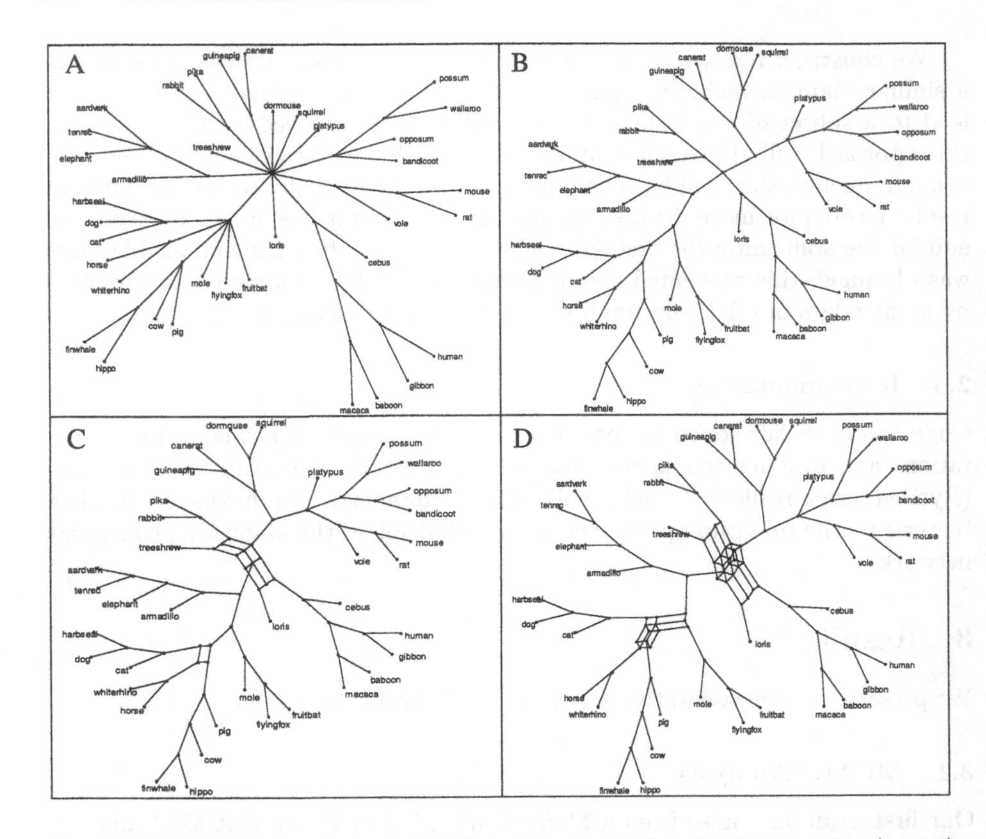

Fig. 2. a) Strict consensus tree for 37 mammal mitochondrial sequences ($x = 1$). b) Majority-rule consensus tree for 37 mammal mitochondrial sequences ($x = 0.5$). c) Consensus network for 37 mammal mitochondrial sequences ($x = 0.25$). This is the smallest value of x for which the associated consensus network contains no 3-cubes. d) Consensus network for 37 mammal mitochondrial sequences ($x = 0.10$). The smallest value of x for which the consensus network contains no 4-cubes is 0.028. However, this network has many tiny edges that detract from the main features.

some idea of the regions of the phylogeny that are uncertain, these regions are displayed either as polytomies, or as edges with weak support, rather than the more informative display of alternative hypotheses in the consensus networks (Figures 2c and 2d). For instance, in Figure 2c there are two possible hypotheses regarding the location of the odd-toed ungulates (horse, white rhino). They could either form a sister group with the carnivores (dog, cat, harbour seal) or with the even-toed ungulates (finwhale, hippo, cow, pig). The relative lengths of edges in the 2-cube indicate that the latter hypothesis is more likely according to this analysis.

3.2 Equally Parsimonious Trees

The second example is a data set consisting of 80 trees. This collection of trees resulted from a heuristic search for the most parsimonious tree for a set of 53 sequences of human mitochondrial DNA [17]. The phylogenetic software package PAUP* [22] was used to search for the maximum parsimony tree, (using the default options Swap=TBR, AddSeq=Simple). All splits appearing in the 80 equally parsimonious trees are shown (Figure 3), this corresponds to $x = 0$, making it unnecessary to compute a greedy network. As we see, rather than sifting through the 80 trees to try and identify similarities and differences, the relevant information is summarized in a single figure.

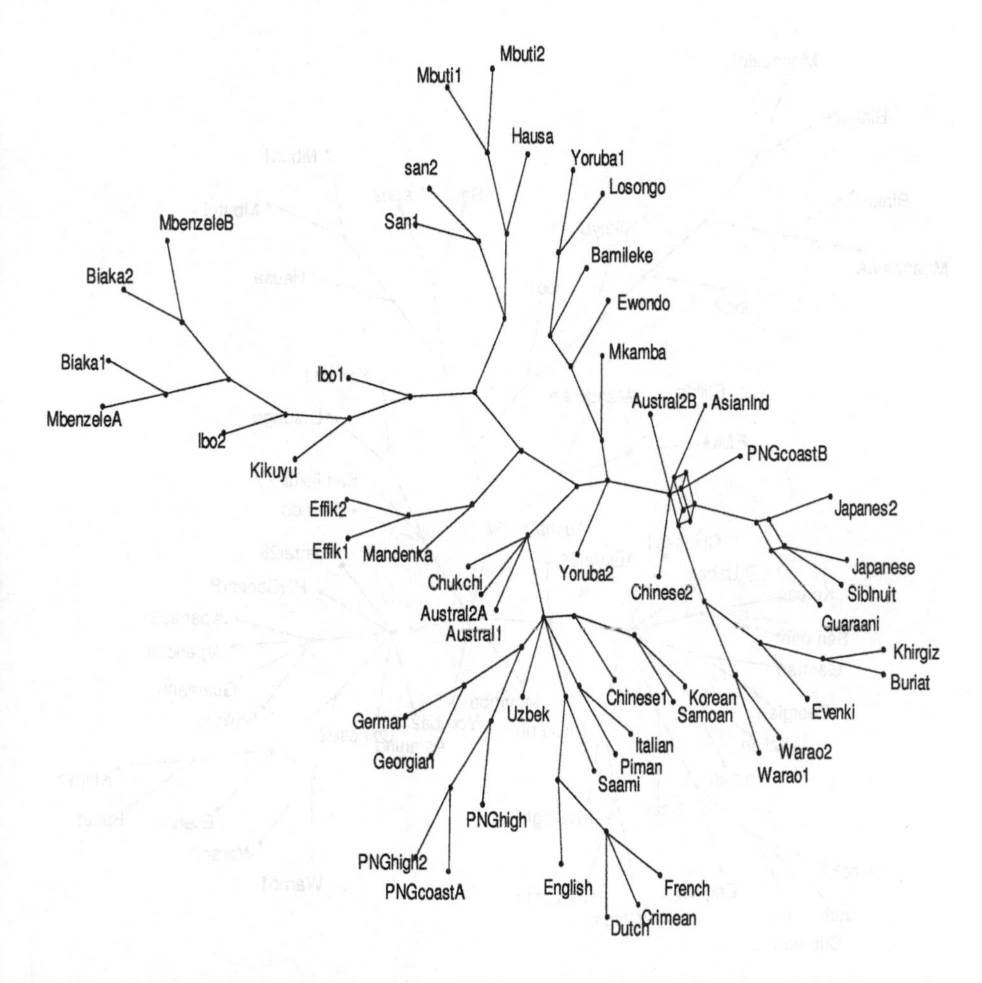

Fig. 3. Consensus network showing all splits in the 80 equally parsimonious trees resulting from a heuristic search on an alignment of 53 human mitochondrial genomes. ($x = 0$).

This data set is typical of intra-species data in that it has many equally likely trees, since taxa are often only separated by a few mutational steps, with a high proportion of the mutations being reversals and parallel changes [4]. As these reversals and parallel changes can lead to conflicting hypotheses about the phylogeny, consensus trees for intra-species data are prone to have many polytomies. This is well illustrated by the majority-rule consensus tree for this data set (Figure 4). There are a large number of resolved trees consistent with the majority-rule tree, only 80 of these are the actual input trees. A greedy consensus tree would provide much greater resolution of the polytomies but would still be unable to display the 3x2 = 6 trees encapsulated by the 3-cube and 2-cube in the network.

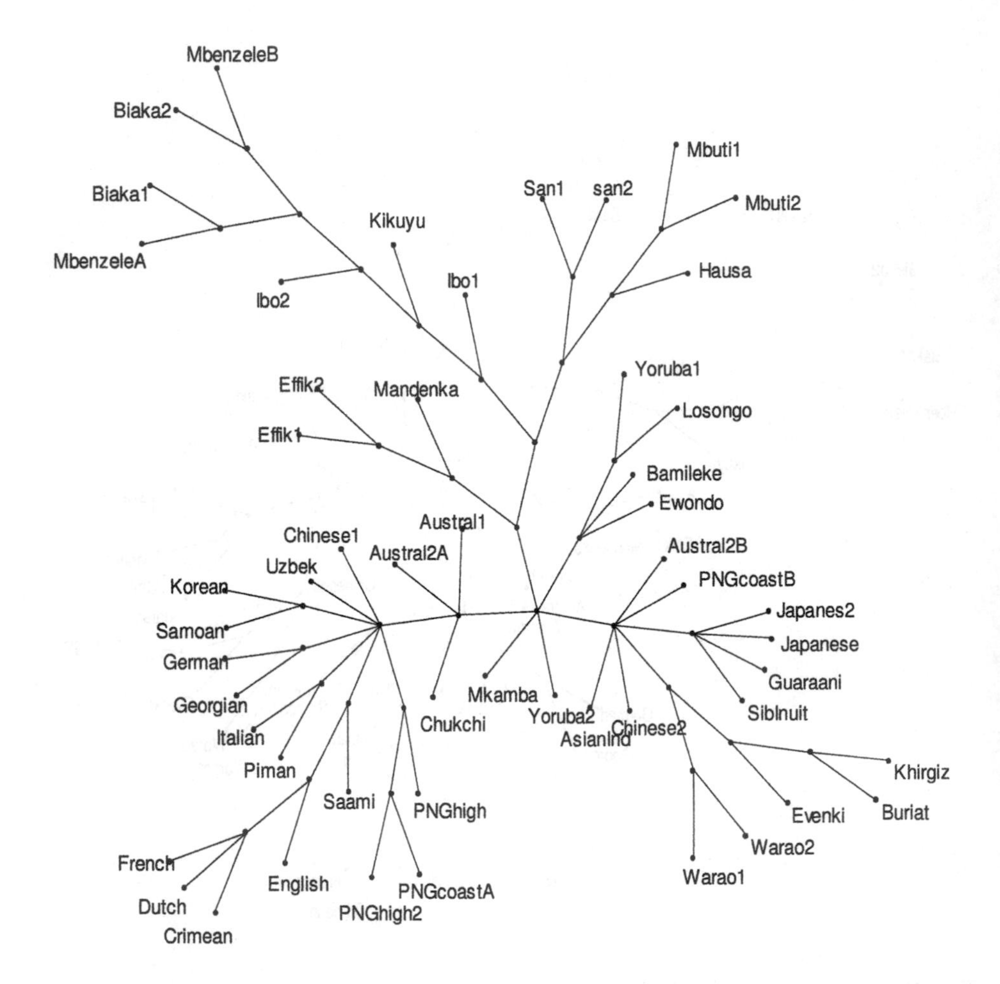

Fig. 4. Majority-rule consensus tree of 80 equally parsimonious trees resulting from a heuristic search on an alignment of 53 human mitochondrial genomes.

The consensus network approach also complements a recent multidimensional scaling method for analyzing collections of trees, TreeViz [18]. This method works by computing a distance between the trees in question (such as the Robinson-Fould's distance), and then using multidimensional scaling to represent the trees as a set of points in a plane (an approach that was also explored in [10]). Using this plot, it is possible to interactively select subcollections of trees and compute consensus trees for these collections.

We show a screenshot of a TreeViz analysis of the 80 equally parsimonious trees (Figure 5a); the multi-dimensional scaling is shown on the left and the consensus tree for 16 selected trees in shown in a panel on the right. In Figure 5b we compare an excerpt from the consensus of the 16 highlighted trees with the corresponding part of the consensus network. Again, where the consensus tree shows a polytomy the network displays the competing hypotheses.

4 Discussion

We have presented a method for generating consensus networks that allows the display of conflicting information within a collection of phylogenetic trees. These networks can be thought of as an extension of strict and majority-rule consensus trees. As with consensus trees, consensus networks can be used as a tool in conjunction with established phylogenetic techniques such as MCMC and bootstrapping. The weights of the edges in consensus networks are open to different interpretations depending on the way in which the input collection of trees is generated. For instance, given a set of trees generated by a MCMC the weights of the splits correspond to posterior probabilities, given bootstrap trees the weights correspond to the confidence level.

One of the main advantages of consensus networks over consensus trees is that they allow conflicting hypotheses within the input collection of trees to be displayed simultaneously in a single diagram. This can be important since a lot of computational effort is usually put into generating large collections of trees, making it somewhat wasteful to only keep a small proportion of this information in the final display. Moreover, it is the conflicts between the trees that are often of interest to biologists and by visual inspection consensus networks allow these to be quickly identified.

Even so, consensus networks still suffer from limitations shared by consensus methods in general. With consensus networks (as with consensus trees) some information may still be lost in order to facilitate display of the network, especially if the data contains many incompatibilities. However, if the data is highly incompatible then it might be questionable in what way a phylogenetic analysis is appropriate.

Another consideration with the consensus networks that we have proposed is that they can still become quite complex, even when restricted to 3-compatible split systems. In practice we found that networks without a distracting level of complexity could be constructed by halting the greedy consensus method

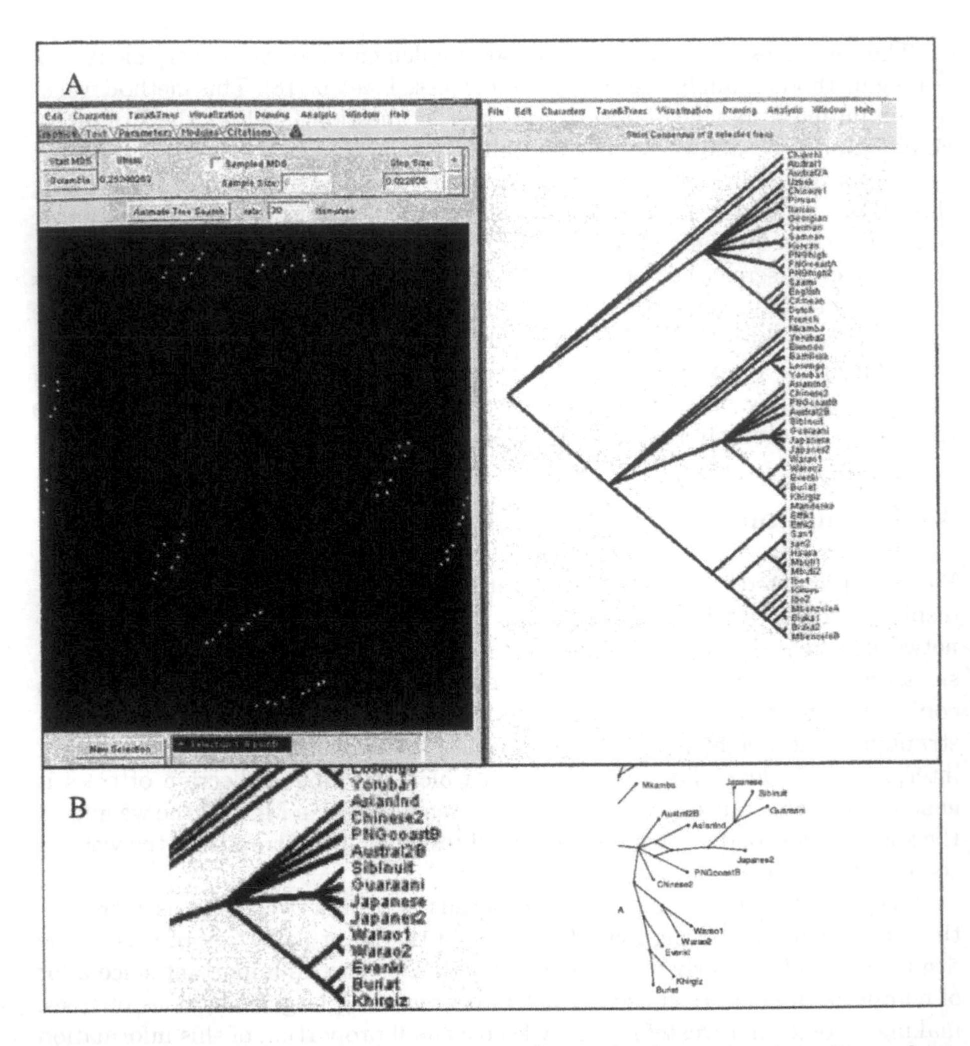

Fig. 5. a) TreeViz screenshot showing the multi-dimensional scaling of 80 equally parsimonious trees resulting from a heuristic search on an alignment of 53 human mitochondrial genomes (left panel), and the consensus tree for 16 selected trees (right panel). b) A comparison of an excerpt from the consensus of the 16 highlighted trees (5a) with the corresponding part of the consensus network.

after the first split which caused a 4-cube was encountered. Another possible approach to controlling the complexity of the networks is to generate *circular split systems* as opposed to k-compatible split systems. These split systems have the advantage that they can be displayed using split-graphs which, as opposed to median networks, are always planar and can be easily computed using the

References

1. Amenta, N., and Klingner, J.: Case Study: Visualizing Sets of Evolutionary Trees. 8th IEEE Symposium on Information Visualization (InfoVIs 2002)
2. Bandelt, H.-J., and Dress, A.: Split decomposition: a new and useful approach to phylogenetic analysis of distance data. Molecular Phylogenetics and Evolution **1**(3) (1992) 242-252
3. Bandelt, H.-J.: 1994. Phylogenetic Networks. Verhandl. Naturwiss. Vereins Hamburg (NF) **34** (1994) 51–71
4. Bandelt, H.-J., Forster, P., Sykes, B.C., Richards, M.B.: Mitochondrial portraits of human populations using median networks. Genetics **14** (1995) 743–753
5. Bandelt, H.-J., Huber, K.T., Moulton, V.: Quasi-median graphs from sets of partitions. Discrete Applied Mathematics **122** (2002) 23–35
6. Bryant, D.: A classification of consensus methods for phylogenetics. in Janowitz, M.. Lapointe, F.J., McMorris, F., Mirkin, B., Roberts, F. Bioconsensus . DIMACS-AMS. 2003. pp 1–21
7. Bryant, D., Moulton, V.: NeighborNet: an agglomerative method for the construction of planar phylogenetic networks. in the proceedings of WABI, 2002. pp 375–391
8. Dress, A., Klucznik, M., Koolen, J., Moulton, V.: A note on extremal combinatorics of cyclic split systems. Seminaire Lotharingien de Combinatoire **47** (2001) (http://www.mat.univie.ac.at/ slc).
9. Felsenstein, J.: Confidence limits on phylogenies: an approach using the bootstrap. Evolution **39** (1985) 783–791
10. Hendy, M.D., Steel, M.A., Penny, D., Henderson, I.M.: Families of trees and consensus. Pp 355–362, in "Classification and Related Methods of Data Analysis" (H.H. Bock ed.) Elsevier Science Publ. 1988 (North Holland).
11. Huber, K.T., Langton, M., Penny, D., Moulton, V., Hendy, M.: Spectronet: A package for computing spectra and median networks. Applied Bioinformatics **1** (2002) 159–161
12. Huber, K.T., Moulton, V., Lockhart, P., Dress, A.: Pruned median networks: a technique for reducing the complexity of median networks. Molecular Phylogenetics and Evolution **19** (2001) 302–310
13. Huelsenbeck, J. P., Larget, B.,Miller, R.E., Ronquist, F.: Potential applications and pitfalls of Bayesian inference of phylogeny. Syst. Biol. **51** (2002) 673-688
14. Huelsenbeck, J. P., Ronquist, F.: MRBAYES: Bayesian inference of phylogenetic trees. Bioinformatics **17** (2001) 754–755
15. Huelsenbeck, J. P., Ronquist, F., Nielsen, R., Bollback, J.P.: Bayesian inference of phylogeny and its impact on evolutionary biology. Science **294** (2001) 2310–2314.
16. Huson, D.: SPLITSTREE: a program for analyzing and visualizing evolutionary data. Bioinformatics **14** (1998) 68–73
`http://bibiserv.techfak.uni-bielefeld.de/intro/seqdept.html`
17. Ingman, M., Kaessmann, H., Paabo, S., Gyllensten, U.: Mitochondrial genome variation and the origin of modern humans. Science **408** (2000) 708-713

18. Klingner, J.: Visualizing Sets of Evolutionary Trees. The University of Texas at Austin, Department of Computer Sciences. Technical Report CS-TR-01-26. (2001)
19. Lin, Y.-H., McLenachan, P.A., Gore, A.R., Phillips, M.J., Ota, R., Hendy, M.D., Penny, D.: Four new mitochondrial genomes and the increased stability of evolutionary trees of mammals from improved taxon sampling. Molecular Biology and Evolution **19** (2002) 2060–2070
20. Lin, Y.-H., Waddell, P.J., Penny, D.: Pika and Vole mitochondrial genomes add support to both rodent monophyly and glires. Gene **294** (2002) 119-129
21. Semple, C., Steel, M.: Phylogenetics, Oxford University Press 2003.
22. Swofford, D.L.: PAUP* - Phylogenetic Analysis Using Parsimony (*and other methods) Version 4. Sinauer Associates, Sunderland, Mass. 1998.

Efficient Generation of Uniform Samples from Phylogenetic Trees*

Paul Kearney[1], J. Ian Munro[2], and Derek Phillips[2**]

[1] Caprion Pharmaceuticals
7150 Alexander-Fleming
Montréal, QC, Canada H4S 2C8
pkearney@caprion.com
[2] School of Computer Science
University of Waterloo
200 University Avenue West
Waterloo, ON, Canada N2L 3G1
{imunro,djphilli}@uwaterloo.ca

Abstract. In this paper, we introduce new algorithms for selecting taxon (leaf) samples from large phylogenetic trees, uniformly at random, under certain biologically relevant constraints on the taxa. All the algorithms run in polynomial time and have been implemented. The algorithms have direct applications to the evaluation of phylogenetic tree and supertree construction methods using biologically curated data. We also relate one of the sampling problems to the well-known clique problem on undirected graphs. From this, we obtain an interesting new class of graphs for which many open problems exist.

1 Introduction

A fundamental problem in computational biology is the accurate reconstruction of evolutionary trees (also called *phylogenetic trees*) from sequence or distance data. The phylogeny reconstruction problem is known to be NP-hard under a variety of measures of tree optimality (see for example [13] and [17]).

This problem has been studied intensely since the 1960s, but due to its complexity, new heuristics continue to be developed. From early distance-based methods such as Fitch-Margoliash [12] and Neighbor-Joining [30], to more recent Quartet [17] and Expectation-Maximization methods [14], there is a wide variety of techniques used to generate phylogenetic topologies and estimate edge lengths.

Due to the difficulty in determining the criteria for an "optimal" tree (see for example the parsimony [8] and maximum-likelihood [11] approaches) experimental methods are used to evaluate the relative performance of each algorithm.

* This work was supported by the Natural Sciences and Engineering Research Council of Canada and the Canada Research Chairs Program.
** Send correspondence to Derek Phillips.

G. Benson and R. Page (Eds.): WABI 2003, LNBI 2812, pp. 177–189, 2003.

Customarily, the performance of an algorithm is judged based on its ability to reconstruct simulated topological and sequence data. The approach involves first generating a topology on n leaves using a topology model (such as the ones described in [2] and [31]), then generating a sequence label for the root, and finally evolving the sequence along the paths in the tree to obtain a label for each leaf. The sequences are then handed to a phylogeny reconstruction algorithm and the algorithms performance is judged based on how closely the inferred tree matches the target tree.

There are a few problems with this approach. First, the topologies generated by the topology models are based on significant assumptions about the process of evolution (some discussion about this problem is found in [22]). It is not clear that topologies generated from these models are representative of actual phylogenetic trees. Second, the sequence evolution models are generally very simple (such as Jukes-Cantor [18] or Kimura [20]) and do not model significant evolutionary changes. For example, insertions, rearrangements, and repeats are ignored by these models. Since many of the phylogeny reconstruction algorithms are built on similar assumptions about evolution, these assumptions could arbitrarily favour certain methods. Finally, there is no reason to believe that good (or bad) performance on simulated phylogeny data is an accurate reflection of the performance on biologically-derived data. To our knowledge, no experimental studies have been conducted which show a correlation between the performance of a phylogeny reconstruction algorithm on simulated data and on real data.

There are several reasons why previous empirical studies have relied on simulated data. For one, it is much easier to generate simulated data than to track down a large number of experimentally derived phylogenetic trees. Since biologically-curated phylogenetic trees are very time-consuming to build, very few accepted trees are available and those that are available can be difficult to acquire. Another reason is that most of the biologically-derived phylogenetic trees are quite large. Since phylogeny reconstruction algorithms are very slow, trees consisting of thousands of taxa are impractical for large studies.

Related to phylogeny reconstruction is another important problem in bioinformatics which concerns the construction of *supertrees* [16] from a set of phylogenetic trees. We refer the reader to the recent review by Bininda-Emonds et al. [5] for a detailed survey of the problem and computational techniques. The basic idea is that we are given a set of phylogenetic trees, each containing a subset of the taxa which are or interest, and we want to construct the "supertree" consisting of all these taxa which is most consistent with the input data. Supertrees have been applied to a variety of different contexts (see for example [4], [21], and [28]).

As in the case of phylogeny reconstruction, there are a wide variety of methods for supertree construction including matrix representation with parsimony (MRP) (proposed in [3] and [29]), BUILD [1] (along with many variants including one described in [24]), and quartet methods [32]. Several studies have been performed on the supertree methods, relying almost exclusively on simulated data (for example [6] and [23]).

In this paper we introduce algorithms for sampling subsets of taxa from a large phylogenetic tree, subject to certain biologically-motivated constraints. Simultaneously satisfying the given constraints and guaranteeing the uniformity of the random samples makes the sampling problem non-trivial. It is easy to show that simple approximations can lead to gross errors in the sampling process; thus, the fact that we can generate the samples in polynomial time is of practical importance.

These algorithms provide a method for systematically testing the performance of phylogeny reconstruction algorithms on real phylogenetic data. Given a phylogenetic tree, a large number of subtrees can be selected uniformly at random (subject to some important constraints), and the performance of the reconstruction algorithms can be assessed based on their ability to reconstruct the subtrees. Since the subtrees are generally much smaller than the original tree, it is feasible to reconstruct these phylogenies, and because they come from a well supported phylogenetic tree, the subtrees represent biologically significant phylogenetic data. We acknowledge here that no proposed phylogenetic tree is without its detractors in the scientific community and so no "known" trees exist for conducting the experiments. Generally, however, the debates involve very small local changes to the topology and thus as long as the experimental methods do not require the reconstructed phylogenies to match the target phylogeny exactly, then these phylogenetic uncertainties should have little effect on the results of the experiments. In addition, we expect the size and certainty of the phylogenetic databases to increase substantially in the next few years, making their use in systematic algorithm evaluation more acceptable.

Similarly, the algorithms can be used to systematically test the performance of supertree algorithms. Given a large phylogenetic tree (preferably consisting of real data, but also applicable in the case of simulated data), several subtrees can be selected uniformly at random, and the performance of the supertree construction algorithms can be measured based on the degree to which the supertree respects the known phylogenetic relationships of the taxa. In particular, experiments on biologically relevant phylogenetic data could provide insight into the debate between the "total evidence" approach to supertree construction, which suggests building large trees directly from the raw data, versus the "taxonomic congruence" approach that favours building small trees first and then combining them. Several surveys concerning the debate have been written, including one by Eernisse and Kluge [7] and one by Page [25].

Our methods have been implemented in a package called *PhyloSample* that can be downloaded freely from `http://monod.uwaterloo.ca/software/`.

2 The Sampling Problems

Given a phylogenetic tree, we would like to select a subset of the leaves, uniformly at random, subject to certain constraints on the subtree induced by the sample. We focus on three simple constraints that specify properties of the induced subtree.

The first constraint, hereafter called the *Leaf Depth Constraint*, specifies a minimum and maximum depth[1] for a leaf in the induced subtree. This roughly translates to a constraint on the amount of evolution that each taxon has experienced since the existence of the least common ancestor of this set of taxa.

The second, which we will refer to as the *Edge Length Constraint*, specifies a minimum and maximum edge length for every edge in the induced subtree. This is a useful constraint since significant disparity in edge lengths can cause the "long branches attract" phenomenon, which creates difficulties for some phylogeny reconstruction algorithms [9]. Varying the parameters of this constraint can help to identify the prevalence of this issue on real-world data and can help to establish the parameters of the "Felsenstein zone" for various reconstruction algorithms.

The final constraint deals with the pairwise distance between any two taxa in the induced subtree. We will refer to this as the *Pairwise Leaf Distance Constraint.* Since very close and very distant sequences can be problematic for phylogeny reconstruction algorithms [9], we can analyze or alleviate these difficulties with the pairwise constraint. As we will see later, there is a close relationship between this problem and the NP-complete clique problem on undirected graphs.

2.1 The Sampling Algorithms

We now present a formal statement of each of the sampling problems and give polynomial-time dynamic programming algorithms to solve each one.

Problem. *(Leaf Depth) Given a phylogenetic tree T with n leaves, a positive integer p, a positive integer $m \geq 2$, and two non-negative real numbers d_{min} and d_{max}, select p samples of m leaves from T, uniformly at random from all sets of size m such that if r is the root of the subtree induced by a sample then for every leaf v in the sample $d_{min} < dist(r, v) < d_{max}$*[2].

Our algorithm for solving this problem begins by enumerating every sample of size m in the tree which satisfies the given constraint and then uses this information to generate uniform samples.

Enumeration can be carried out by constructing a list of leaf descendants, sorted by their depths, at each node u in the tree. Now we can examine every node u in T as the potential root of a sample and determine how many valid samples[3] are rooted there. If there are V_L valid leaves in the left subtree of u

[1] We define the depth of a node u to be the sum of the edge lengths from the root of the tree to u.

[2] For any two nodes u, v in a phylogenetic tree, we define $dist(u, v)$ to be the pairwise distance between these nodes. In other words, it is the sum of the edge lengths on the path from u to v in the tree.

[3] For this problem, we define a *valid leaf* v for a root r as a leaf descendant of r such that $d_{min} < dist(r, v) < d_{max}$. A *valid sample* rooted at r is a subset of m valid leaf descendants of r such that at least one leaf is from the left subtree of r and one is from the right. Similar definitions will be implied in our algorithms for the next two sampling problems.

and V_R in the right, then S_u, the number of valid samples rooted at u, is given by:

$$S_u = \binom{V_L + V_R}{m} - \binom{V_L}{m} - \binom{V_R}{m} \tag{1}$$

Knowing how many valid samples are rooted at each node in T we can establish the probability of a valid sample being rooted at a node u as $\frac{S_u}{S}$, where S is the total number of valid samples in T. Using these probabilities, we pick a root r and then choose m valid leaf descendants of r, uniformly at random, ensuring that at least one is a descendant of the left child of r and one is a descendant of the right child.

Using this algorithm, we can establish the following result:

Theorem 1. *The Leaf Depth problem can be solved in $\Theta(n \lg n + pn)$ time and $\Theta(n)$ space.*

Proof. We first observe that our algorithm requires a list of sorted leaf descendants at each internal node in the tree. This can be done by storing the lists as balanced search tree (for example B-Trees), and merging lists from the leaves up the tree. The following lemma shows that this process can be performed efficiently.

Lemma 1. *Given a phylogenetic tree T with n leaves, the sorted leaf descendant lists for each node in T can be constructed in $\Theta(n \lg n)$ time and space.*

This can be done in a few different ways. See [26] for a simple algorithm which achieves the desired bounds.

With $\Theta(\lg n)$ work, we can count the number of valid descendants of a node in T from the sorted descendant list by searching the descendant list tree to find the closest and farthest valid leaves. This, along with equation (1), is sufficient to enumerate the number of valid samples for each node in T. This means that the enumeration takes $\Theta(n \lg n)$ time. It appears that it also requires $\Theta(n \lg n)$ space, but we note that the sorted leaf descendant list at any node can be destroyed once the parent has finished merging the lists from its two children. Thus, the space required is only $\Theta(n)$.

In order to select a sample, we need to first choose the root of the sample and then pick a set of valid leaf descendants of that node. The following two lemmas show that we can perform these two steps efficiently.

Lemma 2. *Given S_u, the number of valid samples of size m rooted at u for all nodes u in T, the root of a subtree induced by a valid samples can be chosen uniformly at random in $\Theta(n)$ time using $\Theta(1)$ space.*

Proof. Let S be the total number of valid samples in T. We first observe that the probability that the subtree induced by a uniform random sample has u as its root is $\frac{S_u}{S}$. Thus, we choose a root by assigning this probability to each node u in T and randomly choosing a node based on these probabilities. This can be done in $\Theta(n)$ time. □

When p is large, it might be preferable to use an alternate algorithm which uses $\Theta(n)$ space but only $\Theta(n+p\lg n)$ time to generate p samples. This approach involves constructing a vector of running sums of probabilities, where we index each internal node as a number $1, ..., n-1$. Now, to pick r, we choose a number uniformly in the range $[0, ..., 1]$ and do a binary search on the vector to find the chosen root node. This does not affect the asymptotic runtime of the algorithm, however, because the time required by the following procedure dominates the sample generation time.

Lemma 3. *Given a node r, a uniform random valid sample rooted at r can be chosen in $\Theta(n)$ time using $\Theta(1)$ space.*

Proof. Create a set S and add one descendant from the left subtree and one from the right (chosen uniformly at random) to S. Now, traverse both subtrees adding a valid leaf to S with probability $\frac{1}{m'}$ (where m' is the number of valid leaves below r not yet visited). When $m' = |S| - m$, add all the valid leaves not yet visited. At this point S contains the chosen sample. □

The proof of Theorem 1 now follows from Lemmas 1, 2, and 3. □

The second algorithm we present solves the sampling problem under the edge length constraint. We state the problem formally here:

Problem. *(Edge Length) Given a phylogenetic tree T with n leaves, a positive integer p, a positive integer $m \geq 2$, and two non-negative real numbers e_{min} and e_{max}, select p samples of m leaves from T, uniformly at random from all sets of size m such that for every edge e in the subtree induced by the sample, $e_{min} < |e| < e_{max}$.*

The algorithm for this problem is similar to the first one but requires keeping more information about valid samples at each node. The basic idea is to consider each node u in the tree as a potential root and then look at all ways of combining valid samples of size $j = 1, ..., m-1$ rooted at some node in the left subtree of u with valid samples of size $m-j$ rooted at a valid node in the right subtree of u, such that the distance between u and each of these nodes is greater than e_{min} and less than e_{max}.

We will accomplish this by first enumerating and storing $S_u[k]$, the number of valid samples of size $k = 1, ..., m$ rooted at each node u. Next, we choose a root for the subtree by assigning a probability of $\frac{S_u[m]}{S}$ for each node u. Finally, we descend the subtree rooted at u visiting internal nodes to find a valid sample.

The following theorem shows that the algorithm is relatively efficient:

Theorem 2. *The Edge Length problem can be solved in $\Theta(mnh + (p+1)m^2n)$ time and $\Theta(mn)$ space, where h is the height of T.*

Proof. It will be helpful to have sorted descendant lists at each node u in T. The same argument given in Lemma 1, shows that this can be accomplished in $\Theta(n \lg n)$ time. In the previous case, only *leaf* descendants were required, while now we will keep a list of *all* descendants, but this does not affect the asymptotic runtime.

The following lemma shows that we can now perform the required enumeration efficiently:

Lemma 4. *Given a phylogenetic tree T with n leaves, a positive integer $m \geq 2$, and two non-negative real numbers e_{min} and e_{max}, the number of valid samples of size $1, ..., m$ rooted at each node in T can be enumerated in $\Theta(mnh + m^2n)$ time using $\Theta(mn)$ space.*

Proof. If u is a leaf then $S_u[1] = 1$ and $S_u[k] = 0$ for $2 \leq k \leq m$. If u is not a leaf, then we can compute $S_u[k]$ from the values at the valid descendants of u.

First, we traverse the left and right subtrees to obtain the vectors $S^L[k]$ and $S^R[k]$ for $k = 1, ..., m-1$, which represent the number of valid samples of size k rooted at a valid left and right descendant of u, respectively. Since the total number of descendants in a phylogenetic tree on n leaves is at most nh (where h is the height of T), these values can be obtained for each node in total time $\Theta(mnh)$.

Now, $S_u[k]$ can be calculated in $\Theta(m)$ time with this formula:

$$S_u[k] = \sum_{i=1}^{k-1} S^L[i] * S^R[k-i]$$

The time required to obtain sorted descendant lists and to obtain the above $S_u[k]$ values is $\Theta(n \lg n + mnh + m^2n) = \Theta(mnh + m^2n)$. At every node we store a vector of size m so the total space required is $\Theta(mn)$. □

Now that we have all of the $S_u[k]$ values, we can choose a sample uniformly at random. As explained above, selecting the root r of the sample is almost identical to the method used for the leaf depth sampling problem. Once r is obtained, the following lemma shows that we can pick a sample efficiently.

Lemma 5. *Given a node r and an integer $m \geq 1$, a uniform random valid sample of size m rooted at r can be chosen in $\Theta(m^2n)$ time and $\Theta(m)$ space .*

Proof. By induction on m. If $m = 1$ then the only valid samples are rooted at a leaf. Thus, r must be a leaf and so our sample consists simply of r. If $m > 1$ then we traverse the left and right subtrees rooted at r to find the number of valid samples of size $1, ..., m-1$ rooted at all descendants u of r for which $e_{min} < dist(r, u) < e_{max}$. Store these values in the vectors V_L and V_R, respectively.

We want to pick a number i in the range $1 \leq i \leq m-1$ which will represent the number of leaves that will be taken from the left subtree of i, while the rest are taken from the right subtree. There are $V_L[i] * V_R[m-i]$ valid samples of size m with i taken from the left subtree. We use this fact to assign the correct probabilities to each value of i and choose a value for i at random.

Now we traverse the left subtree of r finding the correct left descendant and recursively pick a valid sample rooted there of size i. We do the same to obtain a sample of size $m-i$ from the right subtree.

Clearly, the sample rooted at r is valid since the distance between r and the two descendants we chose must not be too large or too small, and by induction the smaller samples chosen from each of these descendants are also valid.

For each internal node in the induced subtree, we do $\Theta(mn)$ work to traverse the subtrees filling in the V_L and V_R vectors. We only require $\Theta(m)$ time to decide how many leaves to take from the left subtree and then $\Theta(n)$ time to find the correct descendant nodes. There will be exactly $m-1$ internal nodes in the induced subtree, so the total time required is $\Theta((m-1)(mn+m+n)) = \Theta(m^2n)$. The required space stores the V_L and V_R vectors, which have at most $m-1$ entries each. □

The proof of Theorem 2 now follows from Lemmas 4 and 5. □

The final algorithm we present deals with the pairwise leaf distance constraint.

Problem. *(Pairwise Leaf Distance) Given a phylogenetic tree T with n leaves, a positive integer p, a positive integer $m \geq 2$, and two non-negative real numbers d_{min} and d_{max}, select p samples of m leaves from T, uniformly at random from all samples of size m such that the pairwise distance between any two leaves in the sample is greater than d_{min} and smaller than d_{max}.*

This problem is somewhat more difficult because it deals with a constraint on the $\binom{m}{2}$ distances between any two leaves in a sample, rather than just the distance between the root and the leaves or the distance between adjacent nodes (as in the first two problems). While the running-time for the algorithm we present is polynomial, it will be too slow and require too much space to be of much use in practise.

The basic idea is similar to that of the edge length problem, except that much more information must be gathered (and stored) during the enumeration phase. In order to count the number of valid samples rooted at a node u in T, we must now store an $m*m*n*n$ matrix M_u where entry $M_u[k, \ell, v_{min}, v_{max}]$ represents the number of valid samples rooted at u of size k, where ℓ leaves are taken from the left subtree of u, and where v_{min} and v_{max} are the shallowest and deepest leaves in the sample, respectively.

Some of these matrices will be sparse for a number of reasons. For example, if the tree is somewhat balanced then the expected number of descendants in the tree is $\Theta(n \lg n)$. In this case, many choices of v_{min} and v_{max} will be ignored. In addition, unless the d_{min} and d_{max} values are very distant, it is likely that many nodes will exist for which most leaves are not valid, and thus for many (v_{min}, v_{max}) pairs, all entries will be 0.

During the enumeration, we can calculate S_u, the number of valid samples of size m rooted at u, and this value can be used to select the root r of a sample (just as it was in the previous two algorithms). We can now choose a value for ℓ,

v_{min}, and v_{max} by summing over the probabilities of every choice and recursively choose a sample of size ℓ from the left subtree and $m-\ell$ from the right, ensuring that no leaf is shallower than v_{min} and none is deeper than v_{max}.

The following theorem shows that the algorithm runs in polynomial time:

Theorem 3. *The Pairwise Leaf Distance problem can be solved in* $O(m^2n^5 + pm^2n^3)$ *time and* $O(m^2n^3)$ *space.*

Proof. The following lemma implies that the time required for the enumeration of all valid samples is $O(m^2n^5)$.

Lemma 6. *Let u be a node in T. If the enumeration is completed for every descendant of u then the M_u matrix can be computed in $O(m^2n^4)$ time. The M_u matrix will have size $O(m^2n^2)$.*

Proof. (Sketch) If u is a leaf then $M[1,1,u,u] = 1$ and every other entry is 0. If u is an internal node, we need to count all samples of size $k = 2, ..., m$ where $\ell = 1, ..., k-1$ are taken from the left subtree. This canbe done by considering all pairs of leaves (v_L, v'_L) from the left subtree and all pairs (v_R, v'_R) from the right subtree (where $depth(v_L) < depth(v'_L)$ and $depth(v_R) < depth(v'_R)$) as possible values for v_{min} and v_{max}. If $dist(v_L, v_R) > d_{min}$ and $dist(v'_L, v'_R) < d_{max}$, then we want to include samples of this type. If not, we ignore this choice of leaves.

We can now traverse the left and right subtrees of u counting the number of valid samples with these choices of v_{min} and v_{max}, for all choices of k and ℓ. Pick v_{min} to be the shallower of v_L and v_R and v_{max} to be the deeper of v'_L and v'_R. Now, add to $M_u[k, \ell, v_{min}, v_{max}]$ the product of the number of valid samples of size ℓ from the left subtree and of size $k-\ell$ from the right.

It can be shown by induction that this counts all valid samples and never double-counts. We require at most $O(m^2n^4)$ time since we try all choices of k and ℓ (both less than m), for all choices of pairs of leaves from each subtree (fewer than $\binom{n}{4}$ of these). The only space required is for M_u, which contains at most $O(m^2n^2)$ entries since there are $O(n^2)$ choices for v_{min} and v_{max}. □

Having performed the enumeration, we described above how to get the root r of a sample. The following lemma shows how a valid sample can now be generated uniformly at random.

Lemma 7. *Given a root r for the subtree induced by a sample, and the M_u matrix at each node u in T, a valid sample rooted at node r can be generated uniformly at random in $O(m^2n^3)$ time and $O(n^2)$ space.*

Proof. We begin by creating a matrix of size n^2 representing all the possible choices for v_{min} and v_{max}. We calculate exactly how many valid samples there are of size m rooted at r for each possible choice of v_{min} and v_{max} and use the probabilities obtained from these calculations to choose the v_{min} and v_{max} nodes for our sample. This takes $O(m^2n^2)$ time. Next, we select ℓ (the number of leaves that will be taken from the left subtree) by summing up the number of valid samples for all choices of ℓ given the values of v_{min} and v_{max}. This takes $O(mn^2)$ time.

We now recursively perform the sampling at the left and right child asking for samples of size ℓ and $m - \ell$ respectively. We must also ensure that v_{min} and v_{max} both appear in the sample and that they truly are the shallowest and deepest leaves. This is done by ignoring unacceptable choices for the closest and furthest nodes in subsequent samples.

The work at each node takes $O(m^2n^2)$ time, so the total time required is $O(m^2n^3)$. The temporary matrices used to select v_{min}, v_{max}, and ℓ values can be destroyed after each node is completed, so the total space required is $O(n^2+m) = O(n^2)$. □

Since there are $\Theta(n)$ nodes in T, Lemma 6 shows that enumeration takes $O(m^2n^5)$ time and uses $O(m^2n^3)$ space. In Lemma 7 we show that a sample can be generated using $\Theta(n^2)$ space. However, when we wish to generate p samples, again only $\Theta(n^2)$ space is required since the large temporary matrices can be destroyed after generating each sample. The result now follows from Lemmas 2 and 7. □

3 Pairwise Compatibility Graphs and the Clique Problem

There is a natural interpretation of the pairwise leaf distance constraint as a graph. Given a phylogenetic tree T and values for d_{min} and d_{max}, the idea is to construct a graph $PCG(T, d_{min}, d_{max}) = (V, E)$ where each node $u \in V$ is labeled by a leaf of T and the edge $\{u, v\} \in E$ if and only if $d_{min} < dist(u, v) < d_{max}$ in T. We call the class of graphs created in this manner the Pairwise Compatibility Graphs (PCGs). Now, the sampling problem reduces to selecting a clique uniformly at random from $PCG(T, d_{min}, d_{max})$.

3.1 What Are the Pairwise Compatibility Graphs?

Unfortunately, we do not know. They are neither a subset of the chordal graphs (for which the clique problem is known to be polynomial-time solvable [15]) nor are they disjoint from this class. This is easy to see since K_n is both chordal and a PCG, while C_4 is a PCG, but is not chordal.

The relationship between PCGs and perfect graphs is also not known. Again, K_n is both a PCG and a perfect graph, while C_5 is a PCG, but is not perfect.

We have shown that every graph on five vertices or less is a PCG for some choice of topology and values of d_{min} and d_{max} (see [27]) but for larger graphs the super-exponentially increasing number of phylogenetic tree topologies makes the problem difficult [10].

The (seemingly) related "Tree Power" graphs were characterized by Kearney and Corneil in [19], but it does not appear that a similar argument can be applied to PCGs.

3.2 Open Problems on PCGs

Nearly every problem on PCGs remains unsolved. Some interesting questions that remain to be answered include:

1. How many pairwise compatibility graphs on n vertices are there?
2. Is any (or every) cycle of length greater than 5 a pairwise compatibility graph?
3. What is the smallest graph class that encompasses all of the pairwise compatibility graphs?
4. What is the complexity of testing whether a given graph is a pairwise compatibility graph?

With the flexibility afforded to us in the construction of instances of the pairwise leaf distance sampling problem (i.e. choice of topology and values for d_{min} and d_{max}), it is conceivable that the class of PCGs includes all undirected graphs. Since we can solve the sampling problem in polynomial time, this would imply that transforming an instance of the clique problem to an equivalent instance of the sampling problem is NP-hard.

4 Conclusions

We have introduced algorithms for uniform selection of random leaf subsets from a phylogenetic tree. We have also implemented these algorithms in a package called *PhyloSample*. While the time and space requirements for our algorithm for the pairwise leaf distance problem are prohibitive, the other two algorithms are quite useful in practice with relatively large trees (5000-8000 leaves).

Our hope is that researchers will use these algorithms to systematically test new (and existing) phylogeny reconstruction and supertree construction methods on biologically-derived data. This should provide insight into the practical uses of the phylogeny methods and help us to understand the relationship between their performance on simulated data versus their performance on real data.

In addition, we have identified a new class of graphs which link the pairwise leaf distance sampling problem and the clique problem on undirected graphs. It would be interesting to investigate properties of this graph class and look for a method of converting instances of the clique problem to instances of the pairwise leaf distance sampling problem.

5 Acknowledgments

We would like to thank the anonymous reviewers for suggesting the application of our sampling algorithms to supertree evaluation.

References

1. A.V. Aho, Y. Sagiv, T.G. Szymanski, and J.D. Ullman. Inferring a tree from lowest common ancestors with an application to the optimization of relational expressions. *Society of Industrial and Applied Mathematics (SIAM) Journal on Computing*, 10:405–421, 1981.
2. D.J. Aldous. Stochastic models and descriptive statistics for phylogenetic trees, from yule to today. *Statistical Science*, 16:23–34, 2001.
3. B.R. Baum. Combining trees as a way of combining data sets for phylogenetic inference, and the desirability of combining gene trees. *Taxon*, 41:3–10, 1992.
4. O.R.P. Bininda-Emonds, J.L. Gittleman, and A. Purvis. Building large trees by combining phylogenetic information: A complete phylogeny of the extant carnivora (mammalia). *Biological Reviews of the Cambridge Philosophical Society*, 74:143–175, 1999.
5. O.R.P. Bininda-Emonds, J.L. Gittleman, and M.A. Steel. The (super) tree of life. *Annual Review of Ecology and Systematics*, 33:265–289, 2002.
6. O.R.P. Bininda-Emonds and M.J. Sanderson. Assessment of the accuracy of matrix representation with parsimony supertree construction. *Systematic Biology*, 50:565–579, 2001.
7. D.J. Eernisse and A.G. Kluge. Taxonomic congruence versus total evidence, and amniote phylogeny inferred from fossils, molecules, and morphology. *Molecular Biology and Evolution*, 1993.
8. J.S. Farris. Methods for computing Wagner trees. *Systematic Zoology*, 19:83–92, 1970.
9. J. Felsenstein. Cases in which parsimony or compatibility methods will be positively misleading. *Systematic Zoology*, 27:401–410, 1978.
10. J. Felsenstein. The number of evolutionary trees. *Systematic Zoology*, 27:27–33, 1978.
11. J. Felsenstein. Evolutionary trees from DNA sequences: a maximum likelihood approach. *Journal of Molecular Evolution*, 17:368–376, 1981.
12. W.M. Fitch and E. Margoliash. The construction of phylogenetic trees - a generally applicable method utilizing estimates of the mutation distance obtained from cycochrome c sequences. *Science*, 155:279–284, 1967.
13. L.R. Foulds and R.L. Graham. The Steiner problem in phylogeny is NP-complete. *Advances in Applied Mathematics*, 3:43–49, 1982.
14. N. Friedman, M. Ninio, I. Pe'er, and T. Pupko. A structural EM algorithm for phylogenetic inference. In *RECOMB*, pages 132–140, 2001.
15. F. Gavril. Algorithms for minimum coloring, maximum clique, minimum covering by cliques, and maximum independent set of a chordal graph. *Society of Industrial and Applied Mathematics (SIAM) Journal on Computing*, 1(2):180–187, June 1972.
16. A.D. Gordon. Consensus supertrees: The synthesis of rooted trees containing overlapping sets of labelled leaves. *Journal of Classification*, 3:335–348, 1986.
17. T. Jiang, P. Kearney, and M. Li. A Polynomial Time Approximation Scheme for Inferring Evolutionary Trees from Quartet Topologies and Its Application. *Society of Industrial and Applied Mathematics (SIAM) Journal on Computing*, 30(6):1942–1961, 2001.
18. T.H. Jukes and C.R. Cantor. Evolution of protein molecules. In H. N. Munro, editor, *Mammalian Protein Metabolism*, pages 21–132. Academic Press, New York, 1969.

19. P. Kearney and D.G. Corneil. Tree powers. *Journal of Algorithms*, 29:111–131, 1998.
20. M. Kimura. A simple method for estimating evolutionary rates of base substitutions through comparative studies of nucleotide sequences. *Journal of Molecular Evolution*, 10:111–120, 1980.
21. F.-G.R. Liu, M.M. Miyamoto, N.P. Freire, P.Q. Ong, and M.R. Tennant. Molecular and morphological supertrees for eutherian (placental) mammals. *Science*, 291:1786–1789, 2001.
22. J.B. Losos and F.D. Adler. Stumped by trees? A generalized null model for patterns of organismal diversity. *The American Naturalist*, 145(3):329–342, 1995.
23. E.P. Martins. Phylogenies, spatial autoregression, and the comparative method: a computer simulation test. *Evolution*, 50:1750–1765, 1996.
24. M.P. Ng and N.C. Wormald. Reconstruction of rooted trees from subtrees. *Discrete Applied Mathematics*, 69:19–31, 1996.
25. R.D.M. Page. On consensus, confidence, and "total evidence". *Cladistics*, 12:83–92, 1996.
26. C.N.S. Pedersen and J. Stoye. Sorting leaf-lists in a tree. `http://www.techfak.uni-bielefeld.de/~stoye/rpublications/internal_leaflist.ps.gz`, 1998.
27. D. Phillips. Uniform Sampling From Phylogenetics Trees. Master's thesis, University of Waterloo, August 2002.
28. A. Purvis. A composite estimate of primate phylogeny. *Philosophical Transactions of the Royal Society of London Series B*, 348:405–421, 1995.
29. M.A. Ragan. Phylogenetic inference based on matrix representation of trees. *Molecular Phylogenetics and Evolution*, 1:53–58, 1992.
30. N. Saitou and M. Nei. The neighbour-joining method: A new method for reconstructing phylogenetic trees. *Molecular Biology and Evolution*, 4(4):406–425, 1987.
31. J.B. Slowinski and C. Guyer. Testing the stochasticity of patterns of organismal diversity: an improved null model. *The American Naturalist*, 134(6):907–921, 1989.
32. S.J. Willson. An error-correcting map for quartets can improve the signals for phylogenetic trees. *Molecular Biology and Evolution*, 18:344–351, 2001.

New Efficient Algorithm for Detection of Horizontal Gene Transfer Events

Alix Boc[1] and Vladimir Makarenkov[1,2]

[1] Département d'Informatique, Université du Québec à Montréal, C.P. 8888, Succ. Centre-Ville, Montréal (Québec), Canada, H3C 3P8
boc.alix@courrier.uqam.ca and makarenkov.vladimir@uqam.ca
[2] Institute of Control Sciences, 65 Profsoyuznaya, Moscow 117806, Russia

Abstract. This article addresses the problem of detection of horizontal gene transfers (HGT) in evolutionary data. We describe a new method allowing to predict the ways of possible HGT events which may have occurred during the evolution of a group of considered organisms. The proposed method proceeds by establishing differences between topologies of species and gene phylogenetic trees. Then, it uses a least-squares optimization procedure to test the possibility of horizontal gene transfers between any couple of branches of the species tree. In the application section we show how the introduced method can be used to predict eventual transfers of the rubisco *rbcL* gene in the molecular phylogeny including plastids, cyanobacteria, and proteobacteria.

1 Introduction

Evolutionary relationships between species has long been assumed to be a tree-like process that has to be represented by means of a phylogenetic tree. In such a tree, each species can only be linked to its closest ancestor and interspecies relationships are not allowed. However, such important evolutionary mechanism as *horizontal gene transfer* (i.e. *lateral gene transfer*) can be represented appropriately only using a network model. Lateral gene transfer plays an key role in bacterial evolution allowing bacteria to exchange genes across species [6], [21]. Moreover, numerous bacterial sequencing projects reinforced the opinion that evolutionary relationships between species cannot be inferred from the information based on a single gene, i.e. single gene phylogeny, because of existence of such evolutionary events as gene convergence, gene duplication, gene loss, and horizontal gene transfer (see [10], [5], [11], [17]). This article is concerned with studying the possibility of horizontal gene transfers between branches of species trees inferred either from whole species genomes or based on genes that are not supposed to be duplicated, lost or laterally transferred. We will show how the discrepancy between a phylogenetic tree based on a particular gene family and a species tree can be exploited to depict possible scenarios of how this particular gene may have been laterally transferred in course of the evolution.

Several attempts to use network-based evolutionary models to represent lateral gene transfers can be found in the scientific literature (see for example [11],

G. Benson and R. Page (Eds.): WABI 2003, LNBI 2812, pp. 190–201, 2003.

[23]). Furthermore, a number of models based on the subtree transfer operations on leaf labeled trees have also been proposed (see [4], [12], [13]). Recently, a new lateral gene transfer model considering a mapping of a set of gene trees (not necessarily pairwise equal) into a species tree has been proposed [10]. The latter article completed an extensive study of the tree mapping problems considered amongst others in [3], [9], [10], [18], [19].

In this paper we define a mathematically sound model using a least-squares mapping of a gene tree into a species tree. The proposed model is based on the computation of differences between pairwise distances between species in both trees. First, a species phylogeny has to be inferred from available nucleotide or protein sequences using an appropriate tree inferring algorithm. Second, the matrix of evolutionary distances between species with respect to the gene data has to be computed using an appropriate sequence-distance transformation. Third, the length of the species tree should be readjusted with respect to the gene distance matrix (see [1], [15] for more detail). Then, each pair of branches of the species tree (with the original topology and branch lengths adjusted according to the gene distance matrix) has to be evaluated for the possibility of a horizontal gene transfer. A new model introduced in this paper takes into account all different situations sound from the biological point of view when the lateral gene transfer event can explain the discrepancy in positioning two taxa in the gene and species phylogenies.

The new method has been tested with both real and artificial data sets yielding very encouraging results for both types data. In the application section below we show how our method aids to detect possible horizontal gene transfers of the rubisco *rbcL* gene (ribulose-1.5-bisphosphate carboxylase/oxygenase) in the species phylogeny inferred from phylogenetic analysis of 16S rRNA and other evidence (see [5] for more detail on these data). For this data set, the new method provided us with the solution consisting of eight horizontal gene transfers accounting for the conflicts between the *rbcL* and species phylogenies. Among the eight transfers obtained one can find all eventual gene transfers indicated in Delwiche and Palmer (1996). In a latter study, four *rbcL* gene transfers between: cyanobacteria and γ-proteobacteria, α-protcobacteria and red and brown algae, γ-proteobacteria and α-proteobacteria, and γ-proteobacteria and β-proteobacteria, were suggested as the most probable cause (along with the hypothesis of gene duplication) of topological differences between the organismal and gene phylogenies.

2 Description of the New Method

In this section, we describe a new method for detection of lateral gene transfer events obtained by mapping of a gene data into a species phylogeny. The new method allows to incorporate new branches with direction into the species phylogeny to represent gene transfers. Remember that any phylogenetic tree can be associated with a table of pairwise distances between its leaves which are labeled by the names of species; these distances are the minimum path-length distances between the leaves of the tree. All other nodes of the tree are intermediates, they represent unknown ancestors. It has been shown that a distance matrix satisfying the four-point condition (1) defines a unique phylogenetic tree [2].

$$d(i,j) + d(k,l) \leq Max\{d(i,k) + d(j,l); d(i,l) + d(j,k)\}, \text{for any } i,j,k,l. \quad (1)$$

When the four-point condition is not satisfied, what is always the case for real data sets, a tree inferring method has to be applied. There exist a number of efficient methods for inferring phylogenies from distance data; see for example NJ of Saitou and Nei (1987), BioNJ of Gascuel (1997), FITCH of Felsenstein (1997), or MW of Makarenkov and Leclerc (1999).

The main objective of our method is to infer a species tree from sequence or distance data and then test all possible pairs of the tree branches against the hypothesis that a lateral gene transfer could take place between them. Our method consists of the three main steps described below:

Step 1. Let T be a species phylogeny whose leaves are labeled according to the set X of n taxa. T can be inferred from sequence or distance data using an appropriate tree fitting method. Without lost of generality we assume that T is a binary tree, whose internal nodes are all of degree 3 and whose number of branches is $2n$-3. This tree should be explicitly rooted because the position of the root is important in our model.

Step 2. Let T_1 be a gene tree whose leaves are labeled according to the same set X of n taxa used to label the species tree T. Similarly to the species tree, T_1 can be inferred from sequence or distance data characterizing this particular gene. If the topologies of T and T_1 are identical, no horizontal gene transfers between branches of the species tree should be indicated. However, if the two phylogenies are topologically different it may be the result of a horizontal gene transfer. In the latter case the gene tree T_1 can be mapped into the species tree T by fitting by least squares the branch lengths of T to the pairwise distances in T_1 (for an overview of this fitting techniques, see Bryant and Wadell 1998 and Makarenkov and Leclerc 1999). These papers discuss two different ways of computing optimal branch lengths, according to the least-squares criterion, of a phylogenetic tree with fixed topology. After this operation the branch lengths of T will be modified, whereas its original topology will be kept unchanged.

Step 3. The goal of this step is to obtain an ordered list L of all possible HGT connections between pairs of branches in T. This list will comprise ($2n$-3)($2n$-4) entries, which is the possible number of different directed connections (i. e. number of possible HGTs) in a binary phylogenetic tree with n leaves. Each entry of L is associated with the value of the gain in fit obtained after addition of a new HGT branch linking a considered couple of branches. The first entries of L, those contributing the most to decrease the least-squares coefficient, will correspond to the most probable cases of the horizontal gene transfers.

Let us now show how to compute the value of the least-squares coefficient Q for an HGT branch (a,b) added to T to link the branches (x,y) and (z,w). In a phylogenetic tree there exists always a unique path linking any pair of the tree nodes, whereas addition of an HGT branch may create an extra path between them. Fig. 1(a, b, and c) illustrate the three possible cases when the minimum path-length distance between taxa i and j are allowed to be changed after the addition of the new branch (a,b) directed from b to a. Fixing the position of i in the species tree T, these cases differ only by position of j. From the biological point of view it would be plausible to allow the horizontal gene transfer between b and a to affect the evolutionary distance between the pair of taxa i and j if and only if either the node b is an ancestor of j (Fig. 1a) or the attachment point of j on the path (x,z) is located between the node z and Common Ancestor of x and z (Fig. 1b and 1c).

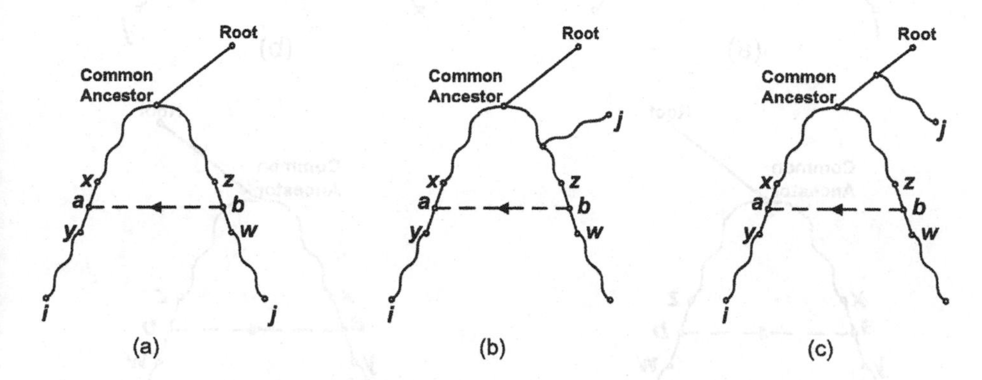

Fig. 1. Three situations when the minimum path-length distance between the taxa i and j can be affected by addition of a new branch (a,b) representing the horizontal gene transfer between branches (z,w) and (x,y) of the species tree. The path between the taxa i and j can now pass by the new branch (a,b).

In all other cases illustrated in Fig. 2 (a to f), the path between the taxa i and j cannot pass by the new branch (a,b) depicting the gene transfer from b to a. To compute the value of the least-squares coefficient Q for a given HGT branch (a,b) the following strategy was adopted: First, we define the set of all pairs of taxa that can be allowed to pass by a new HGT branch (a,b); second, in the latter set we determine all pairs of taxa such that the minimum path-length distance between them may decrease after addition of (a,b); third, we look for an optimal value l of (a,b), according to the least-squares criterion, while keeping fixed the lengths of all other tree branches; and finally, forth, all branch lengths are reassessed one at a time.

Let us define the set $A(a,b)$ of all pairs of taxa ij such that the distances between them may change if an HGT branch (a,b) is added to the tree T. $A(a,b)$ is the set of all pairs of taxa ij such that they are located in T as shown in Fig. 1 (a, b, or c) and:

$$Min\{d(i,a) + d(j,b); d(j,a) + d(i,b)\} < d(i,j), \tag{2}$$

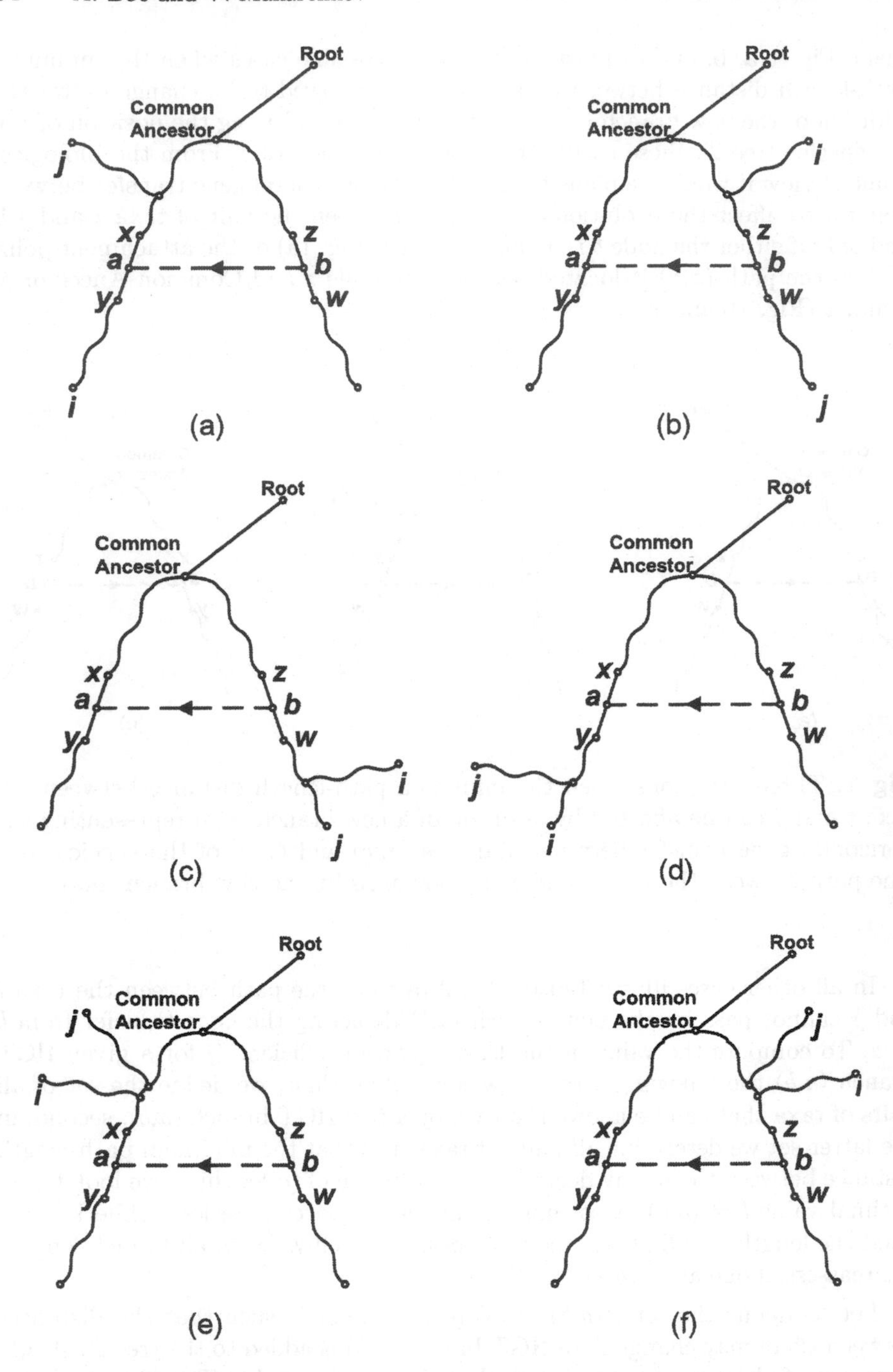

Fig. 2. Six situations when the minimum path-length distance between the taxa i and j is not affected by addition of a new branch (a,b) representing the horizontal gene transfer between branches (z,w) and (x,y) of the species tree. The path between the taxa i and j is not allowed to pass by the new branch (a,b).

where $d(i,j)$ is the minimum path-length distance between the nodes i and j; vertices a and b are located in the middle of the branches (x,y) and (z,w), respectively.

Define the following function:

$$dist(i,j) = d(i,j) - Min\{d(i,a) + d(j,b); d(j,a) + d(i,b)\}, \tag{3}$$

so that $A(a,b)$ is a set of all leaf pairs ij with $dist(i,j) > 0$.

The least-squares loss function to be minimized, with l used as an unknown variable, is formulated as follows:

$$Q(ab,l) = \sum_{dist(i,j)>l} (Min\{d(i,a) + d(j,b); d(j,a) + d(i,b)\} + l - \delta(i,j))^2$$

$$+ \sum_{dist(i,j)\leq l} (d(i,j) - \delta(i,j))^2 \rightarrow min, \tag{4}$$

where $d(i,j)$ is the minimum path-length distance between the taxa i and j in the gene tree T_1. The function $Q(ab,l)$, which is a quadratic polynomial spline, measures the gain in fit when a new HGT branch (a,b) with length l is added to the species tree T. The lower is the value of $Q(ab,l)$, the more likely that a horizontal gene transfer event has occurred between the branches (x,y) and (z,w).

Once the optimal value of a new branch (a,b) is computed, this computation can be followed by an overall polishing procedure for branch lengths reevaluation. In fact, the same calculations can be used to reevaluate the lengths of all other branch in T. To reassess the length of any branch in T, one can use equations (2), (3), and (4) assuming that the lengths of all the other branches are fixed. Thus, the polishing procedure can be carried out for branch number one, then branch number two, and so on, until all branch lengths are optimally reassessed. Then, one can return to the added HGT branch to reassess its length for the second time, and so forth.

These computations are repeated for all pairs of branches in the species tree T. When all pairs of branches in T are tested, an ordered list L providing their classification with respect to the possibility of a horizontal gene transfer can be established. This algorithm takes $O(n^4)$ to compute the optimal value of Q for all possible pairs of branches in T, the algorithmic complexity increases up to $O(n^5)$ when the branch lengths polishing procedure is also carried out.

3 Lateral Gene Transfers of the *rbcL* Gene

The method introduced in the previous section was applied to analyze the plastids, cyanobacteria, and proteobacteria data considered in Delwiche and Palmer (1996). The latter paper discusses the hypotheses of lateral gene transfer events of the rubisco genes between the three above-mentioned groups of organisms. Delwiche and Palmer inferred a maximum parsimony phylogeny of the *rbcL* (large subunit of rubisco) gene for 48 species. This phylogeny is shown in Fig. 3. The latter authors found that the gene classification based on the *rbcL* gene contains a number of conflicts compared to the species classification based on the 16S ribosomal RNA and other evidence. The aligned *rbcL* protein sequences for these species are available at www.life.umd.edu/labs/delwiche/publications.html.

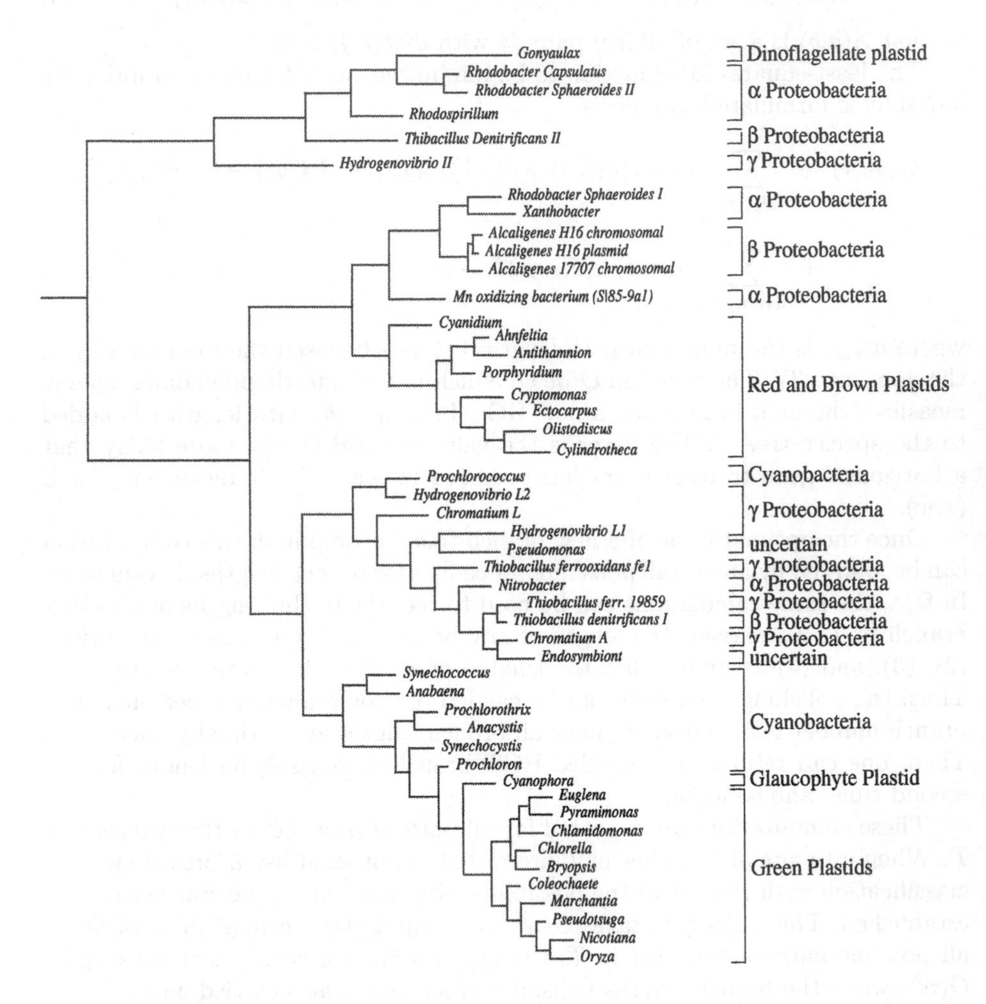

Fig. 3. Maximum parsimony tree of *rbcL* amino acids for 48 bacteria and plastids (from Fig. 2 of Delwiche and Palmer 1996). The tree shown is one of 24 shortest trees of length 2.422, selected arbitrarily. This phylogeny is inferred from 48 *rbcL* amino acid sequences with 497 bases. Classification of taxa based on 16S rRNA and other evidence is indicated to the right.

To apply our method we first attempted to construct the species tree of these 48 organisms based on the sequences data from NCBI [22], Ribosomal Database

Project [14] and other bioinformatics databases. However, the 16S rRNA data were available only for 28 of 48 species considered.

Thus, to carry out the analysis we reduced the number of species to 15 (see trees in Fig. 4a and b). Each species shown in Fig. 4 represents a group of bacteria or plastids from Fig. 3. We decided to conduct our study with three α-proteobacteria, three β-proteobacteria, three γ-proteobacteria, two cyonobacteria, one green plastids, one red and brown plastids, and two single species Gonyaulax and Cyanophora. The species tree in Fig. 4a was built using 16S rRNA sequences and other evidence existing in the scientific literature. The gene tree (Fig. 4b) was derived by contracting nodes of the 48 taxa phylogeny in Fig. 3. While observing the topologies of the species and gene trees one can note an important discrepancy between them.

For instance, the gene tree comprises a cluster regrouping α-proteobacteria3, β-proteobacteria3, and γ-proteobacteria3, as well as cynanobacteria1 is clustering with γ-proteobacteria2, and so on.

These contradictions can be explained either by lateral gene transfers that may have taken place between the species indicated or by ancient gene duplication; two hypotheses which are not mutually exclusive (see [5] for more detail). In this paper, the lateral gene transfer hypothesis is examined to explain the conflicts between the species and gene phylogenies.

The method introduced in this article was applied to the two phylogenies in Fig. 4(a and b) and provided us with a list of lateral gene transfers, ordered according their likelihood, between branches of the species tree. To reduce the algorithmic time complexity the links between adjacent branches were not considered. The solution network depicting the gene tree with eight horizontal gene transfers is shown in Fig. 5. The numbers at the arrows representing HGT events correspond to their position in the ordered list of transfers. Thus, the transfer between α-proteobacteria1 and Gonyaulax was considered as the most significant, then, the transfer between α-proteobacteria1 and β-proteobacteria1, followed by that from α-proteobacteria2 to β-proteobacteria2, and so forth. Delwiche and Palmer (1996, fig. 4) indicated four HGT events of the rubisco genes between cyanobacteria and γ-proteobacteria, γ-proteobacteria and α-proteobacteria, γ-proteobacteria and β-proteobacteria, and finally, between α-proteobacteria and plastids. All these transfers can be found in our model in Fig. 5.

4 Conclusion

We have developed a method for detection of horizontal gene transfers events in evolutionary data. The new method exploits the discrepancy between the species and gene trees built for the same set of observed species to map the gene tree into the species tree and then estimate the possibility of a horizontal gene transfer for each pair of branches of the species tree. As result, the new method gives an ordered list of the horizontal gene transfers between branches of the species tree. Entries of this list should be carefully analyzed using all available information about data in hand to select the gene transfers to be represented

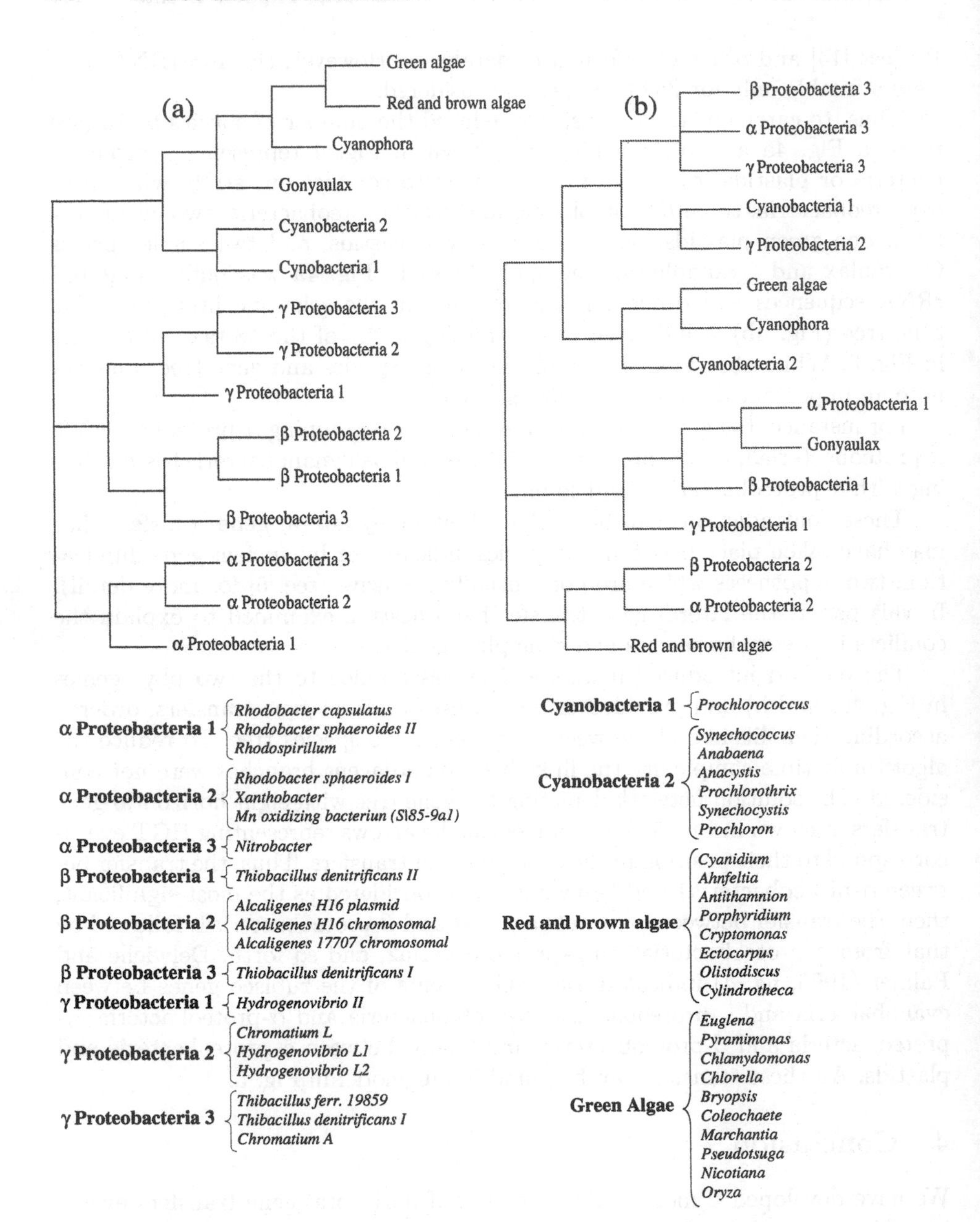

Fig. 4. (a) Species tree for 15 taxa representing different groups of bacteria and plastids from Fig. 3. Each taxon represents a group of organisms reported in the bottom. Species tree is built on the base of 16S rRNA sequences and other evidence. (b) *rbcL* gene tree for 15 taxa representing different groups of bacteria and plastids from Fig. 3. Gene tree is constructed by contracting nodes of the 48 taxa phylogeny in Fig. 3.

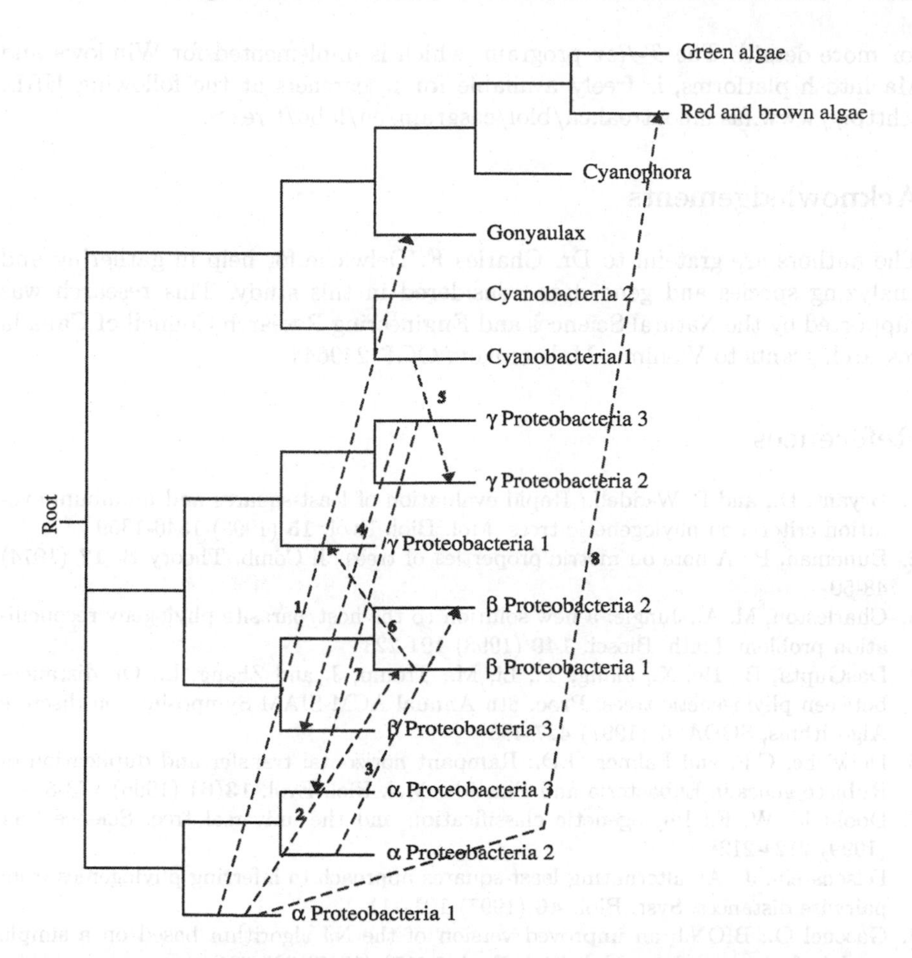

Fig. 5. Species tree from Fig. 4a with 8 dashed, arrow-headed lines representing possible horizontal gene transfers of the *rbcL* gene found by the new algorithm. Numbers on the arrow-headed lines indicate their order of appearance (i.e. order of importance) in the list of all possible HGT transfers found.

as final solution. Any gene transfer branch added to the species phylogeny aids to resolve a discrepancy between it and the gene tree. The example of evolution of the *rbcL* gene considered in the previous section clearly shows that the new method can be useful for prediction of lateral gene transfers in real data sets. In this paper a model based on the least-squares was considered. It would be interesting to extend and test this procedure in the framework of the maximum likelihood and maximum parsimony models. Future developments of this method allowing for different scenarios for addition of several horizontal gene transfers at the same time will be also necessary. The method for detection of horizontal gene transfers events described in this paper will be included (for the May 2003 release) in the *T-Rex* (tree and reticulogram reconstruction) package (see [16] for more detail). The *T-Rex* program, which is implemented for Windows and Macintosh platforms, is freely available for researchers at the following URL: <http://www.fas.umontreal.ca/biol/casgrain/en/labo/t-rex>.

Acknowledgements

The authors are grateful to Dr. Charles F. Delwiche for help in gathering and analyzing species and gene data considered in this study. This research was supported by the Natural Sciences and Engineering Research Council of Canada research grants to Vladimir Makarenkov, OGP 249644.

References

1. Bryant, D., and P. Waddell.: Rapid evaluation of least-squares and minimum evolution criteria on phylogenetic trees. Mol. Biol. Evol. **15** (1998) 1346-1359
2. Buneman, P.: A note on metric properties of trees. Jl Comb. Theory B. **17** (1974) 48-50
3. Charleston, M. A.: Jungle: a new solution to the host/parasite phylogeny reconciliation problem. Math. Biosci. **149** (1998) 191-223
4. DasGupta, B., He, X., Jiang, T., Li, M., Tromp, J. and Zhang, L.: On distances between phylogenetic trees. Proc. 8th Annual ACM-SIAM Symposium on discrete Algorithms, SODA'97 (1997) 427-436
5. Delwiche, C.F. and Palmer, J.D.: Rampant horizontal transfer and duplication of Rubisco genes in Eubacteria and Plastids. Mol. Biol. Evol. **13(6)** (1996) 873-882
6. Doolittle, W. F.: Phylogenetic classification and the universal tree. Science **284** (1999) 2124-2128
7. Felsenstein, J.: An alternating least-squares approach to inferring phylogenies from pairwise distances. Syst. Biol. **46** (1997) 101-111
8. Gascuel O.: BIONJ: an improved version of the NJ algorithm based on a simple model of sequence data, Mol. Biol. Evol. **14(7)** (1997) 685-695
9. Guig, R. I. Muchnik, and T.F. Smith.: Reconstruction of ancient molecular phylogenies. Mol. Phyl. Evol. **6,2** (1996) 189-213
10. Hallet, M., and Lagergreen, J.: Efficient algorithms for lateral gene transfer problems. RECOMB (2001) 149-156
11. Hein, J.: A heuristic method to reconstructing the evolution of sequences subject to recombination using parsimony. Math. Biosci. (1990) 185-200
12. Hein. J., Jiang, T., Wang, L. and Zhang, K.: On the complexity of comparing evolutionnary treees. Combinatorial Pattern Matching (CPM)95, LLNCS **937** (1995) 177-190
13. Hein, J., Jiang, T., Wang, L. and Zhang, K.: On the complexity of comparing evolutionnary trees. Discr. Appl. Math. **71** (1996) 153-169
14. Maidak, B.L., Cole, J.R., Lilburn, T.G., Parker, C.T., Saxman, P.R.,Farris, R.J., Garrity, G.M., Olsen, G.J., Schmidt, T.M., and Tiedje, J.M.: The RDP-II (ribosomal database project). Nucleic Acids Research **29** (2001) 173-174
15. Makarenkov, V. and Leclerc, B.: An algorithm for the fitting of a phylogenetic tree according to a weighted least-squares criterion, J. of Classif. **16,1** (1999) 3-26
16. Makarenkov,V.: T-Rex: reconstructing and visualizing phylogenetic trees and reticulation networks. Bioinformatics, **17** (2001) 664-668

17. Olsen, G. J. and Woese, C. R.: Archael genomics an overview. Cell **89** (1997) 991-994
18. Page, R. D. M.: Maps between trees and cladistic analysis of historical associations among genes, organism and areas. Syst. Biol. **43** (1994) 58-77
19. Page, R. D. M. and Charleston, M. A.: From gene to organismal phylogeny: Reconciled trees. Bioinformatics **14** (1998) 819-820
20. Saitou, N. and Nei, M.: The neighbour-joining method: a new method for reconstructing phylogenetic trees. Mol. Biol. Evol. **4** (1987) 406-425
21. Sneath, P. H.: Reticulate evolution in bacteria and other organisms: How can we study it? J. Classif. **17** (2000) 159-163
22. The NCBI handbook [Internet]. Bethesda (MD): National Library of Medicine (US), National Center for Biotechnology Information; 2002 Oct. Chapter 17, The Reference Sequence (RefSeq) Project
23. von Haseler, A. and Churchill, G. A.: Network models for sequence evolution. J. Mol. Evol. **37** (1993) 77-85

Ancestral Maximum Likelihood of Evolutionary Trees Is Hard

Louigi Addario-Berry[1], Benny Chor[2], Mike Hallett[1], Jens Lagergren[3], Alessandro Panconesi[4], and Todd Wareham[5]

[1] McGill Centre for Bioinformatics, School of Computer Science, McGill University, Montreal, PQ, Canada.
{hallett,laddar}@mcb.mcgill.ca
[2] School of Computer Science, Tel-Aviv University, Israel.
benny@cs.tau.ac.il
[3] Stockholm Bioinformatics Center and Department of Numerical Analysis and Computer Science, KTH, Stockholm, Sweden.
jensl@nada.kth.se
[4] Dipartimento di Informatica, Universitá di Roma "La Sapienza", Rome, Italy.
ale@dsi.uniroma1.it
[5] Department of Computer Science, Memorial University of Newfoundland, St. John's, NL, Canada.
harold@cs.mun.ca

Abstract. Maximum likelihood (ML) (Felsenstein, 1981) is an increasingly popular optimality criterion for selecting evolutionary trees. Finding optimal ML trees appears to be a very hard computational task – in particular, algorithms and heuristics for ML take longer to run than algorithms and heuristics for maximum parsimony (MP). However, while MP has been known to be NP-complete for over 20 years, no such hardness result has been obtained so far for ML.
In this work we make a first step in this direction by proving that ancestral maximum likelihood (AML) is NP-complete. The input to this problem is a set of aligned sequences of equal length and the goal is to find a tree and an assignment of ancestral sequences for all of that tree's internal vertices such that the likelihood of generating both the ancestral and contemporary sequences is maximized. Our NP-hardness proof follows that for MP given in (Day, Johnson and Sankoff, 1986) in that we use the same reduction from VERTEX COVER; however, the proof of correctness for this reduction relative to AML is different and substantially more involved.

1 Introduction

1.1 Background

Most methods for phylogenetic tree reconstruction on n species belong to two major categories – the *distance-based* methods (in which the input is a symmetric

G. Benson and R. Page (Eds.): WABI 2003, LNBI 2812, pp. 202–215, 2003.

n-by-n distance matrix) and *character-based* methods (in which the input is an n-by-m matrix of the values of m characters for each of the n species). Given the increasing availability of genomic sequence data, such character matrices typically consist of an m-length multiple alignment of n homologous sequences, one per species. The two major character-based methods are *maximum parsimony* (MP) and *maximum likelihood* (ML). Each of these methods has many variants, and each has strengths and weaknesses relative to various aspects, *e.g.*, inference consistency (see [16] and references).

An aspect of particular interest is the computational complexity of MP and ML. Each MP and ML variant has a well-defined objective function, and the related decision problems (or at least discretized versions of them) are usually in the complexity class NP. However, the only variants that are known to be solvable in polynomial time are those in which the tree (MP) or the tree and the branch lengths (ML) are given in addition to the data matrix; the goal in those variants is to compute the maximum likelihood of the data relative to that tree (ML) or the optimal maximum parsimony score and ancestral sequence assignments relative to that tree (MP) [4,5,13] (see also [15,16]). The situation for those variants that must determine optimal trees relative to given data matrices is more problematic – though it has been known for over 20 years that all such MP variants are NP-complete [2,6,8] (see also [17] and references), no such results have been found for ML to date. This is particularly frustrating in light of the intuition among practitioners that MP is easier than ML.

In this paper, we make a first step in addressing the complexity of ML by examining the complexity of one of the ML variants, the ANCESTRAL MAXIMUM LIKELIHOOD (AML) problem [9,18]. This variant is "between" MP and ML in that it is a likelihood method (like ML) but it reconstructs sequences for internal vertices (like MP). In this paper, we show that AML is NP-complete using a reduction from VERTEX COVER that is essentially identical to that given for MP by Day, Johnson, and Sankoff [2]. Note, however, that the proof of correctness for this reduction relative to AML is different and substantially more involved than that given for MP in [2].

1.2 Definitions

In this section we briefly describe the ancestral maximum likelihood (AML) problem. The goal of AML is to find the weighted evolutionary tree, together with assignments to all internal vertices, which is most likely to have produced the observed sequence data. To make this notion meaningful, we must have an underlying substitution model for the process of point mutation. Then, we seek the tree(s) T together with the edge probabilities p_e (or weights) and sequence assignments s_v for all internal vertices v of the tree which maximize L, the likelihood of the data.

For a tree T, let $\mathbf{p} = [p_e]_{e \in E(T)}$ be the edge probabilities and $\psi(1), \psi(2), \psi(3), \ldots, \psi(m) \in \{0,1\}^n$ be the observed sequences of length n over m taxa. The edge probability p_e ($p_e \leq 1/2$) is the probability that character states at the two incident vertices of e differ. Given a set $\mathbf{s}$ of sequences of length n labelling the vertices of T, let d_e denote the number of differences between the two sequences labelling the endpoints of the edge $e \in E(T)$. We will assume that we are dealing with a symmetric time-reversible model of character change along edges. In such cases, it is readily seen that the edge probabilities p_e are independent of the position of the root so we can regard T as being unrooted; however, for the purposes of integrating symbol-prior probabilities into the likelihood calculations, we will still designate an arbitrary internal vertex $\mathbf{r}$ as the root. For a specific edge $e \in E(T)$, the probability of generating the d_e differences and $n - d_e$ non-differences equals $p_e^{d_e}(1-p_e)^{n-d_e}$. Given the sequences at the vertices, events across different edges are mutually independent. Therefore the conditional probability (or the *ancestral likelihood*) of observing ψ, given the tree T, the internal sequences $\mathbf{s}$ and the edge probabilities $\mathbf{p}$, equals $L(\psi|T, \mathbf{s}, \mathbf{r}, \mathbf{p}) = (\prod_{e \in E(T)} p_e^{d_e}(1-p_e)^{n-d_e}) \times p(\mathbf{r})$, where $p(\mathbf{r})$ is the term produced by multiplying together all of the prior probabilities of each character-state of the sequence assigned to root-vertex $\mathbf{r}$. This conception of AML is called *joint ancestral likelihood* by Pupko *et al.* [11]. This discussion leads to our first definition of ancestral maximum likelihood as an optimization problem:

Ancestral Maximum Likelihood (Version I)
Input: A set S of m binary sequences, each of length n.
Goal: Find a tree T with m leaves, an assignment $p : E(T) \to [0,1]$ of edge probabilities, and a labelling $\lambda : V(T) \to \{0,1\}^n$ of the vertices such that

1. The m labels of the leaves are exactly the sequences from S, and
2. $(\prod_{e \in E(T)} p_e^{d_e}(1-p_e)^{n-d_e}) \times p(\mathbf{r})$ is maximized.

The AML criterion is usually applied to 4-state (DNA and RNA nucleotide) or 20-state (protein amino acid) sequences. However, to prove hardness of AML, it suffices to consider the simpler case of 2-state characters. We will use the Neyman 2-state model [10]. In this model, each character of the root is assigned a state according to some initial distribution. For each edge e of a tree T, there is a corresponding probability p_e ($p_e \leq 1/2$) that the character states at the two endpoint vertices of e differ. This induce a probability distribution over state assignments to the leaves which is independent of the choice of root. We will assume that both states are equally probable in the initial distribution, which makes the root-prior term a constant that can be ignored. These simplifications yield the second version of the AML optimization problem:

Ancestral Maximum Likelihood (Version II)
Input: A set S of m binary sequences, each of length n.
Goal: Find a tree T with n leaves, an assignment $p : E(T) \to [0,1]$ of edge probabilities, and a labelling $\lambda : V(T) \to \{0,1\}^n$ of the vertices such that

1. The m labels of the leaves are exactly the sequences from S, and
2. $\prod_{e \in E(T)} p_e^{d_e}(1-p_e)^{n-d_e}$ is maximized.

The AML problem may, at first glance, seem like a continuous optimization problem due to the edge probabilities. We now show that this is not the case given the simplifications made above. Consider an edge probability p_e: Given d_e and the length n of the sequences, the value of p_e that maximizes the individual edge likelihood $p_e^{d_e}(1-p_e)^{n-d_e}$ is simply $p_e = d_e/n$. Upon substituting this value and taking the n-th root, the individual edge likelihood now becomes

$$\left(\frac{d_e}{n}\right)^{d_e/n}\left(1-\frac{d_e}{n}\right)^{1-d_e/n}$$

and the tree likelihood expression becomes

$$\sum_{e\in E(T)}\left(\frac{d_e}{n}\log\left(\frac{d_e}{n}\right)+\left(1-\frac{d_e}{n}\right)\log\left(1-\frac{d_e}{n}\right)\right)=\sum_{e\in E(T)}-H_2\left(\frac{d_e}{n}\right),$$

where H_2 is the binary entropy function, $H_2(p) = -p\log_2(p) - (1-p)\log_2(1-p)$. This leads to our third and final formulation (in the sequel we drop the subscript 2 from logarithms and entropies):

Ancestral Maximum Likelihood (Version III)
Input: A set S of m binary strings, each of length n.
Goal: Find a tree T with m leaves and a labelling $\lambda : V(T) \to \{0,1\}^n$ of the vertices such that
1. The m labels of the leaves are exactly the sequences from S, and
2. $\sum_{e\in E(T)} H\left(\frac{d_e}{n}\right)$ is minimized.

The decision version of this formulation is as follows:

Ancestral Maximum Likelihood (Version III / Decision)
Input: A set S of m binary strings, each of length n, and a positive number k.
Question: Is there a tree T with m leaves and a labelling $\lambda : V(T) \to \{0,1\}^n$ of the vertices such that
1. The m labels of the leaves are exactly the sequences from S, and
2. $\sum_{e\in E(T)} H\left(\frac{d_e}{n}\right) \leq k$?

Unless otherwise noted, AML will denote this decision version in the remainder of this paper.

While we show that the decision version of AML-III is NP-complete, simpler versions of AML are tractable [9,11,18]. In these simpler versions, we are given the tree and the edge symbol-change probabilities in addition to the input sequences and the goal is to find sequence assignments to inner vertices as to maximize the likelihood. The dynamic programming algorithm given by Pupko *et al.* [11] is particularly elegant and is built along the same lines as dynamic programming algorithms for MP and ML when the tree is given [4,13,15]. Note that our completeness result is derived for the general version of AML given above in which we have to optimize over trees, assignments, and edge probabilities. This much larger parameter space explains the increase in complexity from polynomial time to NP-complete.

2 The Hardness of Ancestral Maximum Likelihood

In this section we establish that the AMP problem is NP-complete. We begin by recalling the reduction given in [2] which establishes the NP-hardness of maximum parsimony. This reduction is defined relative to the following problems:

VERTEX COVER (VC)
Input: A graph $G = (V, E)$ and a positive integer $k \leq |V|$.
Question: Is there a subset $V' \subseteq V$ such that $|V'| \leq k$ and for each edge $(u, v) \in E$, at least one of u and v belongs in V'?

MAXIMUM PARSIMONY (MP)
Input: A set S of m binary strings, each of length n, and an integer $k \geq 0$.
Question: Is there a tree T with m leaves and a labelling $\lambda : V(T) \rightarrow \{0,1\}^n$ of the vertices such that
1. The m labels of the leaves are exactly the sequences from S, and
2. $\sum_{e \in E(T)} d_e \leq k$?

Given an instance $\langle G = (V, E), k \rangle$ of VC, the reduction constructs an instance $\langle S, k' \rangle$ of MP such that S is a set of $m = |E| + 1$ strings, each of length $n = |V|$, and $k' = k + |E|$. The first string in S consists of all zeros, *i.e.*,

$$\underbrace{000\ldots00}_{n},$$

and then for every edge $e = (i, j) \in E$ there is a string

$$\underbrace{\overbrace{0\ldots0}^{i-1}1\overbrace{0\ldots0}^{j-(i+1)}1\overbrace{0\ldots0}^{n-j}}_{n}$$

where only the ith and jth symbols are set to 1. These latter strings are called *edge strings*. Similarly, a string with only the ith symbol set to 1 will be called a *vertex string*. Where the context is clear, we will refer to a vertex labelled with a string in which symbols $\{i_1, i_2, \ldots, i_t\}$ are set to 1 as $i_1 i_2 \ldots i_t$. For example, a vertex labelled with the edge-string in which symbols i and j are set to 1 would be called ij.

The proof in [2] establishes that optimal solution trees for instances of MP constructed by this reduction have the following form:

Definition 1. *A vertex-labelled tree is* canonical *if it satisfies the following four properties:*

1. *For any edge e in T, $d_e = 1$;*
2. *The sequence labelling the root of T is the all-zero vector;*
3. *The children of the root are internal vertices labelled with vertex strings; and*

4. *The children of the internal vertices in (3) are leaves labelled with edge strings.*

The following observation is crucial in establishing this fact:

> Given a tree T, suppose there are two leaves ij and kl in T that are connected to a common parent. If the numbers i, j, k, and l are all different then there is no increase in tree-cost if we introduce two new vertices i and k, connect these vertices to the root, and then connect ij to i and kl to k.

The proof concludes by noting that each such canonical tree T defines a vertex cover for the graph G in the given instance of VC – namely, the vertices in G corresponding to the vertex-strings labelling the children of the root in T.

Given the similarity of MP and AML as defined here, it is not surprising that we can re-use the reduction above to show that AML is NP-hard. Indeed, our reduction for AML differs from that given above only in that $k' = (k + |E|)H(\frac{1}{n})$. Unfortunately, we cannot immediately re-use the associated proof of correctness because optimal solution trees for instances of AML constructed by this reduction are not necessarily canonical. This is so because for small values of p, the binary entropy function satisfies

$$H(2p) < 2H(p) ,$$

which means that in maximal ancestral likelihood trees, it may be cheaper to connect a vertex ij directly to the root, rather than connecting it to a vertex i that is a child of the root. Hence, in our proof, we will use the following relaxed canonical form:

Definition 2. *A vertex-labelled tree is* weakly canonical *if it satisfies the following four properties:*

1. *For any edge e in T, d_e is either 1 or 2;*
2. *The sequence labelling the root of T is the all-zero vector;*
3. *The children of the root are internal vertices labelled with either vertex strings or edge strings; and*
4. *The children of the internal vertices in (3) are leaves labelled with edge strings.*

Note that internal vertices of weakly canonical trees can be labelled with sequences from S; however, this is not a problem as we can always attach a new leaf with the same label to such vertex, and the resulting edge cost (and thus the entropy contribution $H(d_e/n)$) equals zero.

Given the above, our proof of correctness will be in two parts:

1. Establish that optimal solution trees for instances of AML constructed by the reduction from instances of VC on arbitrary graphs are weakly canonical.

2. Define a class of graphs such that both VC restricted to this class of graphs is NP-hard and AML optimal solution trees for these restricted VC instances are canonical.

The following definitions will be useful. Given a binary string s, the *weight* of s is the number of 1's in s. Given a vertex v labelled by a sequence s, the *weight* of v is the weight of s. Given a vertex-labelled rooted tree T and an edge $e = (u, v)$ in T, the *cost* of e is the Hamming distance between the sequence-labels of the end-point vertices u and v. Given a vertex-labelled tree T and a vertex v in T, v is a *bad vertex* if the weight of the sequence-label of v is greater than 2. Unless otherwise noted, an optimal solution tree for an instance of AML constructed by the reduction will be denoted below as an optimal tree.

Lemma 1. *An optimal tree cannot have bad vertices.*

Proof. Consider an optimal tree T such that T has a minimum number of bad vertices and the sum of the degrees of the bad vertices is minimum. We can assume the following about T:

- *Any leaf in T has weight 2*: Suppose there is a leaf v of weight > 2. One can remove the path in T from v back to either the first vertex with a label from S or the first internal vertex with a descendent-vertex with a label from S; however, this would reduce the number of bad vertices as well as the entropy, violating the choice of T.
- *Any vertex of weight 1 is a child of the root*: One can simply disconnect such a vertex from its parent in T and attach it directly to the root without changing the entropy.
- *Any vertex v with a weight-2 vertex u as a child must have weight 3*: If this is not the case, the Hamming distance between the sequence-labels of v and u is at least 2; however, this would imply that we could attach u directly to the root without increasing the overall entropy, thereby violating the optimality of T.

Let b be a bad vertex in T such that (1) it has maximum distance from the root, (2) it is of minimum weight with respect to (1), and (3) it is of minimum degree with respect to both (1) and (2). By the choice of b, all of its children must have weight 2, which by the initial observation implies that b has weight 3. We will without loss of generality refer to b in the remainder of this proof as ijk.

If ijk has a child attached by an edge of cost 2 or more then that child can be attached directly to the root, violating the choice of T. Therefore ijk can have at most three children, namely ij, jk and ik (see Figure 1). Let us consider the cases of each possible set of children in turn. In what follows let c denote the cost of the edge linking ijk and its parent vertex in T.

1. *ijk has one child*: If ijk has just one child then we can get a tree with lower entropy by removing both ijk and the edges of cost 1 and c and introducing one edge of cost 2 that connects ij directly to the root. This violates the optimality of T, since $H(\frac{c}{n}) + H(\frac{1}{n}) \geq 2H(\frac{1}{n}) > H(\frac{2}{n})$ (see Figure 2).

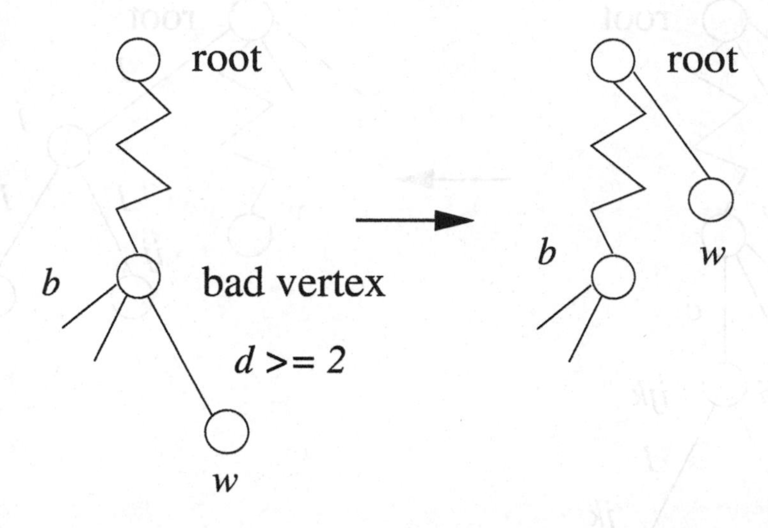

Fig. 1. Proof of Lemma 1: A bad vertex ijk can have at most 3 children.

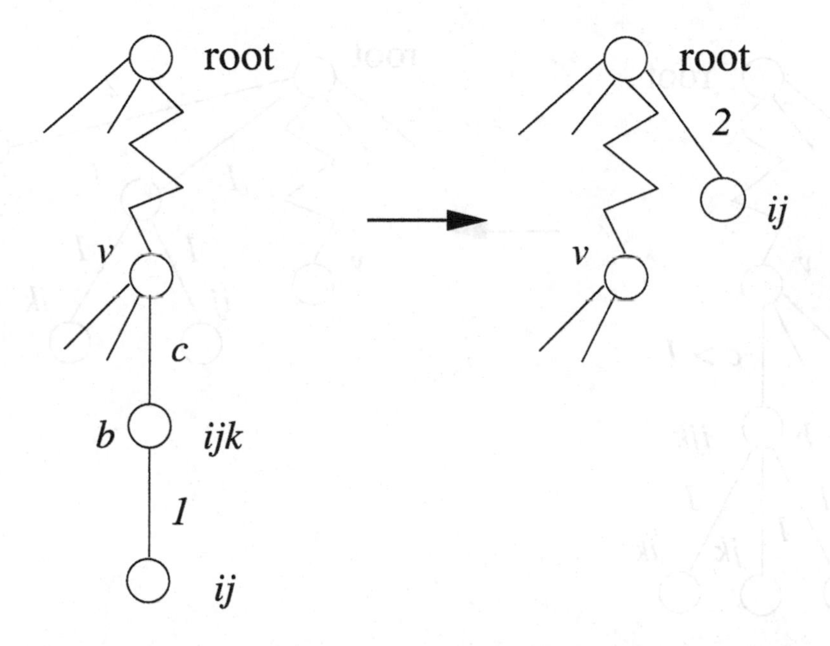

Fig. 2. Proof of Lemma 1, Case 1: Bad vertex ijk has one child.

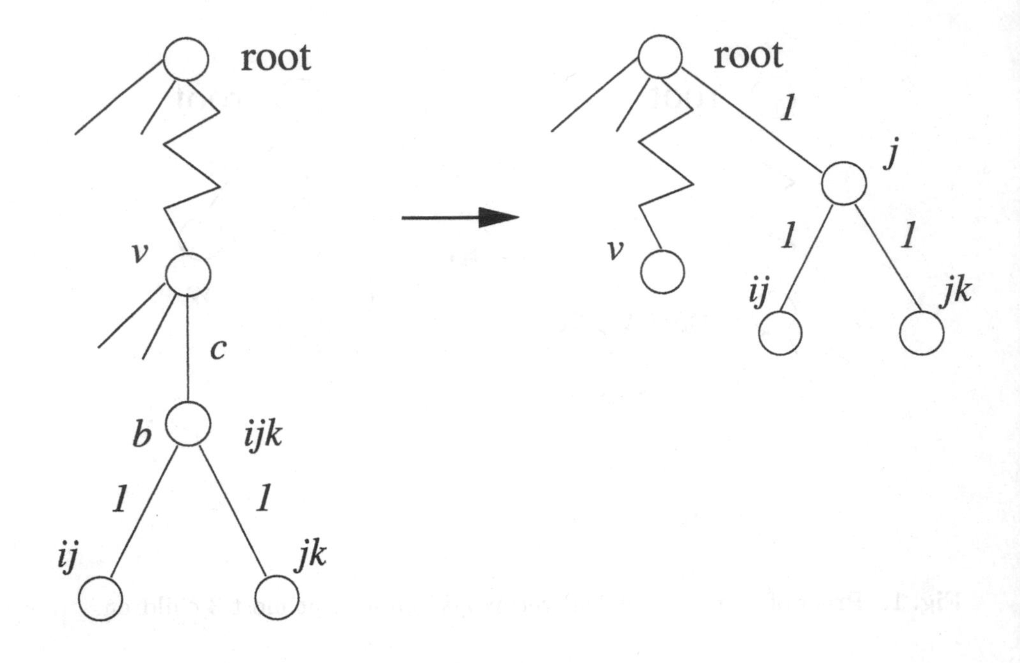

Fig. 3. Proof of Lemma 1, Case 2: Bad vertex ijk has two children.

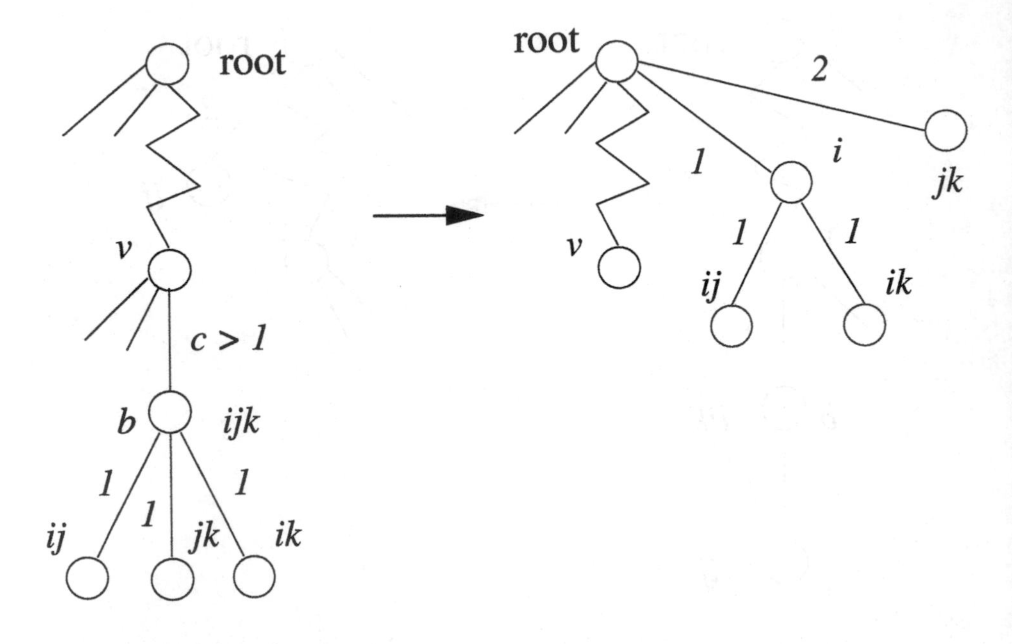

Fig. 4. Proof of Lemma 1, Case 3(a): Bad vertex ijk has three children, cost of edge connecting ijk to parent is > 1.

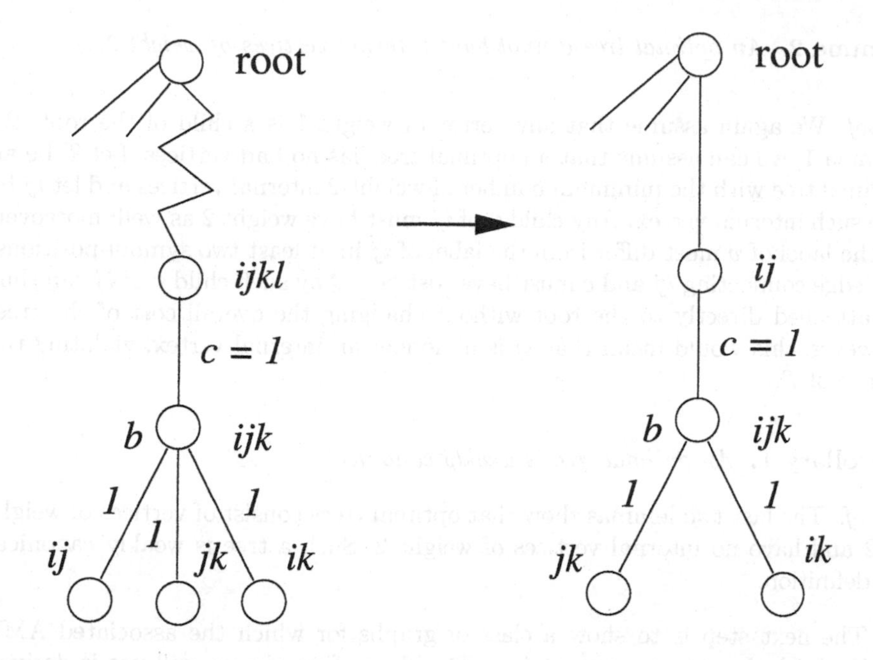

Fig. 5. Proof of Lemma 1, Case 3(b): Bad vertex ijk has three children, cost of edge connecting ijk to parent is 1.

2. *ijk has two children*: Suppose ijk has two children ij and jk. One can attach ij and jk to (a possibly new) vertex j, attach j to the root, and delete ijk. This does not increase the entropy but it eliminates one bad vertex, violating the choice of T (see Figure 3).
3. *ijk has three children*: This has two subcases.
   (a) If $c > 1$, ijk can be eliminated as follows: Vertex jk is attached directly to the root, while ij and ik are attached to (a perhaps new) vertex i that is connected to the root. This does not necessarily increase the entropy since $3H(\frac{1}{n}) + H(\frac{2}{n}) \leq 3H(\frac{1}{n}) + H(\frac{c}{n})$; however, a bad vertex has been removed, violating the choice of T (see Figure 4).
   (b) If $c = 1$, ijk's parent must be of weight 4 (as all suitable weight-2 vertices are already children of ijk). Without loss of generality, let this parent be $ijk\ell$. If ijk is its only child we can eliminate $ijk\ell$ by connecting ij directly to the root and making ijk a child of ij. This lowers the entropy and eliminates one bad vertex, thereby violating the optimality of T (see Figure 5). Thus $ijk\ell$ must have a second child, which we denote by b'. By the initial observation and the choice of b, b' both has weight 3 and has 3 children, all of weight 2. Assume without loss of generality that $ijk\ell$'s second child, b', is $jk\ell$. The children of $jk\ell$ must be jk, $k\ell$ and $j\ell$. This implies that T contains a cycle based on the vertices $ijk\ell$, ijk, $jk\ell$, and jk, which contradicts the acyclicity of T.

Lemma 2. *An optimal tree cannot have internal vertices of weight* 2.

Proof. We again assume that any vertex of weight 1 is a child of the root. By Lemma 1, we can assume that an optimal tree has no bad vertices. Let T be an optimal tree with the minimum number of weight-2 internal vertices and let ij be one such internal vertex. Any child v of ij must have weight 2 as well; moreover, as the label of v must differ from the label of ij in at least two symbol-positions, the edge connecting ij and c must have cost > 2. Any such child v of ij can thus be attached directly to the root without changing the overall cost of the tree. However, this would mean that ij is no longer an internal vertex, violating the choice of T.

Corollary 1. *An optimal tree is weakly canonical.*

Proof. The last two lemmas show that optimal trees consist of vertices of weight ≤ 2 and have no internal vertices of weight 2. Such a tree is weakly canonical by definition.

The next step is to show a class of graphs for which the associated AML optimal solution trees are canonical. The class of graphs we will use is derived using the graph composition operation.

Definition 3. *Given two graphs* $G_1 = (V_1, E_1)$ *and* $G_2 = (V_2, E_2)$, *the* composition graph $G_c = G_1[G_2]$ *is the graph with vertex set* $V = V_1 \times V_2$ *and edge set* E *defined by* $E = \{((u_1, u_2), (v_1, v_2)) :$ *either* $(u_1, v_1) \in E_1$ *or* $u_1 = v_1$ *and* $(u_2, v_2) \in E_2\}$.

If $G_2 = K_h$, the complete graph on $h \geq 1$ vertices, the composition graph G_c consists of h isomorphic copies of G_1. Given a vertex u in G_1, let the *copy* u^i *of* u be the vertex corresponding to u in the ith copy of G_1. Let the set of all copies of u in G_c be called the *column* of u and denote this column by U.

The following general property about vertex covers will be useful in several of the proofs given below.

Property 1. If C is a minimal vertex cover for a graph G, then every $u \in C$ must have a neighbour in G that is not in C.

Indeed, if all neighbours of u were in C then u could be removed from C, violating minimality.

Lemma 3. *For any* h, Vertex Cover *is NP-hard on composition graphs of the form* $G[K_h]$.

Proof. Given a graph G, let $G_c = G[K_h]$. Given a vertex cover C of G consider a vertex cover C' of G_c consisting of all the copies of vertices of C. That is, if $u \in C$ then $u^i \in C'$, for all i. Such vertex covers C' are said to be in *normal form.* Note that if C' is in normal form and U is a column then either U is contained in C' or it is disjoint from it. We will show that optimal vertex covers for G_c must be in normal form.

Let C' be a minimal vertex cover of G_c that is not in normal form. There must therefore be a column U that is neither contained in nor disjoint from C'. Let $u^i \in C' \setminus U$ and let $u^j \in U \setminus C'$. Since C' is minimal, by Property 1, there is a neighbour w of u^i not in C'. However, this would imply that the edge wu^j is not covered, which is a contradiction.

Thus, all minimal vertex covers of G_c must be in normal form, including in particular all optimal vertex covers. NP-hardness follows from the fact that we have established a polynomial-time computable bijection between optimal vertex covers of G and of G_c.

Note that optimal vertex covers of composition graphs of the form $G[K_h]$ have a special structure. In particular, every vertex in any optimal vertex cover of $G[K_h]$ covers at least h edges *uniquely* – that is, it covers at least h edges not covered by any other vertex (this follows from Property 1). This observation will be useful below.

Lemma 4. *An optimal tree associated with an instance of VC based on a composition graph of the form $G[K_h]$ is canonical.*

Proof. Let T be an optimal tree associated with $G_c = G[K_h]$. As T is only known to be weakly canonical by Corollary 1, there is at least one vertex of weight 2 adjacent to the root. Let C be the set of vertices of G_c corresponding to weight-1 vertices adjacent to the root in T. We will now prove that the set of weight-2 vertices directly attached to the root correspond to a matching M in G_c such that $V(M) \cap C = \emptyset$. Suppose that ij and ik are attached to the root. Then, if it is not already there, we introduce i, attach it to the root, and attach ij and ik to it. The new tree has lower entropy since $2H(\frac{2}{n}) > 3H(\frac{1}{n})$. Assume that ij and i are adjacent to the root. We clearly obtain a tree with lower entropy by making ij adjacent to i, instead of the root. It follows that M has the wanted properties. Notice that C is a vertex cover in $G_c \setminus M$.

Let $uv \in M$ and hence $u, v \notin C$. Let $v' \neq v$ be a vertex in the column V. Since $uv \in E(G_c)$, v' must be in the vertex cover C. Let us look at the neighbours of v' apart from u. Let w be such a neighbour. Since $v'w$ is an edge, vw is an edge too. This implies that w must then be in C; otherwise, this edge is not covered. This means that all neighbours $w \neq u$ of v' are in C. In turn this means that the only edge that is covered by v' uniquely is the edge uv' and this holds true for any v' in the column V. However, there is a much more economical way to cover the edges of $G_c \setminus M$, namely

$$C' := (C \setminus V) \cup \{u\} .$$

In terms of entropy of the corresponding trees, this means that if we switch from C to C' (as the weight-1 vertices connected to the root of the tree, and add edges in the natural way) we pay $H(\frac{1}{n})$ but save $H(\frac{2}{n}) + (h-1)H(\frac{1}{n})$. The claim follows.

Theorem 1. *AML is NP-complete.*

Proof. Problem AML is clearly in NP. The result then follows from Lemmas 3 and 4.

3 Concluding Remarks and Future Directions

In this paper, we have shown the NP-completeness of the problem of inferring ancestral sequences under joint maximum likelihood relative to a very simple model of character-change. There are two obvious directions for future research:

1. *Narrow the gap between tractability and intractability with respect to the status of other variants of AML.* Of particular interest is the version of AML with variable rates across sites. Based on work described in [12], it appears that a jump in complexity occurs when mutation-rate variation across sequence sites are allowed.
2. *Extend our results for AML to ML.*

Though the latter remains our ultimate goal, it is not immediately obvious how to extend the present work to attain that goal. A large part of the difficulty is that AML deals with the most likely ancestral-sequence assignment (and hence has that assignment present to be exploited by a reduction) while ML sums likelihoods over all possible ancestral-sequence assignments (leaving only the tree and the likelihood-sum to be exploited by a reduction). It is tempting to believe that the reconstruction of internal vertex sequences is responsible for the computational complexity of MP and AML, and that ML may in fact be easy; however, the NP-hardness of other evolutionary tree inference problems that either do not reconstruct internal states (distance-matrix fitting) [1] or reconstruct internal states in very limited fashions (character compatability, perfect phylogeny) [3,14] (see also [17] and references) suggest that there are more subtle factors at work here. In any case, it may be necessary to resort to reductions very different from that given here for AML to show NP-hardness for ML; one promising source of such reductions is various mathematical arguments showing those cases in which ML and other evolutionary tree inference methods give identical results (see [7] and references).

Acknowledgments

We would like to thank the staff of the Bellairs Research Institute (St. James, Barbados) of McGill University, where much of the research reported here was done, for their hospitality. We would also like to thank the WABI reviewers for their comments, which have been very helpful in making this a better paper. The research reported here was supported by ISF grant 418/00 (BC), the EU thematic network APPOL (AP), and NSERC operating grant 228104 (TW).

References

1. W. Day. Computational Complexity of Inferring Phylogenies from Dissimilarity Matrices. *Bulletin of Mathematical Biology*, 49(4), 461–467, 1987.
2. W. Day, D. Johnson, and D. Sankoff. The Computational Complexity of Inferring Rooted Phylogenies by Parsimony. *Mathematical Biosciences*, 81:33–42, 1986.
3. W. Day and D. Sankoff. Computational Complexity of Inferring Phylogenies by Compatability. *Systematic Zoology*, 35(2), 224–299, 1986.
4. J. Felsenstein. Evolutionary Trees from DNA Sequences: A Maximum Likelihood Approach. *Journal of Molecular Evolution*, 17:368–376, 1981.
5. W. Fitch. Towards defining the course of evolution: minimum change for a specific tree topology. *Systematic Zoology*, 20:406–416, 1971.
6. L. Foulds and R. Graham. The Steiner problem in phylogeny is NP-complete. *Advances in Applied Mathematics*, 3:43–49, 1982.
7. N. Goldman. Maximum likelihood inference of phylogenetic trees, with special reference to a Poisson process model of DNA substitution and to parsimony analyses. *Systematic Zoology*, 39(4):345–361, 1990.
8. R. Graham and L. Foulds. Unlikelihood that minimal phylogenies for a realistic biological study can be constructed in reasonable computational time. *Mathematical Biosciences*, 60:133–142, 1982.
9. M. Koshi and R. Goldstein. Probabilistic reconstruction of ancestral protein sequences. *Journal of Molecular Evolution*, 42, 313–320, 1996.
10. J. Neyman. Molecular studies of evolution: A source of novel statistical problems. In S. Gupta and Y. Jackel (eds.) *Statistical Decision Theory and Related Topics*, pages 1–27. Academic Press, New York, 1971.
11. T. Pupko, I. Pe'er, R. Shamir, and D. Graur. A Fast Algorithm for Joint Reconstruction of Ancestral Amino Acid Sequences. *Molecular Biology and Evolution*, 17(6):890–896, 2000.
12. T. Pupko, I. Pe'er, M. Hasegawa, D. Graur, and N. Friedman. A branch-and-bound algorithm for the inference of ancestral amino-acid sequences when the replacement rate varies among sites: Application to the evolution of five gene families. *Bioinformatics*, 18(8):1116–1123, 2002.
13. D. Sankoff and R. Cedergren. Simultaneous comparison of three or more sequences related by a tree. In D. Sankoff and J. Kruskal (eds.) *Time Warps, String Edits, and Macromolecules: The Theory and Practice of Sequence Comparison*, pages 253–263, Addison-Wesley Publishing Company, Reading, MA, 1983.
14. M. Steel. The complexity of reconstructing trees from qualitative characters and subtrees. *Journal of Classification*, 9, 71–90, 1992.
15. D. Swofford and W. Maddison. Parsimony, Character-State Reconstructions, and Evolutionary Inferences. In R. Mayden (ed.) *Systematics, Historical Ecology, and North American Freshwater Fishes*, pages 186–223, Stanford University Press, 1992.
16. D. Swofford, G. Olsen, P. Waddell, and D. Hillis. Phylogenetic Inference. In D. Hillis, C. Moritz, and B. Mable (eds.) *Molecular Systematics* (Second Edition), pages 407–514, Sinauer Associates, Sunderland, MA, 1996.
17. T. Wareham. *On the Computational Complexity of Inferring Evolutionary Trees.* Technical Report 93-01, Department of Computer Science, Memorial University of Newfoundland, 1993.
18. Z. Yang, S. Kumar, and M. Nei. A new method of inference of ancestral nucleotide and amino acid sequences. *Genetics*, 141:1641–1650, 1995.

A Linear-Time Majority Tree Algorithm

Nina Amenta[1], Frederick Clarke[2], and Katherine St. John[2,3]

[1] Computer Science Department
University of California, 2063 Engineering II
One Sheilds Ave, Davis, CA 95616.
amenta@cs.ucdavis.edu
[2] Dept. of Mathematics & Computer Science
Lehman College– City University of New York
Bronx, NY 12581
fclarke72@aol.com, stjohn@lehman.cuny.edu
[3] Department of Computer Science
CUNY Graduate Center, New York, NY 10016

Abstract. We give a randomized linear-time algorithm for computing the majority rule consensus tree. The majority rule tree is widely used for summarizing a set of phylogenetic trees, which is usually a post-processing step in constructing a phylogeny. We are implementing the algorithm as part of an interactive visualization system for exploring distributions of trees, where speed is a serious concern for real-time interaction. The linear running time is achieved by using succinct representation of the subtrees and efficient methods for the final tree reconstruction.

1 Introduction

Making sense of large quantities of data is a fundamental challenge in computational biology in general and phylogenetics in particular. With the recent explosion in the amount of genomic data available, and exponential increases in computing power, biologists are now able to consider larger scale problems in phylogeny: that is, the construction of evolutionary trees on hundreds or thousands of taxa, and ultimately of the entire "Tree of Life" which would include millions of taxa. One difficulty with this program is that most programs used for phylogeny reconstruction [8,9,17] are based upon heuristics for NP-hard optimization problems, and instead of producing a single optimal tree they generally output hundreds or thousands of likely candidates for the optimal tree. The usual way this large volume of data is summarized is with a consensus tree.

A consensus tree for a set of input trees is a single tree which includes features on which all or most of the input trees agree. There are several kinds of consensus trees. The simplest is the *strict consensus tree*, which includes only nodes that appear in all of the input trees. A node here is identified by the set of taxa in the subtree rooted at the node; the roots of two subtrees with different topologies, but on the same subset of taxa, are considered the same node. For some sets of input trees, the strict consensus tree works well, but for others, it produces

G. Benson and R. Page (Eds.): WABI 2003, LNBI 2812, pp. 216–227, 2003.

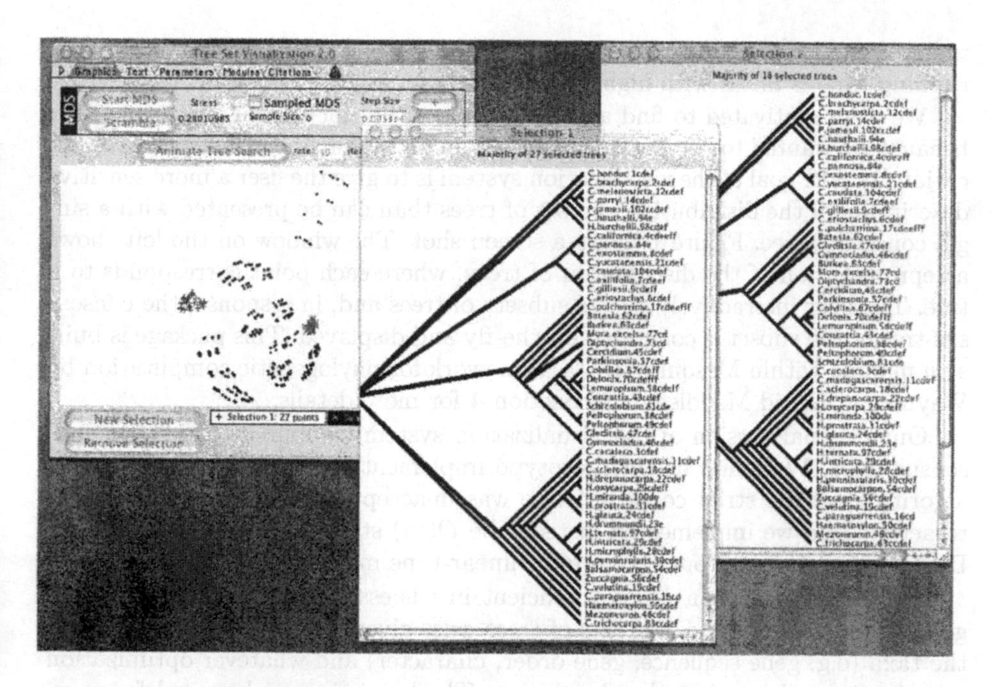

Fig. 1. The tree visualization module in Mesquite. The window on the left shows a projection of the distribution of trees. The user interactively selects subsets of trees with the mouse, and, in response, the consensus tree of the subset is computed on-the-fly and displayed in the window on the right. Two selected subsets and their majority trees are shown.

a tree with very few interior (non-terminal) nodes, since if a node is missing in even one input tree it is not in the strict consensus. The *majority rule consensus tree* includes all nodes that appear in a majority of input trees, rather than all of them. The majority rule tree is interesting for a much broader range of inputs than the strict consensus tree. Other kinds of consensus tree, such as Adams consensus, are also used (see [3], §6.2, for an excellent overview of consensus methods). The maximum agreement subtree, which includes a maximal subset of taxa for which the subtrees induced by the input trees agree, gives meaningful results in some cases in which the majority rule tree does not, but the best algorithm has an $O(tn^3+n^d)$ running time [7] (where d is the maximum outdegree of the trees), which is not as practical for large trees as the majority rule tree. Much recent work has been done on the related question of combining trees on overlapping, but not identical, sets of taxa ([2,13,14,15,16]).

In this paper, we present a randomized algorithm to compute the majority rule consensus tree, where the expected running time is linear both in the number t of trees *and* in the number n of taxa. Earlier algorithms were quadratic in n, which will be problematic for larger phylogenies. Our $O(tn)$ expected running time is optimal, since just reading a set of t trees on n taxa requires $\Omega(tn)$ time. The expectation in the running time is over random choices made during the course of the algorithm, independent of the input; thus, on any input, the running time is linear with high probability.

We were motivated to find an efficient algorithm for the majority rule tree, because we wanted to compute it on-the-fly in an interactive visualization application [1]. The goal of the visualization system is to give the user a more sensitive description of the distribution of a set of trees than can be presented with a single consensus tree. Figure 1 shows a screen shot. The window on the left shows a representation of the distribution of trees, where each point corresponds to a tree. The user interactively selects subsets of trees and, in response, the consensus tree of the subset is computed on-the-fly and displayed. This package is built as a module within Mesquite [10], a framework for phylogenetic computation by Wayne and David Maddison. See Section 4 for more details.

Our original version of the visualization system computed only strict consensus trees. We found in our prototype implementation that a simple $O(tn^2)$ algorithm for the strict consensus tree was unacceptably slow for real-time interaction, and we implemented instead the $O(tn)$ strict consensus algorithm of Day [6]. This inspired our search for a linear-time majority tree algorithm.

Having an algorithm which is efficient in t is essential, and most earlier algorithms focus on this. Large sets of trees arise given any kind of input data on the taxa (e.g. gene sequence, gene order, character) and whatever optimization criterion is used to select the "best" tree. The heuristic searches used for maximizing parsimony often return large sets of trees with equal parsimony scores. Maximum likelihood estimation, also computationally hard, generally produces trees with unique scores. While technically one of these is the optimal tree, there are many others for which the likelihood is only negligibly sub-optimal. So, the output of the computation is again more accurately represented by a consensus tree.

Handling larger sets of taxa is also becoming increasingly important. Maximum parsimony and maximum likelihood have been used on sets of about 500 taxa, while researchers are exploring other methods, including genetic algorithms and super-tree methods, for constructing very large phylogenies, with the ultimate goal of estimating the entire "Tree of Life". Our visualization system is designed to support both kinds of projects. It is also important for the visualization application to have an algorithm which is efficient when $n > t$, so that when a user selects a small subset of trees on many taxa some efficiency can be realized.

1.1 Notation

Let $\mathcal{S}$ represent a set of taxa, with $|\mathcal{S}| = n$. Let $\mathcal{T} = \{T_1, T_2, \ldots, T_t\}$ be the input set of trees, each with n leaves labeled by $\mathcal{S}$, with $|\mathcal{T}| = t$.

Without loss of generality, we assume the input trees are rooted at the branch connecting a distinguished taxon s_0, known as the *outgroup*, to the rest of the tree. If $\mathcal{T}$ is given as unrooted trees, or trees rooted arbitrarily, we choose an arbitrary taxon as s_0 and use it to root (or re-root) the trees.

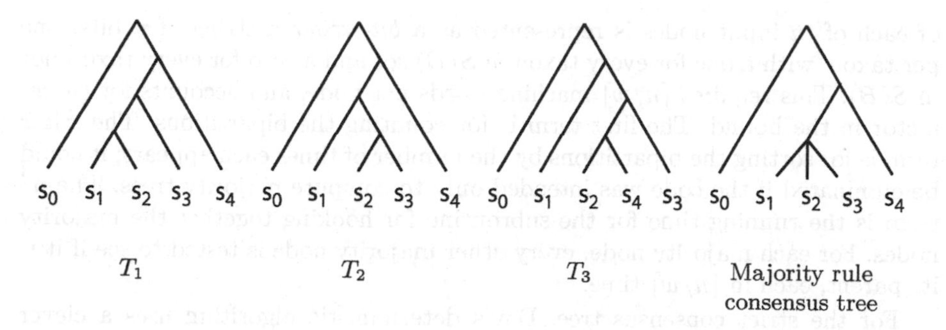

Fig. 2. Three input trees, rooted at the branch connecting s_0, and their majority tree (for a $> 1/2$ majority). The input trees need not be binary.

Consider a node i in an input tree T_j. Removing the branch from i towards the root divides T_j into the subtree below i and the remainder of the tree (including s_0). The induced *bipartition* of the taxa set into two subsets identifies the combinatorial type of node i. We can represent the bipartition by the subset of taxa which does *not* include s_0; that is, by the taxa at the leaves of the subtree rooted at i. If B is the bipartition, this set is $S(B)$. We will says that the *cardinality* of B, and of i, is the cardinality of $S(B)$. For example, in Figure 2, $s_1s_2 \mid s_0s_3s_4s_5$ is a bipartition of tree T_1 and $S(s_1s_2 \mid s_0s_3s_4s_5) = \{s_1s_2\}$. The cardinality of this bipartition is 2.

The *majority rule tree*, or M_l tree, includes nodes for exactly those bipartitions which occur in more than half of the input trees, or more generally in more than some fraction l of the input trees. Margush and McMorris [11] showed that this set of bipartitions does indeed constitute a tree for any $1/2 < l \leq 1$. McMorris, Meronk and Neumann [12] called this family of trees the M_l trees (e.g. the M_1 tree is the strict consensus tree); we shall call them all generically majority rule trees, regardless of the size of the majority.

See Figure 2 for a simple example. While this example shows binary trees, the algorithm also works for input trees with polytomies (internal nodes of degree greater than three).

1.2 Prior Work

Our algorithm follows the same intuitive scheme as most previous algorithms. In the first stage, we read through the input trees and count the occurrences of each bipartition, storing the counts in a table. Then, in the second stage, we create nodes for the bipartitions that occur in a majority of input trees - the *majority nodes* - and "hook them together" into a tree.

An algorithm along these lines is implemented in PHYLIP [8] by Felsenstein *et al.*. The overall running time as implemented seems to be $O((n/w)(tn+x \lg x+n^2))$ where x is the number of bipartitions found ($O(tn)$ in the worst case, but often $O(n)$), and w is the number of bits in a machine word. The bipartition B

of each of tn input nodes is represented as a *bit-string*: a string of n bits, one per taxon, with a one for every taxon in $S(B)$ set and a zero for every taxon not in $S(B)$. This requires $\lceil n/w \rceil$ machine words per node, and accounts for (n/w) factor in the bound. The first term is for counting the bipartitions. The $x \lg x$ term is for sorting the bipartitions by the number of times each appears; it could be eliminated if the code was intended only to compute majority trees. The n^2 term is the running time for the subroutine for hooking together the majority nodes. For each majority node, every other majority node is tested to see if it is its parent, each in $\lceil n/w \rceil$ time.

For the strict consensus tree, Day's deterministic algorithm uses a clever $O((\lg x)/w)$ representation for bipartitions. If we assume that the size of a machine word is $O(\lg x)$, so that for instance we can compare two bipartitions in $O(1)$ time, then we say that Day's algorithm achieves an optimal $O(tn)$ running time. Day's algorithm does not seem to generalize to other M_l trees, however. Wareham, in his undergraduate thesis at the Memorial University of Newfoundland with Day [18], developed an $O(n^2 + t^2 n)$ algorithm, which only uses $O(n)$ space. It uses Day's data structure to test each bipartition encountered separately against all of the other input trees. Majority trees are also computed by PAUP [17], using an unknown (to us) algorithm.

Our algorithm follows the same general scheme, but we introduce a new representation for each bipartition of size $O((\lg x)/w) \approx O(1)$, giving an $O(tn)$ algorithm for the first counting step, and we also give an $O(tn)$ algorithm for hooking together the majority nodes.

2 Majority Rule Tree Algorithm

Our algorithm has two main stages: scanning the trees to find the majority bipartitions (details in Section 2.1) and then constructing the majority rule tree from these bipartitions (details in Section 2.2). It ends by checking the output tree for errors due to (very unlikely) bad random choices. Figure 3 contains pseudo-code for the algorithm.

2.1 Finding Majority Bipartitions

In the first stage of the algorithm, we traverse each input tree in post-order, determining each bipartition as we complete the traversal of its subtree. We count the number of times each bipartition occurs, storing the counts in a table. With the record containing the count, we also store the cardinality of the bipartition, which turns out to be needed as well.

A first thought might be to use the bit-string representation of a bipartition as an address into the table of counts, but this would be very space-inefficient: there are at most $O(tn)$ distinct bipartitions, but 2^n possible bit-strings. A better idea, used in our algorithm and in `PHYLIP`, is to store the counts in a hash-table.

Input:	A set of t trees, $\mathcal{T} = \{T_1, T_2, \ldots, T_t\}$.
Output:	The majority tree, M_l, of $\mathcal{T}$
Algorithm:	
Start-up:	Pick prime numbers m_1 and m_2
	and random integers for the hash functions h_1 and h_2.
1.	For each tree $T \in \mathcal{T}$,
2.	Traverse each node, x, in *post order.*
3.	Compute hashes $h_1(x)$ and $h_2(x)$
4.	If no double collision, insert into hash table (details below).
5.	If double collision, restart algorithm from beginning.
6.	For each tree $T \in \mathcal{T}$,
7.	Let c point to the root of T
8.	Traverse each node, x, in *pre order.*
9.	If x is a majority node,
10.	If not existing in M_l, add it, set its parent to c.
11.	Else x already exists in M_l, update its parent (details below).
12.	In recursive calls, pass x as c.
13.	Check correctness of tree M_l.

Fig. 3. The pseudo-code for the majority tree algorithm

Hash Tables: Our algorithm depends on the details of the hash-table implementation, so we briefly include some background material and terminology (or see, for example, [5], Chapter 11). We use a function, the *hash function*, to compute the address in the table at which to store the data. The address is called the *hash code*, and we say an element *hashes* to its address. In our case, the hash function takes the 2^n possible bipartitions into $O(tn)$ hash table addresses; but since at most tn of the possible bipartitions actually occur in the set of input trees, on average we put only a constant number of bipartitions at each address. More than one bipartition hashing to the same table address is called a *collision.* To handle collisions, we use a standard strategy called *chaining*: instead of storing a count at each table address, we store a linked list of counts, one for each bipartition which has hashed to that address. When a new bipartition hashes to the address, we add a new count to the linked list. See Figure 4.

Universal Hash Functions: As the hash function, we use a *universal hash function* ([5], §11.3.3, or [4]), which we call h_1 (the reader expecting an h_2 later will not be disappointed). As our program starts up, it picks a prime number m_1 which will be the size of the table, and a list $a = (a_1, \ldots, a_n)$ of random integers in $(0, \ldots, m_1 - 1)$. We can select m from a selection of stored primes of different sizes; only the a need to be random. Let $B = (b_1, \cdots, b_n)$ be the bit-string representation of a bipartition. The universal hash function is defined as

$$h_1(B) = \sum_{i=1}^{n} b_i a_i \bmod m_1$$

Notice that $h_1(B)$ is always a number in $0, \ldots, m_1 - 1$.

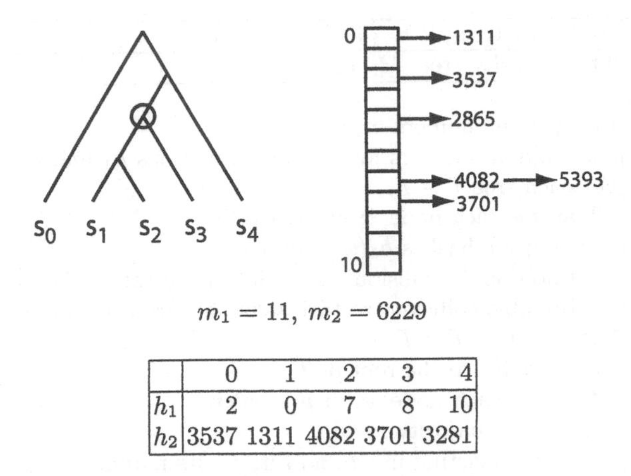

	0	1	2	3	4
h_1	2	0	7	8	10
h_2	3537	1311	4082	3701	3281

Fig. 4. Storing nodes in the hash table: Assume we have the five leaf tree on the left, T_1, and two universal hash functions (and associated prime numbers) given by the table below. The IDs stored in the hash table for the beginning of a post-order traversal of T_1 are shown; the circled node was the last one processed. First s_0 was processed, storing $h_2(s_0) = 3537$ at $h_1(s_0) = 2$. Similarly s_1 and s_2 were processed, and then their parent, storing $h_2 = 1311 + 4082 \bmod 6229$ into $h_1 = 0 + 7 \bmod 11$, and so on.

Using this universal hash function, the probability that any two bipartitions B_1 and B_2 collide (that is, that $h_1(B_1) = h_1(B_2)$) is $1/m_1$ [4], so that if we choose $m_1 > tn$ the expected number of collisions is $O(tn)$.

Collisions: To detect and handle these collisions, we use a second universal hash function h_2 to produce an ID for each bipartition. We represent a bipartition in the hash table by a record which includes the ID as well as the running count of the number of times the bipartition occurs. To increment the count of a bipartition B, we go to the table address $h_1(B)$ and search the linked list for the record with ID $h_2(B)$. If no such record exists, we create one and give it a count of one.

It is possible that a *double collision* will occur; that is, that for two bipartitions B_1, B_2, we have $h_1(B_1) = h_1(B_2)$ and also $h_2(B_1) = h_2(B_2)$. Since these two events are independent, the probability that B_1 and B_2 have a double collision is $1/(m_1 m_2)$. Notice that although we have to choose $m_1 \approx tn$ to avoid wasting space in the table, we can choose m_2 to be very large so as to minimize the probability of double collisions. If we choose $m_2 > ctn$, for any constant c, the probability that no double collisions occur, and hence that the algorithm succeeds, is at least $1-O(1/c)$, and the size of the representation for a bipartition remains $O(\lg tn + \lg c)$.

Nonetheless, we need to detect the unlikely event that a double collision occurs. If so, we abort the computation and start over with new random choices for the parameters a of the hash functions.

If B_1 and B_2 represent bipartitions with different numbers of taxa, we can detect the double collision right away, since we can check to make sure that both the IDs and the cardinality in the record match before incrementing a count. Similarly if B_1 and B_2 are bipartitions corresponding to leaves we can detect the double collision immediately by checking that the two taxa match before incrementing the count. The final remaining case of detecting a double collision for two bipartitions B_1, B_2, both with cardinality $k > 1$, will be done later by a final check of output tree against the hash table.

Implicit Bipartitions: To achieve an $O(tn)$ overall running time, we need to avoid computing the $O(n/w)$-size bit-string representations for each of the $O(tn)$ bipartitions. Instead, we directly compute the hash codes recursively at each node, without explicitly producing the bit-strings.

Fact 1 *Consider a node with a bipartition represented by bit-string B. Let the two children of this node have bipartitions with bit-strings B_L and B_R. Then*

$$h_1(B) = h_1(B_L) + h_1(B_R) \ mod \ m_1$$

This is true because B_l and B_r represent disjoint sets of taxa, so that

$$h_1(B) = \left(\sum_{B_L} b_i a_i \bmod m_1 \right) + \left(\sum_{B_R} b_i a_i \bmod m_1 \right)$$

where a_i is the prime number assigned to i by the universal hash function. A similar statement of course holds for h_2, and when B has more than two children.

We can use this fact to compute the hash code recursively during the post-order traversal. We store the hash codes in the tree nodes as we compute them. At a leaf, we just look up the hash code in the array a; for the leaf containing taxon i, the hash code is a_i. For an internal node, we will have already computed the hash codes of its children $h_1(B_L)$ and $h_1(B_R)$, so we compute $h_1(B)$ in constant time using Fact 1. The reader may wish to go over the example of the computation in Figure 4.

We compute the cardinality of the bipartition in constant time at each node similarly using recursion.

2.2 Constructing the Majority Tree

Once we have all the counts in the table we are ready to compute the majority rule consensus tree. The counts let us identify which are the majority bipartitions that appear in more than lt trees. But since the bipartitions are represented only implicitly, by their hash functions, hooking them up correctly to form the majority rule tree is not totally straightforward. We use three more facts.

Fact 2 *The parent of a majority bipartition B in the majority rule tree is the majority bipartition B' of smallest cardinality such that B is a subset of B'.*

Fact 3 *If majority bipartition B' is an ancestor of majority bipartition B in an input tree T_j, then B' is an ancestor of B in the majority rule tree.*

Fact 4 *For any majority bipartition B and its parent B' in the majority rule tree, B and B' both appear in some tree T_j in the input set. In T_j, B' is an ancestor of B, although it may not be B's parent.*

Fact 4 is true because both B and B' appear in more than $l \geq t/2$ trees, so they have to appear in some tree together, by the pigeon-hole principle.

We do a pre-order traversal of each of the input trees in turn. As we traverse each tree, we keep a pointer to c, the last node corresponding to a majority bipartition which is an ancestor of the current node in the traversal. As we start the traversal of a tree T, we can initialize c to be the root, which always corresponds to a majority bipartition. Let C be the bipartition corresponding to c.

At a node i, we use the stored hash codes to find the record for the bipartition B in the hash table. If B is not a majority node, we ignore it. If B is a majority node and a node for B does not yet exist in the output tree, we create a new node for the output tree and with its parent pointer pointing to C. If, on the other hand, a node in the outut tree does exist for B, we look at its current parent P in the output tree. If the cardinality of P (stored in the hash table record for P) is greater than the cardinality of C, we switch the parent pointer of the node for B to point to the node for C. When we are done, each node B in the output tree, interior or leaf, points to the node of smallest cardinality that was an ancestor in any one of the input trees. Assuming there was no double collision, Facts 2, 3, and 4 imply that the output tree is the correct majority rule consensus tree.

2.3 Final Check

After constructing the majority rule tree, we check it against the hash table in order to detect any occurrence of the final remaining case of a double collision, when two bipartitions B_1, B_2 of the same cardinality $k > 1$ have the same value for both h_1 and h_2. Recall that, if B_1, B_2 are singletons or have different cardinalities, double collisions would already have been detected when putting the data into the hash table.

To check the tree, we do a post-order traversal of the completed majority rule tree, recursively computing the cardinality of the bipartition at each node, and checking that these cardinalities match those in the corresponding records in the hash table. If we find a discrepancy, this indicates a double collision and we discard the majority rule tree and run the algorithm again with new random choices for the parameters a of the hash functions.

Claim. Any remaining double collisions are detected by checking the cardinalities.

Proof: Let us consider the smallest k for which a double collision occurs, and let B_1, B_2 be two of the bipartitions of cardinality k which collide. Since the double collision was undetected in the first stage, there is some record B in the hash table, with cardinality k, representing both B_1 and B_2. Whenever B_1 or B_2 was encountered during the first stage of the algorithm, the count for B was incremented.

In the second reconstruction stage, a node in the output tree is created for B as soon as either B_1 or B_2 is encountered in the traversal of some input tree. Consider a bipartition C which is a child of B_1 (equivalently B_2) in the correct majority rule tree. At the end of stage two, there will be a node for C in the output tree. Its parent pointer will be pointing to B, since in some input tree B_1 (resp. B_2) will be an ancestor of C, with cardinality k, and no majority node of cardinality less than k will be an ancestor of C in any input tree. Thus the nodes for all majority bipartitions which would be children of either B_1 or B_2 in the correct majority rule tree end up as children of B in the incorrect output tree. Notice that since B_1, B_2 have the same cardinality, one cannot be the ancestor of the other; so the two sets $S(B_1), S(B_2)$ are disjoint. Therefore the cardinality of B in the output tree will be $2k$, while the cardinality stored in the hash table for B will be k.

2.4 Analysis Summary

The majority rule consensus tree algorithm runs in $O(tn)$ time. It does two traversals of the input set, and every time it visits a node it does a constant number of operations, each of which requires constant expected time (again, assuming that $w = O(\lg x)$). The final check of the majority tree takes $O(n)$ time.

The probability that any double collision occurs is $1/c$, where c is the constant such that $m_2 > ctn$. Thus the probability that the algorithm succeeds on its first try is $1 - 1/c$, the probability that r attempts will be required decreases exponentially with r, and the expected number of attempts is less than two.

3 Weighted Trees

The majority rule tree has an interesting characterization as the median of the set of input trees, which is useful for extending the definition to weighted trees. When a bipartition is not present in an input tree we consider it to have weight zero, while when a bipartition is present we associate with it the positive weight of the edge determining the bipartition in the rooted tree. Now consider the medial weight of each bipartition over all input trees, including those that do not contain the bipartition. A bipartition which is *not* contained in a majority of the input trees has median weight zero. A bipartition in a majority of input trees has some positive median weight.

Note that it is simple, although space-consuming, to compute this median weight for each majority bipartition in $O(nt)$ time. In the second pass through the set of input trees, we store the weights for each majority edge in a linked list, as they are encountered. Since there are $O(n)$ majority bipartitions and t trees the number of weights stored is $O(nt)$. The median weight for each bipartition can then be computed as the final tree is output, using a textbook randomized median algorithm which runs in $O(t)$ time per edge ([5], §9.2).

4 Implementation

Our majority rule consensus tree algorithm is implemented as part of our tree-set visualization system, which in turn is implemented within Mesquite [10]. Mesquite is a framework for phylogenetic analysis written by Wayne and David Maddison, available for download at their Web site [10]. It is designed to be portable and extensible: it is written in Java and runs on a variety of operating systems (Linux, MacIntosh OS 9 and X, and Windows).

Mesquite is organized into cooperating of modules. Our visualization system has been implemented in such a module, TreeSetVisualization, the first published version of which can be downloaded from our webpage [1]. The module was introduced last summer at Evolution 2002 meeting and has since been downloaded by hundreds of researchers. In the communications we get from users, majority trees are frequently requested. The TreeSetVisualization module includes tools for visualizing, clustering, and analyzing large sets of trees.

The majority tree implementation will be part of the next version of the module. Figure 1 shows our current prototype. We expect to release the new version this summer, including the majority tree code and other new features.

5 Acknowledgments

This project was supported by NSF-ITR 0121651/0121682 and computational support by NSF-MRI 0215942. The first author was also supported by an Alfred P. Sloan Foundation Research Fellowship. We thank Jeff Klingner for the tree set visualization module and Wayne and David Maddison for Mesquite, and for encouraging us to consider the majority tree. The second and third authors would like to thank the Department of Computer Sciences and the Center for Computational Biology and Bioinformatics at University of Texas, and the Computer Science Department at the University of California, Davis for hosting them for several visits during 2002 and 2003.

References

1. Nina Amenta and Jeff Klingner. Case study: Visualizing sets of evolutionary trees. In *8th IEEE Symposium on Information Visualization (InfoVis 2002)*, pages 71–74, 2002. Software available at `www.cs.utexas.edu/users/phylo/`.
2. B.R. Baum. Combining trees as a way of combining data sets for phylogenetic inference, and the desirability of combining gene trees. *Taxon*, 41:3–10, 1992.

3. David Bryant. *Hunting for trees, building trees and comparing trees: theory and method in phylogenetic analysis.* PhD thesis, Dept. of Mathematics, University of Canterbury, 1997.
4. J. Lawrence Carter and Mark N. Wegman. Universal classes of hash functions. *Journal of Computer and Systems Sciences*, 18(2):143–154, 1979.
5. Thomas H. Cormen, Charles E. Leiserson, Ronald L. Rivest, and Clifford Stein. *Introduction to algorithms.* MIT Press, Cambridge, MA, second edition, 2001.
6. William H.E. Day. Optimal algorithms for comparing trees with labeled leaves. *J. Classification*, 2(1):7–28, 1985.
7. Martin Farach, Teresa M. Przytycka, and Mikkel Thorup. On the agreement of many trees. *Information Processing Letters*, 55(6):297–301, 1995.
8. J. Felsenstein. Phylip (phylogeny inference package) version 3.6, 2002. Distributed by the author. Department of Genetics, University of Washington, Seattle. The consensus tree code is in `consense.c` and is co-authored by Hisashi Horino, Akiko Fuseki, Sean Lamont and Andrew Keeffe.
9. John P. Huelsenbeck and Fredrik Ronquist. MrBayes: Bayesian inference of phylogeny, 2001.
10. W.P. Maddison and D.R. Maddison. Mesquite: a modular system for evolutionary analysis. version 0.992, 2002. Available from `http://mesquiteproject.org`.
11. T. Margush and F.R. McMorris. Consensus n-trees. *Bulletin of Mathematical Biology*, 43:239–244, 1981.
12. F.R. McMorris, D.B. Meronk, and D.A. Neumann. A view of some consensus methods for trees. In *Numerical Taxonomy: Proceedings of the NATO Advanced Study Institute on Numerical Taxonomy.* Springer-Verlag, 1983.
13. R.D.M. Page. Modified mincut supertrees. *Lecture Notes in Computer Science (WABI 2002)*, 2452:537–551, 2002.
14. M.A. Ragan. Phylogenetic inference based on matrix representation of trees. *Molecular Phylogenetics and Evolution*, 1:53–58, 1992.
15. M.J. Sanderson, A. Purvis, and C. Henze. Phylogenetic supertrees: assembling the trees of life. *Trends in Ecology and Evolution*, 13:105–109, 1998.
16. Charles Semple and Mike Steel. A supertree method for rooted trees. *Discrete Applied Mathematics*, 105(1-3):147–158, 2000.
17. D.L. Swofford. *PAUP*. Phylogenetic Analysis Using Parsimony (*and Other Methods). Version 4.* Sinauer Associates, Sunderland, Massachusetts, 2002.
18. H. Todd Wareham. An efficient algorithm for computing M_l consensus trees, 1985. BS honors thesis, CS, Memorial University Newfoundland.

Bayesian Phylogenetic Inference under a Statistical Insertion-Deletion Model

Gerton Lunter[1], István Miklós[1], Alexei Drummond[1], Jens Ledet Jensen[2], and Jotun Hein[1]

[1] Department of Statistics, University of Oxford,
1 South Parks Road, Oxford,
OX1 3TG, United Kingdom
{lunter,miklos,drummond,hein}@stats.ox.ac.uk

[2] Department of Mathematical Sciences, University of Aarhus
Ny Munkegade Building 530,
DK-8000 Aarhus C, Denmark
jlj@imf.au.dk

Abstract. A central problem in computational biology is the inference of phylogeny given a set of DNA or protein sequences. Currently, this problem is tackled stepwise, with phylogenetic reconstruction dependent on an initial multiple sequence alignment step. However these two steps are fundamentally interdependent. Whether the main interest is in sequence alignment or phylogeny, a major goal of computational biology is the co-estimation of both. Here we present a first step towards this goal by developing an extension of the Felsenstein peeling algorithm. Given an alignment, our extension analytically integrates out both substitution and insertion–deletion events within a proper statistical model. This new algorithm provides a solution to two important problems in computational biology. Firstly, indel events become informative for phylogenetic reconstruction, and secondly phylogenetic uncertainty can be included in the estimation of insertion-deletion parameters. We illustrate the practicality of this algorithm within a Bayesian Markov chain Monte Carlo framework by demonstrating it on a non-trivial analysis of a multiple alignment of ten globin protein sequences.

Supplementary material: www.stats.ox.ac.uk/~miklos/wabi2003/supp.html

1 Introduction

A fundamental problem in computational biology is the inference of phylogeny given a set of DNA or protein sequences. Traditionally, the problem is split into two sub-problems, namely multiple alignment of the sequences, and inference of a phylogeny based on an alignment. Several methods that deal with one or both of these sub-problems have been developed. ClustalW and T-Coffee are popular sequence alignment packages, while MrBayes [13], PAUP* [25] and Phylip [6] all provide phylogenetic reconstruction.

G. Benson and R. Page (Eds.): WABI 2003, LNBI 2812, pp. 228–244, 2003.

Although these methods can work very well, they share two fundamental problems. First, the division of the phylogenetic inference problem into multiple sequence alignment and alignment-based phylogenetic reconstruction is flawed. For instance, ClustalW computes its alignment based on a 'guide tree', the choice of which will bias any tree inference that is based on the resulting alignment. The solutions of the two sub-problems are interdependent, and ideally phylogenies and alignments should be co-estimated.

The second issue is that heuristic methods are used to deal with insertions and deletions (indels), and sometimes also substitutions. This lack of a proper statistical framework makes it impossible to accurately assess the reliability of the estimated phylogeny. Much biological knowledge and intuition goes into judging the outcomes of these algorithms.

The relevance of statistical approaches to evolutionary inference has long been recognised. Time-continuous Markov models for substitution processes were introduced more than three decades ago [15], and have been considerably improved since then [27]. The first paper on the evolutionary modelling of indel events appeared in the early nineties [26], giving a statistical approach to pairwise sequence alignment, and its extension to an arbitrary numbers of sequences related by a tree has recently been intensively investigated [23, 9, 12, 10, 19, 18]. Such methods are often computationally demanding, and full maximum likelihood approaches are limited to small trees. Markov chain Monte Carlo techniques can extend these methods to practical-sized problems.

While statistical modelling has only recently been used for multiple sequence alignment, it has a long history in population genetic analysis. In particular, coalescent approaches to genealogical inference have been very successful, both in maximum likelihood [16, 7] and Bayesian MCMC frameworks [28, 1]. The MCMC approach is especially promising, as it allows for large data sets to be tackled, as well as allowing for nontrivial extensions of the basic coalescent model, e.g. [21]. Over the short evolutionary time spans considered in population genetics, sequence alignment is generally straightforward, and genealogical inference from a fixed alignment is well-understood [5, 7, 24, 22]. On the other hand, for more divergent sequences, these approaches have difficulty dealing with indels. Not only is the alignment treated as known, but indel events are generally treated as missing data. Treating gaps as unobserved residues [4] renders them phylogenetically uninformative. However, indel events can be highly informative of the phylogeny, because of their relative rarity compared to substitution events.

In this paper, we present an efficient algorithm for computing the likelihood of a multiple sequence alignment given a tree relating the sequences, under the TKF91 model. This model combines probabilistic evolutionary models for substitution events and indel events, allowing consistent treatment of both in a statistical inference framework. The crux of the method is that all missing data (pertaining to the evolutionary history that generated the alignment) is summed out analytically. The algorithm can be seen as an extension of the celebrated peeling algorithm for substitutions [4] to include single residue indels. Summing out missing data eliminates the need for data augmentation of the tree, a methodology referred to in the MCMC literature as *Rao-Blackwellization* [17]. As a result, we can treat indels in a statistically consistent manner with no more than a constant cost over existing methods that ignore indels. Moreover we can utilise existing MCMC kernels for phylogenetic inference, changing only the likelihood calculator. We implemented the new likelihood algorithm in the programming language Java and demonstrated its practicality by interfacing with an existing MCMC kernel for phylogenetics and population genetics [1].

The method presented in this paper represents an important step towards the goal of a statistical inference method that co-estimates phylogeny and alignment. In fact, the only component currently missing is a method for sampling multiple sequence alignments under a fixed tree. Several approaches to sampling alignments employing data augmentation have already been investigated [12, 10]. Such methods may well hold the key to a co-estimation approach, and we are currently investigating the various possibilities.

The rest of the paper is organised as follows. In Section 2 we briefly introduce the TKF91 model of single residue insertion and deletion. Section 3 forms the core of the paper. Here we first derive a recursion (the 'one-state recursion') for the tree likelihood that, unlike in the usual Hidden Markov model formulation of the TKF91 model, does not require states for its computation. We then simplify the computation of transfer coefficients to a dynamic programming algorithm on the phylogenetic tree, similar to Felsenstein's peeling algorithm. Finally we prune the recursion so that only nonzero contributions remain, thereby yielding a linear time algorithm. In Section 4, we apply our method to a set of globin sequences, and estimate their phylogeny. Section 5 concludes with a discussion.

2 The TKF Model

The TKF91 model is a continuous time reversible Markov model for the evolution of nucleotide (or amino acid) sequences. It models three of the main processes in sequence evolution, namely *substitutions*, *insertions* and *deletions* of characters, approximating these as single-character processes. A sequence is represented by an alternating string of *characters* and *links*, connecting the characters, and this string both starts and terminates with a link. We adopt the view that insertions originate from links, and add a character-link pair to the *right* of the original link; deletions originate from characters and have the effect of removing the character and its right link. (This view is slightly different but equivalent to the original description, see [26].) In this way, subsequences evolve independently of each other, and the evolution of a sequence is the sum of the evolutions of individual character-link pairs. The leftmost link of the sequence has no corresponding character to its left, hence it is never deleted, and for this reason it is called the *immortal link*.

Since subsequences evolve independently, it is sufficient to describe the evolution of a single character-link pair. In a given time-span τ, this evolves into a sequence of characters of finite length. Since insertions originate from links, the first character of this sequence may be homologous to the original one, while

Fate:	Probability:	Label:
$C \to C\#^{n-1}$	$e^{-\mu\tau}(1-\lambda\beta(\tau))(\lambda\beta(\tau))^{n-1}$	$H_\tau B_\tau^{n-1}$
$C \to \#^n$	$(1-e^{-\mu\tau}-\mu\beta(\tau))(1-\lambda\beta(\tau))(\lambda\beta(\tau))^{n-1}$	$N_\tau B_\tau^{n-1}$
$C \to -$	$\mu\beta(\tau)$	E_τ
$\star \to \star\#^n$	$(1-\lambda\beta(\tau))(\lambda\beta(\tau))^n$	$(1-B_\tau)B_\tau^n$

Table 1. Possible fates after time τ of a single character (denoted C), and of the immortal link (denoted $\star$), and associated probabilities. The first three lines refer to (1) the ancestral character surviving (with 0 or more newly inserted), (2) the ancestral character dying after giving birth to at least one newly inserted one, and (3) the death of the ancestral character and all of its descendants.

subsequent ones will be inserted characters and therefore non-homologous. Table 1 summarises the corresponding probabilities. On the right-hand side of the arrow in the column labelled "Fate", C denotes a character homologous to the original character, whereas #'s denote non-homologous characters. The immortal link is denoted by $\star$ and other links are suppressed. All final arrangements can be thought of as being built from five basic "processes" which we call Birth, Extinction, Homologous, New (or Non-homologous) and Initial (or Immortal). These processes are labelled by their initials, and each corresponds to a specific probability factor as follows:

$$\begin{aligned} B_\tau &= \lambda\beta(\tau) & E_\tau &= \mu\beta(\tau) \\ H_\tau &= e^{-\mu\tau}(1-\lambda\beta(\tau)) & N_\tau &= (1-e^{-\mu\tau}-\mu\beta(\tau))(1-\lambda\beta(\tau)) \end{aligned} \quad (1)$$

where parameters λ and μ are the birth rate per link and the death rate per character, respectively, and in order to have a finite equilibrium sequence length, we require $\lambda < \mu$. We followed [26] in using the abbreviation

$$\beta(\tau) := \frac{1-e^{(\lambda-\mu)\tau}}{\mu-\lambda e^{(\lambda-\mu)\tau}}. \quad (2)$$

In a tree, time flows forward from the root to the leaves, and to each node of the tree we associate a time parameter τ which is set equal to the length of the incoming branch. For the root, $\tau = \infty$ by assumption of stationarity at the root, and the resulting equilibrium length distribution of the immortal link sequence is geometric with parameter $B_\infty = \lambda/\mu$ (where length 0 is possible); other links will have left no descendants since $H_\infty = N_\infty = 0$.

Because the TKF91 model is time reversible, the root placement does not influence the likelihood (Felsenstein's "Pulley Principle", [4]). Although the algorithms in this paper do not *look* invariant under root placement at all, in fact they are. This follows from the proofs, and we have used it to check the correctness of our implementations.

In the original TKF91 model, a simple substitution process known as the Felsenstein-81 model [4] was used. It is straightforward to replace this by more general models for substitutions of nucleotides or amino acids [11]. In the present paper, when a new non-homologous character appears at a node (as the result of a B or N process), it is always drawn from the equilibrium distribution; if a character at a node is homologous to the character at its immediate ancestral node, then the probability of this event is given by the chosen substitution model.

3 Computing the Likelihood of a Homology Structure

In the previous Monte Carlo statistical alignment papers, the sampled missing data were either unobserved sequences at internal nodes [14], or both internal sequences and alignments between nodes [12]. In both cases the underlying tree was fixed. Here we introduce the concept of *homology structure*, essentially an alignment of sequences at leaves, without reference to the internal tree structure. We present a new algorithm that allows us to compute, under the TKF91 model, the likelihood of observing a set of sequences and their homology structure, given a phylogeny and evolutionary parameters. Missing data in this case are all substitution events and indel events compatible with the observed data, all of which are analytically summed out in linear time. In contrast to previous MCMC approaches [12, 14], we need not store missing data at internal tree nodes, and we can change the tree topology without having to resample missing data. This enables us to consider the *tree* as a parameter, and efficiently sample from tree space.

3.1 Definitions and Statement of the One-State Recursion

Let $A_1, A_2, ...A_m$ be sequences, related to a tree T with vertex set V. Let a_i^j denote the jth character of sequence A_i, and let A_i^k denote its k long prefix.

A *homology structure* $\mathcal{H}$ on $A_1, \ldots, A_m$ is an equivalence relation $\sim$ on the set of all the characters of the sequences, $C = \{a_i^j\}$. It specifies which characters are homologous to which. The evolutionary indel process generating the homology structure on the sequences imposes constraints on the equivalence relations that may occur. More precisely, the equivalence relation $\sim$ has the property that a total ordering, $<_h$, exists on C such that

$$\begin{aligned} a_i^x =_h a_j^y &\iff a_i^x \sim a_j^y, \\ a_i^x <_h a_i^y &\iff x < y \end{aligned} \tag{3}$$

In particular, these imply that characters of a single sequence are nonhomologous. The ordering $<_h$ corresponds to the ordering of columns of homologous characters in an alignment. Note that for a given homology structure, this ordering may not be unique, see Figure 1. This many-to-one relationship of alignment to homology structure is the reason for introducing the concept of homology structure, instead of using the more common concept of alignment.

Below we give an algorithm for calculating the likelihood of the observed data, namely, the sequences with their homology structure. By definition, this

Fig. 1. (Left:) Two alignments representing the same homology structure: residues are homologous if they appear in the same column. These alignments may represent different evolutionary histories, all of which we include in our recursion. Note that this ambiguity is rare in biological sequence alignments. (Right:) Due to the evolutionary process acting on the sequences, homology relationships (arrows) will never 'cross' as depicted. This restriction on the equivalence relation $\sim$ is codified by $<_h$ (see text).

likelihood is the sum of the likelihoods of all evolutionary scenarios resulting in the observed data. In previous works ([12, 14]), it was shown that an evolutionary scenario can be described as a path in a multiple-HMM, so that the likelihood of a homology structure can be calculated using a multiple-HMM. However, this straightforward calculation is infeasible for practical-sized biological problems, since the number of states in the HMM grows exponentially with the number of sequences [18].

In this subsection we show that a so-called *one-state* recursion exists for calculating this likelihood, given evolutionary parameters. In subsequent sections we give an efficient algorithm based on this recursion.

Definitions Let t_r denote the subtree whose root is $r \in V$. Let Ω be the (nucleotide or amino acid) alphabet. An *event* e is a labelling of the nodes of a subtree t_r with $\mathtt{B}_\alpha$, $\mathtt{H}_\alpha$, $\mathtt{N}_\alpha$, $\mathtt{E}$, where $\alpha \in \Omega$ is a character, subject to the following conditions: There is a birth of a character α ($\mathtt{B}_\alpha$) at the root $r = r_e$ of the event, and only there, and if a node is labelled $\mathtt{E}$ then all its descendants are labelled $\mathtt{E}$ as well [18], see Figure 2. In this way, an event codifies the fate (see Table 1) of a single nucleotide born at $r \in T$, but ignoring all subsequent births it may have spawned. We let $e(n)$ denote the symbol labelling the node n. We say that an event e *emits a character* α at a leaf n if $e(n) = \mathtt{H}_\alpha$, $\mathtt{N}_\alpha$ or $\mathtt{B}_\alpha$. The *emission vector* v_e is an m-dimensional vector, and its ith coordinate is 1 if e emits a character into the ith leaf, and 0 otherwise. The *probability factor* $p(e)$ of an event e is the product of probabilities associated to the label of each node (see Table 1), including nucleotide equilibrium probabilities (for $\mathtt{B}_\alpha, \mathtt{N}_\alpha$) or substitution probabilities (for $\mathtt{H}_\alpha$), except that an $\mathtt{E}$ label counts for 1 if its parent is also labelled $\mathtt{E}$. We call these probability *factors* as they do not add up to 1 in an obvious way. However, if we calculate probabilities of observing sequences at leaves using these probability factors, the obtained probabilities do add to 1 (summation is over all observable sequences).

We end this section with a description of which events are allowed given a homology structure:

Definition 1 (Agreement with homology). *An event* e agrees with the homology structure *if the following holds:*

Fig. 2. Three possible events. The first two events emit two characters; in e they are homologous (and at least one mutation event occurred, changing an A into a C), while in e' they are inhomologous by the the N ("new") label; see also Section 2.

1. *If e emits a character $a \in C$, then it emits all $a' \in C$ with $a \sim a'$, and the nodes along the path connecting a and a' are not labelled $\mathtt{N}_\alpha$.*
2. *If e emits characters a and a', and $a \not\sim a'$, then at least one of the nodes along the path connecting them is labelled $\mathtt{N}_\alpha$.*

Statement of the One-State Recursion Let $P(\mathbf{K})$ denote the probability of emitting the prefixes $A^{\mathbf{K}} := (A_1^{K_1}, ..., A_m^{K_m})$ and such that their homology agrees with the given homology structure $\mathcal{H}$. The following equation holds:

$$P(\mathbf{K}) = \left(\prod_{n \in T} (1 - B_n) \right) \sum_{(e_1,\ldots,e_n) \in \mathcal{A}} p(e_1) \cdots p(e_n), \tag{4}$$

where $\mathcal{A}$ is the set of sequences of legal events (see App. A for a definition) that emit $A^{\mathbf{K}}$ and agree with the homology structure, and the factor in front derives from the immortal link. Note that for brevity we wrote B_n instead of $B_{l(n)}$, where $l(n)$ is the length of the incoming branch. Our goal is to find a recursion in terms of $P(\mathbf{K})$. To formulate our result, we need one more definition:

Definition 2. *A set of events $\{e_1, \ldots, e_l\}$ is a* nested set *if, for $i \neq j$, we have that $r_{e_i} \in t_{e_j} \implies e_j(r_{e_i}) = \mathtt{E}$.*

Let $M_l^{\mathbf{K}}$ denote the set of nested sets of l events, where each event is in $\mathcal{E}(\mathbf{K})$. Here $\mathcal{E}(\mathbf{K})$ is the set of events e that emit characters that extend the prefixes to $A^{\mathbf{K}}$ (i.e. from $A^{\mathbf{K}-v_e}$), and that agree with the homology structure $\mathcal{H}$.

Theorem 1 (One-state recursion). *The following equation holds for $P(\mathbf{K})$:*

$$P(\mathbf{K}) = \sum_{l=1}^{2m} \sum_{\{e_1,e_2,\ldots e_l\} \in M_l^{\mathbf{K}}} (-1)^{l-1} P\left(\mathbf{K} - \textstyle\sum_{j=1}^{l} v_{e_j}\right) \prod_{i=1}^{l} p(e_i) \tag{5}$$

Proof: See appendix A ■

3.2 A Reverse Traversal Algorithm for the Transition Factor

Theorem 1 by itself does not give a fast algorithm for calculating the likelihood of a homology structure, as the number of terms in the summation (5) grows exponentially with the number of leaves m. As a first step to arrive at an efficient algorithm, we group terms in (5) as follows:

$$P(\mathbf{K}) = \sum_{v \in \{0,1\}^m} T_v^{\mathbf{K}} P(\mathbf{K} - v) \tag{6}$$

Here v runs over all 2^m length-m vectors with entries 0 or 1 (in section 3.3 we reduce the v-summation). The *transition factors* $T_v^{\mathbf{K}}$ are sums of expressions of the form $(-1)^{l-1} \prod_{i=1}^{l} p(e_i)$. Previously we showed how to calculate these transition factors in linear time when there is no homology structure to be observed [18]. In this section we describe a similar algorithm for the present case.

Note that the zero vector is among the vectors summed over in (6), so that $P(\mathbf{K})$ also appears on the right-hand side. Solving the resulting linear equation results in a proper recursion, and conceptually amounts to summing out all non-emitting events. The relation between this approach and existing ones, see e.g. [3, 23], is discussed in more detail in [18].

To derive the dynamic programming recursion, the main observation is that a nested set of events $\{e_1, \ldots, e_l\}$ can be represented by a labeling of a single tree, since for every node n there is at most one e_i with $e_i(n) \neq \mathtt{E}$. By labeling each node n with the unique non-E label among the $e_i(n)$ (or with $\mathtt{E}$ if none exists), the roots of the events can be recovered as they are precisely the nodes labeled B_α. The term $(-1)^{l-1} p(e_1) \cdots p(e_l)$ that corresponds to a nested set represented in this way, is calculated by multiplying the probabilities for the labels at all nodes according to (1) with the following two caveats: (a) a node labeled $\mathtt{E}$ carries a factor 1 if its parent is also labeled $\mathtt{E}$; (b) a B_α attracts a minus sign to account for the factor $(-1)^{l-1}$, and an additional factor E if its parent is not labeled $\mathtt{E}$. Finally, summing over all possible nested sets is done in linear time with a dynamic programming or 'pruning' algorithm similar to Felsenstein's one [4].

To take the homology structure into account, we need to restrict the tree labelings to those that produce emissions compatible with the given homology, amounting to implementing Definition 1. Input to the algorithm is the total emission vector v, and a numerical vector h that encodes the homology structure of v such that $h_i = h_j \neq 0 \iff a_i^{\mathbf{K}_i} \sim a_j^{\mathbf{K}_j}$, and h_i is zero whenever $v_i = 0$. For example, for $\{e_1, \ldots, e_l\} \in M_l^{\mathbf{K}}$ we have $v = \sum_i v_{e_i}$ and we could set $h = \sum_i i v_{e_i}$. Those leaves with $h_i = h_j$ are said to belong to the same *homology class*.

The algorithm proceeds as follows. First we compute for each homology class the minimum spanning tree of its leaves. If two such spanning trees intersect, no nested sets corresponding to v and satisfying the homology contraints exist, and $T_v^{\mathbf{K}} = 0$. Otherwise, for each node n we set $h(n)$ to be the label of the homology class whose tree contains n, and 0 otherwise:

Algorithm 1 (Computing homology spanning trees)
Input: *Homology vector* $h = (h_1, \ldots, h_m)$, *tree* T.
Output: *Either* "$T_v^{\mathbf{K}} = 0$", *or homology class* $h(n)$ *for each node* n *in* T.
Algorithm:

```
h(n) ← h_i for the index i corresponding to leaf n; 0 for non-leaf nodes.
c(n) ← 1 if h(n) ≠ 0; 0 otherwise.
m(j) ← #{i|h_i = j}      (multiplicity of homology class j)
For all nodes n in postorder traversal (i.e. leaves first):
  If c(n) ≠ m(h(n)), then:
    If h(a(n)) = 0 or h(a(n)) = h(n), then:
      h(a(n)) ← h(n)
      c(a(n)) ← c(a(n)) + c(n)
    Else:
      Return "T_v^K = 0"
    EndIf
  EndIf
EndFor
Return h(·)
```

In this algorithm, $a(n)$ denotes the ancestor of node n. The symbol $\leftarrow$ ('becomes') is the assignment operator.

Our recursion is in terms of quantities $F_H(\alpha, n)$, $F_N(\alpha, n)$ and $F_E(n)$, which are related to (1) character α at node n being homologous to at least one character at the leaves of t_n in the case of F_H; (2) that character being non-homologous to all characters at the leaves of t_n in the case of F_N; (3) no character existing at node n in the case of F_E.

We introduce some notation. For a node n, n_l and n_r denote the left and right descendant, respectively. We abbreviate $\delta_{n,m} := \delta_{h(n),h(m)}$, where the second δ is the usual Kronecker delta, i.e. $\delta_{n,m} = 1$ if $h(n) = h(m)$, 0 otherwise. Let $p_n(\alpha \to \gamma)$ denote the probability that character α evolves into γ in time l_n, which is the length of the incoming branch to node n, and let $\pi(\alpha)$ denote the equilibrium distribution of characters. We finally introduce the following abbreviations, where we again write H_n, N_n, B_n, E_n for $H_{l(n)}, N_{l(n)}, B_{l(n)}, E_{l(n)}$ resp., where $l(n)$ is the length of node n's incoming branch:

$$H(n, \alpha) = \sum_{\gamma \in \Omega} F_H(\gamma, n) H_n p_n(\alpha \to \gamma), \tag{7}$$

$$N(n, \alpha) = F_E(n) E_n + \sum_{\gamma \in \Omega} F_N(\gamma, n) H_n p_n(\alpha \to \gamma) + \sum_{\gamma \in \Omega} \left[F_H(\gamma, n) + F_N(\gamma, n)\right] \left[N_n - E_n B_n\right] \pi(\gamma), \tag{8}$$

$$E(n) = F_E(n) - \sum_{\gamma \in \Omega} \left[F_H(\gamma, n) + F_N(\gamma, n)\right] B_n \pi(\gamma). \tag{9}$$

Algorithm 2 (Computing $T_v^{\mathbf{K}}$) *With the notation as above, and $p(n)$ computed by Algorithm 1, the transition factor in (6) associated to the sequences $A_1, \ldots, A_m$ related by a tree T and homology structure $\mathcal{H}$, is $T_v^{\mathbf{K}} = -E(r)$, where r is the root of T. The terms $F_E(r)$, $F_H(\gamma, r)$ and $F_N(\gamma, r)$ are computed recursively as follows. If $h(n) = 0$ and n is an internal node, then*

$$F_H(\alpha, n) = H(n_l, \alpha)N(n_r, \alpha) + N(n_l, \alpha)H(n_r, \alpha), \tag{10}$$
$$F_N(\alpha, n) = N(n_l, \alpha)N(n_r, \alpha), \tag{11}$$
$$F_E(n) = E(n_l)E(n_r). \tag{12}$$

If $h(n) \neq 0$ and n is an internal node, then

$$F_H(\alpha, n) = [H(n_l, \alpha)\delta_{n,n_l} + N(n_l, \alpha)(1 - \delta_{n,n_l})] \times [H(n_r, \alpha)\delta_{n,n_r} + N(n_r, \alpha)(1 - \delta_{n,n_r})], \tag{13}$$
$$F_N(\alpha, n) = F_E(n) = 0. \tag{14}$$

If n is a leaf node, then

$$F_H(\alpha, n) = 1 \quad \text{if } a_n^{K_n} = \alpha, \quad 0 \text{ otherwise}; \tag{15}$$
$$F_N(\alpha, n) = 0; \tag{16}$$
$$F_E(n) = 1 \quad \text{if } v_n = 0, \quad 0 \text{ otherwise}. \tag{17}$$

Note that we abused notation by confusing (leaf) nodes and sequence indices.

3.3 Finding the Prefix Vectors K

For many vectors $\mathbf{K}$, the quantity $P(\mathbf{K})$ will vanish, since the corresponding sequence prefixes $A^{\mathbf{K}}$ cannot occur while at the same time agreeing with the homology structure. Secondly, for many vectors v the transition factor $T_v^{\mathbf{K}}$ will similarly vanish, for the same reason. Restricting to those v and $\mathbf{K}$ that actually contribute dramatically increases the efficiency of the algorithm.

All this depends on the homology structure; the various paths through $\mathbf{K}$-space that the algorithm traverses correspond to the various possible orderings $\leq_h$ that correspond to the underlying homology structure. In fact, if none of the characters are homologous to any other (corresponding to an alignment with only single-character columns), all of the $\mathbf{K}$-values are valid, and all of the vectors v have to be considered. In practice, however, such alignments will have very low likelihood and can safely be ignored.

As the final part of our algorithm, we give the top-level subroutine that traverses the set of $\mathbf{K}$-vectors for which $P(\mathbf{K}) \neq 0$. Input to the algorithm is a homology structure, which is represented by an alignment, or more precisely, as a sequence of vectors $(c_1, c_2, \ldots, c_n)$, where the jth coordinate of c_i is 1 precisely if the jth character in the ith column in the alignment is a residue, and 0 if it is a gap. The algorithm is independent of the alignment chosen to represent the homology structure.

Algorithm 3 (Traversing contributing prefix vectors)
Input: *Vectors* $(c_1,\ldots,c_n)$*; sequences* $A_1,\ldots,A_m$*; tree* T*; cutoff* N
Output: *Likelihood of sequences under given homology structure.*
Algorithm:

Mark all c_i *'free';* $\mathbf{K}\leftarrow(0,\ldots,0)$; $P(\mathbf{K})\leftarrow\prod_{n\in T}(1-B_n)$
`While` *not all* c_i *marked 'used':*
  Let s_i *be the maximal, increasing subsequence satisfying:* c_{s_i} *'free'* $\forall i$
  Mark c_{s_k} *'possible' iff* $c_{s_k}\cdot\sum_{i=1}^{k-1}c_{s_i}=0$
  `If` *there are no more than* N *'possible' vectors:*
    `For` *all subsets* $C=\{c'_1,\ldots,c'_n\}$ *of 'possible' vectors:*
      Construct $v=\sum_{k=1}^{n}c'_k$ *and* $h=\sum_{k=1}^{n}kc'_k$*; compute* $T_v^{\mathbf{K}+v}$
      $P(\mathbf{K}+v)\leftarrow P(\mathbf{K}+v)+T_v^{\mathbf{K}+v}(1-T_{\mathbf{0}}^{\mathbf{0}})^{-1}P(\mathbf{K})$
    `EndFor`
  `Else:`
    `Return` *"Likelihood too small"*
  `EndIf`
  $k\leftarrow\max\{i|c_i$ *labeled 'possible'*$\}\cup\{\infty\}$
  Mark c_k *'used';* $\mathbf{K}\leftarrow\mathbf{K}+c_k$
  `For` *all* $i>k$ *for which* c_i *is labeled 'used':*
    Mark c_i *'free';* $\mathbf{K}\leftarrow\mathbf{K}-c_i$
  `EndFor`
`EndWhile`
`Return` *"Likelihood=*$P(\mathbf{K})$*"*

In practice, for biologically meaningful alignments and with the cutoff N in place, the algorithm is linear, although it has slower worst-case behaviour.

4 Results

The indel peeling algorithm of Section 3.3 provides a method for calculating the likelihood $L=\Pr\{A,\mathcal{H}|T,Q,\lambda,\mu\}$ of observing the sequences with their homology structure ('alignment') given the tree and model parameters. Here A are the amino acid sequences, $\mathcal{H}$ is their homology structure, T is the tree including branch lengths, Q is the substitution rate matrix, and λ,μ are the amino acid birth and death rates. To demonstrate the practicality of the new algorithm for likelihood calculation we undertook a Bayesian MCMC analysis of ten globin protein sequences. We chose to use the standard Dayhoff rate matrix to describe the substitution of amino acids. For the purpose of this example we generated a homology structure using T-Coffee. Given this homology structure, we co-estimated the parameters of the TKF91 model, and the tree topology and branch lengths. To do this we sampled from the posterior,

$$h(\mu,T)=\frac{1}{Z}\Pr\{A,\mathcal{H}|T,Q,\lambda,\mu\}f(T,\lambda,\mu). \tag{18}$$

Here Z is the unknown normalising constant. We chose the prior distribution on our parameters, $f(T,\lambda,\mu)$, so that T was constrained to a molecular clock, and

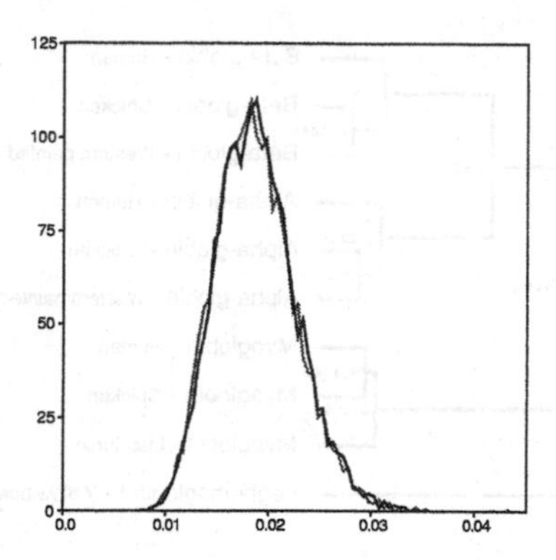

Fig. 3. Estimated posterior densities of the death rate μ sampled according to h (see text), for two independent runs, showing good convergence. Sampled mean is 0.0187; the 95% highest posterior density (HPD) interval was estimated to be $(0.0114, 0.0265)$.

$\lambda = \mu L/(L+1)$ to make the expected sequence length under the TKF91 model agree with the observed lengths. Here L is the geometric average length of the globin sequences, We assume a molecular clock to gain insight into the relative divergence times of the alpha-, beta- and myoglobin families. In doing so we incorporate insertion-deletion events as informative events in the evolutionary analysis of the globin family.

The posterior density h is a complicated function defined on a space of high dimension. We summarise the information it contains by computing the expectations, over h, of various statistics of interest. We estimate these expectations by using MCMC to sample from h. Figure 3 depicts the marginal posterior density of the μ parameter for two independent MCMC runs, showing convergence. Figure 4 depicts the maximum *a posteriori* (MAP) estimate of the phylogenetic relationships of the sequences. This example exhibits only limited uncertainty in the tree topology, however we observed an increased uncertainty for trees that included divergent sequences, such as bacterial and insect globins (results not shown).

The estimated time of the most recent common ancestor of each of the alpha, beta and myoglobin families are all mutually compatible (result not shown), suggesting that the molecular clock hypothesis is at least approximately valid. Analysis of a four sequence dataset demonstrate consistency in μ estimates between MCMC and previous ML analyses [18] (data not shown). Interestingly, the current larger dataset supports a lower value of μ. This is probably due to the fact that no indels are apparent within any of the subfamilies despite a considerable sequence divergence.

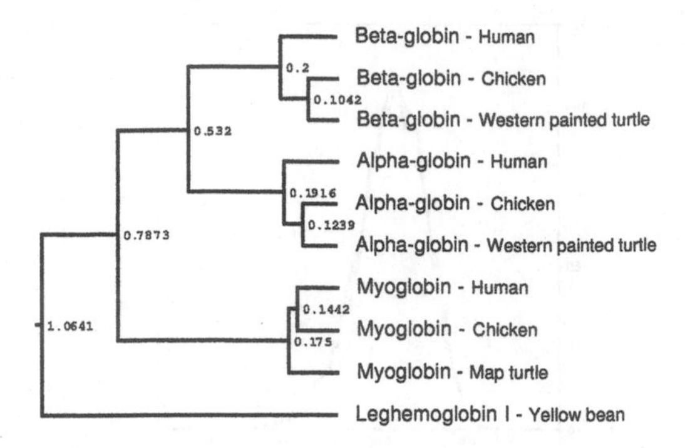

Fig. 4. The maximum *a posteriori* (MAP) estimate of the globin tree. The node heights are given in expected substitutions per site. Notice that the alpha and beta chain sub-families both support the traditional ordering of birds, turtles and mammals, while the three myoglobin sequences support an unconventional phylogeny. This inconsistent signal from myoglobin has been previously observed [8]. The marginal posterior probability (estimated from the MCMC chain) for the monophyly of human and chicken myoglobin was 83.1%, followed by the conventional grouping of turtle and chicken at 11.9%. The third topological arrangement of myoglobin occurred the remaining 5% of the time, suggesting significant homoplasy in this sub-family.

5 Discussion

In this paper we presented a method that extends Felsenstein's peeling algorithm to incorporate insertion and deletion events, under the TKF91 model. This renders indel events informative for phylogenetic inference. Although this incurs considerable algorithmic complications, the resulting algorithm is still linear-time for biological alignments (see also Figure 1).

It should be stressed that the two MCMC analyses of the globin data set were purely illustrative of the practicality of the algorithm described, and no novel biological results were obtained. The two MCMC runs undertaken required only about 3 hours of CPU time each on a 1.25 GHz G4 Apple Macintosh, using an unoptimised implementation of the algorithm, and the estimated number of independent samples (estimated sample size, ESS) obtained for the posterior probability were good at 650 and 820 respectively (see [1] for methods). The estimated ESSs for the death rate, μ, were 4800 and 3900 respectively. We expect analyses of data sets of around 50 sequences to be readily attainable with only a few days computation.

Our method is intended as the first step towards a full co-sampling approach of phylogeny and alignment. The only remaining issue is to combine the current method with a sampling strategy for alignments. Proper sampling algorithms have already been developed [12, 10], varying mainly in the way data augmentation is employed. These or similar methods of data augmentation may prove helpful in certain stages of an MCMC kernel that solves the full problem of phylogeny/alignment co-estimation. However, in the context of the algorithm described herein it may be possible to avoid data augmentation entirely and still achieve efficient co-estimation of phylogeny and alignment through a simple Metropolis-Hastings proposal scheme that perturbs homology structures directly.

As was mentioned in [26], it would be desirable to have a statistical sequence evolution model that deals with 'long' insertions and deletions, which is the statistical counterpart of affine gap penalties in score-based alignment methods. We have made progress on a full likelihood method for statistical sequence alignment under such an evolutionary model [20], but this method seems not to be directly extendable to trees. We believe that here too, Markov chain Monte Carlo approaches, combined with data augmentation, will be the key to practical algorithms.

Acknowledgements

This research is supported by EPSRC (code HAMJW) and MRC (code HAMKA).

A Proof of Theorem 1

Let $\leq_p$ be the partial ordering on V for which $n_1 \leq_p n_2$ iff $n_1 \in t_{n_2}$. We denote $n_1 <>_p n_2$ if n_1 and n_2 are incomparable, and we denote $n_1 \not\leq_p n_2$ if either $n_1 >_p n_2$ or $n_1 <>_p n_2$. Also we introduce a total ordering $\leq$ of the nodes which is an arbitrary refinement of the partial ordering, namely, $n_1 \leq n_2$ implies that $n_1 \leq_p n_2$ or $n_1 <>_p n_2$.

A *state* S is a function $V \to \{0,1\}$, with the property that the root of the tree is labelled with 1, and $S(n_1) = 0 \implies S(n_2) = 0$ for all $n_2 <_p n_1$. The state is used to keep track of where subsequent births may occur (i.e. everywhere except after extinctions E), and to avoid over-counting of independent histories in disjoint subtrees. The *initial state* is the state labelling all nodes with 1. The *action* of an event e on a state S' is defined to be $S = S' * e$, with

$$S(n) = \begin{cases} S'(n) & \text{if } n > r_e \\ 0 & \text{if } n < r_e \text{ and } n <>_p r_e \\ 1 & \text{if } n \leq_p r_e \text{ and } e(n) \neq \mathtt{E} \\ 0 & \text{if } n \leq_p r_e \text{ and } e(n) = \mathtt{E} \end{cases} \tag{19}$$

The first two lines make sure that histories in disjoint subtrees are not counted twice, by disallowing new births in disjoint subtrees at one 'side' of the current event. The last two lines implement the restriction that new births are not allowed after extinction (E) events. If an event e occurs in state S, the state becomes $S*e$ after the event. An event e is a *legal event* in a state S iff $S(r_e) = 1$.

Let $P_S(\mathbf{K})$ denote the probability of emitting the prefixes $A^{\mathbf{K}}$ by legal events agreeing with the homology structure, starting from the initial state, such that the state after the last event is S. A recursion in terms of $P_S(\mathbf{K})$ is easy:

$$P_S(\mathbf{K}) = \sum_{S'} \sum_{\substack{e \in \mathcal{E}(\mathbf{K}): \\ S' * e = S \wedge S'(e)=1}} P_{S'}(\mathbf{K} - v_e) p(e) \tag{20}$$

Here $\mathcal{E}(\mathbf{K})$ is the set of events e that emit characters extending the prefixes from $A^{\mathbf{K}-v_e}$ to $A^{\mathbf{K}}$, and that agree with the homology structure, and as initial condition we have $P_I(\mathbf{0}) = \prod_{n \in T}(1 - B_n)$, with I the initial state (assigning 1 to each node). The correctness of the recursion is discussed in detail in [18]. In the end, the quantity of interest $P(\mathbf{K})$ is obtained by summing over all states:

$$P(\mathbf{K}) = \sum_S P_S(\mathbf{K}) \tag{21}$$

This approach to calculate the likelihood of a set of sequences is very similar to the forward-backward algorithm of HMMs [2], and it is the straightforward extension of the original formulation of the dynamic programming algorithm given in [26]. We now combine (20) and (21) to get:

$$P(\mathbf{K}) = \sum_{\{e_1\} \in M_1^{\mathbf{K}}} \sum_{S: S(r_{e_1})=1} P_S(\mathbf{K} - v_{e_1}) p(e_1) \tag{22}$$

We can rewrite this as

$$P(\mathbf{K}) = \sum_{\{e_1\} \in M_1^{\mathbf{K}}} P(\mathbf{K} - v_{e_1}) p(e_1) - \sum_{\{e_1\} \in M_1^{\mathbf{K}}} \sum_{S: S(r_{e_1})=0} P_S(\mathbf{K} - v_{e_1}) p(e_1) \tag{23}$$

The theorem can be derived from (23) using the following lemma recursively:

$$\begin{aligned}
&\sum_{\{e_1, e_2, \ldots e_l\} \in M_l^{\mathbf{K}}} \sum_{S: S(r_{e_i})=0 \forall i} P_S\left(\mathbf{K} - \textstyle\sum_{i=1}^{l} v_{e_i}\right) \prod_{i=1}^{l} p(e_i) = \\
&\quad = \sum_{\{e_1, e_2, \ldots e_{l+1}\} \in M_{l+1}^{\mathbf{K}}} P\left(\mathbf{K} - \textstyle\sum_{i=1}^{l+1} v_{e_i}\right) \prod_{i=1}^{l+1} p(e_i) - \\
&\quad - \sum_{\{e_1, e_2, \ldots, e_{l+1}\} \in M_{l+1}^{\mathbf{K}}} \sum_{S: S(r_{e_i})=0 \forall i} P_S\left(\mathbf{K} - \textstyle\sum_{i=1}^{l+1} v_{e_i}\right) \prod_{i=1}^{l+1} p(e_i)
\end{aligned} \tag{24}$$

Before we prove (24), we show that it indeed leads to the proof of Theorem 1. When we apply the recursion (24) to (23) $k-1$ times, we end up with the equation

$$\begin{aligned}
P(\mathbf{K}) = &\sum_{l=1}^{k} \sum_{\{e_1, e_2, \ldots, e_l\} \in M_l^{\mathbf{K}}} (-1)^{l-1} P\left(\mathbf{K} - \textstyle\sum_{i=1}^{l} v_{e_i}\right) \prod_{i=1}^{l} p(e_i) \\
&+ (-1)^k \sum_{\{e_1, e_2, \ldots, e_k\} \in M_k^{\mathbf{K}}} \sum_{S: S(r_{e_i})=0 \forall i} P_S\left(\mathbf{K} - \textstyle\sum_{i=1}^{k} v_{e_i}\right) \prod_{i=1}^{k} p(e_i)
\end{aligned} \tag{25}$$

However, $M_k^{\mathbf{K}}$ eventually becomes empty, since no two events in a nested set share a root, and there are $2m-1$ nodes in the tree (recall that m is the number of leaves). This implies the theorem.

To prove the lemma, we first apply (20) to the left-hand side of (24) to get

$$\sum_{\substack{\{e_1,e_2,\dots e_l\}\\ \in M_l^{\mathbf{K}}}} \sum_{\substack{S:\\ S(r_{e_i})=0\forall i}} \sum_{S'} \sum_{\substack{e\in\mathcal{E}(\mathbf{K}-\sum_{i=1}^l v_{e_i}):\\ S'*e=S \wedge S'(r_e)=1}} P_{S'}\Big(\mathbf{K}-v_e-\sum_{i=1}^{l} v_{e_i}\Big)p(e)\prod_{i=1}^{l}p(e_i) \quad (26)$$

The main observation is that for the events $\{e_1,\dots,e_l\}$ and e over which the summation extends, we have that $\{e_1,\dots,e_l,e\}\in M_{l+1}^{\mathbf{K}}$, and $r_e \not\leq_p r_{e_i}$ for all i. To show the latter, suppose that $r_e \leq_p r_{e_i}$ for a particular i, then since $S(r_{e_i})=0$ we have $S(r_e)=0$, contradicting $S=S'*e$, therefore r_e must be a greatest element in the partial ordering. To show the former, note that from the action of e on S' it follows that $e(r_{e_i})$ must be E for all $r_{e_i} <_p r_e$, which implies that $\{e_1,\dots,e_l,e\}$ is a nested set. The events in a nested set have non-overlapping emissions, so that $e\in\mathcal{E}(\mathbf{K}-\sum_{i=1}^l v_{e_i})$ implies that $e\in\mathcal{E}(\mathbf{K})$. From this it follows that $\{e_1,\dots,e_l,e\}\in M_{l+1}^{\mathbf{K}}$.

Since the summand of (26) only involves S' and the events, the above observation implies that we can simplify (26) to a sum of the expression

$$P_{S'}\Big(\mathbf{K}-\textstyle\sum_{i=1}^{l+1} v_{e_i}\Big)\displaystyle\prod_{i=1}^{l+1}p(e_i) \quad (27)$$

over some $\{e_1,\dots,e_{l+1}\}\in M_{l+1}^{\mathbf{K}}$, and some states S'. We claim the sum actually extends over all nested sets in $M_{l+1}^{\mathbf{K}}$, and all states S', except those for which $S'(r_{e_i})=0$ for all i. That these should be excluded is clear as e in (26) satisfies $S'(e)=1$. Conversely, let S' be given and suppose $S'(e_i)=1$ for at least one i, then for e choose the event whose root is maximal in the total ordering among the r_{e_i} for which $S'(r_{e_i})=1$. This event is legal for S', and yields a state $S=S'*e$ for which $S(r_{e_i})=0$ for all remaining e_i; moreover it is the only event among the $\{e_1,\dots,e_{l+1}\}$ that has these two properties. This finishes the proof of the lemma, and of the theorem. ■

References

[1] A. J. Drummond, G. K. Nicholls, A. G. Rodrigo, and W. Solomon. Estimating mutation parameters, population history and genealogy simultaneously from temporally spaced sequence data. *Genetics*, 161(3):1307–1320, 2002.

[2] R. Durbin, S. Eddy, A. Krogh, and G. Mitchison. *Biological sequence analysis.* Cambridge University Press, 1998.

[3] S. Eddy. HMMER: Profile hidden Markov models for biological sequence analysis (http://hmmer.wustl.edu/), 2001.

[4] J. Felsenstein. Evolutionary trees from DNA sequences: a maximum likelihood approach. *J. Mol. Evol.*, 17:368–376, 1981.

[5] J. Felsenstein. Estimating effective population size from samples of sequences: Inefficiency of pairwise and segregating sites as compared to phylogenetic estimates. *Genetical Research Cambridge*, 59:139–147, 1992.

[6] J. Felsenstein. PHYLIP version 3.5c. Dept. of Genetics, Univ. of Washington, Seattle, 1993.
[7] R. C. Griffiths and S. Tavare. Ancestral inference in population genetics. *Statistical Science*, 9:307–319, 1994.
[8] S. B. Hedges and L. L. Poling. A molecular phylogeny of reptiles. *Science*, 283(5404):945–946, Feb 12 1999.
[9] J. Hein. An algorithm for statistical alignment of sequences related by a binary tree. In *Pac. Symp. Biocomp.*, pages 179–190. World Scientific, 2001.
[10] J. Hein, J. L. Jensen, and C. N. S. Pedersen. Recursions for statistical multiple alignment. Technical Report 425, Dept. of Theor. Stat., Univ. of Aarhus, January 2002.
[11] J. Hein, C. Wiuf, B. Knudsen, M. B. Møller, and G. Wibling. Statistical alignment: Computational properties, homology testing and goodness-of-fit. *J. Mol. Biol.*, 302:265–279, 2000.
[12] I. Holmes and W. J. Bruno. Evolutionary HMMs: a Bayesian approach to multiple alignment. *Bioinformatics*, 17(9):803–820, 2001.
[13] J. P. Huelsenbeck and F. Ronquist. MRBAYES: Bayesian inference of phylogenetic trees. *Bioinformatics*, 2001.
[14] J.L. Jensen and J. Hein. Gibbs sampler for statistical multiple alignment. Technical Report 429, Dept. of Theor. Stat., U. Aarhus, September 2002.
[15] T. H.. Jukes and C. R. Cantor. Evolution of protein molecules. In Munro, editor, *Mammalian Protein Metabolism*, pages 21–132. Acad. Press, 1969.
[16] M. K. Kuhner, J. Yamato, and J. Felsenstein. Estimating effective population size and mutation rate from sequence data using Metropolis-Hastings sampling. *Genetics*, 140(4):1421–1430, 1995.
[17] J. S. Liu. *Monte Carlo Strategies in Scientific Computing.* Springer, 2001.
[18] G. A. Lunter, I. Miklós, Y. S. Song, and J. Hein. An efficient algorithm for statistical multiple alignment on arbitrary phylogenetic trees. *J. Comp. Biol.*, 2003. In press.
[19] I. Miklós. An improved algorithm for statistical alignment of sequences related by a star tree. *Bul. Math. Biol.*, 64:771–779, 2002.
[20] I. Miklós, G. A. Lunter, and I. Holmes. A "long indel" model for evolutionary sequence alignment. In preparation.
[21] O. G. Pybus, A. J. Drummond, T. Nakano, B. H. Robertson, and A. Rambaut. The epidemiology and iatrogenic transmission of hepatitis c virus in Egypt: a Bayesian coalescent approach. *Mol Biol Evol*, 20(3):381–387, 2003.
[22] O. G. Pybus, A. Rambaut, and P. H. Harvey. An integrated framework for the inference of viral population history from reconstructed genealogies. *Genetics*, 155(3):1429–1437, 2000.
[23] M. Steel and J. Hein. Applying the Thorne-Kishino-Felsenstein model to sequence evolution on a star-shaped tree. *Appl. Math. Let.*, 14:679–684, 2001.
[24] M. Stephens and P. Donnelly. Inference in molecular population genetics. *J. of the Royal Stat. Soc. B*, 62:605–655, 2000.
[25] D. Swofford. Paup* 4.0. Sinauer Associates, 2001.
[26] J. L. Thorne, H. Kishino, and J. Felsenstein. An evolutionary model for maximum likelihood alignment of DNA sequences. *J. Mol. Evol.*, 33:114–124, 1991.
[27] S. Whelan, P. Lió, and N. Goldman. Molecular phylogenetics: state-of-the-art methods for looking into the past. *Trends in Gen.*, 17:262–272, 2001.
[28] I. J. Wilson and D. J. Balding. Genealogical inference from microsatellite data. *Genetics*, 1998.

Better Hill-Climbing Searches for Parsimony

Ganeshkumar Ganapathy, Vijaya Ramachandran*, and Tandy Warnow**

Department of Computer Sciences, University of Texas, Austin, TX 78712;
{gsgk, vlr, tandy}@cs.utexas.edu

Abstract. The reconstruction of evolutionary trees is a major problem in biology, and many evolutionary trees are estimated using heuristics for the NP-hard optimization problem Maximum Parsimony. The current heuristics for searching through tree space use a particular technique, called "tree-bisection and reconnection", or TBR, to transform one tree into another tree; other less-frequently used transformations, such as SPR and NNI, are special cases of TBR. In this paper, we describe a new tree-rearrangement operation which we call the *p*-ECR move, for *p*-Edge-Contract-and-Refine. Our results include an efficient algorithm for computing the best 2-ECR neighbors of a given tree, based upon a simple data structure which also allows us to efficiently calculate the best neighbors under NNI, SPR, and TBR operations (as well as efficiently running the greedy sequence addition technique for maximum parsimony). More significantly, we show that the 2-ECR neighborhood of a given tree is incomparable to the neighborhood defined by TBR, and properly contains all trees within two NNI moves. Hence, the use of the 2-ECR move, in conjunction with TBR and/or NNI moves, may be a more effective technique for exploring tree space than TBR alone.

1 Introduction

The Maximum Parsimony Problem, also called the Hamming Distance Steiner Tree Problem, is one of the main optimization problems in phylogenetic analysis. Because it is NP-hard [5], heuristics are used to analyze datasets. Most of the favored heuristics operate by hill-climbing through tree space where each move changes a tree using some specific transformation, and then scores the new tree, and the search terminates when no allowed move improves the score. Transformations that are used in standard hill-climbing procedures are NNI, SPR, and TBR, with NNI being a special case of SPR, and SPR being a special case of TBR; thus, TBR searches are the most exhaustive, and also the most preferred [15]. Even TBR searches, however, can get caught in local optima (that is, trees that have no neighbors under TBR moves which are better and yet are not globally optimal)

The main result in this paper is a mathematical analysis of a new transformation, which we call *p*-edge-contract-and-refine, or *p*-ECR. This transformation is similar to other techniques described in other papers [2,9], and has a similar motivation; what is new here is the mathematical analysis. We provide a fast algorithm for computing the

* Supported by NSF grant CCR-9988160
** Supported by NSF grant EIA 01-21680 and by a fellowship from the David and Lucile Packard Foundation

G. Benson and R. Page (Eds.): WABI 2003, LNBI 2812, pp. 245–258, 2003.

optimal 2-ECR neighbors of a given tree, and show that the number of 2-ECR neighbors that are also TBR neighbors is small, namely $O(n)$, where n is the number of leaves. In contrast, we show that the size of the 2-ECR neighborhood is itself $\Theta(n^2)$, and it has been shown that the TBR neighborhood could be $\Theta(n^3)$. Our other main result is a simple algorithm, called *Three-Way-Labels*, which can be used to speed-up exhaustive search for optimal neighbors under these and other transformations on trees. See Section 7 for pointers to related work on these problems.

The rest of the paper is organized as follows. In Section 2 we describe the Maximum Parsimony problem, and describe an algorithm to compute the parsimony score of a given tree. In Section 3, we define the NNI, SPR and TBR moves and present some known properties about the neighborhoods induced by these moves. In Section 4, we describe the p-ECR move, and compare the neighborhoods defined by the 2-ECR, TBR, and NNI operations. In Section 5, we formally define the problem of finding the best neighbor under the p-ECR move and describe a general algorithmic technique that we use to obtain a fast algorithm for the optimal neighbor problem under the 2-ECR move. In Section 6 we describe how our general technique can be used to obtain fast algorithms for the optimal neighbor problem under NNI, SPR and TBR moves, and also for computing the Greedy Sequence Addition algorithm for maximum parsimony. We conclude with Section 7 where discuss related work.

2 Basics

2.1 The Maximum Parsimony Problem

The input to the Maximum Parsimony problem is a collection S of n strings of the same length k over a given alphabet, Σ; these are the "given" nodes. The Steiner nodes (i.e., the nodes which can be used to connect the given nodes together) are drawn from Σ^k, i.e., all strings of length k over Σ. The objective is a tree T, with the given nodes at the leaves, and internal nodes from Σ^k, which minimizes the sum of the Hamming distances on the edges, where the Hamming distance on an edge $e = (x,y)$, denoted $H(x,y)$, is the number of positions in which x and y differ. Informally, this quantity is the minimum number of changes (via point mutations) needed to explain the evolution of the dataset from a common ancestor. We formalize this as follows.

Definition 1. *Parsimony score of a tree*

Let S be a set of sequences of length k over the alphabet Σ. Let T be a binary tree with leaf set S, and let f be an assignment of sequences to the internal nodes of T. The score of T under the assignment f, denoted $score(T,f)$ equals $\sum_{(u,v)\in E(T)} H(f(u),f(v))$. The parsimony score of T, denoted by $pscore(T)$, is the minimum $score(T,f)$ over all possible assignments f.

We now define the Maximum Parsimony (MP) problem.

Definition 2. *The Maximum Parsimony Problem*

Input: *Set S of sequences of length k over an alphabet Σ.*
Output: *A binary tree T whose leaves are bijectively labeled with sequences in S, such that the parsimony score of T, $pscore(T)$, is minimum.*

2.2 Computing the Maximum Parsimony Score of a Fixed Tree

Although finding the most parsimonious tree is NP-hard, we can find the optimal labeling of the internal nodes of a given tree in polynomial time. The standard algorithm for this problem, by Fitch [6], is the basis of our *Three-Way-Labels* algorithm, and so is included here.

The input to the fixed-tree maximum parsimony problem is a set S of n strings over a fixed alphabet Σ; for the typical cases, Σ is either the set of four nucleotides, or the set of amino-acid sequences, and thus is quite small. The elements of Σ are called the "states". We make the typical assumption that the sequences are already aligned, so that all sequences have the same length k. The positions within the sequences are sometimes called "sites."

The algorithm operates as follows. First, the tree is rooted (arbitrarily), either at a leaf, or by subdividing an edge e and rooting the tree at the newly introduced node. The cost of the tree (also called its "length") is then computed using dynamic programming. The algorithm is usually described as having two phases, where the first phase computes the length of the tree as well as a representation of candidate labels (strings over Σ^k that would produce optimal scores) for the root of each subtree of the tree; the second phase then actually produces a specific labeling for each node achieving the optimal score. We are primarily interested in the first phase, which we modify for use in our *Three-Way Labels* algorithm. However, the whole algorithm is of general interest, and so we provide it here.

Note that each position within the strings can be handled separately, so it suffices to describe the fixed-tree maximum parsimony algorithm as though there were only one position to consider. Since we have rooted T (arbitrarily), for every internal node v in T, we can define the rooted subtree T_v, and also the children of v. We let $States_v$ denote the set of state assignments for the node v (i.e., elements from Σ) which are part of an optimal assignment of states to all nodes in T_v so as to minimize the total parsimony score in T_v. We assume that T is binary, and that v's children are x and y (the algorithm can be applied more generally, however), and we similarly define $States_x$ and $States_y$. Then, the following equality holds (see [6]):

$$v \text{ is a leaf}: \quad States_v = \{\text{state of } v\}$$
$$v \text{ has two children } x,\ y: States_v = \begin{cases} States_x \cap States_y \text{ if } States_x \cap States_y \neq \emptyset \\ States_x \cup States_y \text{ otherwise} \end{cases}$$

This allows us to compute $States_v$ for every node v in T, from the bottom up. The optimal cost, i.e. the parsimony score, of T can also be calculated from the bottom-up at the same time: every time $States_x \cap States_y = \emptyset$ we increment the parsimony score of the tree by one. (Since we perform this computation for each site – i.e., position – independently, the sum of these values over all the sites is the parsimony score of the tree.)

In the second phase, we obtain the labeling on the internal nodes using a pre-order traversal. Once again, we can handle the positions (sites) independently. For the root r arbitrarily assign the state for r to be any element of $States_r$. Then visit the remaining nodes in turn, every time assigning a state to the node v from its set $States_v$. When we visit a node v we will have already set the state of its parent, u. If the selected state for u is an element of $States_v$, then we use the same state; otherwise we pick a state arbitrarily from $States_v$.

This algorithm takes $O(nrk)$ time to compute the labeling of every node in T and the optimal length (i.e., maximum parsimony cost) of T, where $r = |\Sigma|$, $n = |S|$, and k is the sequence length.

3 Hill-Climbing Heuristics for MP Analysis

The general structure of a heuristic search is as follows:

- First, an initial tree (or set of trees) is obtained, typically using the Greedy Sequence Addition method (see Section 6.2).
- Then, for each tree in the initial set, a search is initiated in which the given tree is modified (using a transformation that modifies trees), and the new tree is then scored. This process is repeated until a local optimum is found – that is, a tree which has no neighbor that has a better score.
- Finally, of all the local optima found, the set of trees that have the best MP score, or a "consensus" of these trees is returned; sometimes, sub-optimal trees are also returned.

Note that since MP is NP-hard, a local optimum need not be globally optimal, and in general this scheme will not return optimal trees in polynomial time.

We now describe three currently used tree-rearrangement operations and present some properties of the neighborhood induced around a tree by each of the three operations. Our definitions closely follow those in [1].

Nearest Neighbor Interchange (NNI) The NNI move swaps one rooted subtree on one side of an internal edge e with another on the other side; note that this is equivalent to contracting the edge e, and then resolving the resultant tree into a new binary tree. See Figure 1 for an example of this procedure.

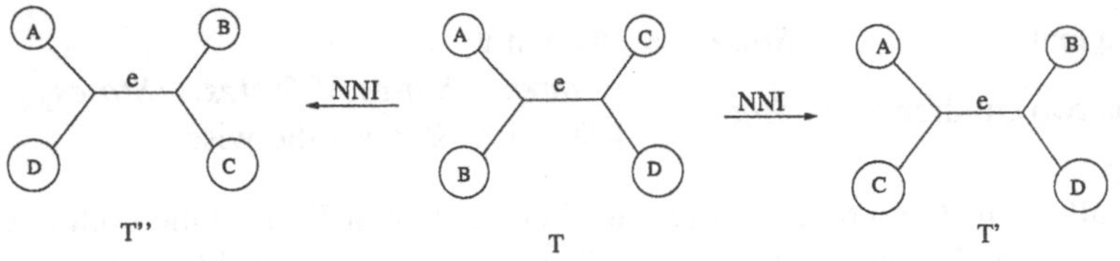

Fig. 1. Tree T can be transformed into either T' or T'' with one NNI move

Subtree Prune and Regraft (SPR) An SPR move on a tree T is defined as cutting any edge and thereby pruning a subtree, t, and then regrafting the subtree by the same cut edge to a new vertex obtained by subdividing a pre-existing edge in $T - t$. Any internal node that might arise that has degree two is suppressed in the resulting tree.

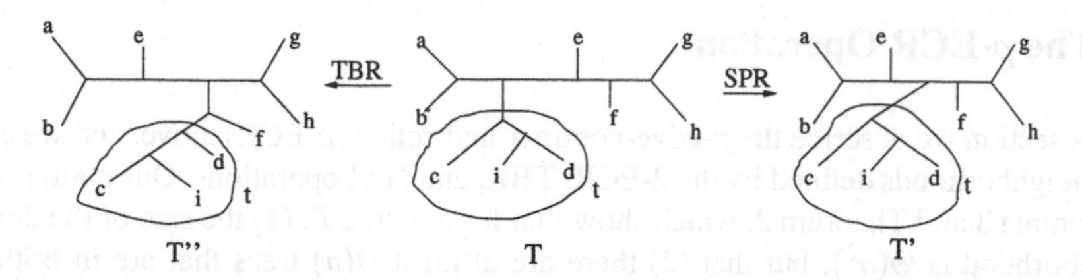

Fig. 2. Tree T' is one SPR move away from T, while T'' is one TBR move away.

Tree Bisection and Reconnection (TBR) In a TBR move an edge is removed from T, creating subtrees t and $T-t$, and then a new edge is added between the midpoints of any two edges in t and $T-t$, creating a new tree. Again, throughout the operation any internal node of degree two is suppressed. The last two operations are illustrated in Figure 2.

Each of the tree rearrangement operations described above naturally induces a distance metric in the space of trees. For instance, the NNI distance between two trees is defined as the minimum number of NNI moves required to transform one tree to another. The metrics induced by NNI, SPR and TBR moves have been discussed in [1]. We will denote the NNI metric by δ_{NNI}, the SPR metric by δ_{SPR}, and the TBR metric by δ_{TBR}. Note that every NNI move is an SPR move, and that every SPR move is a TBR move. Hence we have the following result:

Observation 1 *(From [12]) For any two unrooted leaf-labeled binary trees T and T' on the same set of leaves,*

$$\delta_{TBR}(T,T') \leq \delta_{SPR}(T,T') \leq \delta_{NNI}(T,T').$$

It is known that all of these distances are finite (Robinson showed this for the NNI distance in [14]).

Note that TBR searches explore a *superset* of trees, compared to both SPR and NNI, which is desirable. However, TBR searches are also more expensive, since there are more trees that are TBR neighbors of a given tree.

Induced neighborhoods We define the neighborhood of an unrooted binary leaf-labeled tree T under a tree-rearrangement move to be the set of all trees that can be obtained from T by one move. The following theorem, about the neighborhoods induced by NNI, SPR and TBR moves, is from [1].

Theorem 1. *[1] The size of the neighborhood for T is:*

1. $2n-6$ *for the NNI operation,*
2. $2(n-3)(2n-7)$ *for the SPR operation,*
3. *at most* $(2n-3)(n-3)^2$*, and dependent on the topology of T for the TBR operation.*

See [10,1,4] for results related to computing the distance between trees under these metrics, and [11,1] for results related to the maximum pairwise distance between trees under these metrics.

4 The p-ECR Operation

In this section we describe the p-edge-contract-and-refine (p-ECR) move, and we compare neighborhoods defined by the 2-ECR, TBR, and NNI operations. Our main results are Lemma 3 and Theorem 2, which show that for any tree T, (1) the size of the 2-ECR neighborhood is $\Theta(n^2)$, but that (2) there are at most $O(n)$ trees that are in both the 2-ECR neighborhood and the TBR neighborhood of T. We also show that the 2-ECR neighborhood strictly contains all trees within two NNI moves from T.

The p-ECR move is a generalization of the NNI move in the following sense: Since an NNI move can also be viewed as an edge contraction followed by a refinement at the newly created unresolved node, we can generalize NNI by contracting p edges all at once, creating unresolved nodes in the process, and then refining these unresolved nodes give back a binary tree. Note that this process is *not*, in general, equivalent to contracting and refining each of the p edges in succession. Indeed, in Figure 3 we give an example of two trees T and T' such that $\delta_{p-ECR}(T,T') = 1$, but $\delta_{NNI}(T,T') > p$.

Since NNI is the same as 1-ECR, it follows from [14] that $\delta_{p-ECR}(T,T')$ is finite for all pairs of binary trees on the same leaf set (i.e, we can go from any tree to any other tree through a sequence of p-ECR moves).

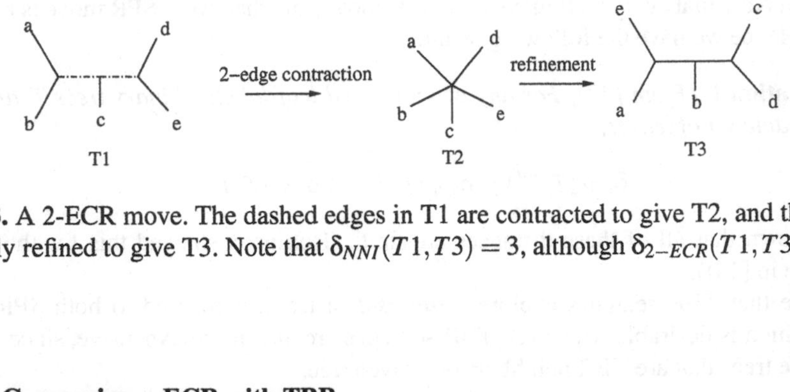

Fig. 3. A 2-ECR move. The dashed edges in T1 are contracted to give T2, and then T2 is fully refined to give T3. Note that $\delta_{NNI}(T1,T3) = 3$, although $\delta_{2-ECR}(T1,T3) = 1$.

4.1 Comparing p-ECR with TBR

In this section we show that the number of 2-ECR neighbors of a tree on n leaves is $\Omega(n^2)$, and that the number of 2-ECR neighbors that are also TBR neighbors is only $O(n)$.

We now define some concepts that will be necessary for our analyses. Every edge in a binary leaf-labeled tree T induces a *bipartition* of the set of leaves. Let the bipartition induced by an edge e be π_e. Then the set $C(T) = \{\pi_e | e \in E(T)\}$ uniquely defines the tree T (see [3,17]). In the subsequent discussion all trees will be assumed to be binary, leaf-labelled trees.

Definition 3. *Robinson-Foulds distance [13].*

The Robinson-Foulds *(RF) distance between two binary leaf-labeled trees T and T′ is defined to be* $|(C(T)\Delta C(T')|$, *i.e,* $|C(T) - C(T')|$ + $|C(T') - C(T)|$.

Neither contraction nor refinement of a set of edges alters bipartitions induced by other edges in a tree, and hence we have the following:

Observation 2 *Let T and T′ be two binary leaf-labeled trees. Then, for any* $1 \leq p \leq n-3$, $\delta_{p\text{-}ECR}(T,T') = 1 \implies RF(T,T') \leq 2p$.

We now define the *Maximum Agreement Forest* [10] between two binary leaf-labeled trees.

Let $F = \{t_1, t_2, \ldots, t_m\}$ be a forest of m trees that results from deleting $m-1$ edges from a tree T. Let F' be a forest of m trees obtained similarly from T'. F (or F') is said to be an *agreement forest* for T and T' iff $F = F'$. A *maximum agreement forest* (MAF) for T and T' is an agreement forest with the *minimum* number of trees.

Lemma 1. *(From [1]) Let T and T′ be two binary leaf-labeled trees. Let F be a maximum agreement forest for T and T′. Then* $\delta_{TBR}(T,T') = |F| - 1$.

We now show that for every and n and every $1 < p < n-3$ there are trees whose p-ECR distance is less than their TBR distance, and vice versa.

Lemma 2. *The following is true for every natural number n and every natural number* $p < n-3$:

1. $\exists T,\ T'$ *s.t* $\delta_{TBR}(T,T') < \delta_{p\text{-}ECR}(T,T')$.
2. $\exists T,\ T'$ *s.t* $\delta_{p\text{-}ECR}(T,T') = 1$ *and* $\delta_{TBR}(T,T') \in \Omega(p)$.

Due to space requirements, we omit the proof; however, see Figure 4 for a pair of trees satisfying the first condition, and Figure 5 for a pair of trees satisfying the second condition.

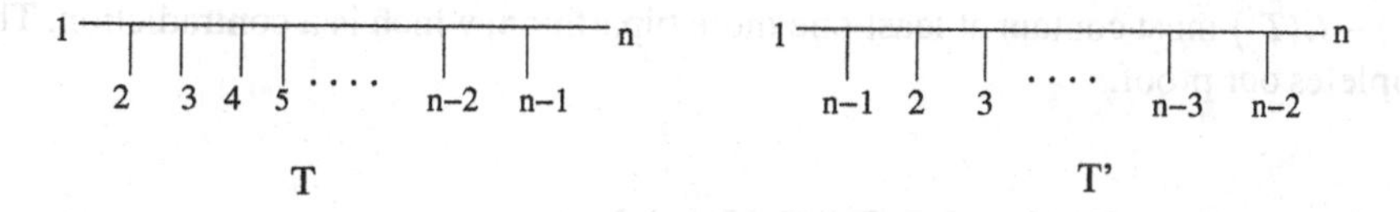

Fig. 4. $\delta_{TBR}(T,T') = 1$ but $RF(T,T') = 2n-6$.

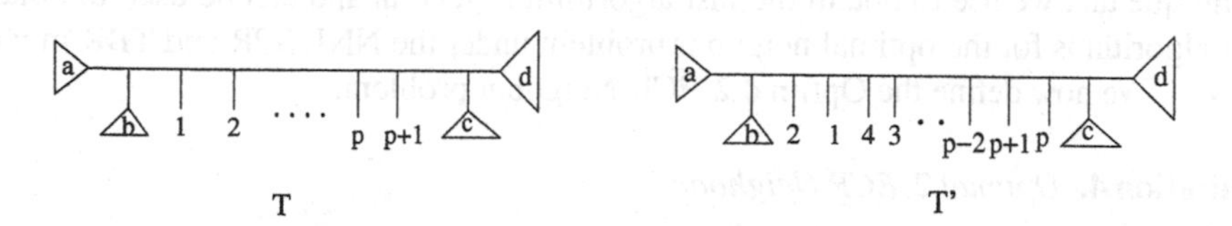

Fig. 5. $|MAF(T,T')| \in \Omega(p)$ but $\delta_{p\text{-}ECR}(T,T') = 1$.

We now prove some results about the neighborhood induced around a tree by the 2-ECR operation, and in particular we show that the neighborhood of a tree induced by the

2-ECR operation is very different from the one induced by the TBR operation. We will denote the 2-ECR neighborhood of a tree T as $\mathcal{N}_{2-ECR}(T)$, and the TBR neighborhood as $\mathcal{N}_{TBR}(T)$.

Lemma 3. *For any binary leaf-labeled tree T, $|\mathcal{N}_{2-ECR}(T)| \in \Theta(n^2)$.*

Proof. Omitted due to space constraints.

We now prove that most of the trees in $\mathcal{N}_{2-ECR}(T)$ are not in $\mathcal{N}_{TBR}(T)$.

Theorem 2. *For a binary leaf-labeled tree T, $|\mathcal{N}_{2-ECR}(T) \cap \mathcal{N}_{TBR}(T)| \in O(n)$.*

Proof. Let $X(T) = \mathcal{N}_{2-ECR}(T) - \mathcal{N}_{NNI}(T)$. Note then that each tree $T' \in X(T)$ can be obtained by contracting two edges e_1 and e_2 in T, and then refining the resultant tree. Consider the set S of all trees T' in $X(T)$ such that the corresponding contracted edges e_1 and e_2 are separated in T by at least two edges. Note that there are only $\Theta(n)$ pairs of edges in T that are either adjacent, or separated by exactly one edge. Consequently, it follows that $|\mathcal{N}_{2-ECR} - S| \in O(n)$.

We now show that $S \cap \mathcal{N}_{TBR}(T) = \emptyset$. Suppose to the contrary that there is a tree $T' \in S \cap \mathcal{N}_{TBR}(T)$. Note that $RF(T,T') = 4$, since $C(T) - C(T') = \{\pi_{e_1}, \pi_{e_2}\}$, where $C(T)$ is the set of bipartitions in T, and e_1 and e_2 are those two edges through whose contraction (and subsequent refinement) T' was obtained from T.

However, T and T' are one TBR move apart. Hence, it can be shown that if $\{\pi_{e_1}, \pi_{e_2}\} \subseteq C(T) - C(T')$, then the bipartitions induced by all edges (except, possibly, the edge that was broken in the TBR move) in the path in T between e_1 and e_2 are in $C(T) - C(T')$. Now, since e_1 and e_2 are separated by at least two edges, the set $C(T) - C(T')$ must contain at least one more bipartition, which is a contradiction. This completes our proof.

5 Computing Optimal 2-ECR Neighbors

In this section we consider the problem of finding an optimal neighbor under the p-ECR tree-rearrangement operation, and present a fast algorithm for solving the problem. The technique that we use to obtain the fast algorithm is general and can be used to obtain fast algorithms for the optimal neighbor problem under the NNI, SPR and TBR moves as well. We now define the Optimal 2-ECR Neighbor problem.

Definition 4. *Optimal 2-ECR Neighbors*

Input *An unrooted binary tree T on n leaves, each bijectively leaf-labeled by a set S of sequences of length k over an alphabet of size r.*

Output*: An unrooted binary tree T' on n leaves, each bijectively labelled by the same set S, such that T' has the minimum MP score among all such trees t for which $\delta_{2-ECR}(T,t) = 1$.*

Henceforth, we will call such a tree an *optimal 2-ECR-neighbor* of T. The *Optimal TBR-neighbor*, *Optimal SPR-neighbor* and *Optimal NNI-neighbor* problems are defined similarly. Note also that there can be more than one optimal neighbor, and that in general the objective is to find all optimal solutions.

At the outset we observe that a brute-force algorithm for the above problem would take $\Theta(n^3rk)$ time, since there are $\Theta(n^2)$ 2-ECR neighbors for any tree, and computing the parsimony score of each tree using Fitch's algorithm would take $\Theta(nrk)$ time. We will obtain a $\Theta(n^2rk)$ time algorithm for the Optimal 2-ECR problem which will return all the optimal neighbors.

As was observed in Section 4, an *NNI* move can be thought of as a 1-ECR move. So, not surprisingly, we use a fast algorithm for computing optimal NNI-neighbors in our algorithm for computing optimal 2-ECR-neighbors. A brute-force optimal NNI neighbors algorithm would run in $\Theta(n^2rk)$ time, but our algorithm runs in $\Theta(nrk)$ time.

5.1 An $O(nrk)$ Algorithm for the Optimal NNI Neighbor Problem

The way we obtain a speed-up over the brute-force techniques for each of the problems we address is by performing a preprocessing step in which we assign three labels to each node in the tree.

In order to understand why we do this preprocessing step, consider an NNI move across an edge, say (u,v), in a given tree T. Let W and X be the rooted subtrees below u, and let Y and Z be the rooted subtrees below v. The NNI move will, e.g, involve swapping W with Y. Let the resulting tree be T'. Supposing that we have the parsimony scores and optimal state assignments for the rooted subtrees W, X, Y and Z, the parsimony score of T' can be computed in $\Theta(rk)$ time, thus: we can subdivide edge (u,v) and root T' at the newly created node, which we shall call x. This produces a binary tree rooted at x, with subtrees off x rooted at u and v. The parsimony score of the subtree of T' rooted at u depends just on the parsimony scores and optimal state assignments of X and Y, and can be computed from them in $\Theta(rk)$ time, as in Fitch's algorithm. Similarly, the parsimony score of the subtree of T' rooted at v can be computed in $\Theta(rk)$ time. Finally, the parsimony score of T' (rooted at x) can be computed in $O(rk)$ time from the parsimony scores and optimal state assignments of the subtrees rooted at u and v.

The above observations suggest that a preprocessing stage that computes the parsimony score and the optimal state assignments for every rooted subtree will let us compute the parsimony score of each NNI neighbor in $\Theta(rk)$ time. The brute-force way of performing this preprocessing step would take $\Theta(n^2rk)$ time, but we will next see how to perform this preprocessing stage in $\Theta(nrk)$ time. We will call the preprocessing step the Three-Way Labels algorithm since it would assign three optimal state-assignment labels to each internal node.

5.2 Three-Way Labels: The Dynamic Programming Algorithm

In a tree T, consider an internal node v with three neighbors a, b and c, as in Figure 6. The node v is the root of three rooted subtrees, one where a and b are its children ($tree(v,a,b)$ in the figure), one where a and c are its children ($tree(v,a,c)$) and one where b and c are its children ($tree(v,b,c)$). The preprocessing step would involve assigning three labels (optimal state assignments) for each such internal node v - namely the optimal state assignments at the roots of $tree(v,a,b)$, $tree(v,b,c)$ and $tree(v,a,c)$.

The parsimony score and the optimal state assignment of, for example, $tree(v,a,b)$ can be computed from the parsimony score and optimal state assignments of the subtrees rooted at a and b. Note that the subtrees of $tree(v,a,b)$ rooted at a and b have fewer leaves than $tree(v,a,b)$. This suggests the following dynamic programming algorithm:

Bucket sort the rooted subtrees in T by the number of leaves in the subtree in $O(n)$ time. For subtrees that contain just a single leaf, the label is just the sequence at the leaf. For subtrees such as $tree(v,a,b)$, the label is computed by in the usual way: for a given site, if the corresponding sets at a and b are disjoint we take the union of the sets, and otherwise we take the intersection. Note the when we compute the label at v corresponding to $tree(v,a,b)$, the necessary labels at a and b are already available since the subtrees rooted a and b are smaller. There are $O(n)$ rooted trees like $tree(v,a,b)$, and for each of them the optimal state assignment at the root can be computed in $\Theta(rk)$ time using the dynamic programming technique. Also, the parsimony score of the rooted subtrees can be computed along side their optimal state assignments.

Therefore, we have the following:

Lemma 4. *The Three-Way Labels algorithm takes $O(nrk)$ time, where n is the number of leaves in the tree T, and each leaf is labeled by a sequence of length k over an alphabet of size r.*

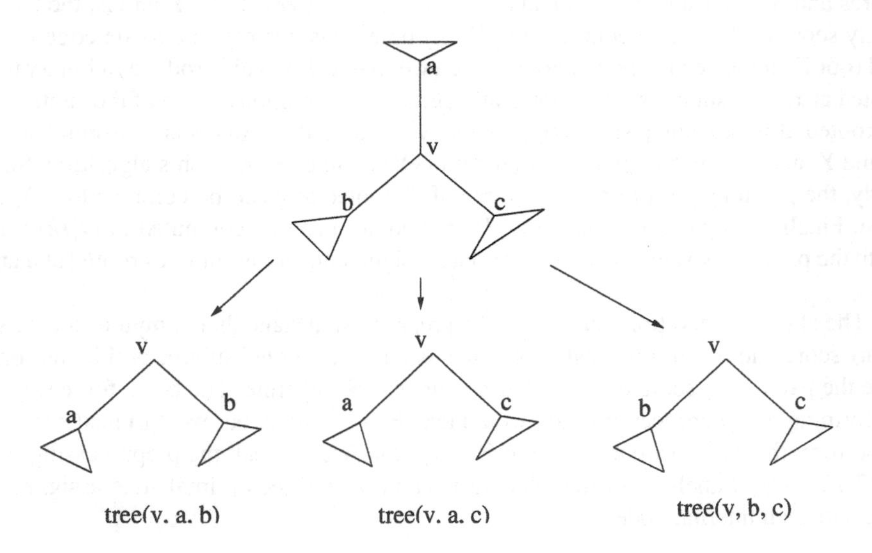

Fig. 6. Internal node v and the three subtrees associated with it.

As we saw in the previous section, the preprocessing stage would let us compute the parsimony score of each NNI-neighbor in $\Theta(rk)$ time. Since there are $2n-6$ NNI-neighbors, the optimal NNI neighbors can be identified in $\Theta(nrk)$ time after the preprocessing step. To summarize,

Theorem 3. *We can solve the Optimal NNI-Neighbors Problem in $\Theta(nrk)$ time.*

5.3 Computing an Optimal 2-ECR Neighbor

We now show how to compute an optimal 2-ECR neighbor of an unrooted binary tree on n labeled leaves, each labeled by a sequence of length k over an alphabet of size r, in $\Theta(n^2rk)$ time, thus spending only $\Theta(rk)$ time per neighbor.

A 2-ECR move on a tree T is specified by the two edges e_1 and e_2 to be contracted, and the refinement of the resulting contracted tree into an unrooted binary tree that differs from T.

Our algorithm will handle the following two cases separately.

1. The edges e_1 and e_2 are not adjacent to each other.
2. The edges e_1 and e_2 are adjacent to each other.

We now show how to handle case (1). We first state a lemma that in this case any 2-ECR move can be "simulated" by two successive NNI moves. We omit the proof because of space limitations.

Lemma 5. *Let T be an unrooted leaf-labeled tree and let T' be a 2-ECR neighbor of T such that the 2-ECR move involves the contraction and refinement of two non-adjacent edges in T. Then T' can be reached from T through two NNI moves.*

We now continue with the discussion of the Optimal 2-ECR neighbors algorithm.
Case 1: the edges are not adjacent By Lemma 5, in this case the optimal 2-ECR neighbors can be obtained by two sequential NNI moves. To compute the optimal 2-ECR neighbors of T in this case, we compute the optimal NNI neighbors of every NNI neighbor of T. There are $\Theta(n)$ NNI neighbors of T, and the optimal NNI neighbors of a given tree can be found in $\Theta(nrk)$ time by Theorem 3. Hence, the set of optimal 2-ECR neighbors can be computed in $\Theta(n^2rk)$ time for this case.
Case 2: the edges are adjacent Note that on a tree with n leaves, there are only $O(n)$ pairs of adjacent edges. For each possible way of contracting a pair of adjacent edges, we create a tree with a single unresolved node (that is, a node of degree more than three), and the unresolved node has degree 5. Hence, there are 15 possible binary trees that resolve each such tree. Furthermore, each of the 15 refinements involves only a rearrangement of the 5 rooted subtrees off the unresolved node around the two new edges that result from the refinement. Therefore, the algorithm operates as follows. First, we compute the optimal labels at the root of all such subtrees in $\Theta(nrk)$ time in a preprocessing step as in Section 5.2. Then, for each of the $O(n)$ pairs of adjacent edges, in $O(rk)$ additional time we can compute the optimal neighbors obtainable by contracting and refining those edges. Hence, for the case of adjacent edges, can compute the set of optimal 2-ECR neighbors in $\Theta(nrk)$ time.

The overall optimal 2-ECR neighbors will be those with the best score, and hence we have the following:

Theorem 4. *Let T be an unrooted binary tree on n leaves, each labeled by a sequence of length k over an alphabet of size r. Then the optimal 2-ECR neighbors of T can be computed in $\Theta(n^2rk)$ time.*

6 Application of Three-Way Labeling to Other Problems

In this section we describe how our Three-Way labeling algorithm can be used to compute the set of optimal SPR and TBR neighbors in $O(n^2rk)$ time and $O(n^3rk)$ time respectively. We then describe how the technique can be used to compute the Greedy Sequence Addition parsimony algorithm in $O(n^2rk)$ time.

6.1 Optimal SPR and TBR Neighbors

An SPR move on a tree T to create a tree T' involves the following three steps:

- Delete an edge $e = (x_1, x_2)$ from T, thus producing two trees T_1 and T_2 with $x_1 \in V(T_1)$ and $x_2 \in V(T_2)$.
- Pick one of the two subtrees (say, T_1). Then, pick an edge e_2 in T_2, and subdivide e_2, thus creating a new node v_2.
- Add the edge (x_1, v_2).

To compute the parsimony score of T', we can root T' at the edge (x_1, v_2). We will then need the parsimony score and optimal state assignments of T_2 rooted at v_2 and those of T_1 rooted at x_1. We perform a Three-Way labeling of the nodes in T_1 and T_2. This takes $O(nrk)$ time, and would allow us to compute the parsimony score and optimal state assignments of T_2 rooted at v_2 and those of T_1 rooted at x_1 in $O(rk)$ time. Once this information is available, the parsimony score of T' can be computed in $O(rk)$ time. Thus, for a fixed way of deleting an edge (x_1, x_2), all SPR neighbors can be evaluated in $O(nrk)$ time. There are $O(n)$ ways of deleting an edge, and thus we can evaluate all SPR neighbors and identify the optimal ones in $O(n^2rk)$ time.

As for TBR, the only difference here is that for every way of deleting an edge (x_1, x_2) in T_1, there can be $O(n^2)$ TBR neighbors. Evaluating all these neighbors can be done in $O(n^2rk)$ time if we Three-Label the two trees (T_1 and T_2) that result from the deletion of (x_1, x_2) from T. There are $O(n)$ ways of deleting an edge from T, and thus we can evaluate all TBR neighbors and identify the optimal ones in $O(n^3rk)$ time.

6.2 A Faster Algorithm for Greedy Sequence Addition Parsimony

We begin by describing a brute-force algorithm for the Greedy-MP algorithm.

Brute-Force Greedy MP Greedy-MP constructs a tree for a set S of sequences based upon a specified (usually random) ordering on S; suppose that ordering is $s_1, s_2, \ldots, s_n$. It begins with the star tree on the first three sequences, s_1, s_2, and s_3, and then sequentially adds each of the remaining sequences into the tree it has constructed so far. Before it attempts to insert the ith sequence, s_i, it has a tree t_{i-1} on the first $i-1$ sequences. In order to insert s_i into t_{i-1}, it computes the length of each possible extension of t_{i-1}, in the obvious way: for each way of adding s_i (by subdividing an edge in t_{i-1}, and making the newly created node the parent of s_i), it uses Fitch's algorithm (see Section 2.2) to score the resultant tree. If there is a tie (more than one way of adding s_i gives a minimal score), then a best tree is selected arbitrarily. When all sequences have been added, the resultant tree is returned.

The running time of this brute-force algorithm follows from the analysis of the cost of adding s_i to the tree t_{i-1}. First, note that there are $O(i)$ ways to add s_i to t_{i-1}, and that scoring the resultant trees costs $O(irk)$ per tree, where r is the alphabet size, and k is the sequence length. Hence, computing t_i, given t_{i-1}, costs $O(i^2rk)$. Since we do this for $i = 4,5,\ldots,n$, the total cost is $O(n^3rk)$.

Faster Greedy-MP If we do a Three-Way Labeling of t_{i-1}, then we can compute the optimal placement of s_i into t_i in only $O(irk)$ time, so that Greedy-MP can be completed in $O(n^2rk)$ time.

7 Related Work

In the "sectorial search" technique used in the parsimony software TNT, developed by Goloboff *et. al* ([9]), repeatedly a set of edges is identified (using some specific technique) to be contracted and then refined. This can be viewed as a p-ECR based search, where the value for p is determined indirectly. The general approach of contracting edges and then finding an optimal resolution has also been suggested in [2]. Empirical comparisons in [9] of sectorial search to other search strategies suggested that this kind of approach would be potentially useful. Our contribution here is theoretical rather than empirical, and our findings are consistent with those positive observations reported in [9].

Our other main contribution, namely the 3-Way-Labels algorithm, and its use in finding optimal neighbors under various tree transformations, is new, but similar techniques have been presented before (see [16,7,8]).

8 Acknowledgments

We thank an anonymous referee for pointing out earlier papers related to this work.

References

1. B. Allen and M. Steel. Subtree Transfer Operations and their Induced Metrics on Evolutionary Trees. *Annals of Combinatorics*, 5:1–15, 2001.
2. M. Bonet, M. Steel, T. Warnow, and S. Yooseph. Better Methods for Solving Parsimony and Compatibility. *Journal of Computational Biology*, 5(3):409–422, 1998.
3. P. Buneman. The Recovery of Trees from Measures of Dissimilarity. *Mathematics in the Archaelogical and Historical Sciences*, pages 387–395, 1971.
4. B. Dasgupta, X. He, T. Jiang, M. Li, J. Tromp, and L. Zhang. On the Distances Between Phylogenetic Trees. In *Proceedings of the 8th Annual ACM-SIAM Symposium on Discrete Algorithms*, pages 427–436. ACM-SIAM, 1997.

5. L. R. Foulds and R. L. Graham. The Steiner problem in Phylogeny is NP-complete. *Advances in Applied Mathematics*, 3:43–49, 1982.
6. W. Fitch. Toward Defining Course of Evolution: Minimum Change for a Specified Tree Topology. *Systematic Zoology*, 20:406–416, 1971.
7. P. A. Goloboff. Character Optimization and Calculation of Tree Lengths. *Cladistics*, 9:433–436, 1994.
8. P. A. Goloboff. Methods for Faster Parsimony Analysis. *Cladistics*, 12:199–220, 1996.
9. P. A. Goloboff. Analyzing Large Datasets in Reasonable Times: Solutions for Composite Optima. *Cladistics*, 15:415–428, 1999.
10. J. Hein, T. Jiang, L. Wang, and K. Zhang. On the Complexity of Comparing Evolutionary trees. *Discrete Applied Mathematics*, 71:153–169, 1996.
11. M. Li, J. Tromp, and L. Zhang. On the Nearest Neighbour Interchange Distance Between Evolutionary Trees. *Journal of Theoretical Biology*, 182:463–467, 1996.
12. D. R. Maddison. The Discovery and Importance of Multiple Islands of Most Parsimonious Trees. *Systematic Zoology*, 43(3):315–328, 1991.
13. D. F. Robinson and L. R. Foulds. Comparison of Phylogenetic Trees. *Mathematical Biosciences*, 53:131–147, 1981.
14. D. F. Robinson. Comparison of Labeled Trees with Valency Three. *Journal of Combinatorial Theory*, 11:105–119, 1971.
15. D. Swofford, G. J. Olson, P. J. Waddell, and D. M. Hillis. *Molecular Systematics*, chapter Phylogenetic Inference, pages 407–425. Sinauer Associates, Sunderland, Massachusetts, second edition, 1996.
16. D. L. Swofford. *Studies in Numerical Cladistics: Phylogentic Inference Under the Principle of Maximum Parsimony*. PhD thesis, University of Illinois at Urbana-Champaign, 1986.
17. T. Warnow. Tree Compatibility and Inferring Evolutionary History. *Journal of Algorithms*, 16:388–407, 1994.

Computing Refined Buneman Trees in Cubic Time

Gerth Stølting Brodal[1,*,**], Rolf Fagerberg[1,*], Anna Östlin[2,*,***], Christian N. S. Pedersen[3,*], and S. Srinivasa Rao[4,***]

[1] BRICS (Basic Research in Computer Science), Department of Computer Science, University of Aarhus, Ny Munkegade, 8000 Århus C, Denmark.
{gerth,rolf}@brics.dk.

[2] IT University of Copenhagen, Glentevej 67, DK-2400 Copenhagen NV.
annao@it-c.dk.

[3] Bioinformatics Research Center (BiRC), Department of Computer Science, University of Aarhus, Ny Munkegade, 8000 Århus C, Denmark.
cstorm@daimi.au.dk.

[4] School of Computer Science, University of Waterloo, 200, University Avenue West, Waterloo, Ontario W2L 3G1.
ssrao@monod.uwaterloo.ca.

Abstract. Reconstructing the evolutionary tree for a set of n species based on pairwise distances between the species is a fundamental problem in bioinformatics. Neighbor joining is a popular distance based tree reconstruction method. It always proposes fully resolved binary trees despite missing evidence in the underlying distance data. Distance based methods based on the theory of Buneman trees and refined Buneman trees avoid this problem by only proposing evolutionary trees whose edges satisfy a number of constraints. These trees might not be fully resolved but there is strong combinatorial evidence for each proposed edge. The currently best algorithm for computing the refined Buneman tree from a given distance measure has a running time of $O(n^5)$ and a space consumption of $O(n^4)$. In this paper, we present an algorithm with running time $O(n^3)$ and space consumption $O(n^2)$. The improved complexity of our algorithm makes the method of refined Buneman trees computational competitive to methods based on neighbor joining.

1 Introduction

The evolutionary relationship for a set of species is commonly described by an evolutionary tree, also called a phylogeny, where leaves correspond to species and internal nodes correspond to points in time where the evolution has diverged in different directions. Reconstructing an unknown evolutionary tree for

* Partially supported by the Future and Emerging Technologies programme of the EU under contract number IST-1999-14186 (ALCOM-FT).

** Supported by the Carlsberg Foundation (contract number ANS-0257/20).

*** Work done while at BRICS.

G. Benson and R. Page (Eds.): WABI 2003, LNBI 2812, pp. 259–270, 2003.

a set of species from obtainable information about the species is a fundamental problem in bioinformatics. A multitude of models and methods for reconstructing evolutionary trees have been proposed in the literature, see e.g. [13] for an overview. A large class of methods use the evolutionary distance between each *pair* of species as their primary source of information for reconstructing the unknown evolutionary relationships. Distance data can e.g. be obtained from sequence data from the species by estimating the evolutionary distance between homologous sequences in a model of sequence evolution.

A widely used distance based method is the neighbor joining method by Saitou and Nei [14], which can be implemented with running time $O(n^3)$ and space consumption $O(n^2)$, where n is the number of species. A common critique of neighbor joining based methods is that they always reconstruct fully resolved evolutionary trees, i.e. unrooted trees where all internal nodes have degree three. A fully resolved tree can be misleading because many of its internal edges can be artifacts of the reconstructing method insisting on a fully resolved tree even though the underlying distance data contains little phylogenetic evidence hereof. To avoid this problem, a number of distance based methods have been studied which only propose evolutionary trees whose edges are well supported by constraints expressed in terms of *quartets*. A quartet is the topological subtree induced by four species. Every edge in an evolutionary tree induces a set of quartets consisting of the quartets with two species in each of the two subtrees induced by removing the edge.

The Q^* method [1,4], which relates to the tree construction method introduced by Buneman in [7], imposes constraints on the proposed edges by requiring that all induced quartets must have positive weight for some given weight function. The running time of the general Q^* method is $O(n^4)$, where n is the number of species, and it has been experimentally shown to introduce very few incorrect edges [4]. If quartets are weighted according to their *Buneman score* the resulting evolutionary tree which satisfies that all induced quartets have positive Buneman score is called a *Buneman tree*. Berry and Bryant [3] show how to compute the Buneman tree for a set of n species in time $O(n^3)$ and space consumption $O(n^2)$. The Buneman tree is a conservative but reliable estimate of the evolutionary tree. However, as discussed in [3,5] and illustrated in [3, Figure 1], the severe constraints of positive Buneman score for all induced quartets often result in a proposed evolutionary tree with few resolved edges. This shortcoming was addressed by Moulton and Steel [12] who proposed the *refined Buneman tree* which loosens the constraints by allowing a limited number of the induced quartets to have negative score. This tree is a refinement of the Buneman tree in the sense that it contains at least the edges in the Buneman tree. The refined Buneman tree thus takes up a useful middle ground between the neighbor joining method and the classic Buneman tree.

Bryant and Moulton [6] presented the first polynomial time algorithm to compute the refined Buneman tree. The running time is $O(n^6)$. Berry and Bryant [3,5] gave an improved algorithm with running time $O(n^5)$ and space consumption $O(n^4)$. In this paper we present a method for constructing the refined Buneman tree for a set of n species in time $O(n^3)$ and space $O(n^2)$.

Our algorithm is based on an incremental approach also used in the algorithms presented in [3,5,6]. The central difference is that we do not construct a sequence of refined Buneman trees, but instead construct a sequence of over-approximations to refined Buneman trees from which we can extract the desired refined Buneman tree at the end.

The improved running time and the simplicity of our algorithm makes the method of refined Buneman trees computational competitive to methods based on neighbor joining and on plain Buneman trees. It will also make it possible to perform comprehensive experiments on biological data to examine the virtues of refined Buneman trees against trees produced by these other methods. An implementation of our algorithm is currently being made, and it is planned to be part of release 4.0 of the well-known SplitsTree package [10].

The rest of this paper is organized as follows. In Section 2 we introduce notation and earlier results related to Buneman and refined Buneman trees. In Section 3 we describe how to maintain a set of compatible splits. In Sections 4 and 5 we present our improved algorithm for computing refined Buneman trees.

2 Preliminaries

In the following we let the set of species be denoted $X = \{x_1, \ldots, x_n\}$, and for an integer $k \in \{1, \ldots, n\}$ we let $X_k = \{x_1, \ldots, x_k\}$.

Evolutionary tree An evolutionary tree (or X-tree) for a set of species X is an unrooted tree $T = (V, E)$ together with an injective labeling of the leaves by members of X.

Dissimilarity measure A dissimilarity (or distance) measure δ on a set of species X is a symmetric function $\delta : X^2 \to \mathbb{R}_+$ where $\delta(x, x) = 0$ for all $x \in X$.

Quartets To every set of four species $a, b, c, d \in X$, there are four ways to associate a leaf-labeled tree, as shown in Figure 1. The three possible binary tree resolutions, *quartets*, are denoted by $ab|cd$, $ac|bd$ and $ad|bc$, indicating how the central edge of the binary tree bipartitions the four species. We say that an edge e in an X-tree induces a quartet $ab|cd$ if e bipartitions the four species in the same way as the central edge of the quartet.

Splits The partition of a finite set into two non-empty parts U and V is denoted a *split* $U|V$. In this paper we represent a split $U|V$ as a bit-vector A such that $x_i \in U$ if and only if $A[i] = 0$. If $|U| = 1$ or $|V| = 1$ the split is called trivial. Removing an edge e from an X-tree T partitions the leaf set of the tree into two parts. This is called the split of T associated with the edge e. The complete set of splits associated with each of its edges is denoted $splits(T)$. The lemma below is proved in [9], but also follows from the construction in Section 3.

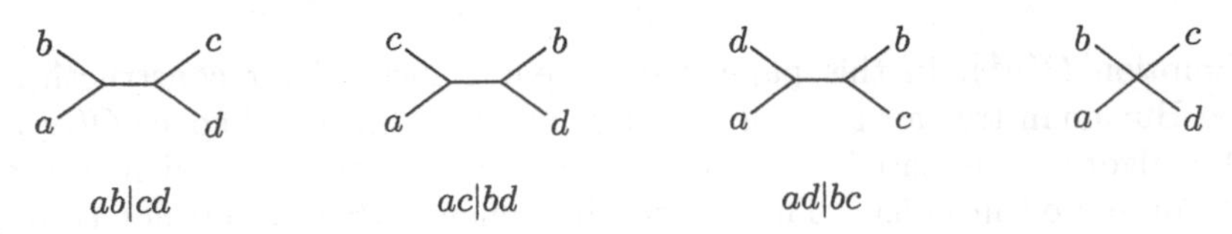

Fig. 1. The possible topologies of four species.

Lemma 1 (Gusfield [9]). *Any unrooted X-tree with n leaves can be constructed from its set of non-trivial splits in time $O(kn)$, where k is the number of non-trivial splits.*

The set of quartets associated with a split $U|V$ is defined by $q(U|V) = \{uu'|vv' : u, u' \in U \ \wedge \ v, v' \in V\}$. Here u and u' (similarly v and v') need not be distinct.

Compatibility A set of splits S is *compatible* if $S \subseteq splits(T)$ for some tree T.

Lemma 2 (Buneman [7]). *Two splits $A|B$ and $C|D$ are compatible if and only if one of $A \cap C$, $A \cap D$, $B \cap C$ or $B \cap D$ is empty. A set of splits is compatible if and only if it is pairwise compatible.*

Buneman trees Buneman [7] shows how to construct a weighted unrooted tree from a dissimilarity measure δ on X by considering quartets. The *Buneman score* of a quartet $q = ab|cd$, where $a, b, c, d \in X$ is defined as:

$$\beta_q = \frac{1}{2}(\min\{ac + bd, ad + bc\} - (ab + cd)) , \tag{1}$$

where ab denotes $\delta(a, b)$ for $a, b \in X$. Two distinct quartets q_1 and q_2 for the same four species satisfy

$$\beta_{q_1} + \beta_{q_2} \leq 0 . \tag{2}$$

The *Buneman index* of a split $\sigma = U|V$ of X is

$$\mu_\sigma(\delta) = \min_{u,u' \in U, v,v' \in V} \beta_{uu'|vv'} .$$

Buneman showed that the set of splits $B(\delta) = \{\sigma : \mu_\sigma(\delta) > 0\}$ is compatible. The *Buneman tree* corresponding to a given dissimilarity measure δ is defined to be the weighted unrooted tree whose edges represent the splits $\sigma \in B(\delta)$ and are weighted according to $\mu_\sigma(\delta)$.

Anchored Buneman tree One relaxation of the condition that $\mu_{U|V} > 0$ is to only look at quartets containing a certain fixed species $x \in X$. For each split $U|V$ with $x \in U$ define

$$\mu^x_{U|V}(\delta) = \min_{u \in U, v,v' \in V} \beta_{xu|vv'} ,$$

and let $B_x(\delta) = \{U|V : \mu^x_{U|V} > 0\}$. Clearly $B(\delta) \subseteq B_x(\delta)$. Bryant and Moulton show that the set of splits $B_x(\delta)$ is compatible [6, Lemma 1]. The weighted unrooted tree representing $B_x(\delta)$ with the edge representing a split $\sigma \in B_x(\delta)$ given the weight $\mu^x_\sigma(\delta)$, is called the *Buneman tree anchored at x*.

Lemma 3 (Bryant and Moulton [6, Proposition 2]). $B(\delta) = \cap_{x \in X} B_x(\delta)$.

Lemma 4 (Berry and Bryant [3, Section 3.2]). *$B_x(\delta)$ can be computed in time (and space) $O(n^2)$.*

Refined Buneman tree Given a split σ for a set size n, let $m = |q(\sigma)|$ and let $q_1, \ldots, q_m$ be an ordering of the elements of $q(\sigma)$ in non-decreasing order of their Buneman scores. The refined Buneman index of the split σ is defined as

$$\bar{\mu}_\sigma(\delta) = \frac{1}{n-3} \sum_{i=1}^{n-3} \beta_{q_i} \,. \tag{3}$$

Moulton and Steel show that the set of splits $\{\sigma : \bar{\mu}_\sigma > 0\}$ is compatible [12, Corollary 5.1]. They define the *refined Buneman tree* as the weighted unrooted tree representing the set $RB(\delta) = \{\sigma : \bar{\mu}_\sigma > 0\}$, with the edge representing the split $\sigma \in RB(\delta)$ given the weight $\bar{\mu}_\sigma(\delta)$.

Lemma 5. *Given two incompatible splits σ_1 and σ_2, there exists an $i \in \{1, 2\}$ such that $\bar{\mu}_{\sigma_i} \leq 0$, and this can be computed in $O(n)$ time.*

Proof. Let $\sigma_1 = U_1|V_1$ and $\sigma_2 = U_2|V_2$. Since σ_1 and σ_2 are incompatible, the sets $A = U_1 \cap U_2$, $B = U_1 \cap V_2$, $C = V_1 \cap U_2$ and $D = V_1 \cap V_2$ are all non-empty. From the bitvector representations of σ_1 and σ_2 these four sets can be computed in time $O(n)$. Since $|A| \cdot |B| \cdot |C| \cdot |D| \geq n - 3$ and $\beta_{ab|cd} + \beta_{ac|bd} \leq 0$ for every $a \in A, b \in B, c \in C$ and $d \in D$ by (2), we can find at least $n-3$ pairs of quartets (q^1_i, q^2_i), where q^1_i and q^2_i contain the same four species and $1 \leq i \leq n-3$, such that $q^1_i \in q(\sigma_1), q^2_i \in q(\sigma_2)$ and $\beta_{q^1_i} + \beta_{q^2_i} \leq 0$. Thus, we have

$$\sum_{i=1}^{n-3} \beta_{q^1_i} + \beta_{q^2_i} \leq 0 \,,$$

which implies

$$\sum_{i=1}^{n-3} \beta_{q^1_i} \leq 0 \quad \text{or} \quad \sum_{i=1}^{n-3} \beta_{q^2_i} \leq 0 \,.$$

It follows that $\bar{\mu}_{\sigma_1} \leq 0$ or $\bar{\mu}_{\sigma_2} \leq 0$. By calculating the two sums $\sum_{i=1}^{n-3} \beta_{q^1_i}$ and $\sum_{i=1}^{n-3} \beta_{q^2_i}$ in time $O(n)$ we get two upper bounds for $\bar{\mu}_{\sigma_1}$ and $\bar{\mu}_{\sigma_2}$ and can discard at least one of the two splits.

The following lemma is due to Bryant and Moulton and forms the basis of the incremental algorithms presented in [3,5,6] as well as the algorithm we present in this paper.

Lemma 6 (Bryant and Moulton [6, Proposition 3]).
Suppose $|X| > 4$, and fix $x \in X$. If $\sigma = U|V$ is a split in $RB(\delta)$ with $x \in U$, and $|U| > 2$, then either $U|V \in B_x(\delta)$ or $U - \{x\}|V \in RB(\delta|_{X-\{x\}})$ or both.

3 Maintaining a Set of Compatible Splits

The running time of our algorithm for computing refined Buneman trees is dominated by the maintenance of a set of compatible splits represented by an X-tree T. In this section, we consider how to support the operations below on X-trees. Recall that we represent a split $U|V$ by a bit-vector A such that $x_i \in U$ if and only if $A[i] = 0$.

- Incompatible(T, σ) Return a split σ' in T that is incompatible with σ. If all splits in T are pairwise compatible with σ then return nil.
- Insert(T, σ) Insert a new split σ into T. It is assumed that σ is pairwise compatible with all existing splits in T.
- Delete(T, σ) Remove the split σ from T.

Theorem 1. *The operations* Incompatible, Insert, *and* Delete *can be supported in time $O(n)$, where $n = |X|$.*

Proof. For the operation Incompatible(T, σ), where $\sigma = U|V$, we root T at an arbitrary leaf, and by a depth first traversal of T for each node v of T compute the number of leaves below v which are in respectively U and V in time $O(n)$. If the parent edge of a node v represents the split $\sigma' = U'|V'$, where U' are the elements below v, then the two counts represent respectively $|U' \cap U|$ and $|U' \cap V|$. From the equalities $|V' \cap U| = |U| - |U' \cap U|$ and $|V' \cap V| = |V| - |U' \cap V|$, we can now in constant time decide if σ' is incompatible with σ, since σ and σ' by definition are incompatible if and only if $|U' \cap U|$, $|U' \cap V|$, $|V' \cap U|$, and $|V' \cap V|$ are all non-zero. We return the first incompatible split found during the traversal of T. If no edge represents a split incompatible with σ, we return nil.

To perform Delete(T, σ) we in linear time find the unique edge (v, u) representing the split σ, by performing a depth first traversal to locate the node v, where the subtree rooted at v contains all elements from U and no element from V or vice versa. Finally we remove the edge (v, u), where u is the parent of v, by contracting v and u into a single node inheriting the incident edges of both nodes.

Finally consider Insert(T, σ), where $\sigma = U|V$. We claim that, since σ is assumed pairwise compatible with all splits in T, there exist a node v, such that removing v and its incident edges leaves us with a set of subtrees where each subtree contains only elements from either U or V. We prove the existence of v below. To locate v we similar to the Incompatible operation root T at an arbitrary leaf and bottom-up calculate for each node the number of leaves below in respectively U and V. We stop when we find the node v described above. We replace v by two nodes v_U and v_V connected by the edge $e = (v_U, v_V)$. Each subtree incident to v containing only elements from respectively U or V is made incident to respectively v_U or v_V. This ensures that e represents the split σ, and that all other edges remain representing the same set of splits.

What remains is to show that such a node v exists. If $|U| = 1$ or $|V| = 1$ the statement is trivially true. Otherwise, assume $|U| > 1$ and $|V| > 1$, and that the root r is a leaf in U. Let a be a leaf in V. We now argue that the lowest node v on the path from a to the root r containing at least one element from U in its subtree is the node required. Let u be the predecessor of v on the path from a to the root r. By definition u only contains elements from V in its subtree. Let u' be a sibling of u that contains at least one leaf b from U in its subtree. Assume now for the sake of contradiction that removing v and its incident edges leaves us with a set of subtrees including a subtree containing elements $c \in U$ and $d \in V$. Consider the case that c and d are not contained in the subtree of u', but in a subtree that was connected to v with an edge representing a split $U'|V'$, where $c \in V'$ and $d \in V'$. Then the splits $U|V$ and $U'|V'$ are incompatible, since $a \in V \cap U'$, $b \in U \cap U'$, $c \in U \cap V'$, and $d \in V \cap V'$. Otherwise if c and d are contained in the subtree of u', then let $U'|V'$ be the split represented by the edge (u', v). The splits $U|V$ and $U'|V'$ are then incompatible by $a \in V \cap U'$, $r \in U \cap U'$, $c \in U \cap V'$, and $d \in V \cap V'$.

4 Computing Refined Buneman Trees

We compute the refined Buneman tree for X by computing a sequence of sets of splits $C_4, \ldots, C_n$, such that each C_k is a set of compatible splits that is an over-approximation of the refined Buneman splits for X_k, i.e. $C_k \supseteq RB(\delta_k)$, where $\delta_k = \delta|_{X_k}$. Each iteration makes essential use of the characterization given by Lemma 6, that enables us to compute C_{k+1} from C_k together with the anchored Buneman tree for X_{k+1} with anchor x_{k+1}. To avoid a blow up in the number of splits, we use the observation that given two incompatible splits, we by Lemma 5 can discard one of the splits as not being a refined Buneman split. By computing the refined Buneman scores for the final set of splits C_n we can exclude all splits with a non-positive refined Buneman score, and obtain the refined Buneman tree $RB(\delta)$ for X. In the following we assume that all sets of compatible splits over X_k are represented by their X_k-tree, i.e. the space usage for storing a compatible set of splits is $O(k)$.

Theorem 2. *Given a dissimilarity measure δ for n species, the refined Buneman tree $RB(\delta)$ can be computed in time $O(n^3)$ and space $O(n^2)$.*

Proof. Pseudo code for the algorithm is contained in Figure 2. The operations Insert, Delete and Incompatible are the operations on a set of compatible splits as described in Section 3. The operation DiscardRight? takes two incompatible splits and returns true/false if the second/first split has been verified not to be a refined Buneman split, c.f. Lemma 5.

In lines 1-11 we compute a sequence of sets of compatible splits $C_4, \ldots, C_n$, such that $C_k \supseteq RB(\delta_k)$. In line 1 we let C_4 be an over-approximation of $B_{x_4}(\delta_4)$,

```
1.   C_4 := B_{x_4}(δ_4)
2.   for k = 5 to n
3.       C_k := B_{x_k}(δ_k)
4.       for U|V ∈ C_{k-1}
5.           for σ ∈ {U ∪ {x_k}|V , U|V ∪ {x_k}}
6.               σ' := Incompatible(C_k, σ)
7.               while σ' ≠ nil and DiscardRight?(σ, σ')
8.                   Delete(C_k, σ')
9.                   σ' := Incompatible(C_k, σ)
10.              if σ' = nil
11.                  Insert(C_k, σ)
12.  Compute refined Buneman index for C_n and discard splits
     with a non-positive score
```

Fig. 2. The overall algorithm for computing the refined Buneman tree

which satisfies $C_4 \supseteq RB(\delta_4)$ since each refined Buneman split for a set of size four must also be contained in any anchored Buneman split. In lines 2-11 we (based on Lemma 6) inductively compute C_k from C_{k-1} by letting C_k be the set of splits

$$B_{x_k}(\delta_k) \;\cup \bigcup_{U|V \in C_{k-1}} \{U \cup \{x_k\}|V \,,\, U|V \cup \{x_k\}\} \,,$$

except for some incompatible splits that we explicitly verify not being in $RB(\delta_k)$ (lines 6-11).

Since C_k and $B_{x_k}(\delta_k)$ are sets of compatible splits, both contain at most $2k-3$ splits. It follows that in an iteration of lines 3 – 11 at most $(2k-3) + 2(2(k-1)-3) \leq 6k$ splits can be inserted and deleted from C_k. The number of calls to DiscardRight? and Incompatible is bounded by the number of insertions and deletions of splits. Since by Lemma 4 computing $B_{x_k}(\delta_k)$ takes time $O(k^2)$ and each operation on a set of compatible splits takes time $O(k)$, it follows that the total time spent in an iteration of lines 3-11 is $O(k^2)$, i.e. for lines 1-11 the total time used is $O(n^3)$. Since we for each iteration of the for loop in line 2 only require access to C_{k-1} and C_k, which are represented by X-trees, it follows that the space usage for lines 1-11 is $O(n)$ (not counting the space usage for the dissimilarity measure), if we discard C_{k-2} at the beginning of iteration k.

In Section 5 we describe how to compute the refined Buneman indexes for C_n in time $O(n^3)$ and space $O(n^2)$, i.e. it follows that the total time and space usage is respectively $O(n^3)$ and $O(n^2)$.

The algorithms in [3,5,6] are based on a similar approach as the algorithm described above, but use the stronger requirement that $C_k = RB(\delta_k)$. A central feature of our relaxed computation is that the number of computations of refined Buneman scores for a set of compatible splits is reduced from $n-3$ to a single computation as the final step of the algorithm.

The algorithm described above in line 3 initializes C_k to be the anchored Buneman tree $B_{x_k}(\delta_k)$ by applying Lemma 4. As a simplification, we note that since the algorithm is based on over-approximation, it remains valid if we in line 3 just require C_k to be a compatible set of splits containing at least all splits of $B_{x_k}(\delta_k)$. Berry and Bryant [3, Theorem 2] prove that the single linkage clustering tree for x_k has this property. Single linkage trees can be found in time $O(n^2)$ using spanning tree based methods [2,8,11].

5 Refined Buneman Indexes

Given an X-tree where the edges E represent a set of compatible splits, we in this section describe how to compute the refined Buneman indexes for the set of splits in time $O(n^3)$ and space $O(n^2)$. The previously best algorithm for this subtask uses time $O(n^4)$ [5, Lemma 3.2] assuming that the scores of the quartets are given in sorted order.

For each edge e, our algorithm finds the $n-3$ quartets of smallest Buneman score induced by e. The refined Buneman indexes for all edges can then be computed according to (3) in time $O(n^2)$.

To identify for each split the quartets with smallest Buneman score, we assume an arbitrary ordering of the species and adopt the following terminology. Let $ab|cd$ be a quartet, where a is the smallest named species among the four species a, b, c and d in the assumed ordering of all species. Motivated by the definition of Buneman scores (1), we consider each quartet $ab|cd$ as two *diagonal* quartets which we denote $ab||cd$ and $ab||dc$. The score of a diagonal quartet $ab||cd$ is defined as $\eta_{ab||cd} = (\delta(b,c) - \delta(a,b) + \delta(a,d) - \delta(d,c))/2$. From the definitions we have $\beta_{ab|cd} = \min\{\eta_{ab||cd}, \eta_{ab||dc}\}$.

Instead of searching for quartets with increasing score we search for diagonal quartets with increasing score. This has the disadvantage that each quartet can be found up to two times (only one time if $c = d$). We say that $ab||cd$ is the minimum diagonal of $ab|cd$, if $\eta_{ab||cd} < \eta_{ab||dc}$ or $\eta_{ab||cd} = \eta_{ab||dc}$ and c is the smallest named species among c and d. Otherwise $ab||dc$ is the minimum diagonal. Note that the Buneman score of $ab|cd$ equals the score of the minimum diagonal. When identifying $ab||cd$ we can by inspecting the quartet check if $ab||cd$ is the minimum diagonal of $ab|cd$; if so we identify $ab|cd$. Otherwise, $ab|cd$ has already been identified by $ab||dc$ since the diagonal quartets are visited in order of increasing score.

The main property of diagonal quartets which we exploit is that for fixed a and c, we can search *independently* for b and d to find the diagonal quartet $ab||cd$ of minimum diagonal score: Find respectively b and d such that respectively $\delta(b,c) - \delta(a,b)$ and $\delta(a,d) - \delta(d,c)$ are minimal.

For an edge e defining the split $U|V$ and $a \in U$ and $c \in V$, where a is the smallest named species among a and c, let $U^e_{ac} = b_1, \ldots, b_{|U^e_{ac}|} \subseteq U$ and $V^e_{ac} = d_1, \ldots, d_{|V^e_{ac}|} \subseteq V$ be the sets of species named at least a, and where b_i and d_j appear in sorted order with respect to increasing $\delta(b_i, c) - \delta(a, b_i)$ and $\delta(a, d_j) - \delta(d_j, c)$ value. We can consider all $ab_i||cd_j$ as entries of a matrix

M^e_{ac}, where $(M^e_{ac})_{i,j} = \eta_{ab_i||cd_j}$. The crucial property of M^e_{ac} is that each row and column is monotonic non-decreasing. This allows us to construct M^e_{ac} in a lazy manner while exploring the diagonal quartets, starting with only computing $(M^e_{ac})_{1,1}$ which we denote the minimal score of the pair (a, c).

For each edge e we will lazily construct a subset Q_e of the diagonal quartets induced by e. We represent each Q_e by a linked list. To identify the $n-3$ quartets with smallest Buneman score it is sufficient to identify the $2(n-3)$ pairs (a, c) with smallest minimum score. Since for a quartet there are at most two diagonal quartets, the $n-3$ quartets induced by e with smallest Buneman score will have minimum diagonal quartets with (a, c) among the $2(n-3)$ pairs found.

```
1.  for e ∈ E
2.     Q_e := ∅
3.  for (a,c) ∈ X^2 and a < c
4.     for each edge e on the path from a to c
5.        find b_e on the same side of e as a with δ(b_e,c) − δ(a,b_e) minimal and b_e ≥ a
6.     for each edge e on the path from c to a
7.        find d_e on the same side of e as c with δ(a,d_e) − δ(d_e,c) minimal and d_e > a
8.     for each edge e on the path from a to c
9.        Q_e := Q_e ∪ {ab_e||cd_e}
10.       if |Q_e| ≥ 3(n − 3)
11.          remove the n − 3 quartets with largest score from Q_e
12. for e ∈ E
13.    S_e := ∅
14.    while |S_e| < n − 3
15.       ab_i||cd_j := DeleteMin(Q_e)
16.       if ab_i||cd_j is a minimum diagonal
17.          Insert(S_e, ab_i|cd_j)
18.       Insert(Q_e, ab_{i+1}||cd_1) provided j = 1 and b_{i+1} exists
19.       Insert(Q_e, ab_i||cd_{j+1}) provided d_{j+1} exists
```

Fig. 3. Algorithm for computing the $n-3$ smallest Buneman scores induced by each edge of an X-tree

The pseudo code for the algorithm to find the $n-3$ quartets for each split is given in Figure 3. In lines 1-11 we identify between $2(n-3)$ and $3(n-3)$ pairs (a, c) with smallest minimal score.

In lines 4-5 and 6-7 we find the b_1 and d_1 species for entries $(M^e_{ac})_{1,1}$. Note that the two loops process the edges between a and c in different directions. Since the set of possible species b (species d) increases along the path from a to c (from c to a), we can compute the species b_e (species d_e) from the minimum found so far for the predecessor edge on the path together with the new species not considered yet. For each pair (a, c) we will then spend a total time of $O(n)$ in lines 4-7.

In lines 10-11 we for an edge e remove the 1/3 of the pairs (a, c) computed with largest minimum score if $|Q_e|$ becomes $3(n-3)$, leaving the $2(n-3)$ pairs with smallest minimum score in Q_e. This ensures that for each of the n edges we at most have to store $3(n-3)$ pairs, in total bounding the space required by $O(n^2)$. Line 11 can be performed in $O(n)$ time using e.g. the selection algorithm in [15], i.e. amortized $O(1)$ time for each element deleted from Q_e. In total we spend time $O(n^3)$ in lines 1-11 and use space $O(n^2)$.

In lines 12-19 we extract for each edge e the $n-3$ quartets S_e with smallest Buneman score in sorted order. In line 15 we delete the next diagonal quartet from Q_e with smallest diagonal score. If Q_e contains several diagonal quartets with the same score we first delete those which are minimum diagonals.

In line 18 we ensure that if the j first entries of row i of M^e_{ac} have been considered, then $(M^e_{ac})_{i,j+1}$ is inserted in Q_e. Similarly in line 19 we ensure that if the i first entries in the first column of M^e_{ac} have been considered then $(M^e_{ac})_{i+1,1}$ is inserted into Q_e. To find the relevant b_{i+1} in line 18 (d_{j+1} in line 19), we make a linear scan of the subtree incident to e which contains a (respectively c). The species $b_{i+1} \geq a$ should have the smallest value $\delta(b_{i+1}, c) - \delta(a, b_{i+1}) \geq \delta(b_i, c) - \delta(a, b_i)$; in case the expressions are equal then the smallest $b_{i+1} > b_i$. Similarly, the species $d_{j+1} > a$ should have the smallest $\delta(a, d_{j+1}) - \delta(d_{j+1}, c) \geq \delta(a, d_j) - \delta(d_j, c)$; and in case the expressions are equal then the smallest $d_{j+1} > d_j$.

The for-loop in lines 12-19 is performed n times, and the while-loop in lines 14-19 is performed at most $2(n-3)$ times for each edge, since each iteration considers one diagonal quartet. Each of the $2(n-3)$ deletions from Q_e inserts at most two diagonal quartets into Q_e, i.e. $|Q_e| \leq 5(n-3)$. It follows that DeleteMin in line 15 takes time $O(n)$. Finally, lines 18 and 19 each require time $O(n)$. The total time used by the algorithm becomes $O(n^3)$ and the space usage is $O(n^2)$.

Theorem 3. *The refined Buneman indexes for all splits in a given X-tree can be computed in time $O(n^3)$ and space $O(n^2)$.*

References

1. H.-J. Bandelt and A. W. Dress. Reconstructing the shape of a tree from observed dissimilarity data. *Advances in Applied Mathematics*, 7:309–343, 1986.
2. J.-P. Barthélémy and A. Guénoche. *Trees and Proximity Representations.* John Wiley & Sons, 1991.
3. V. Berry and D. Bryant. Faster reliable phylogenetic analysis. In *Proc. 3rd International Conference on Computational Molecular Biology (RECOMB)*, pages 69–69, 1999.
4. V. Berry and O. Gascuel. Inferring evolutionary trees with strong combinatorial evidence. *Theoretical Computer Science*, 240:271–298, 2000.
5. D. Bryant and V. Berry. A structured family of clustering and tree construction methods. *Advances in Applied Mathematics*, 27(4):705–732, 2001.
6. D. Bryant and V. Moulton. A polynomial time algorithm for constructing the refined buneman tree. *Applied Mathematics Letters*, 12:51–56, 1999.

7. P. Buneman. The recovery of trees from measures of dissimilarity. In F. Hodson, D. Kendall, and P. Tautu, editors, *Mathematics in Archaeological and Historical Sciences*, pages 387–395. Edinburgh University Press, 1971.
8. J. C. Gower and J. G. S. Ross. Minimum spanning trees and single-linkage cluster analysis. *Applied Statistics*, 18:54–64, 1969.
9. D. Gusfield. Efficient algorithms for inferring evolutionary trees. *Networks*, 21:19–28, 1991.
10. D. Huson. Splitstree: a program for analyzing and visualizing evolutionary data. *Bioinformatics*, 14(1):68–73, 1998. (http://www-ab.informatik.uni-tuebingen.de/software/splits/welcome_en.html).
11. B. Leclerc. Description combinatoire des altramétriqueès. *Math. Sci. Hum.*, 73:5–37, 1981.
12. V. Moulton and M. Steel. Retractions of finite distance functions onto tree metrics. *Discrete Applied Mathematics*, 91:215–233, 1999.
13. M. Nei and S. Kumar. *Molecular Evolution and Phylogenetics.* Oxford University Press, 2000.
14. N. Saitou and M. Nei. The neighbor-joining method: A new method for reconstructing phylogenetic trees. *Molecular Biology Evolution*, 4:406–425, 1987.
15. A. Schönhage, M. S. Paterson, and N. Pippenger. Finding the median. *Journal of Computer and System Sciences*, 13:184–199, 1976.

Distance Corrections on Recombinant Sequences

David Bryant[1], Daniel Huson[2], Tobias Kloepper[2], and Kay Nieselt-Struwe[2]

[1] McGill Centre for Bioinformatics
3775 University
Montréal, Québec, H3A 2B4
Canada
bryant@mcb.mcgill.ca,
[2] University of Tuebingen,
Center for Bioinformatics Tuebingen,
Sand 14
D-72076 Tuebingen, Germany
huson,kloepper,nieselt@informatik.uni-tuebingen.de

Abstract. Sequences that have evolved under recombination have a 'mosaic' structure, with different portions of the alignment having evolved on different trees. In this paper we study the effect of mosaic sequence structure on pairwise distance estimates. If we apply standard distance corrections to sequences that evolved on more than one tree then we are, in effect, correcting according to an incorrect model. We derive tight bounds on the error introduced by this model mis-specification and discuss the ramifications for phylogenetic analysis in the presence of recombination.

1 Introduction

Generally, phylogenetic analysis works under the assumption that the homologous sequences evolved along a single, bifurcating tree. Recombination, gene conversion and hybridisation can all lead to violations of this basic assumption and give rise to 'mosaic' sequences, different parts of which evolved along different trees [12,21].

Simulation experiments have established that a mosaic sequence structure can have a marked effect on phylogenetic reconstruction and evolutionary parameter estimation [13,17]. Our goal in this article is to characterise this effect theoretically. Standard distance corrections assume that the sequences evolved on a single evolutionary tree, so if we correct distance estimates using these methods we are essentially correcting according to an incorrect model. We show that the effect of this model mis-specification is relatively small and derive explicit bounds for the bias introduced by this failure to account for mosaic sequence structure.

The result has important applications in conventional phylogenetic analysis. As we shall discuss in Section 5.1, our characterisation of distance corrections on mosaic sequences provides theoretical explanations for the various forms of bias

G. Benson and R. Page (Eds.): WABI 2003, LNBI 2812, pp. 271–286, 2003.

observed experimentally in [17]. We can also apply the result to discuss the effect of rate heterogeneity on phylogenetic reconstruction. Our observations complement the inconsistency results of [5] by limiting the zone for which distance based methods are inconsistent.

However our principal motivation for this investigation was to better understand the behavior of distance based phylogenetic network algorithms like split decomposition [2] and NeighborNet [3]. We show that recombinant phylogenetic information is indeed retained in corrected distance matrices, and justify the family of network approaches that decompose distance matrices into weighted collections of split metrics. NeighborNet and split decomposition are two members of this family, though there is potential, and perhaps need, for several more.

Perhaps most importantly, we can finally provide a theoretical interpretation of the form and branch lengths of the splits graph. A splits graph is not a reconstruction of evolutionary history: the internal nodes in a splits graph should not be identified with ancestral sequences [18]. These networks had long been justified only in the weak sense that they 'represent' some kind of structure in the data. As we will discuss, we can consistently view a splits graph as an estimation of the splits appearing in the input trees (or, under a Bayesian intepretation, the splits in trees with a high posterior probability).

To illustrate, suppose that we have a collection of sequences that evolved on two different trees. Even so we compute and correct distances over all the sites, giving a 'wrongly corrected' distance matrix d. Suppose that one third of the sites evolved on T_1 and two thirds on T_2. If d_{T_1} is the matrix of path length distances for T_1 and d_{T_2} is the distance matrix for T_2 then, as we will show, the 'wrongly corrected' distance d will closely approximate the weighted sum $1/3d_{T_1} + 2/3d_{T_2}$, as the sequence length increases. Since split decomposition is consistent on distance matrices formed from the sum of two distance matrices, the splits graph produced will exactly represent the splits in T_1 and T_2. Furthermore, the weights of the splits in this graphs will be a weighted sum of the corresponding branch lengths in T_1 and T_2 (where a split has length 0 in a tree that doesn't contain it).

This interpretation of splits graphs ignores some fundamental limitations of the various network methods: existing methods are consistent on particular collections of distance matrices and, as with tree based analysis, are affected by sampling error and model mis-specification. However having the correct theoretical interpretation should enable researchers to better design network methods that overcome these difficulties.

2 Background

2.1 Markov Evolutionary Models

We briefly outline the aspects of Markov processes we need for the paper. For further details, refer to [15,19] or any text book on molecular evolution.

Sequence evolution along a branch is typically modelled using a Markov process. The process is determined by an $n \times n$ *rate matrix* Q, where $Q_{ij} > 0$ for all $i \neq j$ and $Q_{ii} = -\sum_{j \neq i} Q_{ij}$. Nucleotide models have $n = 4$, while amino acid models have $n = 20$. We assume that the process is *time reversible*, which means that there exist positive $\pi_1, \ldots, \pi_n$ such that $\sum_{i=1}^{n} \pi_i = 1$ and $\pi_i Q_{ij} = \pi_j Q_{ji}$ for all i, j. The values π_i correspond to the equilibrium frequency for the process. We assume that the process starts in equilibrium. Let Π denote the diagonal matrix with $\pi_1, \pi_2, \ldots, \pi_n$ on the diagonal.

Suppose that we run the process for time t. The probability of observing state j at time t conditional on being at state i at time 0 equals $P_{ij}(t)$, where $P(t)$ is the *evolutionary matrix*

$$P(t) = e^{Qt} = \sum_{m=0}^{\infty} \frac{Q^m t^m}{m!}.$$

If we assume that the states are in equilibrium at the start then the probability of having i at time 0 and j at time t equals $X_{ij}(t) = \pi_i P_{ij}(t)$. The matrix $X(t) = \Pi P(t)$ is called the *divergence matrix*.

The *mutation rate* r_Q is the expected number of mutations per unit time, and can be shown to equal the sum of the off-diagonal elements of $\frac{d}{dt} X(t)|_{t=0} = \Pi Q$. Hence

$$r_Q = \sum_{i=1}^{n} \sum_{j \neq i} \pi_i Q_{ij} = -\sum_{i=1}^{n} \pi_i Q_{ii} = -\mathrm{tr}(\Pi Q)$$

where $\mathrm{tr}(A)$ denotes the trace of a matrix A. The expected number of changes between time 0 and time t therefore equals $r_Q t$. This is the standard unit for measuring evolutionary divergence.

This general description includes many specific Markov models. The simplest for nucleotide sequences is the Jukes-Cantor model [9]. The rate matrix for this model is

$$Q = \begin{bmatrix} -3\alpha & \alpha & \alpha & \alpha \\ \alpha & -3\alpha & \alpha & \alpha \\ \alpha & \alpha & -3\alpha & \alpha \\ \alpha & \alpha & \alpha & -3\alpha \end{bmatrix}$$

which has only one parameter $\alpha > 0$. Substituting into the above formulae we see that the evolutionary matrix for this model is specified by

$$P_{ij}(t) = \begin{cases} \frac{1}{4} + \frac{3}{4} e^{-4\alpha t} & \text{when } i = j; \\ \frac{1}{4} - \frac{1}{4} e^{-4\alpha t} & \text{when } i \neq j, \end{cases}$$

while the mutation rate is $r_Q = 3\alpha$. Thus letting $\alpha = 1/3$ gives a model with expected mutation rate of 1 per unit time.

We assume that evolution of different sites to be independent. We use $s[i]$ to denote the state at site i in sequence s. The probability of observing sequence s_2 after time t given s_1 at time 0 is then given by

$$P(s_2 | s_1, t) = \prod_i P_{s_1[i] s_2[i]}(t).$$

2.2 Distance Corrections

Rodríguez et al. [15] describe a general method for estimating the evolutionary distance $r_Q t$ between two sequences s_1 and s_2. Let F_{ij} denote the proportion of sites for which s_1 has an i and s_2 has a j (Actually, we can obtain better results using $\frac{1}{2}(F_{ij} + F_{ji})$ since F should be symmetric). The general time reversible (GTR) correction is given by

$$\hat{d} = r_Q \hat{t} = -\mathrm{tr}(\Pi \log(\Pi^{-1} F))$$

where log is the matrix logarithm defined by

$$\log(I + A) = A - \tfrac{1}{2}A^2 + \tfrac{1}{3}A^3 - \tfrac{1}{4}A^4 + \cdots .$$

The correction formula is consistent: if $F = X(t)$ then

$$\begin{aligned} -\mathrm{tr}(\Pi \log(\Pi^{-1} F)) &= -\mathrm{tr}(\Pi \log(e^{Qt})) \\ &= -\mathrm{tr}(\Pi Q)t \\ &= r_Q t. \end{aligned}$$

Most of the standard corrections can be derived from this general formula. For example, under the Jukes-Cantor model, suppose that we have observed that the proportion of changed sites equals p. Our estimate for $X(t)$ would then be the matrix with $p/12$ on the off-diagonal (there are 12 such entries) and $(1-p)/4$ on the diagonal. Substituting into the general formula, we obtain the standard Jukes-Cantor correction $r_Q t = -\frac{3}{4}\log(1 - \frac{4}{3}p)$.

2.3 Trees, Splits, Splits Graphs, and Distance Matrices

A *phylogenetic tree* is a tree with no vertices of degree two and leaves identified with the set of taxa X. A *split* $A|B$ is a partition of the taxa set into two non-empty parts. Removing an edge from a phylogenetic tree T induces a split of the taxa set. The set of splits that can be obtained in this way from T is called the *splits of* T and denoted $\Sigma(T)$. A given set of splits is *compatible* if it is contained within the set of splits of some tree.

A *splits graph* is a bipartite connected graph G with a partition $E(G) = E_1 \cup E_2 \cup \cdots \cup E_k$ of $E(G)$ into disjoint sets such that no shortest path contains more than two edges from the same block and, for each i, $G - E_i$, consists of exactly two components. (This definition is equivalent to the definition of [6]). Some of the vertices are labelled by elements of X so that each edge cut E_i induces a split of X. The set of these splits is denoted $\Sigma(G)$. Note that every set of splits can be represented by a splits graph, but this graph is not necessarily unique. Every phylogenetic tree is a splits graph with every edge in a different block.

Suppose that we assign lengths to the edges of T. The *additive distance* $d_T(x, y)$ between two taxa x, y equals the sum of the edge lengths along the path separating them. We can also assign length to the edges in a splits graph

G, where all edges in the same block are assigned the same length. The distance $d_G(x,y)$ between two taxa in G is then the length of the shortest path connecting them.

Both d_T and d_G have an equivalent formulation in terms of splits. The *split metric* $\delta_{A|B}$ for a split $A|B$ is the (pseudo-)metric

$$\delta_{A|B}(x,y) = \begin{cases} 1 & \text{if } x, y \text{ are on different sides of } A|B; \\ 0 & \text{otherwise.} \end{cases}$$

Let $\lambda_{A|B}$ denote the length of the edge (or in the case of splits graphs, edges) corresponding to $A|B$. Both $d_T(x,y)$ and $d_G(x,y)$ equal the sum of the edge lengths for all of the splits that separate x and y. Hence

$$d_T(x,y) = \sum_{A|B \in \Sigma(T)} \lambda_{A|B}\delta_{A|B}(x,y) \quad \text{and} \quad d_G(x,y) = \sum_{A|B \in \Sigma(G)} \lambda_{A|B}\delta_{A|B}(x,y).$$

Split decomposition [2] and NeighborNet [3] both take a distance metric d and compute a decomposition

$$d = \epsilon + \sum_{A|B} \lambda_{A|B}\delta_{A|B}$$

of d into a positive combination of split metrics and an error term ϵ. Furthermore, both methods are *consistent* over a large class of metrics. If a set of splits $\mathcal{S}$ is *weakly compatible* and

$$\bar{d}(x,y) = \sum_{A|B \in \mathcal{S}} \bar{b}(A|B)\delta_{A|B}$$

then split decomposition will recover the splits $A|B$ as well as the coefficients $\bar{b}$ [2]. NeighborNet will recover this decomposition when the set of splits $\mathcal{S}$ is circular [4].

3 Correcting Distances Estimated from Mosaic Sequences

In a mosaic alignment, different sites evolved along different trees. Correction formulae, such as those described above, make the assumption that the sequences evolved on the same tree. When we apply these corrections to mosaic sequences we are correcting according to an incorrect model.

We show here that the distance correction formulae work just how we would hope, at least up to a small error term. Correcting a heterogeneous collection of sequences using a homogeneous model does introduce error, but the error is quite small compared to the distances themselves.

Suppose that the sequences evolved under the same model on k different trees $T_1, T_2, \ldots, T_k$. Furthermore, for each i, suppose that the proportion of sites coming from T_i is q_i. Let s_1 and s_2 be the sequences for two taxa, and let

$d_1, d_2, \ldots, d_k$ be the expected number of mutations on the path between these taxa on trees $T_1, T_2, \ldots, T_k$. We use

$$\mathbb{E}[d] = \sum_{i=1}^{k} q_i d_i$$

and

$$\mathrm{var}[d] = \sum_{i=1}^{k} q_i (d_i - \mathbb{E}[d])^2$$

to denote the mean and variance of the d_i's.

Let $\hat{F}_{ab}$ denote the proportion of sites with an a in sequence s_1 and a b in sequence s_2. Then $\hat{F}$ will approach

$$F = \sum_{i=1}^{k} q_i e^{Qt_i}$$

as the sequences become sufficiently long. We obtain upper and lower bounds on the distance estimate computed from F.

First, however, we need to prove a small result in matrix analysis.

Lemma 1. *Suppose that all eigenvalues of an $n \times n$ matrix X are real and non-negative, and that there is a diagonal matrix $D > 0$ such that DX is symmetric. Then* $\mathrm{tr}(DX) \geq 0$.

Proof

Since $D > 0$ the inverse D^{-1} and square root of the inverse $D^{-\frac{1}{2}}$ both exist. Define the matrix Y by

$$Y = D^{-\frac{1}{2}}(DX)D^{-\frac{1}{2}} = D^{\frac{1}{2}}XD^{-\frac{1}{2}}.$$

Then Y is symmetric and has the same non-negative eigenvalues as X. It follows that Y is positive semi-definite with non-negative diagonal entries. As $D^{-\frac{1}{2}} > 0$, the matrix DX has non-negative diagonal entries and $\mathrm{tr}(DX) \geq 0$. □

Theorem 1 *Let $F = \sum_{i=1}^{k} q_i e^{Qt_i}$ and let ρ_Q denote the constant $\frac{\mathrm{tr}(\Pi Q^2)}{(r_Q)^2}$. Then*

$$\mathbb{E}[d] - \tfrac{1}{2}\rho_Q \mathrm{var}[d] \leq -\mathrm{tr}(\Pi \log(\Pi^{-1}F)) \leq \mathbb{E}[d].$$

Proof

Let $\lambda_1, \lambda_2, \ldots, \lambda_n$ be the eigenvalues of Q. One of these is zero and all others are negative. Let $v_1, v_2, \ldots, v_n$ be a linearly independent set of eigenvectors, where $Qv_j = \lambda_j v_j$ for all j. Define $t_i = \frac{d_i}{r_Q}$ for all i, and $\bar{t} = \frac{\mathbb{E}[d]}{r_Q}$. Then

$$\begin{aligned}\mathbb{E}[d] - (-\mathrm{tr}(\Pi \log(\Pi^{-1}F))) &= \mathrm{tr}(\Pi(\log(\Pi^{-1}F) - Q\bar{t})) \\ &= \mathrm{tr}(\Pi A)\end{aligned}$$

where $A = (\log(\Pi^{-1}F) - Q\bar{t})$. The matrix A has the same eigenvectors as Q. Let α_j denote the eigenvalue of A corresponding to the eigenvector v_j. We derive lower and upper bounds on α_j.

For the lower bound we have

$$\begin{aligned} \alpha_j &= \log(\sum_{i=1}^{k} q_i e^{\lambda_j t_i}) - \lambda_j \bar{t} \\ &\geq \sum_{i=1}^{k} q_i \log(e^{\lambda_j t_i}) - \lambda_j \bar{t} \\ &= 0 \end{aligned}$$

with the inequality following from the concavity of the logarithm (or Jensen's inequality). It follows that A has only non-negative eigenvalues so, by Lemma 1, $\mathrm{tr}(\Pi A) \geq 0$. Thus

$$0 \leq \mathrm{tr}(\Pi A) = \mathrm{tr}(\Pi \log(\Pi^{-1}F)) - \mathrm{tr}(\Pi Q)\bar{t} = \mathrm{tr}(\Pi \log(\Pi^{-1}F)) + \mathbb{E}[d]$$

For the upper bound, let $x = \sum_{i=1}^{k} q_i(e^{\lambda_j t_i} - 1)$. Then $-1 < x \leq 0$ so $\log(1+x) \leq x - \frac{1}{2}x^2$ and

$$\begin{aligned} \alpha_j &= \log\left(1 + \sum_{i=1}^{k} q_i(e^{\lambda_j t_i} - 1)\right) - \lambda_j \bar{t} \\ &\leq \left(\sum_{i=1}^{k} q_i(e^{\lambda_j t_i} - 1)\right) - \tfrac{1}{2}\left(\sum_{i=1}^{k} q_i(e^{\lambda_j t_i} - 1)\right)^2 - \lambda_j \bar{t} \end{aligned}$$

If we set $y = \lambda_j t_i$ then $y \leq 0$ and $y \leq e^y - 1 \leq y + \frac{1}{2}y^2$. Thus

$$\begin{aligned} \alpha_j &\leq \sum_{i=1}^{k} q_i(\lambda_j t_i + \tfrac{1}{2}\lambda_j^2 t_i^2) - \tfrac{1}{2}\left(\sum_{i=1}^{k} q_i \lambda_j t_i\right)^2 - \lambda_j \bar{t} \\ &= \tfrac{1}{2}\lambda_j^2 \left[\sum_{i=1}^{k} q_i t_i^2 - \left(\sum_{i=1}^{k} q_i t_i\right)^2\right] \\ &= \tfrac{1}{2}\frac{\lambda_j^2}{(r_Q)^2}\mathrm{var}[d] \end{aligned}$$

Thus $(\frac{\mathrm{var}[d]}{2}Q^2/r_Q^2 - A)$ has non-negative eigenvalues, and

$$\begin{aligned} 0 &\leq \mathrm{tr}(\Pi(\tfrac{\mathrm{var}[d]}{2}Q^2/r_Q^2 - A)) \\ &= \tfrac{\mathrm{var}[d]}{2}\mathrm{tr}(\Pi Q^2)/r_Q^2 - \mathrm{tr}(\Pi \log(\Pi^{-1}F)) + \mathrm{tr}(\Pi Q)\bar{t} \\ &= \tfrac{1}{2}\rho_Q \mathrm{var}[d] - \mathrm{tr}(\Pi \log(\Pi^{-1}F)) - \mathbb{E}[d]. \end{aligned}$$

□

This general result implies error bounds for all the standard distance corrections. For example, under Jukes-Cantor, we have $\rho_Q = \frac{4}{3}$ so if the proportion of observed changes approaches p then

$$\mathbb{E}[d] - \frac{2}{3}\mathrm{var}[d_i] \leq -\frac{3}{4}\log(1-\frac{4}{3}p) \leq \mathbb{E}[d]$$

For K2P with parameter $\kappa = 2\frac{\text{expected num. transitions}}{\text{expected num. transversions}}$ we obtain the bound

$$\mathbb{E}[d] - \frac{\kappa^2+2\kappa+3}{(\kappa+2)^2}\mathrm{var}[d] \leq -\mathrm{tr}(\Pi \log(\Pi^{-1}F)) \leq \mathbb{E}[d].$$

The divergences between well aligned molecular sequences are typically small. In this case, the error bound $\frac{1}{2}\rho_Q\mathrm{var}[d]$ comes close to zero. Thus, when distances are small, the corrected distances approximate the convex combination of the distances from the different blocks of the mosaic sequences. We therefore have

Corollary 1. *Let $\hat{\mathbf{d}}$ be the corrected distances estimated from mosaic sequences evolved on trees $T_1, \ldots, T_k$, where for each i, the proportion of sites evolved on T_i is q_i. Let $\mathbf{d}_i$ be the distance matrix estimated from only those sites evolving on T_i. Then*

$$\hat{\mathbf{d}} \approx \sum_{i=1}^{k} q_i \mathbf{d}_i.$$

Even if the bias *is* sufficient to have a significant effect on distance estimates, this bias is well characterised. The lower bound $\mathbb{E}[d] - \frac{1}{2}\mathrm{var}[d]$ is very tight. The distance correction differs from $\mathbb{E}[d] - \frac{1}{2}\mathrm{var}[d]$ only by a term of order $O(d^3)$, as can be easily demonstrated from a Taylor series expansion.

Note also that the error bound $\frac{1}{2}\rho_Q\mathrm{var}[d]$ depends only on the variance of the block distances, and not on the number of different blocks. By taking k to infinity, we see that Theorem 1 extends directly to a continuous distribution on input distances.

4 Experimental Results

We performed two separate experiments to assess the tightness of the approximation established in Theorem 1 for the distance between two sequences. The parameters for each run were

- The number of contiguous blocks k, set to $k = 2$ or 5.
- The height h (in expected mutations) of the root in the coalescent.

The k contiguous segments were determined by randomly selecting $k-1$ breakpoints without repetition. This gave the proportions $q_1, q_2, \ldots, q_k$ of sites from each tree. The distances d_i for each contiguous segment were sampled by constructing a coalescent with 30 leaves and height h, using the protocol outlined

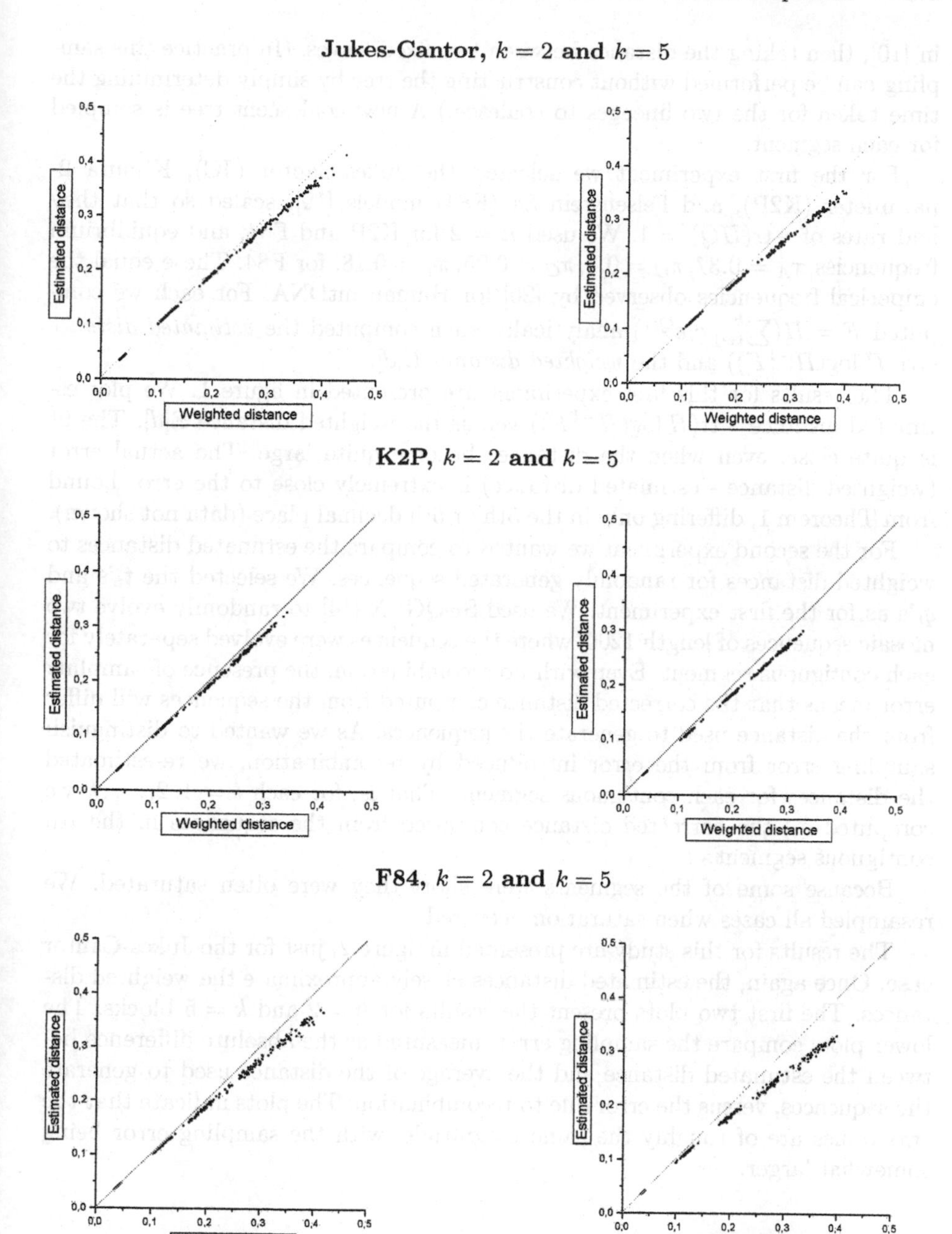

Fig. 1. Results of the first experiment. The estimated distance is computed using a distance correction applied to the whole sequence. The weighted distance is the mean $\mathbb{E}[d]$ of the distances for each contiguous block. Results are presented for Jukes-Cantor, Kimura 2-parameter (K2P) and the Felsenstein 84 model (F84).

in [10], then taking the distance between two fixed leaves. (In practice this sampling can be performed without constructing the tree by simply determining the time taken for the two lineages to coalesce.) A new coalescent tree is sampled for each segment.

For the first experiment we selected the Jukes-Cantor (JC), Kimura 2-parameter (K2P), and Felsenstein 84 (F84) models [19], scaled so that they had rates of $-\mathrm{tr}(\Pi Q) = 1$. We used $\kappa = 2$ for K2P and F84, and equilibrium frequencies $\pi_A = 0.37, \pi_C = 0.4, \pi_G = 0.05, \pi_T = 0.18$. for F84. These equal the emperical frequencies observed by [20] for Human mtDNA. For each we computed $F = \Pi(\sum_{i=1}^{k} q_i e^{Qt_i})$ analytically, then computed the *estimated distance* $-\mathrm{tr}(\Pi \log(\Pi^{-1} F))$ and the *weighted distance* $E[d]$.

The results for this first experiment are presented in figure 1. We plot estimated distance $-\mathrm{tr}(\Pi \log(\Pi^{-1} F))$ versus the weighted distance $\mathbb{E}[d]$. The fit is quite close, even when the distances become quite large. The actual error (weighted distance - estimated distance) is extremely close to the error bound from Theorem 1, differing only in the 5th or 6th decimal place (data not shown).

For the second experiment we wanted to compare the estimated distances to weighted distances for randomly generated sequences. We selected the t_i's and q_i's as for the first experiment. We used SEQGEN [14] to randomly evolve two mosaic sequences of length 1200, where the sequences were evolved separately for each contiguous segment. Even with no recombination, the presence of sampling error means that the corrected distance computed from the sequences will differ from the distance used to generate the sequences. As we wanted to distinguish sampling error from the error introduced by recombination, we re-estimated the distances for each contiguous segment. That is, for each $i = 1, 2, \ldots, k$ we computed a_i, the corrected distance computed from the sequences in the ith contiguous segment.

Because some of the segments were short they were often saturated. We resampled all cases when saturation occurred.

The results for this study are presented in figure 2, just for the Jukes-Cantor case. Once again, the estimated distances closely approximate the weighted distances. The first two plots present the results for $k = 2$ and $k = 5$ blocks. The lower plots compare the sampling error, measured as the absolute difference between the estimated distance and the average of the distance used to generate the sequences, versus the error due to recombination. The plots indicate that the two values are of roughly the same magnitude, with the sampling error being somewhat larger.

5 Applications

5.1 The Consequences of Recombination on Traditional Phylogenetic Analysis

Schierup and Hein [17] conducted an extensive simulation experiment to assess the effect of mosaic sequence structure on features of reconstructed phylogenies.

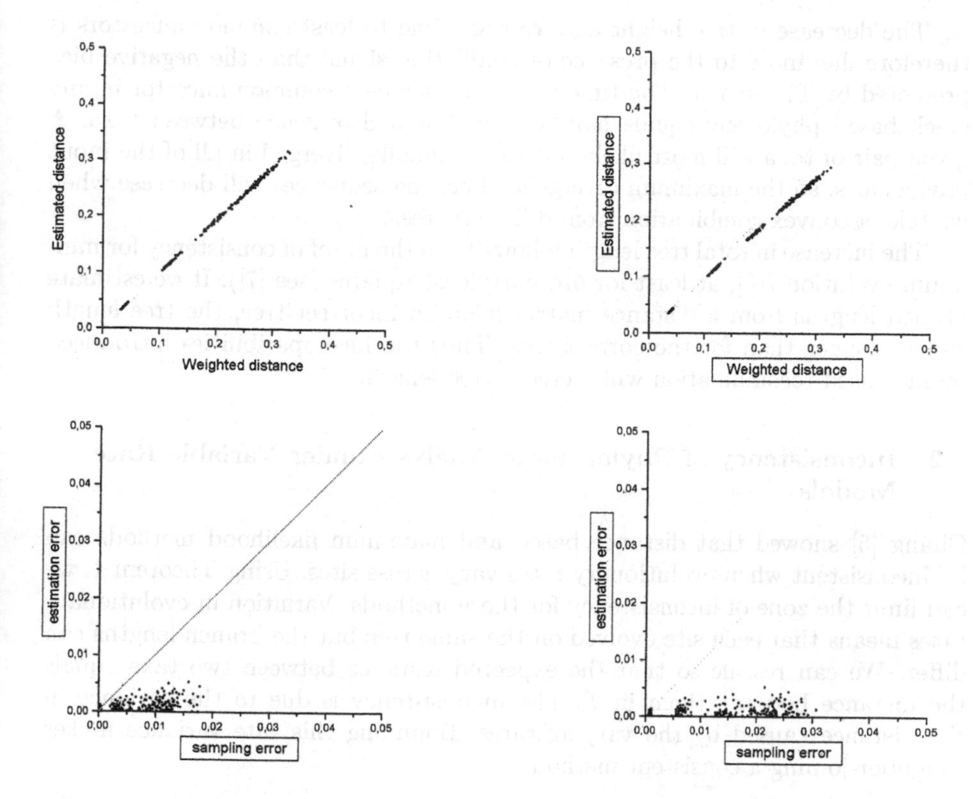

Fig. 2. Results of the second experiment. The top two plots give estimated distance versus weighted distance for JC ($k = 2, 5$). The two lower graphs plot the sampling error versus the error from the approximation .

They used the recombinant coalescent algorithm of [8] to generate genealogies with varying rates of recombination, then evolved simulated DNA sequences along these genealogies. Distances were corrected according to the Jukes-Cantor model, and trees were constructed using a least squares heuristic. As the amount of recombination increased, Schierup and Hein observed

1. a tiny decrease in the average distance between sequences.
2. a decrease in the time to the most recent common ancestor of all taxa, and also in the average time back to the common ancestors of pairs of taxa
3. an increase in the total length (sum of branch lengths) of the topology

All of these observations can be predicted from Theorem 1. We showed that the estimated distances will under-estimate the average distances from the various input trees. This explains the decrease in average pairwise distances. The fact that this decrease is very small indicates that the estimated distances are close to the convex combinations of the input distances.

The decrease in tree height and average time to least common ancestors is therefore due more to the presence of conflicting signal than the negative bias predicted by Theorem 1. The time to the most recent common ancestor in any clock based phylogeny equals half the maximum divergence between taxa. A given pair of taxa will most likely not be maximally diverged in all of the input phylogenies, so the maximum divergence between sequences will decrease when we take a convex combination from different trees.

The increase in total tree length follows from the proof of consistency for minimum evolution [16], at least for ordinary least squares (see [7]). If we estimate branch lengths from a distance matrix from an incorrect tree, the tree length will be longer than for the correct tree. Thus the incompatibilities introduced by increased recombination will increase tree length.

5.2 Inconsistency of Phylogenetic Analysis under Variable Rate Models

Chang [5] showed that distance based and maximum likelihood methods can be inconsistent when evolutionary rates vary across sites. Using Theorem 1, we can limit the zone of inconsistency for these methods. Variation in evolutionary rates means that each site evolved on the same tree but the branch lengths can differ. We can rescale so that the expected distance between two taxa equals the distance between them in T. The inconsistency is due to the variance in the distance caused by the varying rates. Bounding this rate variance makes Neighbor-joining a consistent method.

Theorem 2 *Suppose that sequences are evolved along a phylogeny T under a stochastic model with rate matrix Q and variable evolutionary rates. Let ϵ be the expected number of mutations along the shortest branch of T. If*

$$\mathrm{var}[d] < \frac{\epsilon}{\rho_Q}$$

then Neighbor-Joining (and most other distance based methods) applied to corrected distances will return T with sufficiently long sequences.

Proof
We prove the result for a finite number k of possible evolutionary rate histories, though the result extends immediately to a continuous rate distribution. Each rate history corresponds to an assignment of branch lengths to T. For each i we let T_i denote T with branch lengths modified according to the ith possible rate history. Let d_i denote the additive distance for T_i. Since each of $T_1, \ldots, T_k$ has the same topology, the distance matrix $\mathbb{E}[d]$ formed from their weighted averages is also additive on T.

From Theorem 1 the corrected distance $-\mathrm{tr}(\Pi \log(\Pi^{-1} F))$ differs from $\mathbb{E}[d]$ by at most $\frac{1}{2}\rho_Q \mathrm{var}[d] = \frac{1}{2}\epsilon$. Neighbor-joining therefore returns the correct tree T [1]. □

5.3 Split Decomposition and NeighborNet

Models for generating sequences under recombination combine three parts: the sampling of the recombination history, the sampling of breakpoints, and the evolution of the sites. After the recombination history and breakpoints are determined, each site has an associated phylogenetic tree. While the trees are correlated, we usually assume independence of each site given the trees (e.g. [11,17]). The problem of reconstructing the complete recombination history therefore requires a reconstruction of the contributing phylogenies.

Suppose that we have a mosaic alignment with a proportion of q_1 of the sites evolved on T_1, q_2 of the sites evolved on T_2,...,q_k sites evolved on T_k. For each i let d_i denote the distance matrix for T_i. Since the trees T_i are highly correlated there will be less variation in the individual distances than if the trees had been sampled independently. If the sequences are sufficiently long, the distance estimates computed for the whole sequence will approximate the weighted average

$$\bar{d}(x,y) = \sum_{i=1}^{k} q_i d_i(x,y).$$

Each tree T_i can be decomposed into a non-negative linear combination of split metrics, as we observed in Section 2.3. For each tree T_i, let $b_i(A|B)$ denote the length of the branch corresponding to $A|B$, with $b_i(A|B) = 0$ if $A|B$ is not a split of T_i. We set

$$\bar{b}(A|B) = \sum_{i=1}^{k} q_i b_i(A|B) \geq 0.$$

Let $\mathcal{S}$ equal the union $\Sigma(T_1) \cup \cdots \cup \Sigma(T_k)$ of the splits of $T_1, \ldots, T_k$. Then

$$d_i(x,y) = \sum_{A|B \in \mathcal{S}} b_i(A|B)\delta_{A|B}(x,y)$$

and

$$\begin{aligned}\bar{d}(x,y) &= \sum_{i=1}^{k} \sum_{A|B \in \mathcal{S}} q_i b_i(A|B)\delta_{A|B}(x,y) \\ &= \sum_{A|B \in \mathcal{S}} \bar{b}(A|B)\delta_{A|B}(x,y).\end{aligned}$$

We therefore have

1. The distance matrix $\bar{d}$ is in the positive cone generated by the split metrics for splits in the trees $T_1, \ldots, T_k$
2. The coefficient of $\delta_{A|B}$ in this sum equals the sum of the branch lengths corresponding to $A|B$ in the input tree, where the branch lengths in T_i are weighted by a factor q_i.

Split decomposition and NeighborNet both take a distance metric d and compute a decomposition

$$d = \epsilon + \sum_{A|B} \lambda_{A|B} \delta_{A|B}$$

of d into a positive combination of split metrics and an error term ϵ. Furthermore, both methods are *consistent* over a large class of metrics. If the set of splits $\mathcal{S}$ from the trees $T_1, \ldots, T_k$ is *weakly compatible* and

$$\bar{d}(x, y) = \sum_{A|B \in \mathcal{S}} \bar{b}(A|B) \delta_{A|B}$$

then split decomposition will recover the splits $A|B$ as well as the coefficients $\bar{b}$ [2]. In particular, if $k = 2$ then $\mathcal{S}$ is weakly compatible and split decomposition will recover the splits. If $\mathcal{S}$ is *circular* then NeighborNet will also recover both the splits and coefficients.

The splits in a splits graph therefore represent an estimate of the splits in the trees generating the mosaic sequences. The lengths of the edges represent an estimate of the corresponding branch lengths, weighted by the frequencies.

6 Discussion – Error and the Phylogenetic Analysis of Recombination

We have established a general result for distance corrections on mosaic sequences, and studied applications of this result to phylogenetic tree and network construction. The result characterises the signal present in distance matrices derived from recombinant sequences, a fundamental step towards the design of new methods for recovering this conflicting phylogenetic signal.

Building on the observations made in Theorem 1 we can identify (at least) four sources of error that could cause the splits graph representation to be incorrect, even under the assumption that we have the correct evolutionary model. The first is the (negatively biased) error term introduced by the approximation in Theorem 1. This factor is tiny, however, and is likely to only affect splits with small branch lengths or splits that only contribute to a small fraction of the sites.

A second source of error is sampling. Here the combined network approach seems to have an advantage over sliding window approaches. A consequence of the Theorem is that the variance of the estimate $\mathbb{E}[d]$ is not significantly different than the standard variance estimate for an alignment without recombination. On the other hand, if we knew which sites evolved from which trees and estimated these distances separately, the variances would be far higher.

The third source of error is the systematic error introduced by split decomposition and NeighborNet. For certain classes of splits, both methods are consistent. However if the splits in the trees contributing to the mosaic are not weakly compatible (or in the case of NeighborNet, not circular) then the resulting splits graph could be misleading. There is a need to characterise how both methods respond to these model violations, and scope for developing further methods for decomposing distance matrices.

When the splits in S are not weakly compatible a fourth complication can arise. A distance matrix can have two or more distinct decompositions into split metrics. Consider the three clock-like trees T_1, T_2, T_3 in figure 3. The average of the distance matrices generated by these trees is exactly equal to the distance matrix for T^*. Even if the the proportions were not exact, it would always be possible to decompose the estimated distance matrix into the non-negative sum of six (rather than seven) split metrics [2]. This problem of non-recoverability will become worse the more complicated the recombinations are, and poses a severe challenge for the development, and experimental analysis, of recombination reconstruction algorithms.

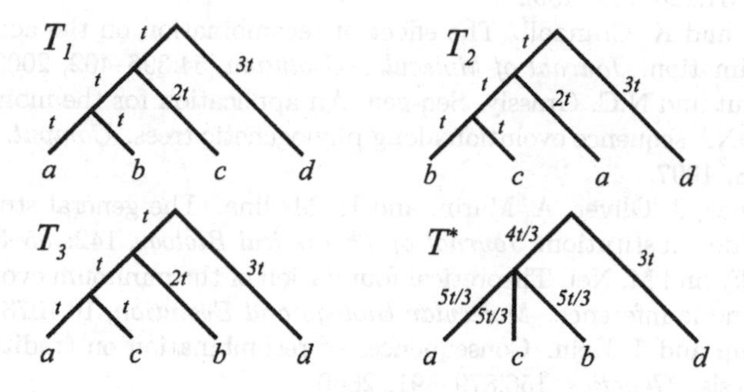

Fig. 3. An example of non-recoverability. The three trees T_1, T_2, T_3 generate the same distance matrix as the single tree T^*.

References

1. K. Atteson. The performance of the neighbor-joining methods of phylogenetic reconstruction. *Algorithmica*, 25:251–278, 1999.
2. H.-J. Bandelt and A.W.M. Dress. A canonical decomposition theory for metrics on a finite set. *Advances in Mathematics*, 92:47–105, 1992.
3. D. Bryant and V. Moulton. Neighbornet: An agglomerative algorithm for the construction of planar phylogenetic networks. In R. Guigo and D. Gusfield, editors, *Workshop in Algorithms for Bioinformatics (WABI)*, number 2452 in LNCS, pages 375–391. Springer-Verlag, 2002.
4. D. Bryant and V. Moulton. Consistency of the neighbornet algorithm for constructing phylogenetic networks. Technical report, School of Computer Science, McGill University, 2003.
5. J. Chang. Inconsistency of evolutionary tree topology reconstruction methods when substitution rates vary across characters. *Mathematical Biosciences*, 134:189–215, 1996.

6. A.W.M. Dress and D. Huson. Computing phylogenetic networks from split systems (manuscript).
7. O. Gascuel, D. Bryant, and F. Denis. Strengths and limitations of the minimum evolution principle. *Systematic Biology*, 50:621–627, 2001.
8. R.R. Hudson. Properties of a neutral allele model with intragenic recombination. *Theoretical Population Biology*, 23:183–201, 1983.
9. T.H. Jukes and C.R. Cantor. Evolution of protein molecules. In H.N. Munro, editor, *Mammalian Protein Metabolism*, pages 21–123. Academic Press, New York, 1969.
10. M.K. Kuhner and J. Felsenstein. A simulation comparison of phylogeny algorithms under equal and unequal evolutionary rates. *Molecular Biology and Evolution*, 11:459–468, 1994.
11. M.K. Kuhner, J. Yamato, and J. Felsenstein. Maximum likelihood estimation of recombination rates from population data. *Genetics*, 156:1393–1401, 2000.
12. J. Maynard-Smith. Analyzing the mosaic structure of genes. *Journal of Molecular Evolution*, 34:126–129, 1992.
13. D. Posada and K. Crandall. The effect of recombination on the accuracy of phylogeny estimation. *Journal of Molecular Evolution*, 54:396–402, 2002.
14. A. Rambaut and N.C. Grassly. Seq-gen: An application for the monte carlo simulation of DNA sequence evolution along phylogenetic trees. *Comput. Appl. Biosci.*, 13:235–238, 1997.
15. F. Rodriguez, J. Oliver, A. Marin, and R. Medina. The general stochastic model of nucleotide substitution. *Journal of Theoretical Biology*, 142:485–501, 1990.
16. A. Rzhetsky and M. Nei. Theoretical foundation of the minimum evolution method of phylogenetic inference. *Molecular Biology and Evolution*, 10:1073–1095, 1993.
17. M. Schierup and J. Hein. Consequences of recombination on traditional phylogenetic analysis. *Genetics*, 156:879–891, 2000.
18. K. Strimmer, C. Wiuf, and V. Moulton. Recombination analysis using directed graphical models. *Molecular Biology and Evolution*, 18:97–99, 2001.
19. D. Swofford, G.J. Olsen, P.J. Waddell, and D.M. Hillis. Phylogenetic inference. In D.M. Hillis, C. Moritz, and B.K. Mable, editors, *Molecular Systematics*, pages 407–514. Sinauer, 2nd edition, 1996.
20. K. Tamura and M. Nei. Estimation of the number of nucleotide substitutions in the control region of mitochondrial DNA in humans and chimpanzees. *Molecular Biology and Evolution*, 10:512–526, 1993.
21. C. Wiuf, T. Christensen, and J. Hein. A simulation study of the reliability of recombination detection methods. *Molecular Biology and Evolution*, 18:1929–1939, 2001.

Parsimonious Reconstruction of Sequence Evolution and Haplotype Blocks:
Finding the Minimum Number of Recombination Events

Yun S. Song* and Jotun Hein

Department of Statistics, University of Oxford,
1 South Parks Road, Oxford, OX1 3TG, UK
song@stats.ox.ac.uk, hein@stats.ox.ac.uk

Abstract. Under the infinite-sites model of mutation, we consider the problem of finding the minimum number of recombination events which must have occurred in the evolutionary history of sampled DNA sequences. Our approach is deterministic and is based on the combinatorics of leaf-labelled rooted trees. In contrast to previously known approaches, which only yield estimated lower bounds, our approach always gives the exact minimum number of recombination events. Furthermore, our method can be used to reconstruct explicitly evolutionary histories with the minimum number of recombination events. As an additional application, we discuss how our work can be used to define haplotype blocks.

1 Introduction

As well as being the major biological process which can destroy linkage disequilibrium (LD), recombination is one of the principal driving forces which generate genetic variations between different individuals of the same population. As such, recombination can have far-reaching consequences on molecular evolution. Knowing how many and where in the sequence recombination events have occurred can thus be a major contributing factor in unravelling many important questions in genetics. Some recombination events do not change the local phylogeny, however, and therefore it is in general impossible to know exactly how many recombination events have occurred in the evolutionary history of sampled sequences. Nevertheless, within the framework of a chosen model, it is meaningful to ask at least how many recombination events must have occurred in the history.

In [6], Hudson and Kaplan considered, within the framework of the infinite-sites model, the problem of finding a lower bound on the number of recombination events which must have occurred in the history of sampled DNA sequences. With the advent of new technologies which allow us to obtain data at an alarming rate, that particular problem of counting recombination events is currently

* Corresponding Author

G. Benson and R. Page (Eds.): WABI 2003, LNBI 2812, pp. 287–302, 2003.

receiving a renewed interest. Myers and Griffiths recently proposed an integer linear programming approach for constructing new lower bounds on the number of recombination events [9], whereas the present authors proposed a set theoretical method to define a new lower bound [12]. Neither of these methods can explicitly reconstruct evolutionary histories. Moreover, a common thread that runs through all currently-existing methods is that, for some data set S, the computed lower bound on the number of recombination events may, in fact, be less than the minimum number $\mathcal{R}_{\min}(S)$. Here, the minimum number $\mathcal{R}_{\min}(S)$ is defined by the property that there exists no evolutionary history with less than $\mathcal{R}_{\min}(S)$ recombination events that can generate S under the infinite-sites model. The goal of our present work is thus twofold; to find the exact minimum number $\mathcal{R}_{\min}(S)$ of recombination events and to reconstruct possible evolutionary histories with exactly $\mathcal{R}_{\min}(S)$ recombination events.

The main idea which underlies our algorithm was first laid out by Hein more than a decade ago [4], and a heuristic implementation of the idea was subsequently carried out [5]. An exact implementation of the idea, however, could not be carried out so far due to several difficulties, the major one being the complexity of the combinatorics involved. For example, as we later elaborate, working with trees with many restrictions and computing the distance between two arbitrary such trees can be very complicated. The approach we take in the present paper is to view recombination events as the so-called subtree-prune-and-regraft (SPR) operations on trees. More precisely, if an appropriate definition of a tree is used, the SPR-distance between two trees – defined as the minimum number of SPR operations required to transform one tree to the other – correctly encodes the number of recombination events.

Perhaps the most interesting recent finding in haplotype analysis is the discovery of the possible existence of haplotype block structures in the human genome [1,7,3], where a block is roughly characterised by the existence of high LD and limited haplotype diversity. In this paper, using our algorithm for detecting recombination events, we propose a new way of defining haplotype blocks. Unlike previous proposals, we explicitly take possible evolutionary histories into account.

Throughout this paper we assume that the data S consists of binary sequences; phased single nucleotide polymorphism (SNP) data satisfies this criterion, for instance. More precisely, we let $S = \{s_\alpha\}$ be a set of n binary sequences $s_\alpha = c_1^\alpha, c_2^\alpha, \ldots, c_\ell^\alpha$, where $c_i^\alpha \in \{0,1\}$ for every $\alpha \in \{1,2,\ldots,n\}$ and $i \in \{1,2,\ldots,\ell\}$. Note that each sequence is of fixed length ℓ. The entry c_i^α is called the ith *character* of sequence s_α, and $\mathbf{c}_i := (c_i^1, c_i^2, \ldots, c_i^n)$ the ith character *column*. A character column $\mathbf{c}_i$ is called *informative* if it contains at least two 0s and two 1s. Otherwise, it is called *non-informative*. We define $\bar{\mathbf{c}}_i := (\bar{c}_i^1, \bar{c}_i^2, \ldots, \bar{c}_i^n)$, where $\bar{c}_i^\alpha = 0$ if $c_i^\alpha = 1$ and $\bar{c}_i^\alpha = 1$ if $c_i^\alpha = 0$. As mentioned before, we assume the infinite-sites model of mutation. That is, we assume that at most one mutation event has occurred at each character column.

The organisation of this paper is as follows. In §2 we discuss the combinatorics of rooted trees relevant to our work, as well as laying out some necessary definitions. Our main ideas and algorithms are described in §3. In §4 we apply our method to analyse Kreitman's 1983 data of the alcohol dehydrogenase locus from 11 chromosomes of *Drosophila melanogaster* [8]. We conclude with some remarks in §5.

2 Trees

In this section, we present some definitions and facts to which we frequently refer in the main part of this paper. The reader is strongly recommended to browse through §2.1 to get familiarised with our notations.

2.1 Definitions

In this paper we consider leaf-labelled rooted binary trees whose branch lengths are not specified. The space of leaf-labelled rooted binary trees with n leaves is denoted by $\mathscr{T}_n^{\mathrm{r}}$. The degree of a vertex v is the number of edges which are incident with v. For $n \geq 2$, a tree in $\mathscr{T}_n^{\mathrm{r}}$ has n labelled degree-1 vertices called *leaves*; $n-2$ unlabelled degree-3 vertices; and a distinguished vertex of degree 2 called the *root*. A 1-leaved tree consists of a single labelled degree-0 vertex which serves as both the root and the leaf. A vertex which is not a leaf is called an *internal* vertex. The leaves of an n-leaved tree are bijectively labelled by a finite set S of n elements. In the remainder of this paper, when we say a tree without any qualification, we shall mean a leaf-labelled rooted binary tree.

A *path* from a vertex v_0 to another vertex v_k is an alternating sequence $v_0, e_1, v_1, e_2, v_2, \ldots, e_k, v_k$ of vertices v_i and edges e_i, such that (1) e_i joins v_{i-1} and v_i, and (2) all e_i's and v_i's are distinct. In a rooted tree, time flows from the root to the leaves. We say that vertex $v \in T$ is a *descendant* of vertex $u \in T$ if there exists a path from u to v which goes strictly forward in time; u is called an *ancestor* of v. A *subtree* t of a tree $T \in \mathscr{T}_n^{\mathrm{r}}$ is a tree in $\mathscr{T}_{n'}^{\mathrm{r}}$, where $n' \leq n$, and is defined by the property that if a vertex $v \in T$ is contained in t, then so are all its descendants. We say that two vertices $u, v \in T$ are *adjacent* if there exists an edge which joins u and v. In this paper, a subtree whose root is adjacent to the root of T is called an R-subtree of T.

In a rooted binary tree, the set $\{v_1, v_2, \ldots, v_{n-2}\}$ of degree-3 vertices is a *partially* ordered set whose binary relation denoted $<$ is given by ancestral relation. More precisely, $v_i < v_j$ if v_i is a descendant of v_j. Note that, if r denotes the root of a tree T, then $v_i < r$, for all degree-3 vertices v_i of T. Two degree-3 vertices v_i and v_j are *incomparable* if v_i is not in the path to the root from v_j and vice versa.

An *ordered tree* is a leaf-labelled rooted binary tree whose corresponding set $\{v_1, v_2, \ldots, v_{n-2}\}$ of degree-3 vertices is a *totally* ordered set; that is, for any two vertices v_i and v_j, either $v_i < v_j$ or $v_j < v_i$. In this case, the binary relation $<$ is given by age ordering. As before, $v_i < v_j$ if v_i is a descendant of v_j. If there exists no ancestral relation between v_i and v_j, then either $v_i < v_j$ or $v_j < v_i$ is allowed. Furthermore, we impose the condition that $v_i \neq v_j$ if $i \neq j$. Two trees equivalent as rooted trees are distinct as ordered trees if the ordering of their degree-3 vertices are different. In a subtree t of an ordered tree T, the ordering of degree-3 vertices in t is determined by their ordering in T. The space of ordered trees with n leaves is denoted by $\mathscr{T}_n^{\text{o}}$.

Let $\{B_i, B_i^c\} = S$ denote the bipartition corresponding to an informative character column $\mathbf{c}_i$, such that s_α and s_β belong to the same subset if and only if $c_i^\alpha = c_i^\beta$. A tree T is said to be *compatible* with the informative column $\mathbf{c}_i$ if there exists an edge in T such that cutting the edge decomposes T into two connected components, one containing the leaves labelled by B_i and the other the leaves labelled by B_i^c. If the column $\mathbf{c}_i$ is not informative, then every n-leaved tree is considered compatible with $\mathbf{c}_i$. In relation to biology, if a tree is compatible with a character column, then at most one mutation event at that column is necessary for the tree to represent the evolutionary history of sampled sequences at that character column, i.e. the tree is consistent with the character column under the infinite-sites model.

NOTE: When we write a symbol (for example, $\mathscr{T}_n$) without a qualifying superscript "r" or "o," it should be understood as referring to both cases.

2.2 SPR Operations

The precise definition of a subtree-prune-and-regraft (SPR) operation depends on the type of tree on which the operation is performed. In general, the more characteristics a tree has, the more restrictive an SPR operation has to be. We begin our discussion with plain leaf-labelled rooted trees. There are three kinds of SPR operations that can be performed on leaf-labelled rooted trees. An illustration of these operations is shown in Figure 1. In what follows, let T (resp. T') denote a tree before (resp. after) an SPR operation. The notation $T \setminus t$ denotes the part of T obtained from removing a subtree t and the edge incident with the root of t but not contained in t. In words the three SPR operations are as follows.

1. An edge e is cut to prune a non-R-subtree t, and t is regrafted onto a pre-existing edge in the remaining part $T \setminus t$ of T, thus creating a new degree-3 vertex. The vertex in $T \setminus t$ where e used to be incident gets removed. The root of T remains the root of T'. (In Figure 1, $T \to T_1$ is an example of this kind. The edge e_b is cut and then regrafted onto the edge e_a.)
2. Let s_1 and s_2 be the two R-subtrees of T, and let e_1 and e_2, respectively, be the edges which join their roots to the root of T. The edge e_1 is cut to prune s_1, and s_1 is regrafted onto a pre-existing edge in s_2. The edge e_2 gets removed and the degree-3 vertex in s_2 where e_2 used to be incident gets replaced by a degree-2 vertex, which becomes the root of T'. (In Figure 1, $T \to T_2$ is an example of this kind. The edge e_c can be cut and regrafted onto e_a. The root of the R-subtree containing t_1, t_2 and t_3 then becomes the root of T_2. In this example, note that T can be transformed into T_2 in several ways. More exactly, the subtree t_1 can be pruned and regrafted onto e_c or

the subtree containing t_2 and t_3 can be pruned and joined onto the root of T; these kinds of SPR operations have already been described above.)

3. An edge e is cut to prune a non-R-subtree t, and t is joined to the root of T. The root of T' is given by creating a new vertex of degree 2 on e. (In Figure 1, $T \to T_3$ is an example of this kind. The edge e_b is cut and then joined to the root of T. A new degree-2 vertex is created on the edge and it serves as the root of T_3.)

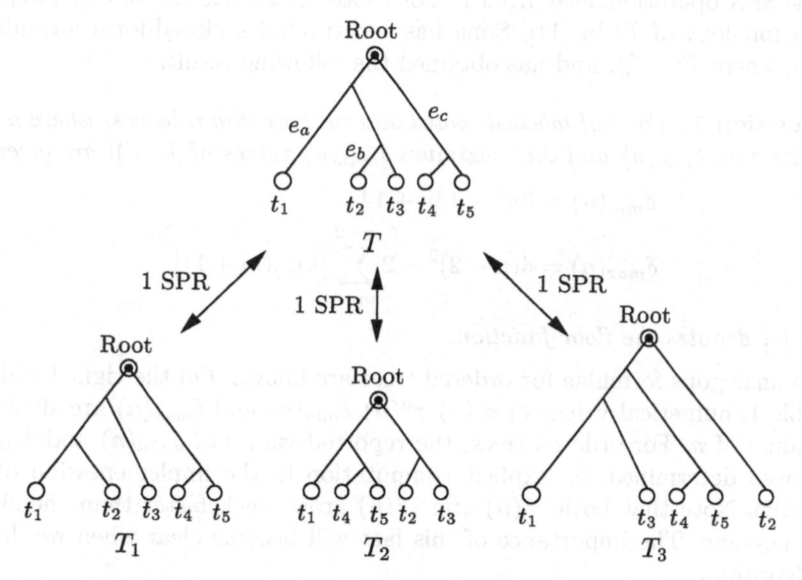

Fig. 1. An illustration of SPR operations. Big open circles ○ labelled by t_j represent subtrees.

For ordered trees, we impose an additional restriction on the definition of SPR operations. Consider a subtree t of an ordered tree $T \in \mathcal{T}_n^{\mathrm{o}}$. An SPR operation of t which transform $T \in \mathcal{T}_n^{\mathrm{o}}$ to $T' \in \mathcal{T}_n^{\mathrm{o}}$ is defined to satisfy the following additional property: Let v (resp. v') denote the vertex in $T \backslash t$ (resp. $T' \backslash t$) which is adjacent to the root of t. Suppose v_i and v_j are two internal vertices other than v and v'. If $v_i < v_j$ before an SPR operation, then $v_i < v_j$ after the SPR operation, and vice versa; that is, a relation between any two internal vertices other than v and v' should be the same before and after an SPR operation.

2.3 Some Enumerations of Trees

It is well-known [10] that the number of inequivalent leaf-labelled rooted binary trees with n leaves is

$$\tau^{\mathrm{r}}(n) := |\mathcal{T}_n^{\mathrm{r}}| = (2n-3)!! = (2n-3) \times (2n-5) \times \cdots \times 3 \times 1 = \frac{(2n-2)!}{2^{n-1}(n-1)!},$$

whereas the number of inequivalent ordered trees with n leaves is

$$\tau^{\mathrm{o}}(n) := |\mathscr{T}_n^{\mathrm{o}}| = \prod_{m=2}^{n} \binom{m}{2} = \frac{n!(n-1)!}{2^{n-1}}.$$

Note that $\tau^{\mathrm{r}}(n) = (2n-3)\tau^{\mathrm{r}}(n-1)$, while $\tau^{\mathrm{o}}(n) = \frac{n(n-1)}{2}\tau^{\mathrm{o}}(n-1)$. The number $\tau^{\mathrm{o}}(n)$ of ordered trees thus grows much faster than the number $\tau^{\mathrm{r}}(n)$ of plain rooted trees.

The adjacency-set $U(T)$ of a tree T is defined as the set of all trees which are one SPR operation away from T. For rooted trees, the size of $U(T)$ depends on the topology of T. In [11], Song has constructed a closed-form formula for $|U(T)|$, where $T \in \mathscr{T}_n^{\mathrm{r}}$, and has obtained the following result:

Proposition 1. *For leaf-labelled rooted binary trees with n leaves, where $n \geq 3$, the minimum $\delta_{min}(n)$ and the maximum $\delta_{max}(n)$ values of $|U(T)|$ are given by*

$$\delta_{min}(n) = 3n^2 - 13n + 14,$$

$$\delta_{max}(n) = 4(n-2)^2 - 2\sum_{m=1}^{n-2} \lfloor \log_2(m+1) \rfloor,$$

where $\lfloor \cdot \rfloor$ denotes the floor function.

No analogous formulae for ordered trees are known. On the right hand side of Table 1, numerical values of $\tau^{\mathrm{r}}(n), \tau^{\mathrm{o}}(n), \delta_{\mathrm{min}}(n)$ and $\delta_{\mathrm{max}}(n)$ are shown for low values of n. For ordered trees, the reported values of $\delta_{\mathrm{min}}(n)$ and $\delta_{\mathrm{max}}(n)$ have been determined via explicit computation in the implementation of our algorithm. Note that both $\tau^{\mathrm{r}}(n)$ and $\tau^{\mathrm{o}}(n)$ grow much faster than the size of adjacency-sets. The importance of this fact will become clear when we discuss our algorithm.

Also of interest to us is the number of trees which are compatible with a given character column. Let $\{B_i, B_i^c\}$ denote the bipartition of S corresponding to a character column $\mathbf{c}_i$, and suppose $|B_i| = k$ and $|B_i^c| = n-k$, where $1 \leq k \leq n-1$. Let $w^{\mathrm{r}}(n,k)$ (resp. $w^{\mathrm{o}}(n,k)$) denote the number of rooted trees (resp. ordered trees) compatible with the bipartition $\{B_i, B_i^c\}$. From [11] we have the following results:

Proposition 2 (Number of rooted trees compatible with a column). *For $n \geq 4$ and $1 \leq k \leq n-1$,*

$$w^{\mathrm{r}}(n,k) := (2n-3)\,\tau^{\mathrm{r}}(k)\,\tau^{\mathrm{r}}(n-k).$$

Proposition 3 (Number of ordered trees compatible with a column). *For $n \geq 4$ and $1 \leq k \leq n-1$,*

$$w^{\mathrm{o}}(n,k) := \tau^{\mathrm{o}}(k)\tau^{\mathrm{o}}(n-k)\left[\binom{n}{k-1} + \binom{n}{n-k-1} - \binom{n-2}{k-1}\right].$$

Numerical values of $w^{\mathrm{r}}(n,k)$ and $w^{\mathrm{o}}(n,k)$ are shown in Table 1 for $n \leq 9$.

Table 1. Numerical summary of tree enumerations. For ordered trees, $\delta_{\min}(n)$ and $\delta_{\max}(n)$ have been determined through explicit computation.

	Rooted Trees			Ordered Trees		
n	$\tau^{r}(n)$	$\delta_{\min}$	$\delta_{\max}$	$\tau^{o}(n)$	$\delta_{\min}$	$\delta_{\max}$
4	15	10	12	18	12	13
5	105	24	28	180	33	37
6	945	44	52	2,700	71	79
7	10,395	70	84	56,700	128	143
8	135,135	102	124	1,587,600	210	233
9	2,027,025	140	170	57,153,600	?	?
10	34,459,425	184	224	2,571,912,000	?	?

n	k	$w^{r}(n,k)$	$w^{o}(n,k)$
4	2	5	6
5	2,3	21	36
6	2,4	135	396
	3	81	216
7	2,5	1,155	6,660
	3,4	495	2,484
8	2,6	12,285	156,600
	3,5	4,095	44,820
	4	2,925	29,808
9	2,7	155,925	4,876,200
	3,6	42,525	1,142,100
	4,5	23,625	567,000

3 Main Ideas

3.1 Recombination Events as SPR Operations

In a graphical representation of evolutionary history, if sampled sequences have been subjected to recombination, different character columns may be described by different trees. The so-called ancestral recombination graph (ARG) is constructed by putting together the set of trees supported at different character columns. For instance, while a column $\mathbf{c}_i$ is described by $T \in \mathscr{T}_n^{o}$, the next column $\mathbf{c}_{i+1}$ may be described by a different tree $T' \in \mathscr{T}_n^{o}$. The tree T can be transformed to the tree T' through SPR operations, and an ARG precisely contains this information; namely, recombinant sequences in the ARG correspond to the leaves in the subtree that gets pruned and regrafted. Furthermore, the number of SPR operations used to transform T to T' corresponds to the number of recombination events occurring between the columns $\mathbf{c}_i$ and $\mathbf{c}_{i+1}$.

3.2 SPR-distance between Trees

For any pair of trees, say $T, T' \in \mathscr{T}_n$, their SPR-distance $d(T,T')$ is a non-negative integer defined as the minimum number of SPR operations necessary to transform T into T'. In practice, determining the SPR-distance between two arbitrary rooted trees can be quite difficult, especially so for ordered trees. It is not very difficult, however, to determine whether two trees are one SPR operation away. Hence, our approach is to determine first which trees are distance one away from each other, and then use that information to compute $d(T,T')$ for arbitrary T and T'.

By the adjacency-set of a tree $T \in \mathscr{T}_n$ we mean the set

$$U(T) := \{T' \in \mathscr{T}_n \mid d(T,T') = 1\}\,.$$

Let $U_0(T) = \{T\}$ and $U_1(T) = U(T)$. Then, for $m \geq 2$, recursively define

$$U_m(T) := \left[\bigcup_{T' \in U_{m-1}(T)} U(T') \right] \setminus [U_{m-1}(T) \cup U_{m-2}(T)] .$$

The distance $d(T, T')$ between T and T' is the value of m which satisfies $T' \in U_m(T)$.

3.3 The Algorithm for Counting Recombination Events

1. To each character column $\mathbf{c}_i$ of S, associate a set W_i° of trees defined as

$$W_i^{\circ} := \{T \in \mathscr{T}_n^{\circ} \mid T \text{ compatible with column } \mathbf{c}_i\} \subseteq \mathscr{T}_n^{\circ} .$$

Let $k(i)$ denote the size of B_i or B_i^c, with $\{B_i, B_i^c\}$ being the bipartition of S corresponding to the column $\mathbf{c}_i$. Then,

$$w_i^{\circ} := |W_i^{\circ}| = \begin{cases} w^{\circ}(n, k(i)), & \text{if } k(i) \geq 2, \\ \tau^{\circ}(n), & \text{otherwise,} \end{cases}$$

where $w^{\circ}(n, k)$ is defined in Proposition 3.

2. Construct a weighted graph G as follows. Introduce ℓ clusters, with the ith cluster containing w_i° vertices labelled by the trees in W_i°.
   (a) For all $T \in W_1^{\circ}$, let $f_1(T) = 0$.
   (b) For all $1 \leq i < \ell$, recursively determine

$$f_{i+1}(T_a) = \min_{T_b \in W_i^{\circ}} [f_i(T_b) + d(T_b, T_a)] \tag{1}$$

   for every tree $T_a \in W_{i+1}^{\circ}$.
   (c) In the weighted graph G, vertices $T_a \in W_{i+1}^{\circ}$ and $T_b \in W_i^{\circ}$ are joined by an edge if $f_{i+1}(T_a) - f_i(T_b) = d(T_a, T_b)$, and the weight of the edge is $d(T_a, T_b)$.

3. The number defined as

$$\mathcal{R}_{\circ}(S) = \min_{T_a \in W_{\ell}^{\circ}} f_{\ell}(T_a) \tag{2}$$

gives the minimum number $\mathcal{R}_{\min}(S)$ of recombination events. A connected path from any tree $T_a \in W_1^{\circ}$ to a tree $T_b \in W_{\ell}^{\circ}$ with $f_{\ell}(T_b) = \mathcal{R}_{\circ}(S)$ is called a *minimal path* in G.

(**Remark:** Why ordered trees, not plain rooted trees, should be used in the above algorithm to obtain $\mathcal{R}_{\min}(S)$ is discussed in §3.6.)

The algorithm described above is a special case of the exact algorithm given in [5]; assuming the infinite-sites model has simplified the algorithm a bit. Because it was not known how the distance $d(T, T')$ for arbitrary ordered trees T and T' could be computed, Hein used unrooted trees in his implementation of the algorithm [5]. Moreover, he assumed that at most one recombination event occurs between any two adjacent character columns. In our present work, we have implemented the above algorithm for ordered trees, without any heuristic assumptions. As computing the distance $d(T, T')$ for arbitrary $T, T' \in \mathscr{T}_n^{\circ}$ is rather non-trivial, step 2 in the above algorithm is computationally intensive. We will return to this point in §3.8, where we present a new, modified algorithm. For now, using the fact that $d(\cdot, \cdot)$ is a proper distance function on $\mathscr{T}_n^{\circ}$, we establish the following result, which allows us to simplify the algorithm further:

Proposition 4. *For all $T_a, T_b \in W_i^{\circ}$, where $i \geq 2$, the function f_i defined in (1) satisfies*

$$|f_i(T_a) - f_i(T_b)| \leq d(T_a, T_b)\,. \tag{3}$$

Proof: We shall prove (3) by induction. Since $f_1(T) = 0$ for all $T \in W_1^{\circ}$, we have $f_2(T') = \min_{T \in W_1^{\circ}} [d(T, T')]$, for all $T' \in W_2^{\circ}$. Let $T_a, T_b \in W_2^{\circ}$. Suppose $T_a, T_b \in W_1^{\circ}$. Then, $f_2(T_a) = 0 = f_2(T_b)$, and therefore (3) is satisfied. Suppose $T_a \notin W_1^{\circ}$ and $T_b \in W_1^{\circ}$. Then, $f_2(T_b) = 0$, whereas $f_2(T_a) = \min_{T \in W_1^{\circ}} [d(T, T_a)]$, which by definition is less than or equal to $d(T_b, T_a)$. Hence, (3) is again satisfied. Lastly, suppose $T_a \notin W_1^{\circ}$ and $T_b \notin W_1^{\circ}$, with $f_2(T_a) \geq f_2(T_b)$. Let $T_s \in W_1^{\circ}$ be a tree which satisfies $d(T_s, T_b) = \min_{T \in W_1^{\circ}} [d(T, T_b)]$. Then,

$$\begin{aligned} f_2(T_a) - f_2(T_b) &= \min_{T \in W_1^{\circ}} [d(T, T_a)] - d(T_s, T_b) \leq d(T_s, T_a) - d(T_s, T_b) \\ &\leq d(T_a, T_b), \end{aligned}$$

where the last inequality follows from triangle inequality of the SPR-distance. Hence, we have shown that (3) holds for $i = 2$.

Assume that (3) holds up to $i = k - 1$. We prove the $i = k$ case by contradiction. Let $T_a, T_b \in W_k^{\circ}$, with $f_k(T_a) \geq f_k(T_b)$. Suppose we have

$$f_k(T_a) - f_k(T_b) > d(T_a, T_b). \tag{4}$$

Further suppose $T_b \in W_{k-1}^{\circ}$. If we had $f_{k-1}(T_b) > f_k(T_b)$, then it would imply that there exists $T \in W_{k-1}^{\circ}$ such that $f_{k-1}(T_b) > f_{k-1}(T) + d(T, T_b)$, which would contradict the induction hypothesis. Also, if we had $f_{k-1}(T_b) < f_k(T_b)$, then it would contradict the definition of f given in (1). Hence, we conclude that $f_k(T_b) = f_{k-1}(T_b)$ if $T_b \in W_{k-1}^{\circ}$. But, then (4) would imply $f_k(T_a) - f_{k-1}(T_b) > d(T_a, T_b)$, thus contradicting definition (1). Therefore, we must have $T_b \notin W_{k-1}^{\circ}$. Now, let $T_c \in W_{k-1}^{\circ}$ satisfy $f_k(T_b) = f_{k-1}(T_c) + d(T_b, T_c)$. Then, (4) implies $f_k(T_a) - f_{k-1}(T_c) > d(T_b, T_c) + d(T_a, T_b)$. But, the triangle inequality $d(T_b, T_c) + d(T_a, T_b) \geq d(T_a, T_c)$ implies $f_k(T_a) - f_{k-1}(T_c) > d(T_a, T_c)$, which contradicts definition (1). Thus, (4) cannot be true and this completes our induction. ■

It now follows from the above proposition that step 2(b) in the algorithm can be modified as follows:
For all $1 \leq i < \ell$ and $T_a \in W_{i+1}^{\circ}$, recursively determine

$$f_{i+1}(T_a) = \begin{cases} f_i(T_a), & \text{if } T_a \in W_i^{\circ}, \\ \min_{T_b \in W_i^{\circ}} [f_i(T_b) + d(T_b, T_a)], & \text{otherwise.} \end{cases}$$

3.4 Reduction of Data

In performing our algorithm, some character columns do not influence the determination of $\mathcal{R}_o(S)$ and therefore can be ignored. It is straightforward to show that, before one carries out any analysis on S, reducing the data as follows does not change the value of $\mathcal{R}_o(S)$; i.e., if S' denotes the reduced data, then $\mathcal{R}_o(S) = \mathcal{R}_o(S')$.

1. Collapse identical sequences into one.
2. Remove all non-informative columns from S. Let $\mathbf{c}'_1, \mathbf{c}'_2, \ldots, \mathbf{c}'_{\ell'}$ denote the character columns in the resulting data.
3. Collapse all consecutive columns $\mathbf{c}'_i, \mathbf{c}'_{i+1}, \ldots, \mathbf{c}'_{i+k}$ where $\mathbf{c}'_{i+j} = \mathbf{c}'_i$ or $\mathbf{c}'_{i+j} = \bar{\mathbf{c}}'_i$ for all $j = 1, 2, \ldots, k$, into a single column $\mathbf{c}'_i$.
4. Sequentially repeat steps $1 \sim 3$ until none of them is possible.

3.5 A Simple Example

To illustrate how our algorithm works, we consider the following very simple example:

$$\begin{matrix} 1\,1\,0\,0\,0\,0\,0\,0\,0 \\ 1\,1\,0\,1\,1\,1\,0\,1\,1 \\ 1\,0\,1\,0\,0\,1\,1\,1\,0 \\ 1\,1\,0\,1\,1\,1\,0\,1\,0 \\ 1\,0\,1\,1\,0\,0\,0\,0\,0 \\ 1\,0\,1\,1\,0\,0\,0\,0\,0 \end{matrix} \quad \longrightarrow \quad \begin{matrix} 1\,0\,0 \\ 1\,1\,1 \\ 0\,0\,1 \\ 0\,1\,0 \end{matrix} \qquad \begin{aligned} W_1^o &= \{T_4, T_9, T_{13}, T_{14}, T_{15}, T_{18}\}\,, \\ W_2^o &= \{T_3, T_7, T_8, T_{10}, T_{12}, T_{17}\}\,, \\ W_3^o &= \{T_1, T_2, T_5, T_6, T_{11}, T_{16}\}\,. \end{aligned}$$

The original data S is shown on the left hand side of the arrow. After the reduction steps described in §3.4, it reduces to the data S' shown on the right hand side of the arrow. Let $\mathbf{c}_1$, $\mathbf{c}_2$ and $\mathbf{c}_3$ denote the 3 columns of S'. As shown in Figure 2(a), there are 18 inequivalent ordered trees with 4 leaves. Trees T_1 and T_2 are inequivalent as ordered trees but equivalent as plain rooted trees. The same goes true for the pairs T_7, T_8 and T_{13}, T_{14}. The ordered trees compatible (c.f. §2.1) with $\mathbf{c}_1, \mathbf{c}_2, \mathbf{c}_3$ are given by W_1^o, W_2^o, W_3^o, respectively.

Applying the algorithm described in §3.3, one can obtain the weighted graph shown in Figure 2(b). In this simple example, all edges have weight 1, and therefore any connected path from a tree $T_a \in W_1^o$ to a tree $T_b \in W_3^o$ is a solution to the problem. In summary, $\mathcal{R}_{\min}(S) = 2$ and there are 132 minimal paths.

3.6 Plain Rooted Trees or Ordered Trees?

In discussing our algorithm so far, we have been careful in using $\mathscr{T}_n^o$ and W_i^o to refer to ordered trees. That is because coalescent events and recombination events occur at specific points in time, and ignoring the time ordering of the events can lead to contradictions. For example, biologically a recombinant cannot be older than its parents.

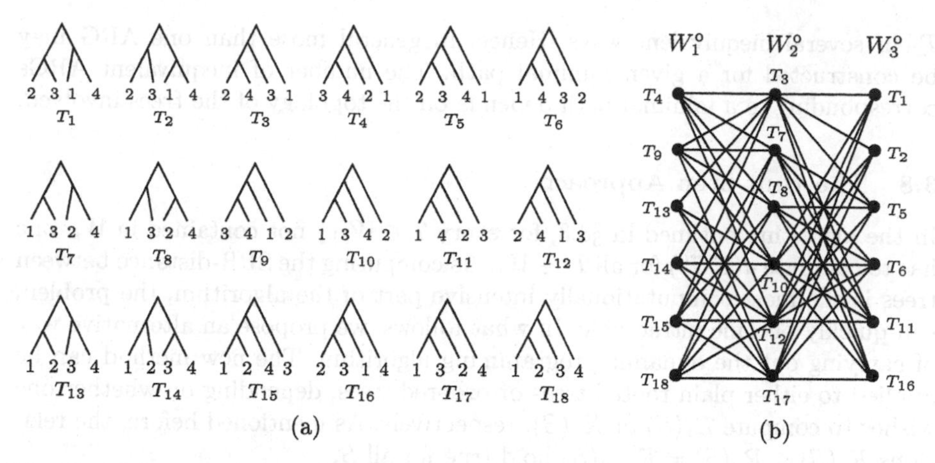

Fig. 2. (a) Inequivalent ordered trees with 4 leaves. (b) A graphical summary of performing our algorithm on the simple example. All edges have weight 1. There are 132 minimal paths, each leading to $\mathcal{R}_{\min}(S) = 2$.

We stress that the algorithm described in §3.3 always gives $\mathcal{R}_{\text{o}}(S) = \mathcal{R}_{\min}(S)$. On the contrary, if plain rooted trees are used in the algorithm – i.e. W_i^{r} instead of W_i^{o} are used – to obtain $\mathcal{R}_{\text{r}}(S) := \min_{T_a \in W_\ell^{\text{r}}} f_\ell(T_a)$, then the number $\mathcal{R}_{\text{r}}(S)$ may or may not be equal to the exact minimum $\mathcal{R}_{\min}(S)$. For less than or equal to 9 sequences, however, our investigation shows that $\mathcal{R}_{\text{r}}(S) = \mathcal{R}_{\min}(S)$ for most cases of S. A possible explanation of this phenomenon is as follows. When there are only a small number of leaves in a tree, SPR operations usually involve subtrees with few leaves. As a 1-leaved subtree contains no internal vertices, there is no restriction on where the 1-leaved subtree can be regrafted, and therefore if $\mathcal{R}_{\text{r}}(S)$ can be obtained through a series of SPR operations each involving a single leaf, then we should have $\mathcal{R}_{\text{r}}(S) = \mathcal{R}_{\min}(S)$.

As shown in Table 1, the number of ordered trees grows much faster than the number of plain rooted trees. For instance, there are over 57 million 9-leaved ordered trees, whereas there are about 2 million plain rooted trees with 9 leaves. So, it would be a good strategy to use first plain rooted trees to compute $\mathcal{R}_{\text{r}}(S)$ and try to reconstruct history as described in §3.7. If a consistent history can be reconstructed using only $\mathcal{R}_{\text{r}}(S)$ recombination events, then we can conclude that $\mathcal{R}_{\text{r}}(S) = \mathcal{R}_{\min}(S)$.

3.7 Reconstruction of History

After performing the algorithm from $\mathbf{c}_1$ to $\mathbf{c}_\ell$, we can find out which trees $T \in W_\ell^{\text{o}}$ have $f_\ell(T) = \mathcal{R}_{\text{o}}(S)$ and obtain minimal paths to those trees. Given a minimal path $T_{i_1}, T_{i_2}, \ldots, T_{i_\ell}$, one can combine the trees in the minimal path into an ARG. More precisely, possible sets of SPR operations that can transform T_{i_m} to $T_{i_{m+1}}$ tell us how the trees can be combined. In general, if $T_{i_m} \neq T_{i_{m+1}}$, there could be more than one way to transform T_{i_m} to $T_{i_{m+1}}$. For instance, as we discussed in §2.2 the tree T shown in Figure 1 can be transformed into

T_2 in several inequivalent ways. Hence, in general more than one ARG may be constructed for a given minimal path. The number of inequivalent ARGs corresponding to a minimal path depends on the topology of the trees involved.

3.8 The Unit-Step Approach

In the algorithm outlined in §3.3, for every $T \in W_{i+1}$ not contained in W_i, one has to compute $d(T, T')$ for all $T' \in W_i$. As computing the SPR-distance between trees is the most computationally intensive part of the algorithm, the problem can quickly become intractable. In what follows, we propose an alternative way of carrying out the dynamic programming algorithm. The new method can be applied to either plain rooted trees or ordered trees, depending on whether one wishes to compute $\mathcal{R}_{\mathrm{r}}(S)$ or $\mathcal{R}_{\mathrm{o}}(S)$, respectively. As mentioned before, the relations $\mathcal{R}_{\mathrm{r}}(S) \leq \mathcal{R}_{\mathrm{o}}(S) = \mathcal{R}_{\min}(S)$ hold true for all S.

For $X \subset \mathscr{T}_n$, let $N_0(X) = X$ and, for $r \geq 1$, define the r-neighbourhood of X as

$$N_r(X) := \left\{ T \in \mathscr{T}_n \;\middle|\; T \in \bigcup_{m=0}^{r} U_m(T') \text{ for some } T' \in X \right\} .$$

In the implementation of our algorithm, we pre-compute the adjacency-set $U(T)$ for all $T \in \mathscr{T}_n$ and store them in a file which can be accessed by our program. Therefore, computing $N_r(X)$ can easily be done. We define the diameter d_n of $\mathscr{T}_n$ as the maximum value of $d(T, T')$ over all trees $T, T' \in \mathscr{T}_n$. As shown in [11], $d_n \leq n-2$. In the following discussion, define $\mathcal{N}(T, m, i) := N_1(\{T\}) \cap N_m(W_i)$.

Let $f_{1,0}(T) = 0$ for all $T \in \mathscr{T}_n$. For all $1 \leq r < d_n$ and $1 \leq i < \ell$, recursively compute the following quantities:
For all $T_a \in N_r(W_i)$, find

$$f_{i,r}(T_a) = \begin{cases} f_{i,r-1}(T_a), & \text{if } T_a \in W_i, \\ \min\limits_{T_b \in \mathcal{N}(T_a, r-1, i)} \left[f_{i,r-1}(T_b) + 1 - \delta_{a,b} \right], & \text{otherwise,} \end{cases}$$

and, for all $T_a \in W_{i+1}$, find

$$f_{i+1,0}(T_a) = \begin{cases} f_{i,d_n-1}(T_a), & \text{if } T_a \in W_i, \\ \min\limits_{T_b \in \mathcal{N}(T_a, d_n-1, i)} \left[f_{i,d_n-1}(T_b) + 1 - \delta_{a,b} \right], & \text{otherwise.} \end{cases}$$

Here, $\delta_{a,b}$ denotes the Kronecker delta, which is 1 if $a = b$ and 0 if $a \neq b$. The minimum number of recombination events is given by $\min_{T \in W_\ell} f_{\ell,0}(T)$, which is equal to the value $\min_{T \in W_\ell} f_\ell(T)$ defined in §3.3.

There are several advantages to the algorithm just described over that in §3.3. First of all, note that for each tree T, we just need to compare a function evaluated at T with a function evaluated at at most $|N_1(T)| = |U(T)| + 1$ trees. As shown in Table 1, the maximum size of $|U(T)|$ does not grow as fast as the number $w(n, k)$ of trees compatible with a character column. Secondly, we do not need to compute the SPR-distance explicitly; the algorithm effectively computes the SPR-distance for us and correctly updates $f_{i+1,0}(T)$, for all $1 \leq i < \ell$.

Our current implementation of the algorithm can analyse up to 8 (resp. 9) sequences in the reduced data if ordered (resp. plain rooted) trees are used in the algorithm.

3.9 Haplotype Blocks

For each minimal path our algorithm finds, in addition to knowing which trees are selected, we know exactly where in the sequence each tree is supported. Hence, we can associate a candidate haplotype block structure to each minimal path. That is, for each minimal path, we define a block as consecutive positions in the sequence where the same tree is supported. As there could be many minimal paths, it could be that there are many inequivalent candidate haplotype block structures predicted by our algorithm. By studying all inequivalent candidate block structures, however, we may be able to learn something useful. For example, many or all structures may share one or more common blocks, thus indicating the robustness of those particular blocks.

Although in general we cannot find a unique haplotype block structure, we can still ask questions to each of which there exists a unique answer for a given data set S. For example, we can ask the following question: If a block is obtained as described above, what is the minimum number of haplotype blocks that can be defined by a minimal path? We address this question in §4, where we consider a specific application of our method.

4 Application

In this section, we apply the methods discussed in this paper to analyse Kreitman's 1983 data of the alcohol dehydrogenase locus from 11 chromosomes of *Drosophila melanogaster* [8]. The data has been taken from 5 geographically distinct populations and the aligned sequence length is 2800 base-pairs. Ignoring insertions and deletions, there are 43 polymorphic columns in the data. We have transformed the polymorphism data into binary sequences as shown in Figure 3.

Sequence	Block 1	Block 2	Block 3	Block 4	Block 5	Block 6
Wa-S =	000	000001100000	0	00110111011110000000	0	0000000
F1-1S =	001	000000000000	0	00110111011110000000	0	0000000
Af-S =	000	000000000000	0	00000000000000000000	1	0000101
Fr-S =	000	000000000000	0	11000000000000000000	1	0011000
F1-2S =	000	110001011001	1	11000000000000000000	0	1000000
Ja-S =	001	000000000000	1	00000000000000010101	1	1000010
F1-F =	001	000000000000	1	00000000000000111111	0	1000000
Fr-F =	111	110001011100	1	00000000000000111111	0	1100000
Wa-F =	111	110001011100	1	00000000000000111111	0	1100000
Af-F =	111	110001011100	1	00000000000000111111	0	1100000
Ja-F =	111	111111000010	1	00001000100001111111	0	1000000

Fig. 3. Kreitman's data in binary form. Also shown is a haplotype block structure with 6 blocks, whose boundaries are indicated by vertical solid lines.

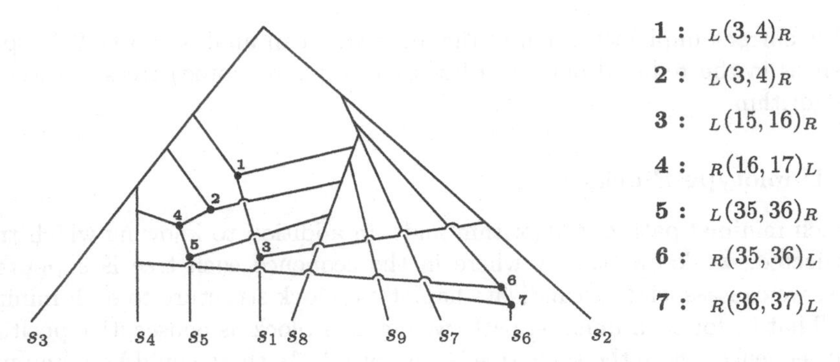

Fig. 4. A minimal ancestral recombination graph for Kreitman's data. Recombination vertices are denoted by •. The notation $_{S_L}(i,j)_{S_R}$ is used to denote the location (i,j) of the break-point and to indicate that S_L part of the recombinant gets descended from the left edge and S_R part from the right edge.

Since the sequences for Fr-F, Wa-F and Af-F are identical, in our analysis we only need to consider 9 distinct sequences. Hence, we relabel the sequences as follows:

$s_1 :=$ Wa-S $\quad s_2 :=$ F1-1S $\quad s_3 :=$ Af-S
$s_4 :=$ Fr-S $\quad s_5 :=$ F1-2S $\quad s_6 :=$ Ja-S
$s_7 :=$ F1-F $\quad s_8 :=$ Fr-F = Wa-F = Af-F $\quad s_9 :=$ Ja-F

As discussed in §3.6, we have performed our analysis on $S = \{s_1, \ldots, s_9\}$ first using rooted trees, obtaining $\mathcal{R}_r(S) = 7$. We have then checked that it is indeed possible to construct an ARG with exactly 7 recombination events. In addition, we have enumerated all minimal paths, each of which leads to $\mathcal{R}_r(S) = 7$. It turns out that there are about 10 million minimal paths for Kreitman's data; this number is not so surprisingly high, since for each recombination event, there could be numerous choices regarding which subtree undergoes an SPR operation and where in the sequence the event occurs.

The minimum number of haplotype blocks that can be defined by a minimal path is 6. A haplotype block structure with 6 blocks is shown in Figure 3, where block boundaries are indicated by vertical solid lines. A minimal ARG associated to a minimal path which generates such a block structure is shown in Figure 4. Note that 2 recombination events occur between columns 3 and 4, as well as between columns 35 and 36. Positions $3, 4, 15, 16, 17, 35, 36, 37$ in S correspond to positions $63, 170, 847, 950, 1030, 1691, 1730, 1827$, respectively, in the actual data. Figure 5 illustrates where in the actual data the recombination events shown in Figure 4 are supposed to occur.

Let us now compare our result on $\mathcal{R}_{\min}(S)$ with that given by some currently-available methods. In [6] Hudson and Kaplan also have examined Kreitman's data. Their algorithm gives 5 as a lower bound on the number of recombination events. If one uses the algorithm developed by Myers and Griffiths [9], one would obtain 6. In [12] the present authors have analysed Kreitman's data using a method based on set theory and have obtained 7 as a lower bound.

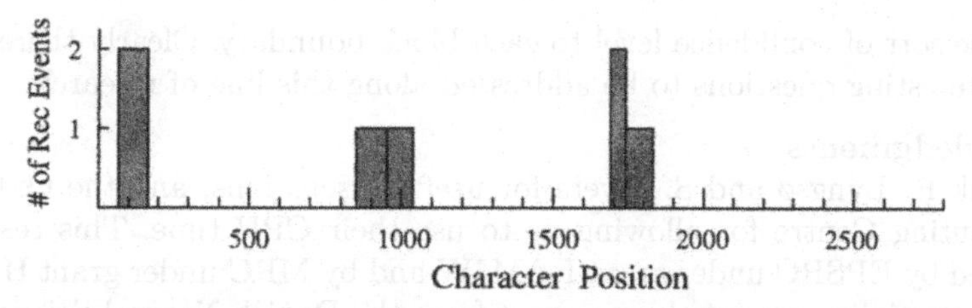

Fig. 5. Recombination locations in the actual data. Shaded areas enclosed by solid lines indicate the regions in which recombination events shown in Figure 4 occur. The number of events for each region is indicated by the height of the shaded area.

5 Concluding Remarks

Under the infinite-sites model of mutation, our algorithm finds the exact minimum number of recombination events in the evolutionary history of sampled sequences. The method introduced in this paper for computing the distance between trees has allowed us to overcome some difficulties which have hitherto prevented an exact implementation of the dynamic programming idea. It is important to note, however, that even our new approach, as it stands, becomes infeasible for more than 9 sequences in the reduced data; when there are many trees, it takes an inordinate amount of memory to store the adjacency-sets. In [13], Wang, Zhang and Zhang have shown that, under the infinite-sites model, the problem of reconstructing the evolutionary history with the minimum number of recombination events is NP-hard. Although they have constructed a polynomial-time algorithm for a restricted version of the problem, no polynomial-time algorithm is known for the general case.

Nevertheless, for more than 9 sequences, we can try the following. The algorithm proposed by Myers and Griffiths uses local bounds for small regions to construct a global bound for the entire data [9]. As there may not be so many distinct haplotypes if small regions are considered, we can use our algorithm to compute exact local bounds and use them in Myers and Griffiths' program to find a global bound. Combining our algorithm with that of Myers and Griffiths as just described should perform quite well.

In our future research, we plan to relax the assumption of the infinite-sites model and extend our investigation to consider gene conversion. Also, for a given data set, it would be interesting to infer mutation and recombination rates – for example, using the method developed by Fearnhead and Donnelly [2] – and estimate the probabilities of the minimal ARGs we obtain, as well as the distribution of the number of recombination events.

In this paper we have only begun to consider defining haplotype blocks based on parsimonious reconstruction of sequence evolution. As the example of Kreitman's data shows, in general our algorithm may find many minimal paths and therefore find many inequivalent candidate haplotype block structures. To be able to determine which structures are more probable, we need to be able to assign some sort of confidence level to each block boundary. Clearly there remain many interesting questions to be addressed along this line of research.

Acknowledgments
We thank R. Lyngsø and S. Myers for useful discussions, and the Oxford Supercomputing Centre for allowing us to use their CPU time. This research is supported by EPSRC under grant HAMJW and by MRC under grant HAMKA. Y.S.S. is partially supported by a grant from the Danish Natural Science Foundation (SNF-5503-13370).

References

1. Daly, M.J. *et al.*, *High-Resolution Haplotype Structure in the Human Genome,* Nat. Genet. **29** (2001) 229-232.
2. Fearnhead, P. and Donnelly, P., *Estimating Recombination Rates from Population Genetic Data,* Genetics **159** (2001) 1299-1318.
3. Gabriel, S.B. *et al.*, *The Structure of Haplotype Blocks in the Human Genome,* Science (2002) **296** 2225-2229.
4. Hein, J., *Reconstructing Evolution of Sequences Subject to Recombination Using Parsimony,* Math. Biosci. **98** (1990) 185-200.
5. Hein, J., *A Heuristic Method to Reconstruct the History of Sequences Subject to Recombination,* J. Mol. Evol. **36** (1993) 396-405.
6. Hudson, R.R. and Kaplan, N.L., *Statistical Properties of the Number of Recombination Events in the History of a Sample of DNA Sequences,* Genetics **11** (1985) 147-164.
7. Johnson, G.C. *et al.*, *Haplotype Tagging for the Identification of common Disease Genes,* Nat. Genet. **29** (2001) 233-237.
8. Kreitman, M., *Nucleotide Polymorphism at the Alcohol Dehydrogenase Locus of Drosophila Melanogaster,* Nature **304** (1983) 412-417.
9. Myers, S.R. and Griffiths, R.C., *Bounds on the Minimum Number of Recombination Events in a Sample History,* Genetics **163** (2003) 375-394.
10. Schröder, E., *Vier Combinatorische Probleme,* Zeit. für. Math. Phys, **15**, (1870), 361-376.
11. Song, Y.S., *Notes on the Combinatorics of Rooted Binary Phylogenetic Trees,* submitted to Annals of Combinatorics for publication.
12. Song, Y.S. and Hein, J., *On the Minimum Number of Recombination Events in the Evolutionary History of DNA Sequences,* to appear in J. Math. Biol.
13. Wang, L., Zhang, K. and Zhang, L., *Perfect Phylogenetic Networks with Recombination,* J. Comp. Biol, **8** (2001) 69-78.

Identifying Blocks and Sub-populations in Noisy SNP Data

Gad Kimmel[1], Roded Sharan[2], and Ron Shamir[1]

[1] School of Computer Science, Tel-Aviv University, Tel-Aviv 69978, Israel.
{kgad,rshamir}@tau.ac.il

[2] International Computer Science Institute, 1947 Center St., Suite 600, Berkeley CA-94704.
roded@icsi.berkeley.edu

Abstract. We study several problems arising in haplotype block partitioning. Our objective function is the total number of distinct haplotypes in blocks. We show that the problem is NP-hard when there are errors or missing data, and provide approximation algorithms for several of its variants. We also give an algorithm that solves the problem with high probability under a probabilistic model that allows noise and missing data. In addition, we study the multi-population case, where one has to partition the haplotypes into populations and seek a different block partition in each one. We provide a heuristic for that problem and use it to analyze simulated and real data. On simulated data, our blocks resemble the true partition more than the blocks generated by the LD-based algorithm of Gabriel et al. [7]. On single-population real data, we generate a more concise block description than extant approaches, with better average LD within blocks. The algorithm also gives promising results on real 2-population genotype data.

Keywords: haplotype, block, genotype, SNP, sub-population, stratification, algorithm, complexity.

1 Introduction

The availability of a nearly complete human genome sequence makes it possible to look for telltale differences between DNA sequences of different individuals on a genome-wide scale, and to associate genetic variation with medical conditions. The main source of such information is single nucleotide polymorphisms (SNPs). Millions of SNPs have already been detected [17, 18], out of an estimated total of 10 millions common SNPs [9]. This abundance is a blessing, as it provides very dense markers for association studies. Yet, it is also a curse, as the cost of typing every individual SNP becomes prohibitive. Haplotype blocks allow researchers to use the plethora of SNPs at a substantially reduced cost.

The sequence of alleles in contiguous SNP positions along a chromosomal region is called a *haplotype*. A major recent discovery is that haplotypes tend to be preserved along relatively long genomic stretches, with recombination occurring

G. Benson and R. Page (Eds.): WABI 2003, LNBI 2812, pp. 303–319, 2003.

primarily in narrow regions called *hot spots* [7, 16]. The regions between two neighboring hot spots are called *blocks*, and the number of distinct haplotypes within each block that are observed in a population is very limited: typically, some 70-90% of the haplotypes within a block belong to very few (2-5) *common haplotypes* [16]. The remaining haplotypes are called *rare haplotypes.* This finding is very important to disease association studies, since once the blocks and common haplotypes are identified, one can hopefully obtain a much stronger association between a haplotype and a disease phenotype. Moreover, rather than typing every individual SNP, one can choose few representative SNPs from each block that suffice to determine the haplotype. Using such *tag SNPs* allows a major saving in typing costs.

Due to their importance, blocks have been studied quite intensively recently. Daly et al. [5] and Patil et al. [16] used a greedy algorithm to find a partition into blocks that minimizes the total number of SNPs that distinguish a prescribed fraction of the haplotypes in each block. Zhang et al. [20] provided a dynamic programming algorithm for the same purpose. Koivisto et al. [14] provided a method based on Minimum Description Length to find haplotype blocks. Bafna et al. [2] proposed a combinatorial measure for comparing block partitions and suggested a different approach to find tag SNPs, that avoids the partition into blocks. For a recent review on computational aspects of haplotype analysis, see [12].

In this paper we address several problems that arise in haplotype studies. Our starting point is a very natural optimization criterion: We wish to find a block partition that minimizes *the total number of distinct haplotypes* that are observed in all the blocks. This criterion for evaluating a block partition follows naturally from the above mentioned observation, that within blocks in the human genome, only a few common haplotypes are observed [16, 5, 7]. The same criterion is used in the pure parsimony approach for haplotype inference, where the problem is to resolve genotypes into haplotypes, using a minimum number of distinct haplotypes [11]. In this case, the problem was shown to be NP-hard [13]. This criterion was also proposed by Gusfield [10] as a secondary criterion in refinements to Clark's inference method [3]. Minimizing the total number of haplotypes in blocks can be done in polynomial time if there are no data errors, using a dynamic programming algorithm. The problem becomes hard when errors are present or some of the data are missing. In fact, the problem of scoring a single given block turns out to be the bottleneck. Note that in practice, one has to account for rare haplotypes and hence minimize the total number of *common* haplotypes.

The input to all the problems we address is a binary *haplotype matrix* A with columns corresponding to SNPs in their order along the chromosome and rows corresponding to individual chromosomal segments typed. A_{ij} is the allele type of chromosome i in SNP j. The first set of problems that we study concerns the scoring of a single block in the presence of errors or missing data. In one problem variant, we wish to find a minimum number of haplotypes such that by making at most E changes in the matrix, each row vector is transformed into one of them.

We call this problem *Total Block Errors (TBE)*. We show that the problem in NP-hard, and provide a polynomial 2-approximation algorithm when the number of haplotypes is bounded. In a second variant, we wish to minimize the number of haplotypes when the *maximum* number of errors between a given row and its (closest) haplotype is bounded by e. We call this problem *Local Block Errors (LBE)*. This problem is shown to be NP-hard too, and we provide a polynomial algorithm (for fixed e), which guarantees a logarithmic approximation factor. In a third variant, some of the data entries are missing (manifested as "question marks" in the block matrix), and we wish to replace each of them by zero or one, so that the total number of haplotypes is minimum. Again, we show that this *Incomplete Haplotypes (IH)* problem is NP-hard. To overcome the hardness we resort to a probabilistic approach. We define a probabilistic model for generating haplotype data, including errors, missing data and rare haplotypes, and provide an algorithm that scores a block correctly with high probability under this model.

Another problem that we address is stratifying the haplotype populations. It has been shown that the block structure in different populations is different [7]. When the partition of the sample haplotypes into sub-populations is unknown, determining a single block structure for all the haplotypes can create artificial solutions with far too many haplotypes. We define the *Minimum Block Haplotypes (MBH)* problem, where one has to partition the haplotyped individuals into sub-populations and provide a block structure for each one, so that the total number of distinct haplotypes over all sub-populations and their blocks is minimum. We show that MBH is NP-hard, but provide a heuristic for solving it in the presence of errors, missing data and rare haplotypes. The algorithm uses ideas from the probabilistic analysis.

We applied our algorithm to several synthetic and real datasets. We show that the algorithm can identify the correct number of sub-populations in simulated data, and is robust to noise sources. When compared to the LD-based algorithm of Gabriel et al. [7], we show that our algorithm forms a partition into blocks that is much more faithful to the true one. On a real dataset of Daly et al. [5] we generate a more concise block description than extant approaches, with a better average value of the high LD-confidence fraction within blocks. As a final test, we applied our MBH algorithm to the two largest sub-populations reported in Gabriel et al. [7]. As this was genotype data, we treated heterozygotes as missing data. Nevertheless, the algorithm was able to determine that there are two sub-populations and correctly classified over 95% of the haplotypes.

The paper is organized as follows: In Section 2 we study the complexity of scoring a block under various noise sources and present our probabilistic scoring algorithm. In Section 3 we study the complexity of the MBH problem and describe a practical algorithm for solving it. Section 4 contains our results on simulated and real data.

2 Scoring Noisy Blocks

In this section we study the problem of minimizing the number of distinct haplotypes in a block under various noise sources. This number will be called the *score* of the block. The scoring problem arises as a key component in block partitioning in single- and multiple-population situations.

The input is a haplotype matrix A with n rows (haplotypes) and m columns (SNPs). A may contain errors (where '0' is replaced by '1' and vice versa), resulting from point mutations or measurement errors, and missing entries, denoted by '?'. Clearly, if there are no errors or missing data then a block can be scored in time proportional to its size by a hashing algorithm. Below we define and analyze several versions of the scoring problem which incorporate errors into the model. We assume until Section 2.4 that there are no rare haplotypes. In the following we denote by v_i the i-th row vector (haplotype) of A, and by $V = \{v_1, \ldots, v_n\}$ the set of all n row vectors.

2.1 Minimizing the Total Number of Errors

First we study the following problem: We are given an integer E, and wish to determine the minimum number of (possibly new) haplotypes, called *centroids*, such that by changing at most E entries in A, every row vector is transformed into one of the centroids. Formally, let $h(\cdot, \cdot)$ denote the Hamming distance between two vectors. Define the following problem:

Problem 1 (Total Block Errors (TBE)). Given a block matrix A and an integer E, find a minimum number k of centroids $v_1, \ldots, v_k$, such that $\sum_{u \in V} \min_i h(u, v_i) \leq E$.

Determining if $k = 1$ can be done trivially in $O(nm)$ time by observing that the minimum number of errors is obtained when choosing v_1 to be the consensus vector of the rows of A. The general problem, however, is NP-hard, as shown below:

Theorem 1. *TBE is NP-hard.*

Proof. We provide a reduction from VERTEX COVER. Given an instance $(G = (V = \{1, \ldots, m\}, F = \{e_1, \ldots, e_n\}), k)$ of VERTEX COVER, where w.l.o.g. $k < m - 1$, we form an instance $(A, k+1, E)$ of TBE. A is an $(n + mn^2) \times m$ matrix, whose rows are constructed as follows:

1. For each of edge $e_i = (s, t) \in F$, we form a binary vector v_{e_i} with '1' in positions s and t, and '0' in all other places.
2. For vertex $i \in V$ define the *vertex vector* u_i as the vector with '1' in its i-th position, and '0' otherwise. For each $i \in V$ we form a set U_i of n^2 identical copies of u_i.

We shall prove that G has a vertex cover of size at most k iff there is a solution to TBE on A with at most $k+1$ subsets and $E = n + n^2(m-k)$ errors.

($\Rightarrow$) Suppose that G has a vertex cover $\{v_1, \ldots, v_t\}$ with $t \le k$. Partition the rows of A into the following subsets: For $1 \le i \le t$ the i-th subset will contain all vectors corresponding to edges that are covered by v_i (if an edge is covered by two vertices, choose one arbitrarily), along with the n^2 vectors in U_i. Its centroid will be v_i. The $(t+1)$-st subset will contain all vectors corresponding to vertices of G that are not members of the vertex cover, with its centroid being the all-0 vector. It is easy to verify that the number of errors induced by this partition is exactly $n + n^2(m-t) \le E$.

($\Leftarrow$) Suppose that A can be partitioned into at most $t+1$ subsets (with $t \le k$) such that the number E^* of induced errors is at most E. W.l.o.g. we can assume that for each i all vectors in U_i belong to the same set in the partition. For each vertex $i \in V$, the set U_i induces at least n^2 errors, unless u_i is one of the centroids. Let l be the number of centroids that correspond to vertex vectors. Then the number of errors induced by the rest $(m-l)$ sets of vertex vectors is $E' = (m-l)n^2 \le (m-k)n^2 + n$. Hence, $k \le l \le t+1 \le k+1$. Suppose to the contrary that $l = k+1$. Since the Hamming distance of any two distinct vertex vectors is 2, we get $E' \ge 2(m-k-1)n^2 > E$ (since $m > k+1$), a contradiction. Thus, $l = k$. We claim that these k vertices form a vertex cover of G. By the argument above each other vertex vector must belong to the $(k+1)$-st subset and, moreover, its centroid must be the all-0 vector. Consider a vector w corresponding to an edge (u, w). If w is assigned to the $(k+1)$-st subset it adds 2 to E^*. Similarly, if w is assigned to one of the first k subsets corresponding to a vertex v and $u, w \ne v$ then w adds 2 to E^*. Since there are n edges and the assignment of vertex vectors induced $E' = n^2(m-k) \ge E - n$ errors, each edge can induce at most one error. Hence, each edge induces exactly one error, implying that every edge is incident to one of the k vertices. □

Thus, we study enumerative approaches to TBE. A straightforward approach is to enumerate the centroids in the solution and assign each row vector of A to its closest centroid. Suppose there are k centroids in an optimum solution. Then the complexity of this approach is $O(kmn2^{mk})$, which is feasible only for very small m and k. In the following we present an alternative approach to a variant of TBE, in which we wish to minimize the total number of errors induced by the solution. We devise a $(2 - \frac{2}{n})$-approximation algorithm for this variant, which takes $O(n^2m + kn^{k+1})$ time.

To describe the algorithm and prove its correctness we use the following lemma, that focuses on the problem of seeking a single centroid $v \in V$ to the n vectors $v_1, \ldots, v_n$. Denote $\widetilde{v}_b = \arg\min_{v \in \{0,1\}^m} \sum_{i=1}^n h(v, v_i)$, and let E be $\max_{v \in V} h(v, \widetilde{v}_b)$.

Lemma 1. *Let* $v_b = \arg\min_{v \in V} \sum_{i=1}^n h(v, v_i)$. *Then* $\sum_{i=1}^n h(v_b, v_i) \le (2 - \frac{2}{n})E$.

Proof. Define $s \equiv \sum_{1 \le i < j \le n} h(v_i, v_j)$. We first claim that $s \le E(n-1)$. Then,

$$s = \sum_{i<j} h(v_i, v_j) \le \sum_{i<j} [h(v_i, \widetilde{v}_b) + h(\widetilde{v}_b, v_j)] = (n-1) \sum_i h(v_i, \widetilde{v}_b) = (n-1)E .$$

The first inequality follows since the Hamming distance satisfies the triangle inequality. The last equality follows by using $\widetilde{v}_b$ as the centroid. This proves the claim.

By the definition of v_b, for every $v_c \neq v_b$ we have

$$\sum_{v_i \in V} h(v_b, v_i) \leq \sum_{v_i \in V} h(v_c, v_i)$$

Summing the above inequality for all n vectors, noting that $h(v, v) = 0$, we get

$$n \sum_{v_i \in V} h(v_b, v_i) \leq 2 \sum_{1 \leq i < j \leq n} h(v_i, v_j) = 2s \leq 2E(n-1)$$

□

Theorem 2. *TBE can be $(2 - \frac{2}{n})$-approximated in $O(n^2 m + kn^{k+1})$ time.*

Proof. **Algorithm:** Our algorithm enumerates all possible subsets of k rows in A as centroids, assigns each other row to its closest centroid and computes the total number of errors in the resulting solution.

Approximation factor: Consider two (possibly equal) partitions of the rows of A: $P_{alg} = (A_1, \ldots, A_k)$, the one returned by our algorithm; and $P_{best} = (\hat{A}_1, \ldots, \hat{A}_k)$, a partition that induces a minimum number of errors. For $1 \leq i \leq k$ denote $v_b^i = \arg\min_{v \in A_i} \sum_{v_j \in A_i} h(v, v_j)$ and $\hat{v}_b^i = \arg\min_{v \in \hat{A}_i} \sum_{v_j \in \hat{A}_i} h(v, v_j)$. The number of errors induced by P_{alg} and P_{best} are $E_{alg} = \sum_{i=1}^k \sum_{v \in A_i} h(v_b^i, v)$ and $E_{best} = \sum_{i=1}^k \sum_{v \in \hat{A}_i} h(\hat{v}_b^i, v)$, respectively. Finally, let $n_i = |\hat{A}_i|$ and denote by e_i the minimum number of errors induced in subset $\hat{A}_i$, by the optimal solution. In particular, $\sum_{i=1}^k n_i = n$ and $\sum_{i=1}^k e_i = E_{best}$.

Since our algorithm checks all possible solutions that use k of the original haplotypes as centroids and chooses a solution that induces a minimal number of errors, $E_{alg} \leq E_{best}$. By Lemma 1, $\sum_{v \in \hat{A}_i} h(\hat{v}_b^i, v) \leq (2 - \frac{2}{n_i})e_i$ for every $1 \leq i \leq k$. Summing this inequality over all $1 \leq i \leq k$ we get

$$E_{alg} \leq E_{best} = \sum_{i=1}^{k} \sum_{v \in \hat{A}_i} h(\hat{v}_b^i, v) \leq \sum_{i=1}^{k} (2 - \frac{2}{n_i}) e_i \leq \sum_{i=1}^{k} (2 - \frac{2}{n}) e_i = (2 - \frac{2}{n}) E \, .$$

Complexity: As a preprocessing step we compute the Hamming distance between every two rows in $O(n^2 m)$ time. There are $O(n^k)$ possible sets of centroids. For each centroid set, assigning rows to centroids and computing the total number of errors takes $O(kn)$ time. The complexity follows. □

2.2 Handling Local Data Errors

In this section we treat the question of scoring a block when the *maximum* number of errors between a haplotype and its centroid is bounded. Formally, we study the following problem:

Problem 2 (Local Block Errors (LBE)). Given a block matrix A and an integer e, find a minimum number k of centroids $v_1, \ldots, v_k$ and a partition $P = (V_1, \ldots, V_k)$ of the rows of A, such that $h(u, v_i) \leq e$ for every i and every $u \in V_i$.

Theorem 3. *LBE is NP-hard even when* $e = 1$.

Proof. We use the same construction as in the proof of Theorem 1. We claim that the VERTEX COVER instance has a solution of cardinality at most k iff the LBE instance has a solution of cardinality at most $k + 1$ such that at most one error is allowed in each row. The 'only if' part is immediate from the proof of Theorem 1. For the 'if' part observe that any two vectors corresponding to a pair of independent edges cannot belong to the same subset in the partition, and so is the case for a vertex vector and any vector corresponding to an edge that is not incident on that vertex. This already implies a vertex cover of size at most $k+1$. Since $m > k+1$ there must be a subset in the partition that contains at least two vectors corresponding to distinct vertices. But then either it contains no edge vector, or it contains exactly one edge vector and the vectors corresponding to its endpoints. In any case we obtain a vertex cover of the required size. □

In the following we present an $O(\log n)$ approximation algorithm for the problem.

Theorem 4. *There is an* $O(\log n)$ *approximation algorithm for LBE that takes* $O(n^2 m^e)$ *time.*

Proof. Our approximation algorithm for LBE is based on a reduction to SET COVER. Let V be the set of row vectors of A. Define the *e-set* of a vector v with respect to a matrix A as the set of row vectors in A that have Hamming distance at most e to v. Denote this e-set by $e(v)$. Let U be the union of all e-sets corresponding to row vectors of A. We reduce the LBE instance to a SET COVER instance $(V, \mathcal{S})$, where $\mathcal{S} \equiv \{e(v) \cap V : v \in U\}$. Clearly, there is a 1-1 mapping between solutions for the LBE instance and solutions for the SET COVER instance, and that mapping preserves the cardinality of the solutions. We now apply an $O(\log n)$-approximation algorithm for SET COVER (see, e.g., [4]) to $(V, \mathcal{S})$ and derive a solution to the LBE instance, which is within a factor of $O(\log n)$ of optimal. The complexity follows by observing that $|U| = O(nm^e)$. □

2.3 Handling Missing Data

In this section we study the problem of scoring an *incomplete matrix*, i.e., a matrix in which some of the entries may be missing. The problem is formally stated as follows:

Problem 3 (Incomplete Haplotypes (IH)). Given an incomplete haplotype matrix A, complete the missing entries so that the number of haplotypes in the resulting matrix is minimum.

Theorem 5. *IH is NP-hard.*

Proof. By reduction from GRAPH COLORING [8]. Given an instance $(G = (V, E), k)$ of GRAPH COLORING we build an instance (A, k) of IH as follows: Let $V = \{1, \ldots, n\}$. Each $i \in V$ is assigned an n-dimensional row vector v_i in A with '1' in the i-th position, '0' in the j-th position for every $(i, j) \in E$ and '?' in all other positions.

Given a k-coloring of G, let $V_1, \ldots, V_k$ be the corresponding color classes. For each class $V_i = \{v_{j_1}, \ldots, v_{j_i}\}$ we complete the '?'-s in the vectors corresponding to its vertices as follows: Each '?' in one of the columns $j_1, \ldots, j_i$ is completed to 1, and all other are completed to 0. The resulting matrix contains exactly k distinct haplotypes: Each haplotype corresponds to a color class, and has '1' in position i iff i is a member of the color class.

Conversely, given a solution to IH of cardinality at most k, each of the solution haplotypes corresponds to a color class in G. This follows since any two vectors corresponding to adjacent vertices must have a column with both '0' and '1' and, thus, represent two different haplotypes. □

2.4 A Probabilistic Algorithm

In this section we define a probabilistic model for the generation of haplotype block data. The model is admittedly naive, in that it assumes equal allele frequencies and independence between different SNPs and distinct haplotypes. However, as we shall see in Sections 3 and 4, it provides useful insights towards an effective heuristic, that performs well on real data. We give a polynomial algorithm that computes the optimal score of a block under this model with high probability (w.h.p.). Our model allows for all three types of confusing signals mentioned earlier: Rare haplotypes, errors and missing data.

Denote by T the hidden true haplotype matrix, and by A the observed one. Let T' be a submatrix of T, which contains one representative of each haplotype in T (common and rare). We assume that the entries of T' are drawn independently according to a Bernoulli distribution with parameter 0.5. T is generated by duplicating each row in T' an arbitrary number of times. This completes the description of the probabilistic model for T. Note that we do not make any assumption on the relative frequencies of the haplotypes. We now introduce errors to T by independently flipping each entry of T with probability $\alpha < 0.5$. Finally, each entry is independently replaced with a '?' with probability p. Let A be the resulting matrix, and let A' be the submatrix of A induced by the rows in T'. Under these assumptions, the entries of A' are independently identically distributed as follows: $A'_{ij} = 0$ with probability $\frac{1-p}{2}$, $A'_{ij} = 1$ with probability $\frac{1-p}{2}$ and $A'_{ij} =?$ with probability p.

We say that two vectors x and y have a *conflict* in position i if one has value 1 and the other 0 in that position. Define the dissimilarity $d(x, y)$ of x and y as the number of their conflicting positions (in the absence of '?'s, this is just the Hamming distance). We say that x is *independent* of y and denote it by $x \parallel y$, if x and y originate from two different haplotypes in T. Otherwise, we say that

x and y are *mates* and denote it by $x \approx y$. Intuitively, independent vectors will have higher dissimilarity compared to mates. In particular, for any i:

$$p_I \equiv Prob(x_i = y_i | x \parallel y;\ x_i, y_i \in \{0,1\}) = 0.5, \tag{1}$$
$$p_M \equiv Prob(x_i = y_i | x \approx y;\ x_i, y_i \in \{0,1\}) = \alpha^2 + (1-\alpha)^2 > 0.5\,.$$

Problem 4 (Probabilistic Model Block Scoring (PMBS)). Given an incomplete haplotype block matrix A, find a minimum number k of centroids $v_1, \ldots, v_k$, such that under the above probabilistic model, w.h.p., each vector $u \in A$ is a mate of some centroid.

Our algorithm for scoring a block A under the above probabilistic model is described in Figure 1. It uses a threshold t^* on the dissimilarity between vectors, to decide on mate relations. t^* is set to be the average of the expected dissimilarity between mates and the expected dissimilarity between independent vectors (see proof of Theorem 6). The algorithm produces a partition of the rows into mate classes of cardinalities $s_1 \geq s_2 \geq \ldots \geq s_l$. Given any lower bound γ on the fraction of rows that need to be covered by the common haplotypes, we give A the score $h = \arg\min_j \sum_{i=1}^{j} s_i \geq \gamma n$. We prove below that w.h.p. h is the correct score of A.

Score(A):

1. Let V be the set of rows in A.
2. Initialize a heap S.
3. **While** $V \neq \emptyset$ **do:**
   (a) Choose some $v \in V$.
   (b) $H \leftarrow \{v\}$.
   (c) **For** every $v' \in V \setminus \{v\}$ **do:**
   If $d(v, v') < t^*$ **then** $H \leftarrow H \cup \{v'\}$.
   (d) $V \leftarrow V \setminus H$.
   (e) Insert(S,$|H|$).
4. Output S.

Fig. 1. An algorithm for scoring a block under a probabilistic model of the data. Procedure Insert(S,s) inserts a number s into a heap S.

Theorem 6. *If* $m = \omega(\log n)$ *then w.h.p. the algorithm computes the correct score of* A.

Proof. We prove that w.h.p. each mate relation decided by the algorithm is correct. Applying a union bound over all such decisions will give the required result. Fix an iteration of the algorithm at which v is the chosen vertex and let $v' \neq v$ be some row vector in A. Let X_i be a binary random variable which is 1 iff v_i and v'_i are in conflict. Clearly, all X_i are independent identically distributed Bernoulli random variables. Define $X \equiv d(v, v') = \sum_{i=1}^{m} X_i$ and $f \equiv (1-p)^2$. Using Equation 1 we conclude:

$$(X|v' \parallel v) \sim Binom(m, f(1-p_I)) ,$$
$$(X|v' \approx v) \sim Binom(m, f(1-p_M)) .$$

We now require the following Chernoff bound (cf. [1]): If $Y \sim Binom(n, s)$ then for every $\epsilon > 0$ there exists $c_\epsilon > 0$ that depends only on ϵ, satisfying:

$$Prob[|Y - ns| \geq \epsilon ns] \leq 2e^{-c_\epsilon ns}.$$

Let $\mu = mf(1-p_M)$. Define $\epsilon \equiv \frac{(1-p_I)-(1-p_M)}{2(1-P_M)}$ and $t^* \equiv \epsilon\mu$. Applying Chernoff bound we have that for all $c > 0$:

$$Prob(X > t^*|v' \approx v) \leq 2e^{-c_\epsilon \mu m} < \frac{1}{n^c} , Prob(X \leq t^*|v' \parallel v) < \frac{1}{n^c} .$$

Since we check whether $d(v, v') < t^*$ a total of $O(n^2)$ times, applying a union bound we conclude that the probability that throughout the algorithm some implied mate relation is incorrect, is bounded by a polynomial in $\frac{1}{n}$. □

When using the algorithm as part of a practical heuristic (see Section 3), we do not report the rare haplotypes. Instead, we report only the smallest number of most abundant haplotypes as computed by the algorithm that together capture a fraction γ of all haplotypes. In applications in which the error rate α is not known, t^* cannot be directly computed. Instead, we calculate the ratio $\frac{d(v_1,v_2)}{fm}$ for any two row vectors, and keep these values in a sorted array: $d_1 \leq \ldots \leq d_{\binom{n}{2}}$. Next we find $(a, b) = \arg\max_{a=d_i, b=d_{i+1},\ 1 \leq i < \binom{n}{2}} [b - a]$. Then we set $t^* = \frac{a+b}{2}$. It can be shown that using this strategy the algorithm solves PMBS with high probability.

3 Minimum Block Haplotypes

Suppose that the matrix A contains haplotypes from several homogeneous populations. The partitioning into blocks can differ among populations [7]. Here, we study how to reconstruct the partitioning of the rows of A into sets called *sub-populations*, and the columns in each set into blocks, such that the sum of the scores of the submatrices corresponding to these blocks is minimized. Formally:

Problem 5 (Minimum Block Haplotypes (MBH)). Given a haplotype matrix A, find a partition of its rows into sub-populations so that the total number of block haplotypes is minimized.

We usually know which populations the haplotypes came from, however, in certain situations, there may be a hidden stratification of the population, that can dramatically change the conclusions of association studies.

Given a partition of the rows, one can compute the score in the noiseless case using a simple adaptation of the dynamic programming algorithm of [20]. However, the general MBH problem is NP-hard.

Theorem 7. *MBH is NP-hard.*

For lack of space, the proof is omitted here. Interestingly, the problem can be solved in polynomial time if each sub-population is required to be a contiguous set of rows. This may be useful for designing heuristics that permute the matrix rows for local improvement.

We now present an efficient heuristic for MBH. The algorithm has three components: A block scoring procedure; a dynamic programming algorithm to find the optimum block structure for a single sub-population; and a simulated annealing algorithm to find an optimum partition into homogeneous sub-populations. We describe these components below.

The dynamic programming component computes the score for a given sub-population in a straightforward manner, similar to [20]. Let T_i, $0 \leq i \leq m$, be the minimum number of block haplotypes in the submatrix of A induced on the columns $1, \ldots, i$, where $T_0 = 0$. For a pair of columns i, j let B_{ij} be the score of the block induced by the row in S and the columns in $\{i, \ldots, j\}$. Then the following recursive formula can be used to compute T_m:

$$T_i = \min_{0 \leq j \leq i-1} T_j + B_{ji} \ .$$

For scoring a block within the dynamic programming, we use the probabilistic algorithm described in Section 2.4 with a small modification. Instead of using a fixed threshold t^*, we compute a different threshold $t^*_{v,v'}$ for every two vectors v, v'. This is done by counting the number l of positions, in which none of the vectors has '?', and setting $t^*_{v,v'} = \frac{l((1-p_M)+(1-P_I))}{2}$. Scoring an $n \times t$ block takes $O(tnk)$ time, where k is a bound on the number of common haplotypes. Hence, the dynamic programming takes $O(mb^2nk)$ total time, where b is an upper bound on the allowed block size. Additional saving may be possible by precomputing the pairwise distances of rows in contiguous matrix segments of size up to b.

The goal of the annealing process is to optimize the partition of the haplotypes into sub-populations. We define a *neighboring partition* as any partition that can be obtained from the current one by moving one haplotype from one group to another. A crucial factor in obtaining a good solution is the initialization of the annealing process. We perform the initialization as follows: We compute pairwise similarities between every two haplotypes. The similarity S_{uv} of vectors u and v is calculated as follows: Initially we set $S_{uv} = 0$. We then slide a window of size $w = 20$ along u and v. For each position i we check whether $d((u_i, \ldots, u_{i+w-1}), (v_i, \ldots, v_{i+w-1})) \leq w\alpha$. If this is the case, we increment S_{uv} and jump to $i + w$ for the next iteration. Otherwise, we jump to $i + 1$. The intuition is that rows from the same sub-population should be more similar in blocks in which they share the same haplotypes and, thus, have better chance to hit good windows, and accumulate higher score in the scan. We next cluster the haplotypes based on their similarity values, using the K-means algorithm [15]. The resulting partition is taken to be the starting point for the process. To determine the number of sub-populations K, we try several choices and pick the one that results in the lowest score.

The running time of the practical algorithm is dominated by the cost of each annealing step. Since an annealing step changes the haplotypes of two sub-populations only, it suffices to recompute the scores of these sub-populations only.

4 Experimental Results

4.1 Simulations

We applied our algorithm to simulated and real haplotype data. First, we conducted extensive simulations to check the ability of our algorithm to detect sub-populations and recognize their block structure. Our simulation setup was as follows: Each simulated haplotype matrix contained 100 haplotypes and 300 SNPs. The number of sub-populations varied in the simulations. Sub-populations were of equal sizes. For each sub-population we generated block boundaries using a Poisson process with rate 20. Each block within a sub-population contained 2-5 common haplotypes covering 90% of the block's rows (with the rest 10% being rare haplotypes). Errors and missing data were introduced with varying rates up to 30%. The haplotype matrix was created according to the probabilistic model described in Section 2.4.

As a first test we simulated several matrices with 1-4 sub-populations and applied our algorithm with K ranging from 1 to 8. For each K we computed the score of the partition obtained, as described in Section 3. In each of the simulations the correct number got the lowest score (Figure 2.A). Next, we simulated several matrices with 3 sub-populations and different levels of errors and missing data. Figure 2.B summarizes our results in correctly assigning a haplotype to a sub-population (the set with the largest overlap with the true one was declared as correct). It can be seen that the MBH algorithm gives highly accurate results for missing data and error levels up to 10%.

For comparison, we also implemented the LD-based algorithm of Gabriel et al. [7] for finding blocks. We compared the block structures output by our algorithm and the LD-based algorithm to the correct one, using an alignment score similar to the one used in comparison of two DNA restriction enzyme maps [19, Sec. 9.10]. The score of two partitions P_1 and P_2 of m SNPs is computed as follows: We form two vectors of size $m-1$, in which '1' in position i denotes a block boundary between SNPs i and $i+1$, and '0' denotes that the two SNPs belong to the same block. We then compute an alignment score of these vectors using an affine gap penalty model with penalties 3, 2 and 0.5 for mismatch, gap open and gap extension, respectively, and a match score of zero.

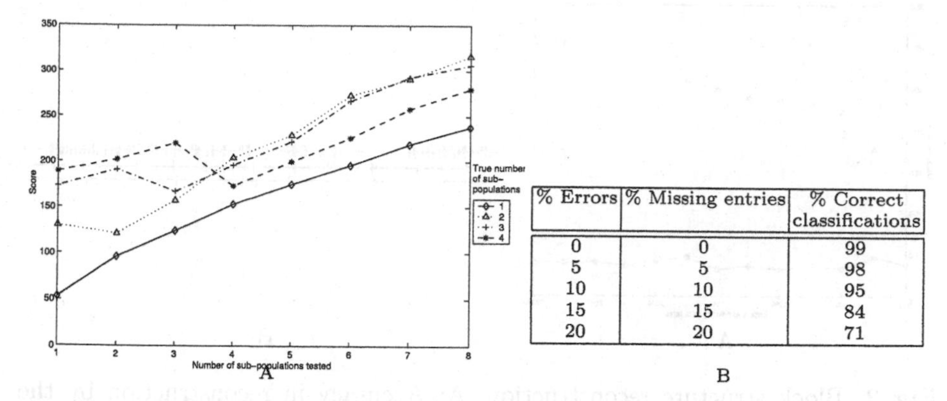

% Errors	% Missing entries	% Correct classifications
0	0	99
5	5	98
10	10	95
15	15	84
20	20	71

Fig. 2. Simulation results. A: Determining the number of sub-populations. For each simulated matrix, containing 1-4 sub-populations, the figure shows the score assigned by the algorithm to partitions (y-axis) with different number of sub-populations (x-axis). Simulations were performed with 1% errors and no missing entries. B: Accuracy of haplotype classification by the MBH algorithm for different noise levels. Data are for 3 sub-populations.

We simulated one population with 3000 haplotypes, computed its block structure with both algorithms and compared them to the true one. We repeated this experiment with different error and missing data rates. The results are shown in Figure 3.A. It can be observed that our algorithm yields partitions that are closer to the true ones, particularly as the rate of errors and missing data rises. An example of the actual block structures produced is shown in Figure 3.B.

4.2 Real Data

We applied our algorithm to two published datasets. The first dataset of Daly et al. [5] consists of 258 haplotypes and 103 SNPs. We applied our block partitioning algorithm with the following parameters: The maximal allowed error ratio between two vectors, to be considered as resulting from a single haplotype, was 0.02. In addition, we allowed 5% of rare haplotypes, i.e., in scoring a block we sought the minimum number of different haplotypes that together cover 95% of the rows.

In order to assess our block partitioning and compare it to the one reported by Daly et al. [5], we calculated LD-based measures for both partitions. Specifically, we calculated the LD-confidence values between every pair of SNPs inside the same block, using a χ^2-test. For each block, we calculated the fraction of SNP pairs in the block whose LD-confidence value exceeded 95% (*high LD pairs*). The average fraction over all blocks was computed as the ratio of the total number of high LD pairs inside blocks to the total number of SNP pairs within blocks.

A comparison between our block partition to the one obtained by Daly et al. is presented in Table 1. Overall, the two block partitions have similar bound-

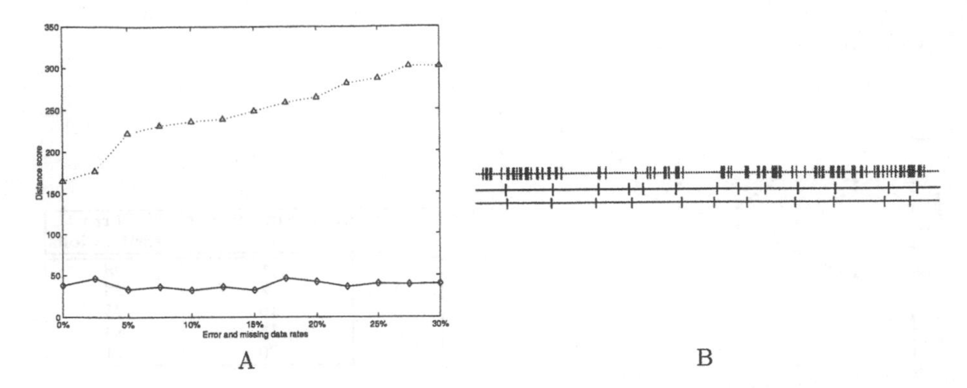

Fig. 3. Block structure reconstruction. A: Accuracy in reconstruction by the PMBS algorithm (solid line) and the algorithm of Gabriel et al. [7] (dashed line). y-axis: the score of aligning the reconstructed structure with the correct one. x-axis: the noise rate. B: An example of the block structures produced for an error rate of 1% by our algorithm (bottom), the LD-based algorithm of [7] (top) and the true solution (middle). Each block boundary is denoted by a vertical line.

aries and similar scores. The average fraction of high LD pairs in blocks for our partition was 0.823. For the partition of Daly et al. the average fraction was 0.796. Another available partition for this data by Eskin et al. [6], was based on minimizing the number of representative SNPs. Their partition contained 11 blocks and its average fraction of high LD pairs was 0.814.

The second dataset we analyzed, due to Gabriel et al. [7], contains unresolved genotype data. In order to apply our algorithm to this data, we transformed it into haplotype data by treating heterozygous SNPs as missing data. Notably, the fraction of heterozygous sites was relatively small, so the loss in information was moderate. We considered the two largest populations in the data, A (European) and D (individuals from Yoruba), consisting of 93 and 90 samples, respectively. Each population was genotyped in ~60 different regions in the genome. We analyzed 6 of those regions that contained over 70 SNPs. In all cases we were able to detect two different populations in the data and classify correctly over 95% of the haplotypes.

The results are shown in Table 2. The results with three populations were poorer, due to the smaller size of the third population.

5 Concluding Remarks

We have introduced a simple and intuitive measure for scoring and detecting blocks in a haplotype matrix: The total number of distinct haplotypes in blocks. Using this measure along with several error models, we have studied the computational problems of scoring of a block, and of finding an optimal block structure.

Daly et al. blocks	Fraction of high LD pairs	Our blocks	Fraction of high LD pairs
1: 1-9	0.78	1: 1-15	0.81
2: 10-15	1		
3: 16-24	0.78	2: 16-24	0.78
4: 25-35	0.95	3: 25-36	0.94
5: 36-40	0.70	4: 37-44	0.68
6: 41-45	1		
7: 46-77	0.77	5: 45-67	0.84
		6: 68-78	0.71
8: 78-85	0.50	7: 79-81	0.33
9: 86-91	0.93	8: 82-90	0.89
10: 92-98	0.95	9: 91-95	1
11: 99-103	1	10: 96-103	0.75
Average	**0.796**		**0.822**

Table 1. Comparison between the blocks of Daly et al. [5] and the blocks generated by our algorithm.

Most versions of the scoring problem that address imperfect data are shown to be NP-hard. A similar situation occurred with the f score function of Zhang et al. [20]. We devised several algorithms for different variants of the problem. In particular, we gave a simple algorithm, which, under an appropriate probabilistic model, scores a block correctly with high probability, in the presence of errors, missing data and rare haplotypes.

Note that our measure is adequate only when the ratio n/m of the data matrix is not too extreme: When the number of typed individuals n is very small and the number of SNPs m is large, our measure might be optimized by the trivial solution of a single block.

In simulations, our score leads to more accurate block detection than the LD-based method of Gabriel et al. [7]. While the simulation setup is quite naive, it seems to act just as favorably for the LD-based methods. The latter methods apparently tend to over-partition the data into blocks, as they demand a very stringent criterion between every pair of SNPs in the same block. This criterion is very hard to satisfy as block size increases, and the number of pairwise comparisons grows quadratically. On the data of Daly et al. [5] we generated a slightly more concise block description than extant approaches, with a somewhat better fraction of high LD pairs.

We also treated the question of partitioning a set of haplotypes into sub-populations based on their different block structures, and devised a practical heuristic for the problem. On a genotype dataset of Gabriel et al. [7] we were able to identify sub-populations correctly, in spite of ignoring all heterozygous types. A principled method of dealing with genotype data remains a computational challenge. While in some studies the partition into sub-populations is

Chromosome: Region	#SNPs	Discovered blocks	% Correct classifications
1: 3a	119	1: 1-35, 36-119 2: 1-46, 47-119	95
2: 8a	73	1: 1-73 2: 1-73	99
8: 29a	104	1: 1-27, 28-104 2: 1-40, 41-104	100
14: 41a	141	1: 1-48, 49-63, 64-141 2: 1-12, 13-63, 64-141	100
6: 24a	121	1: 1-52, 53-121 2: 1-44, 45-121	98
9: 32a	110	1: 1-25, 26-110 2: 1-38, 39-110	99

Table 2. Separation to populations and block finding on different regions in part of the data of [7], which includes populations A and D.

known, others may not have this information, or further, finer partition may be detectable using our algorithm. In our model we implicitly assumed that block boundaries in different sub-populations are independent. In practice, some boundaries may be common due to the common lineage of the sub-populations. A more detailed treatment of the block boundaries in sub-populations should be considered when additional haplotype data reveal the correct way to model this situation.

Acknowledgments

R. Sharan was supported by a Fulbright grant. R. Shamir was supported by a grant from the Israel Science Foundation (grant 309/02). We thank Chaim Linhart and Dekel Tsur for their comments on the manuscript.

References

[1] N. Alon and J. H. Spencer. *The Probabilistic Method.* John Wiley and Sons, Inc., 2000.

[2] V. Bafna, B. V. Halldorsson, R. Schwartz, A. Clark, and S. Istrail. Haplotyles and informative SNP selection algorithms: Don't block out information. *Proc. of RECOMB*, pages 19–27, 2003.

[3] A. Clark. Inference of haplotypes from PCR-amplified samples of diploid populations. *Molecular Biology and Evolution*, 7(2):111–22, 1990.

[4] T. H. Cormen, C. E. Leiserson, and R. L. Rivest. *Introduction to Algorithms.* MIT Press, Cambridge, Mass., 1990.

[5] M.J. Daly et al. High-resolution haplotype structure in the human genome. *Nature Genetics*, 29(2):229–232, 2001.

[6] E. Eskin, E. Halperin, and R. M. Karp. Large scale reconstruction of haplotypes from genotype data. *Proc. of RECOMB*, pages 104–113, 2003.
[7] S. B. Gabriel et al. The structure of haplotype blocks in the human genome. *Science*, 296:2225–2229, 2002.
[8] M. R. Garey and D. S. Johnson. *Computers and Intractability: A Guide to the Theory of NP-Completeness.* W. H. Freeman and Co., San Francisco, 1979.
[9] L. Grugliyak and D. A. Nickerson. Variation is the spice of life. *Nature Genetics*, 27:234–236, 2001.
[10] D. Gusfield. Inference of haplotypes in samples of diploid populations: Complexity and algorithms. *Journal of Computational Biology*, 8(3):305–323, 2001.
[11] D. Gusfield. Haplotyping by pure parsimony. *Technical Report UCDavis CSE-2003-2, To appear in the Proceedings of the 2003 Combinatorial Pattern Matching Conference*, 2003.
[12] B. V. Halldorsson et al. Combinatorial problems arising in SNP. *DMTCS '03 Conference.*
[13] E. Hubbell. Finding a parsimony solution to haplotype phase is NP-hard. *Personal's communication.*
[14] M. Koivisto et al. An MDL method for finding haplotype blocks and for estimating the strength of haplotype block boundaries. Proc. PSB 2003.
[15] J. MacQueen. Some methods for classification and analysis of multivariate observations. In *Proceedings of the 5th Berkeley Symposium on Mathematical Statistics and Probability*, pages 281–297, 1965.
[16] N. Patil et al. Blocks of limited haplotype diversity revealed by high-resolution scanning of human chromosome 21. *Science*, 294:1719–1723, 2001.
[17] R. Sachidanandam et al. A map of human genome sequence variation containing 1.42 million single nucleotide polymorphisms. *Nature*, 291:1298–2302, 2001.
[18] C. Venter et al. The sequence of the human genome. *Science*, 291:1304–51, 2001.
[19] M.S. Waterman. *Introduction to Computational Biology: Maps, Sequences and Genomes.* Chapman and Hall, 1995.
[20] K. Zhang, M. Deng, T. Chen, M.S. Waterman, and F. Sun. A dynamic programming algorithm for haplotype block partitioning. *Proc. Natl. Acad. Sci. USA*, 99(11):7335–9, 2002.

Designing Optimally Multiplexed SNP Genotyping Assays

Yonatan Aumann[1], Efrat Manisterski[1], and Zohar Yakhini[2]

[1] Dept. of Computer Science, Bar-Ilan University, Ramat Gan, Israel
aumann@cs.biu.ac.il
[2] Agilent Laboratories and Departmemt of Computer Science, the Technion, Haifa, Israel
zohar_yakhini@agilent.com

Abstract. We consider the task of SNP (Single Nucleotide Polymorphism) genotyping. In many studies, genotyping of a large number of SNP must be performed. Multiple SNPs can be genotyped together in the same assay (a process called *multiplexed genotyping*), provided they adhere to some constraints. We address the optimization problem of designing assays that maximize the number of SNPs genotyped, subject to the multiplexing constraints. We focus on the SNP genotyping method based on *primer extension* and *mass-spectrometry* (PEA/MS). We translate the optimization problem to a graph coloring problem, and provide an essentially optimal heuristics for solving the corresponding coloring problem. In addition, we present a method that enables a dramatic increase in the multiplexing rate by modifying primer masses. In this case, the multiplexing design problem can be modelled as a matching problem in hypergraphs. We analyze the problem from both theoretical and practical aspects, providing theoretical hardness results and practical heuristics. The heuristics are tested using simulation methods, and prove to be close to optimal in practice.

1 Introduction

1.1 Background

SNP Genotyping. The genetic makeup of any two given individuals, as determined by their genomic DNA sequences, differ in a variety of ways. Some of these variations, or *polymorphisms*, occur in coding regions and may thus have phenotypic manifestations in cellular and larger scale functions, such as disease susceptibility, metabolism, protein production, etc. Variations in other regions of the genomic sequence are useful in studies aimed at finding genomic regions linked to phenotypic variations. Such studies are performed by seeking correlations between the phenotypic inheritance patterns and the polymorphic genetic variations (see [11] for a more detailed background). In this work we focus on a specific type of polymorphism called *single nucleotide polymorphism* (SNP). Single Nucleotide Polymorphism is characterized by a differences, across the population, in a single base within an otherwise conserved genomic sequence [14].

G. Benson and R. Page (Eds.): WABI 2003, LNBI 2812, pp. 320–338, 2003.

SNPs have become extremely useful both as indicators of variations in coding regions and as markers used in linkage, association and linkage disequilibrium studies ([12,3]).

Given a known SNP and a sample of genetic material containing the locus of this SNP a *genotyping assay* is aimed at determining the specific variation of the SNP present in the sample (see [6]). Association and linkage analysis studies require genotyping of multiple SNPs sites over multiple individuals. That is, for a relatively large number of individuals, M, one needs to genotype a large number, n, of SNPs for each of the individuals. The range of parameters n and M that will be practical in the near future is a matter of scientific and policy discussions. $n \approx 60000$ and $M \approx 1000$ are examples of the order of magnitude.

Multiplexing. SNP genotyping is a time-consuming and expensive procedure. Thus, we are interested in minimizing the number of times the genotyping assay must be performed in a given study. Under certain circumstances, genotyping of multiple SNP sites can be performed simultaneously, in a single genotyping assay; a process called *multiplexed genotyping.* However, not all SNPs can be genotyped together. Each genotyping method imposes a different set of constraints on which SNPs can and cannot be assayed together. Thus, in order to achieve high multiplexing rates it is necessary to carefully plan the genotyping assays, in order to allow simultaneous genotyping of as many SNPs as possible. In this paper we present methods for achieving high multiplexing rates when genotyping using the *primer extension and mass spectrometry* method.

In association studies certain technologies allow for pooling individuals and for estimating allele frequencies from the pooled measurements ([6]). We note that this pooling process does not effect the multiplexing design questions considered in this paper. Multiplexing schemes that work for the single individual case can be extended to the pooled case, assuming that an adequate quantitative reading is afforded in the mass-spectrometry stage.

The SNP Genotyping Process. A genotyping assay is typically preceded by a step in which the relevant regions of the genome are isolated and amplified (typically using PCR). Therefore, the input to the genotyping assay is a set of n sequences, spanning the polymorphic sites. The output of the assay is a set of pairs of letters over the alphabet $\Sigma = \{A, C, G, T\}$, representing the SNP variations present in the sample, one pair for each SNP locus. Note that the SNP polymorphism can be bi-allelic, hence the for each SNP site there are two possible outcomes. Hence, the pair of letters for each SNP site.

Because of the importance seen for SNPs as components of genetic studies and as clinically meaningful indicators, we are witnessing the development of innovative approaches to high throughput SNP genotyping. In this work we focus on the method based on *primer extension and mass-spectrometry (PEA/MS)*, which works as follows. For each SNP, a primer composing the Watson-Crick complement of the downstream sequence immediately following the variation site is designed. The primer is put in contact with the sampled DNA amplicon,

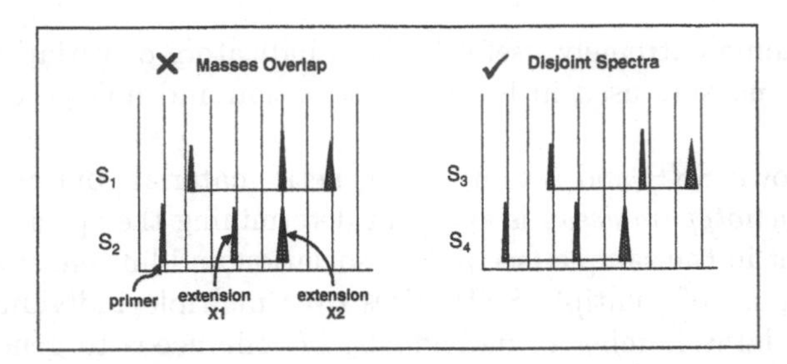

Fig. 1. Multiplexing Criteria: A schematic representation of the mass peaks that can be obtained for four different SNP sites. SNP sites s_1 and s_2 cannot be assayed together since their mass spectra overlap. SNP sites s_3 and s_4 can be assayed together since their spectra are disjoint.

in conditions that favor hybridization. The primer thus hybridizes to the amplicon at the location immediately following the SNP locus. Next, the primer is extended by one base by polymerase-extension on its 3'-side (which corresponds to the upstream direction on the amplicon)[1]. The properties of DNA binding provide that the additional base that extends the primer is the WC-complement of the base present at the SNP locus. To complete the genotyping process, the extended primer is separated from the original amplicon, and its mass is measured using mass-spectrometry. The mass of the extended primer is sum of the mass of the original primer plus the mass of the additional base (minus 18 amu). Thus, since the four bases have different masses, the total mass of the extended primer directly indicates what base is present at the SNP locus on the amplicon.

Multiplexing Criteria and Problem Definition. For a single biallelic SNP, the mass spectrum obtained in this genotyping measurement will show two or three peaks - one at the mass of the unextended primer, and one or two at the masses of the two possible extensions, corresponding to the two SNP alleles. Thus, two distinct SNP sites can be jointly measured in the same assay if the corresponding triplets of peaks are disjoint (see Figure 1)[2]. This observation provides a basis for allowing high multiplexing schemes: a plurality of SNP sites can be jointly measured if the corresponding triplets of masses are pairwise disjoint. Thus, given a set of SNPs to be genotyped, we seek to partition the set into a minimal number of subsets such that each subset can be jointly measured. This is the computational problem we address in this paper.

Our Results. We model the multiplexing problem as a graph coloring problem on graphs of a special type, which we call *tuple graphs.* We show that the general problem of coloring tuple graphs is NP-hard, but heuristics provide essentially optimal results for our problem in practice. However, the actual multiplexing

[1] A polymerase and a mixture of the four ddNTPs are added to the mixture, in conditions that favor extension.

[2] Theoretically, primers with the same mass do not obstruct unique calling. For purposes of assay control we do, however, want the primers to also not mass overlap.

rates obtained, while optimal, do not provide for sufficiently high multiplexing. Thus, we present a method that enables a dramatic increase in the multiplexing rate by modifying primer masses. In this case, the multiplexing design problem can be modelled as a matching problem in hypergraphs. We show that the corresponding approximation problem is NP-hard, and provide heuristics methods which achieve close to optimal results in practice, when tested on simulated data.

Related Work. [13] describe a multiplexed SNP genotyping assay utilizing primer extension and MALDI-TOF mass spectrometry, relying on the natural masses of the extended primers. The results of calling 12 SNP sites and the methods that enable this performance are discussed. The authors use additional non-complimentary bases at the 5' end of primers to resolve conflicting spectra. The total length of the resulting primers is between 15 and 23 bases. The current work provides a general algorithmic and statistical framework for a systematic design process that obtains optimal multiplexing rates especially useful in much larger scale assays, where manual design is not possible.

Kivioja et al. [8] consider the problem of optimization in multiplexed transcription profiling. In this case the aim is to measure transcriptional expression level of multiple genes, using hybridization probes. Two genes can be measured together iff their respective probes have different lengths (for electrophoresis). Each gene can be measured by any one of a number of probes, and the optimization problem is to choose the probes so as to minimize the total number of measurements necessary. Kivioja et al. provide a 2-approximation algorithm for this optimization problem. This problem is of a similar flavor, but different from the one considered in this paper.

In [15] the authors describe methods that utilize graph coloring techniques to obtain optimal design for multiplexed genotyping of microsattelite markers (a different kind of genetic variation). Here we take a similar approach and translate several variants of our design optimization problem into graph coloring problems. We analyze the complexity of the latter problems and the performance of heuristic approaches on appropriate stochastic models.

1.2 Mathematical Formalism

Consider a set of n SNP genomic loci, or *sites*, $S = s_1, \ldots, s_\eta$. To each SNP site, s, we associate a *triplet*, $(\mathrm{P}_s, \mathrm{X1}_s, \mathrm{X2}_s)$, where:

- P_s is the primer used for s, and
- $\mathrm{X1}_s$ and $\mathrm{X2}_s$ are the two potential extended primers for s.

For a oligonucleotide molecule x, we denote by $m(x)$ the mass of x. The three mass peaks that can possibly be obtained when genotyping an SNP s are: $m(\mathrm{P}_s)$, $m(\mathrm{X1}_s)$, $m(\mathrm{X2}_s)$.

A pair of sites, s_i and s_j (with their associated primers) are said to be *non-conflicting* if there is no coincidence between their mass spectra. Namely, the sets $\{m(\mathrm{P}_{s_i}), m(\mathrm{X1}_{s_i}), m(\mathrm{X2}_{s_i})\}$ and $\{m(\mathrm{P}_{s_j}), m(\mathrm{X1}_{s_j}), m(\mathrm{X2}_{s_j})\}$ are disjoint.

A subset $U \subseteq S$ is said to be *conflict free* if any two members of U are non-conflicting. By definition, a set of SNP sites can be jointly assayed iff it is conflict free. Thus, we are interested to partition S into a minimal number of conflict free sets.

2 Optimal Multiplexing and Coloring 3-Tuple Graphs

In this section we study the situation where the set S of (bi-allelic) SNP sites is given and for each site the primer is predetermined. We are interested in a partition of S into a minimal number k of conflict free sets. Define a graph $G(S) = (V, E)$ with $V = S$ and $(s_i, s_j) \in E$ iff the intersection of s_1 and s_2 is non-empty (i.e. s_1 and s_2 conflict). The graph $G(S)$ is called the *interference graph* of S. Note that an independent set in $G(S)$ corresponds to a conflict-free set in S. Thus, a coloring of $G(S)$ corresponds to a partition of S into conflict-free sets. Thus, it remains to color $G(S)$. In general, coloring is known to be NP-hard, even to approximate. However, the graph $G(S)$ is of special type, which we now analyze.

2.1 *k*-Tuple Graphs and Their Coloring

Each SNP is represented by a triplet of masses. This motivates the following definition.

Definition 1 *Let $G = (V, E)$ be a graph with $|V| = n$. We say that G is a* k-tuple graph *if there are sets $T_1, T_2, \ldots, T_n$ such that:*

- *For each i, $|T_i| = k$.*
- *G is the intersection graph of this collection: $(v_i, v_j) \in E$ iff $T_i \bigcap T_j \neq \emptyset$.*

Thus, the graph $G(S)$ is a 3-tuple graph. It is easy to see that any graph is a k-tuple graph for some k.

Theorem 1. *For any $k \geq 2$, coloring k-tuple graphs is NP hard.*

Proof. For the case $k = 2$, by reduction from edge coloring (which is NP hard by [7]). Consider a graph $G = (V, E)$ for which we seek an edge coloring. Define a graph $G' = (V', E')$, where $V' = E$ and $(e_1, e_2) \in E'$ iff $e_1 \cap e_2 \neq \emptyset$. Then, by definition G' is a 2-tuple graph, and a coloring of G' is an edge coloring of G.

For $k > 2$ reduce from the case $k = 2$ by padding each tuple T_i with $k - 2$ unique elements (different elements for each i).

For a graph G we denote by $\chi(G)$ its chromatic number.

Theorem 2. *There is a polynomial time algorithm that colors any k-tuple graph G in less than $k \cdot \chi(G)$ colors.*

Proof. For a graph G, we denote by $\Delta(G)$ the maximal degree of G and by $\kappa(G)$ the maximum clique size. Any graph can be colored with $\Delta(G)+1$ colors, in quadratic time. We show that this provides the desired approximation.

Let G be a k-tuple graph, and let $T_1, \ldots, T_n$ be the corresponding k-tuples. For each element x in any of the tuples, let $f(x)$ be the number of distinct tuples containing x. Note that for each element x, the set of tuples containing x constitutes a clique. Let v be the vertex with the highest degree in G and let T_v be the corresponding k-tuple. Then,

$$\Delta(G) = d(v) \leq \sum_{x \in T_v} (f(x) - 1) \leq k \cdot \max_{x \in T_v} f(x) - k \leq k \cdot \kappa(G) - k \leq k \cdot \chi(G) - k.$$

Thus, $\Delta(G) + 1 \leq k \cdot \chi(G)$.

2.2 A Heuristic Approach

We tested several coloring heuristic approaches to the task of coloring the interference graph that corresponds to the multiplexing problem. The most successful approach, as measured by performance on synthetic simulated data, was SLO-coloring (Smallest Last Order coloring, [10]). SLO-coloring proceeds as follows. First we define an order on the vertices of the graph G, as follows. Vertex $v_{\pi(1)}$ is the the vertex of minimal degree in G. Inductively $v_{\pi(i+1)}$ is the vertex of minimal degree, in the subgraph of G induced by $V - \{v_{\pi(1)}, \ldots, v_{\pi(i)}\}$. The coloring itself is then effected by greedily assigning colors according to the above order, reversed. That is: assign a random color to $v_{\pi(n)}$; having colored $v_{\pi(n)}, \ldots, v_{\pi(n-i)}$ assign one of the used colors to $v_{\pi(n-i-1)}$; if this is impossible, open a new color for $v_{\pi(n-i-1)}$.

2.3 Simulation Results

We tested the heuristics on synthetically generated data using primers of length 20, and for varying values of n - the numbers of SNPs. For each value of n, 100 independent experiments were conducted. For each experiment, we randomly generated n different SNP tuples, as follows. First we generated the primer by choosing uniformly at random a sequence of 20 bases. Then, we chose two distinct bases, uniformly at random, for the two possible extensions. As mentioned, the SLO heuristic provided the best results. The average results are summarized in Figure 2.

In order to gauge the quality of the results, we compared the results provided by the SLO heuristic to the following lower bound. For any element x, the set $\{T_i : x \in T_i\}$ is a clique. Thus, $\chi(G) \geq \kappa(G) \geq \max_x f(x)$ (where $f(x)$ is the number of tuples containing x). SLO matched the lower bound throughout.

3 Mass Modification

In the previous section we provided a heuristic for the multiplexing problem, and showed that this heuristic is essentially optimal. Specifically, given the set of

n - number of SNP sites (n)	k - average no. of reactions (lower bound and SLO
100	5.23
500	15.26
1000	26.56
2000	47.84
5000	110.88
10000	216.63

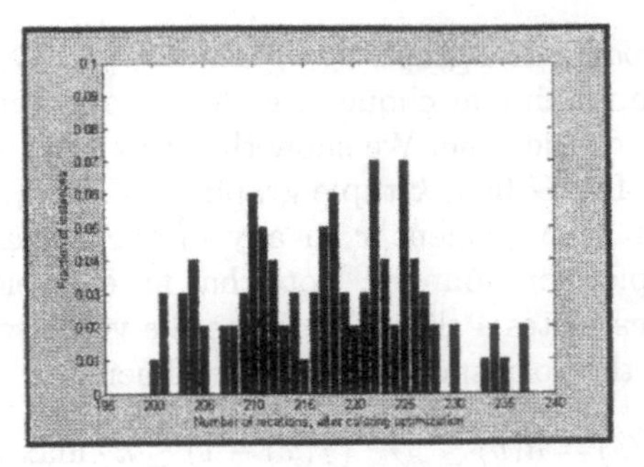

Fig. 2. Number of conflict free sets obtained by SLO and lower bound for synthetically generated sets of SNPs of various sizes, with primers of size 20. Table provides average number computed over 100 instances. Results obtained by SLO were equal to the lower bound throughout. Histogram: distribution of number of conflict free sets for $n = 10000$

SNPs to be genotyped, and the masses of the associated tuples, it is impossible to subdivide the SNP set into a smaller number of conflict-free sets. Thus, if we seek to further increase the multiplexing rate, we must somehow modify the masses of the tuples. In this section we suggest a simple technique for mass modification and show how to use this technique to achieve dramatic improvements in the multiplexing rate.

3.1 Primer Elongation

Consider a specific SNP. The SNP appears at a specific site on the genome and the primer for this SNP is constructed as the WC complement of the bases downstream of this location. Specifically, a primer of length ℓ is the sequence of bases that complement the ℓ bases immediately following the SNP site. Note however, that if we further extend the primer on its 5'-end (which corresponds to the amplicon's 3'-end) with a small number of additional bases, the resulting sequence will still function as a primer for the given SNP site. The reason is that the original sequence will still hybridize at the same location, the enzymatic reaction takes place on the 3'-end of the sequence, and the additional bases will simply hang as a "tail" at the other (5') end of the primer (see Figure 3). This process of adding a "tail" on the 5'-end is a standard practice in PCR reactions (see, for example, [1] page 319). The mass of the elongated primer is the mass of the original primer, plus the mass of the additional bases. Thus, by elongating the primer on the 5'-end we can modify its mass without affecting its function. It is important to note that the entire elongated primer can be synthesized at once, in the same way ordinary primers are synthesized.

As mentioned, the additional bases of the elongated primer need not be the complement of the base sequence of the original genom. Thus, there are multiple possible elongations for any given primer. If we restrict elongations to a maximum length k, then any sequence of k or less bases is a valid elongation.

Furthermore, different primers, for different SNP sites, may be elongated using different base sequences. This provides the opportunity to modify the masses of the primers in a way that will allow for higher multiplexing rates. In general, in order to achieve a good multiplexing rate, we must:

- for each SNP site s, choose an elongation sequence, and
- given the elongations, partition the set of resulting SNP tuples into conflict-free sets.

Clearly, these two steps are related, and have to be addressed together. We call the combination of both a *multiplexing strategy.* In this section we discuss the hardness of obtaining an optimal multiplexing strategy, and provide efficient heuristics for the problem.

Terminology and Notation Let s_1 and s_2 be sequences of bases. We denote by $s_1 + s_2$ the sequence of bases obtained by elongating s_1 by s_2 on the 5'-end. Let $s = (\mathrm{P}_s, \mathrm{X1}_s, \mathrm{X2}_s)$ be an SNP tuple, and let t be an elongation (i.e. a sequence of bases). We denote by $s + t = \{\mathrm{P}_s + t, \mathrm{X1}_s + t, \mathrm{X2}_s + t\}$ the tuple resulting by elongating the primer s by t. We call this tuple an *elongated tuple.*

Definition 2 *Let S be a set of SNPs, and let T be a set of possible elongations. A* multiplexing strategy *for (S, T) is a pair (f, C), where:*

- *$f : S \rightarrow T$ is a function that assigns an elongation $f(s) \in T$ to each $s \in S$,*
- *C is a partition of the set $\{s + f(s) : s \in S\}$ of elongated tuples into conflict free sets.*

A multiplexing strategy is optimal *if the number of conflict free sets is minimal over all possible multiplexing strategies.*

The *multiplexing strategy problem* is: given a pair (S, T) find an optimal multiplexing strategy.

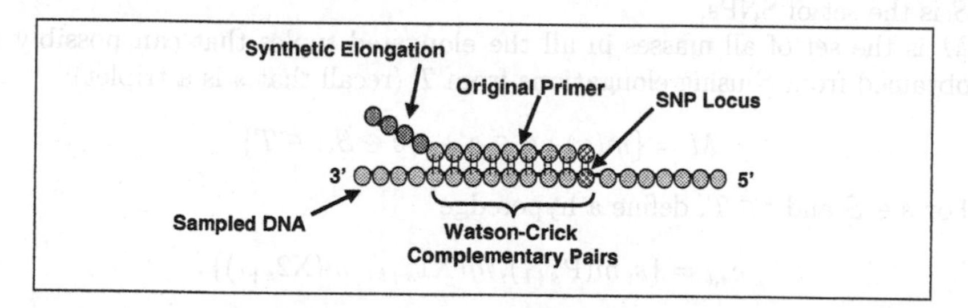

Fig. 3. Primer Elongation. The primer is elongated on the 5'-end with bases that do not necessarily form the WC-complement of the original DNA amplicon. The original primer hybridizes with the amplicon, and extension at the SNP locus takes place on the other end of the primer. The mass spectra is shifted by the mass of the elongation.

3.2 Mathematical Modelling

Consider a multiplexing strategy problem on (S,T). We model the problem as a matching problem in hypergraphs. A hypergraph is a graph where edges connect between 2 or more vertices (see [2] for more details on hypergraphs). For example, an edge in a hypergraph can connect 4 vertices. In our construction, we use a specific type of hypergraphs, which we call *extension graphs*.

Extension Graphs The following is a formal definition of extension graphs.

Definition 3 *Let $G=(A,B,E)$ be hypergraph, where A and B are two disjoint sets of vertices and E is the set of hyperedges. We say that G is an* extension graph *of A if in each hyperedge $e \in E$ there is exactly one vertex from A.*

Next, we define the notion of matchings in extension graphs.

Definition 4 *Let $G=(A,B,E)$ be an extension graph of A. For edges e, e', we say that e and e'* conflict *if $e \cap e' \neq \emptyset$. For an edge e and a set $C \subseteq E$ we say that e conflicts with C if there exists $e' \in C$ such that e and e' conflict.*

A matching *in G is a subset $C \subseteq E$ such that for any $e \in C$, e does not conflict with $C - \{e\}$. For a vertex v and matching C, we say that C* covers *v if there is an edge $e \in C$, such that $v \in e$. We denote by Cover(C) the set of nodes covered by C. For a collection of matchings $\mathcal{P} = \{C_1, C_2, \dots, C_\ell\}$, we denote Cover($\mathcal{P}$) $= \bigcup_{C \in \mathcal{P}}$ Cover(C).*

A matchings partition *for G is a collection $\mathcal{P} = \{C_1, C_2, \dots, C_\ell\}$, such that Cover($\mathcal{P}$) $\supseteq A$ (i.e. each vertex $v \in A$ is covered by at least one of the matchings in $\mathcal{P}$). A matching partition is* minimal *if there is no matching partition with fewer matchings. For an extension graph G, we denote by $\gamma(G)$ the number of matchings in the minimal matching partition for G.*

The Graph $G_{S,T}$ Given a multiplexing problem pair (S,T) we construct an extension graph $G_{S,T} = (S, M, E)$.

- S is the set of SNPs,
- M is the set of all masses in all the elongated tuples that can possibly be obtained from S using elongations from T (recall that s is a triplet):

$$M = \{m(x) : x \in s + t, s \in S, t \in T\},$$

- For $s \in S$ and $t \in T$, define a hyperedge

$$e_{s,t} = \{s, m(\mathrm{P}_{s+t}), m(\mathrm{X1}_{s+t}), m(\mathrm{X2}_{s+t})\}.$$

  In words, $e_{s,t}$ connects an SNP s with the triplet of masses that this SNP gets given the elongation t. The edges of $G_{S,T}$ are:

$$E = \{e_{s,t} : s \in S, t \in T\}.$$

For an edge $e_{s,t}$, we denote $s(e_{s,t}) = s$ and $t(e_{s,t}) = t$.

We now establish a one-to-one correspondence between matchings in $G_{S,T}$ and conflict-free sets for S.

Claim. Let C be a matching in $G_{S,T}$, and let $S' = \{s(e) + t(e) : e \in C\}$ be the induced set of elongated SNP's. Then, S' is a conflict free set.

Proof. Consider $e_1, e_2 \in C$. Since C is a matching, e_1 and e_2 are disjoint. Thus, in particular, the masses of the tuples $s_1 + t_1$ and $s_2 + t_2$ are disjoint. Hence, they are non-conflicting as SNPs.

The reverse also holds:

Claim. Let $S' \subset S$ be a set of SNP sites. For each $s \in S'$, let $t(s)$ be an elongation for s. Suppose that the set $\{s + t(s) : s \in S'\}$ is a conflict free set. Then $C = \{e_{s,t(s)} : s \in S'\}$ is a matching in $G_{S,T}$.

Proof. Similar.

Corollary 1. *Let S be a set of SNP sites, and let T be a set of possible elongations. Let $G_{S,T}$ be the hypergraph as defined above. Then, the number of non-conflicting sets in the optimal multiplexing strategy for (S,T) is equal to $\gamma(G_{S,T})$.*

We therefore translated the multiplexing optimization task to finding a minimal matching partition in $G_{S,T}$.

3.3 Hardness Result

The minimum matching partition problem is hard:

Theorem 3. *The problem of finding the minimal matching partition in an extension graph is NP-Hard. Furthermore, for any $\epsilon > 0$ there is no polynomial approximation algorithm for the problem that guarantees an* additive *approximation factor of $n^{1-\epsilon}$, unless P = NP.*

The proof is by reduction to the 3D-matching problem, which is known to be NP-Hard[4]. The 3D-matching problem is defined as follows. Let X, Y, Z be sets and let $M \subseteq X \times Y \times Z$. A *matching* for X, Y, Z, M is a subset $C \subseteq M$ such for each $c_1, c_2 \in C, c_1 \neq c_2$, c_1 and c_2 differ in all three coordinates. The 3D-matching problem is, given X, Y, Z, M and a natural number t, determine if there is a matching in X, Y, Z, M of cardinality t. The 3-matching problem is NP-Hard even in the special case that $|X| = |Y| = |Z| = t$ [4].

Let X, Y, Z, M be a 3D-matching problem such that $|X| = |Y| = |Z| = t$. Suppose that there is an algorithm A, which approximates the matching partition problem to within an additive factor of $n^{1-\epsilon}$ for some $\epsilon > 0$. We reduce the 3D-matching problem to a matching partition problem in extension graphs, as follows.

Let $\delta = \epsilon/4$ and $n = t^{1/\delta}$. Set $k = n/|M|$. W.l.o.g. X, Y and Z are disjoint. We define an extension graph $G = (A, B, E)$, as follows:

- $A = \{x^{(i)} : x \in X, 1 \le i \le k\}$, i.e. for each x in X, we create k copies of x in A,
- $B = Y \cup Z$,
- $E = \{(x^{(i)}, y, z) : (x, y, z) \in M, 1 \le i \le k\}$, i.e. for each triplet (x, y, z) in M, we add k hyperedges to G, corresponding to the k copies of x.

Note that $|G| = \Theta(|M|k) = \Theta(n)$.

Claim. If for X, Y, Z, M, there exists a matching $C \subseteq M$ such that $|C| = t$, then in G there exists a matching partition of size k.

Proof. Let C be a matching such that $|C| = t$. We define a set of k matchings in G. For $i = 1, \ldots, k$, set $C_i = \{(x^{(i)}, y, z) : (x, y, z) \in C\}$. The set $C_1, \ldots, C_k$ is a matching partition for G.

Conversely,

Claim. Suppose that there is no matching C for X, Y, Z, M such that $|C| = t$. Then, $\gamma(G) \ge k + n^{1-\epsilon}$.

Proof. Let $\mathcal{P} = C_1, \ldots, C_\ell$ be a minimal matching partition for G. First we show that $|C_i| < t$ for all i. In contradiction suppose that $|C_i| \ge t$. Define $C = \{(x, y, z) : \exists j, (x^{(j)}, y, z) \in C_i\}$. Then, C is a matching of cardinality t for the 3-matching problem, in contradiction.

Since $\mathcal{P}$ is a matching partition, it fully covers A. Thus,

$$|A| = tk \le \sum_{i=1}^{\ell} |C_i| \le \ell(t-1).$$

Note that since $M \subseteq X \times Y \times Z$, we have $|M| \le t^3$. Hence

$$\gamma(G) = \ell \ge \frac{tk}{t-1} = k + \frac{k}{t-1} = k + \frac{n}{|M|(t-1)} \ge k + \frac{n}{t^4} = k + n^{1-4\delta} = k + n^{1-\epsilon}.$$

Combining Claims 3.3 and 3.3 we obtain:

Proof. (of Theorem 3) Let A be the algorithm that approximates the matching partition problem to within an additive factor of $n^{1-\epsilon}$. Given a 3D-matching problem, use the above polynomial reduction to reduce the problem to a matching partition problem, as described above, and provide the resulting graph to A. By Claim 3.3 if there is a matching of cardinality t in X, Y, Z, M then $\gamma(G) = k$ and hence $A(G) < k + n^{1-\epsilon}$. On the other hand, if there is no matching of cardinality t in X, Y, Z, M then by Claim 3.3 $A(G) \ge \gamma(G) \ge k + n^{1-\epsilon}$. Hence, by the output of $A(G)$ we can decide on the 3-matching problem.

```
Greedy(G_{S,T})
Input: G_{S,T} = (S, M, E) - Extension graph
Output: P partition matching for G_{S,T}
1  P ← ∅
2  foreach s ∈ S one by one do
3      foreach C ∈ P do
4          e ← ChooseEdge(s, C)
5          if e ≠ "Nil" then
6              C ← C ∪ {e}
7              break (goto line 2 with next s)
8      end foreach
9      e ← ChooseEdge(s, ∅)
10     P ← P ∪ {{e}}
11 end foreach

ChooseEdge(s, C)
Input: s - vertex to cover, C - matching
Return Value: Edge e that covers s or "Nil"

1  if exits e ∈ E that covers s and does not conflict with C then
2      return e (any such e is ok)
3  else
4      return "Nil"
```

Fig. 4. Greedy Procedure

3.4 The Greedy Algorithm

Given the hardness result, we must seek heuristics to find efficient multiplexing strategies. In the next sections we describe three such heuristics. We start with a simple *greedy algorithm*. Pseudocode of the algorithm is provided in Figure 4. A high level description follows.

Let $G_{S,T}$ be an extension graph of S. We seek a matching partition $\mathcal{P} = \{C_1, C_2, \ldots\}$. The greedy algorithm starts with $\mathcal{P}$ being the empty set. As the algorithm progresses, additional matchings are added to $\mathcal{P}$. The algorithm iterates through the vertexes of S, one by one, according to some arbitrary order. For each vertex $s \in S$, the algorithm considers all currently available matchings $C \in \mathcal{P}$, one by one. For each matching the algorithm checks if it can be extended to also cover s. Extending a matching is performed by adding to it an edge that, on the one hand covers s, and on the other hand does not conflict with the other edges in the matching. If none of the matchings can be extended, a new matching is added to $\mathcal{P}$.

3.5 Refined Greedy Algorithm

The greedy heuristic is greedy in the sense that it considers the vertices one by one, chooses an edge to cover the vertex, and never revisits its choices. However, even in this greedy process the algorithm does not make any attempt to optimize its choices in the first place. The *Refined Greedy* algorithm, described next, attempts at making better choices within the greedy process.

The key idea of the Refined Greedy algorithm is to try and choose the edge to cover a vertex as the edge that will be least limiting for future choices. Specifically, suppose we wish to cover vertex s, and that e covers s. By choosing to cover s with e we rule out the possibility to use in the future any edge e' that conflicts with e. Accordingly, the Refined Greedy algorithm chooses the edge that eliminates the least number of other edges from future use. However, we only count edges which can be used to cover currently uncovered edges. This more refined choice of edges is reflected in a refined ChooseEdge procedure, as provided in Figure 5.

ChooseEdge(s, C)
Input: s - vertex to cover, C - matching
Return Value: Edge e that covers s or "Nil"

1 $A_s \leftarrow \{e \in E : e$ does not conflict with C and e covers $s\}$
2 **if** $A_s \neq \emptyset$ **then**
3 $\quad S' \leftarrow S - \text{Cover}(\mathcal{P})$
4 $\quad$ **foreach** $e \in A_s$ set $\text{Conflict}(e) = |\{e' : e$ and e' conflict and $\exists s' \in S', s' \in e\}|$
5 $\quad$ let e_0 be the $e \in A_s$ with smallest Conflict(e)
6 $\quad$ **return** e
7 **else** (A_s is empty)
8 $\quad$ **return** "Nil"

Fig. 5. Refined Greedy ChooseEdge Procedure

3.6 Rematch Algorithm

The Rematch Algorithm, which we present next, differs from the greedy algorithm in that, if necessary, it backtracks on previous choices. An overview of the algorithm follows. A detailed pseudo-code is provided in Figure 6.

The algorithm starts with the matchings partition $\mathcal{P}$ being the empty set, and adds more matching to $\mathcal{P}$ if and when necessary. In its outer most loop (lines 2-13), the algorithm loops through all uncovered vertices, in search for a vertex that can covered without the need to add another matching to the set $\mathcal{P}$. The main procedure in this loop is the recursive RecursiveVertexRematch$(s, \mathcal{P})$ procedure.

ReMatch($G_{S,T}$)
Input: $G_{S,T} = (S, M, E)$ - Extension graph
Output: $\mathcal{P}$ - a covering partition for $G_{S,T}$
Global Variables: $G_{S,T}$, *Visited*

1 $\mathcal{P} \leftarrow \emptyset$
2 **while** exists $s \in S$ not yet covered by $\mathcal{P}$ **do**
3 $\quad \mathcal{P} \leftarrow \mathcal{P} \cup \{\emptyset\}$
4 $\quad$ **repeat**
5 $\quad\quad$ *Visited* $\leftarrow \emptyset$
6 $\quad\quad$ **foreach** $s \in S$ not covered by $\mathcal{P}$ one by one **do**
7 $\quad\quad\quad$ *Visited* $\leftarrow$ *Visited* $\cup \{s\}$
8 $\quad\quad\quad$ $\mathcal{P}' \leftarrow$ RecursiveVertexRematch($s, \mathcal{P}$)
9 $\quad\quad\quad$ **if** $\mathcal{P}' \neq$ False **then**
10 $\quad\quad\quad\quad$ $\mathcal{P} \leftarrow \mathcal{P}'$
11 $\quad\quad\quad\quad$ **break** (goto line 4)
12 $\quad$ **until** $V \cup \text{Cover}(\mathcal{P}) \supseteq S$
13 **end while**

RecursiveVertexRematch($s, \mathcal{P}$)
Input: s - vertex to cover, $\mathcal{P}$ - current set of matchings
Return Value: $\mathcal{P}'$ - a valid set of matching covering s if found, False- if no such matchings found

1 **foreach** $e \in E$ such that $s \in e$ one by one **do**
2 $\quad$ **foreach** $C \in \mathcal{P}$ one by one **do**
3 $\quad\quad$ **if** e does not conflict with C **then**
4 $\quad\quad\quad$ $C' \leftarrow C \cup \{e\}$
5 $\quad\quad\quad$ $\mathcal{P}' \leftarrow \mathcal{P} - \{C\} \cup \{C'\}$
6 $\quad\quad\quad$ **return** $\mathcal{P}'$
7 **foreach** $e \in E$ such that $s \in e$ one by one **do**
8 $\quad$ **foreach** $C \in \mathcal{P}$ one by one **do**
9 $\quad\quad$ **if** exists only one $e' \in C$ such that e and e' conflict **then**
10 $\quad\quad\quad$ Let $s' \in S$ be the vertex such that e' covers s'
11 $\quad\quad\quad$ **if** $s' \notin$ *Visited* **then**
12 $\quad\quad\quad\quad$ $D \leftarrow C - \{e'\} \cup \{e\}$
13 $\quad\quad\quad\quad$ $\mathcal{D} \leftarrow \mathcal{P} - \{C\} \cup \{D\}$
14 $\quad\quad\quad\quad$ $\mathcal{P}' \leftarrow$ RecursiveVertexRematch($s', \mathcal{D}$)
15 $\quad\quad\quad\quad$ **if** $\mathcal{P}' \neq$ False **return** $\mathcal{P}'$
16 **return** False

Fig. 6. ReMatch Algorithm

RecursiveVertexRematch($s, \mathcal{P}$) accepts as input a vertex s and a set of matchings $\mathcal{P}$, such that s is not covered by $\mathcal{P}$. Its goal is to output a new set of covers $\mathcal{P}'$ which also covers s. In the course of doing so, it may omit edges from $\mathcal{P}$ and try to cover the corresponding vertices with other edges. However, in the end of the process, any vertex covered by $\mathcal{P}$ must also be covered by $\mathcal{P}'$, in addition to s.

RecursiveVertexRematch($s, \mathcal{P}$) operates as follows. First (lines 1-6) it tries to cover s without changing the covering of any other vertex. If this is not possible, the procedure tries to cover s by uncovering another node s', covering s and then recursively trying to cover s' (lines 7-15). The vertex s' which is uncovered must have the following properties:

1. uncovering it alone must allow to cover s (lines 9-10),
2. there was no attempt to be cover (or re-cover) it since the last time the main algorithm successfully managed to add a new vertex to set of matchings. The set *Visited* is the set of all such covered vertices.

Property (1) guarantees no branching in the recursive process, thus avoiding exponential blowup. Property (2) guarantees termination.

3.7 Simulation Results

We tested the heuristics using simulation on synthetically generated data. We conducted the tests on a wide range of parameters, both in the number of SNPs and in the length of the elongation. In order to measure the quality of the algorithms we would have liked to compare their output to the true optimum in each instance. However, computing the optimum is NP hard, and thus infeasible in the sizes we tested. Instead, for each instance we provide a *lower bound* on the optimum, and compare the output of the heuristics to this lower bound. As we shall see, the Rematch algorithm provides results very close to the lower bound, which shows that the lower bound is very close to the true optimum, on one hand, and that the Rematch algorithm is close to optimal, on the other.

A Lower Bound Given an extension graph $G = (S, M, E)$ we define a new extension graph $G' = (S', M', E')$ as follows.

- $S' = \{u^{(1)}, u^{(2)}, u^{(3)} : u \in S\}$, i.e. each vertex of S has three copies in S',
- $M' = M$,
- $E' = \{(u^{(1)}, x), (u^{(2)}, y), (u^{(3)}, z) : (u, x, y, z) \in E\}$, i.e. each hyperedge (u, x, y, z) in E is split into three separate simple edges, each connecting one member of M (x, y or z) to one of the copies of u.

Claim. Let $G = (S, M, E)$ be an extension graph and let $G' = (S', M', E')$ be the resulting graph, as described above. Then, $\gamma(G') \leq \gamma(G)$.

Proof. Let $C_1, \ldots, C_k$ be a minimal matching partition for G. We construct a matching partition $C'_1, \ldots, C'_k$ for G' as follows. For $j = 1, \ldots, k$,

$$C'_j = \left\{(u^{(1)}, x), (u^{(2)}, y), (u^{(3)}, z) : (u, x, y, z) \in C\right\}$$

That is, for each edge $e \in C_j$, the set C'_j contains the three edges of E' obtained by splitting e. It is easy to see that $C'_1, \ldots, C'_k$ is a matching partition for G'.

G' is a simple graph, not a hypergraph. Furthermore, it is bi-partite. Finding the minimal matching partition in a bi-partite graph is polynomial, as shown below.

Let $H = (A, B, E)$ be a bi-partite graph. Suppose we want to check if there is a minimal matching partition for H with at most k matchings. Construct a new bi-partite graph $H_{(k)} = (A', B', E')$, as follows:

- $A' = A$,
- $B' = \{v^{(1)}, \ldots, v^{(k)} : v \in B\}$, i.e. each vertex of B has k copies in B',
- $E' = \{(u, v^{(i)}) : (u, v) \in E, i = 1, \ldots, k\}$, i.e. each edge is duplicated for each of the copies of the vertexes of B.

Claim. There exists a matching partition of size $\leq k$ in H iff there exists a (regular) matching in $H_{(k)}$ which covers all vertexes of A'.

Proof. Let $C_1, \ldots, C_k$ be a matching partition in H. Then

$$C = \bigcup_{i=1}^{k} \left\{(u, v^{(i)}) : (u, v) \in C_i\right\}$$

is a matching in $H_{(k)}$ covering all of A'. Conversely, let C be a matching in $H_{(k)}$ covering all of A'. For $i = 1, \ldots, k$, let $C_i = \{(u, v) : (u, v^{(i)}) \in C\}$. Then $C_1, \ldots, C_k$ is a matching partition in H.

Finding a maximal matching in a bi-partite graph is polynomial [5], and we can therefor calculate the number of matching partitions in G' by applying Claim 3.7 for increasing values of k.

The Test Data We tested our algorithms for varying values of n - the number of SNPs, and k - the maximum length of elongation. For a given k, we allowed all elongations up to length k (thus, T is the set of all elongations up to length k). For each pair of n and k, 100 independent experiments were conducted. For each experiment, we randomly generated n different SNP tuples, as follows. First we generated the primer by choosing uniformly at random a sequence of 20 bases. Then, we chose two distinct bases, uniformly at random, for the two possible extensions.

Results Tests were conducted for $n = 500, 1000, 2000, 5000$, and 10000, and k ranging from 1 to 5. For each instance we recorded: (i) the size of the matching partition produced by the Refined Greedy algorithm, (ii) the size of the matching partition produced by the Rematch algorithm, and (iii) the lower bound. The results (averages) are provided in Table 1. Figure 7 depicts a summary. The average number of matchings is plotted as a function of the length of the elongation, for the $n = 500$ and $n = 10000$. The figure clearly demonstrates the power of the mass modification technique. With an elongation of length 5, the number of matchings is reduced by a factor of more than 10. Both heuristics provide close approximation of the lower bound. The Rematch algorithm, in particular, exhibits performance close to optimal throughout the entire range, as seen in Figure 7.

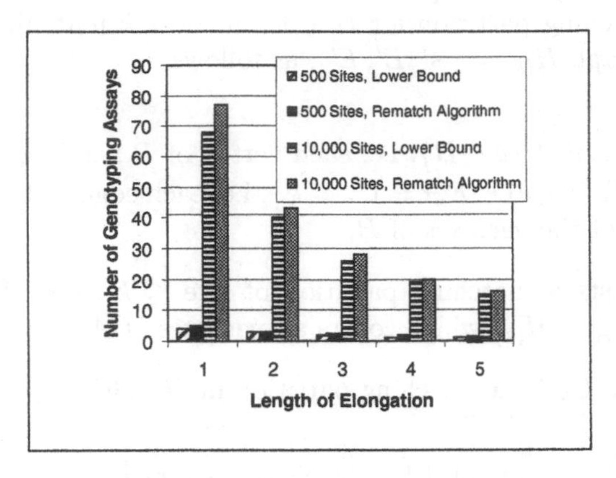

Fig. 7. Simulation Results.

Num of SNPs	100					500					1,000				
Elongation	1	2	3	4	5	1	2	3	4	5	1	2	3	4	5
Lower Bound	2	1	1	1	1	4	3	2	1	1	8	5	3	2	2
Refined Greedy	2	1	1	1	1	6	3	2	2	2	11	6	4	3	2
Rematch	2	1	1	1	1	5	3	2	2	1	9	5	3	3	2

Num of SNPs	2,000					5,000					10,000				
Elongation	1	2	3	4	5	1	2	3	4	5	1	2	3	4	5
Lower Bound	14	8	6	4	3	35	20	13	10	8	68	40	26	19	15
Refined Greedy	20	11	7	5	4	56	25	16	12	8	85	46	31	23	18
Rematch	17	9	6	5	4	39	22	14	10	8	77	43	28	20	16

Table 1. Simulation results. Averages of 100 simulation for each entry.

4 Discussion

We presented an analysis of optimizing the multiplexing rate when using the primer-extension mass-spectrometry genotyping method. There are several variants of the problems that arise in this context, some of mathematical interest and some of more practical interest. For example, suppose that the primers corresponding to S all have distinct masses (i.e. $s_i \neq s_j \Longrightarrow m(\mathrm{P}_{s_i}) \neq m(\mathrm{P}_{s_j})$). In this case, the corresponding conflict graph can be colored in at most 8 colors.

This is achieved by an SLO-like process working with the $m(P_s)$ ordered left to right. See [9] for more details and further variants.

Resolution. The results we presented assume 1-amu resolution of the mass-spectrometry process. In some cases mass-spectrometry may only provide lesser resolution. If this is the case then additional restrictions on the multiplexing may apply. Specifically, suppose that the mass-spectrometer can only distinguish between masses that are d apart. If this is the case, SNPs *conflict* with each other if in their combined mass-spectrums there are two masses that are less than d apart. We seek to measure the entire set of SNPs in a minimal number of non-conflicting sets. The algorithms presented in this paper can be readily extended to handle this case as well. The only difference is in the definition of *conflict* (Definition 4). Given the altered definition, the algorithms need no modification. Simulation results for the reduced resolution case will be provided in the full version of the paper.

PCR Multiplexing. An issue of great practical importance is the compatibility of the multiplexing schemes provided here with PCR multiplexing schemes, which precede the extension and mass-spectrometry stages. We note that our results can be used to allow greater flexibility in the PCR multiplexing stage. Specifically, suppose that there is a set S of SNPs to be genotyped, first by undergoing PCR amplification and then by primer extension and mass spectrometry. Then, any set of SNP loci that undergo PCR amplification together must also undergo mass spectrometry together. Thus, only non-conflicting SNPs can be considered for joint PCR multiplexing. Accordingly, larger non-conflicting sets allow greater flexibility in multiplexing the PCR stage. Our results provide a method to produce large non conflicting sets.

References

1. B. Alberts, D. Bray, J. Lewis, M. Raff, K. Roberts, and J. D. Watson. *Molecular Biology of the Cell.* Garland Publishing, 3rd edition, 1994.
2. Claude Berge. *Hypergraphes.* Gauthier-Villars, 1987.
3. L. Cardon and J. Bell. Association study design for complex diseases. *Nat. Rev. Genet.*, 2:91–99, 2001.
4. M. R. Garey and D. S. Johnson. *Computers and Intractability: A Guide to the Theory of NP-Completeness.* W. H. Freeman and Co., San Francisco, 1979.
5. Allan Gibbons. *Algorithmic Graph Theory*, chapter 5. Cambridge University Press, 1985.
6. I. G. Gut. Automation of genotyping of single nucleotide polymorphisms. *Human Mutation*, 17:475–492, 2001.
7. I. Holyer. The np-completeness of edge colorings. *SIAM J. Comput.*, 10:718–720, 1981.
8. Teemu Kivioja, Mikko Arvas, Kari Kataja, Merja Penttila, Hans Soberlund, and Esko Ukkonen. Assigning probes into a small number of pools separable by electrophoresis. *Bioinformatics*, 1(1):1–8, 2002.

9. Efrat Manisterski. Optimal multiplexing schemes for snp genotyping mass-spectrometry. Master's thesis, Bar Ilan University, Department of Computer Science, 2001.
10. D.W. Matula and L.L. Beck. Smallest-last ordering and clustering and graph coloring algorithms. *J. ACM*, 30:417–427, 1983.
11. J. Ott. *Analysis of Human Genetic Linkage*. Johns Hopkins University Press, 1991.
12. J. Pritchard and M. Przeworski. Linkage disequilibrium in humans: Models and data. *Am. J. Hum. Genet.*, (69):1–14, 2001.
13. P. Ross, L. Hall, I. Smirnov, and L. Haff. High level multiplex genotyping by maldi-tof mass spectrometry. *Nature Biotechnolgy*, 16:1347–1351, 1998.
14. D. G. Wang, J. B. Fan, C. J. Siao, A. Berno, P. P. Young, et al. Large-scale identification, mapping, and genotyping of single- nucleotide polymorphisms in the human genome. *Science*, 280(5366):1077–82., 1998.
15. Z. Yakhini, P. Webb, and R. Roth. *Partitioning of Polymorphic DNAs*. US Patent 6,074,831, 2000.

Minimum Recombinant Haplotype Configuration on Tree Pedigrees

Koichiro Doi[1], Jing Li[2], and Tao Jiang[2]

[1] Department of Computer Science
Graduate School of Information Science and Technology, University of Tokyo
7–3–1 Hongo, Bunkyo-ku, Tokyo 113-0033, Japan
doi@is.s.u-tokyo.ac.jp
[2] Department of Computer Science and Engineering, University of California
Riverside, CA 92521, USA
{jili,jiang}@cs.ucr.edu

Abstract. We study the problem of reconstructing haplotype configurations from genotypes on pedigree data under the Mendelian law of inheritance and the minimum recombination principle, which is very important for the construction of haplotype maps and genetic linkage/association analysis. Li and Jiang [9,10] recently proved that the *Minimum Recombinant Haplotype Configuration* (MRHC) problem is NP-hard, even if the number of marker loci is 2. However, the proof uses pedigrees that contain complex mating loop structures that are not common in practice. The complexity of MRHC in the loopless case was left as an open problem. In this paper, we show that loopless MRHC is NP-hard. We also present two dynamic programming algorithms that can be useful for solving loopless MRHC (and general MRHC) in practice. The first algorithm performs dynamic programming on the members of the input pedigree and is efficient when the number of marker loci is bounded by a small constant. It takes advantage of the tree structure in a loopless pedigree. The second algorithm performs dynamic programming on the marker loci and is efficient when the number of the members of the input pedigree is small. This algorithm also works for the general MRHC problem. We have implemented both algorithms and applied the first one to both simulated and real data. Our preliminary experiments demonstrate that the algorithm is often able to solve MRHC efficiently in practice.

1 Introduction

In order to understand and study the genetic basis of complex diseases, the modeling of human variation is very important. *Single nucleotide polymorphisms* (SNPs), which are mutations at single nucleotide positions, are typical variations central to ongoing development of high-resolution maps of genetic variation and haplotypes for human [8]. While a dense SNP haplotype map is being built, various new methods [11,14,19,23] have been developed to use haplotype information in linkage disequilibrium mapping. Some existing statistical methods for gene mapping have also shown increased power by incorporating SNP haplotype information [18,25]. But, the use of haplotype maps has been limited due to the fact that the human genome is diploid, and in practice, genotype data instead of haplotype data are collected directly, especially in large-scale

G. Benson and R. Page (Eds.): WABI 2003, LNBI 2812, pp. 339–353, 2003.

sequencing projects, because of cost considerations. Although recently developed experimental techniques [3] give the hope of deriving haplotype information directly with affordable costs, efficient and accurate computational methods for haplotype reconstruction from genotype data are still highly demanded.

The existing computational methods for haplotyping can be divided into two categories: statistical methods and rule-based (*i.e. combinatorial*) methods. Both methods can be applied to population data and pedigree data. Statistical methods [4,12,15,20,22] often estimate the haplotype frequencies in addition to the haplotype configuration for each individual, but their algorithms are usually very time consuming and not suitable for large data sets. On the other hand, rule-based methods are usually very fast although they normally do not provide any numerical assessment of the reliability of their results. Nevertheless, by utilizing some reasonable biological assumptions, such as the minimum recombination principle, rule-based methods have proven to be powerful and practical[7,13,16,17,21,24].

The minimum recombination principle states that the genetic recombinants are rare and thus haplotypes with fewer recombinants should be preferred in a haplotype reconstruction [7,16,17]. The principle is well supported by real data from the practice. For example, recently published experimental results [2,5,8] showed that the human genomic DNA can be partitioned into long *blocks* such that recombinants within each block are rare or even non-existent. Thus, within a single block, the true haplotype configuration is most likely one of the configurations with the minimum number of recombinants.

1.1 The Minimum Recombinant Haplotype Configuration Problem and Previous Work

Qian and Beckman [17] formulated the problem of how to reconstruct haplotype configurations from genotype data on a pedigree under the Mendelian law of inheritance such that the resulting haplotype configurations require the minimum number of recombinants (*i.e.* recombination events). The problem is called *Minimum Recombinant Haplotype Configuration* (MRHC). They proposed a rule-based heuristic algorithm for MRHC that seems to work well on small pedigrees but is very slow for medium-sized pedigrees, especially in the case of biallelic genotype data.

Recently, Li and Jiang [9,10] proved that MRHC is NP-hard, even if the number of marker loci is 2. They also devised an efficient (iterative) heuristic algorithm for MRHC, which is more efficient than the algorithm in [17] and can handle pedigrees of any practical sizes. However, their NP-hardness proof uses pedigrees with complex mating loop structures that may not ever rise in practice. The complexity of MRHC on tree pedigrees (*i.e.* loopless pedigrees, or pedigrees without mating loops) was left as an open question in [9,10].

1.2 Our Results

This paper is concerned with the MRHC problem on tree pedigrees (called *loopless MRHC*). First, we show that loopless MRHC is NP-hard, answering an open question in [9,10]. Therefore, loopless MRHC does not have a polynomial time (exact) algorithm

Table 1. Computational complexities of MRHC and loopless MRHC on pedigrees with n members, m marker loci, and at most m_0 heterozygous loci in each member.

	$m = 2$	m =constant	n =constant	no restriction
MRHC	NP-hard [9,10]	NP-hard [9,10]	$O(nm2^{4n})$	NP-hard [9,10]
loopless MRHC	$O(n)$	$O(nm_0 2^{3m_0})$	$O(nm2^{4n})$	NP-hard

unless P = NP. The proof is based on a simple reduction from the well-known MAX CUT problem. Then we observe that the complexity of an instance of MRHC is defined by two independent parameters, namely, the number of members in the pedigree (*i.e.* the size of the pedigree) and the number of marker loci in each member, and we can construct a polynomial time algorithm for MRHC if one of these parameter is bounded by a constant (which is often the case in practice). This gives rises to two dynamic programming algorithms for loopless MRHC. The first algorithm assumes that the number of marker loci is bounded by a small constant and performs dynamic programming on the members of the input pedigree. The algorithm, called the *locus-based* algorithm, takes advantage of the tree structure of the input pedigree and has a running time linear in the size of the pedigree. Recall that the general MRHC problem is NP-hard even if the number of marker loci is 2 [9,10]. This algorithm will be very useful for solving MRHC in practice because most real pedigrees are loopless and involve a small number of marker loci in each block. For example, the pedigrees studied in [5] are all loopless and usually contain four to six marker loci. The second algorithm assumes that the input pedigree is small and performs dynamic programming on the marker loci in each member of the pedigree simultaneously. The algorithm, called the *member-based* algorithm, works in fact for any input pedigree and has a running time linear in the number of marker loci. This algorithm will be useful as a subroutine for solving MRHC on small pedigrees, which could be nuclear families from a large input pedigree (*i.e.* components of the pedigree consisting of parents and children) or independent nuclear families from a (semi-)population data.

Table 1 exhibits the difference between the computational complexities of (general) MRHC and loopless MRHC. We have implemented both algorithms and applied the locus-based algorithm to both simulated and real datasets. Our preliminary experiments demonstrate that the algorithm is often able to solve MRHC efficiently in practice.

The rest of the paper is organized as follows. Some necessary definitions and notations used in MRHC are reviewed in Section 2. Section 3 presents the NP-hardness proof for loopless MRHC and Section 4 describes the two dynamic programming algorithms. The experimental results are summarized in Section 5.

2 Preliminaries

This section presents some concepts, definitions, and notations required for the MRHC problem. First, we define pedigrees.

Definition 1. *A pedigree is a connected directed acyclic graph* $G = \{V, E\}$, *where* $V = M \cup F \cup N$, *M stands for the male nodes, F stands for the female nodes, N*

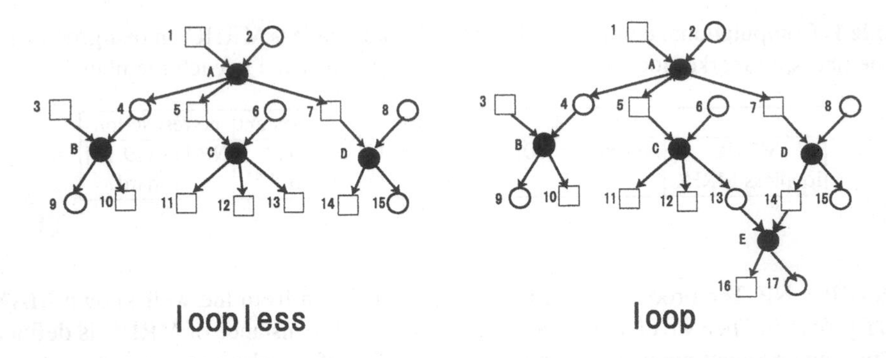

Fig. 1. Example pedigrees without mating loops and with mating loops. Here, a square box represents a male node, a circle represents female node, and a solid circle represents a mating node. The pedigree on the left has 15 members and 4 nuclear families.

stands for the mating nodes, and $E = \{e = (u, v) : (u \in M \cup F$ *and* $v \in N)$ *or* $(u \in N$ *and* $v \in M \cup F)\}$. $M \cup F$ *are called the individual nodes or members of the pedigree. The in-degree of each individual node is at most 1. The in-degree of a mating node must be 2, with one edge from a female node (called mother), and the out-degree of a mating node must be larger than zero.*

In a pedigree, the individual nodes adjacent *to* a mating node are called the children of the two individual nodes adjacent *from* the mating node (*i.e.* the father and mother nodes, which have edges to the mating node). For each mating node, the induced subgraph containing the father, mother, mating, and child nodes is called a *nuclear family*. A parents-offspring *trio* consists of two parents and one of their children. The individual nodes that have no parents are called the *founders*. A *mating loop* is a cycle in the pedigree graph when the directions of edges are ignored. A loopless pedigree is also called a *tree pedigree*. Fig. 1 shows two example pedigrees, one without mating loops and one with mating loops.

The genome of an organism consists of chromosomes that are double strand DNA. Locations on a chromosome can be identified using markers, which are small segments of DNA with some specific features. The position of a marker on a chromosome is called a *marker locus* and the state of a marker locus is an *allele*. A set of markers and their positions define a genetic map of chromosomes. There are many types of markers. The two most commonly used markers are microsatellite markers and SNP markers.

In a *diploid* organism such as human, the status of two alleles at a particular marker locus of a pair of *homologous* chromosomes is called a marker *genotype*. The genotype information of a locus will be denoted using a set, *e.g.* $\{a, b\}$. If the two alleles are the same, the genotype is *homozygous*. Otherwise it is *heterozygous*. A *haplotype* consists of all alleles, one in each locus, that are on the same chromosome.

The Mendelian law of inheritance states that the genotype of a child must come from the genotypes of its parents at each maker locus. In other words, the two alleles at each locus of the child have different origins: one is from its father (which is called the *paternal* allele) and the other from its mother (which is called the *maternal* allele).

Usually, a child inherits a complete haplotype from each parent. However, recombination may occur, where the two haplotypes of a parent get shuffled due to crossover of chromosomes and one of the shuffled copies is passed on to the child. Such an event is called a *recombination event* and its result is called a *recombinant*. Since markers are usually very short DNA sequences, we assume that recombination only occurs between markers.

Following [9,10], we use PS (*parental source*) to indicate which allele comes from which parent at each locus. The PS value at a heterozygous locus can be $-1, 0, 1$, where -1 means that the parental source is unknown, 0 means that the allele with the smaller identification number is from the father and the allele with the larger identification number is from the mother, and 1 means the opposite. The PS value will always be set as 0 for homozygous locus. A locus is *PS-resolved* if its PS value is 0 or 1. For convenience, we will use GS (*grand-parental source*) to indicate if an allele at a PS-resolved locus comes from a grand parental allele or a grand-material allele. Similar to a PS value, a GS value can also be -1, 0, or 1. An allele GS-resolved if its value is 0 or 1. A locus is GS-resolved if both of its alleles are GS-resolved. The PS and GS information can be used to count the number of recombinants as follows. For any two alleles that are at adjacent loci and from the same haplotype, they induce a recombinant if their GS values are 0 and 1.

Definition 2. *A haplotype configuration of a pedigree is an assignment of nonnegative values to the PS of each locus and the GS of each allele for each member of the pedigree that is consistent with the Mendelian law of inheritance.*

Now, the MRHC problem [9,10,17,21] can be defined precisely as follows:

Definition 3 (Minimum Recombinant Haplotype Configuration (MRHC)). *Given a pedigree and genotype information for each member of the pedigree, find a haplotype configuration for the pedigree that requires the minimum number of the recombinants.*

It is known that MRHC is NP-hard even when the number of loci is 2 [9,10]. In this paper, we are interested in the restricted case of MRHC on tree pedigrees like the left one in Fig. 1.

Definition 4. *The loopless MRHC problem is MRHC on pedigrees without mating loops.*

3 Loopless MRHC Is NP-hard

In this section, we show the NP-hardness of the loopless MRHC problem, thus answering an open question in [9,10]. The proof is via a reduction from the well-known MAX CUT problem.

Theorem 1. *Loopless MRHC problem is NP-hard.*

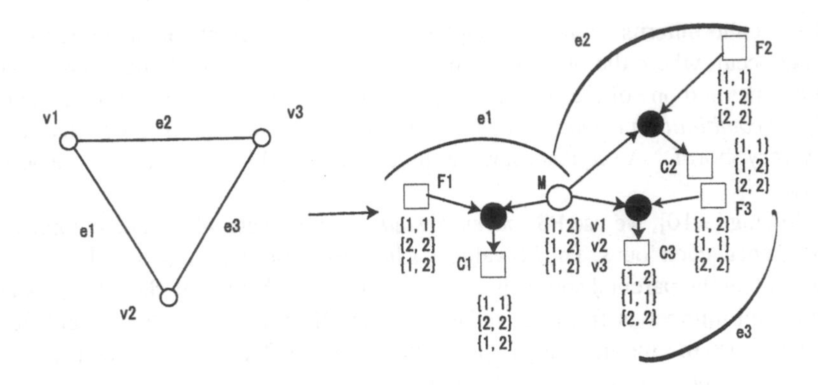

Fig. 2. An example reduction from MAX CUT to loopless MRHC.

Proof. We give a polynomial time reduction from MAX CUT, which is known NP-hard [6]. Recall that an instance of MAX CUT is a graph $G = (V, E)$ and the objective is to partition V into two disjoint subsets V_1 and V_2 such that the number of edges of E that have one endpoint in V_1 and other endpoint in V_2 is maximized.

Assume that $V = \{v_1, v_2, \ldots, v_n\}$, $E = \{e_1, e_2, \ldots, e_m\}$, and the value of an optimal solution of MAX CUT on G is OPT. We construct a pedigree with n loci. First, we introduce an individual node M_0. The memberM_0 has n loci. Each locus corresponds to a vertex v_i with a heterozygous genotype $\{1, 2\}$. Observe that an assignment of PS values to the loci (in M_0) naturally divide the loci into two groups, which also bipartitions the corresponding vertices of G. Therefore, the PS values of the loci in M_0 encode a solution of the MAX CUT problem. We construct a nuclear family with mother M_0, father F_j, and child C_j for each edge $e_j = (v_{j1}, v_{j2})$, as shown in Fig. 2. The loci in F_j and C_j corresponding to v_{j1} are set to have a homozygous genotype $\{1, 1\}$, and the loci in F_j and C_j corresponding to v_{j2} are set to have a homozygous genotype $\{2, 2\}$. Other loci in F_j and C_j are all of heterozygous genotype $\{1, 2\}$.

Consider an assignment of PS values to the loci in M_0, which specifies a solution to the MAX CUT problem on G. Recall that the assignment partitions V into two groups. If v_{j1} and v_{j2} are in the same group (*i.e.* the loci in M_0 corresponding to v_{j1} and v_{j2} have the same PS value), the parents-child trio consisting of M_0, F_j, and C_j requires at least 1 recombinant, no matter how the PS values at other loci in M_0 and the PS values at the loci in F_j and C_j are assigned. Moreover, the parents-child trio can be made to require exactly 1 recombinant if the PS values at the loci in F_j and C_j are assigned appropriately On the other hand, if v_{j1} and v_{j2} are in different groups, the parents-child trio of M_0, F_j, and C_j requires no recombinants, no matter how the PS values at other loci in M_0 are assigned and if the PS values at the loci in F_j and C_j are assigned appropriately. Therefore, the number of the minimum recombinants required for the above instance of loopless MRHC is equal to $m - OPT$. Hence, MAX CUT reduces to loopless MRHC and loopless MRHC is NP-hard. □

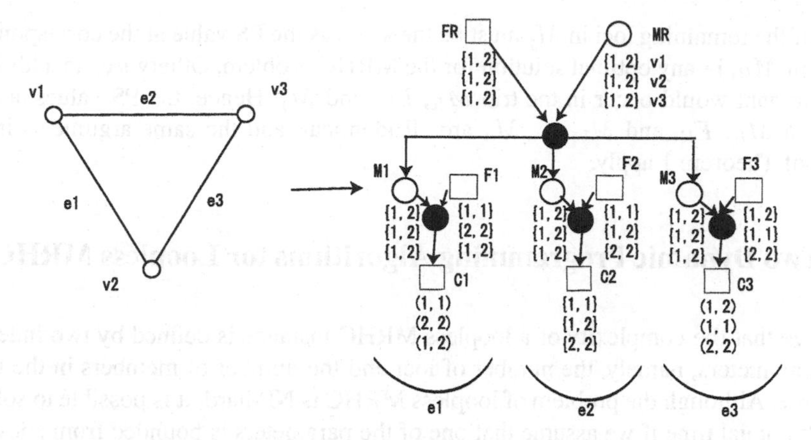

Fig. 3. An example reduction from MAX CUT to restricted loopless MRHC where each member of the pedigree has at most one mating partner.

In the above construction, the member M_0 has many mating partners. This does not happen very often in real datasets. We can convert the above construction to one where each member of the input pedigree has exactly one mating partner. This shows that loopless MRHC is still NP-hard if we further require that each member of the pedigree to have at most one mating partner.

Corollary 1. *Loopless MRHC is still NP-hard even if each member of the pedigree has at most one mating partner.*

Proof. This corollary can be proved by a simple modification of the proof of Theorem 1 as follows. We replace member M_0 by a nuclear family consisting father F_R, mother M_R, and female children $M_1, \ldots, M_m$, whose loci are of heterozygous genotype $\{1, 2\}$. We construct a nuclear family with mother M_j, father F_j, and child C_j for each edge $e_j = (v_{j1}, v_{j2})$. The loci in F_j and C_j are defined in the same way as in the proof of Theorem 1. Fig. 3 illustrates the new complete construction.

We would like to show that the minimum number of recombinants required for the above constructed instance of MRHC is also equal to $m - OPT$, and the PS values of the loci in M_R define an optimal solution of the MAX CUT problem. If the PS values at each locus in M_R, F_R, and $M_1, \ldots, M_m$ are the same, this would follow from the above proof. To show that the PS values at each locus in M_R, F_R, and $M_1, \ldots, M_m$ are the same, we first observe that we can always fix the PS value at the one of the loci (*e.g.* the first locus) in F_R to be the same as the PS value at the corresponding locus in M_R, because F_R is a founder. Moreover, we claim that the PS values at each of the remaining loci in M_R and F_R must be the same. Otherwise, there would be recombinants in each parents-offspring trio consisting of M_R, F_R, and M_j no matter how the PS values in M_j are assigned, for all j, and the total number of recombinants would be at least m. Similarly, we can assume, without loss of generality, that the PS value at the first locus in M_j is the same as the PS value at the first locus in M_R (because both parents M_R and F_R of M_j have the identical haplotype configurations), and claim that the PS value at each of the remaining loci in M_j must be the same as the PS value at the corresponding locus in M_R, in any optimal solution for the MRHC problem. Otherwise, an additional recombinant would occur in the trio M_R, F_R, and M_j. Hence, the PS values at each locus in M_R, F_R, and $M_1, \ldots, M_n$ are all identical, and the same arguments in the proof of Theorem 1 apply. □

4 Two Dynamic Programming Algorithms for Loopless MRHC

Observe that the complexity of a loopless MRHC instance is defined by two independent parameters, namely, the number of loci and the number of members in the input pedigree. Although the problem of loopless MRHC is NP-hard, it is possible to solve it in polynomial time if we assume that one of the parameters is bounded from above by a constant. Here, we present two such dynamic programming algorithms. One assumes that the number of loci is bounded and is called the *locus-based* algorithm and the other assumes that the number of members is bounded and is called the *member-based* algorithm. As mentioned in Section 1, both algorithms are useful in practice.

4.1 The Locus-Based Dynamic Programming Algorithm

The algorithm starts by converting the input tree pedigree into a rooted tree (at an arbitrary member). An example conversion is shown in Fig. 4. Then we traverse the tree in postorder and solve MRHC for the subtree rooted at each member when the member is required to have a certain haplotype configuration (*i.e.* PS assignment). A key step in the computation is the processing of a nuclear family. First, observe that if the PS values of all members in a nuclear family are known, we can easily compute the GS values of the children in the family to minimize the number of recombinants required in the nuclear family. Our second observation is that, once the PS values of the parents are fixed, the PS values for each child can be assigned independently. In other words, we can think of a nuclear family as a collection of (independent) parents-offspring trios and process the trios accordingly.

A more detailed description of the algorithm is given below. Let $numtrio(p, q, c)$ denote the minimum number of recombinants required for a parents-offspring trio consisting of a father, a mother, and a child with PS assignments p, q, and c, respectively.

Step 1: Root the pedigree at an arbitrary member R.

Step 2: Traverse the tree in postorder. For each individual node (*i.e.* a member) r and each PS assignment s at r, compute the minimum number of recombinants required in the subtree rooted at r. If r has more than 2 children (*i.e.* mating nodes), we do the above computation for each child mating node separately. Each child mating node of r defines a unique nuclear family, which may contain r as a parent or a child. Suppose that the nuclear family consists of father F, mother M, and children $C_1, \ldots, C_l$. We construct an array $num[node][ps]$, where $node$ denotes an individual node and ps denotes a PS assignment at the node.

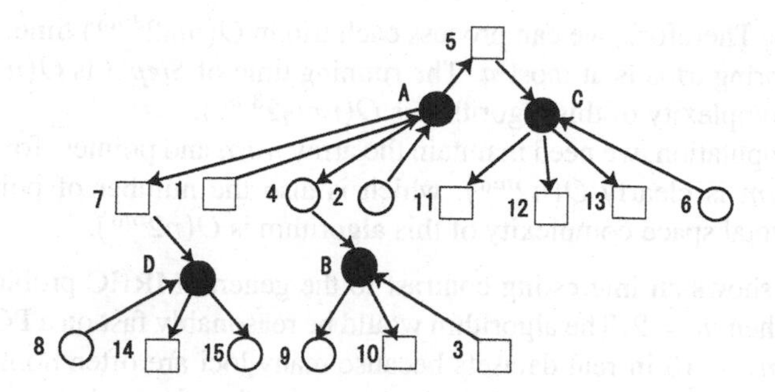

Fig. 4. A rooted tree for the tree pedigree in Fig. 1, where node 5 is selected as the root.

If r is M (or F), then for each PS assignment s at r, we do the following.

1. For each PS assignment p (or q) at F (or M, respectively), compute

$$num[r][s] = \min_{p} \sum_{1 \le i \le l} \min_{c}(num[F][p] + num[C_i][c] + numtrio(p, s, c))$$

$$(\text{or} = \min_{q} \sum_{1 \le i \le l} \min_{c}(num[M][m] + num[C_i][c] + numtrio(s, q, c)))$$

2. Keep pointers to the corresponding PS assignments at F (or M) and C_i ($1 \le i \le l$) in an optimal solution.

On the other hand, if r is C_j for some j, then for each PS assignment s at r, we do the following.

1. For each PS assignments p and q at F and M, respectively, compute

$$num[r][s] = \min_{p,q}(numtrio(p, q, s) + \sum_{1 \le i \le l, i \ne j} \min_{c}(num[F][p] + num[M][q] + num[C_i][c] + numtrio(p, q, c)))$$

2. Keep pointers to the corresponding PS assignments at F, M, and C_i ($1 \le i \le l$) in an optimal solution.

Step 3: At root R, we compute the minimum number of recombinants required for the whole tree for any PS assignment at R, and the corresponding PS assignments at all members by backtracing the pointers.

The time and space complexities of this algorithm are given below.

Theorem 2. *Let n denote the number of members and m_0 the maximum number of heterozygous loci in any member. The above locus-based dynamic programming algorithm runs in $O(nm_0 2^{3m_0})$ time and $O(n2^{m_0})$ space.*

Proof. The rooted tree can be constructed in $O(n)$ time. In *Step 2*, we have to consider all combinations of PS assignments at the members of a parents-offspring trio, which is at most 2^{3m_0}. The computation of *numtrio* is $O(m_0)$ for each trio and combination of

PS assignments. Therefore, we can process each trio in $O(m_0 2^{3m_0})$ time. The number of parents-offspring trios is at most n. The running time of *Step 3* is $O(n)$. Therefore, the total time complexity of this algorithm is $O(nm_0 2^{3m_0})$.

For this computation, we need maintain the array num and pointers for backtracing. The size of num is clearly $O(n2^{m_0})$, which is also the number of pointer needed. Therefore, the total space complexity of this algorithm is $O(n2^{m_0})$. □

This result shows an interesting contrast to the general MRHC problem, which is still NP-hard when $m = 2$. The algorithm would be reasonably fast on a PC for $m_0 \leq 8$ (which means $m \leq 15$ in real datasets because many loci are often homozygous). In practice, the algorithm can be sped up by preprocessing the pedigree and fixing as many PS values as possible using the Mendelian law of inheritance. This could reduce the number of "free" heterozygous loci significantly. Moreover, if each nuclear family in the input pedigree has only one child, the algorithm can be made much faster.

Theorem 3. *If each nuclear family in the input pedigree has only one child, we can solve loopless MRHC in* $O(nm_0 2^{2m_0})$ *time and* $O(n2^{m_0})$ *space.*

Proof. We need only modify *Step 2* in the above algorithm. For parents-offspring trio, instead of considering all combinations of PS assignments at both the father and mother for each PS assignment at the child, we can consider each parent independently and optimize the number of recombinants required from the parent to the child. More precisely, if r is M (or F), we compute the minimum number of recombinants from F (or M) to C_1 for all PS assignments at F (or M) and C_1 and memorize the minimum number of recombinants for each PS assignment at C_1. Next, we compute the minimum number of recombinants in the parents-offspring trio for all PS assignments at M (or F) and C_1. Similarly, if r is C_1, we compute the minimum number of recombinants from M to C_1 for all PS assignments at M and C_1 and the minimum number of recombinants from F and C_1 for all PS assignments at F and C_1, separately. Then, we add up the corresponding numbers for each PS assignment at C_1. The running time of the modified algorithm is $O(m_0 2^{2m_0})$ for each parents-offspring trio and $O(nm_0 2^{2m_0})$ for the whole pedigree. □

4.2 The Member-Based Dynamic Programming Algorithm

Many real pedigrees in practice are often of small or moderate sizes. For example, the dataset studied in [2] consists of a collection of parents-offspring trios and the dataset in [5] consists of pedigrees of sizes between 7 and 8. Here, we present an algorithm that is efficient when the size of the input pedigree is bounded from above by a small constant. The algorithm considers the PS and GS values at each locus across all members of the pedigree, and performs dynamic programming along the loci. A detailed description of the algorithm is given below. The algorithm in fact works for MRHC on general pedigrees.

We construct an array $num[i][t]$ that denotes the minimum number of recombinants required in the pedigree from locus 1 to locus i, if the PS/GS assignment at locus i in all members is t. Let $num1(s,t)$ denote the number of recombinants between any two adjacent loci with PS/GS assignment s and t in the whole pedigree. The algorithm works as follows. For $i = 1$ to $m - 1$,

1. Compute the minimum number of recombinants required in the pedigree from locus 1 to $i+1$ for every possible PS/GS assignment t at locus $i+1$, by considering all possible PS/GS assignment s at locus i, using the recurrence

$$num[i+1][t] = \min_{s}(num[i][s] + num1(s,t))$$

2. Keep track of the PS/GS assignment s achieving the minimum in the above.

Finally, we find the minimum number of the recombinants from locus 1 to m and the corresponding PS/GS assignments at all loci by a standard backtracing procedure.

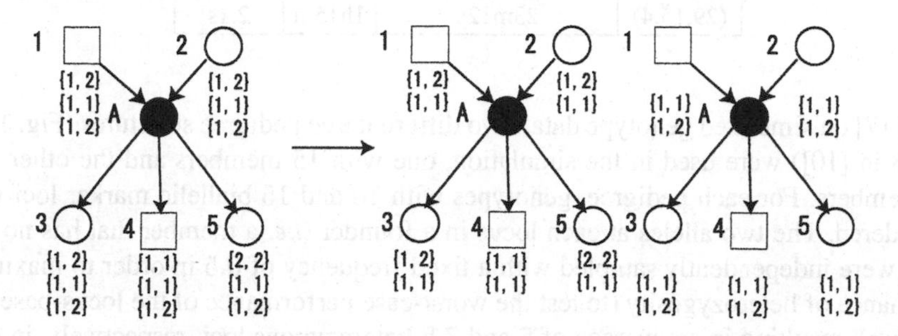

Fig. 5. An example pedigree with genotype information. The figure shows that the PS/GS assignments at locus 3 can be computed by considering all possible PS/GS assignments at locus 2 (independently from the PS/GS assignments at locus 1).

Theorem 4. *Let n denote the number of members in the pedigree and m the number of loci. The time complexity of the above member-based dynamic programming algorithm is $O(nm2^{4n})$. Its space complexity is $O(m2^{2n})$.*

Proof. The number of all possible PS/GS assignments at a single locus is trivially bounded 2^{3n}. However, this number includes many impossible assignments, *i.e.* assignments inconsistent with the Mendelian law of inheritance. It is easy to prove that the number of all possible PS/GS assignments at a locus is maximized if all genotypes at this locus are homozygous, such as locus 2 in Fig. 5. Therefore, the number of all possible (feasible) PS/GS assignments at a locus is $O(2^{2n})$, and the running time of the above algorithm is $O(mn2^{4n})$. The space for keeping the pointers used in backtracing is clearly $O(m2^{2n})$. □

In practice, the member-based algorithm runs reasonably fast when $n \leq 6$.

5 Experimental Results

We have implemented the above two algorithms in C++. To evaluate the performance of our program, we compare the locus-based algorithm with PedPhase [10] and MRH

Table 2. Running times of the locus-based algorithm, MRH and PedPhase on biallelic markers. The parameters in the first column are the number of members, number of marker loci, and number of recombinants.

Parameters	locus-based algorithm	MRH	PedPhase
(15,10,0)	6.4s	16s	1.3s
(15,10,4)	3.4s	3m13s	1.7s
(29,10,0)	13.6s	16m49s	1.8s
(29,10,4)	5.1s	14m2s	1.4s
(15,15,0)	35m15s	14m29s	1.3s
(15,15,4)	8m7s	10m9s	1.8s
(29,15,0)	15m10s	1h11m	1.9s
(29,15,4)	23m12s	1h15m	2.1s

v0.1 [17] on simulated genotype data. Two different tree pedigree structures (Fig. 1 and Fig. 8 in [10]) were used in the simulation, one with 15 members and the other with 29 members. For each pedigree, genotypes with 10 and 15 biallelic marker loci were considered. The two alleles at each locus in a founder (*i.e.* a member that has no parents) were independently sampled with a fixed frequency of 0.5 in order to maximize the chance of heterozygosity (to test the worst-case performance of the locus-based algorithm), resulting in an average of 5 and 7.5 heterozygous loci, respectively, in each member of the pedigree. The number of recombinants used in generating each pedigree ranged from 0 to 4. For each combination of the above parameters, 100 sets of random genotype data were generated and the average performance of the programs was recorded, as shown in Table 2. The experiments were done on a Pentium IV PC with 1.7GHz CPU and 512MB RAM. In terms of the quality of solutions, the locus-based algorithm always output a haplotyping solution with the minimum number of recombinants, while PedPhase and MRH do not guarantee optimal solutions. When the data was generated with zero or few recombinants, the optimal solution was often unique and both PedPhase and MRH could recover it. However, the performance of PedPhase and MRH decreases when the number of recombinants increases. In terms of efficiency, PedPhase was the fastest among the three programs in all cases. But, surprisingly, the locus-based algorithm outperformed MRH in many cases.

We further compared the performance of the locus-based algorithm, PedPhase, MRH, and the EM algorithm used in [5] on a real dataset consisting of 12 multi-generation tree pedigrees, each with 7-8 members, from [5]. The comparison was based on the inference of *common* haplotypes (*i.e.* haplotypes with frequencies > 5%). We focused on a randomly selected autosome (*i.e.* chromosome 3). There are 10 blocks (*i.e.* regions with few recombinants) in the chromosome 3 data, of which one block consists of 16 marker loci and all the others have only 4-6 marker loci each. The test results of the locus-based algorithm, MRH and the EM algorithm are summarized in Table 3. (We obtained the results of the EM algorithm directly form the authors [5]. The results of PedPhase are very similar to those of MRH's [10], and are thus not shown here due to space limit.) The results show that both rule-based haplotyping methods could discover almost all the common haplotypes that were inferred by the EM algorithm [5]. The haplotype frequencies estimated from the haplotype configuration results by simple

Table 3. Common haplotypes and their frequencies obtained by the locus-based algorithm, MRH and the EM method. In the haplotypes, the alleles are encoded as 1=A, 2=C, 3=G, and 4=T.

Block	EM		locus-based		MRH	
	Common haplotypes	Frequencies	Common haplotypes	Frequencies	Common haplotypes	Frequencies
16a-1	4 2 2 2 2	0.4232	4 2 2 2 2	0.3817	4 2 2 2 2	0.3779
	3 4 3 4 4	0.2187	3 4 3 4 4	0.1720	3 4 3 4 4	0.1744
	4 2 2 2 4	0.2018	4 2 2 2 4	0.1989	4 2 2 2 4	0.1802
	3 4 2 2 4	0.1432	3 4 2 2 4	0.1667	3 4 2 2 4	0.1802
16b-1	3 2 4 1 1 2	0.8014	3 2 4 1 1 2	0.7634	3 2 4 1 1 2	0.7849
	1 3 2 3 3 4	0.0833	1 3 2 3 3 4	0.0753	1 3 2 3 3 4	0.0753
16b-2	4 1 2 2	0.5410	4 1 1 2	0.5000	4 1 1 2	0.4826
	2 3 3 4	0.2812	2 3 3 4	0.2634	2 3 3 4	0.2616
	2 3 3 2	0.1562	2 3 3 2	0.1398	2 3 3 2	0.1512
17a-1	3 1 3 4 4 4	0.3403	3 1 3 4 4 4	0.3172	3 1 3 4 4 4	0.3226
	1 3 3 2 4 2	0.3021	1 3 3 2 4 2	0.2527	1 3 3 2 4 2	0.2473
	3 3 2 4 2 4	0.1354	3 3 2 4 2 4	0.0968	3 3 2 4 2 4	0.0914
	3 3 3 4 4 4	0.1021	3 3 3 4 4 4	0.1129	3 3 3 4 4 4	0.1183
	3 3 2 4 4 4	0.0681	3 3 2 4 4 4	0.0806	3 3 2 4 4 4	0.0806
	1 3 3 2 4 4	0.0521				
17a-2	2 3 2 4 2	0.3542	2 3 2 4 2	0.3065	2 3 2 4 2	0.2903
	3 3 4 2 4	0.3333	3 3 4 2 4	0.3118	3 3 4 2 4	0.3118
	3 3 4 4 2	0.1458	3 3 4 4 2	0.1505	3 3 4 4 2	0.1237
	3 4 4 4 4	0.1250	3 4 4 4 4	0.1452	3 4 4 4 4	0.1452
17a-3	4 4 3 1	0.4129	4 4 3 1	0.4408	4 4 3 1	0.4167
	3 1 1 2	0.2813	3 1 1 2	0.2312	3 1 1 2	0.2051
	4 1 3 1	0.2363	4 1 3 1	0.1935	4 1 3 1	0.2115
	4 1 3 2	0.0696	4 1 3 2	0.0753	4 1 3 2	0.0705
17a-4	3 4 4 1 2 4	0.3854	3 4 4 1 2 4	0.3656	3 4 4 1 2 4	0.4429
	2 3 2 4 3 2	0.3333	2 3 2 4 3 2	0.3065	2 3 2 4 3 2	0.2357
	3 4 2 4 2 4	0.2500	3 4 2 4 2 4	0.1935	3 4 2 4 2 4	0.1857
18a-1	1444231214144132	0.2697	1444231214144132	0.2688	1444231214144132	0.1706
	1444111214144132	0.2396	1444111214144132	0.2097	1444111214144132	0.2357
	1444131214144132	0.1887	1444131214144132	0.1989	1444131214144132	0.2176
	4222133313412211	0.1250				
	1444231234144132	0.0833	1444231234144132	0.0699	1444231234144132	0.0764
18a-2	3 1 2 4 4 2	0.4967	3 1 2 4 4 2	0.4839	3 1 2 4 4 2	0.4765
	1 3 2 4 3 4	0.2604	1 3 2 4 3 4	0.1989	1 3 2 4 3 4	0.1765
	3 1 2 2 4 2	0.1271	3 1 2 2 4 2	0.0806	3 1 2 2 4 2	0.0765
	1 3 4 4 4 4	0.0938	1 3 4 4 4 4	0.0860	1 3 4 4 4 4	0.0941
			1 3 2 4 3 2	0.0538	1 3 2 4 3 2	0.0588
18a-3	2 2 1 1	0.4186	2 2 1 1	0.4140	2 2 1 1	0.4214
	4 3 3 3	0.2188	4 3 3 3	0.1935	4 3 3 3	0.1714
	2 3 1 1	0.2064	2 3 1 1	0.2097	2 3 1 1	0.1928
	4 3 1 3	0.1250	4 3 1 3	0.1613	4 3 1 3	0.1857

counting are also similar to the estimations by the EM algorithm. Most of the blocks (> 85%) were found by the locus-based algorithm and MRH to have involved 0 recombinants. The speeds of all three programs on this real dataset were very fast (less 1 minute), even though one of the blocks has 16 marker loci.

6 Concluding Remarks

Missing data usually occur in real data set due to many reasons. Unfortunately, Aceto *et al.* [1] recently showed that checking consistency (thus imputing missing data) for pedigree is in general NP-hard. The two algorithms above take data with no missing values as input. In our experiment on the real data set, we had to impute the missing values first before feeding the data to the locus-based algorithm. Our current imputation algorithm simply uses the Mendelian law and allele frequencies. It would be desirable to combine missing data imputation and haplotype inference in a unified framework.

Acknowledgements

KD's work was partially supported by a Grant-in-Aid for Scientific Research on Priority Areas (C) for "Genome Information Science" from the Ministry of Education, Culture, Sports, Science and Technology, Japan. JL is supported by NSF grant CCR-9988353. TJ is supported by NSF Grants CCR-9988353, ITR-0085910, DBI-0133265, and National Key Project for Basic Research (973).

References

1. L. Aceto, J. A. Hansen, A. Ingólfsdóttir, J. Johnsen, and J. Knudsen. The complexity of checking consistency of pedigree information and related problems. *Manuscript*, 2003.
2. M. Daly, J. Rioux, S. Schaffner, T. Hudson, and E. Lander. High-resolution haplotype structure in the human genome. *Nat Genet*, 29(2):229–232, 2001.
3. J. A. Douglas, M. Boehnke, E. Gillanders, J. Trent, and S. Gruber. Experimentally-derived haplotypes substantially increase the efficiency of linkage disequilibrium studies. *Nat Genet*, 28(4):361–364, 2001.
4. L. Excoffier and M. Slatkin. Maximum-likelihood estimation of molecular haplotype frequencies in a diploid population. *Mol Biol Evol*, 12:921–927, 1995.
5. S. B. Gabriel, *et al.* The structure of haplotype blocks in the human genome. *Science*, 296(5576):2225–29, 2002.
6. M. R. Gary, D. S. Johnson, and L. Stockmeyer. Some simplified NP-complete graph problems. *Theor. Comput. Sci.*, 1, 237–267, 1976.
7. D. Gusfield. Haplotyping as perfect phylogeny: conceptual framework and efficient solutions. *Proc. RECOMB*, 166–175, 2002.
8. L. Helmuth. Genome research: Map of the human genome 3.0 *Science*, 293(5530):583–585, 2001.
9. J. Li and T. Jiang, Efficient rule-based haplotyping algorithms for pedigree data. *Proc. RECOMB'03*, pages 197–206, 2003.
10. J. Li and T. Jiang, Efficient inference of haplotypes from genotypes on a pedigree. *J. Bioinfo. and Comp. Biol.* 1(1):41-69, 2003.
11. J. C. Lam, K. Roeder, and B. Devlin. Haplotype fine mapping by evolutionary trees. *Am J Hum Genet*, 66(2):659–673, 2000.
12. S. Lin and T. P. Speed. An algorithm for haplotype analysis. J Comput Biol, 4(4):535–546, 1997.
13. R. Lippert, R. Schwartz, G. Lancia, and S. Istrail. Algorithmic strategies for the single nucleotide polymorphism haplotype assembly problem. Briefings in Bioinformatics, 3(1):23–31, 2002.
14. J. S. Liu, C. Sabatti, J. Teng, B. J. Keats, and N. Risch. Bayesian analysis of haplotypes for linkage disequilibrium mapping. *Genome Res*, 11(10):1716–24, 2001.
15. T. Niu, Z. S. Qin, X. Xu, and J. S. Liu. Bayesian haplotyping interface for multiple linked single-nucleotide polymorphisms. *Am J Hum Genet*, 70(1):157–169, 2002.
16. J. R. O'Connell. Zero-recombinant haplotyping: applications to fine mapping using snps. *Genet Epidemiol*, 19 Suppl 1:S64–70, 2000.

17. D. Qian and L. Beckman. Minimum-recombinant haplotyping in pedigrees. *Am J Hum Genet*, 70(6):1434–1445, 2002.
18. H. Seltman, K. Roeder, and B. Delvin. Transmission/disequilibrium test meets measured haplotype analysis: family-based association analysis guided by evolution of haplotypes. *Am J Hum Genet*, 68(5):1250–1263, 2001.
19. S. K. Service, D. W. Lang, N. B. Freimer, and L. A. Sandkuijl. Linkage-disequilibrium mapping of disease genes by reconstruction of ancestral haplotypes in founder populations. *Am J Hum Genet*, 64(6):1728-1738, 1999.
20. M. Stephens, N. J. Smith, and P. Donnelly. A new statistical method for haplotype reconstruction from population data. *Am J Hum Genet*, 68(4):978-989, 2001.
21. P. Tapadar, S. Ghosh, and P. P. Majumder. Haplotyping in pedigrees via a genetic algorithm. *Hum Hered*, 50(1):43–56, 2000.
22. A. Thomas, A. Gutin, V. Abkevich, and A. Bansal. Multilocus linkage analysis by blocked gibbs sampling. *Stat Comput*, 259–269, 2000.
23. H. T. Toivonen, P. Onkamo, K. Vasko, V. Ollikainen, P. Sevon, H. Mannila, M. Herr, and J. Kere. Data mining applied to linkage disequilibrium mapping. *Am J Hum Genet*, 67(1):133–145, 2000.
24. E. M. Wijsman. A deductive method of haplotype analysis in pedigrees. *Am J Hum Genet*, 41(3):356–373, 1987.
25. S. Zhang, K. Zhang, J. Li and H. Zhao. On a family-based haplotype pattern mining method for linkage disequilibrium mapping. *Pac Symp Biocomput*, 100–111, 2002.

Efficient Energy Computation for Monte Carlo Simulation of Proteins

Itay Lotan, Fabian Schwarzer, and Jean-Claude Latombe

Dept. of Computer Science, Stanford University, Stanford, CA 94305
{itayl, schwarzf, latombe}@cs.stanford.edu

Abstract. Monte Carlo simulation (MCS) is a common methodology to compute pathways and thermodynamic properties of proteins. A simulation run is a series of random steps in conformation space, each perturbing some degrees of freedom of the molecule. A step is accepted with a probability that depends on the change in value of an energy function. Typical energy functions sum many terms. The most costly ones to compute are contributed by atom pairs closer than some cutoff distance. This paper introduces a new method that speeds up MCS by efficiently computing the energy at each step. The method exploits the facts that proteins are long kinematic chains and that few degrees of freedom are changed at each step. A novel data structure, called the ChainTree, captures both the kinematics and the shape of a protein at successive levels of detail. It is used to find all atom pairs contributing to the energy. It also makes it possible to identify partial energy sums left unchanged by a perturbation, thus allowing the energy value to be incrementally updated. Computational tests on four proteins of sizes ranging from 68 to 755 amino acids show that MCS with the ChainTree method is significantly faster (as much as 12 times faster for the largest protein) than with the widely used grid method. They also indicate that speed-up increases with larger proteins.

1 Introduction

1.1 Monte Carlo Simulation (MCS)

The study of the conformations adopted by proteins is an important topic in structural biology. MCS [1] is one common methodology for this study. In this context, it has been used for two purposes: (1) estimating thermodynamic quantities over a protein's conformation space [2, 3, 4] and, in some cases, even kinetic properties [5, 6]; and (2) searching for low-energy conformations of a protein, including its native structure [7, 8, 9]. The approach was originally proposed in [10], but many variants and improvements have later been suggested [11].

MCS is a series of randomly generated *trial steps* in the conformation space of the studied molecule. Each such step consists of perturbing some degrees of freedom (DOFs) of the molecule [4, 5, 6, 9, 12], in general torsion (dihedral) angles around bonds (see Section 1.2). Classically, a trial step is *accepted* – i.e., the simulation actually moves to the new conformation – with probability

G. Benson and R. Page (Eds.): WABI 2003, LNBI 2812, pp. 354–373, 2003.

$\min\{1, e^{-\Delta E/k_b T}\}$ (the so-called Metropolis criterion [10]), where E is an energy function defined over the conformation space, ΔE is the difference in energy between the new and previous conformations, k_b is the Boltzmann constant, and T is the temperature of the system. So, a downhill step to a lower-energy conformation is always accepted, while an uphill step is accepted with a probability that goes to zero as the energy barrier grows large. It has been shown that a long MCS with the Metropolis criterion and an appropriate step generator produces a distribution of accepted conformations that converges to the Boltzmann distribution.

The need for general algorithms to speed-up MCS has often been mentioned in the biology literature, most recently in [4]. In this paper, we propose a new algorithm that achieves this goal, independent of a specific energy function, step generator, and acceptance criterion. More precisely, our algorithm reduces the average time needed to decide whether a trial step is accepted, or not, without affecting which steps are attempted, nor the outcome of the acceptance test. It achieves this result by incorporating efficient techniques to incrementally update the value of the energy function during simulation. Although we will describe this algorithm for its application to classic MCS, it could also be used to speed up other kinds of MCS methods, as well as other optimization and sampling techniques. Several such applications will be discussed in Section 8.2.

1.2 Kinematic Structure of a Protein

A protein is the concatenation of small molecules (the amino acids) forming a long backbone chain with small side chains. Since bond lengths and angles between any two successive bonds are almost constant across all conformations at room temperature [13], it is common practice to assume that the only DOFs of a protein are its torsion angles, also called the internal coordinates. Each amino acid contributes two torsion DOFs to the backbone – the so-called ϕ and ψ angles. See Figure 1 for illustration. Thus, the backbone is commonly modelled as a long chain of links separated by torsion joints (the backbone's DOFs). A link, which designates a rigid part of a kinematic chain, is a group of atoms with no DOFs between them. For example, in the model of Figure 1a, the C and O atoms of amino acid $i-1$ together with the N and H atoms of amino acid i form a link of the protein's backbone, since none of the bonds between them is rotatable. While a backbone may have many DOFs (between 136 and 1510 in the proteins used for the tests reported in this paper), each side-chain has between 0 and 4 torsion DOFs (known as the χ angles). In Figure 1a, these DOFs are hidden inside the ball marked R in each amino acid.

The model of Figure 1a is the most common torsion-DOF representation used in the literature, and is also the one we use in this paper. However, it is possible to apply our algorithm to models that include additional DOFs, such as: ω angles (rotations about the peptide bonds C–N between adjacent amino acids), bond lengths, and bond angles. At the limit, one can make each link a single atom and each joint a rigid-body transform. However, while it is theoretically possible to perform MCS in the Cartesian coordinate space, where each atom

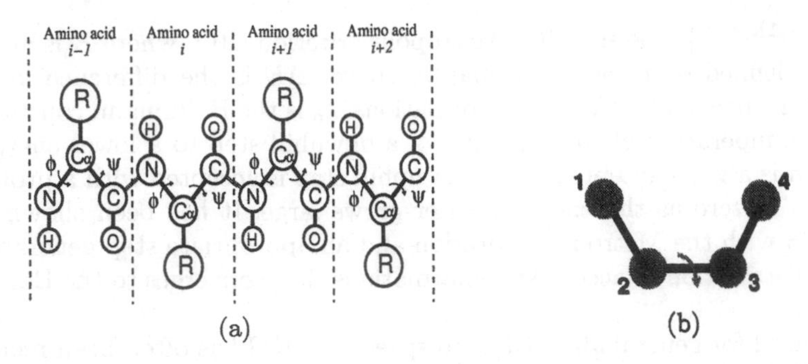

Fig. 1. (a) An illustration of a protein fragment with its backbone DOFs. R represents any side-chain (b) A torsional DOF: it is the angle made by the two planes containing the centers of atoms 1, 2, and 3, and 2, 3, and 4, respectively.

has 3 DOFs, it is more efficient to run it in the torsion-DOF space [14]. Hence, the vast majority of MCS are run in this space [4, 5, 6, 9, 12].

Due to the chain kinematics of the protein, a small change in one DOF of the backbone may cause large displacements of some atoms. Thus, in an MCS, a high percentage of steps are rejected because they lead to high-energy conformations, in particular conformations with steric clashes (self-collisions). In fact, the rejection rate tends to grow quickly with the number k of DOFs randomly changed in a single step. This fact is well-known in the literature [12, 15] and as a result it is common practice in MCS to change few DOFs (picked at random) at each trial step [4, 5, 6, 9, 12, 16, 17].

1.3 Computing the Energy

Various energy functions have been proposed for proteins [16, 18, 19, 20, 21]. For all of them, the dominant computation is the evaluation of non-bonded terms, namely energy terms that depend on distances between pairs of non-bonded atoms. These may be physical terms (e.g., van der Waals and electrostatic potentials [20]), heuristic terms (e.g., potentials between atoms that should end up in proximity to each other [18]) and/or statistical potentials derived from a structural database (e.g. [16]).

To avoid the quadratic cost of computing and summing up the contributions from all pairs, cutoff distances are usually introduced, exploiting the fact that physical and heuristic potentials drop off quickly toward 0 as the distance between atoms increases. We refer to the pairs of atoms that are close enough to contribute to the energy function as the *interacting pairs*. Because van der Waals forces prevent atom centers from getting very close, the number of interacting pairs in a protein is often less than quadratic in practice [22].

Hence, one may try to reduce computation by finding interacting pairs without enumerating all atom pairs. A classical method to do this is the grid algorithm (see Section 2), which indexes the position of each atom in a regular three-dimensional grid. This method takes time linear in the number of atoms, which is asymptotically optimal in the worst case. However, it does not exploit an important property of proteins, namely that they form long kinematic chains. It also does not take advantage of the common practice in MCS to change only a few DOFs at each time-step. Moreover, it does not address the remaining problem of efficiently summing up the contributions of the interacting pairs. These issues are addressed in this paper.

1.4 Contributions

A key consequence of making only a small number of DOF changes in a single MCS step is that at every step large fragments of the protein remain rigid. Hence, at each step, many partial energy sums are unaffected. The grid method re-computes all interacting pairs at each step and cannot directly identify partial sums that have remained constant. Instead, the method proposed in this paper finds the new interacting pairs and retrieves unaffected partial sums without enumerating all interacting pairs. It uses a novel hierarchical data structure – the *ChainTree* – that captures both the chain kinematics and shape of a protein at successive levels of detail. At each step, the ChainTree can be maintained and queried efficiently to find new interacting pairs. It also enables the identification of unchanged partial energy sums stored in a companion data structure — the *EnergyTree* — thus allowing for efficient energy updates throughout the simulation.

Our test results (see Section 6) show that MCS with the ChainTree method is significantly faster than with the grid method when the number k of DOF changes at each step is sufficiently small. We observed speed-ups by factors up to 12 for the largest of the four proteins. Therefore, not only does a small k sharply increases the step acceptance ratio, it also makes it possible to expedite the evaluation of the acceptance criterion. Simulation methodologies other than classical MCS may also benefit from our algorithm (see Section 8.2).

1.5 Outline of Paper

The rest of this paper is organized as follows. Section 2 discusses related work. Section 3 presents the ChainTree data structure. Section 4 introduces the algorithm for finding new interacting atom pairs and Section 5 describes how to efficiently update the energy at each simulation step. Section 6 gives experimental results comparing our algorithm with the grid algorithm in MCS. A downloadable version of our algorithm is described in Section 7. Section 8 discusses applications of our algorithm to other MCS methods as well as to other types of molecular simulation methods and points to possible extensions and future directions of research. The application of the ChainTree to testing a long kinematic chain for self-collision was previously presented in [23].

Throughout this paper, we always use n to denote the number of links of a kinematic chain (e.g., a protein's backbone) and k to denote the number of DOF changes per simulation step. Since the number of atoms in any amino acid is bounded by a constant, the number of atoms in a protein is always $O(n)$. Although k can be as big as $O(n)$, it is much smaller in practice [4, 5, 6, 9, 12, 15].

2 Related Work

Because biologists are more interested in simulation results than in the computational methods they use to achieve these results, the literature does not extensively describe algorithms for MCS.

A prevailing algorithm – referred to as the *grid algorithm* in this paper – reduces the complexity of finding all interacting pairs in a molecule to asymptotically linear time by indexing the atoms in a regular grid. This approach exploits the fact that van der Waals potentials prevent atom centers from coming very close to one another. In [22] it is formally shown that in a collection B of n possibly overlapping balls of similar radii, such that no two sphere centers are closer than a small fixed distance, the number of balls that intersect any given ball of B is bounded by a constant. This result yields the grid algorithm, which subdivides the 3D space into cubes whose sides are set to the maximum diameter of the balls in B, computes the cubes intersected by each ball, and stores the results in a hash-table. This data structure is re-computed after each step in $\Theta(n)$ time. Determining which balls intersect any given ball of B then takes $O(1)$ time. Hence, finding all pairs of intersecting balls takes $\Theta(n)$ time. The grid method can be used to find all pairs of atoms within some cutoff distance, by growing each atom by half this distance. The method is asymptotically optimal in the worst case, but updating the data structure always takes linear time. This is too costly for very large proteins, making it impractical to perform MCS in this case.

A variant of this method mostly used for Molecular Dynamics simulation maintains, for each atom, a list of atoms within a distance d somewhat larger than the cutoff distance d_c by updating it every s steps [24]. The idea is that atoms further apart than d will not come closer than d_c in less than s steps. There is a tradeoff between s and $d - d_c$ since the larger this difference, the larger the value of s that can be used. However, choosing d big causes the neighbor lists to become too large to be efficient. A method for updating neighbor lists based on monitoring the displacement of each atom is described in [25].

The ChainTree includes a bounding volume hierarchy (BVH) to represent a protein at successive levels of detail. BVHs have been extensively used to detect collisions and compute distances between rigid objects [26, 27, 28, 29, 30, 31]. They have been extended in [26, 31, 32] to handle deformable objects by exploiting the facts that topological proximity in the meshed surface of an object is invariant when the object deforms and implies spatial proximity. However, BVH techniques alone lose efficiency when applied to testing a deformable object for self-collision, because they cannot avoid detecting the trivial interaction of each object component with itself. They also lose efficiency when many components move independently.

Finding interacting pairs in a protein is equivalent to finding self-collision in a deformable chain after having grown all links by the cutoff distance. The ChainTree borrows from previous work on BVHs. It uses a BVH based on the invariance of topological proximity along a chain. But it combines it with a transform hierarchy that makes it possible to efficiently prune the search for interacting pairs, when few DOFs change simultaneously.

3 The ChainTree

In this section we describe the ChainTree, the data structure we use to represent a protein. We begin by stating the key properties of proteins and MCS that motivated this data structure (Subsection 3.1). Then follows a description of the two hierarchies that make up the ChainTree. The *transform* hierarchy that approximates the kinematics of the backbone is introduced in Subsections 3.2 and the *bounding-volume* hierarchy that approximates the geometry of the protein is presented in Subsection 3.3. Next, we discuss the representation of the side-chains (Subsection 3.4). Finally, we describe how the two aforementioned hierarchies are combined to form a single balanced binary tree (Subsection 3.5) and the way it is updated (Subsection 3.6).

In the following we refer to the algorithm that updates the ChainTree as the *updating* algorithm and to the algorithm that finds interacting pairs as the *testing* algorithm.

3.1 Properties of Proteins and MCS

A protein backbone is commonly modelled as a kinematic chain made up of a sequence of n links (atoms or rigid groups of atoms) connected by torsional DOFs. The ChainTree is motivated by three key properties derived from this model:

Local changes have global effects: Changing a single DOF causes all links beyond this DOF, all the way to the end of the chain, to move. Any testing algorithm that requires knowing the absolute position of every link at each step must perform $O(n)$ work at each step even when the number k of DOF changes is $O(1)$.

Small changes may cause large motions: The displacement of a link caused by a DOF change depends not only on the angular variation, but also on the distance between the DOF axis and the link (radius of rotation). So, at each step, any link with a large radius of rotation undergoes a large displacement.

Large sub-chains remain rigid at each step: If we only perturb few DOFs at each step, as is the case during MCS, then large contiguous fragments of the chain remain rigid between steps. So, there cannot be any new interacting pairs inside each of these fragments.

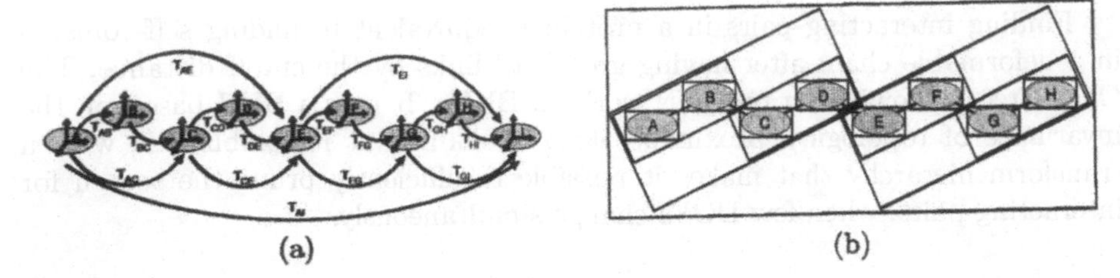

Fig. 2. The two hierarchies. (a) The transform hierarchy: grey ovals depict links; $T_{\alpha\beta}$ denotes the rigid-body transform between the reference frames of links α and β. (b) The bounding volume hierarchy: each BV approximates the geometry of a chain-contiguous sequence of links

3.2 Transform Hierarchy

We attach a reference frame to each link of the protein's backbone and map each DOF to the rigid-body transform between the frames of the two links it connects. The transform hierarchy is a balanced binary tree of transforms. See Figure 2a, where ovals and labelled arrows depict links and transforms, respectively. At the lowest level of the tree, each transform represents a DOF of the chain. Products of pairs of consecutive transforms give the transform at the next level. For instance, in Figure 2a, T_{AC} is the product of T_{AB} and T_{BC}. Similarly, each transform at every level is the product of two consecutive transforms at the level just below. The root of the tree is the transform between the frames of the first and last links in the chain (T_{AI} in the figure). Each of the $\log n$ levels of the tree can be seen as a chain that has half the links and DOFs of the chain at the level just below it. In total, $O(n)$ transforms are cached in the hierarchy. We say that each intermediate transform $T_{\alpha\beta}$ *shortcuts* all the transforms that are in the subtree rooted at $T_{\alpha\beta}$.

The transform hierarchy is used from the top down by the testing algorithm to propagate transforms defining the relative positions of bounding volumes (from the other hierarchy) that need to be tested for overlap.

3.3 Bounding-Volume Hierarchy

The bounding-volume (BV) hierarchy is similar to those used by prior collision checkers (see Section 2). As spatial proximity in a deformable chain is not invariant, our BVH is based on the proximity of links along the chain. See Figure 2b. Like the transform hierarchy, the BVH is a balanced binary tree. It is constructed bottom up in a "chain-aligned" fashion. At the lowest level, one BV bounds each link. Then, pairs of neighboring BVs at each level are bounded by new BVs to form the next level. The root BV encloses the entire chain. So, at each level, we have a chain with half the number of BVs as the chain at the level below it. This chain of BVs encloses the geometry of the chains of BVs at all lower levels.

The BV type we use is called RSS (for rectangle swept sphere). It was introduced in [29] and is defined as the Minkowski sum of a sphere and a rectangle.

The RSS bounding a set of points in 3D is created as follows. The two principal directions spanned by the points are computed and a rectangle is constructed along these directions to enclose the projection of all points onto the plane defined by these directions. The RSS is the Minkowski sum of this rectangle and the sphere whose radius is half the length of the interval spanned by the point set along the dimension perpendicular to the rectangle. To compute the distance between two RSSs, one simply computes the distance between the two underlying rectangles minus the radii of the swept spheres. RSSs offer a good compromise between tightness and efficiency of distance computation. They bound well both globular objects (single atoms, small groups of atoms) and elongated objects (chain fragments). In addition, RSSs are invariant to a rigid-body transform of the geometry they bound.

We construct each intermediate RSS to enclose its two children, thus creating what we term a *not-so-tight* hierarchy (in contrast to a *tight* hierarchy where each BV tightly bounds the links of the sub-chain it encloses). In [23] we show that the size of these BVs does not deteriorate too much as one climbs up the hierarchy. As a result, the shape of the BV stored at each intermediate node depends only on the two BVs held by this node's children.

3.4 Side-Chain Representation

Side-chains, one per amino acid, are short chains with up to 20 atoms, that protrude from the backbone [33]. A side-chain may have some internal torsional DOFs (between 0 and 4). The biology literature proposes different ways to model side-chains ranging from a single sphere approximating the entire side-chain, to a full atomistic model [20, 33]. The choice depends on both the physical accuracy one wishes to achieve and the amount of computation one is willing to pay per simulation step. One may choose to make the side-chains completely rigid, or allow their DOFs to change during the simulation. In both cases we expect the overhead of using a sub-hierarchy for the side-chain atoms to exceed any benefit it may provide. Therefore, we allow each link of the protein backbone to be an aggregate of atoms represented in a single coordinate frame and contained in a BV that is a leaf of the BVH. Each such aggregate includes one or several backbone atoms forming a rigid piece of the backbone and the atoms of the side-chain stemming from it (contained in the circle marked R in Figure 1a).

3.5 Combined Data Structure

The ChainTree combines both the transform and the BV hierarchies into a single binary tree as the one depicted in Figure 3a. The leaves of the tree (labelled A through H in the figure) correspond to the links of the protein's backbone with their attached side-chains (using any of the representations discussed above). Each leaf holds both the BV of the corresponding link and the transform (symbolized by a horizontal arrow in the figure) to the reference frame of the next link. Each internal node (nodes J through P) has the frame of the leftmost link in its sub-tree associated with it. It holds both the BV of the BVs of its two

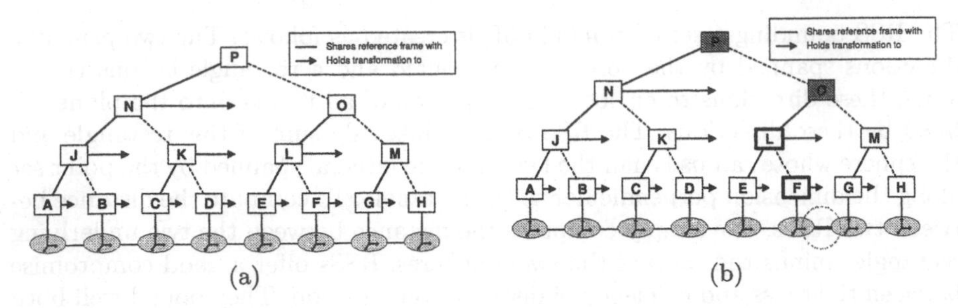

Fig. 3. The ChainTree: (a) a binary tree that combines the transform and BV hierarchies, and (b) after applying a 1-DOF perturbation. The transforms of the nodes with bold contours (F and L) were updated. The BVs of the nodes in grey (O and P) were re-computed.

children and the transform to the frame of the next node at the same level, if any. The ChainTree contains both pointers from children to parents, which are used by the updating algorithm to propagate updates from the bottom up, as described below, and pointers from parents to children, which are used by the testing algorithm (Section 4).

3.6 Updating the ChainTree

When a change is applied to a single arbitrary DOF in the backbone, the updating algorithm re-computes all transforms that shortcut this DOF and all BVs that enclose the two links connected by this DOF. It does this in bottom-up fashion, by tracing the path from the leaf node immediately to the left of the changed DOF up to the root. A single node is affected at each level. If this node holds a transform, this transform is updated. If it holds a BV that contains the changed DOF, then the BV is re-computed. For example, see Figure 3a. Since the shape of an RSS is invariant to a rigid-body transform of the objects it bounds, all other BVs remain unchanged.

If a DOF is changed in a side-chain, the BV stored at the corresponding leaf node of the ChainTree and the BVs of all the ancestors of this node are re-computed, but *all transforms in the hierarchy remain unchanged.* By updating from the bottom up, each affected transform is re-computed in $O(1)$ time. By using not-so-tight BVs – thus, trading tightness for speed – re-computing each BV is also done in $O(1)$ time. Since the ChainTree has $O(\log n)$ levels, and at each level at most one transform and one BV are updated, the total cost of the update process is $O(\log n)$.

When multiple DOFs are changed simultaneously (in the backbone and the side-chains), the ChainTree is updated one level at a time, starting with the lowest level. Hence, all affected transforms and BVs at each level are updated at most once before proceeding to the next level above it. The total updating time is then $O(k \log (n/k))$. When k grows this bound never exceeds $O(n)$.

The updating algorithm marks every node whose BV and/or transform is recomputed. This mark will be used later by the testing algorithm. See Figure 3b.

4 Finding Interacting Pairs

BVHs have been widely used to detect collision or compute separation distance between pairs of rigid objects, each described by its own hierarchy [27, 28, 29, 30, 31, 34]. If each object is divided into small fragments (e.g. links of a chain) the hierarchies are easily applied to finding all pairs of fragments that are closer than some threshold. A simple variant of this algorithm can detect pairs of fragments of the same object that are closer than a threshold by testing the BVH of the object against itself. This variant skips the test of a BV against itself and proceeds directly to testing the BV's children. However, it takes $\Omega(n)$ time, since all leaves will inevitably be visited as each leaf is at distance 0 from itself. The ChainTree allows us to avoid this lower bound by exploiting the third property stated in Section 3.1 — large sub-chains remain rigid between steps.

Since each leaf BV may contain a number of atoms, when two leaf BVs are within the cutoff distance, the interacting atom pairs are found by examining all pairs of atoms, one from each leaf.

When only a small number k of DOFs are changed simultaneously, long sub-chains remain rigid at each step. These sub-chains cannot contain new interacting pairs. So, when we test the BVH contained in the ChainTree against itself, we prune the branches of the search that would look for interacting pairs within rigid sub-chains.

There are two distinct situations where pruning occurs:

1. If the algorithm is about to test a BV against itself and this BV was not updated after the last DOF changes, then the test is pruned.
2. If the algorithm is about to test two different BVs, and neither BV was updated after the last DOF changes, and no backbonc DOF between those two BVs was changed, then the test is pruned.

The last condition in this second situation – that no backbone DOF between the two BVs was changed – is slightly more delicate to recognize efficiently. We say that two nodes at the same level in the ChainTree are *separated* if there exists another node between them at the same level that holds a transform that was modified after the last DOF changes. This node will be dubbed *separator*. Hence, if two nodes are separated, a DOF between them has changed. We remark that:

– If two nodes at any level are separated, then any pair consisting of a child of one and a child of the other is also separated.
– If two nodes at any level are *not* separated, then a child of one and a child of the other are separated if and only if they are separated by another child of either parent.

Hence, by pushing separation information downward, the testing algorithm can know in constant time whether a DOF has changed between any two BVs it is about to test. The algorithm also propagates transforms from the transform tree downward to compute the relative position of any two separated BVs in constant time before performing the overlap test.

To illustrate how the testing algorithm works, consider the ChainTree of Figure 3b obtained after a change of the DOF between F and G. F and L are the only separators. The algorithm first tests the BV stored in the root P against itself. Since this BV has changed, the algorithm examines all pairs of its children, (N, N), (N, O) and (O, O). The BV held in N was not changed, so (N, N) is discarded (i.e., the search along this path is pruned). (N, O) is not discarded since the BV of O has changed, leading the algorithm to consider the four pairs of children (J, L), (J, M), (K, L), and (K, M). Both (J, L) and (K, L) satisfy the conditions in the second situation described above; thus, they are discarded. (J, M) is not discarded because J and M are separated by L. The same is true for (K, M), and so on.

In [23] we proved that the worst-case complexity of the testing algorithm to detect self-collision is $\Theta(n^{\frac{4}{3}})$. This bound holds unchanged when using RSSs to compute all interacting pairs. Note, however, that it is an asymptotic bound, which relies on the fact that the number of atoms that may interact with any given atom is bounded by a constant as the size n of the protein *grows arbitrarily large*. For many proteins, this constant may correspond to a significant fraction of n.

The worst-case bound on finding all interacting pairs using the ChainTree hides the practical speed-up allowed by search pruning. We evaluate this speed-up, in a set of benchmarks in Section 6.

5 Energy Maintenance

A typical energy function is of the form $E = E_1 + E_2$, where E_1 sums terms depending on a single parameter commonly referred to as *bonded* terms (e.g., torsion-angle and bond-stretching potentials) and E_2 sums terms commonly known as *non-bonded* terms, which account for interactions between pairs of atoms or atom groups closer than a cutoff distance [5, 6, 12, 16, 17, 19]. Updating E_1 after a conformational change is straightforward. This is done by computing the sum of the differences in energy in the terms affected by the change and adding it to the previous value of E_1. After a k-DOF change, there are only $O(k)$ affected single-parameter terms. So, in what follows we focus on the maintenance of E_2.

At each simulation step we must find the interacting pairs of atoms and change E_2 accordingly. When k is small, many interacting pairs are unaffected by a k-DOF change. The number of affected interacting pairs, though still $O(n)$ in the worst case, is usually much smaller than the total number of interacting pairs at the new conformation. Therefore, an algorithm like the grid algorithm that computes all interacting pairs at each step is not optimal in practice. Moreover,

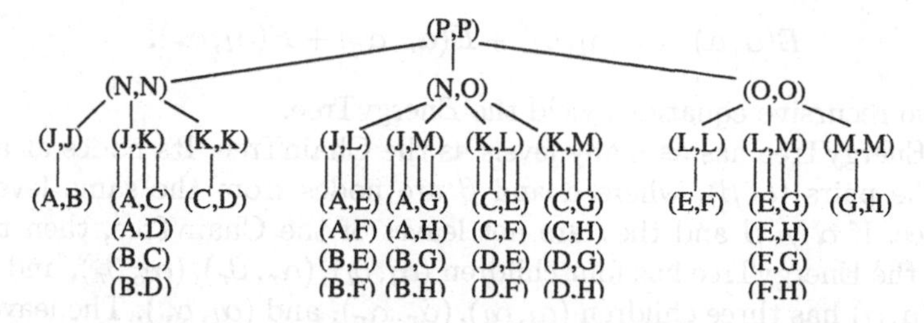

Fig. 4. EnergyTree for the ChainTree of Figure 3a. For simplification, leaves of the form (α, α) are not shown.

after having computed the new set of interacting pairs, we still have to update E_2, either by re-computing it from scratch, or by scanning the old and new sets of interacting pairs to determine which terms should be subtracted from the old value of E_2 and which terms should be added to get the new value. In either case, we perform again a computation at least proportional to the total number of interacting pairs. Instead, our method detects partial energy sums unaffected by the DOF change (these sums correspond to interacting pairs where both atoms belong to the same rigid sub-chains). The energy terms contributed by the new pairs are then added to the unaffected partial sums to obtain the new value of E_2. In practice, the total cost of this computation is roughly proportional to the number of *changing* interacting pairs.

Recall that when the testing algorithm examines a pair of sub-chains (including the case of two copies of the same sub-chain), it first tests whether these sub-chains have not been affected by the DOF change and are contained in the same rigid sub-chain. If this is the case, the two sub-chains cannot contribute new interacting pairs, and the algorithm prunes this search path. But, for this same reason, the partial sum of energy terms contributed by the interacting pairs from these sub-chains is also unchanged. So, we would like to be able to retrieve it. To this end, we introduce another data structure, the *EnergyTree*, in which we cache the partial sums corresponding to all pairs of sub-chains that the testing algorithm may possibly examine. Figure 4 shows the EnergyTree for the ChainTree of Figure 3a.

Let α and β be any two nodes (not necessarily distinct) from the same level of the ChainTree. If they are not leaf nodes, let α_l and α_r (resp., β_l and β_r) be the left and right children of α (β). Let $E(\alpha, \beta)$ denote the partial energy sum contributed by all interacting pairs in which one atom belongs to the sub-chain corresponding to α and the other atom belongs to the sub-chain corresponding to β. If $\alpha \neq \beta$, we have:

$$E(\alpha, \beta) = E(\alpha_l, \beta_l) + E(\alpha_r, \beta_r) + E(\alpha_l, \beta_r) + E(\alpha_r, \beta_l). \quad (1)$$

Similarly, the partial energy sum $E(\alpha, \alpha)$ contributed by the interacting pairs inside the sub-chain corresponding to α can be decomposed as follows:

$$E(\alpha, \alpha) = E(\alpha_l, \alpha_l) + E(\alpha_r, \alpha_r) + E(\alpha_l, \alpha_r). \tag{2}$$

These two recursive equations yield the EnergyTree.

The EnergyTree has as many levels as the ChainTree. Its nodes at any level are all the pairs (α, β), where α and β are nodes from the same level of the ChainTree. If $\alpha \neq \beta$ and they are not leaves of the ChainTree, then the node (α, β) of the EnergyTree has four children $(\alpha_l, \beta_l), (\alpha_r, \beta_r), (\alpha_l, \beta_r)$, and (α_r, β_l). A node (α, α) has three children $(\alpha_l, \alpha_l), (\alpha_r, \alpha_r)$, and (α_l, α_r). The leaves of the EnergyTree are all pairs of leaves of the ChainTree (hence, correspond to pairs of links of the protein chain). For simplification, Figure 4 does not show the leaves of the form (α, α). Each node (α, β) of the EnergyTree holds the partial energy sum $E(\alpha, \beta)$ after the last accepted simulation step. The root holds the total sum.

At each step, the testing algorithm is called to find new interacting pairs. During this process, whenever the algorithm prunes a search path, it marks the corresponding node of the EnergyTree to indicate that the energy sum stored at this node is unaffected. The energy sums stored in the EnergyTree are updated next. This is done by performing a recursive traversal of the tree. The recursion along each path ends when it reaches a marked node or when it reaches an unmarked leaf. In the second case, the sum held by the leaf is re-computed by adding all the energy terms corresponding to the interacting pairs previously found by the testing algorithm. When the recursion unwinds, the intermediate sums are updated using Equations (1) and (2). In practice, the testing algorithm and the updating of the EnergyTree are run concurrently, rather than sequentially.

The size of the EnergyTree grows quadratically with the number n of links. For most proteins this is not a critical issue. For example, in our experiments, the memory used by the EnergyTree ranges from 0.4 MB for 1CTF ($n = 137$) to 50 MB for 1JB0 ($n = 1511$). If needed, however, memory could be saved by representing only those nodes of the EnergyTree which correspond to pairs of RSSs closer than the cutoff distance.

6 Experimental Results for MCS

6.1 Experimental Setup

We implemented ChainTree as described in Section 3. Since each step of an MCS may be rejected, we keep two copies of the ChainTree and the EnergyTree. RSS and distance computation routines were borrowed from the PQP library [27, 29]. We also implemented the grid method (henceforth called Grid) to find interacting pairs by setting the side length of the grid cubes to the cutoff distance. As we mentioned in Section 5, Grid finds all interacting pairs at each step, not just the new ones, and does not cache partial energy sums. So, it computes the new energy value by summing the terms contributed by all the interacting pairs.

Tests were run on a 400 MHz UltraSPARC-II CPU of a Sun Ultra Enterprise 5500 machine with 4.0 GB of RAM.

We performed MCS with the ChainTree and Grid on the four proteins 1CTF, 1LE2, 1HTB and 1JB0 of length 68, 144, 374 and 755 amino-acids respectively, which represent the different sizes of known proteins. The total number of atoms in the MCS was between 487 (1CTF) and 5878 (1JB0). The side-chains were included in the models, as rigid groups of atoms (no internal DOF) and no sub-hierarchies was used to represent each link with its side-chain (see Subsection 3.4). So, if two leaf RSSs are within the cutoff distance, ChainTree finds the interacting pairs from the two corresponding links by examining all pairs of atoms. The energy function we used for these tests includes a van der Waals (vdW) potential with a cutoff distance of 6Å, an electrostatic potential with a cutoff of 10Å, and a native-contact attractive quadratic-well potential with a cutoff of 12Å. Hence, the cutoff distance for both ChainTree and Grid was set to 12Å.

Each simulation run consisted of 300,000 trial steps. The number k of DOFs changed at each step was constant throughout a run. We performed runs with $k = 1, 5$ and 10. Each change was generated by picking k backbone DOFs at random and changing each DOF independently with a magnitude picked uniformly at random between 0° and 12°. Each run started with a random, partially extended conformation of the protein. Since the vdW term for a pair of atoms grows as $O(d^{-12})$ where d is the distance between the atom centers, it quickly approaches infinity as d becomes small (steric clash). When a vdW term was detected to cross a very large threshold, the energy computation was halted (in both ChainTree and Grid), and the step was rejected.

ChainTree and Grid compute the same energy values for the same protein conformations. Hence, to better compare their performance, we ran the same MCS with both of them on each protein, by starting at the same initial conformation and using the same seed of the random-number generator.

6.2 Results

The results for all the experiments are found in Table 1. Illustrations of the average time results for $k = 1$ and $k = 5$ are presented in Figures 5a and 5b respectively. As expected, ChainTree gave its best results for $k = 1$, requiring on average one quarter of the time of Grid per step for the smallest protein (1CTF) and one twelfth of the time for the largest protein (1JB0). The average number of interacting pairs for which energy terms were evaluated at each step was almost 5 times smaller with ChainTree than with Grid for 1CTF and 30 times smaller for 1JB0.

We see similar results when $k = 5$. In this case, ChainTree was only twice as fast as Grid for 1CTF and 6 times faster for 1JB0. The average number of interacting pairs for which energy terms were evaluated was twice smaller with ChainTree for 1CTF and 14 times smaller for 1JB0. When $k = 10$, the relative effectiveness of ChainTree declined further, being only 1.2 times faster than Grid for 1CTF and 4 times faster for 1JB0. The average number of interacting pairs for which energy terms were evaluated using ChainTree was 60% of the number evaluated using Grid for 1CTF and 10 times smaller for 1JB0.

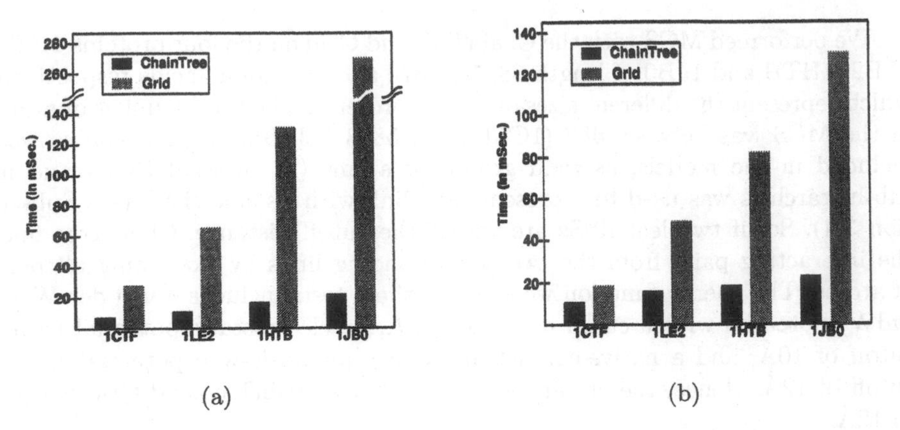

Fig. 5. Comparing the average time per MCS step of ChainTree and Grid (a) when $k = 1$ and (b) when $k = 5$.

	$k = 1$		$k = 5$		$k = 10$	
	CT	Grid	CT	Grid	CT	Grid
1CTF	7.82	27.7	8.34	18.22	12.57	15.07
1LE2	11.16	65.05	14.31	48.84	14.29	27.12
1HTB	16.72	130.9	18.2	81.86	21.75	60.33
1JB0	21.71	271.4	22.18	130.5	29.88	133.8

(a)

	$k = 1$		$k = 5$		$k = 10$	
	CT	Grid	CT	Grid	CT	Grid
1CTF	5,100	25,100	7,400	16,900	8,000	13,500
1LE2	5,100	48,500	6,000	36,600	7,700	23,400
1HTB	5,400	100,000	7,000	56,800	8,200	43,100
1JB0	5,900	200,000	7,000	95,600	10,300	102,000

(b)

Table 1. MCS results: (a) average time per simulation step (in milliseconds) and (b) number of interacting pairs for which energy terms were evaluated per step, when $k = 1$, 5 and 10. (CT stands for ChainTree.)

The larger k, the less effective our algorithm compared with Grid. When k is small, there are few new interacting pairs at each step, and ChainTree is very effective in exploiting this fact. For both ChainTree and Grid the average time per step decreases when k increases. This stems from the fact that a larger k is more likely to yield over-threshold vdW terms and so to terminate energy computation sooner.

In order to examine the full effect of reusing partial energy sums, we re-ran the simulations for the four proteins without the vdW threshold for $k = 1$ and 5. The results are presented in Tables 2a and 2b. Removing the vdW threshold does not significantly alter the behavior of the algorithms. The average time per step is of course larger, since no energy computation is cut short by a threshold crossing. The relative speed-up of ChainTree over Grid is only slightly smaller without the threshold.

6.3 Two-Pass ChainTree

In the previous MCS the percentage of steps that were rejected before energy computation completed, due to an above-threshold vdW term for 1CTF, for

	$k = 1$		$k = 5$	
	CT	Grid	CT	Grid
1CTF	12.8	37.2	29.6	37.7
1LE2	20.9	86.5	24.6	65.4
1HTB	26.6	185	51.8	173
1JB0	40.0	401	89.1	348

(a)

	$k = 1$		$k = 5$	
	ChainTree	Grid	ChainTree	Grid
1CTF	8,600	33,300	21,000	34,700
1LE2	9,900	61,900	11,400	47,500
1HTB	9,900	134,000	21,500	129,000
1JB0	12,000	280,000	30,300	248,000

(b)

	unfolded		folded	
	CT	CT+	CT	CT+
1CTF	8.34	2.6	15.74	6.2
1LE2	14.31	6.4	32.37	9.06
1HTB	18.2	9.23	68.92	11.35
1JB0	22.18	6.33	81.15	15.51

(c)

Table 2. (a) Average time (in milliseconds) per step when running an MCS without a threshold on the vdW terms. (b) Average number of interacting pairs evaluated per step for the same simulation. (c) Average running times (in milliseconds) of ChainTree and ChainTree+ per step when the simulations start at unfolded conformations and when they start at the folded conformation of the proteins. (CT stands for ChainTree.)

example, rose from 60% when $k = 1$ to 98% when $k = 10$. This observation not only motivates choosing a small k. It also suggests the following two-pass approach. In the first pass, ChainTree uses a very small cutoff distance chosen such that atom pairs closer than this cutoff yield above-threshold vdW terms. In this pass, the algorithm stops as soon as it finds an interacting pair, and then the step is rejected. In the second pass the cutoff distance is set to the largest cutoff over all energy terms and ChainTree computes the new energy value. We refer to the implementation of this two-pass approach as ChainTree+.

We compared ChainTree and ChainTree+ by running an MCS of 300,000 trial steps with $k = 5$ and measuring the average time per step. The results for the four proteins are given in Table 2c. We ran two different simulations for each protein. One that started at a partially extended conformation and another that started at the folded state of the protein. Hence, the conformations reached in the first case were less compact than in the second case. Consequently, the rejection rate due to self-collision was higher in the second case. While ChainTree+ is faster in both cases, speed-up factors are greater (as much as 5) when starting from the folded state.

7 MCS Software

We have extended our implementation of the ChainTree algorithm to include a physical, full-atomic energy function. We chose to implement the force-field *EEF1* [19]. This force field is based on the CHARMM19 potential energy function [35] with an added implicit solvent term. We chose *EEF1* because it has been shown to discriminate well between folded and misfolded structures [36]. It is well suited for ChainTree because its implicit solvent term is pairwise and thus our algorithm can compute it efficiently. We have packaged our software into a program that runs fast MCS. It can be downloaded from `http://robotics.stanford.edu/~itayl/mcs`.

This software loads an initial structure that is described in terms of its amino acid sequence and the corresponding backbone angles of each residue. It then performs a classical MCS. The user can control some parameters of the simulation (e.g. the number of angles to change, the length of the simulation, the temperature ...) by specifying them on the command line.

8 Conclusion

8.1 Summary of Contribution

This paper presents a novel algorithm based on the ChainTree and EnergyTree data structures to reduce the average step time of MCS of proteins, independent of the energy function, step generator, and acceptance criterion used by the simulator. Tests show that, when the number of simultaneous DOF changes at each step is small (as is usually the case in MCS), the new method is significantly faster than previous general methods — including the worst-case optimal grid method — especially for large proteins. This increased efficiency stems from the treatment of proteins as long kinematic chains and the hierarchical representation of their kinematics and shape. This representation — the ChainTree — allows us to exploit the fact that long sub-chains stay rigid at each step, by systematically re-using unaffected partial energy sums cached in a companion data structure — the EnergyTree. Our tests also demonstrate the advantage of using the ChainTree to detect steric clash before computing the energy function.

8.2 Other Applications

Although we have presented the application of our algorithm to classical Metropolis MCS, it can also be used to speed up other MCS methods as well as other optimization and simulation methodologies.

For example, MCS methods that use a different acceptance criterion can benefit from the same kind of speed-up as reported in Section 6, since the speed-up only derives from the faster maintenance of the energy function when relatively few DOFs are changed simultaneously, and is independent of the actual acceptance criterion. Such methods include Entropic Sampling MC [3], Parallel Hyperbolic MC [4], and Parallel-hat Tempering MC [9]. MCS methods that use Parallel Tempering [2] (also known as Replica Exchange) such as [4, 9], which require running a number of replicas in parallel, could also benefit by using a separate ChainTree and EnergyTree for each replica.

Some MCS methods use more sophisticated move sets (trial step generators). Again, our algorithm can be applied when the move sets do not change many DOFs simultaneously, which is in particular the case of the moves sets proposed in [7, 8] (biasing the random torsion changes), and in [37] and [38] (moves based on fragment replacement). More computationally intensive step generators use the internal forces (the gradient of the energy function) to bias the choice of the next conformation (e.g., Force-Biased MC [39], Smart MC [40] and MC plus minimization [12]). For such step generators, the advantage of using our algorithm is questionable, since they may change all DOFs at each step.

Some optimization approaches could also benefit from our algorithm. For instance, a popular one uses genetic algorithms with crossover and mutation operators [41, 42, 43]. The crossover operator generates a new conformation by combining two halves, each extracted from a previously created conformation. Most mutation operators also reuse long fragments from one or several previous conformations. For both types of operators, our algorithm would allow partial sums of energy terms computed in each fragment to be re-used, hence saving considerable amounts of computation.

8.3 Current and Future Work

We are currently using our algorithm to run MCS of proteins on the order of 100 residues and larger using a full-atomic model and a physical energy function (*EEF1* [19]). To the best of our knowledge this has not been attempted so far. We also intend to perform MCS of systems of several small proteins, in order to study protein misfolding, which is known to cause diseases such as Alzheimer. Each protein in the system will have its own ChainTree, which will be used to detect interaction both within each molecule and between molecules.

One possible extension of our work would be to use the ChainTree to help select "better" simulation steps. Indeed, the rejection rate in MCS becomes so high for compact conformations that simulation comes to a quasi standstill. This is a known weakness of MCS, which makes it less useful around the native conformation. This happens because almost any DOF change causes a steric clash. To select DOF changes less likely to create such clashes, one could pre-compute the radius of rotation of every link relative to each DOF. These radii and the distances between interacting atom pairs at each conformation would allow computing the range of change for each DOF such that no steric clash will occur.

Another natural extension, which exploits the hierarchical nature of the ChainTree, is to vary the resolution of the molecular representation as MCS progresses. This could be accomplished by changing on the fly the level of the ChainTree that is considered the leaf level (the bottom level). For full atomic resolution, searches in the ChainTree would continue until reaching the absolute bottom level. If a coarser resolution can be tolerated, the search could be stopped at a higher level in the hierarchy, where each node represents one or some amino acids. A different energy function could then be used for each level of resolution. This scheme could entail large savings in CPU time in regions of the conformation space where the protein structure is not very compact, while not compromising precision in other regions when it is needed.

Acknowledgements: This work was partially funded by NSF ITR grant CCR-0086013 and a Stanford BioX Initiative grant.

References

[1] Binder, K., Heerman, D.: Monte Carlo Simulation in Statistical Physics. 2nd edn. Springer Verlag, Berlin (1992)
[2] Hansmann, U.: Parallel tempering algorithm for conformational studies of biological molecules. Chemical Physics Letters **281** (1997) 140–150
[3] Lee, J.: New Monte Carlo algorithm: entropic sampling. Physical Review Letters **71** (1993) 211–214
[4] Zhang, Y., Kihara, D., Skolnick, J.: Local energy landscape flattening: Parallel hyperbolic Monte Carlo sampling of protein folding. Proteins **48** (2002) 192–201
[5] Shimada, J., Kussell, E., Shakhnovich, E.: The folding thermodynamics and kinetics of crambin using an all-atom Monte Carlo simulation. J. Mol. Bio. **308** (2001) 79–95
[6] Shimada, J., Shakhnovich, E.: The ensemble folding kinetics of protein G from an all-atom Monte Carlo simulation. Proc. Natl. Acad. Sci. **99** (2002) 11175–80
[7] Abagyan, R., Totrov, M.: Biased probability Monte Carlo conformational seraches and electrostatic calculations for peptides and proteins. J. Mol. Bio. **235** (1994) 983–1002
[8] Abagyan, R., Totrov, M.: Ab initio folding of peptides by the optimal-bias Monte Carlo minimization procedure. J. of Computational Physics **151** (1999) 402–421
[9] Zhang, Y., Skolnick, J.: Parallel-hat tempering: A Monte Carlo search scheme for the identification of low-energy structures. J. Chem. Phys. **115** (2001) 5027–32
[10] Metropolis, N., Rosenbluth, A., Rosenbluth, M., Teller, A., Teller, E.: Equation of state calculations by fast computing machines. J. Chem Phys **21** (1953) 1087–1092
[11] Hansmann, H., Okamoto, Y.: New Monte Carlo algorithms for protein folding. Current Opinion in Structural Biology **9** (1999) 177–183
[12] Li, Z., Scheraga, H.: Monte Carlo-minimization approach to the multiple-minima problem in protein folding. Proc. National Academy of Science. **84** (1987) 6611–15
[13] Grosberg, A., Khokhlov, A.: Statistical physics of macromolecules. AIP Press, New York (1994)
[14] Northrup, S., McCammon, J.: Simulation methods for protein-structure fluctuations. Biopolymers **19** (1980) 1001–1016
[15] Abagyan, R., Argos, P.: Optimal protocol and trajectory visualization for conformational searches of peptides and proteins. J. Mol. Bio. **225** (1992) 519–532
[16] Kikuchi, T.: Inter-Ca atomic potentials derived from the statistics of average interresidue distances in proteins: Application to bovine pancreatic trypsin inhibitor. J. of Comp. Chem. **17** (1996) 226–237
[17] Kussell, E., Shimada, J., Shakhnovich, E.: A structure-based method for derivation of all-atom potentials for protein folding. Proc. Natl. Acad. Sci. **99** (2002) 5343–8
[18] Gō, N., Abe, H.: Noninteracting local-structure model of folding and unfloding transition in globular proteins. Biopolymers **20** (1981) 991–1011
[19] Lazaridis, T., Karplus, M.: Effective energy function for proteins in solution. Proteins **35** (1999) 133–152
[20] Leach, A.: Molecular Modelling: Principles and Applications. Longman, Essex, England (1996)
[21] Sun, S., Thomas, P., Dill, K.: A simple protein folding algorithm using a binary code and secondary structure constraints. Protein Engineering **8** (1995) 769–778
[22] Halperin, D., Overmars, M.H.: Spheres, molecules and hidden surface removal. Comp. Geom.: Theory and App. **11** (1998) 83–102

[23] Lotan, I., Schwarzer, F., Halperin, D., Latombe, J.C.: Efficient maintenance and self-collision testing for kinematic chains. In: Symp. Comp. Geo. (2002) 43–52
[24] Thompson, S.: Use of neighbor lists in molecular dynamics. Information Quaterly, CCP5 **8** (1983) 20–28
[25] Mezei, M.: A near-neighbor algorithm for metropolis Monte Carlo simulation. Molecular Simulations **1** (1988) 169–171
[26] Brown, J., Sorkin, S., Latombe, J.C., Montgomery, K., Stephanides, M.: Algorithmic tools for real time microsurgery simulation. Med. Im. Ana. **6** (2002) 289–300
[27] Gottschalk, S., Lin, M.C., Manocha, D.: OBBTree: A hierarchical structure for rapid interference detection. Comp. Graphics **30** (1996) 171–180
[28] Klosowski, J.T., Mitchell, J.S.B., Sowizral, H., Zikan, K.: Efficient collision detection using bounding volume hierarchies of k-DOPs. IEEE Tr. on Visualization and Comp. Graphics **4** (1998) 21–36
[29] Larsen, E., Gottschalk, S., Lin, M.C., Manocha, D.: Fast distance queries with rectangular swept sphere volumes. In: IEEE Conf. on Rob. and Auto. (2000)
[30] Quinlan, S.: Efficient distance computation between non-convex objects. In: IEEE Intern. Conf. on Rob. and Auto. (1994) 3324–29
[31] van den Bergen, G.: Efficient collision detection of complex deformable models using AABB trees. J. of Graphics Tools **2** (1997) 1–13
[32] Guibas, L.J., Nguyen, A., Russel, D., Zhang, L.: Deforming necklaces. In: Symp. Comp. Geo. (2002) 33–42
[33] Creighton, T.E.: Proteins : Structures and Molecular Properties. 2nd edn. W. H. Freeman and Company, New York (1993)
[34] Hubbard, P.M.: Approximating polyhedra with spheres for time-critical collision detection. ACM Tr. on Graphics **15** (1996) 179–210
[35] Brooks, B., Bruccoleri, R., Olafson, B., States, D., Swaminathan, S., Karplus, M.: CHARMM: a program for macromolecular energy minimizationand dynamics calculations. J. of Computational Chemistry **4** (1983) 187–217
[36] Lazaridis, T., Karplus, M.: Discrimination of the native from misfolded protein models with an energy funbction including implicit solvation. J. Mol. Bio. **288** (1998) 477–487
[37] Elofsson, A., LeGrand, S., Eisenberg, D.: Local moves, an efficient method for protein folding simulations. Proteins **23** (1995) 73–82
[38] Simons, K., Kooperberg, C., Huang, E., Baker, D.: Assembly of protein tertiary structure from fragments with similar local sequences using simulated annealing and bayesian scoring functions. J. Mol. Bio. **268** (1997) 209–225
[39] Pangali, C., Rao, M., Berne, B.J.: On a novel Monte Carlo scheme for simulating water and aqueous solutions. Chemical Physics Letters **55** (1978) 413–417
[40] Kidera, A.: Smart Monte Carlo simulation of a globular protein. Int. J. of Quantum Chemistry **75** (1999) 207–214
[41] Pedersen, J., Moult, J.: Protein folding simulations with genetic algorithms and a detailed molecular description. J. Mol. Bio. **269** (1997) 240–259
[42] Sun, S.: Reduced representation model of protein structure prediction: statistical potential and genetic algorithms. Protein Science **2** (1993) 762–785
[43] Unger, R., Moult, J.: Genetic algorithm for protein folding simulations. J. Mol. Bio. **231** (1993) 75–81

Speedup LP Approach to Protein Threading via Graph Reduction

Jinbo Xu

Department of Computer Science, University of Waterloo,
Waterloo, Ontario N2L 3G1, Canada.
j3xu@math.uwaterloo.ca

Abstract. In our protein structure prediction computer program RAPTOR, we have implemented a linear programming (LP) approach to protein threading based on the template contact map graph. Our protein threading model considers pairwise contact potential rigorously. In order to further improve the computational efficiency of our LP approach, this paper proposes a graph reduction technique to reduce a template contact graph to a new one with fewer vertices and edges. Our graph reduction operation is formalized to minimize the number of variables, constraints and non-zero elements in the constraint matrix of our linear programs. These three factors are key to the computational time of solving linear programs by both the Simplex method and the Interior-Point method. This graph reduction technique does not impact the quality of protein threading (i.e., the energy function is still globally optimized). Experiments show that the more a template contact graph can be reduced, the more computational efficiency improvement can be attained and that in average, the computational efficiency of threading any long sequence to the whole template database can be improved by 30%.

1 Introduction

In our previous papers [1, 2], we have introduced a novel integer linear programming approach to protein threading. The algorithm is implemented in our protein structure prediction package RAPTOR which is ranked first among all automatic individual servers in the latest public and blind test CAFASP3 (The Third Critical Assessment of Fully Automated Structure Prediction) [3]. Protein threading makes a structure prediction through aligning a target sequence to each template protein structure in the template database and choosing the best-fit template as a basis to build a structural model for the target. The quality of sequence-template alignments is measured by an energy function. When the energy function contains pairwise contact (interaction) potential and variable gaps are allowed in the alignment, the protein threading problem is NP-hard [4]. Besides many approximation algorithms [5, 6], several exact algorithms are also proposed such as the divide-and-conquer method by Xu et al. [7], the branch-and-bound algorithm by Lathrop et al. [8], as well as the LP approach by us. A French research group also proposed a similar LP approach to protein threading

G. Benson and R. Page (Eds.): WABI 2003, LNBI 2812, pp. 374–388, 2003.

in their technical report [9] on October 2002, although our first paper [1] had already appeared on the PSB (Pacific Symposium on Biocomputing) 2003 website since late August 2002 and RAPTOR had participated CAFASP3 [3] since late May 2002. Xu et al's divide-and-conquer method works very well with the templates having simple contact (interaction) topology but is very inefficient for threading a long sequence to the templates with complex interaction topology. As reported in Lathrop et al.'s paper [10], the branch-and-bound algorithm can not converge within 2 hours for 10% of some test data sets. Our LP approach works very well for all protein threading instances.

In our paper[1, 2], we have formulated our integer linear program based on the following several basic assumptions:

- Each template is parsed as a sequence of cores with the connecting loops between the adjacent cores. Each core is the most conserved segment of an α-helix or β-sheet secondary structure among its homologs.
- When aligning a query protein sequence with a template, alignment gaps are confined to only loops. The biological justification is that cores are conserved so that the chance of insertion or deletion within them is very scarce.
- We consider only contacts (interactions) between core residues. It is generally believed that interactions involving loop residues can be ignored as their contribution to fold recognition is relatively insignificant. We say that an interaction exists between two residues if the spatial distance between their C_β atoms is within $7\AA$ and they are at least 4 positions apart in the template sequence. *We say that an interaction exists between two cores if there exists at least one residue-residue interaction between the two cores.*

Let c_i ($i = 1, 2, \ldots, M$) denote all cores of one template. We use an undirected graph $CMG = (V(CMG), E(CMG))$ to denote the (simplified) contact map graph of a protein template structure. Here, $V(CMG) = \{c_1, c_2, ..., c_M\}$, and $E(CMG) = \{(c_i, c_j)|$ there are interactions between c_i and c_j, or $|i - j| = 1\}$. We call the edges between two adjacent cores *gap edges* and the other edges *arc edges*. Figure 1 gives an example of a contact graph and an alignment between a template and a sequence.

Definition 1. *Given a template with its (simplified) contact graph CMG and a sequence $s_1 s_2 ... s_l ... s_n$, and assume $D[i]$ be the set of candidate alignment positions of the first residue of core c_i, an alignment between them is defined to be a set of ordered pairs (c_i, l_i) ($i = 1, 2, ..., M$, $l_i \in \{1, 2, ..., n\}$) such that (i) each core c_i is aligned to a sequence segment from position l_i to $l_i + len_i - 1$ where len_i is the length of core c_i and $l_i \in D[i]$; (ii) there is no overlap and crossover between the alignments of any two cores, that is, $l_{i+1} - l_i$ must be no less than len_i. An optimal alignment is one that minimizes the energy function $\sum_i E_{single}(c_i, l_i) + \sum_{(c_i,c_j)\in E(CMG)} E_{pairwise}(c_i, c_j, l_i, l_j)$. The algorithmic problem of protein threading is to find the optimal alignment between sequences and templates given an energy function.*

Where $E_{single}(c_i, l_i)$ refers to the singleton score when core c_i is aligned to position l_i. $E_{pairwise}$ refers to the pairwise score between two sequence segments

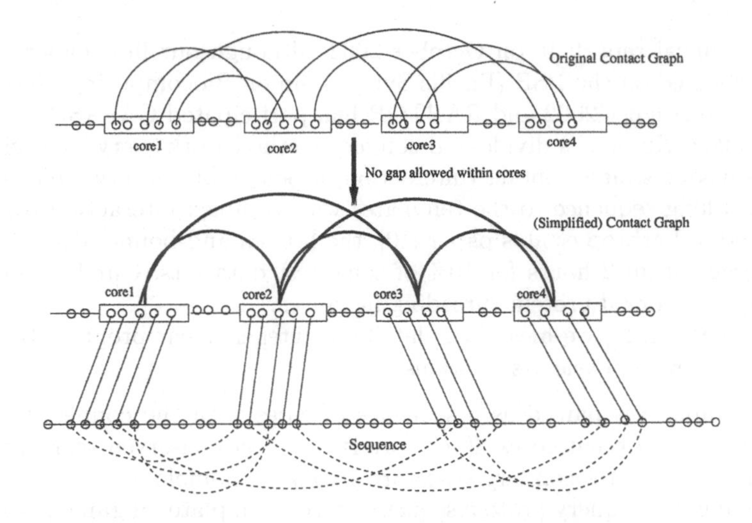

Fig. 1. Template contact graph and an example alignment between the template and the sequence. A small circle represents one residue. One solid arc in the original contact graph means that its two end residues have an interaction. A dashed arc means that if two sequence residues are aligned to two template residues having interaction to each other, then the interaction score of these two sequence residues must be counted in the energy function. The interaction score between two sequence segments is the sum of the interaction scores between two residues which are aligned by two interacted template residues.

(starting from l_i and l_j respectively) when c_i and c_j are aligned to sequence position l_i and l_j respectively [1].

We have already formulated the optimal alignment problem as a large scale integer program problem in [1, 2]. The linear program contains $O(|E|n^2)$ variables, $O(|E|n)$ constraints and $O(|E|n^2)$ non-zero elements in the constraint matrix where $E = E(CMG)$ and n is the sequence size. The CPU time of current methods solving linear programs is closely related to the number of variables, constraints and the number of non-zero elements. In this paper, we will introduce a contact graph reduction operation to minimize the number of variables and non-zero elements. The contact graph reduction technique enables us to reduce some "regular" subgraphs of a template contact graph to an edge and hence reduce the size of our linear programs. Experiments on a large scale data set show that the computational efficiency of threading a long sequence to the whole template database can be improved by 20% to 50% and that the more a template contact graph is reduced, the more computational efficiency improvement can be attained.

[1] The gap score between two adjacent cores could be treated as one special kind of pairwise score.

This paper is organized as follows. Section 1 introduces protein threading and some basic assumptions. Section 2 analyzes and compares the divide-and-conquer method and our LP approach. Section 3 describes our contact graph reduction technique which leads to our improved LP approach to protein threading problem. In this section, we also presents some theoretical results on the reducibility of a contact graph. Section 4 presents the contact graph reduction algorithm to search for the best reduction of a contact graph. Section 5 gives the experimental results on the effectiveness of our graph reduction technique. Finally, Section 6 draws a conclusion and discusses the scope of the graph reduction technique. In this paper, for the convenience of readers, we also attach an appendix to present some details on our LP formulation, the divide-and-conquer method and the proof of one lemma.

2 LP Approach Vs. Divide-and-Conquer Method

Based on the observation that many template contact graphs are "regular", PROSPECT [7] uses a divide-and-conquer algorithm to search for the globally optimal alignment between the target sequence and each template. The basic idea is to split a template into two subsegments such that each subsegment is connected to as few external vertices as possible, recursively align each subsegment to the target sequence, and finally merge the alignments of two subsegments to form an alignment for the whole segment. The key point is how to split the template to render the number of external vertices, which the alignment of each subsegment to the target sequence depends on, as small as possible. The computational time is dominated by the alignment of one subsegment with the maximum number of external vertices. The best splitting method would minimize the maximum number of external vertices of all generated subsegments. PROSPECT uses topological complexity (TC) to describe this value, which will be further described in the appendix. The value of TC can be regarded as the quantification of the "regularity" of contact graphs. The computational complexity of the divide-and-conquer method is $O(Mn^{\frac{3}{2}TC+1})$ and the memory usage is $O(Mn^{TC+1})$ where n is sequence size and M the number of template cores. Therefore, PROSPECT is only efficient for those templates with contact graph having low topological complexity. It meets difficulty in both memory and CPU usage when threading a long sequence to a template with topological complex contact graph ($TC \geq 4$). If the contact distance cutoff is 7A, there are about 25% of templates with contact graph having $TC \geq 4$. Nonetheless, these templates dominate the computational time of threading one sequence to the template database.

As reported in our paper [2], for almost all threading instances, our LP formulation produces integral solutions directly. Therefore, we can use the computational complexity of LP to approximate the computational time of our integer linear program. A linear program can be solved within $O(\hat{n}^3L)$ by the Interior-Point method where $\hat{n}$ is the number of variables and L is the input size[2].

[2] In most of our cases, the Simplex method is faster than the Interior-Point method.

Our LP formulation has $O(|E|n^2)$ variables, $O(|E|n)$ constraints and $O(|E|n^2)$ non-zero elements in the constraint matrix where $E = E(CMG)$. As such, the computational complexity of our LP is $O(|E|^3n^6L)$ if the Interior-Point method is used. Another advantage of our LP approach is that if the Simplex method is used, the memory used is roughly $O(|E|n^2)$. Therefore, our LP approach can deal very well with those templates with "non-regular" contact graph. However, $O(|E|^3n^6L)$ is a little expensive for those templates with very "regular" contact graph. A straightforward idea is to combine our LP approach and the divide-and-conquer method together. For those templates with "regular" contact graphs, we use divide-and-conquer method, whereas for those with "non-regular" contact graphs, we use LP approach instead.

Can we go further beyond the abovementioned simple idea? The answer is positive. A further observation shows that even a template with topologically complex contact graph often contains some subsegments inducing topologically simple subgraphs. We can exploit this kind of partial regularity by employing the graph reduction technique which can reduce such a subsegment into a single edge. The subsegment can be aligned to the sequence by the divide-and-conquer method because of its low topological complexity. The original contact graph is reduced into a new "non-regular" contact graph with fewer vertices and edges. Our LP approach can be used to align the template with the reduced contact graph to the sequence. If the cost of aligning the subsegments to the sequence is relatively small, then the computational efficiency of our LP approach can be improved due to the smaller number of variables, constraints and non-zero elements in the constraint matrix. Please notice that the quality of the protein threading is not impacted by the graph reduction technique because the same energy function is still globally optimized within the same search space. In the next section, we will detailedly describe how to formalize this idea as a graph reduction problem which enables us to combine the LP approach and the divide-and-conquer method together to speedup our structure prediction computer program RAPTOR.

3 Contact Graph Reduction

For the purpose of simplicity, we first give a definition as follows.

Definition 2. *Given a template contact graph CMG and a segment (c_i, c_j), a segment subgraph derived from segment (c_i, c_j) is a subgraph $SG(c_i, c_j) = (V(SG), E(SG))$, where $V(SG) = \{c_i, c_{i+1}, ..., c_j\}$, and $E(SG)$ is a subset of $E(CMG)$. Each edge in $E(SG)$ has both ends in $V(SG)$.*

Let $N(c_i)$ denote the set of vertices (cores) adjacent to core c_i in the template contact graph. If there are two cores $c_i, c_k (i < k)$ such that $\forall j(i < j < k)$, $N(c_j) \in \{c_i, c_{i+1}, ..., c_k\}$ and the segment subgraph $SG(c_i, c_k)$ has low topological complexity, then we can first align segment (c_i, c_k) to the sequence by the divide-and-conquer method within low-degree polynomial time and calculate the new pairwise scores $\hat{P}(c_i, c_k, l_i, l_k)$ for all $l_i \in D[i]$ and $l_k \in D[k]$. Given a pair

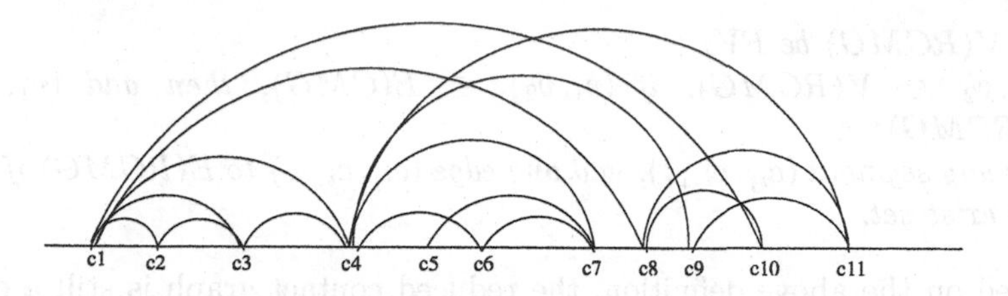

Fig. 2. A template contact graph.

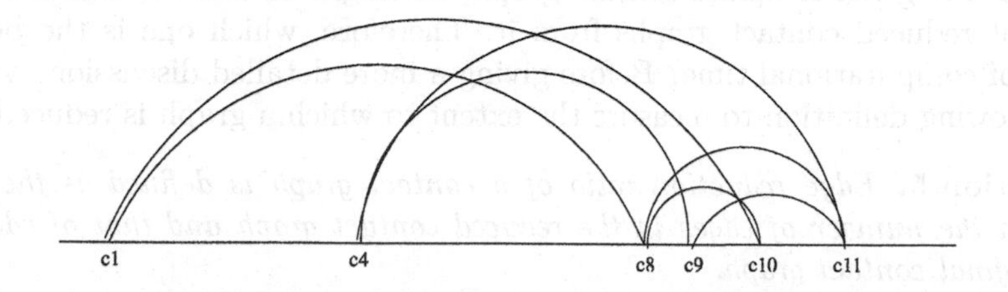

Fig. 3. A reduced template contact graph.

of l_i and l_k, the optimal alignment positions of $c_{i+1}, ..., c_{k-1}$ are completely determined by this process because they have no interaction relationship with the cores not within segment (c_i, c_k). Therefore, segment (c_i, c_k) can be reduced to a single edge and the pairwise scores related to this new edge are $\hat{P}(c_i, c_k, l_i, l_k)$. The original contact graph is reduced to a smaller graph. As shown in Figure 2 and Figure 3, segment (c_1, c_4) and segment (c_4, c_8) are reduced to two edges respectively cause the segment subgraphs $SG(c_1, c_4)$ and $SG(c_4, c_8)$ are topologically simple. Both segments can be aligned to the target sequence by the divide-and-conquer method within $O(L_{seg}n^3)$ and $O(L_{seg}n^4)$ where L_{seg} is the length of the segment.

We formally define this kind of graph reduction through Definition 3 and Definition 4.

Definition 3. *Given a template contact graph $CMG = (V, E)$, $|V| \geq 2$, a subset $RV = \{c_{i_1}, c_{i_2}, ..., c_{i_k}\}$ $(i_1 < i_2, ..., < i_k,\ k \geq 2)$ of V is called* Reduced Vertex Set *if $\forall e = (c_l, c_p) \in E$, then $c_l, c_p \in RV$ or there exists one j $(1 \leq j \leq k)$ such that $l, p \in \{i_j, i_j + 1, ..., i_{j+1}\}$.*[3]

Definition 4. *Given a template contact graph $CMG = (V(CMG), E(CMG))$, and its* Reduced Vertex Set *RV, We construct its* Reduced Contact Graph *$RCMG$ according to the following procedures:*

[3] In the context of segment (c_j, c_{j+1}), $j + 1$ should be interpreted as 1 if j equals to the size of Reduced Vertex Set, that is, $\{i_j, i_j + 1, ..., i_{j+1}\} = \{i_j, ..., M, 1, ..., i_1\}$.

1. *Let $V(RCMG)$ be RV;*
2. *$\forall v_1, v_2 \in V(RCMG)$, if $(v_1, v_2) \in E(CMG)$, then add (v_1, v_2) to $E(RCMG)$;*
3. *For any segment $(c_{i_j}, c_{i_{j+1}})$, add one edge $(c_{i_j}, c_{i_{j+1}})$ to $E(RCMG)$ if it does not exist yet.*

Based on the above definition, the reduced contact graph is still a contact graph. Therefore, our LP approach to protein threading can be used on the reduced contact graph. According to Definition 3 and Definition 4, we can see that for any given template contact graph, we might be able to induce several different reduced contact graphs from it. Therefore, which one is the best in terms of computational time? Before giving a more detailed discussion, we give the following definition to measure the extent to which a graph is reduced.

Definition 5. *Edge reduction ratio of a contact graph is defined as the ratio between the number of edges in the reduced contact graph and that of edges in the original contact graph.*

According to Definition 5, the smaller the edge reduction ratio, the more the contact graph is reduced.

Let $RV = \{c_{i_1}, c_{i_2}, ..., c_{i_p}\}$ denote the reduced vertex set of the contact graph after graph reduction operation. Let T_{seg} denote the computational complexity of threading the target sequence to all segments $(c_{i_r}, c_{i_{r+1}})$ $(r = 1, 2, .., p)$ by the divide-and-conquer method and $T_r(LP)$ that of threading the sequence to the template with the reduced contact graph by our LP approach. Ideally, the best graph reduction operation should minimize the sum of T_{seg} and $T_r(LP)$. If we only consider the template contact graph with high topological complexity (i.e., $TC \geq 4$), because LP approach is more efficient than the divide-and-conquer method on these templates, then our objective is to minimize $T_r(LP)$ such that $T_{seg} = o(T_r(LP))$. As mentioned before, our LP formulation has $O(|E|n^2)$ variables, $(O|E|n)$ constraints and $O(|E|n^2)$ non-zero elements in the constraint matrix. In order to minimize $T_r(LP)$, we should make the number of variables, constraints and non-zero elements as small as possible. Therefore, our graph reduction problem can be formalized as follows.

Given a template contact graph, find an optimal reduced vertex set to minimize its edge graph ratio such that $T_{seg} = o(T_r(LP))$.

In a summary, our improved LP approach consists of the following three major steps:

- Find the optimal reduced vertex set $RV = \{c_{i_1}, c_{i_2}, ..., c_{i_k}\}$. Its algorithm will be discussed in the next section;
- Align each subsegment $(c_{i_j}, c_{i_{j+1}})$ to the sequence by the divide-and-conquer method;
- Employ our LP approach to align the template to the target sequence based on the reduced contact graph.

An interesting question is that if T_{seg} is given as a low-degree polynomial time, is there certain limit to the edge reduction ratio for all (mathematically) possible contact graphs? Lemma 1, 3 and 4 give some theoretical results about the edge reduction ratio. Please refer to the appendix for the proof of Lemma 1.

Lemma 1. *Given a template with a contact graph having only $\frac{M}{2}$ arc edges $(i, i+\frac{M}{2})$ $(i = 1, 2, ..., \frac{M}{2})$, the computational complexity of its optimal alignment to a sequence of length n by the divide-and-conquer method is $\Omega(n^{\frac{M}{2}})$.*

Lemma 1 gives an extreme example of "non-regular" contact graph. The divide-and-conquer algorithm will encounter difficulty in aligning the template with such a sparse contact graph to the sequence. This is the reason that we say the divide-and-conquer method is good for the templates with "regular" contact graphs rather than with sparse contact graphs. Our LP approach for such a template is very efficient because the number of variables and non-zero elements in the constraint matrix is relatively small. Should every contact graph be reduced to such a case or close to one, the computational efficiency of our LP approach would be improved greatly.

Definition 6. *A template contact graph is called a nested contact graph if it contains no more than three vertices or for any two different arc edges (i_1, j_1) and (i_2, j_2), either $i_1 \leq i_2 < j_2 \leq j_1$ or $i_2 \leq i_1 < j_1 \leq j_2$.*

Lemma 2. *There exists a nested contact graph with M vertices and $M - 2$ arc edges.*

Proof. Obvious.

Lemma 3. *The optimal alignment between a sequence of length n and a template with a nested contact graph can be solved by the divide-and-conquer method within $O(Mn^3)$ time.*

Proof. For segment subgraph $SG(c_i, c_j)$, if it is a nested contact graph, then we can always find a core c_k $(i < k < j)$ such that segment subgraphs $SG(c_i, c_k)$ and $SG(c_{k+1}, c_j)$ are nested contact graphs. Divide segment (c_i, c_j) at position k. The computational complexity of merging these two subsegments is $O(n^3)$ through enumerating the alignment positions of c_i, c_k and c_j. Therefore, by mathematical induction, we can prove that the alignment of this template to any sequence can be solved within $O(Mn^3)$ time.

Based on the above lemmas, we can show that there exists a template contact graph which can be reduced greatly, but the divide-and-conquer method will encounter difficulty in threading such a template to a long sequence.

Lemma 4. *Given a ratio $\epsilon > 0$ and an integer $m > 0$, mathematically, there exists a template such that (i) the computational complexity of its alignment to a sequence of length n by the divide-and-conquer method is $\Omega(n^m)$; (ii) its contact* *graph can be reduced such that its edge reduction ratio is no more than ϵ and the computational complexity of aligning the reduced segments to the sequence by the divide-and-conquer method is $O(L_{seg}n^3)$ where L_{seg} is the length of all reduced segments.*

Proof. Construct a contact graph G_1 with $2m$ vertices and m arc edges $(i, i+m)$ $(i = 1, 2, ..., m)$. Then construct a new graph G by expanding each gap edge in G_1 into a nested contact graph with d vertices and $d-2$ edges. The computational complexity of aligning the template with contact graph G to the sequence by the divide-and-conquer method is $\Omega(n^m)$ based on Lemma 1. G can be reduced into graph G_1 by letting the reduced vertex set be $V(G_1)$. The computational complexity of aligning the reduced segments to the sequence is $O(L_{seg}n^3)$. A proper choice of d can make the edge reduction ratio smaller than ϵ.

Based on Lemma 4, we can see that mathematically, even if T_{seg} is given as $O(L_{seg}n^3)$, there still exists some contact graphs such that the graph reduction technique can make their edge reduction ratios very small and as such improve the computational efficiency of our LP approach greatly.

The left problem is to design a graph reduction algorithm to minimize the edge reduction ratio given $T_{seg} = O(L_{seg}n^k)$, $k \geq 3$, which will be described in the next section.

4 Graph Reduction Algorithm

In this section, we will discuss the algorithm for graph reduction operation. Given a template contact graph CMG and a small constant k $(k \geq 3\ldots)$, our goal is to find an optimal reduced vertex set RV such that the edge reduction ratio is minimized and $T_{seg} = O(L_{seg}n^k)$ where L_{seg} is the number of cores of all reduced segments. It can be fulfilled by Algorithm 1. Notice that for each template, we only need to search for its optimal reduced vertex set once given a T_{seg} regardless of how many sequences being threaded to it, therefore, we do not care too much about the computational efficiency of our graph reduction algorithm. Our graph reduction algorithm is straightforward but far from efficient. A simple pruning strategy can improve its efficiency greatly. Let MIN_MAX_CUT denote the algorithm to output the computational complexity of threading a target sequence to a template segment by the divide-and-conquer method. MIN_MAX_CUT is detailedly discussed in Xu et al.'s paper [7].

Algorithm 1 enumerates all feasible reduced vertex sets and then calculate the computational complexity of aligning each segment to the sequence by calling the MIN_MAX_CUT algorithm and finally to check if $T_{seg} = O(L_{seg}n^k)$ and if the edge reduction ratio size is minimal or not.

5 Experimental Results

In order to test the effectiveness of graph reduction operation, we compare the efficiency of our original LP algorithm and the algorithm proposed in this

Algorithm 1 Calculate RV subject to $T_{seg} = O(L_{seg}n^k)$, $k \geq 3$

REDUCE-k(CMG, RV, k)
1: $RV \leftarrow V(CMG)$
2: $EdgeReductionRatio \leftarrow 1$
3: **for all** feasible reduced vertex sets rv **do**
4: $T_{seg} \leftarrow 0$
5: **for all** segments $(c_{i_r}, c_{i_{r+1}})$ with $_{r+1} - r_r \geq 2$ **do**
6: $T_{seg} \leftarrow T_{seg} + MIN\_MAX\_CUT(c_{i_r}, c_{i_{r+1}})$ {Calculate the computational complexity of threading a sequence to all segments}
7: **end for**
8: **if** $T_{seg} \leq 2L_{seg}n^k$ **then**
9: calculate new edge reduction ratio ERR
10: **if** $EdgeReductionRatio > ERR$ **then**
11: $EdgeReductionRatio \leftarrow ERR$
12: $RV \leftarrow rv$
13: **end if**
14: **end if**
15: **end for**

paper by threading approximately 60 sequences to all templates in our template database. The 60 sequences are chosen from CASP5 target sequences [11], LiveBench target sequences [12] and the sequences submitted by Genesilico group (http://www.genesilico.pl) to the RAPTOR server, with size ranging from 330 to 450 residues. We do not experiment on short sequences because RAPTOR runs faster on them [4]. We run our test on Flexor at the University of Waterloo, which is a Silicon Graphics Origin 3800 system, with 40 400 MHz MIPS R12000 CPUs and 20 GB of RAM. For the sake of quantified comparison of the two algorithms, we define *CPU time ratio* as follows.

Definition 7. *For a template, CPU time ratio refers to the ratio between the total CPU time of our improved LP approach for threading all test sequences to this template and that of our original LP algorithm. For a sequence, CPU time ratio refers to the ratio between the total CPU time of our improved LP approach for threading this sequence to the whole template database* [5] *and that of our original LP algorithm.*

In our experiments, all template contact graphs are reduced to minimal edge reduction ratio such that $T_{seg} = O(L_{seg}n^4)$, which gives a best overall performance. We examine the following two relationships: (1) The relationship between the computational efficiency improvement and contact graph edge reduction ratio. As shown in Figure 4, the CPU time ratio is approximately square of the edge reduction ratio, which means several edges reduction can improve the computational efficiency of our LP approach a lot. For almost all templates with contact

[4] The average size of protein sequences is 250 $\sim$ 300 AAs.
[5] Our template database contains about 3200 templates.

graph being reduced, the computational efficiency is improved. For those templates with an irreducible contact graph, the CPU time ratio ranges from 0.95 to 1.05. Therefore, the measurement error in our experiment can be estimated to be 5%. All templates used for this figure have a contact graph with topological complexity no less than 4. (2) The relationship between the computational efficiency improvement and the target sequence size. As shown in Figure 5, the computational efficiency of threading all sequences are improved to some extent. The CPU time ratio ranges from 50% to 80%, that is, the computational efficiency is improved by 20% to 50%. The average CPU time ratio is 70% and its standard deviation is 5%, which means that graph reduction operation can improve computational efficiency by 30% in average for threading a sequence to the whole template database.

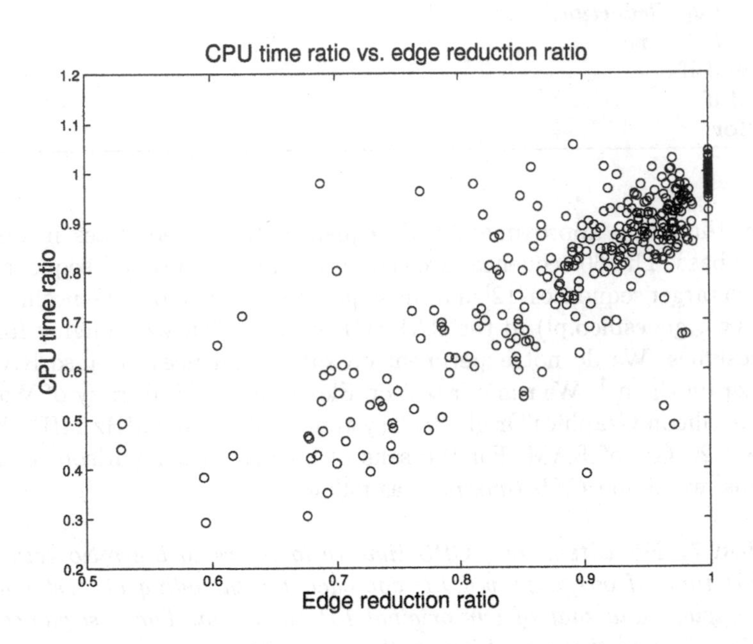

Fig. 4. The relationship between the computational efficiency improvement and the edge reduction ratio.

6 Conclusion

In this paper, we have discussed how to reduce template contact graphs to minimize the number of variables, constraints and non-zero elements contained in our linear programs formulating the protein threading problem. This technique can improve computational efficiency by about 30% for any long sequence. The computational efficiency improvement is highly related to how much the template

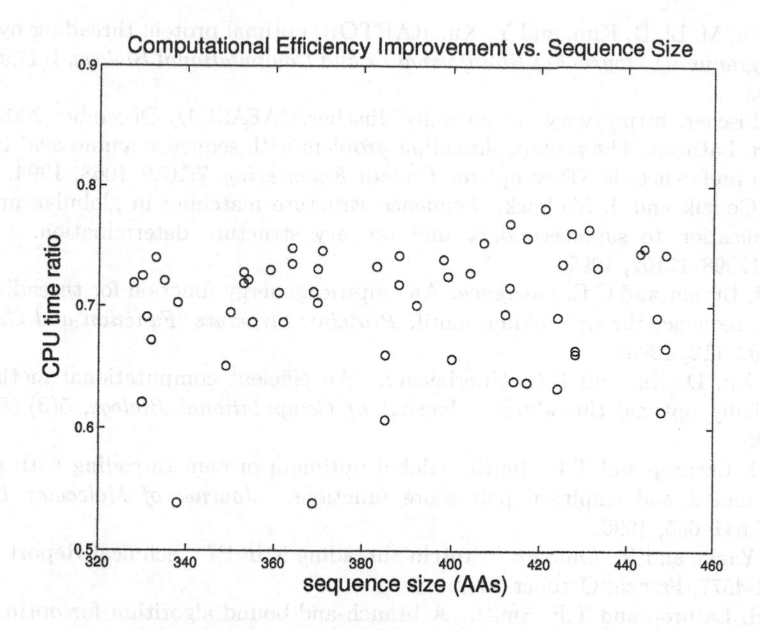

Fig. 5. The relationship between the computational efficiency improvement and the sequence size.

contact graph can be reduced. Given a template, its contact graph reduction ratio is determined by how its contacts are defined. In our threading model, two residues have a contact if and only if their spatial distance is within $7\dot{A}$. If this distance limit is increased to a bigger one which often occurs in a threading model considering distance-dependent pairwise contacts, then the number of contacts within a template will also be increased. Consequently, the edge reduction ratio of template contact graphs will be bigger and the gain of graph reduction technique will be less. Conversely, if certain biological grounds can be used to prune some meaningless contacts, then the computational efficiency of our LP approach can be improved more.

7 Acknowledgements

The author would like to thank Ming Li, Ying Xu, Dong Xu, Dongsup Kim, Guohui Lin for collaborating with me on the RAPTOR project.

References

[1] J. Xu, M. Li, G. Lin, D. Kim, and Y. Xu. Protein threading by linear programming. pages 264–275, Hawaii, USA, 2003. Biocomputing: Proceedings of the 2003 Pacific Symposium.

[2] J. Xu, M. Li, D. Kim, and Y. Xu. RAPTOR: optimal protein threading by linear programming. *Journal of Bioinformatics and Computational Biology*, 1(1):95–117, 2003.
[3] D. Fischer. http://www.cs.bgu.ac.il/~dfischer/CAFASP3/, December 2002.
[4] R.H. Lathrop. The protein threading problem with sequence amino acid interaction preferences is NP-complete. *Protein Engineering*, 7:1059–1068, 1994.
[5] A. Godzik and J. Skolnick. Sequence-structure matching in globular proteins: application to supersecondary and tertiary structure determination. *PNAS*, 89:12098–12102, 1992.
[6] S.H. Bryant and C.E. Lawrence. An empirical energy function for threading protein sequence through folding motif. *Proteins: Structure, Function and Genetics*, 16:92–112, 1993.
[7] Y. Xu, D. Xu, and E.C. Uberbacher. An efficient computational method for globally optimal threadings. *Journal of Computational Biology*, 5(3):597–614, 1998.
[8] R.H. Lathrop and T.F. Smith. Global optimum protein threading with gapped alignment and empirical pair score functions. *Journal of Molecular Biology*, 255:641–665, 1996.
[9] N. Yanev and R. Andonov. Protein threading is in P? Technical Report INRIA RR-4577, French, October 2002.
[10] R.H. Lathrop and T.F. Smith. A branch-and-bound algorithm for optimal protein threading with pairwise (contact potential) amino acid interactions. IEEE Computer Society Press, 1994.
[11] CASP5. http://predictioncenter.llnl.gov/casp5/Casp5.html, December 2002.
[12] L. Rychlewski. http://www.bioinfo.pl/livebench/6, 2002.

8 Appendix

8.1 IP Formulation to Protein Threading

This subsection gives a brief description of our IP formulation to the protein threading problem based on the template contact graph. Let $x_{i,l}$ be a 0-1 variable indicating that core c_i is aligned to the lth residue of the sequence. For any $(c_{i_1}, c_{i_2}) \in E(CMG)$, let $y_{(i_1,l_1),(i_2,l_2)}$ indicate the pairwise interactions between two x variables x_{i_1,l_1} and x_{i_2,l_2}. $y_{(i_1,l_1),(i_2,l_2)} = 1$ if and only if $x_{i_1,l_1} = 1$ and $x_{i_2,l_2} = 1$. Let $R[i,j,l]$ denote the set of feasible alignment positions of core j given that core i is aligned to sequence position l. The constraint set of our IP formulation is as follows.

$$\sum_{l \in D[i]} x_{i,l} = 1, i = 1, 2, \ldots, M; \tag{1}$$

$$\sum_{k \in R[i,j,l]} y_{(i,l)(j,k)} = x_{i,l}, (c_i, c_j) \in E(CMG); \tag{2}$$

$$\sum_{l \in R[j,i,k]} y_{(i,l)(j,k)} = x_{j,k}, (c_i, c_j) \in E(CMG); \tag{3}$$

$$x, y \in \{0, 1\}. \tag{4}$$

Constraint 1 says that one core can be aligned to a unique sequence position. Constraints 2 and 3 imply that one y variable equals to 1 if and only if its two generating x variables are 1, i.e., if two cores have interactions and are aligned to two sequence positions respectively, then the interaction energy between the sequence residues aligned by these two cores must be calculated.

8.2 Divide-and-Conquer Method

In this subsection, we outline the divide-and-conquer algorithm to align one segment to a target sequence. For a more detailed exposition, please refer to Xu et al's paper [7]. Our description is slightly different. Given a template contact graph CMG, for any two cores c_i and c_j $(i < j)$, let $In(c_i, c_j)$ denote $\{c_i, c_{i+2}, ..., c_j\}$ and $N(c_i, c_j) = V(CMG) - In(c_i, c_j)$. Let $E(c_i, c_j)$ denote the set of all edges with both ends in set $In(c_i, c_j)$ and $XE(c_i, c_j)$ the set of all edges with one end in $In(c_i, c_j)$ and the other end in $N(c_i, c_j)$.

Definition 8. *Given a segment (c_i, c_j), a vertex set cover of edge set $XE(c_i, c_j)$ is called an anchor set of this segment.*

Given a segment (c_i, c_j), we split it into two subsegments (c_i, c_k) and (c_{k+1}, c_j) at position k. Let $Anchor(c_i, c_j)$ denote one anchor set of segment (c_i, c_j). Let $CUT(k) = \{(c_{i_1}, c_{i_2}) | i \le i_1 \le k \le i_2 \le j\}$ and $CutCover(k)$ the set of vertices which covers all edges in $CUT(k)$. The cut at position k also splits $Anchor(c_i, c_j)$ into two disjoint subsets $X^1(k)$ and $X^2(k)$, where $X^1(k)$ contains the cores in $\{c_i, c_{i+1}, ..., c_k\}$ or adjacent to cores in $\{c_i, c_{i+1}, ..., c_k\}$ and $X^2(k)$ contains the cores in $\{c_{k+1}, ..., c_j\}$ or adjacent to cores in $\{c_{k+1}, ..., c_j\}$. Obviously, $X^1(k) \cup CutCover(k)$ and $X^2(k) \cup CutCover(k)$ are the anchor sets of segment (c_i, c_k) and segment (c_{k+1}, c_j) respectively. Let $SCORE((c_i, c_j), Anchor(c_i, c_j))$ denote the optimal alignment score of aligning the sequence to segment (c_i, c_j) given the alignment positions of all cores in $Anchor(c_i, c_j)$. Let $D(C)$ denote all feasible combinations of alignment positions of vertices in set C, where C is a subset of $\{c_1, c_2, ..., c_M\}$. If we divide the segment (c_i, c_j) at position k, then we have the following equation.

$$\begin{aligned} & SCORE((c_i, c_j), Anchor(c_i, c_j)) \\ = & \min_{D(CutCover(k))} \{SCORE((c_i, c_k), X^1(k) \cup CutCover(k)) \\ & + SCORE((c_{k+1}, c_j), X^2(k) \cup CutCover(k))\} \end{aligned} \tag{5}$$

Then the optimal alignment between a template and a sequence can be calculated by calling $SCORE((c_1, c_M), \phi)$. Now let's consider the computational complexity of the recursive algorithm based on Eq. 5. Let $T((c_i, c_j), Anchor(c_i, c_j))$ denote the computational time of $SCORE((c_i, c_j), Anchor(c_i, c_j))$, then we have

$$\begin{aligned} T((c_i, c_j), Anchor(c_i, c_j)) = & \, T((c_i, c_k), X^1(k) \cup CutCover(k)) \\ & + T((c_{k+1}, c_j), X^2(k) \cup CutCover(k)) \\ & + O(n^{(|CutCover(k)| + |Anchor(c_i, c_j)|)}) \end{aligned} \tag{6}$$

Where, the third item of the right hand side of Eq. 6 is the computational time needed for merging the two subsegments. For a given segment and its anchor set in order to minimize its alignment computational time, we need to choose a proper cut position k and a proper cut cover $CutCover(k)$ to minimize the maximum of three items in the right hand side of Eq. 6 during the recursive process. In their paper [7], Xu et al. has proposed an algorithm MIN_MAX_CUT to search for the optimal partition scheme of the whole template and the optimal set cover of cut edges at each cut position. The topological complexity of a template contact graph TC is defined as the maximum among all $X^1(k) \cup CutCover(k))$ and $X^2(k) \cup CutCover(k))$. MIN_MAX_CUT is designed to minimize TC.

8.3 Proof of Lemma 1

Proof. For any parition scheme adopted by the divide-and-conquer method, we can always find a segment (c_i, c_j) of length no less than $\frac{M}{2}$ such that its alignment to the sequence is constructed from the alignments of two subsegments (c_i, c_k) and (c_{k+1}, c_j), each of which has length less than $\frac{M}{2}$. There are totally $\frac{M}{2}$ arc edges with at least one end in segment (c_i, c_j). These arc edges can be grouped into two disjoint subsets A and B. A is the set of arc edges with only one end in this segment and B the set of arc edges with both ends in this segment. Obviously, any arc edge in B has one end in segment (c_i, c_k) and the other end in segment (c_{k+1}, c_j). Therefore, in merging the alignments of these two subsegments, the divide-and-conquer method has to enumerate (i) the alignment positions of at least one end of each arc edge in B; and (ii) the alignment positions of at least one end of each arc edge in A. Because any two arcs in set $A \cup B$ are not incident to the same vertex, the algorithm has to enumerate totally $O(n^{|A|+|B|})$ possible combinations of alignment positions in order to merge these two subsegments to segment (c_i, c_j). Therefore, the computational complexity of aligning this template to the sequence of length n is $\Omega(n^{\frac{M}{2}})$.

Homology Modeling of Proteins Using Multiple Models and Consensus Sequence Alignment

Jahnavi C. Prasad[1], Michael Silberstein[1], Carlos J. Camacho[2], and Sandor Vajda[2]

[1] Program in Bioinformatics, Boston University, Boston MA 02215
[2] Department of Biomedical Engineering, Boston University, Boston MA 02215
vajda@bu.edu

Abstract. Homology modeling predicts the three-dimensional structure of a protein (target), given its sequence, on the basis of sequence similarity to a protein of known structure (template). The main factor determining the accuracy of the model is the alignment of template and target sequences. Two methods are described to improve the reliability of this step. First, multiple alignment are produced, converted into models, and then the structure with the lowest free energy is chosen. The method performs remarkably well for targets for which a good template is available. In the second approach, the alignment is based on the consensus of five popular methods. It provides reliable prediction of the structurally conserved framework region, but the alignment length is reduced. A homology modeling tool combining the two methods is in preparation.

1 Introduction

Over the last decade the exponential growth of sequenced genes has prompted the development of several methods for the prediction of protein structures. The most successful prediction method to date is homology modeling (also known as comparative modeling), which predicts the three-dimensional structure of a protein (target), given its sequence, on the basis of sequence similarity to proteins of known structure (templates) [1,2]. The approach is based on the structural conservations of the framework regions between the members of a protein family. Since the 3D structures are more conserved in evolution than sequence, even the best sequence alignment methods frequently fail to correctly identify the regions that possess the desired level of structural similarity, and the quality of alignment remains the single most important factor determining the accuracy of the 3D model [3]. Therefore, it is of substantial interest to develop methods that can provide highly accurate sequence alignment, and possibly identify regions were the similarity is too low for building a meaningful model on the basis of the template structure [4].

In this paper we describe two approaches to reduce the uncertainty of the alignment. The first approach to dealing with this uncertainty is based on the use of multiple models [5]. A number of pairwise alignments (using simple dynamic programming with variation of parameters), is generated. This is followed

G. Benson and R. Page (Eds.): WABI 2003, LNBI 2812, pp. 389–401, 2003.

by energy based discrimination of the generated models [6,7]. The approach was tested at the CASP4 (Comparative Assessment of Structure Prediction) competition in 2000 (see http://predictioncenter.llnl.gov/casp4/). As we will show, in view of its relative simplicity the approach provides surprisingly good result for the easy targets, i.e., for targets with a good template available, but the dynamic programming is too rudimentary to obtain any good alignment for difficult targets. Thus, the free energy ranking algorithm had too choose one among such inferior models, and the method was unable to compete with approaches based on more sophisticated sequence alignment algorithms which employed evolutionary relationships between all homologues, and accounted for the known structure of the template.

The second approach involves a consensus alignment algorithm for the prediction of the framework regions that are structurally conserved between two proteins [8]. The target and template sequences are aligned by the five best algorithms currently available, and each position is assigned a confidence level (consensus strength) based on the consensus of the five methods. The regions reliable for homology modeling are predicted by applying criteria involving secondary structure and solvent exposure profile of the template, predicted secondary structure of the target, consensus confidence level, template domain boundaries and structural continuity of the predicted region with other predicted regions. The methodology was developed based on a diverse set of 79 pairs of homologues with an average sequence identity of 18.5%, and was validated using a different set of 48 target-template pairs. On the average, our method predicts structures that deviate from the native structures by about 2.5 Å, and the predictions extend to almost 80% of the regions that are structurally aligned in the FSSP database [9]. The approach was tested at the as an automatic server, participating in the CAFASP3 competition of such servers, described on the webpage http://www.cs.bgu.ac.il/~dfischer/CAFASP3/.

2 Methods

2.1 Multiple Model Approach to Homology Modeling

The basic idea of the method is to generate a large number of alignments, construct a homology model for each, and rank the models according to their free energies. The current implementation of the procedure starts with traditional template selection using BLAST and PSI-BLAST [10]. The Domain Profile Analysis developed in Temple Smith's lab (http://bmerc-www.bu.edu/bioinformatics/profile_request.html) has also been consulted. One or (infrequently) several proteins have been selected as templates for the comparative modeling. In the second step of the algorithm, we generate multiple alignments between target and template sequences by varying the alignment parameters (gap-opening, gap-extension, and scoring matrix) for producing semi-global alignments by standard dynamic programming. The blosum62 and gonnet matrices were used with gap opening penalty values 5, 6, 7, 8, 9, 10, 12, 14, 17, 20, 25, and gap extension penalty values 0.1, 0.2, 0.3, 0.5, 0.75, 1.0, 1.25, 1.6, 2.0,

2.5, 3, 4, 5, 7, 10. We produced only one alignment for each set of parameters using a single trace-back path in the dynamic programming matrix, thus resulting in 330 alignments for each template-target pair. Any alignment was deleted if it was a duplicate, or less than 75% of the target residues were aligned to the template, generally resulting in 80 to 150 retained alignments.

In the third step, all alignments are used for model construction via the MODELER program developed by Sali and co-workers [2,11]. The resulting models were minimized for 200 steps using the Charmm potential [12], and ranked by using an empirical free energy function [6,7]. The function combines molecular mechanics with empirical solvation/entropic terms to approximate the free energy G of the system consisting of the protein and the solvent, the latter averaged over its own degrees of freedom. The free energy is given by $G = E_{conf} + G_{solv}$. The conformational energy E_{conf} is calculated by Version 19 of the Charmm potential, $E_{conf} = E_{elec} + E_{int}$, where the internal (bonded) energy, E_{int}, is the sum of bond stretching, angle bending, torsional, and improper terms, $E_{int} = E_{bond} + E_{angle} + E_{dihedral} + E_{improper}$[12]. The electrostatic energy, E_{elec}, is calculated using neutral side chains and the distance-dependent dielectric $\varepsilon g = 4r$. G_{solv} is the solvation free energy, obtained by the atomic solvation parameter model of Eisenberg and McLachlan [13].

Notice that the function does not include the van der Waals energy term [6,7]. This approximation is based on the concept of van der Waals cancellation which assumes that the solute-solute and solute-solvent interfaces are equally well packed, and hence the van der Waals contacts lost between solvent and solute are balanced by new solute-solute contacts formed upon protein folding. This cancellation is promoted by a procedure called van der Waals normalization, prior to the free energy calculations. Van der Waals normalization impliesthat all conformations are minimized for a moderate number of steps, the structure with the lowest van der Waals energy is selected, and all other structures are further minimized to attain the same van der Waals energy value. The van der Waals cancellation implies that we can remove both the solute-solvent and the solute-solute van der Waals terms from the free energy function.

2.2 Consensus Alignment

In a benchmarking analysis [8], we have tested ten widely used methods and selected five of them in a hierarchical manner so that we cover a broad range of alignments. The five methods are as follows:

(1) BLAST-Pairwise: Target sequence is blasted against the template sequence to get an alignment.

(2) T99-BLAST: A PSI-BLAST [10] alignment of the target and template hits is supplied as the 'seed alignment' to Target99 script. It is tuned up, an HMM is built using this alignment, and target and template are aligned to it [14].

(3) HMMER-BLAST: The PSI-BLAST generated alignment of target and template hits is used to build a model. Target and template sequences are then aligned to it [15].

(4) T99-HSSP: Family alignment around the template (downloaded from the FSSP database) is used to build the initial model. The combined hits of target and template sequences are then aligned to this model and "tuned up" using the target99 script. A model is then constructed from this alignment and target and template are aligned to it.

(5) HMMER-HSSP: A model is built using the template family alignment. The combined hits of target and template are aligned to it to get a multiple sequence alignment. Another model is then constructed from this alignment and used to get a target-template pairwise alignment.

For our training set targets [8], at least one of the above methods was able to produce an alignment resulting in a low RMSD model, but it was not possible to predict which of the methods would perform well for a particular problem. Therefore we developed a selection procedure for determining whether or not two aligned residues can be included for accurate homology modeling. This process also invokes structural considerations, e.g., secondary structure and solvent exposure information, on the template, and to a lesser degree secondary structure prediction of the target. A flow chart of the overall algorithm is depicted in Fig. 1.

As shown in Fig. 2, the consensus strength (CS) is a measure of the agreement between the five alignment methods calculated for all target-template residue pairs. If all three methods T99-BLAST, HMMER-BLAST and BLAST-PW align target residue X_{tar} to template residue X_{tem} then CS = 9. If only two of the above three methods concur in aligning X_{tar} to X_{tem}, then CS = 6 for the $X_{tar}X_{tem}$ pair. If any three out of the five methods concur then CS = 7, and concurrence of only any 2 out of 5 means CS = 5. Consensus strengths between 4 and 0 are assigned to the residue pairs aligned by only one method, the methods respectively being T99-BLAST, HMMER-BLAST, BLAST-PW, T99-HSSP and HMMER-HSSP. Obviously, the methods will differ in certain regions. Consensus among certain alignment methods for a certain region may be incompatible with consensus among other methods for a different region. In such a case, the region with higher consensus strength receives priority.

Since consensus strength does not eliminate all the regions of potential structural dissimilarity, the following selection method is applied. If CS is 9 and template residue X_{tem} is buried, $X_{tar}X_{tem}$ pair is selected. If there are no pairs with a CS of 9, then pairs with a CS of 7 that are buried are selected. This forms the core of the selection. Subsequently the selected regions are extended towards the N and C termini as long as neighboring residues have CS of 7, or until a misaligned GLY residue occurs. Moreover, alignment regions where the template has long helices and sheets are selected subject to their CS, solvent exposure and percentage match of the predicted secondary structure of the target (using JNET of Cuff and Barton [16]) with the actual secondary structure of the corresponding template region. Other structural criteria such as single beta-sheet pairing, taut regions in template with limited potential for conformational variation are also used for selection. Regions corresponding to potentially loose termini, and

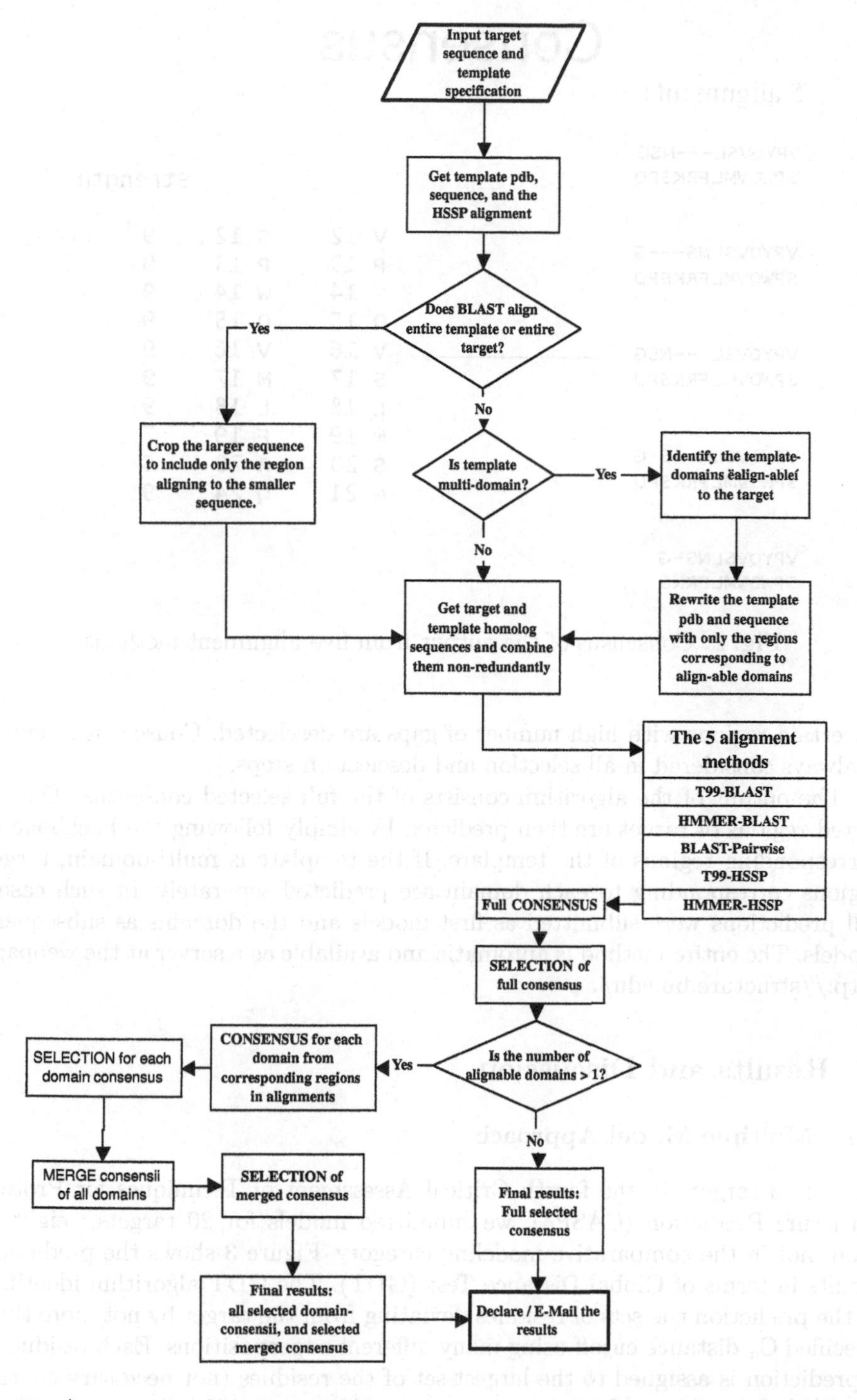

Fig. 1. A summary flow chart of the method, as implemented in the Consensus server

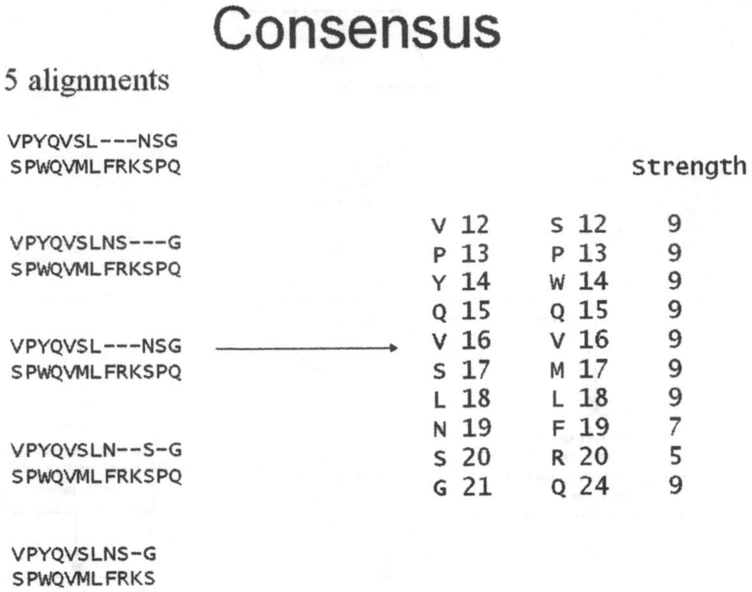

Consensus

5 alignments

```
VPYQVSL---NSG
SPWQVMLFRKSPQ                                            Strength

                              V 12      S 12       9
VPYQVSLNS---G                 P 13      P 13       9
SPWQVMLFRKSPQ                 Y 14      W 14       9
                              Q 15      Q 15       9
VPYQVSL---NSG   ---------->   V 16      V 16       9
SPWQVMLFRKSPQ                 S 17      M 17       9
                              L 18      L 18       9
                              N 19      F 19       7
VPYQVSLN--S-G                 S 20      R 20       5
SPWQVMLFRKSPQ                 G 21      Q 24       9

VPYQVSLNS-G
SPWQVMLFRKS
```

Fig. 2. Consensus of the output from five alignment methods

uncertain regions with high number of gaps are deselected. Consensus strength is always considered in all selection and deselection steps.

The output of the algorithm consists of the full selected consensus. The selected regions of target are then predicted by simply following the backbone of corresponding regions of the template. If the template is multi-domain, target regions corresponding to each domain are predicted separately. In such cases, full predictions were submitted as first models and the domains as subsequent models. The entire method is automatic and available as a server at the webpage http://structure.bu.edu/.

3 Results and Discussion

3.1 Multiple Model Approach

Out of 43 targets in the fourth Critical Assessment of Techniques for Protein Structure Prediction (CASP4), we submitted models for 20 targets, eight of them not in the comparative modeling category. Figure 3 shows the prediction results in terms of Global Distance Test (GDT). The GDT algorithm identifies in the prediction the sets of residues deviating from the target by not more than specified C_a distance cutoff using many different superpositions. Each residue in a prediction is assigned to the largest set of the residues (not necessary continuous), deviating from the target by no more than a specified distance cutoff on the x axis of the plot. The figures show submissions by all groups, participating

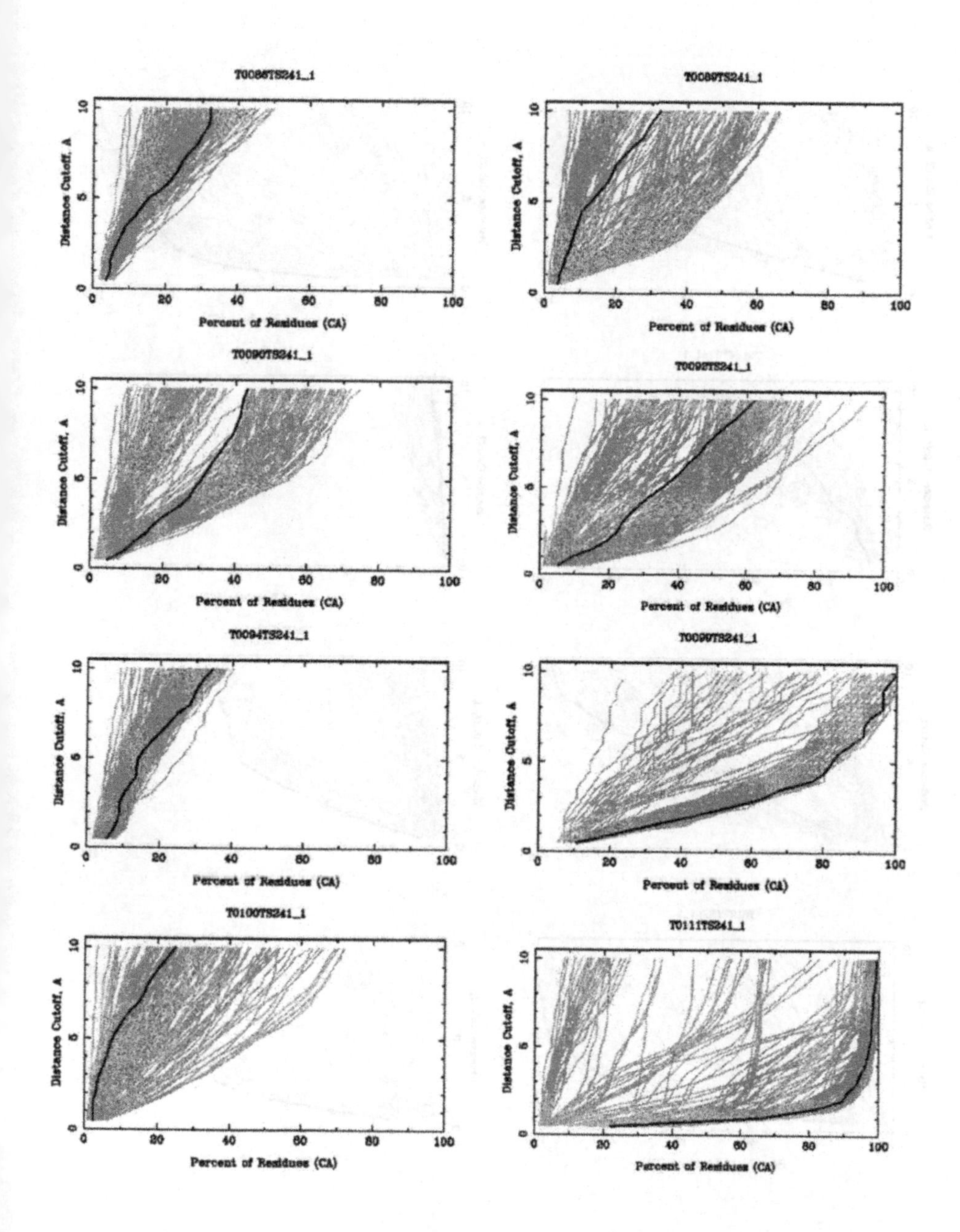
T0086TS241_1
Distance Cutoff, A
Percent of Residues (CA)
T0089TS241_1
Distance Cutoff, A
Percent of Residues (CA)
T0090TS241_1
Distance Cutoff, A
Percent of Residues (CA)
T0092TS241_1
Distance Cutoff, A
Percent of Residues (CA)
T0094TS241_1
Distance Cutoff, A
Percent of Residues (CA)
T0099TS241_1
Distance Cutoff, A
Percent of Residues (CA)
T0100TS241_1
Distance Cutoff, A
Percent of Residues (CA)
T0111TS241_1
Distance Cutoff, A
Percent of Residues (CA)

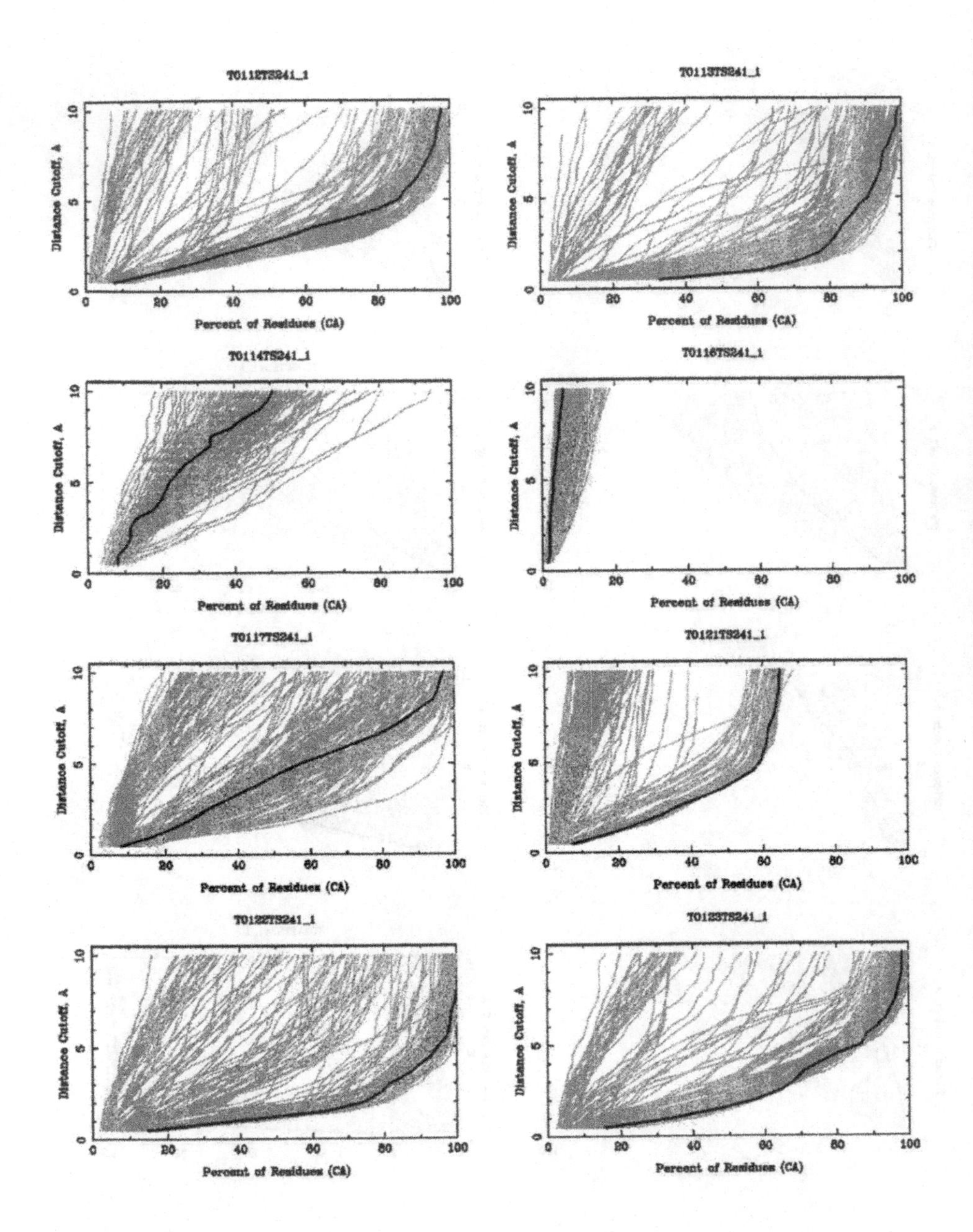
T0112TS241_1
T0113TS241_1
T0114TS241_1
T0116TS241_1
T0117TS241_1
T0121TS241_1
T0122TS241_1
T0123TS241_1
Distance Cutoff, A
Percent of Residues (CA)
0
5
10
0
20
40
60
80
100

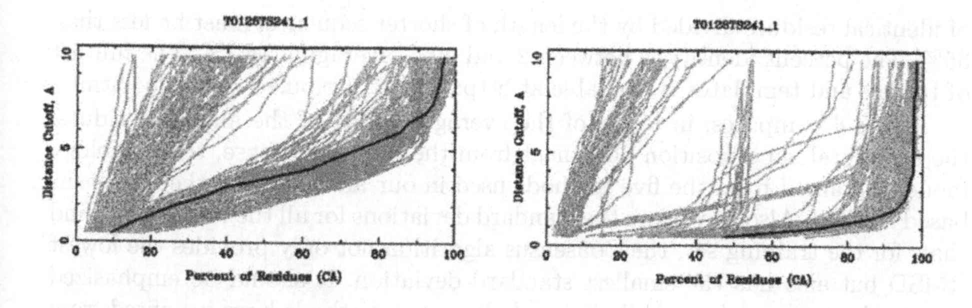

Fig. 3. Results of homology modelling at CASP4 in terms of the Global Distance Test (GDT). The GDT algorithm identifies in the prediction the set of residues deviating from the target by not more than a specified distance cutoff. The figures show the percent of such residues vs. the distance cutoff. All predictions for each target are shown, our prediction is indicated by the dark blue curve. For Target 114 we also submitted a second model, shown in light blue.

in CASP4, our submission being shown in dark blue. For target 114 there were two submissions from us, shown in dark and light blue. Results for targets 88, 93 and 119 have not been published, and hence are not shown here.

According to the above Figure 3, the multiple model approach provides good result for the relatively easy targets, i.e., for the targets were at least one groups predicted over 80 % of the residues below 5 Å distance cutoff (targets 99, 111, 112, 113, 117, 122, 123, 125, and 128). Indeed, with the exception of targets 117 and 125, for these easy cases our predictions are among the bests, but they are above average even for targets 117 and 125. Our prediction is also very good for target 121, a more difficult case. In addition, we had one of the largest prediction lengths for these targets. However, for most of the really difficult targets for which no group was able to predict at least 80% of residues below 5 Å distance cutoff (targets 86, 89, 90, 92, 94, 100, 114, 116, and 117) our method yields average or below average results. The main reason is that for these targets any available template had low sequence similarity, and the simple alignment by dynamic programming produced poor results for any of the parameters. Thus, the free energy ranking algorithm had too choose one among inferior models, and the method was unable to compete with approaches based on more sophisticated sequence alignment algorithms, utilizing the evolutionary relationships between all available homologues, and accounting for the known structure of the template.

3.2 Homology Modeling Using Consensus Alignment

For the validation of the method, 48 target-template pairs were selected from the FSSP database. The selection was governed by following criteria: (a) each target belongs to a different family in FSSP; (b) the length of the structural alignment must be greater than 100 residues; (c) the percent identity, defined as the number of identical residues divided by the length of shorter sequence, must be less than 35%. The percent identity is between 2 and 29%, averaging 16.8%. The full list of targets and templates is available at http://structure.bu.edu/consdoc.html.

Figure 4 compares, in terms of the average RMSD of the aligned residues, the structural superposition alignment from the DALI database, the homology models obtained from the five methods used in our analysis, and the consensus based method. Also shown are the standard deviations for all the models. We find that, for the training set, the Consensus algorithm not only provides the lowest RMSD but also has the smallest standard deviation. It should be emphasized that in these comparison the individual alignment methods have benefited from an automatic splitting and cropping of domains which do not align with the target sequence [8], otherwise the average RMSD for, say, T99-BLAST would be higher than 6 Å. The main result to consider here is that the RMSD has been brought down to about 2.2 Å (lower than the average DALI RMSD) while keeping the alignment length to about 75% of the DALI alignment.

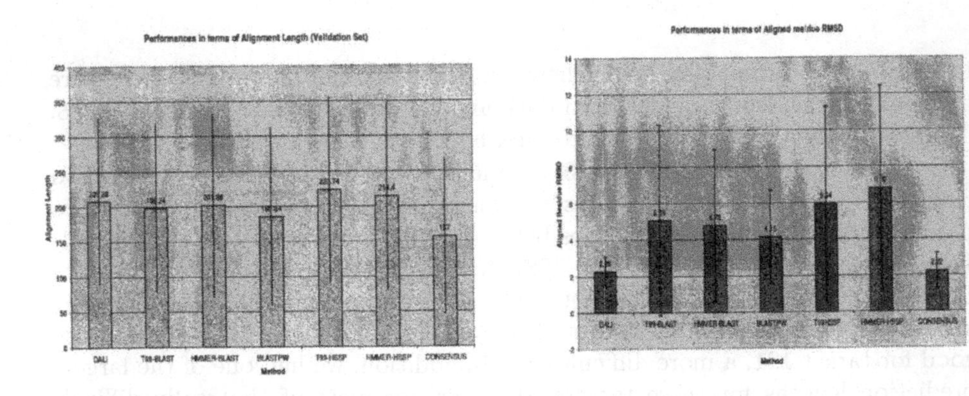

Fig. 4. Comparison of the models generated from five alignment methods and their selected consensus with respect to DAL1 in terms of RMSD and the number of aligned residues. Bars indicate one standard deviation from the average.

Figure 5 compares the CASP5 predictions we have obtained using the automatic consensus server (server #98) to the best result for each target, as well as to the output of server #45, that was deemed to produce the best overall results in the homology modeling competition. Since we restrict consideration to the highly reliable regions of the proteins, the direct comparison is of somewhat limited value, because the reliable regions, selected by the consensus method, constitute about 60 % of the total length, thus substantially smaller than for the competing methods. However, the average RMSD on the reliable regions is 2.65 Å, much lower than for the other methods. Thus, homology modeling based on consensus alignment is a reasonable first step, provided the alignment can be extended to the less reliable regions of the target (see below).

CAFASP COMPARISON		CONSENSUS SERVER #98			Best case		PMODEL3 SERVER #45	
Target	Type	NT	NP	CA	NP	CA	NP	CA
133	CM	293	148	3.26	263	5.01	271	7.26
137	CM	133	105	1.09	132	0.96	130	0.96
140_1	CM	87	24	2.23	40	2.74	71	7.87
141	CM	187	55	2.96	130	4.95	162	8.56
142	CM	280	212	3.33	280	3.47	280	3.68
143_1	CM	121	99	3.68	100	1.89	120	5.79
143_2	CM	95	86	3.18	95	1.61	95	3.46
149_1	CM	201	49	11.11	191	7.82	182	8.79
150	CM	96	85	2.14	94	1.86	94	1.86
151	CM	106	74	2.07	99	2.8	106	4.75
152	CM	198	68	2.47	166	5.24	166	5.24
153	CM	134	94	1.19	129	1.35	134	4.95
154_2	CM	103	84	1.78	100	2.29	100	2.72
155	CM	117	103	0.79	116	0.8	117	0.89
160	CM	125	84	1.45	118	2.01	124	2.49
165	CM	318	123	1.94	196	3.74	278	6.31
167	CM	180	131	1.47	168	2.84	180	7.33
169	CM	156	96	2.78	146	3.85	128	3.75
172_1	CM	192	34	4.2	144	3.56	97	4.56
176	CM	100	68	4.51	81	4.33	99	5.83
177_1	CM	57	55	1.4	56	1.24	57	1.92
177_2	CM	88	86	1.57	87	1.29	88	1.54
177_3	CM	75	68	1.79	70	1.78	75	2.81
178	CM	219	170	1.48	219	1.68	216	2.8
179_1	CM	56	50	0.8	54	0.81	56	1.12
179_2	CM	218	211	3.06	208	1.8	218	3.15
182	CM	249	229	1.01	247	1.27	249	1.39
183	CM	247	185	1.31	212	2.22	218	2.46
184_2	CM	72	46	5.24	72	2.37	12	3.26
185_1	CM	101	47	2.22	97	2.66	101	2.94
185_2	CM	197	134	2	197	3.92	197	6.61
185_3	CM	130	23	1.97	130	4.21	130	5.29
186_1	CM	77	35	0.77	72	2.96	77	3.85
186_2	CM	250	36	4.69	250	9.25	250	13.38
188	CM	107	68	1.62	106	2.18	107	2.26
189	CM	319	70	3.03	318	4.29	313	4.73
190	CM	111	97	2.09	105	1.52	107	1.52
191_2	CM	143	95	3.88	143	4.87	136	5.58
192	CM	170	37	1.33	135	2.06	148	3.6
195	CM	290	130	2.8	240	3.55	232	5.67
130	CM/FR	100	36	2.33	47	2.84	77	7.54
168_2	CM/FR	141	27	8.86	140	12.9	141	16.76
193_2	CM/FR	130	26	1.52	66	2.15	95	3.97
148_2	FRA	91	33	10.63	71	5.89	91	16.28
156	FRH	156	47	18.52	90	6.86	104	13.54
181	NF	111	29	7.09	37	5.68	71	12.45
146_3	NF/FR	56	10	3.65	52	10.4	56	11.71
146_4	NF/FR	47	15	4.29	47	7.93	47	9.19
Mean length		151	81.6		132.4		137.6	
RMSD per al. res.				2.65		3.65		5.41

Fig. 5. CAFASP3 results of the consensus server, http://structure.bu.edu. NT – total number of nucleotides, NP – number of aligned residues, CA – C_αRMSD from the x-ray structure of the target.

4 Conclusions

The multiple model approach, using a simple dynamic programming alignment with variation of the parameters, yields very good models for the relatively easy homology modeling targets, but its performance deteriorates if good template is not available. The consensus method keeps the RMSD values low (2.65 Å), but models are constructed only for 60 % of all residues. We are in the process of developing a homology modeling procedure that will integrate the two methods, and will perform the following steps:

1. The consensus alignment method is used to identify and align regions on which the alignment is highly reliable.

2. Multiple alignments are generated with the reliable regions constrained. This would result in far fewer alignments than without constraints, and will also reduce the false positive problem.

3. The models are ranked using a simple free energy evaluation expression, and the model with the lowest free energy is used as the prediction. Alternatively, the generated models are clustered on the basis of the pairwise RMSD, and the largest clusters are retained as predictions.

Acknowledgement This research has been supported by grants DBI-0213832 from the National Science Foundation, and P42 ES07381 from the National Institute of Environmental Health.

References

1. Marti-Renom,M.A., Stuart,A.C., Fiser,A., Sanchez,R., Melo,F., Sali,A.: Comparative protein structure modeling of genes and genomes. Ann. Rev. Biophys. Biomol. Struct. **29** (2000) 291-325
2. Sanchez, R., Sali, A.: Advances in protein-structure comparative modeling. Curr. Opinion in Struct. Biol. **7** (1997) 206-214
3. Fiser, A., Sanchez, R., Melo, F., Sali, A. Comparative protein structure modeling. In: M. Watanabe, M., Roux, B., MacKerell, A., Becker, O (eds.): Computational Biochemistry and Biophys. Marcel Dekker . (2001) 275-312
4. Cline,M., Hughey,R. and Karplus,K.: Predicting reliable regions in protein sequence alignments. Bioinformatics **18** (2000) 306-314.
5. Jaroszewski, L., Rychlewski, L., Godzik, A.: Improving the quality of twilight-zone alignments. Protein Science **9** (2000) 1487-1496
6. Janardhan, A., Vajda, S.: Selecting near-native conformations in homologymodeling: The role of molecular mechanics and solvation terms. Protein Science **7** (1997) 1772-1780, 1997.
7. Gatchell, D., Dennis, S., Vajda, S.: Discrimination of near-native protein structures from misfolded models by empirical free energy functions. Proteins **41** (2000) 518-534
8. Prasad, J.C., Comeau, S.R., Vajda, S. and Camacho, C.J.: Consensus alignment for reliable framework prediction in homology modeling. Bioinformatics, in press.
9. Holm,L., Sander,C.: Mapping the protein universe. Science **273** (1996) 595-602
10. Altschul,S.F., Madden,T.L., Schaffer,A.A., Zhang,J., Zhang,Z., Miller,W., Lipman,D.J.: Gapped BLAST and PSI-BLAST: a new generation of protein database search programs. Nucleic Acid Res. **25** (1997) 3389-3402

11. Sanchez,R., Sali,A.: Evaluation of comparative protein structure modeling by MODELLER-3. PROTEINS: Structure, Function and Genetics, **Suppl. 1** (1997) 50-58
12. Brooks, B.R., Bruccoleri, R.E., Olafson, B.D., States, D. J., Swaminathan, S., Karplus. M.: CHARMM: A Program for Macromolecular Energy, Minimization, andDynamics Calculations, J. Comp. Chem. **4** (1983) 187-217
13. Eisenberg, D., McLachlan, A.D.: Solvation energy in protein folding and binding. Nature **319** (1986) 199-203
14. Karplus, K., Barrett, C., Hughey, R.: Hidden Markov models for detecting remote protein homologies. Bioinformatics **14** (1998) 846-856
15. Eddy S. R. : Profile hidden Markov models. Bioinformatics **14** (2001) 755-763
16. Cuff, J.A., Barton, G.J.: Application of enhanced multiple sequence alignment profiles to improve protein secondary structure prediction. Proteins 40 (2000) 502-511

Side-Chain Structure Prediction Based on Dead-End Elimination: Single Split DEE-criterion Implementation and Elimination Power

Jan A. Spriet

Katholieke Universiteit Leuven Campus Kortrijk, 8500 Kortrijk, Belgium.
Jan.Spriet@kulak.ac.be

Abstract. The three-dimensional structure of a protein is a very fundamental piece of knowledge. Part of the problem of obtaining that knowledge is modeling the side-chain spatial positions. Dead-End Elimination (DEE) is a powerful philosophy concerning search in the abstract conformational space, meant to yield the optimal side-chain orientations, fixed on the given protein backbone. The approach permits the formulation of several DEE-criteria, which differ in quality and computational efficiency for the abstract optimization problem that is rooted in and framed by the macromolecular architectural aspects of the protein universe. The present work investigates time complexity, elimination power and optimized implementation of the recently proposed Single Split DEE- criterion for protein side-chain placement or prediction. Properly inserted in a suitable DEE-cycle, the criterion is found to follow the classic cost bound and is proven to be worthwhile. On a test set of sixty proteins, the elimination power is 17.7%, in addition to the combined pruning effect of the standard criteria. It is also found that algorithm implementation can be optimized using the efficacy – "Magic Bullet Character", of the side-chain residue types.

1 Introduction

Proteins are one of the most important classes of macromolecules, playing major roles in a large variety of cellular mechanisms, which constitute the activity of living systems. Because those macromolecules are essentially linear chains of amino acids, one important perspective on their three-dimensional architecture consists in distinguishing the backbone structure from the side-chain or residue elements, that "stuff" the peptide chain. The more rigid backbone has less chemical and atomic conformational diversity than the twenty amino acid side-chains. Because a protein is a more or less thermodynamically stable atomic dense aggregate, side-chains play a major role in the required proper packing state by taking the adequate 3D-orientations. In addition, the enzymatic function of many proteins requires an "active site", which is made up of a piece of backbone with a number of well-positioned side-chains, providing in that manner among other

G. Benson and R. Page (Eds.): WABI 2003, LNBI 2812, pp. 402–416, 2003.

things, the three-dimensional organization for eventual chemical reactive action. In the medical field, for instance, that molecular activity is basic to drug effects and consequently also to their design. Because protein folding is not yet completely understood and because the theoretical protein structure prediction problem is also not yet sufficiently solved, many problems in protein science and engineering require approximate approaches. The approach of dividing the structure in a backbone and side-chain elements, has proved to be a partly theoretical and certainly practical valid one. In many cases, trying to predict side-chain orientations starting from some given backbone structure or model has been useful. Side-chain packing and structure prediction, consequently, are essential components of protein science, modeling and engineering [1]. The determination of optimal side-chain conformations, given a particular protein backbone structure (or template), emerges in many sub-disciplines of protein structure analysis and prediction: homology modeling, protein design, loop structure prediction, peptide docking and so forth. The problem mentioned, however, requires substantial mathematical and computational sophistication, at least for the more useful or mature issues. Consequently, it belongs to the more advanced sub-fields of computational molecular biology.

2 Problem Formulation and General Approaches

2.1 The Problem Perspectives

Simply stated, in this paper we investigate, relative to previously known "pruning" criteria, the value of a specific, recently discovered "bounding function" – the Single Spit DEE-criterion [2][3] – for a class of integer programming problems, whose properties are rooted in the biophysical structure of protein biomolecules. Although the side-chains of real proteins adopt an equilibrium conformation out of a continuum of possible orientations, the underlying biophysical laws seem to yield a finite number of dominant orientations, depending on the type of amino acid [4][5][6]. The major theoretical entity involved is the free energy of the molecules within the environment of interest. Full theoretical and mathematical correctness in building a proof for the conjecture is not possible. Approximate reasoning and experimental investigation indicate, at the present, that side-chains of naturally occurring proteins are tightly clustered around a finite number of preferential orientations. The discretization of the allowed conformational space into "rotamers", representing the statistically dominant orientations, is thus partly well-founded, serving also the computational requirements in respect to continuous variables. A schematic representation of a short peptide is given on Fig. 1. For a specific system, those side-chain positions that minimize the free energy, are the actual occurring orientations.

The related mathematical problem, thus, has discrete variables and an objective or cost function, that models the free energy. The solution approaches will belong to the field of integer programming [7]. The mathematical formulation is as follows: there are p variables, one for each residue in the protein: $1, 2, \ldots, i, j, \ldots, p$.

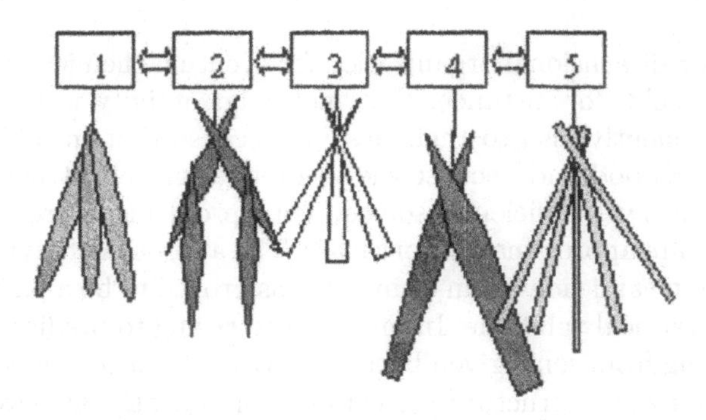

Fig. 1. Chain of five amino acids ($p = 5$), each having a specific number of rotamers for the residue: ($n_1 = 3, n_2 = 2, n_3 = 3, n_4 = 2, n_5 = 5$).

Each variable : e.g. i, is discrete and can take only a finite number of integer values: $1, 2, \ldots, r, s, t, u, v, \ldots, n_i$, which code for specific rotamers. The number of integer values that a discrete variable can take, is different from variable to variable: $n_1, n_2, \ldots, n_i, n_j, \ldots, n_p$. In physical terms, it simply means that the number of rotamers may differ depending on the residue type. The average number of rotamers per residue is $n = (n_1, n_2, \ldots, n_i, n_j, \ldots, n_p)/p$. A candidate protein conformation C is mathematically specified by a vector of p integer values: $C = [\ldots, i_r, j_u, \ldots]$. Such a conformation has an approximate free energy value – cost function value – of the following type as explained in [8]:

$$E_C = E_{template} + \sum_i E(i_r) + \sum_i \sum_{j \neq k} E(i_r, j_u). \qquad (1)$$

This type of cost function is defined as being pair wise additive and the property is sometimes exploited in designing a search strategy. The solution of the theoretical side-chain prediction problem – TSCPP, is the conformation C_{min} that minimizes E_C.

It was soon clear that with growing size of the peptide chain, search space dimensionality encompasses a combinatorial explosion, while the cost functions, being approximations of the free energy, are misleading and treacherous due to the rugged nature of the "cost function landscape" and the many local extrema [9]. Because of the importance of the underlying biological issues, those dismal integer programming conditions, did not deter researchers in computational molecular biology. In the course of the last decades, many algorithmic strategies have been proposed to place side-chains. Often the molecular biochemists and protein scientists have been disappointed with the prediction capabilities and results when trying to use them for their biochemical or biological system problems. Unbelief concerning the value of gigantic computational projects has resulted is studies ascertaining that the "practical" actual side-chain placement problem is in fact almost trivial [10]. The latter claim was and still is [11] justified if one incorporates as "boundary conditions" the many approximations and limitations of molecular science, if one chooses for "hybrid" solution approaches,

where alternative biochemical methods are included as often as necessary to solve the overall problem and if one judges the quality of the solution according to practical standards, having the major focus on the applied biochemical or biological result. However, those three framing conditions are too stringently limiting the complete problem perspective. The diversity in the types of theoretical and applied molecular biology problems is recognized to be so pronounced that side-chain placement with a more trivial method – roughly speaking working with a database- rule- or knowledge-based approach, or using "hybrid" approaches with limited precision, has its too restricted scope so that interesting results could be missed or overlooked. The steady trend towards higher computational power, which is still unavoidable, challenges the economy argument involved when preferring alternative solution routes. In addition, advances in many fields, may easily invalidate the framing conditions mentioned.
The proper perspective for the research done here, has to be chosen on the basis of the following considerations. The mathematical integer programming problem, described above continues to serve the computational molecular biology problem of side-chain structure prediction at the more sophisticated level – whenever the knowledge-based or simple hybrid procedures are unsatisfactory, the complex approach becomes important. The abstract formulation mentioned is not only instrumental in solving side-chain packing problems, it is also representative for other physical entities: e.g. spin glasses [1][12]. Furthermore, in the course of the last fifteen years, the TSCPP has attracted a considerable amount of effort. The results constitute a valuable amount of knowledge concerning the power of algorithmic approaches to integer programming problems, having the properties of being high dimensional in the search space and being related to "rugged" cost functions with multiple local extrema. New work adds to this existing sophisticated level.
Theoretically, the problem has been shown to be NP-complete [13]. In practical terms, the meaning is that for every algorithm there are problems instances that will take an infinite amount of computation time to get solved. Actual algorithmic experience in those cases can throw light on the nature of the limitations, that the brutal fact of NP-completeness entails. For instance, in case a specific problem instance turns out to be drowned in excessive computation time requirements, it may be appealing to discover to what degree the algorithm reorganization has to be pushed to obtain a solution anyhow.

2.2 The Different General Approaches

It is beyond the scope of the paper to present, explain and compare the different algorithmic approaches to the TSCPP. However to situate the Dead-End Elimination philosophy, it is important to give a major overview of the algorithm classes. The different broad classes to be mentioned are *the stochastic class* – comprising algorithms to be subdivided under the headings: Mont-Carlo schemes, Simulated Annealing techniques, Genetic Algorithm approaches, *the deterministic class* – comprising pruning strategies like Dead-End Elimination, and network techniques as artificial neural networks, *the variational class* with

as typical representatives the Mean-Field algorithms, and *the standard class*, referring to the general mathematical programming algorithms and algorithm adaptations, e.g. simplex techniques in various problem set-up's.
All approaches have there limitations for the theoretical side-chain prediction problem. In general, the problem solution requires a number of "preprocessing" and "post-processing" steps, that sandwich the basic algorithm. Implementing the latter requires often hierarchical and loop program structures and parametrization approaches that are resolved at run time. The different classes have their own strengths and weaknesses as discussed in [7]. Apart from that, at the level of sophistication of the comparisons performed, few of those algorithmic classes can be found to be really inferior. However, this level is not extremely high because of differences in test sets, objective functions, pre- and post-processing steps, implementation details etc. Much work remains to be done on that issue.

3 Dead-End Elimination Philosophy

3.1 Algorithmic Approach and Criteria

The Dead-End Elimination (DEE) solution philosophy is the only one that sets the goal of finding the global minimum energy conformation (GMEC) in a reasonable amount of computation time for almost all problems [14][16] and DEE has survived most practical tests and problem instances [14]. The basic idea is that instead of searching for the best solution in discrete conformational space, one eliminates rotamers, that can be proven not to belong to the GMEC. The final aim is to be able to continue pruning the set of all candidate rotamers and thus also all candidate conformations until one is left with a single conformation, which then is by result the theoretical right answer. This viewpoint often turned out to be too ambitious. If, however, the necessary bounding functions as rejection criteria can be found to reduce the size of the conformational space to a value, were exhaustive search is not prohibitive in computation time, the GMEC is found too
The high appeal for DEE stems from the fact that the problems of "rugged landscape" and local extrema is completely circumvented. The original Dead-End Elimination criterion [8] states that a rotamer i_r can be eliminated if an alternative rotamer i_t, at the same position, can be found satisfying:

$$E(i_r) + \sum_{j,j \neq i} \mathrm{Min_u}\, E(i_r, j_u) > E(i_t) + \sum_{j,j \neq i} \mathrm{Max_u}\, E(i_t, j_u). \quad (2)$$

In words, the superiority of the second rotamer i_t is checked by verifying that the free energy of a protein structure with it, supplemented with side-chain orientations for the other residues j_u, that are, on a mutual basis, chosen to be poorest, *is <u>more</u> adequate* than the free energy of a protein structure with the original rotamer, supplemented with side-chain orientations for the other residues, that are, on a mutual basis, chosen to be most advantageous. With time, the philosophy has resulted in quite a number of DEE-criteria [2][3]. A mature overview was obtained after the development of the "Bottom Line" DEE-criterion [3].

3.2 DEE-cycle Design

In practice, any single DEE-criterion on its own is grossly unsatisfactory for the chosen TSCPP problems. A complete program requires, preprocessing, a DEE-cycle and postprocessing [14].The actual program part of main interest, termed a DEE-cycle [2], is a judicious sequence of sub-algorithms, each built on the basis of a different but specific DEE-criterion, used in iteration and structured in more complex nested loops. An example of a DEE-cycle is shown on Fig. 2. Three DEE-criteria are used, each of them is in iteration until exhaustion of their effects. Suppose that the second criterion has been implemented for reasons of computational speed in an approximate manner. A nested loop after the use of the third criterion maý be useful. In practice, the third criterion often has changed the space of remaining conformations in a substantial manner so that the second approximated criterion may yield new eliminations.
Nowadays, research often focuses on DEE-cycle analysis, design and evaluation

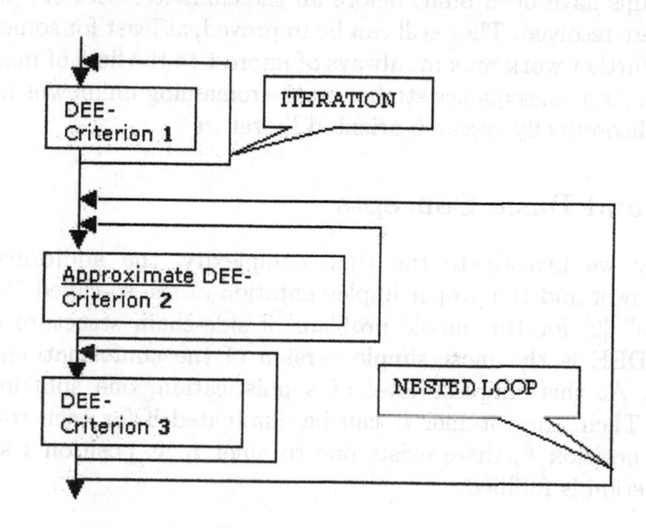

Fig. 2. Example of a DEE-cycle structure.

over the range of all TSCPP's, limited to natural occurring proteins. A recent paper [2] gave a well-summarized overview of DEE-criteria and proposed the strategy of conformational space splitting, giving rise to more powerful but more computational intensive DEE-criteria. Due to the huge search space, which growths dramatically with the size of the protein, it is now well-established [15][16] that in the very beginning of the computation, the simplest DEE-criteria, though being less powerful, have to be used intensively in order to reduce the gigantic solution space as fast as possible. Only at that stage in the DEE-cycle is the use of a more powerful, but time-consuming DEE-criterion worthwhile. The work here has focused on such a criterion, placed further downstream in the DEE-cycle.

3.3 Practical Hybrid Programs

Referring to the discussion in Sect. 2, the use of a single algorithm type for solving many TSCPP's, whose solutions most often yield intermediate but necessary results for the actual global biological problem under investigation, is inadequate. Indeed such a "pure breed" program often remains really too slow for practical issues. In that case, hybrid programs, that combine more than one algorithm philosophy, are used. The gain in computation time may be impressive – e.g. FASTER [17]. However, most hybrid programs do not yield the GMEC. The study of "pure breed" algorithms, therefore, remains worthwhile.
Theoretical issues often require the GMEC. Unsatisfactory solution results for the actual total problem often suggest that the GMEC itself, instead of a close relative, is of interest. At the present, there remains still the danger that a specific problem instance gives rise to an unstable, non-robust, or ill-defined solution model. Then, the GMEC is the major rock to cling to. Furthermore, hybrid programs have been built, before all the characteristics of "pure breed" algorithms were resolved. They still can be improved, at least for some problems. Finally, such further work remains always of interest to the field of mathematical programming. As a consequence studies on the remaining unknowns have moved to the more theoretically research oriented literature.

4 Issues and Basic Concepts

In this study we investigate the time complexity, the additional rotamer elimination power and the proper implementation of the so-called "Single Split DEE-criterion" [2] for the classic problem of side-chain structure prediction. Single Split DEE is the most simple version of the conformational splitting DEE-criteria. At this simplest level of sophistication, one splitting position k_i is chosen. Then, the rotamer i_r can be eliminated if for each rotamer v at the splitting position k, there exists one rotamer i_t at position i so that the following criterion is fulfilled:

$$E(i_r) - E(i_t) + \sum_{j,j \neq i \neq k} \{\mathrm{Min_u}\ [E(i_r, j_u) - E(i_t, j_u)]\} + [E(i_r, k_v) - E(i_t, k_v)] > 0. \qquad (3)$$

For a proof of the validity of (3) as a bounding function, one is referred to the original papers [2][3].

4.1 Time Complexity

The first concept required is that of problem complexity. Basically, given a specific rotamer library and force field, it is related to the number and the types of residues and the nature of the protein fold. A priori, it is not clear which quantitative measure of problem complexity is most adequate for the present study. Protein and rotamer library size parameters give indications. Ideally, problem complexity could be specified to be the computation time required to solve the specific side chain packing problem with the best choice and ordering of criteria and with their optimal algorithmic implementations. A next concept, then, is that of DEE-criterion complexity, its measure being the computation time required to solve the specific problem with that criterion in its most efficient algorithmic implementation. Clearly, DEE-criterion complexity will depend on problem complexity. Finally, algorithmic complexity is measured here as the computation time resulting from executing the actual program algorithm parts for the problem to be solved. Having chosen those measures of complexity, the solution cost is directly related to the complexity and cost bounds can be used as complexity bounds. Specification of a cost bound estimate, based on the size and the number of nested loops, such as $O(p^2n^3)$ for instance, relates thus algorithmic complexity to some protein problem size parameters, that certainly are involved in problem complexity.

4.2 Elimination Power

In case of protein prediction problems using the DEE-methodology, the theoretical worth of a specific criterion is its rotamer elimination power. Quantification of the elimination power can be done in different manners. The abstract view on predicting side-chain positions as an optimization of the force field in combinatorial space, suggest as a power measure for a criterion, the reduction in the size of the conformational space it is able to achieve. Unfortunately, such a measure scales poorly and a mathematical transformation of it – e.g. taking the logarithm of it, is sometimes open to some arbitrariness. In this work, the percentage reduction in the *number of rotamers* is chosen as measure for the Elimination Power. This measure is dimensionless. The scaling is no problem as the number of rotamers is in linear proportion to the number of residues. It directly relates to the number of rotamers pruned and is better focused on the aim of getting a single rotamer combination as GMEC conformation. As a drawback, one may mention that the measure does not expose the multiplicative aspects of conformational space; however the latter is more readily investigated by the time complexity.

4.3 The "Magic Bullet" Character of Residue Types

When a criterion is used, that is derived from splitting conformational space at certain residue locations, the nature of the residue, being it its specific position, its type or another characteristic, may make a difference in elimination performance. The adequacy of a residue property, when the residues, having that property, are used as conformation splitters, is called its "magic bullet" character. Because the proper choice of adequate measures is better understood when seeing the data, partial and overall measures for it will be defined later on.

5 Experimental Set-Up and Implementation Details

The Test Set The test set of sixty non-homologous (< 25%) proteins were chosen from the Protein Data Bank [20], using the results of Hobohm et al. [18][19]. Resolution was kept below $1.85 A^o$ and the R-value below 20%, while the protein was to be of the single chain type with as few heterogens as possible. The PDB-identifiers of the chosen proteins were: 1crn, 5pti, 1zia, 1jer, 1thx, 1lz1, 1rie, 1whi, 1rpg, 1lit, 1jcv, 2end, 1tfe, 1rbu, 2cpl, 1vhh, 1ilk, 1l68, 2eng, 1wba, 1lbu, 1thv, 1hbq, 1chd, 153l, 1amm, 1xnb, 1sgt, 1knb, 9pap, 1dts, 1din, 1iae, 1akz, 2hft, 1arb, 1bec, 1xjo, 1mml, 1tml, 2prk, 2dri, 1mla, 2cba, 1cnv, 1amp, 8abp, 1cvl, 3tln , 1nar, 3app, 1nif, 2ctc, 1eur, 3pte, 1cem, 1omp, 1edg, 2sil, 1gnd.

The Potential Function The CHARMM force field of Brooks et al. [21] was used throughout this work to calculate adequate approximate free energy values. No modifications were made to the energy function nor to the parameters. No specific solvent terms were taken into account. Coulombic interactions were calculated by using a distance-dependent dielectric constant ($\epsilon = r$).

The Rotamer Library The rotamer library was the one proposed by De Maeyer et al. [22]. Being quite detailed, it has 859 elements, covering all twenty amino acids except Glycine, Alanine and Proline. The side-chain rotamers were modeled in standard geometry (i.e., they were generated with ideal bond lengths and angles).

Problem Preparation Each PDB-structure was completed with the hydrogen atoms using the Brugel modeling package [23]. All structures were then refined by 100 steps steepest descent minimization to remove clashes. For each structure the protein backbone template was composed of the main-chain supplemented with the C_β-atoms and the full Gly, Ala, Pro and disulfide bond forming Cys residues. This structural template was kept fixed during side-chain prediction.

The Computer Facility The algorithms were programmed in FORTRAN. The programs have been running on a SGI Indy 4600 workstation.

The Simulation Experiments In a first step the complexity of the side-chain placement or packing problem for the proteins of the test set was quantified by two numbers: the number of side-chains to be predicted: p, and the average number of rotamers per side-chain n. The ranges of those numbers over the test set were: p between 26 and 346 and n between 9 and 21. In a second step, the Single Split DEE-algorithmic structure was placed after a first elimination round based on the so-called fast "Singles" algorithmic structure [2][3] and a second elimination round based on the so-called still fast "Doubles Assisted Singles" algorithmic structure [2][3]. As in the DEE-cycle the "Single Split DEE-criterion" is only applied in a third instance, after two DEE-criteria that are faster, the actual problem complexity for the criterion under investigation is smaller because the faster algorithms have already eliminated a substantial amount of rotamers. An overview of the problem complexity (n, p) of the test set instances is shown on

Fig. 3 below. The sixty packing problems were fed in the DEE-cycle mentioned and all relevant results were gathered.

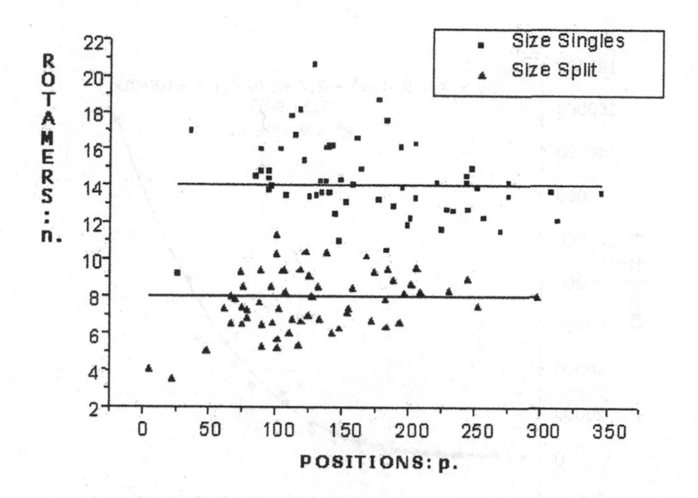

Fig. 3. Overview of the spread in problem complexity of the test set proteins for two DEE-criteria in terms of a (p, n)-tuple.

6 Results

6.1 Time Complexity

For the algorithmic implementation of the Single Split DEE-criterion used on the problem set in this work, the cost bound is $O(n^3p^3)$ and thus $O(N_{rot}^3)$, where $N_{rot} = n * p$: the total number of rotamers at the start of the Single Split DEE elimination process. Concerning the quantitative precision of the cost bound, the following results could be gathered. Proper fit to the data could only be obtained using a full third order polynomial, indicating the existence of quadratic and linear terms that contribute to algorithmic complexity. In that manner the fit is adequate as illustrated on Fig. 4. Even over the limited range of problem complexities in the data set, the quadratic and linear terms cannot be neglected: fitting a pure power law results in a power coefficient of 3.77, the value being statistically significant different from 3.0. The, on the cost bound, based empirical law – see Fig. 4, permits computation time predictions precise to within 30% of the estimated value. Higher prediction precision probably requires more elaborate measures for problem complexity, including e.g. the amino acid type distribution within the chain and/or certain fold descriptors. The classic approach to design DEE-cycles consists in ordering criteria according to the simple cost bounds of the type mentioned above. Based on the results discussed, it can be concluded that those cost bounds are largely adequate. However, the actual values of the power coefficients could, under certain circumstances, allow for more detailed design.

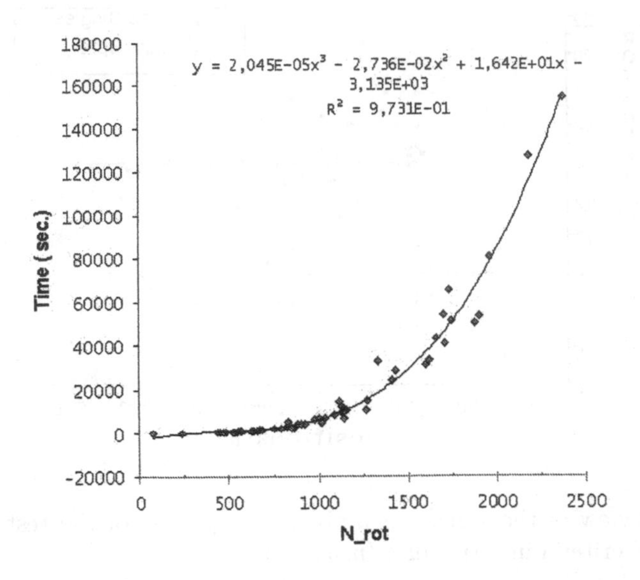

Fig. 4. Single Split DEE algorithm computation times versus the total number of rotamers making up conformational space at start of the algorithm.

6.2 Single Split Dead-End Elimination Efficiency

Elimination power was quantified as the percentage reduction in the number of protein rotamers. Standard statistical analysis of the Single Split DEE additional elimination power on the test set mentioned, yields an average of 17.7% – the 95% confidence interval indicates an estimation precision to 1%. The standard deviation is 4%. Clearly, splitting conformational space encompasses quite substantial additional dead-end elimination. The approach is thus really worthwhile. If one makes a histogram of the numbers, it is found that the shape of it is skewed – Fig. 5, above, at the left. Such a finding is indicative for further needed analysis. The first issue that comes to mind is the eventual impact of problem complexity. A check on the hypothesis that problem complexity has an impact on elimination power, leads to the counter-intuitive result that additional elimination power seems to degrade with increasing complexity of the packing problem. A scatter plot of percentage of rotamers eliminated versus the logarithm of the number of rotamers in the problem – N_{rot}, shows a linear trend – Fig. 5. After removing that trend, the skewness of the histogram disappears – Fig. 5, above, at the right. The result is unfortunate: computational time demands already grow according to a power law in problem size and less elimination power for the larger problems will exacerbate computer loads.

Because in this work the Single Split DEE-criterion is under study, it is im-

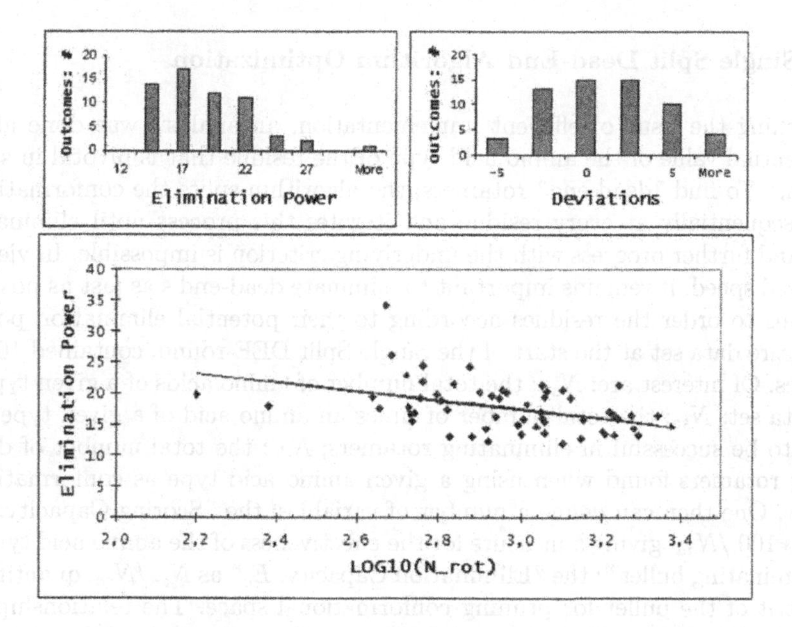

Fig. 5. Discovery of elimination power dependency on problem complexity when analyzing sample histogram skewness.

portant to check whether elimination efficiency degradation with complexity is a characteristic of the conformational splitting approach or not. To clarify the issue, the elimination capacity of the DEE-Singles pass – the first elimination round based on the most elementary DEE-criteria that are completely independent from space splitting, was analyzed and a similar trend, being even more pronounced, was observed – see data in Table 1. Due to the type of DEE-cycle

Table 1. Numeric data concerning the dependency of the elimination power on problem complexity.

Elimination Power $= \mathbf{a} - \mathbf{b} * Log_{10}(N_{rot})$

Criterion	**a**	**b**	R^2
DEE-Singles	119.98	21.86	0.35
Single Split DEE	39.14	7.24	0.18

used in the computational experiments, the Single Split DEE-criterion is under more difficult test conditions than the Singles DEE. Efficiency deterioration with problem size can better be attributed to the DEE problem solution approach in general until another study discovers a proper explanation.

6.3 Single Split Dead-End Algorithm Optimization

Concerning the issue of efficient implementation, an analysis was done about the eventual value of the amino acid type of the residue that is pivotal in space splitting. To find "dead-end" rotamers, the algorithm splits the conformational space sequentially at every residue and iterates this process until elimination stops and further progress with the underlying criterion is impossible. In view of increased speed, it remains important to eliminate dead-end's as fast as possible and thus to order the residues according to their potential elimination power. The entire data set at the start of the Single Split DEE-round, contained 10,117 residues. Of interest are: N_{aa}: the total number of amino acids of a given type in the data set; N_{sc}: the total number of times an amino acid of a given type was found to be successful in eliminating rotamers; N_{de}: the total number of dead-ending rotamers found when using a given amino acid type as conformational splitter. One then can define a number of variables: the "Scoring Capacity: S_c" as $N_{sc} * 100./N_{aa}$, giving a measure for the effectiveness of the amino acid type as an "eliminating bullet"; the "Elimination Capacity: E_c" as N_{de}/N_{sc}, quantifying the merit of the bullet for pruning conformational space. The relationship between the Elimination Capacity and the Scoring Capacity of the different amino acids is shown in Fig. 6. Amino acids that are more often valuable for eliminating rotamers have also the property of being able to eliminate more rotamers on the average. The "Magic Bullet Character: MB_{ch}" as $S_c * E_c$, is thus indicative for the overall performance value of the amino acid type in eliminating dead-end's. The range of MB_{ch} is between 24.02 for Serine and 240.75 for Arginine. The

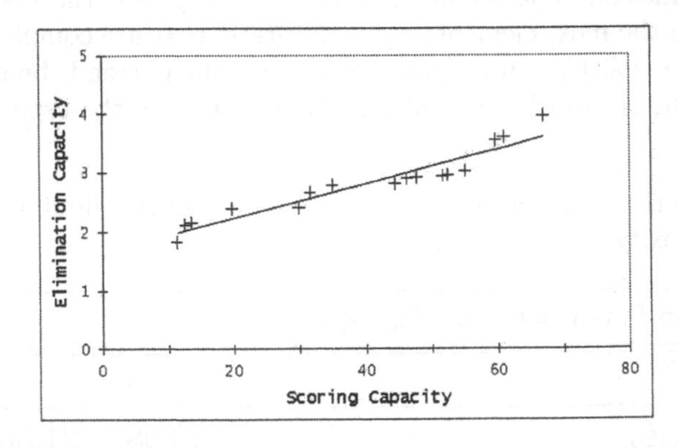

Fig. 6. Relationship between Elimination Capacity and Scoring Capacity of the different amino acid types.

Magic Bullet Character of the latter is ten times that of the former. Arginine is ten times more effective in splitting conformational space than Serine. The results for the complete set of amino acid types to be placed were the following: Arg – 240.75; Trp – 217.62; Tyr – 185.64; His – 180.14; Glu – 179.45; Gln – 151.82; Met – 138.22; Phe – 129.80; Asp – 128.03; Lys – 96.6; Leu – 79.42; Asn – 75.25; Ile – 47.18; Val – 27.03; Thr – 24.72; Ser – 24.02. For the purpose of algorithm optimization, it is best to order the residues according to their Magic Bullet Character and in that manner the space of candidate conformations is likely to reduce at the fastest pace.

7 Conclusions

The Dead-End Elimination philosophy exists already for more than a decade. From the beginning on, specific DEE-criteria have been discovered on a regular basis while DEE-cycle development has evolved, tuned to the specific nature of the computational molecular biology problem involved in side-chain packing. Conformational Splitting has been incorporated with success in DEE-cycles developed for protein design [2][3]. In this work a begin is made of quantifying time complexity and elimination capacity of sub-algorithms based on different DEE-criteria within a DEE-cycle in a more systematic and detailed manner. Here, the Single Split DEE criterion in a DEE-cycle for protein side-chain prediction was further evaluated and its merit put in a more elaborate perspective. At the same time, based on gathered efficiency data, the program algorithm could be optimized.

References

1. Vasquez, M.: Modeling side-chain conformation. Curr. Opin. Struct. Biol. **6** (1996) 217–221
2. Desmet, J., De Maeyer, M., Lasters, I.: Theoretical and algorithmical optimization of the dead-end elimination theorem. In: Altman, R., et al., (Eds.): Proceedings of the Pacific Symposium on Biocomputing '97, World Scientific New Jersey (1997) 122–133
3. Pierce, N., Spriet J., Desmet, J., Mayo, S.: Conformational Splitting: A More Powerful Criterion for Dead-End Elimination. J. Comput. Chem. **11** (2000) 999–1009
4. Janin, J., Wodak, S., Levitt, M., Maigret, D.: Conformation of amino acid side-chains in proteins. J. Mol. Biol. **125** (1978) 357–386
5. Ponder, P., Richards, F.: Tertiary templates for proteins. Use of packing criteria in the enumeration of allowed sequences for different structural classes. J. Mol. Biol. **193** (1987) 775–791
6. Dunbrack, R., Karplus, M.: Backbone-dependent rotamer library for proteins: application to side-chain prediction. J. Mol. Biol. **230** (1993) 543–574
7. Eriksson, O., Zhou, Y., Elofsson, A.: Side-Chain Positioning as an Integer Programming Problem. In: Proceeding of WABI'01. Springer Verlag, LNCS **2149** (2001) 128–141

8. Desmet, J., De Maeyer, M., Hazes, B., Lasters, I.: The Dead-End Elimination Theorem and its Use in Protein Side-Chain Positioning. Nature **356** (1992) 539–542
9. Lee, E., Subbiah, S.: Prediction of Protein Side-chain Conformation by Packing Optimization. J. Mol. Biol. **217** (1991) 373–388
10. Eisenmenger, F., Argos, P., Abagyan R.: A Method to Configure Protein Side-chains from the Main-chain Trace in Homology Modelling. J. Mol. Biol. **231** (1993) 849–860
11. Xiang, Z., Honig, B.: Extending the Accuracy Limits of Prediction for Side-chain Conformations. J. Mol. Biol. **311** (2001) 421–430
12. Goldstein, R.: Efficient rotamer elimination applied to protein side-chains and related spin glasses. Biophysic. Journ. **66** (1994) 1335–1340
13. Fraenkel, A.: Protein folding, spin glass and computational complexity. In: Third annual DIMACS workshop on DNA based computers. Proceedings ,Philadelphia (1997) 23–25
14. Desmet, J., De Maeyer, M., Lasters, I.: The Dead-End Elimination Theorem: a New Approach to the Side-Chain Packing Problem. In: Merz, K., Le Grand, S. (Eds.): The Protein Folding Problem and Tertiary Structure Prediction, Birkhuser, Boston, (1994) 307–337
15. Dahiyat, B., Mayo, S.: De novo protein design: fully automated sequence selection. Science **278** (1997) 82–87
16. Lasters, I., De Mayer, M., Desmet, J.: Enhanced dead-end elimination in the search for the global minimum energy conformation of a collection of protein side-chains. Prot. Eng. **8** (1995) 815–822
17. Desmet, J., Spriet, J., Lasters, I.: Fast and Accurate Side-Chain Topology and Energy Refinement (FASTER) as a New Method for Protein Structure Optimization. PROTEINS: Struct. Funct. Genet. **48** (2002) 31–4
18. Hobohm, U., Scharf, M., Schneider, R., Sander, C.: Selection of representative protein data sets. Prot. Sci. **1** (1992) 409–417
19. Hobohm, U., Sander, C.: (1994): Enlarged representative set of protein structures. Prot. Sci. **3** (1994) 522–524
20. Bernstein, F.; Koetzle, T., Williams, G., Meyer Jr, E., Brice, M., Rodgers, J., Kennard, O., Shimanouchi, T., Tasumi, M.: The Protein Data Dank: a computer-based archival file for macromolecular structures. J. Mol. Biol. **112** (1977) 535–542
21. Brooks, B., Bruccoleri, R., Olafson, D., States, D., Swaminathan, S., Karplus, M.: CHARMM: a program for macromolecular energy minimization and dynamics calculations. J. Comput. Chem. **4** (1983) 187–217
22. De Maeyer, M., Desmet, J., Lasters, I.: All in One: a Highly Detailed Rotamer Library Improves both Accuracy and Speed in the Modelling of Sidechains by Dead-End Elimination. Folding Design **2** (1997) 53–66
23. Delhaise, P., Bardiaux, M., Wodak, S.: Interactive computer animation of macromolecules. J. Mol. Graph. **2** (1984) 103–106

A Large Version of the Small Parsimony Problem

Jakob Fredslund[1], Jotun Hein[2], and Tejs Scharling[1]

[1] Bioinformatics Research Center, Department of Computer Science, University of Aarhus, Denmark,
{jakobf|,tejs}@birc.dk

[2] Department of Statistics, University of Oxford, United Kingdom,
hein@stats.ox.ac.uk

Abstract. Given a multiple alignment over k sequences, an evolutionary tree relating the sequences, and a subadditive gap penalty function (e.g. an affine function), we reconstruct the internal nodes of the tree optimally: we find the optimal explanation in terms of indels of the observed gaps and find the most parsimonious assignment of nucleotides. The gaps of the alignment are represented in a so-called gap graph, and through theoretically sound preprocessing the graph is reduced to pave the way for a running time which in all but the most pathological examples is far better than the exponential worst case time. E.g. for a tree with nine leaves and a random alignment of length 10.000 with 60% gaps, the running time is on average around 45 seconds. For a real alignment of length 9868 of nine HIV-1 sequences, the running time is less than one second.

1 Introduction

The relationship of biological sequences such as proteins and DNA can be described by a rooted tree, with a series of bifurcations corresponding to duplications in the past. Because of fluctuations in rates of evolution of such molecules it is not possible to determine the position of the root and the relative order of ancient duplications. When a rooted bifurcating tree is stripped of this information it becomes an unrooted tree, where all internal vertices have three incident edges. Protein and DNA sequencing techniques allow the determination of the sequences at the leaves of such a tree (the present). The sequences that represent ancestral molecules are unobservable and must be reconstructed according to some principle.

Usually the principle of parsimony is invoked: given a function that measures the distance between two sequences, choose ancestral sequences that minimize the total distance between neighbors in the tree, i.e. the overall amount of evolution. The distance function defines the penalty, or cost, of basic events such as substitutions and insertions/deletions (indels), and thus the distance between two sequences s_1 and s_2 is the "cheapest" process of evolution in terms of substitutions and indels which transforms s_1 into s_2. Application of the parsimony

G. Benson and R. Page (Eds.): WABI 2003, LNBI 2812, pp. 417–432, 2003.

principle falls in two categories, depending on whether the phylogeny is known or not. If the phylogeny is known, it is the small parsimony problem; otherwise it is the large parsimony problem. This article will only address the small parsimony problem, but since it is a hard version involving complete sequences, it is called a large version of the small parsimony problem.

The simplest case where only substitutions have occurred and they all have cost 1 was first addressed by Fitch [1] and independently by Hartigan [3]. This was generalized to allow substitutions with arbitrary costs by Sankoff [5]. The computing time used by these algorithms was of the order $O(n \cdot k)$, where n is the length of the sequences and k the number of sequences. The situation with two sequences subject to substitutions and indels of length 1 was solved by Sellers [4] and Sankoff [2]. This is called the *approximate string matching problem* and uses $O(n^2)$ computing time. Sankoff [5] combined the string matching and Fitch-Hartigan-Sankoff algorithms to devise an algorithm that reconstructs the history of a set of sequences if their phylogenetic tree is known. Its running time was $O(n^k \cdot 2^k \cdot k)$, which is prohibitive.

Waterman et al. [6] introduced sequence comparisons of two sequences with indels longer than one. Their algorithm uses $O(n^3)$ computing time. Gotoh [7] gave an algorithm that uses time $O(n^2)$ if the gap penalty function is of the form $g(l) = \alpha + \beta \cdot l$, where l is the length of the gap. Fredman [8] generalized Gotoh's algorithm to three sequences if the gap penalty function was of the form $g(l) = \alpha$, i.e. the cost of an indel is independent of its length.

Wang et al. [10] studied the *tree alignment* problem where a multiple alignment is found by reconstructing the sequences at the internal nodes of a phylogeny minimizing the sum of all pairwise alignments between neighbors in the tree. In their approach gap costs are linear and the overall goal is to find a multiple alignment of the given sequences. The algorithm was improved in [11] where a polynomium time approximation scheme is given which yields a solution with a cost of at most 1.583 times the optimum, for sequences of length up to 200.

Finally, much work exists on finding the best multiple alignment of k sequences with no phylogeny given, e.g. [12,13]. Hein [9] devised an algorithm which finds an approximation to the optimal multiple alignment of k sequences and simultaneously reconstructs their ancestors, given the tree. However, multiple alignment is not our aim in this article: we give an algorithm which *optimally* reconstructs the sequences at the internal nodes of a phylogeny, given an alignment of the sequences at the leaves. The algorithm allows indels of any length and any subadditive gap penalty function (e.g. an affine function on the form $g(l) = \alpha + \beta \cdot l$ with $\alpha, \beta \geq 0$). Its worst-case time complexity is exponential, but in practice it is very fast and has been succesfully tested for nine HIV sequences of length 9868.

2 Indels of Length Greater than One

Given a multiple alignment the problem arises of how to interpret gap symbols as indels. When only indels of length one are allowed each column can be treated separately. When longer indels are allowed the problem becomes non-trivial as illustrated in Figure 1.

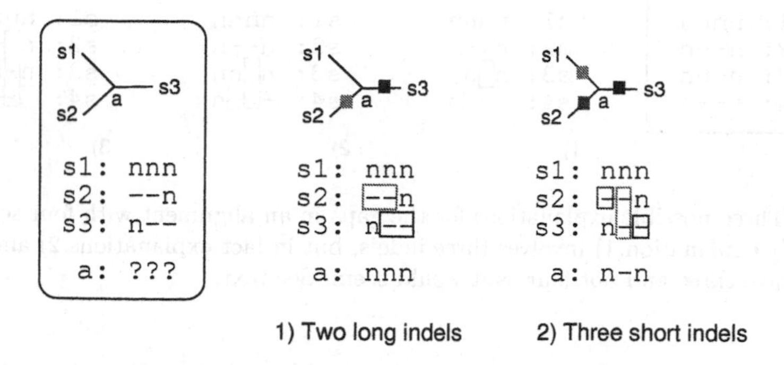

Fig. 1. Two possible explanations for the gaps in an alignment with three sequences (in rounded rectangle). Small boxes represent indels. For clarity we focus on indels and ignore substitutions, thus denoting all nucleotides simply by 'n'.

There are gap symbols in sequences s_2 and s_3. The gaps partly overlap and they can be explained in two ways:

(1) Two long indels have occurred: one on the evolutionary path between a and s_2 causing s_2's gaps, and one on the path between a and s_3 causing s_3's gaps. Consequently, a, the closest common relative of s_2 and s_3, has no gaps.
(2) Three short indels have occurred: one on the path between s_1 and a causing gaps in the middle interval in both a, s_2 and s_3 in one go; one between a and s_2 accounting for s_2's gap in the first column, and one between a and s_3 accounting for s_3's gap in the last column. With this explanation, a has a gap in the middle.

If we solve the problem the same way as when indels have length one, we would place the gaps as high (i.e., as far from the leaves) as possible in the tree, and so we would choose explanation 2) in Figure 1 — but that means that three indels have occurred whereas choosing explanation 1 means only two. On the other hand, since now both the *number* and *length* of the indels may vary between possible explanations, it is no longer obvious that the most parsimonious explanation is simply the one with the fewest indels. Whether two long indels are cheaper than three short ones depends on the chosen gap penalty function. With the function $g(l) = \alpha + \beta l$, where l is the length of the gap, the costs of the two explanations are $2(\alpha + 2\beta)$ versus $3(\alpha + \beta)$, respectively.

Not only is it non-trivial to exhaustively list all possible explanations; it may also be non-trivial to find the cost of a given explanation. Consider the example in Figure 2: in explanation 1), the gaps are caused by three indels; one of length 2 on the edge between s_2 and a causing s_2's gaps, one of length 1 on the edge

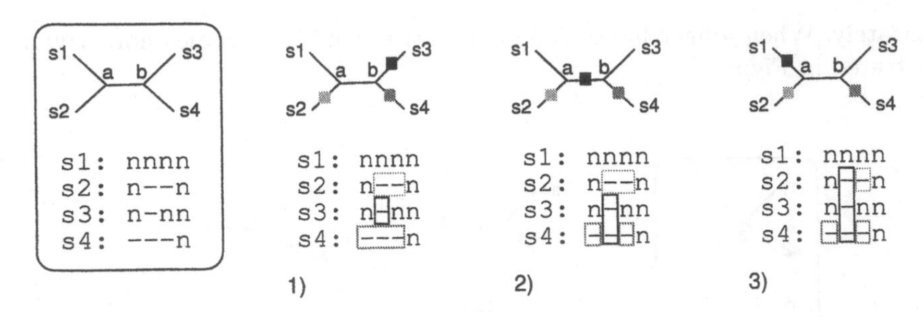

Fig. 2. Three possible explanations for the gaps in an alignment with four sequences. Obviously explanation 1) involves three indels, but in fact explanations 2) and 3) also only involve three and not four as it would seem. See text.

between s_3 and b causing s_3's gaps, and one of length 3 on the edge between s_4 and b causing s_4's gaps.

In explanation 2), a single indel occurring between a and b causes gaps in both s_3 and s_4 (and b) in the second column of the alignment. In explanation 3) this indel is pushed further back to the edge between s_1 and a now causing gaps in s_2, s_3 and s_4 (and a and b) all at once. Depending on the direction of evolution, this indel is either an insertion happening *after*, or a deletion happening *before* the emergence of b, s_3, and s_4. In both cases, the first and third alignment columns are in fact consecutive in b, s_3, and s_4. Therefore, though it might seem we need two indels to account for the two remaining gaps in s_4, they may be the result of only one. Thus, in this particular case, one indel can explain gaps in two non-adjacent alignment columns because the intermediate symbols have been deleted earlier or inserted later in the evolution. Concluding the example, we end up with the following costs for the three explanations: 1) $3\alpha + 6\beta$, 2) $3\alpha+5\beta$, and 3) $3\alpha+4\beta$. In passing, we observe that 3) is the optimal explanation for any (positive) choices of α and β.

These examples hopefully illustrate the complexity of handling indels of length more than 1. In general two questions arise: "What are the consistent explanations of a given multiple alignment in terms of indels?" and second: "Given some gap penalty function, which explanation is optimal?" In the following we give an algorithm that solves the problem.

3 Our Algorithm

We assume we are given a multiple alignment $\mathcal{M}$ of $k > 2$ sequences $S_1 \ldots S_k$, and a phylogeny in the form of a binary evolutionary tree $\mathcal{T}$ relating these sequences. Thus each S_i is a leaf in $\mathcal{T}$, and all other nodes in $\mathcal{T}$ have three neighbors. We do not take branch lengths into account. The length of $\mathcal{M}$ is n with the columns numbered 1 to n. Each sequence is a string over some alphabet Σ extended with a gap symbol '$-$'. Further, we are given a gap penalty function $g(l)$ defining the cost of a gap of length l. We assume g is subadditive: $g(l_1+l_2) \leq g(l_1)+g(l_2)$. In the following, for brevity we will let Σ denote the set of nucleotides $\{A, C, G, T\}$, and we will use the affine gap cost function $g(l) = \alpha + \beta l$, although any alphabet applies as well as, e.g., logarithmic or root gap cost functions. We will write $[a:b]$ to denote the sequence of integers from a to b, including the endpoints a and b.

No sequence has all gaps, and neither does any column. We assume that aligned nucleotides are related (call it the R assumption): if sequences s_i and s_j both have a nucleotide in some column, the closest common ancestor of s_i and s_j also has a nucleotide in this position. It follows that *all* sequences on the path from s_i to s_j in $\mathcal{T}$ have nucleotides in this position. We make no assumption about the direction of time in the tree $\mathcal{T}$. In other words, $\mathcal{T}$ is not *rooted.* For a leaf, we can talk about the unique *closest relative* since it has only one edge; for an internal node, the closest relative is not well-defined since internal nodes have three edges.

If an edge connecting two nodes a and b is removed from $\mathcal{T}$, the result is two disjoint *rooted* subtrees $\mathcal{T}_a$ and $\mathcal{T}_b$ rooted at a and b, respectively. An insertion that occurs on this edge inserts nucleotides in all nodes in $\mathcal{T}_a$ and gaps in all nodes in $\mathcal{T}_b$, or vice versa. Similarly, a deletion converts nucleotides to gaps in all nodes in $\mathcal{T}_a$ and leaves the nucleotides in all nodes in $\mathcal{T}_b$, or vice versa. In other words, *an indel affects a whole subtree* of $\mathcal{T}$. Thus we can and will identify an indel by the subtree in which it inserts gaps. Note that the closest relative of a subtree $\mathcal{T}' \subset \mathcal{T}$ is well-defined: it is the unique node $n \notin \mathcal{T}'$ which has an edge to the root of $\mathcal{T}'$.

It is easily shown that a tree with $k > 2$ leaves has $4k - 5$ different non-empty subtrees. We can represent a subtree $\mathcal{T}'$ of $\mathcal{T}$ unambiguously by its leaves; of course not all leaf subsets represent subtrees. Consider Figure 3: this tree with $k = 4$ leaves has $4k - 5 = 11$ different subtrees (counting $\mathcal{T}$ itself); e.g., $\{s_1\}$, $\{s_3, s_4\}$, and $\{s_1, s_2, s_4\}$. Note that, for example, $\{s_2, s_3\}$ is *not* a subtree: one cannot cut one edge and end up with a subtree containing exactly the leaves $\{s_2, s_3\}$.

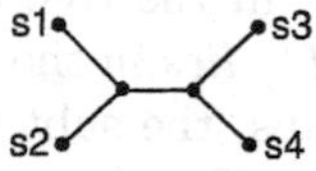

s1 s3
s2 s4

```
s1:nn-----
s2:nn--nnn
s3:----nnn
s4:n-nn---
```

Fig. 3.

3.1 Construction of a Gap Graph

An alignment column has a *gap configuration* given by the sequences that have gap symbols in this column. A maximal interval $I \subseteq [1:n]$ where alle columns have the same gap configuration is called a *gap interval.* E.g., the alignment in Figure 3 has four gap intervals: $I_1 = [1:1]$, $I_2 = [2:2]$, $I_3 = [3:4]$, and $I_4 = [5:7]$. Gap intervals are important since the indels used to explain the gaps in one column of a gap interval can be extended to explain the remaining columns without the cost of introducing extra indels (by subadditivity of the gap cost function this is cheaper, so indels are never broken off in the middle of a gap interval). Recall that each leaf in the evolutionary tree $\mathcal{T}$ corresponds to a sequence. The *tree covering* of a gap interval I is the minimal forest $\mathcal{F}$ of subtrees of $\mathcal{T}$ where

(C1) All leaves of $\mathcal{T}$ with gap symbols in I belong to a subtree in $\mathcal{F}$
(C2) All leaves in each subtree in $\mathcal{F}$ have gap symbols in I

Thus, a tree covering of I "covers" exactly the sequences that have gaps in I and not the ones that have nucleotides. Note that a tree covering is minimal in the *number* of subtrees that satisfy the above conditions. In other words, each subtree is maximal: it can not be extended without including leaves with nucleotides in I. In the following, we will represent a gap interval by a pair $(\mathcal{F}, I)$, where I is the gap interval itself and $\mathcal{F}$ is its associated tree covering.

Look again at Figure 3. In I_1, to satisfy condition C1 we need to cover the sequence s_3 only. Therefore, the tree covering $\mathcal{F}_1$ associated with I_1 is $\{\{s_3\}\}$, a forest of only one subtree. I_2 and I_3 are also covered by single subtree forests: $\mathcal{F}_2 = \{\{s_3, s_4\}\}$, and $\mathcal{F}_3 = \{\{s_1, s_2, s_3\}\}$. Finally, to cover I_4 we need two subtrees: $\mathcal{F}_4 = \{\{s_1\}, \{s_4\}\}$. This is a minimal forest: any subtree with more than one leaf would include s_2 or s_3 which have nucleotides in I_4, and thus the subtree would violate condition C2.

For now we focus on the gaps, and so we simply represent the alignment by its constituent gap intervals, i.e. the set $\{(\mathcal{F}_i, I_i)\}$, where, for all i, interval I_i precedes interval I_{i+1} in the alignment and $\bigcup I_i$ is a disjoint partition of $[1:n]$. We call such a set a *gap division.*

A gap division $\mathcal{D}$ now induces a *gap graph* in the following manner. Each gap interval $(\mathcal{F}_i, I_i) \in \mathcal{D}$ gives rise to a set of vertices in the graph: for each subtree $\mathcal{T}'$ in the tree covering $\mathcal{F}_i$, a vertex $(\mathcal{T}')_{I_i}$ is created. We say that the vertex $(\mathcal{T}')$ *lies in* the interval I_i. We observe that since no alignment columns have all gaps, the subtree $\mathcal{T}'$ of any vertex is a true subtree of the whole evolutionary tree $\mathcal{T}$ and so $\mathcal{T}'$ has a closest relative. As we shall see, a vertex $(\mathcal{T}')_{I_i}$ represents a *potential* indel: an indel occurring on the edge between the root of $\mathcal{T}'$ and its closest relative *could* be the cause of all the gaps in the sequences in $\mathcal{T}'$ in the gap interval I_i.

Consider a gap graph with two vertices $(\mathcal{T}_1)_{I_1}$ and $(\mathcal{T}_2)_{I_2}$ that lie in consecutive intervals I_1 and I_2. Then the subtree $\mathcal{T}_1$ has one of five possible relationships with $\mathcal{T}_2$:

- (*twin*) $\mathcal{T}_1 = \mathcal{T}_2$
- (*son*) $\mathcal{T}_1 \subset \mathcal{T}_2$
- (*father*) $\mathcal{T}_1 \supset \mathcal{T}_2$
- (*cousin*) $\mathcal{T}_1 \cap \mathcal{T}_2 \neq \emptyset$, but $\mathcal{T}_1 \not\subseteq \mathcal{T}_2$ and $\mathcal{T}_1 \not\supseteq \mathcal{T}_2$
- (*unrelated*) $\mathcal{T}_1 \cap \mathcal{T}_2 = \emptyset$

These five relationships are manifested in the gap graph as follows:

- If $\mathcal{T}_1$ and $\mathcal{T}_2$ are twins the two vertices are *merged* into a vertex $(\mathcal{T}_1)_{I_1 \cup I_2}$ that lies in the interval $I_1 \cup I_2$ (note that this new interval is not a gap interval since it does not have the same gap configuration everywhere).
- If $\mathcal{T}_1$ is the son of $\mathcal{T}_2$ a *directed edge* is created from the vertex $(\mathcal{T}_1)_{I_1}$ to the vertex $(\mathcal{T}_2)_{I_2}$.

– If $\mathcal{T}_1$ is the father of $\mathcal{T}_2$ a *directed edge* is created from the vertex $(\mathcal{T}_2)_{I_2}$ to the vertex $(\mathcal{T}_1)_{I_1}$.
– If $\mathcal{T}_1$ and $\mathcal{T}_2$ are cousins, an *undirected zigzag edge* is created between the vertices $(\mathcal{T}_1)_{I_1}$ and $(\mathcal{T}_2)_{I_2}$.
– If $\mathcal{T}_1$ and $\mathcal{T}_2$ are unrelated, no edges are created.

Thus, to construct a gap graph from a gap division $\mathcal{D} = \{(\mathcal{F}_i, I_i)\}$, we first create a vertex for each subtree in each tree covering $\mathcal{F}_i$. Then we traverse the gap intervals I_i looking at two consecutive intervals at a time and merge any twin vertices, create directed edges between fathers and sons, and create undirected zigzag edges between cousins. Note that twin vertices may be extended to more than two gap intervals. We next proceed to a full example.

Figure 4 shows an alignment of five sequences. It has four gap intervals $I_1, \ldots I_4$, each with its own tree covering $\mathcal{F}_1, \ldots \mathcal{F}_4$. Each subtree in each tree covering induces a vertex in the gap graph, so initially we get the six vertices $\{2\}_{I_1}$, $\{2\}_{I_2}$, $\{4,5\}_{I_2}$, $\{1,2,3,4\}_{I_3}$, $\{1,2\}_{I_4}$, and $\{4\}_{I_4}$. Traversing the gap intervals, first we merge the two $\{2\}$-vertices since they lie in consecutive intervals, and then we create a directed edge from this new vertex to $\{1,2,3,4\}_{I_3}$ since $\{2\} \subset \{1,2,3,4\}$. Next we create an undirected zigzag edge between $\{4,5\}_{I_2}$ and $\{1,2,3,4\}_{I_3}$ since neither is contained in the other while they still share the leaf 4. Finally, we create directed edges going from $\{1,2\}_{I_4}$ and $\{4\}_{I_4}$ to $\{1,2,3,4\}_{I_3}$. For clarity we write the subtree of a gap graph vertex as either a variable in parentheses, like in $(\mathcal{T}')_I$, or a specific list of leaves with no parentheses, like in $\{1,2\}_I$.

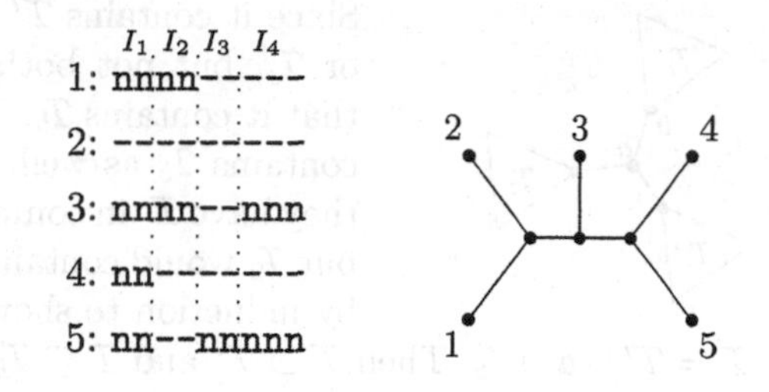

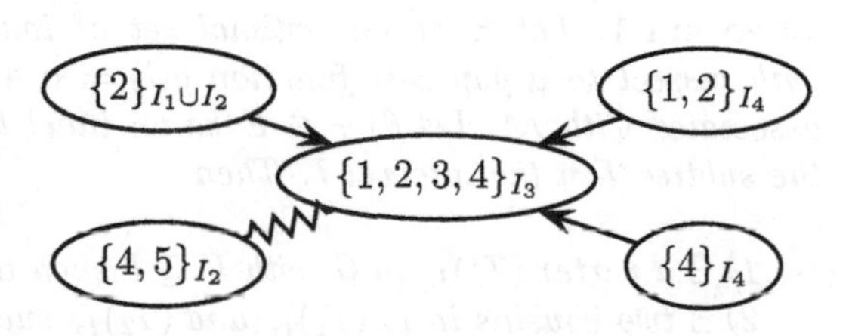

Fig. 4. Alignment with five sequences 1–5 and four gap intervals I_1–I_4, evolutionary tree, and induced gap graph. The tree coverings for the four gap intervals are $\mathcal{F}_1 = \{\{2\}\}$, $\mathcal{F}_2 = \{\{2\},\{4,5\}\}$, $\mathcal{F}_3 = \{\{1,2,3,4\}\}$, and $\mathcal{F}_4 = \{\{1,2\},\{4\}\}$.

Our algorithm now works with the gap graph to find the most parsimonious set of indels that explains the gaps in the alignment, given a gap penalty function $g(l) = \alpha + \beta l$. We first go through a preprocessing phase in which we reduce the potentially very large and complex gap graph. In the second phase, we resolve the reduced graph to find the optimal solution.

3.2 Preprocessing the Gap Graph

As already said, an indel $\theta_{I,\mathcal{T}}$ results in gaps in all nodes of a subtree $\mathcal{T}$ in some (set of) interval(s) I of the alignment. The following theorem establishes a strong connection between indels and gap graph vertices. Before we can prove the theorem, we need a small lemma.

Lemma 1. *Given a subtree $\mathcal{T}'$ and a sequence S of gap graph vertices $(\mathcal{T}_k)_{I_k}$ such that $\forall k$, (1) $\mathcal{T}' \subset \mathcal{T}_k$, and (2) $\mathcal{T}_k \subseteq \mathcal{T}_{k+1}$ or $\mathcal{T}_k \supseteq \mathcal{T}_{k+1}$ (i.e., no neighbors are cousins). Then there exists a subtree $\mathcal{T} \supset \mathcal{T}'$ such that $\mathcal{T} \subseteq \mathcal{T}_k$ for all k.*

Proof. Since $\mathcal{T}'$ is a real subtree of the $\mathcal{T}_k$'s, $\mathcal{T}'$ has a closest relative a, and each $\mathcal{T}_k$ consists of at least $\mathcal{T} \cup a$. Now a cannot be a leaf since the $\mathcal{T}_k$'s are real subtrees of the whole evolutionary tree (if some $\mathcal{T}_w$ were the whole tree, all sequences would have gaps in the associated interval I_w). Thus, a has relatives b and c with associated (possibly empty) subtrees $\mathcal{T}_b$ and $\mathcal{T}_c$, and the evolutionary tree looks like the figure below. Consider $\mathcal{T}_1$, the first subtree in the sequence S.

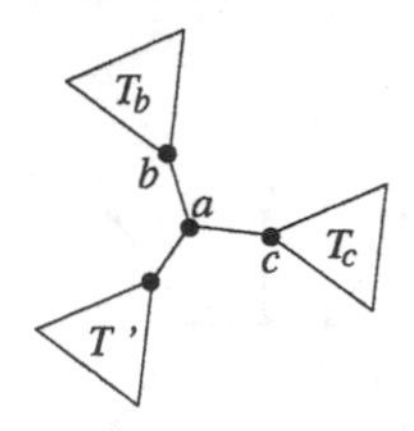

Since it contains $\mathcal{T}'$ and a, it must also contain either $\mathcal{T}_b$ or $\mathcal{T}_c$, but not both. Assume without loss of generality that it contains $\mathcal{T}_b$. Then its neighbor in the sequence, $\mathcal{T}_2$ contains $\mathcal{T}_b$ as well. If not, $\mathcal{T}_1$ and $\mathcal{T}_2$ would be cousins: they have $\mathcal{T}'$ in common, $\mathcal{T}_1$ contains $\mathcal{T}_b$ (and not all of $\mathcal{T}_c$), but $\mathcal{T}_2$ would contain $\mathcal{T}_c$. This argument is easily extended by induction to show that *all* subtrees $\mathcal{T}_k$ contain $\mathcal{T}_b$. Let $\mathcal{T} = \mathcal{T}' \cup a \cup \mathcal{T}_b$. Then $\mathcal{T} \supset \mathcal{T}'$ and $\mathcal{T} \subseteq \mathcal{T}_k$ for all $\mathcal{T}_k \in S$. □

Theorem 1. *Let E be an optimal set of indels explaining an alignment $\mathcal{M}$ with respect to a gap cost function $g(l) = \alpha + \beta l$, and let $\mathcal{G}$ be the gap graph associated with $\mathcal{M}$. Let $\theta_{I,\mathcal{T}} \in E$ be an indel that creates gaps in the nodes of the subtree $\mathcal{T}$ in the interval I. Then*

1) $\exists$ a vertex $(\mathcal{T}')_{I'}$ in $\mathcal{G}$ with $I' \subseteq I$ such that $\mathcal{T}' = \mathcal{T}$, or
2) $\exists$ two cousins in I, $(\mathcal{T}_1)_{I_1}$ and $(\mathcal{T}_2)_{I_2}$ such that $\mathcal{T} \subseteq \mathcal{T}_1$ and $\mathcal{T} \subseteq \mathcal{T}_2$.

Thus, either the indel corresponds to (the subtree of) an exisiting vertex that lies somewhere in the indel's interval I, or it "crosses" a cousin edge in I.

Proof. Let $I = \bigcup I_k$ where the I_k are are disjoint gap intervals in the gap graph $\mathcal{G}$. All leaves of $\mathcal{T}$ have gaps in all of I, and so these leaves are covered in each gap interval I_k in $\mathcal{G}$. Thus, in each I_k there exists a vertex $(\mathcal{T}_k)_{I_k}$ such that $\mathcal{T} \subseteq \mathcal{T}_k$. Let S be the set of subtrees of these vertices. If two neighbor subtrees in S, $\mathcal{T}_k$ and $\mathcal{T}_{k+1}$, are cousins, we are done. Otherwise, we have that $\mathcal{T}_k \subseteq \mathcal{T}_{k+1}$ or $\mathcal{T}_k \supseteq \mathcal{T}_{k+1}$ for all k. Assume that $\forall \mathcal{T}_k : \mathcal{T} \subset \mathcal{T}_k$.

Lemma 1 ensures the existence of a subtree $\mathcal{T}^*$ such that $\mathcal{T} \subset \mathcal{T}^* \subseteq \mathcal{T}_k$ for all $\mathcal{T}_k \in S$. Recalling that each $\mathcal{T}_k$ comes from a vertex $(\mathcal{T}_k)_{I_k}$ in the gap graph, all nodes/sequences in $\mathcal{T}^*$ must have gaps in I. By assumption, the optimal explanation E contains one indel $\theta_{I,\mathcal{T}}$ that accounts for the gaps in

$\mathcal{T}$ in interval I. The cost of this indel is $\alpha + \beta\ell$, if we let ℓ be the length of I. Further, some number of indels (say, $m > 0$) must account for the gaps in the nodes in $\mathcal{T}^* \setminus \mathcal{T}$ in interval I. Note that these m indels may stretch beyond I, so their total cost is $m\alpha + \beta(m\ell + d)$, where we sum the extra lengths beyond I in d. Thus, we can write the total cost g of the optimal explanation E as

$$g = (\alpha + \beta\ell) + m\alpha + \beta(m\ell + d) + q$$

where q is the cost of all indels accounting for other gaps. Consider the explanation E^* obtained from E by replacing $\theta_{I,\mathcal{T}}$ with the indel $\theta_{I,\mathcal{T}^*}$ which creates gaps in interval I in all nodes in the subtree $\mathcal{T}^*$. The cost of this new indel is still $\alpha + \beta\ell$, while we no longer have to let the m extended indels affect interval I. Thus we save $\beta m\ell$ on the cost, and the total cost g^* of the explanation E^* is

$$g^* = (\alpha + \beta\ell) + m\alpha + \beta d + q < g$$

But this contradicts the optimality of E. Therefore, $\mathcal{T}$ must be equal to at least one of the $\mathcal{T}_k$'s, and so the gap graph $\mathcal{G}$ must have a vertex $(\mathcal{T}_k)_{I_k}$ that corresponds to the indel $\theta_{I,\mathcal{T}}$. □

In general, given a vertex $(\mathcal{T}')_I$, the question is whether it corresponds directly to an indel happening on the edge between the root of $\mathcal{T}'$ and its closest relative, or whether the gaps of $\mathcal{T}'$ resulted from several indels happening inside $\mathcal{T}'$. In the latter case we say that the vertex is *decomposed*, otherwise it is *confirmed*. Theorem 1 tells us where to look when searching for the indels in the optimal explanation: Candidates are found among the (subtrees of the) gap graph vertices, and the *intersection between cousins*. In a (part of a) gap graph with no cousin edges all indels in the optimal explanation correspond directly to subtrees of existing vertices. If a cousin edge is present, however, the optimal explanation may contain indels not represented directly as vertices; such an indel must then correspond to a subtree which is contained in both of the cousins' subtrees and thus lies in their intersection.

Thus, Theorem 1 serves two causes. First it pins out cousin vertices as special cases, and second it justifies the perhaps intuitively reasonable notion that one should not place indels "lower" than necessary in the tree: if a vertex $\{1,2\}_I$ is not connected to leaf vertices $\{1\}_{I_j}$ and $\{2\}_{I_k}$ anywhere (and no cousins either), the corresponding gaps in sequences s_1 and s_2 do not both extend beyond I, and so there is no reason not to make the gaps in s_1 and s_2 in I in one go, i.e. on the edge between the root of the subtree $\{1,2\}$ and its closest relative. In fact we can say that if a vertex is decomposed, *all* of the decomposing indels continue and help explain other vertices in the gap graph. We formalize this in a lemma (which follows easily from Theorem 1 so we omit the proof) and elaborate it in a theorem.

Lemma 2. *If a gap graph vertex $(\mathcal{T})_I$ is decomposed in the optimal explanation of the corresponding alignment, then all of the indels explaining gaps in $\mathcal{T}$ extend beyond interval I.*

Theorem 2. *If a gap graph vertex $(T')_I$ is decomposed in the optimal explanation of the corresponding alignment, the decomposing indels do not all extend from I in the same direction.*

Proof. Let $(T')_I$ be decomposed into m indels. By Lemma 2, all decomposing indels extend beyond interval I. Assume they all extend in the same direction.

Let r be the root of T' and let a be the closest relative of T'. Let ℓ be the length of I. Let $g(l) = \alpha + \beta l$ be the cost of an indel of length l, and let g^* be the cost of the optimal explanation E^* of all gaps in the alignment, (i.e. the explanation that minimizes the total cost over all indels). In this explanation, the gaps in the leaves of T' are explained by the m indels that extend beyond I. The total cost of these indels can be written as $m\alpha + \beta(m\ell + d)$ (each such indel has some length greater than ℓ and we sum these extra lengths in the constant d). We can then write

$$g^* = m\alpha + \beta(m\ell + d) + q$$

where q is the total cost of all other indels. Since T' is decomposed into at least two indels, $m \geq 2$.

These m indels all start in interval I and by assumption they extend in the same direction. If they all end in exactly the same column c_e in the alignment, we replace them with only one indel occurring between r and a and extending from I to c_e. Such an explanation would trivially be cheaper than g^*, contradicting the optimality of E^*. Thus, the m indels do not have equal lengths. Let c be the first alignment column where one of the m indels ends (call this indel θ_1). The column c is outside of I since θ_1 extends beyond I. Define I_x to be the extended interval that starts with and includes I and ends in column c; let its length be ℓ_x.

Consider the explanation E_1 obtained from E^* by replacing θ_1 with the indel $\theta_{I_x,T'}$ occurring between r and a in the evolutionary tree and extending all through the interval I_x. This new indel explains the gaps in all T' in all of I_x, including I, while $m-1$ of the original indels (all except θ_1) still have to be made since they go beyond I_x; however, as we have covered all gaps in I_x with $\theta_{I_x,T'}$ we save $\beta m\ell_x$ on the cost. Therefore, the cost of explanation E_1 is

$$\begin{aligned} g_1 &= (\alpha + \beta\ell_x) + (m-1)\alpha + \beta(m\ell + d) + q - \beta m\ell_x = \\ & m\alpha + \beta(m\ell + d) + q - \beta(m\ell_x - \ell_x) \qquad = g^* - \beta\ell_x(m-1) \end{aligned}$$

Since $m \geq 2$ we have that $\beta\ell_x(m-1) > 0$ and thus $g_1 < g^*$. But this contradicts the optimality of E^* and so our assumption must be wrong. Therefore, the decomposing indels do not all extend in the same direction. □

Recall that a gap graph vertex represents gaps in a subtree of sequences in some region of an alignment. The vertex may have edges to other vertices, and these edges can have three different types (in-, out-, and cousin edges). It turns out that vertices with certain configurations of edges can immediately be confirmed to correspond to optimal indels. In other words certain types of vertices can be "decided" immediately and locally. Using the two strong Theorems 1 and 2 we can characterize formally exactly which types of vertices we can immediately decide. We will not go into the details here but simply give the results.[1]

A vertex is called a *leaf vertex* if its subtree consists of one leaf node only. Such a vertex can be confirmed right away and thus corresponds to an optimal indel occurring between this node and its closest relative in the tree. Likewise, *orphans* and *end vertices* can be confirmed: they are vertices with no edges and vertices with edges on one side only, respectively. All sequences in the subtree of an orphan have nucleotides on either side of its interval.

Consider a vertex $(\mathcal{T}')_I$ where *some but not all* of the leaves of its subtree $\mathcal{T}'$ have gaps in an adjacent interval. Such a vertex is called a *patriarch*. A patriarch has at least one edge (zigzag or in-going), but its subtree also has at least one leaf which is not included in a subtree in any vertex in an adjacent interval. We say that such a leaf has *solo gaps*: it has gaps in interval I but nucleotides on both sides. Patriarchs may also be confirmed immediately (using Lemma 2); intuitively, since it is necessary to spend an indel on the solo gaps, this indel may as well create all other gaps of the vertex' subtree also. A patriarch has another important property: after its confirmation, its edges are removed and new edges are created directly between its former neighbors on either side (if any are needed). A patriarch has no out-edges, and its subtree contains the subtrees of its in-edge neighbors; thus, by the same argument explained in connection with Figure 2, the intervals associated with the patriarch's neighbors are in fact consecutive and so the neighbors may be connected directly if they share leaves. A special case is when the patriarch has two cousins; in this situation the edges cannot be removed since we have to keep the information that no indel with a larger subtree than that of the patriarch can "pass" the patriarch and cause gaps on both of its sides in one go.

A few other reduction rules help reduce the gap graph further. The preprocessing phase makes use of these theoretical results by going over the gap graph in a series of passes, each pass reducing the graph by local applications of the reduction rules. Since the removal of edges following a patriarch confirmation may turn previously undecidable vertices decidable, several passes are performed.

3.3 Resolving the Reduced Gap Graph

Once the preprocessing is done, the gap graph has been reduced significantly (as illustrated by Figure 6 below). In other words, most of the gaps have been (optimally) explained. We now turn to the *chains* of the gap graph: two vertices belong to the same chain if and only if there exists a path connecting the two which does not cross a leaf vertex. Thus, chains often end in leaf vertices, some of which may belong to two chains (see Figure 5). The chains are important

[1] Proofs are found at `http://www.daimi.au.dk/~chili/BiRC/gapgraphAppendix.pdf`.

since they can be dealt with independently: indels causing gaps in the vertices of one chain could not have caused gaps in vertices of another chain.

Say that after preprocessing, (a part of) the gap graph looks like Figure 5. In the second phase of our algorithm, we analyze each chain in turn. Looking first at chain a, we see that only the vertex (123) is still undecided (the others being leaf vertices). Referring to Theorem 1, we note that since there are no cousin edges in this chain, the only indels we need to consider for the optimal explanation are the ones already represented by vertices in the chain. That means there are only two possible explanations for the gaps represented by the vertex (123): 1) Either they are the result of one indel occurring between the root of the subtree $\{1, 2, 3\}$ and its closest relative (in which case we confirm the vertex), or 2) they are the result of three indels occurring between 2 and its closest relative, 1 and its closest relative, and 3 and its closest relative, respectively, extending the already confirmed neighboring indels/vertices (in which case we decompose the vertex). If we let l_{123} be the length of its associated alignment interval, the extra cost for resolving the vertex (123) in each of these two explanations is $\alpha + \beta l$ versus $3\beta l$. For the optimal explanation we simply choose the cheaper one (in general, if several options are optimal, we choose one arbitrarily).

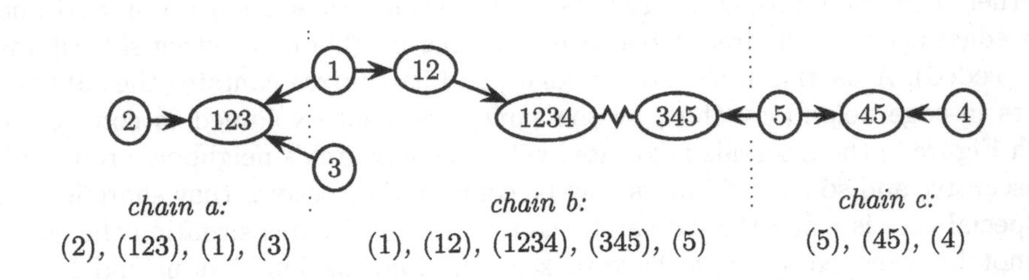

Fig. 5. Gap graph with three chains (same tree as in Figure 4). We omit interval labels on the vertices, and we write, e.g., (2345) to denote the subtree $\{2, 3, 4, 5\}$.

Chain b has three undecided vertices: (12), (1234), and (345). Moreover, the latter two are cousins and so by Theorem 1 we need to consider indels whose subtrees lie in their intersection. That gives two additional potential indels not already represented in the chain by vertices, namely $\{3\}$ and $\{4\}$. Interestingly, the vertex (12) can therefore be confirmed right away: its only possible decomposition is in subtrees $\{1\}$ and $\{2\}$, but since $\{2\}$ is neither present as a vertex nor lies in the cousin intersection, this indel could never be part of an optimal explanation. For the same reason, the vertex (1234) may either be confirmed or decomposed in one way only, namely with three indels with the subtrees $\{1, 2\}$, $\{3\}$, and $\{4\}$. And finally the vertex (345) may either be confirmed or decomposed in three indels with the subtrees $\{3\}$, $\{4\}$, and $\{5\}$ (this time, Theorem 1 dictates that the indel $\{4, 5\}$ is not an option in chain b). In total therefore we have four combinations which we need to check in this chain. Lastly, in chain c we may either confirm the vertex (45) or decompose it in $\{4\}$ and $\{5\}$.

Checking all combinations we find the optimal indels. Each causes gaps in some set of columns in a full subtree, and by the R assumption (see Section 3) we place anonymous nucleotides in the internal nodes in the rest of the tree in the same columns. It remains to optimally name these nucleotides. For each column we ignore the subtrees that have all gaps and get a still connected tree with all nucleotides, and on this tree we now perform the Finch-Hartigan-Sankoff algorithm as a final step to optimally assign nucleotides to the internal nodes.

4 Performance and Future Work

Checking the cost of each combination of confirmation/decomposition of all the vertices in a chain cannot be done in constant time. E.g., looking again at Figure 5 and the combination in chain b of decomposing both (1234) and (345) we have to look at both decompositions to see that the decomposing indels $\{3\}$ and $\{4\}$ can extend to both vertices and do not have to be opened anew. Note also that since we have to check "all with all", the number of combinations to check is exponential in the number of undecided vertices in the chain. I.e., it is critical that as many edges as possible are removed in the preprocessing phase. If the tree is large, some vertices may have many possible decompositions; if the alignment has many "crisscrossing" gaps, the chains may become very long. Encountering, e.g., a chain with 11 undecided vertices each with five possible decompositions (including not decomposing) except two which have 10 decompositions, we have in total $5^9 \cdot 10^2 = 195312500$, close to 200 million, combinations to check (each taking some hard to characterize but more than constant amount of time) for this chain alone (see Figure 6). Thus, the worstcase time complexity is exponential in the length of the alignment.

On the other hand, if the gaps mostly fall in straight blocks delimited by full columns of nucleotides, chains will be short and the running time will be very fast. Thus, the running time of our algorithm is extremely dependent on the particular problem instance. To demonstrate this, we have tested the algorithm on a huge number of different alignments and phylogenies. The main type of data was randomly generated alignments with some ratio of gaps scattered individually across all sequences. We wanted to challenge our algorithm with very hard data, and a long such alignment with a high gap ratio of 60% is indeed a very hard problem to solve. Real alignments do not look like this (see Figure 7 for a screenshot of our program running a 60% gap alignment): first they do not have that many gaps unless the aligned sequences are only very distantly related. And second, since gaps appear as the result of indels, they do appear in connected blocks and not as much as independent single gaps. For this reason we also did some experiments on alignments with only 40% and 50% gaps, and finally we ran the algorithm on an alignment of nine full HIV-1 subtype genomes.

In the left plot of Figure 6 we show the average performance for random alignments with 60% gaps over different alignment lengths with 1000 trials for each, on a tree with nine leaves. For an alignment of length 10000 the average time was around 40 seconds. For lengths below 6000, the time is less than 10

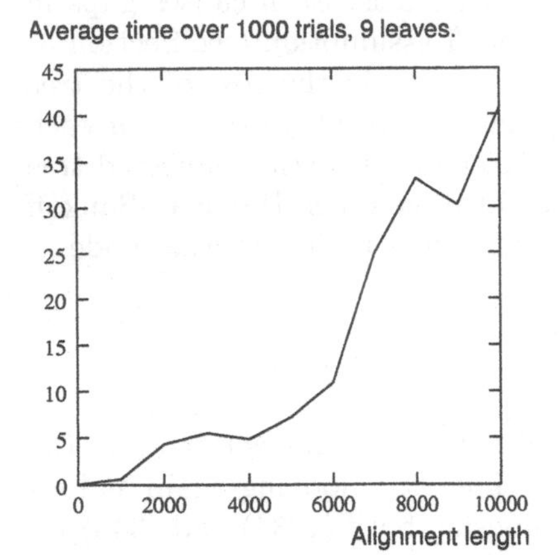

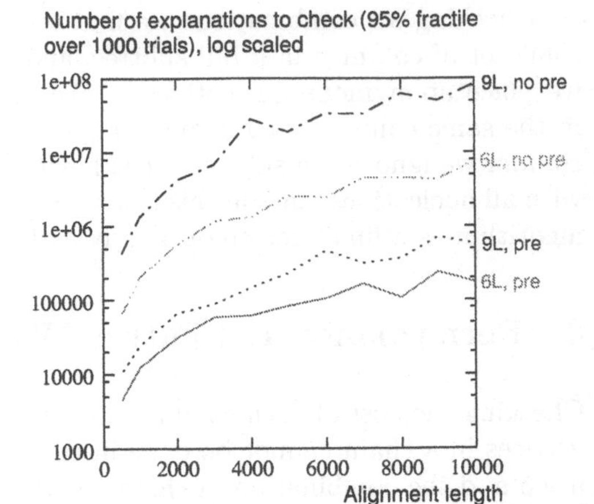

Fig. 6. Performance on random alignments with 60% gaps. Left plot: average times for different lengths, on tree with nine leaves. Right plot: 95% fractiles of the number of combinations to check for different alignment lengths, on trees with six and nine leaves (6L/9L) with and without preprocessing (pre/no pre). The curve '9L, pre' comes from an experiment similar to the one underlying the left plot.

seconds.[2] The curve does not appear smooth; this is due to the abovementioned dependency on the particular problem instance resulting in a huge variance.
The right plot of Figure 6 is an attempt to show the effect of the preprocessing phase. Since it would have been too time consuming to actually find the optimal explanation in 1000 trials with no preprocessing, we instead report the number of combinations it would have been necessary to check. Again, the variance is huge so we report the 95% fractiles (i.e., a value of 10000 means that in 95% of the trials, the number of explanations to check was at most 10000). We did experiments with and without preprocessing, and to also get an idea of the performance for different tree sizes we used both a tree with nine leaves and one with six leaves. For example, for alignment length 10000 and a tree with nine leaves, the numbers are about 500.000 with preprocessing versus about 100 million without preprocessing; i.e. preprocessing reduces the number of combinations to check by a factor of about 200 in this case. We also observe that for the smaller tree, the number of combinations to check is substantially smaller.

This performance is contrasted by our results for a tree with nine leaves and alignments with only 40% or 50% gaps: for all lengths up to 8000, the average running time was less than one second. We also analyzed a full genome alignment of B.ES.89.S61K15, B.FR.83.HXB2, B.GA.88.OYI, B.GB.83.CAM1, B.NL.86.3202A21, B.TW.94.TWCYS, B.US.86.AD87, B.US.84.NY5CG, and B.US.83.SF2, nine HIV-1 subtypes retrieved from the Los Alamos HIV database (`http://hiv-web.lanl.gov/content/index`). After constructing a phylogeny using the neighbor-joining Sanger Institute Quicktree software (`http://www.sanger.ac.uk/Software/analysis/quicktree/`) we had a tree with nine leaves

[2] All experiments were done on a 2.4 GHz Pentium 4 machine with 512 MB RAM running Linux.

and an alignment of length 9868 (this example only serves to demonstrate our algorithm, the methods chosen to obtain phylogeny and alignment were arbitrary). Solving this problem again took less than one second. Because of the great similarity of the sequences, the resulting gap graph was small and simple.

We are interested in running our program on more, real data sets with alignments and trees of various sizes. Much larger trees with shorter alignments (100–500 leaves, length < 500) is an interesting application which we have not yet looked at. Also, we intend to combine the algorithm with some heuristic search method so as to *find* an approximately optimal alignment which fits the tree, rather than assume it is given (a variant of the tree alignment problem). One might consider our algorithm a way of ranking a given combination of tree and alignment using a gap cost function and a substitution matrix, and so trying different alignments, or even trees, might actually improve the data.

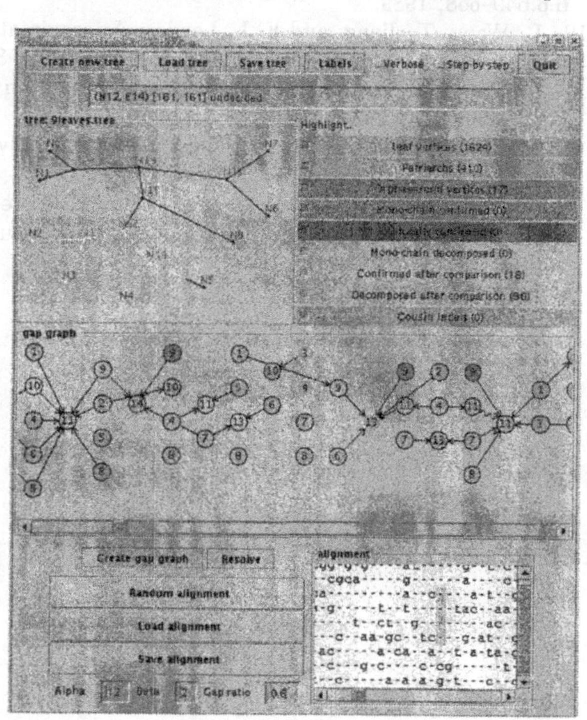

Fig. 7. Screenshot with phylogeny, alignment, and induced gap graph. Note what a random alignment with 60% gap looks like.

References

1. W. M. Fitch: Towards defining the course of evolution: minimum change for a specific tree topology. Systematic Zoology 20.406–416, 1971.
2. D. Sankoff: Matching sequences under deletion/insertion constraints. Proc. Natl. Acad. Sci. USA 69.4–6, 1972.
3. J. A. Hartigan: Miminum mutation fits to a given tree. Biometrics 20.53–65, 1973.
4. P. Sellers: An algorithm for the distance between two finite sequences. J. Comb. Theory 16.253–258, 1974.
5. D. Sankoff: Minimal mutation trees of sequences. SIAM J. appl. Math 78.35–42, 1975.
6. M. S. Waterman, T. F. Smith, and W. A. Beyer: Some biological sequence metrics. Advances in Mathematics 20.367–387, 1976.
7. O. Gotoh: An improved algorithm for matching biological sequences. J. Mol. Biol. 162.705–708, 1981.

8. M. L. Fredman: Algorithms for computing evolutionary similarity measures with length independent gap penalties. Bull. Math. Biol. 46.4.545–563, 1984.
9. J. J. Hein: A method that simultaneously aligns and reconstructs ancestral sequences for any number of homologous sequences, when the phylogeny is given. Mol.Biol.Evol. 6.6.649–668, 1989.
10. L. Wang, T. Jiang, and E. L. Lawler: Approximation Algorithms for Tree Alignment with a Given Phylogeny. Algorithmica 16:302–315, 1996.
11. L. Wang and D. Gusfield: Improved Approximation Algorithms for Tree Alignment. CPM 1996: 220–233.
12. J. Stoye: Multiple sequence alignment with the divide-and-conquer method. Gene 211 (2), GC45–GC56, 1998.
13. E. Althaus, A. Caprara, H.-P. Lenhof, and K.Reinert: Multiple sequence alignment with arbitrary gap costs: Computing an optimal solution using polyhedral combinatorics. ECCB 2002: 4–16.

Optimal Multiple Parsimony Alignment with Affine Gap Cost Using a Phylogenetic Tree

Bjarne Knudsen

Department of Zoology, University of Florida, Gainesville, FL 32611-8525, USA
bk@birc.dk

Abstract. Many methods in bioinformatics rely on evolutionary relationships between protein, DNA, or RNA sequences. Alignment is a crucial first step in most analyses, since it yields information about which regions of the sequences are related to each other. Here, a new method for multiple parsimony alignment over a tree is presented. The novelty is that an affine gap cost is used rather than a simple linear gap cost. Affine gap costs have been used with great success for pairwise alignments and should prove useful in the multiple alignment scenario. The algorithmic challenge of using an affine gap cost in multiple alignment is the introduction of dependence between different columns in the alignment. The utility of the new method is illustrated by a number of protein sequences where increased alignment accuracy is obtained by using multiple sequences.

1 Introduction

An algorithm for pairwise parsimony alignment of biological sequences was devised by Needleman and Wunsch in 1970 [1]. The minimum cost of an alignment is found by a recursive algorithm in the prefixes of the two sequences. Once the matrix of optimal prefix alignment costs has been found, the optimal alignment can be found by going back through the matrix (backtracking). The cost of gaps in this algorithm is a linear function of the gap length.

The original pairwise alignment algorithm had a time and memory complexity of $O(L^2)$ (for sequences of length L). In 1975, however, Hirschberg described an algorithm for pairwise alignment that reduced the memory complexity to $O(L)$, while only increasing time consumption by a factor of two [2,3].

In 1982, Gotoh published an algorithm for pairwise sequence alignment that can use a gap cost which is an affine function of the gap length, rather than linear [4]. This was done in a way resembling the Needleman-Wunsch algorithm while keeping track of three different types of residue configurations of the last alignment step: 1) residues in both sequences, 2) a residue in sequence one aligned to a gap in the other, and 3) a residue in sequence two aligned to a gap in sequence one. The use of an affine gap cost improves the alignments of biological sequences significantly [5].

At close to the same time, Sankoff and Cedergren described an algorithm for multiple parsimony alignment over a tree with linear gap cost [6]. Other

G. Benson and R. Page (Eds.): WABI 2003, LNBI 2812, pp. 433–446, 2003.

methods have used the sum of pairs scores for multiple alignments, rather than scores based on a tree [7]. This scoring scheme does not have an evolutionary explanation, which gives it some drawbacks [5]. There has also been various approximation methods proposed for multiple alignment [8,9,10].

This work gives an algorithm for optimal multiple parsimony alignment with affine gap cost for any number of sequences related by a tree. It combines ideas from pairwise alignment with affine gap cost, and ideas from multiple alignment with linear gap cost [4,6]. The memory reducing technique of Hirschberg is also applied [2]. The resulting algorithm has a memory complexity of $O(7.442^N L^{N-1})$ and a time complexity of $O(16.81^N L^N)$ (for N sequences). These complexities should be compared to $O(L^{N-1})$ and $O(2^N L^N)$, respectively, for the linear gap cost method [2,6].

Some examples are given to show the utility of the algorithm. Due to the large time and memory complexities, it is not practical to use for more than three sequences or maybe four short sequences. The method could, however, prove useful in a number of other situations, such as alignment quality evaluation and pair-HMM based probabilistic alignment, particularly when using sampling based approaches.

2 The Problem

2.1 Pairwise Parsimony Alignment

Gaps in pairwise alignments are the result of insertions and deletions (indels) of sequence pieces. The cost of indels in the affine gap cost setting is based on a gap introduction cost ($g_i > 0$) and a gap extension cost ($g_e < g_i$), giving a total cost of $g_i + (l-1)g_e$ for an indel of length l. For a sequence alphabet Σ of size n, a symmetric $n \times n$ matrix, $\{m_{ij} \geq 0\}$, gives the cost of matches for all possible pairs of residues. The cost of matching two identical residues is zero. Here is an example:

```
G C G G G T -
C C - - G T A
```

$$\text{Cost: } m_{GC} + m_{CC} + g_i + g_e + m_{GG} + m_{TT} + g_i = m_{GC} + 2g_i + g_e$$

Notice that the total indel cost is independent of the nature of the residues, given the alignment. The optimal alignment of two sequences is the one that minimizes the total cost.

2.2 Multiple Parsimony Alignment

Assume that N sequences are given, as well as an unrooted binary tree relating them. The N sequences are located at the leaves of the tree, while the ancestral sequences are at the internal vertices. Assume for a moment that the ancestral sequences are known, so an alignment of all these sequences implies a pairwise alignment of the two sequences at the ends of each edge in the tree. For these pairwise alignments, columns with gaps in both sequences are removed. Now, the

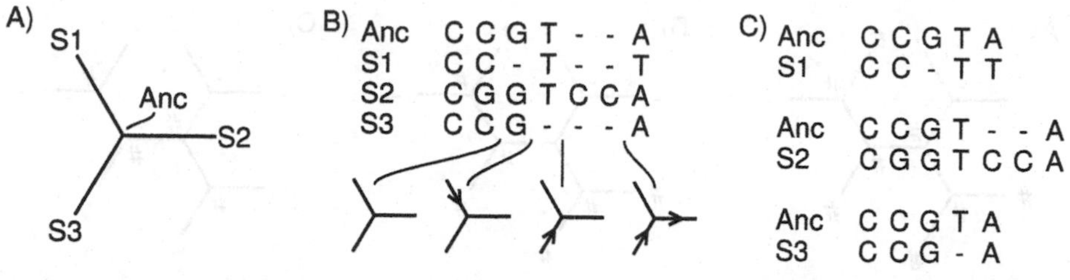

Fig. 1. A) The phylogenetic tree relating the three sequences S1, S2, and S3, and their common ancestor, Anc. B) The multiple alignment of the four sequences and indel configurations at various points (see Sect. 3.3). C) The implied pairwise alignments. The cost of the multiple alignment is $3g_i + g_e + 5m_{CC} + 2m_{GG} + 2m_{AA} + 2m_{TT} + m_{AT} + m_{CG} = 3g_i + g_e + m_{AT} + m_{CG}$.

price of the whole alignment is the sum of the costs of all the implied pairwise alignments of neighboring sequences (see Fig. 1).

If the ancestral sequences are not given, the cost of the multiple alignment of the N sequences is the minimum cost over all possible ancestral sequences and all possible alignments of the sequences. An optimal multiple alignment of the N sequences is one with minimal cost.

2.3 The Objective

Given N sequences, a phylogenetic tree relating them, a matrix of match costs, a gap introduction cost, and a gap extension cost, we seek the optimal multiple parsimony alignment of the sequences.

3 Algorithm

3.1 Residue Configurations

For a given tree, label each vertex with either a residue (#) or a gap (-), and denote that a residue configuration for the tree. Call a residue configuration *acceptable* if 1) the vertices labeled by residues form a connected graph and 2) at least one leaf has a residue (see Fig. 2).

A residue configuration will be used to define what occurs in a single column of a multiple alignment. When aligning two residues, the biological interpretation is that they are related to each other, which is the reasoning behind condition one of an acceptable residue configuration. A column with no residues in any of the sequences being aligned can not be part of an optimal alignment (since $g_i > 0$), leading to condition two for an acceptable configuration.

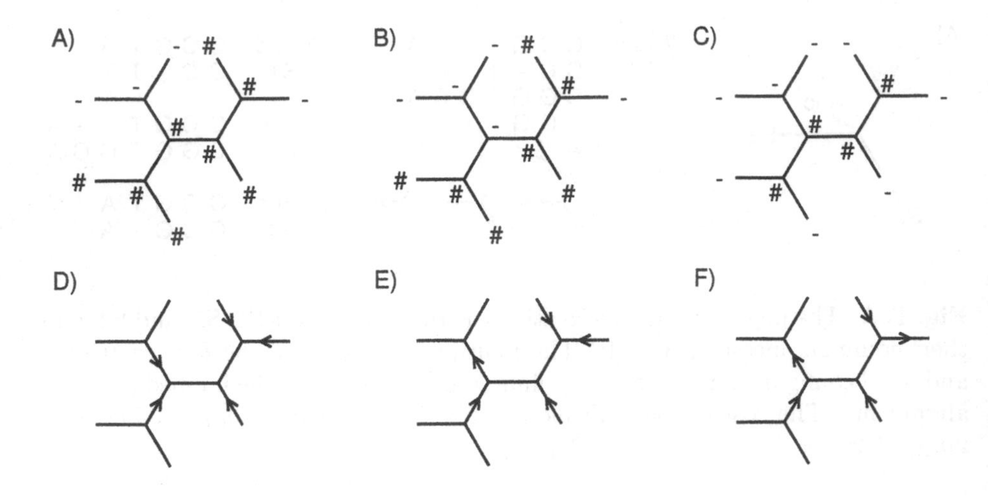

Fig. 2. The three top trees show various residue configurations, while the bottom three shows indel configurations. A) An acceptable residue configuration, since the vertices with residues form a connected subtree and some leaves have residues. B) and C) Unacceptable residue configurations. D) An unacceptable indel configuration, since some vertices are isolated from all the leaves. E) Another unacceptable indel configuration, where changing the direction of one arrow isolates some vertices from all the leaves. F) An acceptable indel configuration.

3.2 Indel Configurations

Given a tree, we can choose to label each edge in one of three ways: 1) an arrow pointing from one end to the other, 2) an arrow pointing the opposite direction, and 3) just by a line. We will call such a labeling an indel configuration for the tree.

For an indel configuration on a tree, a vertex, a, is isolated from another, b, if the path from a to b has an arrow pointing opposite to the direction of the path. An indel configuration is *acceptable* if it satisfies the following constraints: 1) no vertex is isolated from all the leaves and 2) this still holds if the direction of any single arrow is reversed (see Fig. 2).

The indel configurations are used for keeping track of ongoing indels in a multiple alignment. There is no immediate biological reasoning for the unacceptable indel configurations, but it turns out that they can never be part of an optimal alignment, see Sect. 3.4.

3.3 The Recursion

Define a prefix alignment of the N sequences as the optimal alignment of certain prefixes of the sequences. An N dimensional integer vector $\boldsymbol{v}$ defines the number of residues from each sequence that is included in the prefix alignment.

The algorithm starts with the empty prefix alignment (no residues from any sequences) and builds it up column by column, while keeping track of the minimal cost. Using this approach, the optimal cost of the entire alignment can be found, and by backtracking through the costs an optimal alignment may also be found. Notice that there could be more than one optimal alignment.

For a prefix alignment, the indel cost of an additional column will depend on which pairwise indels have been initiated, since the gap cost is affine. The indel configurations are used to keep track of ongoing indels between two sequences that are neighbors on the tree. An arrow indicates that an indel was present between the sequences at the ends of the corresponding edge in the last column involving any of these two sequences. The arrow points in the direction of the sequence that had a residue and away from the one with a gap. A line with no arrow indicates a match in the last column involving the two sequences (see Fig. 1).

Because of the dependence on past indels, we will keep track of each possible ending indel configuration for the prefix alignments. Let $E_{\boldsymbol{v}}^{(d)}$ denote the optimal cost of the prefix alignment of the sequences corresponding to the sequence vector $\boldsymbol{v}$ and indel configuration d.

The addition of a column in the alignment may change the indel configuration depending on which matches are made and which gaps are introduced and extended. For a given indel configuration, let us say that we apply a residue configuration when adding a column with the corresponding residues to give a new indel configuration (see Fig. 3). The change in indel configuration is given by one of the following three possibilities. 1) An edge with residues at both ends changes the indel configuration to a line at that edge (at no cost). 2) An edge with a residue at one end, and a gap at the other, changes the indel configuration to an arrow pointing toward the residue (at a cost of g_e or g_i if the same arrow was there already or not, respectively). 3) An edge with gaps at both ends does not change the indel configuration (and has no cost). Thus, the gap cost of applying a residue configuration to an indel configuration is the gap cost of the new alignment column, given the existing prefix alignment.

For a vector, $\boldsymbol{e}$, with coordinate values of one or zero for each sequence (including ancestors), the corresponding residue configuration is the one with residues for ones and gaps for zeros. For an indel configuration d', applying the residue configuration corresponding to the vector yields another indel configuration, d. We write this as $d' \xrightarrow{\boldsymbol{e}} d$. The gap cost of applying the residue configuration is written $G_{\boldsymbol{e}}^{(d')}$.

When going from prefix alignment $\boldsymbol{v} - \hat{\boldsymbol{e}}$ to $\boldsymbol{v}$ (with $\hat{\boldsymbol{e}}$ denoting the vector with only the N entries of $\boldsymbol{e}$ corresponding to the leaves), the residues of the sequences are known, and the optimal ancestral residues are found by a post order tree traversal giving the minimal residue match cost [11]. The match cost is written as $M_{(\boldsymbol{v}-\hat{\boldsymbol{e}})\rightarrow\boldsymbol{v}}$, which is not dependent on which ancestral sequences has residues, since matching two identical residues is free ($m_{ii} = 0$). For the alphabet, Σ, the match costs can be found for all possible residue configurations

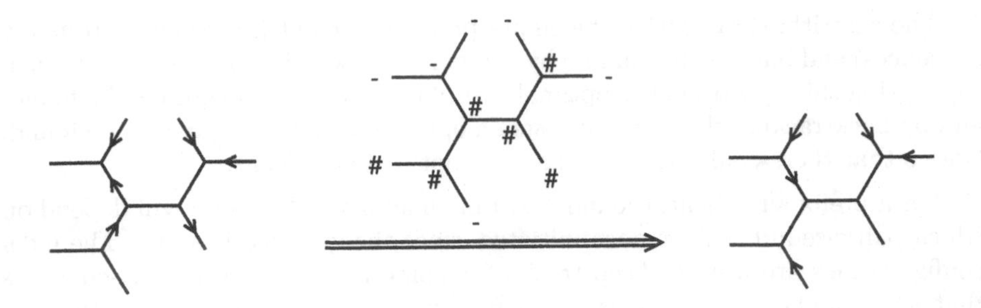

Fig. 3. A residue configuration (over double arrow) is applied to an indel configuration (left) giving a new indel configuration (right). The associated cost is $2g_i + 2g_e$.

before starting the recursion, which will have no effect on the time or memory complexity for large L.

Each acceptable residue configuration defines a possible column in the alignment, and thus a step from one prefix alignment to the next. We can write the algorithm for finding the optimal cost as this recursion:

$$E_{\mathbf{v}}^{(d)} = \min_{\mathbf{e}, d' \xrightarrow{\mathbf{e}} d} \{E_{\mathbf{v}-\hat{\mathbf{e}}}^{(d')} + G_{\mathbf{e}}^{(d')} + M_{\mathbf{v}-\hat{\mathbf{e}} \rightarrow \mathbf{v}}\} , \tag{1}$$

with $E_{\mathbf{0}}^{(d)} = 0$ for the indel configuration, d, where all edges are labeled with a line, and $E_{\mathbf{0}}^{(d')} = \infty$ for all other indel configurations, d'. Keeping all the optimal prefix alignment scores, we can see which shorter prefix alignment each one came from. Doing this from the full alignment to the empty alignment, we obtain an optimal multiple alignment.

3.4 Reduced Resources Recursion

We saw in Sect. 3.1 that only acceptable residue configurations can be part of an optimal alignment. It will be shown here that the same applies to indel configurations.

Assume that we have an indel configuration where some vertices are isolated from all the leaves. Let a be the last of these vertices to have a residue (or one of the last if there is more than one), and consider the column in the prefix alignment where that occurred. Since only acceptable residue configurations are used, this column must have had a residue at a leaf and there would be a connected set of vertices with residues between a and the leaf. Since this is the case, a can not be isolated from that leaf immediately after this column. The only way to isolate a from the leaf is by having a later column with a residue at a and none at the leaf, which is a contradiction.

Now, assume that we have an indel configuration where no vertices are isolated from all leaves, but the reversal of an arrow at the edge α isolates some vertices from all leaves. The last column involving residues at either end of α must have had a residue at one end and a gap at the other end. The vertex, a, that had the gap must become isolated upon the reversal of the arrow at α. Thus, the last column, C_i, with a residue at a could not have had a residue at any of the leaves on the same side of the tree relative to α. Let C_j be the last column before C_i, where a leaf on this side of the tree had a residue.

The prefix alignment of all sequences (including ancestral), leading to this indel configuration, could be changed to have a gap at a, and all other vertices on that side of the tree, in the columns C_{j+1} to C_i. This would not change the alignment of the leaf sequences. Furthermore, the cost of this prefix alignment would be lower, since the match cost would be the same and it has less gap introductions or less gap extensions, or both. This means that the assumed indel configuration could never be part of an optimal alignment if $g_e > 0$. If $g_e = 0$, both configurations could be optimal, so the assumed indel configuration is again not necessary to ensure finding an optimal alignment.

With this, it has been shown that only acceptable indel configurations need to be taken into account in the recursion. Furthermore, for certain indel configurations only a subset of the acceptable residue configurations (the ones leading to acceptable indel configurations) needs to be applied.

3.5 Hirschberg's Memory Reduction

In (1), each prefix alignment cost, $E_{\boldsymbol{v}}^{(d)}$, only depends on the costs for $\boldsymbol{v}'$, where any coordinate is the same or one lower than $\boldsymbol{v}$. Denote a row of costs as a set of costs where the coordinate of $\boldsymbol{v}$ corresponding to sequence one is the same. To find the cost of the optimal alignment, we only need to keep the present and the previous row of costs in memory. This reduces the memory complexity by a factor of L. With this we can not, however, find the optimal alignment since we lost the information we need to backtrack.

Using a backwards recursion for finding the prices from the full alignment to shorter prefixes, an optimal prefix in the middle of the alignment (relative to the first sequence) can be chosen and the alignment problem can be split into two problems of smaller size. The two problems together will take at most half as much time as the original problem, so the total time will be no more than a factor $1 + \frac{1}{2} + \frac{1}{4} + \cdots = 2$ longer [2]. With this, the memory complexity has been reduced by a factor of L, while the time complexity has only increased by a factor of two.

4 Algorithm Analysis

Here, it is assumed that the tree is binary. For trees with higher vertex orders, the time and memory limits will still hold.

4.1 Memory Complexity

The memory complexity of the algorithm is in the order of L^{N-1} times the number of acceptable indel configurations for N sequences.

The indel configurations for a rooted tree can be represented by a three dimensional vector, $\boldsymbol{f}$. The first coordinate is the number of indel configurations where the root is not isolated from all leaves and reversing a single arrow does not change that. The second coordinate is the number of indel configurations where the root is not isolated from all leaves, but reversing a single arrow changes that. The third coordinate is the number of indel configurations where the root is isolated from all leaves.

The configurations for a rooted tree can be calculated from the two rooted subtrees (a and b) connected to the root as:

$$\boldsymbol{f}_{ab} = \left[\boldsymbol{f}_a \boldsymbol{A} \boldsymbol{f}_b^T \ \boldsymbol{f}_a \boldsymbol{B} \boldsymbol{f}_b^T \ \boldsymbol{f}_a \boldsymbol{C} \boldsymbol{f}_b^T\right] ,$$

where:

$$\boldsymbol{A} = \begin{bmatrix} 6\,3\,1 \\ 3\,1\,0 \\ 1\,0\,0 \end{bmatrix} \qquad \boldsymbol{B} = \begin{bmatrix} 2\,2\,1 \\ 2\,2\,1 \\ 1\,1\,0 \end{bmatrix} \qquad \boldsymbol{C} = \begin{bmatrix} 1\,1\,1 \\ 1\,1\,1 \\ 1\,1\,1 \end{bmatrix} .$$

These relationships can be found by connecting the roots of the two subtrees to a vertex (the new root) with all nine combinations of labels for the two new edges.

Two rooted trees (a and b) can be connected by a single edge to give an unrooted tree. There are three configurations for the connecting edge, but only some of these lead to acceptable indel configurations for the whole unrooted tree, depending on the configurations of a and b. By going through these possibilities, the total number of acceptable indel configurations for the unrooted tree can be determined as:

$$\boldsymbol{f}_a \begin{bmatrix} 3\,2\,1 \\ 2\,1\,0 \\ 1\,0\,0 \end{bmatrix} \boldsymbol{f}_b^T , \tag{2}$$

where $\boldsymbol{f}_a$ and $\boldsymbol{f}_b$ are the configurations for the two subtrees.

Define a linear tree as one where any vertex is at most one edge away from a leaf. Here, a proof will be sketched that for a given number of leaves, the linear tree has the highest number of acceptable indel configurations. Consider the following rearrangement of a tree, where T_1 and T_2 represent subtrees with at least two leaves and the dashed lines represent a linear connecting tree:

T_1, T_2 ⟹ T_1, T_2 .

The number of acceptable indel configurations never decreases by such a rearrangement. This can be proven by showing that it applies for all symmetric (in the two ends) pairs of configurations for the connecting linear tree (proof left out because of space limitations). By doing a finite number of these changes, the linear tree will always be reached and it therefore has the maximal number of acceptable configurations.

For the linear rooted tree with n leaves, we can write the number of configurations of the different kinds as $\boldsymbol{f}_n$, with $\boldsymbol{f}_1 = [1\ 0\ 0]$:

$$\boldsymbol{f}_n = \left[\boldsymbol{f}_{n-1}\boldsymbol{A}\boldsymbol{f}_1^T\ \boldsymbol{f}_{n-1}\boldsymbol{B}\boldsymbol{f}_1^T\ \boldsymbol{f}_{n-1}\boldsymbol{C}\boldsymbol{f}_1^T\right] = \boldsymbol{f}_{n-1}\boldsymbol{X} = \boldsymbol{f}_1\boldsymbol{X}^{n-1}$$

$$\text{with } \boldsymbol{X} = \begin{bmatrix} 6\,2\,1 \\ 3\,2\,1 \\ 1\,1\,1 \end{bmatrix} .$$

By connecting the rooted tree with $n-1$ leaves to the rooted tree with one leaf, the maximal number of acceptable configurations for the unrooted linear tree with n leaves (g_n) can be found by using (2):

$$g_n = \boldsymbol{f}_{n-1} \begin{bmatrix} 3\,2\,1 \\ 2\,1\,0 \\ 1\,0\,0 \end{bmatrix} \boldsymbol{f}_1^T = [1\ 0\ 0]\boldsymbol{X}^{n-2} \begin{bmatrix} 3 \\ 2 \\ 1 \end{bmatrix} .$$

$[3\ 2\ 1]$ is a linear combination of all three eigenvectors of $\boldsymbol{X}$, so g_n grows exponentially as $O(m^n)$, where m is the maximal eigenvalue of $\boldsymbol{X}$:

$$m = 3 + \frac{1}{\sqrt[3]{2}}\left(\sqrt[3]{21 + i\sqrt{59}} + \sqrt[3]{21 - i\sqrt{59}}\right) = 7.441622\ldots$$

It follows that the memory complexity of the algorithm is $O(m^N L^{N-1})$. For specific numbers of acceptable indel configurations, see Table 1.

4.2 Time Complexity

Since each acceptable residue configuration is applied to each acceptable indel configuration, the time complexity of the algorithm is at most proportional to the memory complexity times the number of acceptable residue configurations.

Here, the growth rate of the numbers of acceptable indel configurations as a function of the number of leaves will be found. First, let us look at rooted trees that has a residue at the root. For a rooted tree, a, let x_a denote the number of residue configurations with a residue at the root and where the vertices labeled with a residue form a connected graph. These configurations are acceptable, except for the possibility that no leaf has a residue.

Combining two rooted trees, a and b, to a new root vertex gives:

$$x_{ab} = (x_a + 1)(x_b + 1) ,$$

Table 1. The number of the various configurations as a function of the topology (from the top): 1) total indel configurations, 2) acceptable indel configurations, 3) total residue configurations, and 4) acceptable residue configurations

3 3 4 3	19,683 9,534 1,024 124	1,594,323 527,985 16,384 543
27 23 16 10	19,683 9,494 1,024 132	1,594,323 525,342 16,384 603
243 172 64 25	177,147 70,950 4,096 263	1,594,323 524,975 16,384 620
2,187 1,281 256 57	177,147 70,603 4,096 287	1,594,323 522,336 16,384 671

where the addition of one comes from having no residues in the corresponding subtree. Define:

$$y = \log(x+1) .$$

Let x_i and y_i be the maximal x and y values, respectively, over the rooted tree topologies with i leaves. Also let $w_1 = 1$, and:

$$w_{2^{k+1}} = (w_{2^k}+1)^2 \quad \text{and} \quad z_{2^k} = \log(w_{2^k}+1) \quad \text{for } k \geq 0 .$$

By recursively forming balanced trees, it is clear that $x_i \geq w_i$ for $i = 2^k$. It follows that the growth rate of x_i is at least equal to the growth rate of w_i (when defined, i.e. $i = 2^k$). To see that the growth rate is not higher, we will show that:

$$y_i \leq \frac{2^{k+1}-i}{2^k} z_{2^k} + \frac{i-2^k}{2^k} z_{2^{k+1}} \quad \text{for } 2^k \leq i < 2^{k+1} . \tag{3}$$

The right side is an affine function going through z_{2^k} for $i = 2^k$ and $z_{2^{k+1}}$ for $i = 2^{k+1}$. By showing this, we have a strict limit for the growth of x_n between the values where w_n is defined. We have:

$$\frac{w_{2^{k+1}}+1}{w_{2^k}+1} < \frac{(w_{2^k}+1)^2}{(w_{2^{k-1}}+1)^2} \quad \Rightarrow \quad \frac{z_{2^{k+1}}-z_{2^k}}{2^{k+1}-2^k} < \frac{z_{2^k}-z_{2^{k-1}}}{2^k-2^{k-1}} .$$

This means that the slope of the affine function in (3) decreases as k increases. If (3) holds, we then also have:

$$y_i \leq \frac{2^{k+1} - i}{2^k} z_{2^k} + \frac{i - 2^k}{2^k} z_{2^{k+1}} \quad \text{for all } k \geq 0 \ . \tag{4}$$

Equation (3) holds for $i = 1$ since $y_1 = z_1 = \log(2)$. Assume that (3) also holds for $i \leq n$ and choose k so that $2^k \leq n + 1 < 2^{k+1}$. By using (4) we have:

$$y_i \leq \frac{2^k - i}{2^{k-1}} z_{2^{k-1}} + \frac{i - 2^{k-1}}{2^{k-1}} z_{2^k} \quad \text{for } i \leq n \ .$$

It follows that:

$$\begin{aligned} y_{n+1} &\leq \max_{1 \leq i \leq n} \log(\exp(y_i + y_{n+1-i}) + 1) \\ &= \log\left(\exp\left(\frac{2^{k+1} - (n+1)}{2^{k-1}} z_{2^{k-1}} + \frac{(n+1) - 2^k}{2^{k-1}} z_{2^k}\right) + 1\right) \qquad (5) \\ &\leq \frac{2^{k+1} - (n+1)}{2^k} z_{2^k} + \frac{(n+1) - 2^k}{2^k} z_{2^{k+1}} \ . \qquad (6) \end{aligned}$$

The fact that (5) is less than or equal to (6) for $2^k \leq n + 1 < 2^{k+1}$ can be seen by realizing that A) the two expressions are identical for $n + 1 = 2^k$ and for $n + 1 = 2^{k+1}$, and B) (6) is affine in n, while (5) is convex in n, since $\frac{d^2}{dx^2} \log(\exp(ax + b) + 1) > 0$ for $a > 0$.

Induction leads us to conclude that (3) holds for all i. Aho and Sloane [12] have shown that the recursion $q_{i+1} = q_i^2 + 1$ with $q_0 = 1$ grows as $O(a^{2^i})$ with a given by:

$$\begin{aligned} a &= \exp\left(\frac{1}{2}\log 2 + \frac{1}{4}\log\frac{5}{4} + \frac{1}{8}\log\frac{26}{25} + \frac{1}{16}\log\frac{677}{676} + \ldots\right) \\ &= 1.502837\ldots \end{aligned}$$

Because of the affine relationship described in (3) and because $w_{2^k} = q_{k-1}$, we see that x_n grows as $O(a^{2n})$, where $a^2 = 2.258518\ldots$.

These calculations were for rooted trees with a residue at the root and the possibility of no residue at any leaves. Due to the exponential growth rate of x_n, it can easily be shown that the contributions from the configurations with no residues at the root, and from those with no residues at any leaves, are negligible. Furthermore, the growth rate for unrooted trees is the same as for rooted trees.

This means that the time complexity of the algorithm is $O(t^N L^N)$, with $t = ma^2 = 16.807040\ldots$. A slightly smaller exponential value than t is likely to apply for two reasons. 1) For a given indel configuration, only the residue configurations leading to acceptable indel configurations need to be considered. 2) The worst case was the linear tree for the residue configurations and the balanced tree for the indel configurations. It is uncertain whether there is any type of tree topology that is worst case for both types of configurations.

Much like for the indel configurations, specific numbers of acceptable residue configurations can be found by considering vectors representing different cases for rooted trees and then joining two rooted trees to form an unrooted one, see Table 1.

5 Results

The HOMSTRAD database consists of 1033 protein alignments (April 2003) which were made using structural information [13]. These alignments will be considered correct in the following analyses. Three data sets of triple alignments were made by choosing sequences with pairwise identities in the ranges [15%; 25%), [25%; 35%), and [35%; 50%) (see Table 2). Since the starting and ending points of the sequences varied, indels in the ends of the alignments were removed. As many triples as possible (16 to 18) within the identity ranges and with an average sequence length between 100 and 150 amino acids (after removing the ends) were randomly chosen from different HOMSTRAD families to form the data sets.

Table 2. The percentages of correctly aligned positions for pairwise and triple alignments of three data sets. Each set has a number of triple alignments from different families in the HOMSTRAD database [13]. The minimum and maximum pairwise identities between each pair in the alignments define each data set. The size refers to the number of alignments each set contains. Gap introduction and extension costs were chosen to optimize the number of correctly aligned positions. The percentage of correctly aligned positions by Clustal W is included for comparison [10].

		Identity	Pairwise				Triple			
Set	Size	range	g_i	g_e	Correct	Clustal	g_i	g_e	Correct	Clustal
A	18	35% – 50%	3.6	1.3	93.9%	94.1%	3.2	1.6	93.7%	94.2%
B	16	25% – 35%	4.4	2.0	82.1%	83.2%	4.5	2.0	83.8%	85.2%
C	18	15% – 25%	4.4	1.2	60.9%	66.6%	4.1	1.3	64.0%	66.5%

For each triple alignment in the data sets, two analyses were made. 1) All three possible pairwise alignments were performed and the percentage of correct matches was found. 2) The triple alignment was performed and the percentage of pairwise correct matches was found. For each data set, the gap introduction and extension costs were chosen to maximize the percentage of correct matches, which are recorded in Table 2. Triple alignments are better than pairwise alignments for the less similar sequences. For the most similar sequences the performance is very close to identical.

Alignments made by the program Clustal W are a little better than with the algorithm presented here. This is not surprising given the more sophisticated position specific gap penalties of Clustal W and the simple distance matrix used here.

6 Conclusion

It has been shown how an optimal multiple parsimony alignment with affine gap cost over a tree can be found. The algorithm has a memory complexity of $O(7.442^N L^{N-1})$ and time complexity of $O(16.81^N L^N)$. This should be compared to the memory complexity of $O(L^{N-1})$ and time complexity of $O(2^N L^N)$ of the linear gap cost algorithm of [6] combined with the Hirschberg approach [2].

Triple alignment has been shown to outperform pairwise alignment on protein sequences. It is likely that this trend will continue for more sequences, but time and memory become limiting factors in the calculations. However, it may be worthwhile to spend the extra computation time to do triple rather than use pairwise alignments in progressive alignment methods.

Given a multiple alignment, the present method may be used to find its cost by limiting the dynamic programming to the prefixes defined by the alignment. The time and memory complexity for this would be quite small. This may be useful in heuristic alignment methods that seek to improve alignments by local iterative adjustments.

The advances made in this work may form the basis of more probabilistic methods. Statistical alignment methods may be developed that can utilize some of the same approaches for keeping track of gap introductions and extensions. This could also be useful in the context of analyses of multiple alignments using stochastic grammars, e.g., in RNA and protein structure prediction, and gene finding.

7 Acknowledgments

I would like to thank the University of Florida and the Carlsberg Foundation for support. I would also like to thank G. A. Lunter, R. Lyngsø, I. Miklós, and Y. S. Song for helpful comments.

References

1. Needleman, S.B., Wunsch, C.D.: A general method applicable to the search for similarities in the amino acid sequence of two proteins. J. Mol. Biol. **48** (1970) 443–453
2. Hirschberg, D.S.: A linear space algorithm for computing maximal common subsequences. Comm. Assoc. Comp. Mach. **18** (1975) 341–343
3. Myers, E.W., Miller, W.: Optimal alignments in linear space. Comput. Appl. Biosci. **4** (1988) 11–17
4. Gotoh, O.: An improved algorithm for matching biological sequences. J. Mol. Biol. **162** (1982) 705–708

5. Durbin, R., Eddy, S., Krogh, A., Mitchison, G.: Biological sequence analysis: Probabilistic models of proteins and nucleic acids. Cambridge University Press, Cambridge, UK (1998)
6. Sankoff, D., Cedergren, R.J.: Simultaneous comparison of three or more sequences related by a tree. In Sankoff, D., Kruskal, J.B., eds.: Time Warps, String Edits, and Macromolecules: the Theory and Practice of Sequence Comparison, Reading, MA, Addison-Wesley (1983) 253–264
7. Carrillo, H., Lipmann, D.: The multiple sequence alignment problem in biology. SIAM J. Appl. Math. **48** (1988) 1073–1082
8. Feng, D.F., Doolittle, R.F.: Progressive sequence alignment as a prerequisite to correct phylogenetic trees. J. Mol. Evol. **25** (1987) 351–360
9. Hein, J.: A new method that simultaneously aligns and reconstructs ancestral sequences for any number of homologous sequences, when the phylogeny is given. Mol. Biol. Evol. **6** (1989) 649–668
10. Thompson, J.D., Higgins, D.G., Gibson, T.J.: CLUSTAL W: improving the sensitivity of progressive multiple sequence alignment through sequence weighting, position specific gap penalties and weight matrix choice. Nucleic Acids Res. **22** (1994) 4673–4680
11. Fitch, W.M.: Toward defining the course of evolution: minimal change for a specific tree topology. Syst. Zool. **20** (1971) 406–416
12. Aho, A.V., Sloane, N.J.A.: Some doubly exponential sequences. Fib. Quart. **11** (1973) 429–437
13. Mizuguchi, K., Deane, C.M., Blundell, T.L., Overington, J.P.: HOMSTRAD: a database of protein structure alignments for homologous families. Protein Sci. **7** (1998) 2469–2471

Composition Alignment

Gary Benson*

Department of Biomathematical Sciences
The Mount Sinai School of Medicine
New York, NY 10029-6574
benson@ecology.biomath.mssm.edu

Abstract. In this paper, we develop a new approach for analyzing DNA sequences in order to detect regions with similar nucleotide *composition*. Our algorithm, which we call **composition alignment** or, more whimsically, **scrambled alignment**, employs the mechanisms of string matching and string comparison yet avoids the overdependence of those methods on position-by-position matching. In composition alignment, we extend the matching concept to *composition matching*. Two strings have a composition match if their lengths are equal and they have the same nucleotide content.
We define the composition alignment problem and give a dynamic programming solution. We explore several composition match weighting functions and show that composition alignment with one class of these can be computed in $O(nm)$ time, the same as for standard alignment. We discuss statistical properties of composition alignment scores and demonstrate the ability of the algorithm to detect regions of similar composition in eukaryotic promoter sequences in the absence of detectable similarity through standard alignment.

1 Introduction

Most algorithms which characterize DNA functional sites concentrate on identifying *position-specific patterns*, such as consensus patterns or weight-matrix-based profiles. Each is a string $P = p_1p_2 \cdots p_k$ in which p_i is specified as either a single character or a choice (weighted or unweighted) of characters. The data for a pattern are typically collected from multiple known occurrences of the functional site. Well known examples include the TATA box, a frequent component of the eukaryotic gene promoter, where transcription from DNA to RNA starts [9] and the weight matrices used to describe transcription binding sites in the TRANSFAC database [16].

Often, position-specific patterns have low selectivity. That is, when used to search for unknown occurrences, they yield many non-functional sites. This, in part, is due to the nature of the patterns which are frequently short and degenerate (more than one letter choice at some positions) making random matching

* Partially supported by NSF grants CCR-0073081 and DBI-0090789.

G. Benson and R. Page (Eds.): WABI 2003, LNBI 2812, pp. 447–461, 2003.

a relatively common event. A less obvious reason is that functional sites contain multiple features working synergistically and accurate recognition requires identifying several of those features. What hampers current detection methods is the likelihood that some sequence properties which contribute to function *are not reducible to position-specific patterns* and thus are not easily incorporated into current search techniques.

An example is double helical strand separation, a necessary antecedent for several important DNA functions. Rather than exhibiting position specific properties, the propensity for strand separation appears to be spread, in a complex way, over thousands of nucleotides. Programs have been developed to calculate regions of conjectured low energy requirement for strand separation [5, 6, 30, 31] and these programs have had some success in identifying sites which are functionally dependent on this process.

In this paper, we develop a new approach for DNA sequence analysis which employs the mechanisms of string matching and string comparison yet avoids the overdependence of these methods on position-by-position matching. We initiate the study of what we call the *composition property*, which manifests as similarity in *composition*, rather than in position-specific patterns.

Definition: *Composition* is a vector quantity describing the frequency of occurrence of each alphabet letter in a particular string. Let S be a string over Σ. Then, $C(S) = (f_{\sigma_1}, f_{\sigma_2}, \dots, f_{\sigma_{|\Sigma|}})$ is the composition of S, where for $\sigma_i \in \Sigma$, f_{σ_i} is the fraction of the characters in S that are σ_i. Note that the *order* of letters in S is irrelevant as it has no effect on the composition of S. Two strings S and T have a *composition match* if their lengths are equal and $C(S) = C(T)$.

There is an accumulating body of evidence that variation in composition along the DNA strands contributes to function. There are a number of important DNA features, at both large and small scales, whose unifying characteristic is composition bias including:

- *Isochores.* These multi-megabase regions of genomic sequence are specifically GC-rich or GC-poor. GC-rich isochores exhibit greater gene density. Human ALU and L1 retrotransposons appear preferentially in isochores with composition that approaches their own [7, 8, 25].
- *CpG islands.* These regions of several hundred nucleotides are rich in the dinucleotide CpG which is generally underrepresented (relative to overall GC content) in eukaryotic genomes. The level of methylation of the cystine (C) in these dinucleotide clusters has been associated with gene expression in nearby genes [12, 11, 13].
- *Protein binding regions.* Within these domains, tens of nucleotides long, dinucleotide, or base-step composition, can contribute to DNA flexibility, allowing the helix to change physical conformation, a common property of protein-DNA interactions [24, 19, 14, 18].

The springboard for our study of the composition property is a new alignment algorithm which we call **composition alignment** or, more whimsically,

scrambled alignment. Standard sequence alignment is based on single character matching. In composition alignment, we extend the matching concept to substrings which have the same composition *i.e.*, composition matching. This allows us to identify subsequences that share regions of similar composition. More specifically, composition alignment is a pairing of substrings of *exactly matching* composition separated by insertions, deletions or mismatches.

As an example, let

$$X = AACGTCTTTGAGCTC$$
$$Y = AGCCTGACTGCCTA$$

One possible composition alignment for X and Y is

```
AACGTCTTTGAGCTC
|  |<-> | <--->
AGCCTGACT-GCCTA
```

where symbols between the letters are used to indicate single character matches (|) and substring composition matches $(< - - - >)$. Note that composition matches (either single or multicharacter) can occur consecutively in an alignment.

An idea related to composition alignment is the "swap" operation in string comparison, first discussed by Lowrance and Wagner [21, 28]. A swap or transposition is the exchange of two letters that are side-by-side. Matching two letters and their transposition is equivalent to a composition match for substrings of length two. Lowrance and Wagner gave a $O(nm)$ time algorithm for alignment including the swap operation. More recently, Amir et al. [1] gave an algorithm for finding all swapped matchings of a pattern of length m in a text of length n with time complexity $O(nm^{1/3} \log m \log \sigma)$. This was later improved to $O(n \log m \log \sigma)$ [2] and followed by an algorithm to count the number of swaps in each swapped match within the same time complexity [3].

The remainder of the paper is organized as follows. In section 2 we present a formal description of the composition alignment problem. In section 3 we present our composition alignment algorithm and discuss scoring functions for composition matches. In section 4 we discuss the match length *limit* parameter and its effect on the statistical relevance of scores. Finally, in section 5, we demonstrate the use of composition alignment on real biological sequences.

2 Problem Description

The problem we address is the following:

Composition Alignment

Given: Two sequences, S of length m, and T of length n, over an alphabet Σ, and a scoring function $cm(s,t)$ for the score of a composition match between substrings s and t.

Find: The best scoring alignment (global or local) of S with T such that the allowed scoring options include composition match between substrings of S and T as well as the standard options of single character match, single character mismatch, insertion and deletion.

For the remainder of this paper, we will assume 1) that the alphabet Σ is fixed and 2) that similarity scoring is used. The latter means that matches are given positive weight; mismatches, insertions and deletions are given negative weight; and the best alignment has the highest score.

3 Composition Alignment Algorithm

Our composition alignment algorithm is similar to standard alignment algorithms [22, 26, 15] and is computed using dynamic programming. A single step in filling the dynamic programming array, W, can be analyzed in the following way. Given two sequences $X = x_1 \cdots x_n$ and $Y = y_1 \cdots y_m$, the best composition alignment of the two prefix strings $X[1,i] = x_1 \cdots x_i$ and $Y[1,j] = y_1 \cdots y_j$ ends in one of the following four ways:

New:

1. A *composition match* between suffixes of length $l, 1 \leq l \leq \min(i, j, limit)$, *i.e.*, $x_{i-l+1} \cdots x_i$ and $y_{j-l+1} \cdots y_j$, where *limit* is an upper bound on the length of a substring that can participate in a composition match. The necessity of *limit* will be explored further in section 4.

Standard:

1. A *mismatch* between x_i and y_j.
2. A *deletion* of x_i.
3. A *deletion* of y_j.

The score in cell (i, j) is the maximum obtained by these four possibilities (global alignment) or the maximum among these four and a score of zero (local alignment). Note that the score of alternative 1 is

$$W[i-l, j-l] + cm(x_{i-l+1} \cdots x_i, y_{j-l+1} \cdots y_j),$$

where W is the alignment score matrix and cm() is the score function for composition matches. The time complexity of the algorithm is $O(nmZ)$ where Z is the time required, per (i, j) pair, to find the best length l in alternative 1 and compute its score. In the next section, we show how to precompute the *length* of the *shortest* composition match for every (i, j) pair in constant time per pair.

3.1 Finding Composition Matches

Our goal here is to find the length, l, of the shortest suffixes of the strings $X[1, i]$ and $Y[1, j]$ which have a composition match, and to do this for every (i, j) pair,

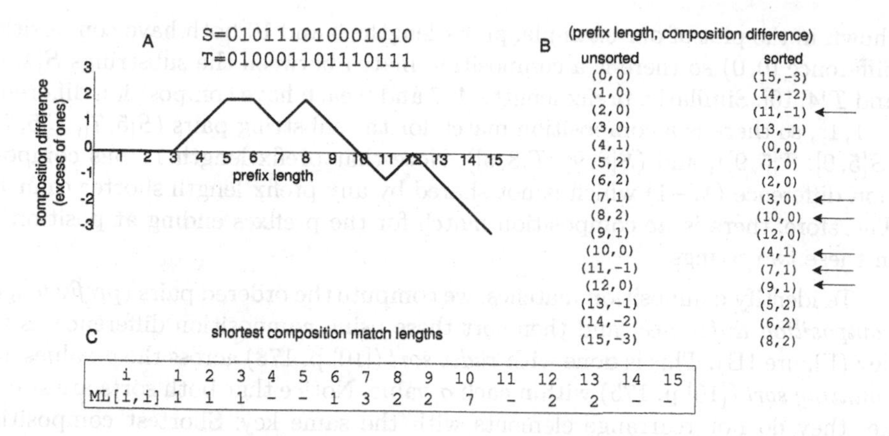

i	1	2	3	4	5	6	7	8	9	10	11	12	13	14	15
ML[i,i]	1	1	1	-	-	1	3	2	2	7	-	2	2	-	-

Fig. 1. A) Graph of composition differences (excess of 1's in prefixes of S relative to T). B) Ordered pairs (prefix length, composition difference) unsorted and sorted using composition difference as the key. Arrows mark prefixes mentioned in text. C) The $ML[i,j]$ array for diagonal zero ($i = j$).

in constant time per pair. For example, if $X = AACGTCTTTGAGCTC$ and $Y = AGCCTGACTGCCTA$, then for the pair $(4,8)$, the shortest suffixes that have a composition match are $X[2,4] = ACG$ and $Y[6,8] = GAC$ each with a length of 3. To find these matches, we use composition difference.

Definition: *Composition difference* is a vector quantity for two strings. Let x and y be strings over an alphabet Σ. Then $CD(x,y) = (c_{\sigma_1}, \ldots, c_{\sigma_{|\Sigma|}})$ is the composition difference of x and y, where $c_{\sigma_r} = c^x_{\sigma_r} - c^y_{\sigma_r}$ is the difference between the number of times σ_r occurs in x ($c^x_{\sigma_r}$) and the number of times it occurs in y ($c^y_{\sigma_r}$). For example, let $\Sigma = \{0,1\}$, $x = 010111010001000$ and $y = 010001101110111$. Then $CD(x,y) = (3,-3)$ because x has three more zeros and three fewer ones than y.

We compute composition match lengths for one diagonal of the alignment matrix at a time. Each diagonal is defined by a value d, $-n \leq d \leq m$ and contains the index pairs (i,j) such that $j - i = d$. The substrings processed in a diagonal for $d \geq 0$ are $X[1,k]$ and $Y[1+d, k+d]$ and for $d \leq 0$ are $X[1-d, k-d]$ and $Y[i,k]$ for all values of $k = 1, \ldots, min(m,n)$ such that both strings are non-null.

To illustrate, let $S = 010111010001000$ and $T = 010001101110111$. We show how to process diagonal zero. Figure 1A is a plot of the composition differences for successive prefixes of S and T. Since the components of any composition difference vector must sum to zero, we plot only the excess of ones in S relative to T (the second number in the composition difference vector).

The *key observation* is that two identical composition differences at prefix lengths g and h with $g < h$, indicate a composition match between the substrings of S and T from position $g+1$ to position h, *i.e.*, of length $h - g$. As shown in the plot of our example, prefix lengths 3 and 10 both have composition difference $(0, 0)$ so there is a composition match between the substrings $S[4, 10]$ and $T[4, 10]$. Similarly, prefix lengths 4, 7 and 9 each have composition difference $(-1, 1)$, so there is a composition match for the substring pairs $(S[5, 7], T[5, 7])$, $(S[5, 9], T[5, 9])$, and $(S[8, 9], T[8, 9])$. Note that prefix length 11 has composition difference $(1, -1)$ which is not shared by any prefix length shorter than 11. Therefore, there is *no* composition match for the prefixes ending at position 11 in these two strings.

To identify composition matches, we compute the ordered pairs (*prefix length*, *composition difference*) and then sort these using composition difference as the key (Figure 1B). This is done with *radix sort* ([10] p. 178) across the σ values and *counting sort* ([10] p. 175) within each σ value. Notice that both sorts are stable, i.e. they do not rearrange elements with the same key. Shortest composition matches are determined by scanning the sorted list to find *adjacent* elements with the same key (composition difference). The difference between prefix lengths is then stored in an array $ML[i, j]$. In our example, since prefix lengths 10 and 3 are adjacent with the same composition difference, we store $7(= 10 - 3)$ as the shortest composition match for prefixes of S and T of length 10.

Time complexity. The sorting is linear in the number of elements when the alphabet is fixed. For a single diagonal of the alignment matrix, the number of elements is the diagonal length. Over all diagonals, the number of elements is $(n + 1)(m + 1)$. Therefore the time complexity to preprocess the strings to find all shortest composition match lengths is $O(nm)$.

3.2 Scoring Functions

The overall complexity of our composition alignment algorithm depends on the complexity of computing the best length l for a composition match. This in turn depends on the scoring function, $cm()$, for composition matches. Below, we discuss several scoring functions we have tested.

Functions based on match length. In this group, the score of a composition match depends on the *length*, k, of the match. We have tested

- Function 1: $cm(k) = ck$
- Function 2: $cm(k) = c\sqrt{k}$
- Function 3: $cm(k) = c\log(k + 1)$

where c is a constant. These functions are additive or subadditive, meaning that $cm(i + j) \leq cm(i) + cm(j)$. Function 1 (additive) treats matches of different lengths equally. Functions 2 and 3 (subadditive) give less weight, per character, to long composition matches, than to short composition matches.

A convenient property of additive or subadditive functions is that, when computing alternative 1 of the alignment score, for any (i, j) pair, it is sufficient to find the length of the *shortest* suffixes of $X[1, i]$ and $Y[1, j]$ which have a composition match.

Lemma 1. *For an index pair (i,j), let $l = l_1 < l_2 < \ldots < l_k$, $1 \leq l \leq min(i,j)$, be the lengths for which there is a composition match between the suffixes $X[i-l+1,i]$ and $Y[j-l+1,j]$. Then, the score for the best alignment which ends in a composition match between suffixes of $X[1,i]$ and $Y[1,j]$ is equal to the score when the suffixes have length l_1. That is, $\forall h, 2 \leq h \leq k, W(i-l_1,j-l_1)+cm(l_1) \geq W(i-l_h,j-l_h)+cm(l_h)$.*

Proof. Assume by way of contradiction that there is an $l_h > l_1$ such that

$$W(i-l_h, j-l_h) + cm(l_h) > W(i-l_1, j-l_1) + cm(l_1).$$

Let $l_h = l' + l_1$. Then

$$W(i-l_h, j-l_h) + cm(l'+l_1) > W(i-l_1, j-l_1) + cm(l_1)$$
$$W(i-l_h, j-l_h) + cm(l'+l_1) - cm(l_1) > W(i-l_1, j-l_1)$$

but by additivity or subadditivity,

$$cm(l') \geq cm(l'+l_1) - cm(l_1)$$

so

$$W(i-l_h, j-l_h) + cm(l') > W(i-l_1, j-l_1)$$

which is a contradiction because $W(i-l_1, j-l_1)$ is assumed to be optimal including the possibility that the alignment which yields this score ends with the composition match of length l'. □

This means that breaking up a long composition match into shorter matches (if possible) will leave the score the same (function 1) or increase the score (functions 2 and 3). The alignment shown in the introduction contains a 4 character composition match which is broken into a single character match and a 3 character match.

Theorem 2. *Composition alignment with an additive or subadditive composition match scoring function has time complexity $O(nm)$.*

Proof: Follows from the discussion in section 3.1. □

Functions based on substring composition. Here, the score of a composition match depends not just on length, but on the composition of the matching substrings. We have tested:

– Function 4: $cm(x, y, k) = ck \cdot H(C, B)$

where x and y are substrings with common composition C, k is their length, and c is a constant. $H(C, B)$ is the relative entropy of composition C given a background composition B. *Relative entropy* is defined as

$$H(C,B) = -\sum_{\sigma \in \Sigma} f_\sigma \log(f_\sigma / b_\sigma)$$

where f_σ is a frequency in the composition vector C and b_σ is the corresponding frequency in the background composition B. This function can only be used if for every non-zero f_σ in C there is a corresponding non-zero b_σ in B, else there will be a divide-by-zero problem. In our studies this has not arisen because we use the overall frequency of letters in the sequences to be aligned as the background. If divide-by-zero is possible, H can be replaced with the unweighted Jensen-Shannon divergence [20]. We have not yet tested this function extensively.

Function 4 favors composition matches where the substrings differ significantly from the background. This could, for example, be a long repetition of a single letter, assuming the background is relatively balanced. This function is not additive or subadditive, so finding the shortest composition match for any (i, j) pair does not always yield the optimal match length. Since a longer match may yield a higher alignment score, we must test all substring match lengths per (i, j) pair. This can easily be done in time linear in the number of match lengths by stepping through the ML array which stores the shortest match lengths, but requires at most $\min(m, n)$ tests per (i, j) pair. In practice though, the *limit* parameter explained below restricts the number of tests to at most *limit* per (i, j) pair.

Theorem 3. *Composition alignment using function 4 for compositon match scoring and the limit parameter has time complexity $O(nm \cdot limit)$.*

Proof. Follows from discussion above. □

Retrieving the alignment. After the alignment score array W has been computed, the optimal alignment is retrieved by tracing back as in standard alignment. When a score $W[i, j]$ is derived from a composition match, we need to reference the length of the matching substring. For Functions 1, 2 and 3, this is done by querying the $ML[i, j]$ value. For Function 4, we must store the optimal match lengths separately as the scores are being computed and then refer to these values when tracing back. In either case, retrieving the alignment requires $O(n + m)$ time.

4 Alphabet Size and the Limit Parameter

The study of alignment using similarity scoring has shown that for ungapped local alignments of randomly generated sequences, the parameter space for match and mismatch weights is divided into logarithmic and linear regions. In the logarithmic region, the parameters produce alignment scores proportional to the logarithm of sequence lengths whereas in the linear region, the scores are directly proportional to the sequence lengths [29]. It is generally accepted that weight combinations which fall within the logarithmic region are useful for detecting *biologically related* sequences, whereas those in the linear region do not distinguish between related and unrelated sequences. The same general features have been observed in gapped local alignments where gap weight is an additional parameter [27, 4]. The rubric for determining if parameters fall within the logarithmic or linear regions is to look at the expected score per aligned letter pair (ungapped alignments [17]) or the expected global alignment score (gapped alignments [4]). In either case, if the expected score is negative, and assuming that positive scores are possible, then the parameters fall within the logarithmic region.

Here we are primarily interested in how the *limit* parameter, the length of the longest allowed composition match, fits into this framework. When $limit = 1$, composition alignment is equivalent to standard alignment. When $limit = 2$, any pair of adjacent letters in one sequence is allowed to match its transposition in the other sequence. This corresponds to the swap operation mentioned earlier [21]. For $limit = 3$, both scrambled triplets and transposed doublets are allowed to match, etc. Intuitively, allowing scrambled letters to match should increase the amount of matching. If too much matching occurs, then the average score will be positive and the alignments will not be meaningful.

4.1 Expected Fraction of Matching Characters in Alignments

We have examined both ungapped and gapped composition alignments to determine the expected fraction of aligned letter pairs that are counted as matches.

Ungapped alignments. Suppose we have a binary alphabet and we examine ungapped aligned strings of length 2 where the characters are generated iid with probability 0.5. Under single character matching, the expected fraction of characters counted as matching is 0.5, *i.e.* on average half the characters will be counted as matches. When we allow composition matches with $limit = 2$, the expected number of characters counted as matches increases to 0.625. For aligned strings of length 3, the results are similar. For single character matching the expected fraction is still 0.5, but for composition matching with $limit = 3$ (substring pairs of length 2 or 3 can match as long as they have the same composition), the expected fraction of matches is 0.6875. As the sequence length grows, calculating the fraction of matches becomes complicated, so we turn to simulation. When the string length reaches 10, and we allow composition matching with $limit = 10$, the fraction of matches is above 0.82 (table 1).

For the four letter DNA alphabet, when the letters are generated iid with probability 0.25, the fraction of characters matching grows similarly but more slowly. For the sixteen letter dinucleotide alphabet, the probability grows until reaching an apparent asymtotic upper bound around 0.075. For dinucleotides, we first generate an iid DNA sequence and then convert it to a dinucleotide sequence. Notice that consecutive letters in the dinucleotide sequence are not independent (i.e. if the first dinucleotide is AC, then the next must start with a C).

To investigate the fraction of matches in longer sequences where the *limit* is smaller than the sequence length, we use global composition alignment to count the matches. Here, insertions and deletions are not allowed, all matches are weighted 1 and all mismatches weighted zero. Results for DNA sequences of length 100 and dinucleotide sequences of length 400 are shown in Table 2. As

Sequence length	1	2	3	4	5	6	7	8	9	10
Binary (%)	50.0	62.5	68.75	72.7	75.6	77.3	78.9	80.3	81.3	82.4
DNA (%)	25.0	30.0	32.3	35.3	37.5	39.7	40.7	42.4	43.3	44.2
Dinucleotide (%)	6.2	6.5	6.7	6.9	7.1	7.3	7.3	7.3	7.4	7.5

Table 1. Fraction of characters counted as matching in randomly generated ungapped alignments where *limit* equals alignment length, for three alphabets. In each case, all letters in an alphabet have equal probability. Results are derived from simulations except for sequence length 1 in all alphabets and lengths 2 and 3 in the binary alphabet.

can be seen in the table, the fraction matching in DNA sequences is nearly 45% with $limit = 5$ and reaches 50% when $limit = 9$. For dinucleotides, the fraction levels off at 7.78% for $limit \geq 20$.

For *local, ungapped* composition alignments with DNA sequences using alignment parameters (composition match constant, single character match, mismatch) $= (1, 1, -1)$, alignment scores grow in proportion to the log of the sequence length for $limit \leq 5$ (data not shown). For *limit* between 6 and 10, growth is proportional to the square root of the sequence length. Note that with these alignment parameters, the expected score of an aligned pair is negative until $limit = 9$ (table 2). Thus negative expected score per aligned letter pair is an inaccurate predictor of logarithmic score growth for ungapped composition alignments.

Gapped alignments. For gapped alignments, we use simulations with actual parameter values and composition match scoring functions because the interaction of these determines the number of gapped positions. The results are useful for several purposes:

1. To define values for the limit parameter that fall within the logarithmic region.
2. To test for concurrence between 1) the change from negative to positive average global alignment score and 2) the change from logarithmic to lin-

DNA: sequence lengths $= 100$; iid; $p = 0.25$										
limit	1	2	3	4	5	6	7	8	9	10
fraction matching (%)	25.0	33.7	38.6	42	44.4	46.3	47.8	49.0	50.0	51.0

dinucleotide: sequence lengths $= 400$; iid; $p = 0.25$								
limit	1	2	5	10	20	30	40	50
fraction matching (%)	6.25	6.81	7.66	7.76	7.78	7.78	7.78	7.78

Table 2. Fraction of characters counted as matching in longer randomly generated DNA and dinucleotide sequences, after composition alignment without gaps, for various *limit* values.

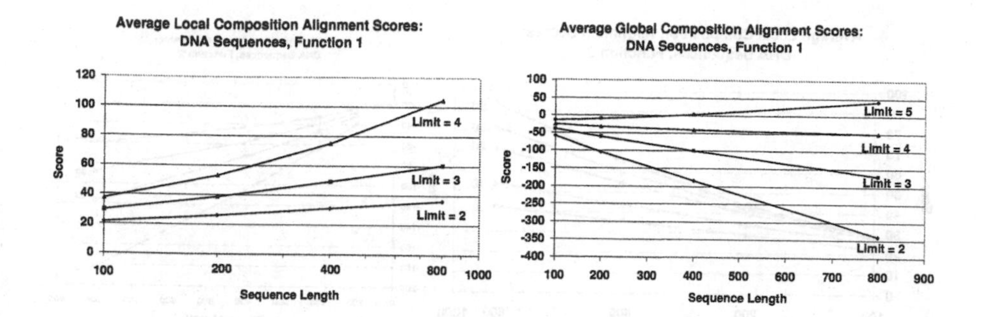

Fig. 2. Function 1 results from alignment of randomly generated DNA sequences (iid, $p = 0.25$) when *limit* is allowed to vary. Alignment parameters: (2,2,-3,-5). Left) Average local scores proportional to log of sequence lengths (note log scale) for $limit \leq 3$. Right) Average global scores become positive at $limit = 5$.

ear behavior in the local alignment score. This is not obvious because the assumptions that underly the theory for alignment scores do not include scrambled alignment.

3. To estimate local alignment score distributions so that the composition alignment algorithm can be used to search for statistically significant alignments in real biological sequences.

We used DNA sequences generated iid with probability 0.25 for each letter and two sets of alignment parameters (composition match constant, single character match, mismatch, indel), $(2, 2, -3, -5)$ and $(2, 2, -7, -7)$. Here we summarize some of the more important results.

Function 1. Local alignment scores grow in proportion to the logarithm of sequence length for $limit \leq 3$. At $limit = 4$, scores are proportional to the square root of sequence length. Note though, that the average global alignment score does not become positive until $limit = 5$. See figure 2. These results indicate that function 1 should be used with *limit* set to 3 and not higher. Also, positive or negative expected global score is inaccurate as a predictor of the parameter values that yield logarithmic growth in local alignment scores.

Functions 2 and 3. Local alignment scores for function 2 are logarithmically related to sequence length below $limit = 10$. Average global scores do not become positive with *limit* as high as 50. See figure 3. Function 3 behaves similarly. Again, average global score is an inaccurate predictor of local alignment score behavior.

Function 4. Local alignment scores are proportional to the log of the sequence lengths up to $limit = 50$.

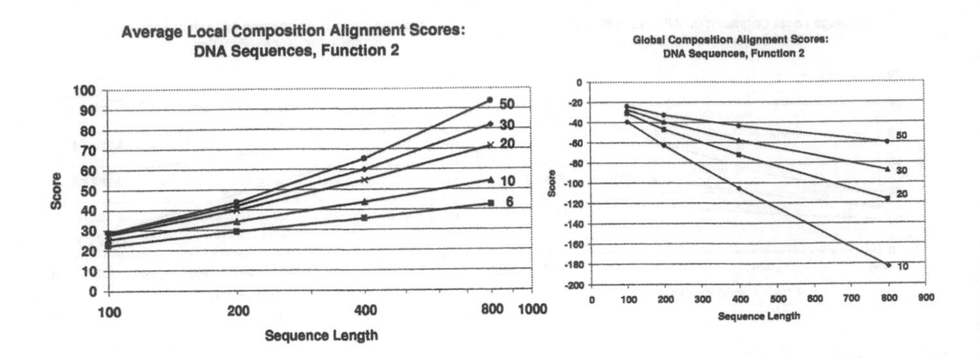

Fig. 3. Function 2 results from alignment of randomly generated DNA sequences. Same conditions as in figure 2. Limit values shown to right of curves. Left) Average local scores proportional to log of sequence lengths (note log scale) for *limit* < 10. Right) Average global scores do not become positive at *limit* as high as 50.

5 Examples

We tested our composition alignment algorithm on a set of 1796 human promoter sequences from the Eukaryotic Promoter Database (EPD) [23] maintained by the Bioinformatics Group of the Swiss Institute for Experimental Cancer Research. The database contains a collection of roughly 3000 annotated non-redundant eukaryotic RNA polymerase II promoters for which the transcription start site has been experimentally determined. Each sequence is 600 bases long and consists of 500 bases upstream and 100 bases downstream of the transcription initiation point. The sequences are non-redundant as selected from the database, which means that no two share greater than 50% sequence identity.

The sequences were aligned pairwise using composition match scoring function 1 with alignment parameters (composition match constant, single character match, mismatch, indel) of $(2, 2, -7, -7)$. This produced a score W. Each pair

```
Composition Alignment:
GCCCGCCCGCCGCGCTCCCGCCCGCCGCTCTCCGTGGCCC-CGCCG-CGCTGCCGCCGCCGCCGCTGC
<->||||<>|<>||<>| ||||<>||<> |<-> |||||| <>|<> ||||<><> |<>| ||<->||
CCGCGCCGCCGCCGTCCGCGCCGCCCCG-CCCT-TGGCCCAGCCGCTCGCTCGGCTCCGCTCCCTGGC

Standard Alignment:
CGCCGCCGCCG
CGCCGCCGCCG
```

Fig. 4. Composition alignment and standard alignment of promoters EP27006 and EP73975, positions 474-539 and 430-495 respectively. Composition of aligned subsequences (top alignment) is $(0.01, 0.59, 0.30, 0.11)$. Background composition of these promoters is $(0.11, 0.44, 0.34, 0.11)$. Standard alignment is not statistically significant.

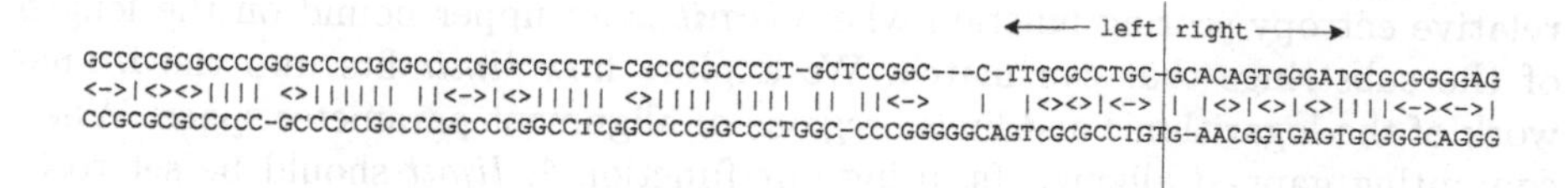

Fig. 5. Composition alignment of promoters EP73298 and EP11149, positions 323-409 and 444-534 respectively. Composition of left two thirds, $(0.01, 0.61, 0.30, 0.08)$, is dramatically different from composition of right third, $(0.19, 0.16, 0.56, 0.09)$.

was also aligned with a standard alignment algorithm using the same parameter values, producing a score S. Those pairs for which 1) W was above the statistical significance cutoff score for composition alignment for a set of sequences this large (as determined by simulation) and 2) $W \geq 3 \cdot S$ were retained. The second criterion was used to exclude composition alignments that scored highly because they were redetecting good standard alignments. Two high scoring alignments are shown here.

The first example was obtained with the promoter pair EP27006 and EP73975 (Figure 4). The standard local alignment which is not statistically significant is shown for comparison. The composition alignment is characterized by high GC content (89%), an enrichment over the background frequency of these sequences (78%). The number of CpG dinucleotides found in the aligned regions is more than expected given either the background composition of these subsequences or the entire sequences. This suggests that the aligned regions are part of CpG islands which are defined [13] as being 200 bp subsequences with a C+G content exceeding 50% and a ratio of observed CpG to expected CpG in excess of 0.6. CpG islands are known to occur in the 5' region of many genes. This alignment is typical of many obtained with the promoter set.

A second example involves promoters EP73298 and EP11149 (Figure 5). Again, the standard local alignment score for this pair is not statistically significant. An interesting feature of the composition alignment is the change in composition of the subsequences from left to right. The left two thirds is GC rich with C dominant, and a single A: $(0.01, 0.61, 0.30, 0.08)$. The situation changes at the right which is G dominant with the fraction of As, and Cs equivalent: $(0.19, 0.16, 0.56, 0.09)$. The fraction of Ts is roughly the same throughout. Notice that the background composition for these sequences is typical, GC rich with the complementary nucleotides balanced: $(0.15, 0.36, 0.34, 0.15)$.

6 Conclusion

We define a new type of alignment problem, *composition alignment* which extends the matching concept to substrings of equal length and the same nucleotide composition. We give an algorithm for composition alignment which has time complexity $O(nm)$ for a fixed alphabet when the composition match scoring function is additive or subadditive. The time complexity is $O(nm \cdot limit)$ for a relative entropy scoring function where *limit* is an upper bound on the length of the substrings that can match. We explore how *limit* fits into the framework of the logarithmic and linear regions of alignment parameter space. When computing gapped alignments, using our function 1, *limit* should be set to 3. For functions 2 and 3 *limit* should be under 10 and for function 4, *limit* can be as high as 50. We give two examples of composition alignments for human RNA polymerase II promoters where the composition alignment scores are statistically significant even though there is no detectable similarity with standard alignment.

References

[1] A. Amir, Y. Aumann, G. Landau, M. Lewenstein, and N. Lewenstein. Pattern matching with swaps. *J. Algorithms*, 37:247–266, 2000.

[2] A. Amir, R. Cole, R. Hariharan, M. Lewenstein, and E. Porat. Overlap matching. In *Proc. 12th ACM-SIAM Sym. on Discrete Algorithms*, pages 279–288, 2001.

[3] A. Amir, M. Lewenstein, and E. Porat. Approximate swapped matching. *Information Processing Letters*, 83:33–39, 2002.

[4] R. Arratia and M. Waterman. A phase transition for the score in matching random sequences allowing deletions. *Ann. Appl. Prob.*, 4:200–225, 1994.

[5] C.J. Benham. Duplex destabilization in superhelical DNA is predicted to occur at specific transcriptional regulatory regions. *J. Mol. Biol.*, 255:425–434, 1996.

[6] C.J. Benham. The topologically driven strand separation transition in DNA-methods of analysis and biological significance. *DIMACS Series in Discrete Mathematics and Theoretical Computer Science*, 47:173–198, 1999.

[7] G. Bernardi. The isochore organization of the human genome. *Annu. Rev. Genet.*, 23:637–661, 1989.

[8] G. Bernardi. The human genome: Organization and evolutionary history. *Annu. Rev. Genet.*, 29:445–476, 1995.

[9] P. Bucher. Weight matrix descriptions of four eukaryotic RNA polymerase II promoter elements derived from 502 unrelated promoter sequences. *J. Mol. Biol.*, 212:563–578, 1990.

[10] T. Cormen, C. Leiserson, and R. Rivest. *Introduction to Algorithms.* MIT Press, 1990.

[11] W. Doerfler. DNA methylation and gene activity. *Ann. Rev. Biochem.*, 52:93–124, 1983.

[12] G. Felsenfeld and J. McGhee. Methylation and gene activity, 1982.

[13] M.G. Garden and M. Frommer. CpG islands in vertebrate genomes. *J.Mol. Biol.*, 196:261–282, 1987.

[14] D.S. Goodsell and R.E. Dickerson. Bending and curvature calculations in B-DNA. *Nucleic Acids Research*, 22:5497–5503, 1994.

[15] O. Gotoh. An improved algorithm for matching biological sequences. *J. Mol. Biol.*, 162:705–708, 1982.

[16] T. Heinemeyer, X. Chen, H. Karas, A. Kel, O. Kel, I. Liebich, T. Meinhardt, I. Reuter, F. Schacherer, and E. Wingender. Expanding the TRANSFAC database towards an expert system of regulatory molecular mechanisms. *Nucleic Acids Res.*, 27:318–322, 1999.

[17] S. Karlin and S. Altschul. Methods for assessing the statistical significance of molecular sequence features by using general scoring schemes. *Proc. Natl. Acad. Sci. USA*, 87:2264–2268, 1990.

[18] H-S. Koo, H-M. Wu, and D.M. Crothers. DNA bending at adenine - thymine tracts. *Nature*, 320:501–506, 1986.

[19] M. Lewis, G.Chang, N.C. Horton, M.A. Kercher, H.C. Pace, M.A. Schumacher, R.G. Brennan, and P. Lu. Crystal structure of the lactose operon repressor and its complexes with DNA and inducer. *Science*, 271:1247–1254, 1996.

[20] J. Lin. Divergence measures based on the Shannon entropy. *IEEE Trans. Inf. Theor.*, 37:145–151, 1991.

[21] R. Lowrance and R.A. Wagner. An extension of the string-to-string correction problem. *JACM*, 22:177–183, 1975.

[22] S. Needleman and C. Wunch. A general method applicable to the search for similarities in the amino acid sequence of two proteins. *J. Mol. Biol.*, 48:443–453, 1970.

[23] R. Périer, V. Praz, T. Junier, C. Bonnard, and P. Bucher. The Eukaryotic Promoter Database (EPD). *Nucleic Acids Research*, 28:302–303, 2000.

[24] S.C. Schultz, G.C. Shields, and T.A. Steitz. Crystal structure of a CAP-DNA complex: The DNA is bent by 90 degrees. *Science*, 253:1001–1007, 1991.

[25] A. Smit. The origin of interspersed repeats in the human genome. *Curr. Opin. Genet. Dev.*, 6:743–748, 1996.

[26] T. Smith and M. Waterman. Identification of common molecular subsequences. *J. Mol. Biol.*, 147:195–197, 1981.

[27] M. Vingron and M. Waterman. Sequence alignment and penalty choice: review of concepts, case studies and implications. *J. Mol. Biol.*, 235:1–12, 1994.

[28] R.A. Wagner. On the complexity of the extended string-to-string correction problem. In *Proceedings 7th ACM STOC*, pages 218–223, 1975.

[29] M. Waterman, L. Gordon, and R. Arratia. Phase transitions in sequence matches and nucleic acid structure. *Proc. Natl. Acad. Sci. USA*, 84:1239–1243, 1987.

[30] E. Yeramian. Genes and the physics of the DNA double-helix. *Gene*, 255:139–50, 2000.

[31] E. Yeraminan, S. Bonnefoy, and G. Langsley. Physics-based gene identification:proof of concept for *Plasmodium falciparum*. *Bioinformatics*, 18:190–193, 2002.

Match Chaining Algorithms for cDNA Mapping

Tetsuo Shibuya[1] and Igor Kurochkin[2]

[1] IBM Tokyo Research Laboratory, 1623-14, Shimo-tsuruma, Yamato, Kanagawa, 242-8502, Japan.
tshibuya@jp.ibm.com
[2] RIKEN Genomic Sciences Center,
1-7-22, Suehiro-cho, Tsurumi, Yokohama, Kanagawa 230-0045, Japan.
igork@gsc.riken.go.jp

Abstract. We propose a new algorithm called the MCCM (Match Chaining-based cDNA Mapping) algorithm that allows mapping cDNAs to the genomes efficiently and accurately, utilizing local matches called MUMs (maximal unique matches) or MRMs (maximal rare matches) obtained with suffix trees. From the MUMs (or MRMs), our algorithm selects appropriate matches which are related to the cDNA mapping. We call the selection the match chaining problem. Several $O(k \log k)$-time algorithms are known where k is the number of the input matches, but they do not permit overlaps of the matches. We propose a new $O(k \log k)$-time algorithm for the problem with provision for overlaps. Previously, only an $O(k^2)$-time algorithm existed. Furthermore, we also incorporate a restriction on the distances between matches for accurate cDNA mapping. We examine the performance of our algorithm through computational experiments using sequences of the FANTOM mouse cDNA database and the mouse genome. According to the experiments, the MCCM algorithm is not only very fast, but also very accurate: We achieved $> 95\%$ specificity and $> 97\%$ sensitivity at the same time against the mapping results of the FANTOM annotators.

1 Introduction

Since the complete and accurate human genome sequence has just been released, the interpretation and analysis of this large dataset has become a major task for the scientific community. One of the most fundamental analyses of a genome sequence is mapping a large number of cDNAs (mRNAs) or ESTs (expression sequence tags) to the regions of the genome they are transcribed from. Figure 1 illustrates the mapping. Precise mapping allows further analysis of the gene regulatory elements (promoters, transcription factor binding sites, etc.) to be performed. It also provides a basis for understanding how genomes evolve. However, the large size of the mammalian genomes, the existence of splice sites, and the numerous repetitive regions pose a serious problem for the alignment between cDNA and genome sequences. Alignments taking into account splice sites are called *spliced alignments.* Several spliced alignment algorithms based on $O(nm)$-time dynamic programming (DP) (where n and m are the lengths of

G. Benson and R. Page (Eds.): WABI 2003, LNBI 2812, pp. 462–475, 2003.

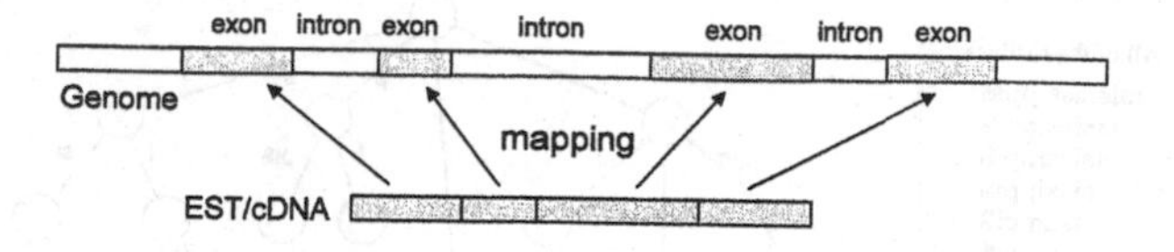

Fig. 1. Mapping a cDNA to the genome.

the two sequences to align) have been developed [11,18,21]. However, in practice, their time bound is too large for aligning large numbers of cDNAs to genome sequences of enormous size. Several other algorithms that heuristically map cDNAs to the genome have been proposed recently. These include sim4 [10], BLAT [14], and Squall [20], which are based on a hash or hash-like tools like BLAST [2].

Suffix trees [9,12,17,22,24] are known to be very powerful and flexible data structures for pattern matching. For example, exactly matching substrings can be found very efficiently using suffix trees, while an ordinary hash structure can deal with only fixed-length matches. Suffix trees require memory space that is several times larger than many of the other indexing structures, but the memory cost is going down at a remarkable rate these days and is becoming a less serious problem. Consequently many bioinformatics tools use suffix trees. For example, MUMmer [6,7] uses suffix trees for computing alignments of bacteria-size long sequences. They at first enumerate short exact matches called MUMs (Maximal Unique Matches) and align the sequences by chaining some of the MUMs. In this paper, we propose a new match chaining algorithm with better time complexity, and we also extend the strategy of the MUMmer to make it more suitable for cDNA mapping. We call the new algorithm the MCCM (Match Chaining-based cDNA Mapping) algorithm.

2 Algorithms

2.1 Preliminaries

We describe in this section several basic algorithms and data structures that will be used to describe our MCCM algorithm.

Suffix Trees The suffix tree [9,12,17,22,24] of a string $S \in \Sigma^n$ is the compacted trie of all the suffixes of $S^+ = S\$$ where \$ is a character such that $\$ \notin \Sigma$. This data structure is known to be buildable in $O(n)$ time. Figure 2 shows an example of this data structure. Each leaf represents a suffix of the string S^+, and each node represents some substring. This data structure is very useful for various problems in sequence pattern matching. Using it, we can query a substring of length m in $O(m)$ time, we can find frequently appearing substrings in a given sequence in linear time, we can find a common substring of many sequences in linear time, and so on [12]. The list of leaves corresponds to a data structure called the suffix array [16] that is a list of the indices of the lexicographically

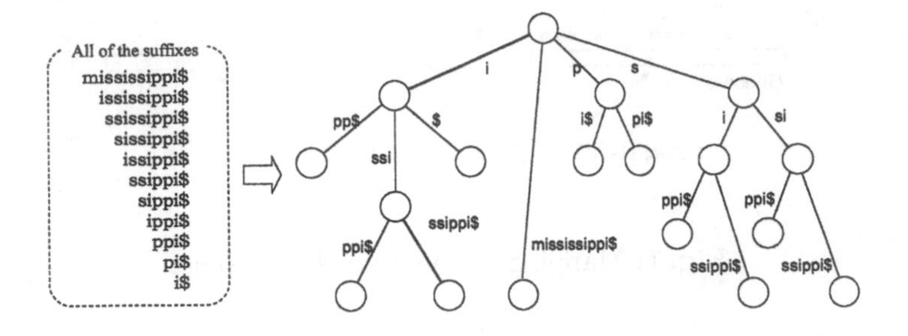

Fig. 2. The suffix tree of a string 'mississippi'.

sorted suffixes. Note that the suffix array is also a very useful data structure that can be an alternative to the suffix tree for many purposes, especially when memory is limited.

Maximal Rare Matching We can easily find the longest prefix of a query string that matches somewhere in a given text sequence in a time linear to the query size if we have the suffix tree of the text sequence. Moreover, with the suffix tree, we can also find all the longest prefixes of all of the suffixes of the query string that matches somewhere in a given text sequence in the same time bound [7,12]. This is done by tracing the nodes of the suffix tree that correspond to the suffixes using suffix links. Note that a suffix link is a pointer from a suffix tree's node that represents some substring T to another node that represents the suffix of T which is shorter than T by 1.

A 'maximal matching substring' of two sequences (a pattern and a text) is an exactly matching substring that occurs in a text sequence r times, but any other string that includes the substring appears less than r times in the text. The MUMmer [6,7] uses to compute the list of what they call MUMs (Maximal Unique Matching substrings), which is a list of the maximal matching substrings where $r = 1$. Using the algorithm above, we can compute all the MUMs in linear time relative to the query size if we already have the suffix tree of the text [7,12]. We will use them for cDNA mapping, considering the cDNA sequences to be pattern sequences and the genome sequence to be a text sequence.

If the genomic sequences contain many repetitive substrings, there can be cases that corresponding matching substrings are not 'unique', and it could be better to use a larger value of r. We call a maximal matching substring a maximal rare matching, or an MRM for short, if $r \leq l$ for some fixed l. All the MRMs for some l can be easily computed with suffix trees in $O(l \cdot m)$ time where m is the length of the query sequences, by using a minor extension of the above MUM algorithm. But in the case of the FANTOM cDNAs and the mouse genome, we do not need to use large ls according to our experiments.

Match Chaining Problem If we select appropriate matches from the large MUM or MRM list, we can build a sort of an alignment of the two sequences [6,13,19]. We call this the match chaining problem. To form an actual alignment, we must postprocess these selected matches, but we do not consider this in this paper. Let $M = \{m_1, m_2, \ldots, m_k\}$ be a list of input matches sorted by their start positions in the pattern sequence. Note that the MUM or MRM results are already sorted in this order if we use typical algorithms based on suffix trees. Let match m_i's start positions in the pattern sequence and in the text sequence be p_i and t_i respectively, and its length be l_i. An M's subset $\{m_{c_1}, m_{c_2}, \ldots, m_{c_f}\}$ is called a match chain of M if $p_{c_i} < p_{c_j}$ and $t_{c_i} < t_{c_j}$ for any i and j such that $i < j$. A match m_i is said to have an overlap with a match m_j on the pattern sequence, if the two regions $[p_i, p_i + l_i - 1]$ and $[p_j, p_j + l_j - 1]$ overlap with each other. Similarly, a match m_i is said to have an overlap with a match m_j on the text sequence, if the two regions $[t_i, t_i + l_i - 1]$ and $[t_j, t_j + l_j - 1]$ overlap with each other.

The match chaining problem is a problem to find a match chain that represents an appropriate alignment of the two sequences, and is defined as follows, letting $score(C)$ be some measure for the 'goodness' of a chain C, which we will discuss later.

Problem 1. Find the match chain $C = \{m_{c_1}, m_{c_2}, \ldots, m_{c_f}\}$ with the largest $score()$ value among all the possible match chains.

There can be several strategies of defining $score()$ values. The simplest measure is the number of matches, *i.e.*, $score(C) = f$. In this case, the best chain is known to be obtained in $O(k \log k)$ time using the longest increasing subsequence (LIS) algorithm [7,12]. We can also use as the $score()$ values the total lengths of the match chains instead of just the number of matches, *i.e.*, $score(C) = \Sigma_{1 \le i \le f} l_{c_i}$. If we do not permit the overlaps of the adjacent matches of the chain, it can also be computed in the same time bound $O(k \log k)$ using a tree structure called a range tree [5,19]. Let C_i be the chain that has the largest $score()$ value among the match chains that ends with m_i, and $prev(m_i)$ be the match previous to m_i in the chain C_i. Most previous algorithms for this problem can be described in the following form.

Algorithm 1 (Basic Match Chaining Algorithm).

1. For all i from 1 to k in this order, do the following.
   – Find $prev(m_i)$ for m_i.
2. Select the match chain with the largest $score()$ value among the match chains obtained in step 1 for all the matches. The actual match chain is constructed by tracing $prev()$ entries.

If $prev(m)$ can be obtained in $O(g(k))$-time in step 1 for a match m, the total computation time is $O(k \cdot g(k))$ time, as step 2 can be done in linear time.

However, no algorithm has been known to exist that considers overlaps and runs in $O(k \log k)$ time. Let $pattern\_overlap(m_i, m_j) = \max\{0, p_i + l_i - p_j\}$. When $p_i \le p_j$ and $p_i + l_i \le p_j + l_j$, this represents the actual overlap length of the two matches on the pattern sequence. Similarly, we let $text\_overlap(m_i, m_j) = \max\{0, t_i + l_i - t_j\}$ and $overlap\_length(m_i, m_j) =$

$\max\{text\_overlap(m_i, m_j), pattern\_overlap(m_i, m_j)\}$. Then the total length of a match chain $\{m_{c_1}, m_{c_2}, \ldots, m_{c_f}\}$ is described as $\sum_{1 \le i \le f} l_{c_i} - \sum_{1 \le i < f} overlap\_length(m_{c_i}, m_{c_{i+1}})$ if we consider overlaps. In this paper, we use this value as the *score*() measure. The MUMmer uses the same measure, but their algorithm requires $O(k^2)$ time, which is inefficient for large k's. We will improve this bound later.

2.2 Dynamic Range Maximum Query

In this section, we deal with yet another problem, the range maximum query (RMQ) problem [4], which plays an important role in our match chaining algorithm. Given a length n array A of numbers, the RMQ problem is the problem of finding the index of the maximum value in the subarray $A[i \ldots j]$ for any query of i and j. For a static array, we can find the index in a constant time with linear-time preprocessing [4]. We consider the dynamic version of this problem as follows.

Problem 2 (Dynamic RMQ). Let S be a set of items that is a null set ϕ at first. Items that are not known in advance are inserted to or deleted from S one by one. An item I has two given values $a(I)$ and $b(I)$, and the problem is to find the item with the maximum b value among the items $I \in S$ whose a value is within the range $[p, q]$ for any query of p and q at any time.

We present a dynamic RMQ algorithm which allows $O(\log k)$ query time with $O(\log k)$ update time for both insertions and deletions, where k is the size of S at the time of the query or the update. In our dynamic RMQ algorithm, we maintain a queue of items sorted by the a values. Such queues are often implemented with balanced binary tree data structures. The height of a balanced tree data structure is always $O(\log k)$, where k is the current number of nodes or the current queue length. For any node v in these balanced trees, the left child of a node v always has a key (the a value in this case) that is no larger than the key of v, while the right child of v always has a key that is no smaller than the key of v. We use the AVL tree [3] for this purpose. The details of the AVL tree are described in many textbooks such as [1]. The AVL tree can maintain such a queue implicitly with the update (insertion and deletion) time of $O(\log k)$. Each node of the AVL tree corresponds to an item and the in-order traversal of the tree corresponds to the sorted list of the items. Each item can be accessed in $O(\log k)$ time using its a value as its key.

In our dynamic RMQ algorithm, we maintain a pointer from each node v to the item with the maximum b value among the items whose corresponding nodes are in the subtree under v (including v). We call this a max pointer. We show that maintenance of the max pointers for insertion and deletion of an item can also be done in $O(\log k)$ time as follows. The update procedure of the AVL tree consists of five procedures: position lookup, leaf deletion, leaf addition, node replacement, and a procedure called rotation, each of which can be done in a constant time except for the lookup procedure. The lookup procedure finds the node position to add, delete or replace, and it can be done in $O(\log k)$ time. This

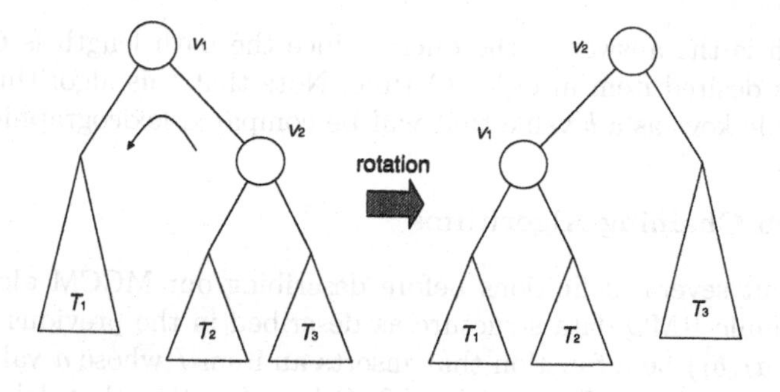

Fig. 3. The rotation procedure for the AVL tree.

procedure does not change the data structure, so we do not have to change any of the max pointers related to the lookup procedure. The next three procedures are executed at most once for an update, while the rotation procedure is executed $O(\log k)$ times. If we add or delete a leaf, or replace a node, we must also update the max pointers of the ancestors of the updated node. As the number of the ancestors is $O(\log k)$, we can do it in $O(\log k)$ time. Figure 3 shows the rotation procedure. Notice that the max pointers of nodes except for nodes v_1 and v_2 in the figure do not change at all. Furthermore, the new max pointers for the two nodes can be obtained in a constant time using the max pointers of the root nodes of the subtrees T_1, T_2, and T_3. Consequently, we need only a constant time for one rotation procedure to maintain the max pointers. In this way, we can maintain the max pointers in $O(\log k)$ time for each update.

We can compute the dynamic RMQ in $O(\log k)$ time using this data structure as follows. Using the AVL tree, we can easily find the item with the smallest a value that is not smaller than p in $O(\log k)$ time. Let the item be I_{left} and the corresponding node be v_{left}. Similarly, we can also find the item with the largest a value that is not larger than q in the same time bound $O(\log k)$. Let the item be I_{right} and the corresponding node be v_{right}. By checking the max pointers of the children of the nodes on the path from v_{left} to v_{right} as follows, we can find the item with the maximum b value among the items whose a value is not smaller than p and not larger than q. Let v_{lca} be the lowest common ancestor of the two nodes v_{left} and v_{right}. A constant-time query for the lowest common ancestor is possible with linear-time preprocessing [4], but we do not have to use the algorithm, as it can be easily computed in a time linear to the size of the path from v_{left} to v_{right}, that is $O(\log k)$. For each node v on the path P from v_{left} to v_{lca} including v_{left} but not including v_{lca}, we check the item pointed to by the max pointer of the right child of v and also the item of node v, if the right child of v is not on the path P. Similarly, for each node v on the path P' from v_{right} to v_{lca} including v_{right} but not including v_{lca}, we check the item pointed to by the max pointer of the left child of v and also the item of node v, if the left child of v is not on the path P'. Add to these, we check the item of v_{lca}. Then we choose the item with the maximum b value among these checked items, which is the answer to the query. Since the path length is $O(\log k)$, we can find the desired item in $O(\log k)$ time. Note that this algorithm can use a set of multiple keys as a b value that will be compared lexicographically.

2.3 Match Chaining Algorithms

Let us set out several definitions before describing our MCCM algorithm. Let R be a dynamic RMQ data structure as described in the previous section. Let $insert(R, I, a_I, b_I)$ be a function that inserts an item I whose a value is a_I and whose b value is b_I into R. Let $delete(R, I)$ be a function that deletes the item I from R. Let $rmq(R, p, q)$ be a function that returns the item that has the maximum b value among the items in R such that $p \leq a \leq q$. If such an item does not exist, it returns nil. Each function can be executed in $O(\log k)$ time, where k is the number of current items in R.

Maximum Inter-match Region Length The genome sequence has many repetitive regions, and copies of a gene subset can be seen in the genome. As a result, an ordinary match chaining algorithm described in the preliminary section often chains matches that are too far away from each other (several hundred millions bp apart in some cases). This can be avoided easily if we set a maximum inter-match region length and we do not permit chaining matches whose distance in the text sequence is larger than that. It can easily be incorporated without increasing the time complexity by using the following algorithm. Note that this algorithm does not permit overlaps of the matches. Let max_len be the maximum inter-match region length.

Algorithm 2 (Match Chaining Algorithm with Maximum Inter-match Region Length).

1. Let R be an empty dynamic RMQ data structure.
2. For all i from 1 to k in order, do the following.
   (a) For all matches m_j such that $j < i$, $p_j + l_j \leq p_i$ and $m_j \notin R$, execute the function $insert(R, m_j, t_j + l_j - 1, score(m_j))$. Note that $score(m_j)$ is described in the next step.
   (b) Let $prev(m_i) = rmq(R, t_i - max\_len - 1, t_i - 1)$. If $prev(m_i) = nil$, let $score(m_i) = l_i$. Otherwise let $score(m_i) = score(prev(m_i)) + l_i$.
3. Find the match m with the largest value of $score(m)$, and construct a chain from m by tracing back the $prev()$ information.

This algorithm runs in $O(k \log k)$ time in total, as $insert()$ and $rmq()$ routines are executed k times at most.

Incorporating Overlaps We next incorporate overlaps without increasing the time complexity, as maximal matching substrings often overlap with each other. Our problem is to find the match chain with the largest total length considering overlaps. The MUMmer uses the same metric, but their algorithm requires $O(k^2)$ time. We break this barrier by proposing an $O(k \log k)$ algorithm. Moreover, at

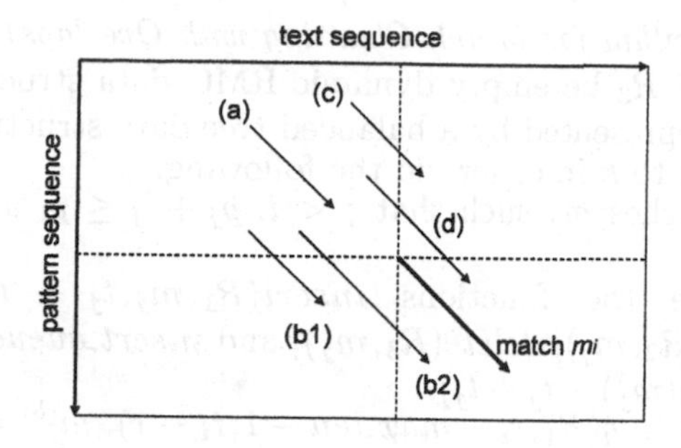

Fig. 4. Types of overlapping matches.

the same time, we use the restriction of the maximum inter-match region as stated above.

In our algorithm, we classify the match candidates for $prev(m_i)$ into 4 groups as follows. (a) A match that does not have an overlap with m_i. (b) A match m_j that has an overlap with m_i on the pattern sequence and satisfies $pattern\_overlap(m_j, m_i) > text\_overlap(m_j, m_i)$. (c) A match that has an overlap on the text sequence but not on the pattern sequence. (d) A match m_j that has overlaps with m_i both on the pattern and on the text and satisfies $pattern\_overlap(m_j, m_i) < text\_overlap(m_j, m_i)$. Figure 4 shows these match types graphically. In the figure, Type (b) is divided into the following two subtypes: (b1) A match of Type (b) that has no overlap with m_i on the text sequence. (b2) A match of Type (b) that also has an overlap with m_i on the text sequence. Note that we have to consider only the case (a) if we do not permit overlaps.

Our algorithm finds the best match candidates among the matches of Type (a), among those of Types (b), among those of Type (c), and among those of Type (d) separately, each of which is computed in $O(\log k)$ time. After that, it determines the actual best match from the four candidates. As for the matches of Type (a), we do not have to consider about any overlaps, so it is easy to deal with them with the dynamic RMQ data structure. To compute scores for the matches of Type (b), we must take the overlap lengths on the pattern sequence into account. Thus we construct a dynamic RMQ data structure based on the scores minus the pattern positions instead of the scores. As for those of Types (c) and (d), we must concisder the overlap lengths on the text sequence. Thus we use the scores minus the text positions instead of the scores as the measure. For the matches of Type (c), we use a different data structure for queuing and two functions for it, *insert_queue*() and *find_best*(), which we will describe later. Letting $t(m)$ denote the start position of a match m (*i.e.* t_i for m_i), our match chaining algorithm is described as follows. In the algorithm, R_1, R_2, R_3, and Q maintain the data for querying matches of types (a), (b), (d), and (c) respectively.

Algorithm 3 (Algorithm for Match Chaining with Overlaps).

1. Let R_1, R_2 and R_3 be empty dynamic RMQ data structures, and Q be an empty queue represented by a balanced tree data structure.
2. For all i from 1 to k in order, do the following.
   (a) For all matches m_j such that $j < i$, $p_j + l_j \leq p_i$ and $m_j \notin R$, do the following.
      – Execute the functions $insert(R_1, m_j, t_j + l_j - 1, score(m_j))$, $delete(R_2, m_j)$, $delete(R_3, m_j)$, and $insert\_queue(Q, m_j, t_j, t_j + l_j - 1, score(m_j) - t_j - l_j)$.
   (b) Let $m^{(a)} = rmq(R_1, t_i - max\_len - 1, t_i - 1)$, $m^{(b)} = rmq(R_2, t_i - p_i - max\_len - 1, t_i - p_i - 1)$, $m^{(c)} = find\_best(Q, t_i)$, and $m^{(d)} = rmq(R_3, t_i - p_i, t_i - 1)$.
   (c) If $m^{(d)} \neq nil$ and $t(m^{(d)}) > t_i$, let $m^{(d)} = nil$.
   (d) If the four candidates $m^{(a)}$, $m^{(b)}$, $m^{(c)}$ and $m^{(d)}$ are all nil, let $prev(m_i)$ be nil and $score(m_i) = l_i$. Otherwise, let $prev(m_i)$ be the match m that is selected from the four (or less when some of them are nil) matches so that $score(m_i) = score(m) + l_i - overlap(m, m_i)$ is maximized.
   (e) Execute the functions $insert(R_2, m_i, t_i - p_i, score(m_i) - p_i - l_i)$ and $insert(R_3, m_i, t_i - p_i, key(score(m_i) - t_i - l_i, t_i - p_i))$, where $key(x, y)$ denotes a multiple key that is compared lexicographically, *i.e.*, x value is compared first and y is used only when x is same.
3. Find the match m with the largest value of $score(m)$, and construct a chain from m by tracing back the $prev()$ information.

In step 3(b), RMQ against R_3 sometimes returns a match whose text position is larger than t_i and we dismiss it in step 3(c). There is no problem in doing so because we always have a candidate of another type that is not worse than any candidates of Type (d) in this case. Next we explain the two functions for the matches of Type (c). Letting $\{I_i\}$ be a set of items for each of which we will give three values a_i, b_i, and c_i, Q be a queue of the items sorted by their b_i values, the function $insert\_queue(Q, I_i, a_i, b_i, c_i)$ is described as follows.

Algorithm 4 ($insert\_queue(Q, I_i, a_i, b_i, c_i)$).

1. Find the item I_e in Q that has the smallest b value such that $b > b_i$.
2. Insert the item I_i into Q if $b_i < a_e$ or $c_i > c_e$. Otherwise, the algorithm is finished.
3. For the items I_j such that $a_i \leq b_j \leq b_i$ from the items that have larger b_j, do the following.
   – If $c_j > c_i$, the algorithm is finished. Otherwise delete I_j from Q.

This procedure runs in $O((d + 1) \log k)$ time where d is the number of items deleted in step 3, and k is the current size of Q, if we implement the queue Q with a balanced tree data structure like the AVL trees. The total number of deletions can be bounded by the number of matches k, thus the computation time for this function in total is $O(k \log k)$. Using this data structure, we can maintain the match candidates of Type (c). Note that this data structure does not maintain some of the matches of Type (c) if a better match candidate or that with the same score exists. We can obtain the best candidate with the following function, $find\_best(Q, b)$.

Table 1. Computation time for mapping all the 60,770 FANTOM cDNAs to three mouse chromosomes.

Chromosome #	1	15	19
Genome size (bp)	196,842,934	104,633,288	61,356,199
Query time in total (sec)	1565.48	1431.37	1333.87
Suffix tree construction time (sec)	2294.28	1159.37	665.88

Algorithm 5 (find_best(Q, b)).

1. Find the item I_e in Q that has the smallest b value such that $b_j > b$.
2. If $b > a_e$, return the item I_e. Otherwise, return *nil*.

This can also be done in $O(\log k)$ time. Thus our match chaining algorithm runs in $O(k \log k)$ time in total.

There are several possible extensions of the MCCM algorithm. We can easily extend our algorithm to use different scores for different bases instead of just using the total length of the match chain. In our match chaining algorithm, we find the *previous*(m) for a match m from the matches only from the region specified by the maximum inter-match region length on the same strand. By selecting *previous*(m) in our algorithm from both strands, without increasing the time complexity, we can extend our algorithm for finding irregular mapping sites when a protein is encoded from both DNA strands [15]. An extension to multiple alignments as in [13] is also interesting, but the analysis of overlaps will be very difficult and remains as a future task.

3 Computational Experiments

In this section, we examine the performance of the MCCM algorithm through computational experiments in which we mapped cDNA sequences of the FANTOM 2.0 database [8] to the mouse chromosome 1 genome sequence and several others taken from the whole genome sequence assembly version 3 of Mouse Genome Sequencing Consortium (MGSC) (ftp://wolfram.wi.mit.edu/pub/mouse_contigs/MGSC_V3). We used this data because the FANTOM annotators used it for mapping the cDNAs. The FANTOM 2.0 database consists of 60,770 full-length cDNA clones whose total size is around 120 Mbp. Chromosome 1 is the largest chromosome of the mouse with a length of 196,842,934 bp. In the following experiments, we used a Power4 CPU running at 1.3 GHz.

Table 1 shows the speed of the MCCM algorithm for the genome sequences of three mouse chromosomes. In the experiments, we let $l = 1$ for the MRM computation (*i.e.* we use MUMs) and set the maximum inter-match to 3 Mbp, which is larger than any intron size as far as the authors know. We did not use a larger l because using a larger l leads to decrease of specificity according to our preliminary experiments. It may be because the mouse genome does

Table 2. Numbers of cDNAs mapped to the positive strand of Chromosome 1.

Algorithm	MIML = 3 Mbp					MIML not used				
Ratio (%)	90	80	70	60	50	90	80	70	60	50
#mapped	1804	1907	1964	2032	2112	1808	1914	1983	2057	2153
#annotated	1773	1824	1846	1860	1864	1774	1824	1846	1861	1864
Specificity (%)	98.3	95.6	94.0	91.5	88.3	98.1	95.3	93.1	90.5	86.6
Sensitivity (%)	94.6	97.3	98.5	99.3	99.5	94.7	97.3	98.5	99.3	99.5

not have many long copies of genes in the same chromosome. We set the minimum MRM length threshold to 20, because extremely short matches can appear randomly and they also cause the decrease of specificity. The table shows our results are much faster than previous tools like BLAT [14] and sim4 [10], which require about 2 to 20 seconds per query for a problem of the same size with the same CPU. The MCCM algorithm requires only 0.026 seconds per query on average. Note that the query time is roughly proportional to the query size. The Squall [20] algorithm runs at a speed similar to ours, but the query time of Squall is proportional to the genome size. On the other hand, the query time of our algorithm is not much influenced by the the genome size. Table 1 reveals that the query time increases only about 17% even if the genome size becomes more than 3 times larger. This means that the query time will be still reasonable even if we apply our algorithm to an entire genome with a size of 2 Gbp size or larger. A suffix tree for a sequence of that size is difficult to construct in the main memory of today's typical machines, but we believe it will be easy to do so in the very near future and then our algorithm will work very efficiently. The time for constructing the suffix tree is proportional to the genome size, but we only have to build it once.

Next, we examine the accuracy of the MCCM algorithm. We compared the results of our algorithms with the FANTOM annotations. We also did experiments with the same algorithm but without setting the maximum inter-match length (MIML) to see the importance of the MIML. Table 2 shows the results. In the table, the 'MIML = 3 Mbp' columns show the results of our algorithm where we set the maximum inter-match length to 3Mbp, while the 'MIML not used' columns show the results of the algorithm without MIML. Note that we did experiments only against the positive strand of the chromosome. In the experiment, we accept that the cDNAs are mapped to the chromosome if the total match chain length exceeds ratio r of the cDNA length for some r. The 'Ratio (%)' row shows the ratio r, which we adjusted from 90% to 50%. The '#mapped' row shows the numbers of cDNAs that are determined to be mapped to the positive strand of the chromosome by our algorithms. FANTOM annotators mapped 1,874 cDNAs to the positive strand of this chromosome. We call these 1,874 cDNAs 'annotated cDNAs'. The '#annotated' row shows the numbers of annotated cDNAs among our results such that the annotated regions are the same as ours or overlap with ours. The 'Specificity (%)' row shows the ratio of them among our results, while

Table 3. An example of a cDNA mapping result.

Exon	Annotated positions		Maximal match positions	
	cDNA	Genome	cDNA	Genome
1	1 : 220	65168568 : 65168788	2 : 224	65168570 : 65168792
2	221 : 342	65186935 : 65187056	220 : 346	65186934 : 65187060
3	343 : 489	65194686 : 65194832	341 : 489	65194684 : 65194832
4	490 : 590	65202019 : 65202119	485 : 591	65202014 : 65202120
5	591 : 731	65205936 : 65206078	589 : 628	65205934 : 65205973
			633 : 702	65205978 : 65206047
			712 : 731	65206059 : 65206078
6	732 : 915	65209875 : 65210057	732 : 779	65209875 : 65209922
			848 : 886	65209990 : 65210028
			887 : 914	65210028 : 65210055
7	916 : 1067	65211981 : 65212131	925 : 1068	65211989 : 65212132
8	1068 : 1768	65240140 : 65240840	1066 : 1768	65240138 : 65240840

the 'Sensitivity (%)' row shows the ratio of them among the annotated 1,874 cDNAs. Both the sensitivity and the specificity of our algorithm are remarkably high. For $r = 80\%$, we achieved 95.6% specificity and 97.3% sensitivity for example. The table also shows that we succeeded in increasing the specificity without decreasing the sensitivity by setting the maximal inter-match length, especially in the case of low ratio thresholds.

Finally, we show an example of the output of the MCCM algorithm in Table 3. This table shows the result for the cDNA whose ID is '1110063D23' in the FANTOM database. It is annotated as 'Camp-response element binding protein-homolog' and mapped to the chromosome 1. Our algorithm also maps this cDNA to the chromosome 1. The 'Annotated positions' columns show the positions given by the FANTOM annotators, while the 'Maximal match positions' show the corresponding maximal match positions. We can see that several matches overlap with each other, which means consideration of overlaps is mandatory, though the overlaps are very short. Our algorithms do not consider biological information like donor and acceptor signals. As a result, the boundaries are not correct in many cases as seen in this example. Add to these, two of the exons are divided into three matches, as our algorithm only outputs exact matches, which is caused by indels or substitutions. But, all in all, the match boundaries are quite close to the actual annotated boundaries, and we succeeded in mapping this cDNA to the correct site of the genome.

4 Concluding Remarks

We proposed a new cDNA mapping algorithm called the MCCM (Match Chaining-based cDNA Mapping) algorithm based on suffix trees and a match chaining technique. For it, we proposed a new $O(k \log k)$ match chaining algorithm which considers overlaps and distances between matches. We also examined the speed and accuracy of our algorithm through computational experiments using FANTOM cDNA sequences and the mouse genome. According to the experiments, our algorithm runs very fast, and we succeeded in finding almost the same (> 95%) set of cDNAs that were given by the FANTOM annotators.

Several tasks remain as future work. The suffix tree is a very large data structure. An implementation with suffix arrays or compressed suffix arrays is very attractive. As shown in the experiments, the match boundaries obtained with our algorithms are a little bit different from the actual boundaries of the exons. Correcting the boundaries does not seem to be a complex problem and this will be addressed in our future work.

Acknowledgment

The authors thank Prof. Kunihiko Sadakane of Kyushu University and Prof. Akihiko Konagaya of RIKEN for various interesting discussions on related problems.

References

1. Aho, A.V., Hopcroft, J.E. and Ullman, J.D. *Data Structures and Algorithms*, (1983) Addison-Wesley.
2. Altschul, S.F., Gish, W., Miller, W., Myers, E.W. and Lipman, D.J. (1990) Basic local alignment search tool. *J. Mol. Biol.*, 215, 403-410.
3. Adelson-Velskil, G.M. and Landis, E.M. (1962) *Soviet Math. (Dokl.)* 3, 1259-1263.
4. Bender, M.A. and Farach, M. (2000) The LCA problem revisited. *Proc. Latin American Theoretical Informatics*, LNCS 1776, 88-94.
5. Bentley, J. and Maurer, H. (1980) Efficient worst-case data structures for range searching. *Acta Informatica*, 13, 155-168.
6. Delcher, A.L., Kasif, S., Fleischmann, D., Paterson, J., White, O. and Salzberg, S.L. (1999) Alignment of whole genomes. *Nucleic Acids Res.*, 27(11), 2369-2376.
7. Delcher, A.L., Phillippy, A., Carlton, J. and Salzberg, L. (2002) Fast algorithms for large-scale genome alignment and comparison. *Nucleic Acids Res.*, 30(11), 2478-2483.
8. FANTOM Consortium and the RIKEN Genome Exploration Research Group Phase I & II Team. (2002) Analysis of the mouse transcriptome based on functional annotation of 60,770 full-length cDNAs. *Nature*, 420, 563-573.
9. Farach, M. (1997) Optimal suffix tree construction with large alphabets. *Proc. 38th IEEE Symp. Foundations of Computer Science,* 137-143.
10. Florea, L., Hartzell, G., Zhang, Z., Rubin, G.M. and Miller, W. (1998) A computer program for aligning a cDNA Sequence with a Genomic DNA Sequence. *Genome Res.*, 8, 967-974.
11. Gelfand, M.S., Mironov, A.A. and Pevzner, P.A. (1996) Gene recognition via spliced sequence alignment. *Proc. Natl. Acad. Sci. USA*, 93, 9061-9066.
12. Gusfield, D. (1997) *Algorithms on strings, trees, and sequences: computer science and computational biology,* Cambridge University Press.
13. Hoehl, M., Kurtz, S. and Ohlebusch, E. (2002) Efficient multiple genome alignment. *Bioinformatics*, 18(Suppl. 1), S312-320.

14. Kent, W. J. (2002) The BLAST-like alignment tool. *Genome Res.*, 12, 656-664.
15. Labrador, M., Mongelard, F., Plata-Rengifo, P., Bacter, E.M., Corces, V.G. and Gerasimova, T.I. (2001) Protein encoding by both DNA strands. *Nature*, 409, 1000.
16. Manber, U. and Myers, G. (1993) Suffix arrays: a new method for on-line string searches. *SIAM J. Comput.*, 22(5), 935-948.
17. McCreight, E.M. (1976) A space-economical suffix tree construction algorithm. *J. ACM,* 23, 262-272.
18. Mott, R. (1997) EST_GENOME: A program to align spliced DNA sequences to unspliced genomic DNA. *Comput. Applic. Biosci.*, 13(4), 477-478.
19. Myers, E and Miller, W. (1995) Chaining multiple-alignment fragments in sub-quadratic time. *Proc. ACM-SIAM Symp. on Discrete Algorithms*, 38-47.
20. Ogasawara, J. and Morishita, S. (2002) Fast and sensitive algorithm for aligning ESTs to Human Genome. *Proc. 1st IEEE Computer Society Bioinformatics Conference*, Palo Alto, CA, 43-53.
21. Sze, S-H. and Pevzner, P.A. (1997) Las Vegas algorithms for gene recognition: suboptimal and error-tolerant spliced alignment. *J. Comp. Biol.*, 4(3), 297-309.
22. Ukkonen, E. (1995) On-line construction of suffix-trees. *Algorithmica,* 14, 249-260.
23. Usuka, J., Zhu, W. and Brendel, V. (2000) Optimal spliced alignment of homologous cDNA to a genomic DNA template. *Bioinformatics*, 16(3), 203-211.
24. Weiner, P. (1973) Linear pattern matching algorithms. *Proc. 14th Symposium on Switching and Automata Theory,* 1-11.

Sequencing from Compomers: Using Mass Spectrometry for DNA De-Novo Sequencing of 200+ nt

Sebastian Böcker*

AG Genominformatik, Technische Fakultät, Universität Bielefeld,
PF 100 131, 33501 Bielefeld, Germany
boecker@CeBiTec.uni-bielefeld.de

Abstract. One of the main endeavors in today's Life Science remains the efficient sequencing of long DNA molecules. Today, most de-novo sequencing of DNA is still performed using electrophoresis-based Sanger Sequencing, based on the Sanger concept of 1977. Methods using mass spectrometry to acquire the Sanger Sequencing data are limited by short sequencing lengths of 15–25 nt.
We propose a new method for DNA sequencing using *base-specific cleavage* and *mass spectrometry*, that appears to be a promising alternative to classical DNA sequencing approaches. A single stranded DNA or RNA molecule is cleaved by a base-specific (bio-)chemical reaction using, for example, RNAses. The cleavage reaction is modified such that not all, but only a certain percentage of those bases are cleaved. The resulting mixture of fragments is then analyzed using MALDI-TOF mass spectrometry, whereby we acquire the molecular masses of fragments. For every peak in the mass spectrum, we calculate those base compositions that will potentially create a peak of the observed mass and, repeating the cleavage reaction for all four bases, finally try to uniquely reconstruct the underlying sequence from these observed spectra. This leads us to the combinatorial problem of Sequencing From Compomers and, finally, to the graph-theoretical problem of finding a walk in a subgraph of the de Bruijn graph. Application of this method to simulated data indicates that it might be capable of sequencing DNA molecules with 200+ nt.

1 Introduction

Suppose we want to reconstruct an (unknown) string s over the alphabet Σ. Multiple copies of s are cleaved with a certain probability whenever a specific character $x \in \Sigma$ appears, comparable to the Partial Digestion Problem [24]. Then, every resulting fragment y is scrambled by a random permutation so that the only information we are left with is how many times y contains each

---

* Sebastian Böcker is currently supported by "Deutsche Forschungsgemeinschaft" (BO 1910/1-1) within the Computer Science Action Program. This work was carried out in part while Sebastian Böcker was employed by SEQUENOM GmbH., Hamburg, Germany.

G. Benson and R. Page (Eds.): WABI 2003, LNBI 2812, pp. 476–497, 2003.

character $\sigma \in \Sigma$. In addition, we discard all fragments y that contain the cleavage character more than k times for a fixed threshold k. This threshold is usually chosen very small, for example $k \in \{2, 3, 4\}$. If we are given such reduced and scrambled fragment sets for every character $x \in \Sigma$, can we uniquely reconstruct the string s from this information? The main challenge for such reconstruction is not scrambling the fragments, but discarding fragments containing too many cleavage characters. Nevertheless, it is often possible to reconstruct the string.

The above problem arises in the context of sequencing DNA by the use of mass spectrometry. Today, most de-novo sequencing of DNA without any *a priori* information regarding the amplicon sequence under examination, is still performed based on the Sanger concept of 1977 [19]. Maxam and Gilbert [12] proposed a method utilizing base-specific chemical cleavage, but this method has not been viable for the dramatically increased demand in DNA sequencing. Both sequencing technologies use gel or capillary electrophoresis to acquire the experimental data. Other approaches like combining the Sanger concept with mass spectrometry for data acquisition [10], or PyroSequencing [18] are limited by the short sequencing length of 15–25 nt, while Sequencing by Hybridization (SBH) [7,2,11] never became practical due to the high number of false reads as well as the current costs of SBH chips.

Here we propose a new approach to DNA sequencing that is *not* based on the Sanger concept, using MALDI-TOF mass spectrometry to acquire the experimental data. Since MALDI-TOF mass spectrometry reads can be obtained in milliseconds to seconds, compared to hours for electrophoresis reads, and mass spectrometry generally provides reliable and reproducible results even under high throughput conditions, our approach seems to be a promising alternative to traditional electrophoresis-based de-novo sequencing. We have applied our method to simulated mass spectra generated from random as well as biological sequences, and simulation results indicate high chances of successful reconstruction even when sequencing 200 and more nucleotides. The reconstruction accuracy, however, highly depends on the underlying sample sequence.

The main focus of this paper is to give a suitable mathematical formulation for the problem of reconstructing the sample sequence from compomers — that represent the randomly scrambled fragments — and to propose a branch-and-bound algorithm that is usually sufficient to reconstruct the sequence in reasonable runtime.

2 Experimental Setup and Data Acquisition

Suppose we are given a target DNA molecule (or *sample DNA*) of length 100–500 nt. Using polymerase chain reaction (PCR) or other amplification methods we amplify the sample DNA. We assume that we have a way of generating a single stranded target, either by transcription or other methods,[1] and we talk about sample DNA even though the cleavage reaction might force us to transcribe the

[1] The method can easily be extended to deal with double stranded data, but we will concentrate in the following on single stranded data.

sample to RNA. We cleave the single stranded sequence with a base-specific chemical or biochemical cleavage reaction: Such reactions cleave the amplicon sequence at exactly those positions where a specific base can be found. Such base-specific cleavage can be achieved using endonucleases RNAse A [17] and RNAse T1 [8], uracil-DNA-glycosylase (UDG) [23], pn-bond cleavage [20], and others.

We modify the cleavage reaction by offering a mixture of cleavable versus non-cleavable "cut bases," such that not all cut bases but only a certain percentage of them will be cleaved. The resulting mixture contains in principle all fragments that can be obtained from the sample DNA by removing two cut bases, cf. Fig. 1 for an example. We call such cleavage reactions *partial.*

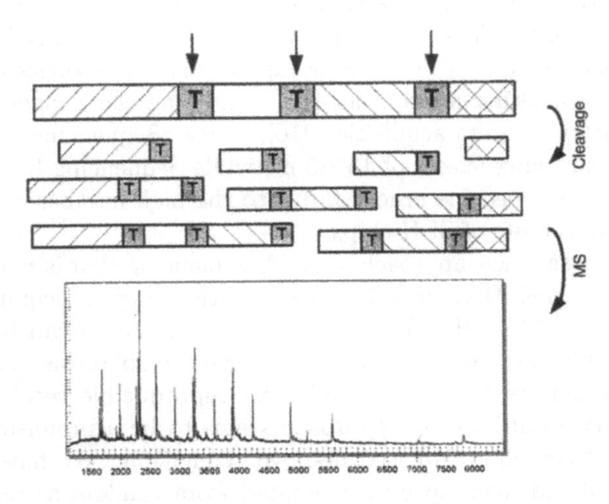

Fig. 1. Partial cleavage using RNAse A with dCTP, rUTP, and dTTP.

MALDI (matrix assisted laser desorption ionization) TOF (time-of-flight) mass spectrometry (MS for short) is then applied to the products of the cleavage reaction, resulting in a sample spectrum that correlates mass and signal intensity of sample particles [9].[2] The sample spectrum is analyzed to extract a list of signal peaks with masses and intensities.

We can repeat the above analysis steps using cleavage reactions specific to all four bases — alternatively, we can apply two suitably chosen cleavage reactions twice, to forward and reverse strands. So, we obtain up to four mass spectra, each corresponding to a base-specific cleavage reaction. We repeat the following steps of the analysis for every cleavage reaction.

[2] More precisely, MALDI-TOF mass spectrometers measure "mass per charge" instead of "mass" of sample particles. To simplify matters, we speak of "mass" instead of "mass per charge" because most particles in a MALDI mass spectrum will be single charged. Even more precisely, MALDI-TOF MS does not provide us with masses but only with time-of-flight of sample particles, so calibration (correlation of time-of-flight and mass) has to be determined beforehand.

If the sample sequence is known, then exact chemical results of the used cleavage reactions and, in particular, the masses of all resulting fragments are known in advance and can be simulated by an *in silico* experiment. Clearly, this holds up to a certain extent only, and measured spectra often differ significantly from the *in silico* predicted spectrum. Compared to other mass spectrometry applications, though, there is only a comparatively small number of differences between the simulated spectrum and the measured one.

Having said that, we can also solve the inverse problem: For every peak detected in the measured mass spectrum, we can calculate one or more base compositions (that is, DNA molecules with unknown order but known multiplicity of bases) that could have created the detected peak, taking into account the inaccuracy of the mass spectrometry read. Therefore, we obtain a list of base compositions and their intensities, depending on the sample DNA *and* the incorporated cleavage method. We want to stress that this calculation is simple when we are dealing with DNA or RNA, because the alphabet size is small and the average mass of a base (about 300 Dalton[3]) is much higher than the maximal mass difference between any two bases (about 50 Da). A simple algorithm based on searching $X+Y$ [4] can compute all base compositions in time $O(m^3)$, where m is the mass of the detected peak. Every base composition with mass sufficiently close to the detected peak must be seen as a potential explanation of the peak, and we use all such base compositions independently in the following.

Clearly, we cannot use the trivial approach of de-novo sequencing because of the exponential number of sequences: In this approach, we would (a) simulate the mass spectra for every potential sequence, for example $s \in \Sigma^l$ for some given length l, and (b) compare the resulting simulated spectra against the measured mass spectrum, finding the one that gives a best fit of the measured spectrum. Here a sequencing length of 200 nt results in about $2.6 \cdot 10^{120}$ mass spectra that we have to test for every cleavage reaction.

2.1 Limitations

The experimental setup described above has been successfully applied to problems such as Pathogen Identification [23] or SNP discovery [3, 17]. There, information on the sample sequence(s) under consideration is known beforehand, so that the requirements to the mass spectrometer (with regards to calibration accuracy and resolution) are comparatively small. Furthermore, we can use the additional information provided by the known reference sequence to reduce the algorithmic complexity of answering such questions. In the setting of this paper, though, almost no information but the mass spectrometry data itself is available.

In real life, several limitations characteristic for mass spectrometry and the experimental setup make the problem of de-novo sequencing from mass spectrometry data more challenging:

[3] Dalton (Da), a unit of mass equal to $\frac{1}{12}$ the mass of a carbon-12 nucleus, about 0.992 times the mass of a single H atom.

i. Current mass spectrometers limit the mass range in which particles can be detected: Signals above 8 000 Dalton ($\approx$ 25 nt) tend to get lost in the spectrum.
ii. Using MS, we can determine masses up to some inaccuracy only. Novel MS technologies like OTOF (orthogonal time of flight) MS allow us to measure particle masses with an inaccuracy of less than 0.3 Da, while current (ATOF, axial time of flight) mass spectrometers can show an inaccuracy of 1-2 Da under high throughput conditions.
iii. Because MS spectra are noisy, it is often impossible to distinguish between signal peaks with low intensities and noise peaks randomly found in the spectrum. Henceforth, one has to choose between minimizing either the number of false positive, or the number of false negative detected peaks.
iv. For a fixed cleavage reaction, several potential base compositions can have nearly identical masses. In the following, we independently use every potential explanation of a mass signal as a base composition. Therefore we transform a single mass signal found in the mass spectrum, into a list of base compositions with masses sufficiently close to the signal mass, depending on the sample sequence and the incorporated cleavage method.
v. Using partial cleavage results in an *exponential decay* (in the number of uncleaved cut bases) of signal intensities in the mass spectrum, so peaks from fragments containing many uncleaved cut bases will be difficult or impossible to detect.
vi. We often know 3–20 terminal bases of the sample string in advance. This can be due to primer or promoter regions used for amplification, transcription, or the like. Furthermore, the masses of terminal fragments located at beginning or end of the sample sequence in general differ from those of non-terminal fragments, and it is often possible to uniquely identify such fragments via their masses.

Depending on the underlying model of fragment ionization, we can calculate useful ratios of cleaved vs. uncleaved cut bases: Let $r \in [0, 1]$ denote the portion of cleaved cut bases and $(1 - r)$ the portion of uncleaved cut bases, so that the ratio equals $r : (1 - r)$. Useful choices for r are $r = \frac{2}{3}$, $\frac{1}{2}$, and $\frac{1}{3}$, because these choices maximize peak intensities of certain types of fragments in the mass spectrum. In addition, the use of small $r \ll \frac{1}{2}$ is not recommended because then, it becomes difficult to discriminate between so-called noise peaks and *any* type of signal peaks in the mass spectrum.

3 Methods

3.1 The Compomer Spectrum

Let $s = s_1 \dots s_n$ be a string over the alphabet Σ where $|s| = n$ denotes the *length* of s. We denote the concatenation of strings a, b by ab, the empty string of length 0 by ϵ.

If $s = axb$ holds for some strings a, x, b then x is called a *substring* of s, a is called a *prefix* of s, and b is called an *suffix* of s. We define the *number of occurrences* of x in s by:

$$\mathrm{ord}_x(s) := \max\{k : \text{there exist } s_0, \ldots, s_k \in \Sigma^* \text{ with } s = s_0 x s_1 x \ldots x s_k\}$$

Hence, x is a substring of s if and only if $\mathrm{ord}_x(s) \geq 1$. For $x \in \Sigma^1$, $\mathrm{ord}_x(s)$ simply counts the number of appearances of x in s. For general x, this is not necessarily the case, because $\mathrm{ord}_x(s)$ counts non-overlapping occurrences only.

For strings $s, x \in \Sigma^*$ we define the *string spectrum* $\mathcal{S}(s, x)$ of s by:

$$\mathcal{S}(s, x) := \{y \in \Sigma^* : \text{there exist } a, b \in \Sigma^* \text{ with } s \in \{yxb, axyxb, axy\}\} \cup \{s\} \tag{1}$$

So, the string spectrum $\mathcal{S}(s, x)$ consists of those substrings of s that are bounded by x or by the ends of s. In this context, we call s *sample string* and x *cut string*, while the elements $y \in \mathcal{S}(s, x)$ will be called *fragments* of s (under x).

Example 1. Consider the alphabet $\Sigma := \{\mathbf{0}, \mathrm{A}, \mathrm{C}, \mathrm{G}, \mathrm{T}, \mathbf{1}\}$ where the characters $\mathbf{0}$, $\mathbf{1}$ are exclusively used to denote start and end of the sample string. Let $s := \mathbf{0}\mathrm{ACATGTG}\mathbf{1}$ and $x := \mathrm{T}$, then:

$$\mathcal{S}(s, x) = \{\mathbf{0}\mathrm{ACA}, \mathrm{G}, \mathrm{G}\mathbf{1}, \mathbf{0}\mathrm{ACATG}, \mathrm{GTG}\mathbf{1}, \mathbf{0}\mathrm{ACATGTG}\mathbf{1}\}$$

The use of special characters $\mathbf{0}$, $\mathbf{1}$ to uniquely denote start and end of the sample sequence is motivated by the observation that terminal fragments in general differ in mass from inner fragments with otherwise identical sequence, see Lim. (vi). We make use of these characters throughout this paper to reduce the symmetry of the problem, see Example 2 below.

Following [3], we introduce a mathematical representation of base compositions: We define a *compomer* to be a map $c : \Sigma \to \mathbb{Z}$, where $\mathbb{Z}$ denotes the set of integers. We say that c is a *natural* compomer if $c(\sigma) \geq 0$ holds for all $\sigma \in \Sigma$. For the rest of this paper, we assume that all compomers are natural compomers, unless explicitly stated otherwise. Let $\mathcal{C}_+(\Sigma)$ denote the set of all natural compomers over the alphabet Σ. Clearly, $\mathcal{C}_+(\Sigma)$ is closed with respect to addition, as well as multiplication with a scalar $n \in \mathbb{N}$, where $\mathbb{N}$ denotes the set of natural numbers *including* 0. For finite Σ, in particular, $\mathcal{C}_+(\Sigma)$ is isomorphic to the set $\mathbb{N}^{|\Sigma|}$. We denote the canonical partial order on the set of compomers over Σ by $\preceq$, that is, $c \preceq c'$ if and only if $c(\sigma) \leq c'(\sigma)$ for all $\sigma \in \Sigma$. Furthermore, we denote the *empty compomer* $c \equiv 0$ by 0.

Suppose that $\Sigma = \{\sigma_1, \ldots, \sigma_k\}$, then we use the notation $c = (\sigma_1)_{i_1} \ldots (\sigma_k)_{i_k}$ to represent the compomer $c : \sigma_j \mapsto i_j$ omitting those characters σ_j with $i_j = 0$. For DNA, c represents the number of adenine, cytosine, guanine, and thymine bases in the compomer, and $c = \mathrm{A}_i\mathrm{C}_j\mathrm{G}_k\mathrm{T}_l$ denotes the compomer with $c(\mathrm{A}) = i$, $\ldots$, $c(\mathrm{T}) = l$. Since the characters $\mathbf{0}$, $\mathbf{1}$ appear at most once in any fragment, we usually omit the indices for these two characters.

The function $\mathrm{comp} : \Sigma^* \to \mathcal{C}_+(\Sigma)$ maps a string $s = s_1 \ldots s_n \in \Sigma^*$ to the compomer of s by counting the number of characters of each type in s:

$$\mathrm{comp}(s) : \Sigma \to \mathbb{N}, \quad \sigma \mapsto \big|\{1 \leq i \leq |s| : s_i = \sigma\}\big|$$

Note that compomers comp(·) are also called Parikh-vectors, see [1]. The *compomer spectrum* $\mathcal{C}(s,x)$ of s consists of the compomers of all fragments in the string spectrum:

$$\mathcal{C}(s,x) := \text{comp}\big(\mathcal{S}(s,x)\big) = \big\{\text{comp}(y) : y \in \mathcal{S}(s,x)\big\} \tag{2}$$

For Example 1 we can compute:

$$\mathcal{C}(s,\mathrm{T}) = \{\mathbf{0}\,\mathrm{A}_2\mathrm{C}_1,\ \mathrm{G}_1,\ \mathrm{G}_1\mathbf{1},\ \mathbf{0}\,\mathrm{A}_2\mathrm{C}_1\mathrm{G}_1\mathrm{T}_1,\ \mathrm{G}_2\mathrm{T}_1\mathbf{1},\ \mathbf{0}\,\mathrm{A}_2\mathrm{C}_1\mathrm{G}_2\mathrm{T}_2\mathbf{1}\}$$

Now, the following question arises: For an unknown string s and a known set of cleavage strings $\mathcal{X}$, can we uniquely reconstruct s from its compomer spectra $\mathcal{C}(s,x)$, $x \in \mathcal{X}$? One can easily see that this problem becomes trivial if there exist characters $\mathbf{0}$, $\mathbf{1}$ that uniquely denote the start and end of the sample string — then, for suitable $\mathcal{X}$ like $\mathcal{X} = \Sigma^1 \setminus \{\mathbf{0},\mathbf{1}\}$, the subsets $\{c \in \mathcal{C}(s,x) : c(\mathbf{0}) = 1\}$ are sufficient to reconstruct s. This fact was exploited in the Maxam-Gilbert approach [12]. Furthermore, this problem is related to, and appears to be computationally at most as hard as, the well-known Partial Digestion Problem (PDP) [24]: There, one cleaves a sample sequence using restriction enzymes, and measures the lengths of the resulting fragments. It seems likely that we can use algorithms efficiently tackling PDP [21, 22], to solve the above problem in reasonable runtime.

Unfortunately, this approach must fail when applied to experimental MS data, because our theoretical approach of compomer spectra does not take into account the limitations of mass spectrometry and partial cleavage mentioned in the previous section. As we have seen there, Lim. (v) suggests that the probability that some fragment y cannot be detected, strongly depends on the multiplicity of the cut string x as a substring of y. In fact, signals from fragments with $\text{ord}_x(y)$ above a certain threshold will most probably be lost in the noise of the mass spectrum and, due to Lim. (iii) and (iv), this threshold will be rather small — say, $k \le 4$ — in real-life applications. This leads us to the following two definitions: For strings s, x and $k \in \mathbb{N} \cup \{\infty\}$, we define the k-string spectrum of s, where k is called the *order* of the string spectrum, by:

$$\mathcal{S}_k(s,x) := \{y \in \mathcal{S}(s,x) : \text{ord}_x(y) \le k\} \tag{3}$$

The k-compomer spectrum of s is, in analogy to above, defined by:

$$\mathcal{C}_k(s,x) := \text{comp}\left(\mathcal{S}_k(s,x)\right) = \big\{\text{comp}(y) : y \in \mathcal{S}(s,x),\ \text{ord}_x(y) \le k\big\} \tag{4}$$

If the cut string is a single character $x \in \Sigma$, we infer $\mathcal{C}_k(s,x) = \{c \in \mathcal{C}(s,x) : c(x) \le k\}$.

For Example 1 we calculate $\mathcal{C}_0(s,\mathrm{T}) = \{\mathbf{0}\,\mathrm{A}_2\mathrm{C}_1,\ \mathrm{G}_1,\ \mathrm{G}_1\mathbf{1}\}$, $\mathcal{C}_1(s,\mathrm{T}) = \mathcal{C}_0(s,\mathrm{T}) \cup \{\mathbf{0}\,\mathrm{A}_2\mathrm{C}_1\mathrm{G}_1\mathrm{T}_1,\ \mathrm{G}_2\mathrm{T}_1\mathbf{1}\}$, and $\mathcal{C}_2(s,\mathrm{T}) = \mathcal{C}_1(s,\mathrm{T}) \cup \{\mathbf{0}\,\mathrm{A}_2\mathrm{C}_1\mathrm{G}_2\mathrm{T}_2\mathbf{1}\} = \mathcal{C}(s,\mathrm{T})$.

Under what conditions can we uniquely reconstruct a sample string s from its compomer spectra $\mathcal{C}_k(s,x)$, $x \in \mathcal{X}$? One can easily see that different strings can share the same k-compomer spectra:

Example 2. Let $\Sigma := \{\mathbf{0}, \mathrm{A}, \mathrm{B}, \mathbf{1}\}$. Then, we cannot uniquely reconstruct the sample string $s = \mathbf{0}\mathrm{BABAAB}\mathbf{1}$ from its complete cleavage compomer spectra $\mathcal{C}_0(s, \mathrm{A})$ and $\mathcal{C}_0(s, \mathrm{B})$, because the string $\mathbf{0}\mathrm{BAABAB}\mathbf{1}$ leads to the same spectra. Analogously, the string $s = \mathbf{0}\mathrm{BABABAABAB}\mathbf{1}$ cannot be reconstructed from its compomer spectra $\mathcal{C}_1(s, x)$ for $x \in \{\mathrm{A}, \mathrm{B}\}$, and we can create such examples for every order k. Furthermore, every string s and its reverse string have identical compomer spectra if all cut strings have length one or — more generally — if all cut strings are symmetric.

Yet, the question stated above does not take into account the problem of false positives: That is, compomers in the set $\mathcal{C}_x$ that do not correspond to actual fragments of the sample sequence. Due to Lim. (ii) and (iv), transforming a mass spectrum into a set of compomers will in general create huge numbers of false positive compomers, since there is usually only one sample fragment corresponding to a peak, but there may be many more compomers with almost identical mass. This number is potentially further increased in view of false positive peaks, see Lim. (iii).

To address this issue we can formulate an optimization problem as follows: For a fixed order $k \in \mathbb{N} \cup \{\infty\}$ let $\mathcal{X} \subseteq \Sigma^*$ be a set of cut strings, and let $\mathcal{C}_x \subseteq \mathcal{C}_+(\Sigma)$ be compomer sets for $x \in \mathcal{X}$. Let $S \subseteq \Sigma^*$ be the set of sample string candidates. Now, find a string $s \in S$ with $\mathcal{C}_k(s, x) \subseteq \mathcal{C}_x$ for all $x \in \mathcal{X}$ that minimizes $\sum_{x \in \mathcal{X}} |\mathcal{C}_x \setminus \mathcal{C}_k(s, x)|$. A potential choice of $S \subseteq \Sigma^*$ are all strings s such that $|s|$ lies in some given interval I, or strings with prefix $\mathbf{0}$ and suffix $\mathbf{1}$. In addition, we may want to search for those strings s of minimal length.

We cannot offer a solution to this problem, but note that the (purely combinatorial) optimization formula $f(s) := \sum_{x \in \mathcal{X}} |\mathcal{C}_x \setminus \mathcal{C}_k(s, x)|$ does not adequately reproduce the experimental "truth": In applications, a tiny peak detected in a mass spectrum can account for several compomers in the corresponding compomer set due to Lim. (iv), and trying to minimize the number of "unused" compomers contradicts the experimental observation. To this end, it makes more sense to find all "good" strings that satisfy at least the inclusion condition:

Sequencing From Compomers (SFC) Problem. For a fixed order $k \in \mathbb{N} \cup \{\infty\}$, let $\mathcal{X} \subseteq \Sigma^*$ be the set of cut strings and, for all $x \in \mathcal{X}$, let $\mathcal{C}_x \subseteq \mathcal{C}_+(\Sigma)$ be a compomer set. Finally, let $S \subseteq \Sigma^*$ be the set of sample string candidates. Now, find all strings $s \in S$ that satisfy $\mathcal{C}_k(s, x) \subseteq \mathcal{C}_x$ for all $x \in \mathcal{X}$.

Note first that this problem differs substantially from PDP and related problems, and approaches for solving PDP cannot be modified to tackle SFC: Such approaches [21,22] rely on the fact that fragments of s where exactly one base in s has been cleaved, can be detected. Unfortunately, these are precisely the fragments that will always be missing from the compomer sets $\mathcal{C}_x$ due to Lim. (v)! So, SFC is somewhat "in-between" the Partial Digestion Problem and the Double Digestion Problem (DDP) [24,14].

Second, it is trivial to find solutions to the SFC Problem in case $S = \Sigma^*$, because $s = \epsilon$ always satisfies the inclusion conditions. So, S should be chosen to exclude such trivial solutions.

Note that even small compomer sets may lead to a huge number of solutions, that is, exponentially many in the fixed length of the reconstructed strings:

Example 3. Let $\Sigma := \{A, B\}$, $k := 0$, and $S := \Sigma^n$ for some $n \in \mathbb{N}$; furthermore $\mathcal{X} := \Sigma^1$, $\mathcal{C}_A := \{B_1\}$, and $\mathcal{C}_B := \{A_1, A_2\}$. Every string $s \in S$ that is an arbitrary concatenation $s_0 s_1 s_2 \dots s_k$ where $s_0 \in \{A, AA\}$ and $s_j \in \{BA, BAA\}$ for $j = 1, \dots, k$ satisfies the conditions $\mathcal{C}_0(s, A) \subseteq \mathcal{C}_A$ and $\mathcal{C}_0(s, B) \subseteq \mathcal{C}_B$.

In applications, SFC is of interest because there hopefully are fewer solutions for experimental data and, as mentioned above, the optimization problem we introduced, does not capture all aspects of the connection between sample sequences and mass spectra. A reasonable approach here is to judge every solution $s \in S$ of SFC by, say, an adequate probability measure.

3.2 The Undirected Sequencing Graph

In this section, we introduce undirected sequencing graphs to tackle the SFC Problem of order $k = 1$. We shall see in the following section that this concept can be seen as a special case of the more elaborate directed sequencing graphs. For the sake of lucidity, we concentrate on the undirected case first.

In the following, we often limit our attention to cut strings x of length 1 to simplify our constructions. Doing so, we still cover all (bio-)chemical cleavage reactions mentioned in Section 2. We indicate in Section 6 how to extend our constructions to arbitrary cut strings $x \in \Sigma^*$. Note that we do not distinguish between the character $x \in \Sigma$ and the corresponding string of length 1.

An (undirected) *graph* consists of a set V of vertices, and a set $E \subseteq \binom{V}{2} \cup V = \{\{u, v\} : u, v \in V\}$ of edges. An edge $\{v\}$ for $v \in V$ is called a *loop*. We suppose that such graphs are *finite*, that is, have finite vertex set. A *walk* in G is a finite sequence $p = (p_0, p_1, \dots, p_n)$ of elements from V with $\{p_{i-1}, p_i\} \in E$ for all $i = 1, \dots, n$. Note that p is in general not a path because $p_0, \dots, p_n$ do not have to be pairwise distinct. We still use the letter p to denote a walk for convenience. The number $|p| := n$ is defined to be the *length* of p.

Let $\mathcal{C} \subseteq \mathcal{C}_+(\Sigma)$ be an arbitrary set of compomers, and let $x \in \Sigma$ be a single cut string of length one. We define the *undirected sequencing graph* $G_u(\mathcal{C}, x) = (V, E)$ as follows: The vertex set V consists of all compomers $c \in \mathcal{C}$ such that $c(x) = 0$ holds. The edge set E consists of those $\{u, v\}$ with $u, v \in V$ that satisfy:

$$u + \text{comp}(x) + v \quad \in \mathcal{C} \tag{5}$$

The vertices u, v are not required to be distinct in this equation.

Example 4. For $\Sigma := \{\mathbf{0}, A, C, G, T, \mathbf{1}\}$, $s := \mathbf{0}$CTAATCATAGTGCTG$\mathbf{1}$, and $x := T$ we can calculate the compomer spectrum of order 1:

$$\mathcal{C} := \mathcal{C}_1(s, T) = \{\mathbf{0}C_1, \mathbf{0}A_2C_1T_1, A_2, A_3C_1T_1, A_1C_1, A_2C_1G_1T_1, A_1G_1, A_1C_1G_2T_1, C_1G_1, C_1G_2T_1\mathbf{1}, G_1\mathbf{1}\}$$

We have depicted the corresponding sequencing graph $G_u(\mathcal{C}, T)$ in Figure 2.

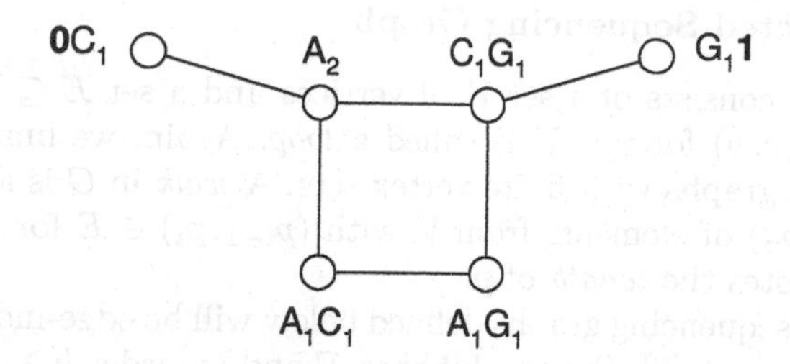

Fig. 2. The sequencing graph $G_u(\mathcal{C}, \mathrm{T})$ from Example 4.

How are sequencing graphs related to the SFC Problem? To this end, we say that a string $s \in \Sigma^*$ is 1-*compatible* with a compomer set $\mathcal{C} \subseteq \mathcal{C}_+(\Sigma)$ under $x \in \Sigma^1$ if $\mathcal{C}_1(s, x) \subseteq \mathcal{C}$ holds. Then s is a solution to the SFC Problem of order 1 with respect to x. And, we say that s is *compatible* with a walk $p = p_0 \dots p_l$ in the sequencing graph $G_u(\mathcal{C}, x)$ if there exist strings $s_0, \dots, s_l \in \Sigma^*$ such that

$$s = s_0 x s_1 x s_2 x \dots x s_l \tag{6}$$

satisfying $l = |p|$ and $\mathrm{comp}(s_j) = p_j$ for $j = 0, \dots, l$. This definition implies $\mathrm{ord}_x(s_j) = 0$ for $j = 0, \dots, l$. We call strings $s_0, \dots, s_l$ satisfying (6) and $\mathrm{ord}_x(s_j) = 0$ for all $j = 0, \dots, l$ an *x-partitioning* of s. For $x \in \Sigma^1$, there exists exactly one x-partitioning of s.

In Example 4, the walk ($\mathbf{0}\mathrm{C}_1$, A_2, $\mathrm{A}_1\mathrm{C}_1$, $\mathrm{A}_1\mathrm{G}_1$, $\mathrm{G}_1\mathbf{1}$) is compatible with our input sequence s, but other sequences like **0**CTAATCGTG**1** or **0**CTAATCGTGATGCTG**1** are also compatible with walks in $G_u(\mathcal{C}, \mathrm{T})$.

The next lemma follows from the above definitions, see Lemma 2 for a proof:

Lemma 1. *Let $s \in \Sigma^*$ be a string and $\mathcal{C} \subseteq \mathcal{C}_+(\Sigma)$ a set of compomers. Then, s is* 1*-compatible with $\mathcal{C}$ under $x \in \Sigma^1$ if and only if there exists a walk p in $G_u(\mathcal{C}, x)$ such that s is compatible with p. Furthermore, this walk p is unique.*

Proposition 1. *For every walk p of a sequencing graph $G_u(\mathcal{C}, x)$ there exist one or more sequences $s \in \Sigma^*$ that are compatible with p and, hence,* 1*-compatible with $\mathcal{C}$.*

Although basic, the above lemma allows us to search for all strings 1-compatible with a compomer set by simply building all walks in a graph if our set of cut strings $\mathcal{X}$ equals $\Sigma^1 \setminus \{\mathbf{0}, \mathbf{1}\}$: Let $\mathcal{C}_x$ be compomer sets and let p_x be walks in $G_u(\mathcal{C}_x, x)$ for all $x \in \mathcal{X}$. If a sample string $s \in \Sigma^*$ is compatible with p_x for every $x \in \mathcal{X}$, then s is uniquely determined by this property: For the prefix x of s of length 1, we infer from (6) that for every string s' compatible with p_x, x is also a prefix of s'. Repeating this argument leads to $s = s'$ as claimed.

We do not provide an algorithm here to build 1-compatible strings based on the above observation but refer the reader to the next section where we tackle the more general case of directed sequencing graphs.

3.3 The Directed Sequencing Graph

A *directed graph* consists of a set V of vertices and a set $E \subseteq V^2 = V \times V$ of edges. An edge (v, v) for $v \in V$ is called a *loop*. Again, we limit our attention to finite directed graphs with finite vertex sets. A *walk* in G is a finite sequence $p = (p_0, p_1, \dots, p_n)$ of elements from V with $(p_{i-1}, p_i) \in E$ for all $i = 1, \dots, n$, and $|p| := n$ denotes the *length* of p.

The directed sequencing graphs defined below will be edge-induced subgraphs of the de Bruijn graph [6]: For an alphabet Σ and an order $k \geq 1$, the *de Bruijn graph* $B_k(\Sigma)$ is a directed graph with vertex set $V = \Sigma^k$ and edge set

$$E = \{(u, v) \in V^2 : u_{j+1} = v_j \text{ for all } j = 1, \dots, k-1\}$$

where $u = (u_1, \dots, u_k)$ and $v = (v_1, \dots, v_k)$. In the following, we denote an edge $((e_1, \dots, e_k), (e_2, \dots, e_{k+1}))$ of $B_k(\Sigma)$ by $(e_1, \dots, e_{k+1})$ for short.

For an arbitrary set of compomers $\mathcal{C} \subseteq \mathcal{C}_+(\Sigma)$ and a cut string $x \in \Sigma$ of length one, we define the *directed sequencing graph* $G_k(\mathcal{C}, x)$ of order $k \geq 1$ as follows: $G_k(\mathcal{C}, x)$ is an edge-induced sub-graph of $B_k(\Sigma_x)$ where

$$\Sigma_x := \{c \in \mathcal{C} : c(x) = 0\}, \tag{7}$$

and an edge $e = (e_1, \dots, e_{k+1})$ of $B_k(\Sigma_x)$ belongs to $G_k(\mathcal{C}, x)$ if and only if the following condition holds:

$$e_i + c_x + e_{i+1} + c_x + \dots + c_x + e_{j-1} + c_x + e_j \in \mathcal{C} \quad \text{for all } 1 \leq i \leq j \leq k+1 \tag{8}$$

where $c_x := \text{comp}(x)$. By definition, the vertex set of $G_k(\mathcal{C}, x)$ is a subset of $(\Sigma_x)^k$.

Example 5. Let $\mathcal{C} := \mathcal{C}_2(s, \text{T})$, where $s = \mathbf{0}\text{CTAATCATAGTGCTG}\mathbf{1}$ was defined in Example 4. We have depicted the directed sequencing graph $G_2(\mathcal{C}, \text{T})$ in Figure 3. Note that there exist two paths connecting $\mathbf{0}\text{C}_1$ and $\text{G}_1\mathbf{1}$ in $G_\text{u}(\mathcal{C}, \text{T})$, but only one directed walk from $(\mathbf{0}\text{C}_1, \text{A}_2)$ to $(\text{C}_1\text{G}_1, \text{G}_1\mathbf{1})$ in $G_2(\mathcal{C}, \text{T})$: The ambiguity of the compomer $\text{A}_2\text{C}_1\text{G}_1\text{T}_1$ is resolved in $\mathcal{C}_2(s, \text{T})$ by the existence of compomers $\text{A}_4\text{C}_1\text{G}_1\text{T}_2$ and $\text{A}_2\text{C}_2\text{G}_2\text{T}_2$, and the non-existence of compomers $\mathbf{0}\text{A}_2\text{C}_2\text{G}_1\text{T}_2$ and $\text{A}_2\text{C}_1\text{G}_2\text{T}_2\mathbf{1}$.

We want to point out that the alphabet Σ_x of the underlying de Bruijn graph does not coincide with the sequence alphabet Σ, unlike the Sequencing By Hybridization approach introduced by Pevzner [15], but instead consists of those compomers $c \in \mathcal{C}$ with $c(x) = 0$. The main distinction between SFC and SBH, though, is that we search the de Bruijn graph for walks instead of Eulerian paths:

- We have to deal with many false positive edges here, because of "noise" peaks, misinterpreted peaks, and misinterpreted compomers.
- We do not know the multiplicity of compomers in $\mathcal{C}_x$, and compomers $c \in \mathcal{C}_x$ of small "order" will regularly correspond to two or more fragments of the sample sequence.

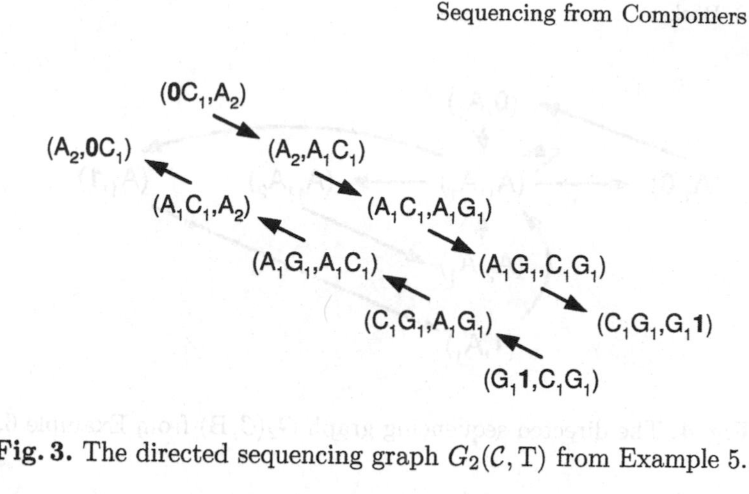

Fig. 3. The directed sequencing graph $G_2(\mathcal{C}, \mathrm{T})$ from Example 5.

Analogously to the previous section, we say that a string $s \in \Sigma^*$ is *k-compatible* with a compomer set $\mathcal{C} \subseteq \mathcal{C}_+(\Sigma)$ under $x \in \Sigma^1$ if $\mathcal{C}_k(s,x) \subseteq \mathcal{C}$ holds. Note that by definition, such a string s satisfies the condition of the Sequencing From Compomers Problem of order k. The string s is called *compatible* with a walk $p = p_0 \dots p_{|p|}$ in the sequencing graph $G_k(\mathcal{C}, x)$ if the x-partitioning $s_0, \dots, s_l \in \Sigma^*$ of s from (6) satisfies $l = |p| + k - 1$ and

$$p_j = \left(c_j, c_{j+1}, \dots, c_{j+k-1}\right) \quad \text{for } j = 0, \dots, |p| , \tag{9}$$

where $c_j := \mathrm{comp}(s_j)$ for $j = 0, \dots, l$. Recall that $\mathrm{ord}_x(s_j) = 0$ must hold for $j = 0, \dots, |p|$. We note that the definitions of "1-compatible" in the previous section, and "k-compatible" for $k = 1$ are equivalent, and so are the graphs $G_\mathrm{u}(\mathcal{C}, x)$ and $G_1(\mathcal{C}, x)$: For every edge $\{u, v\}$ with $u \neq v$ in $G_\mathrm{u}(\mathcal{C}, x)$ there exist two edges (u, v) and (v, u) in $G_1(\mathcal{C}, x)$, and for every loop $\{v\}$ in $G_\mathrm{u}(\mathcal{C}, x)$ there exists a loop (v, v) in $G_1(\mathcal{C}, x)$.

Example 6. For $s :=$ **0**BABABABAABABAB**1** created analogously to Example 2, the graph $G_2(\mathcal{C}, \mathrm{B})$ for $\mathcal{C} := \mathcal{C}_2(s, \mathrm{B})$ is depicted in Figure 4. If we remove the (superfluous) vertices $(\mathrm{A}_1, \mathbf{0})$ and $(\mathbf{1}, \mathrm{A}_1)$, there still exist two walks of length 6 from $(\mathbf{0}\mathrm{A}_1, \mathrm{A}_1)$ to $(\mathrm{A}_1, \mathrm{A}_1\mathbf{1})$ that traverse all edges of the resulting graph; the two sequences compatible with these two walks are our initial sequence, plus the "stutter" sequence **0**BABABAABABABAB**1**, respectively: The positions 7 and 8 of s are exchanged in this string.

Lemma 2. *Let $s \in \Sigma^*$ be a string with $\mathrm{ord}_x(s) \geq k$ for $x \in \Sigma^1$, and let $\mathcal{C} \subseteq \mathcal{C}_+(\Sigma)$ be a set of compomers. Then, s is k-compatible with $\mathcal{C}$ under x if and only if there exists a walk p in the sequencing graph $G_k(\mathcal{C}, x)$ such that s is compatible with p. Furthermore, this walk p is unique.*

Proposition 2. *For every walk p of a sequencing graph $G_k(\mathcal{C}, x)$ there exist one or more sequences $s \in \Sigma^*$ compatible with p and, hence, k-compatible with $\mathcal{C}$.*

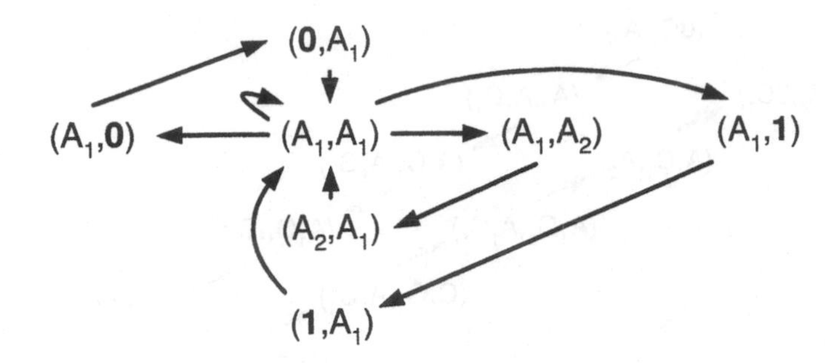

Fig. 4. The directed sequencing graph $G_2(\mathcal{C}, \mathrm{B})$ from Example 6.

It is straightforward to derive the proof of Lemma 2 from the definitions:

Proof (Lemma 2). Suppose that s is k-compatible with $\mathcal{C}$, and let $s_0, \ldots, s_l \in \Sigma^*$ be the x-partitioning of s. Using equation (9) we can define $l - k + 1$ compomers $p_0, \ldots, p_{l-k+1} \in \mathcal{C}^k$, then the definition of $\mathcal{C}_k(s, x)$ implies $\{\mathrm{comp}(s_j) : j = 0, \ldots, l\} \subseteq \Sigma_x$ and, hence, that $p = (p_0, \ldots, p_{l-k+1})$ is a walk in the de Bruijn graph $B_k(\Sigma_x)$. By definition of $\mathcal{C}_k(s, x)$, (8) must hold for every edge $(e_1, \ldots, e_{k+1})$ of p, so p is also a walk in $G_k(\mathcal{C}, x)$ as claimed. If s is compatible with paths p, p' in $G_k(\mathcal{C}, x)$ then (9) implies $p = p'$.

Suppose now that s is compatible with a walk p in $G_k(\mathcal{C}, x)$. Let $s_0, \ldots, s_l \in \Sigma^*$ be the x-partitioning of s, then $l = |p| - k + 1$ must hold. Let $y \in \mathcal{S}(s, x)$ be a fragment of s with $\mathrm{ord}_x(y) \leq k$, then there exist indices i_0, j_0 with $j_0 - i_0 \leq k$ such that $y = s_{i_0} x \ldots x s_{j_0}$. Let $j := \min\{i_0, |p| - k\}$ then by (9), the $(j+1)$-th edge of p is

$$e = \Big(\big(\mathrm{comp}(s_j), \ldots, \mathrm{comp}(s_{j+k-1})\big), \big(\mathrm{comp}(s_{j+1}), \ldots, \mathrm{comp}(s_{j+k})\big) \Big)$$

and, since e is an edge of $G_k(\mathcal{C}, x)$, we infer from (8) that

$$\mathrm{comp}(s_{i_0}) + \mathrm{comp}(x) + \cdots + \mathrm{comp}(x) + \mathrm{comp}(s_{j_0}) \in \mathcal{C}$$

must hold and, hence, $\mathrm{comp}(y) \in \mathcal{C}$. We conclude $\mathcal{C}_k(s, x) \subseteq \mathcal{C}$ as claimed. □

For a given set of compomers, how sparse is the corresponding sequencing graph in general? Clearly, the de Bruijn graph of order k over the alphabet Σ_x has $|\Sigma_x|^k$ vertices and $|\Sigma_x|^{k+1}$ edges. Unfortunately, the number of vertices and edges of a sequencing graph may be of the same order as those of the de Bruijn graph itself even for small compomer sets of size $O(k\,|\Sigma_x|)$ and "short" strings of length $O(n^2 + kn)$, as the following two lemmata show:

Lemma 3. *Let Σ be an alphabet of size $|\Sigma| \geq 2$, $x \in \Sigma^1$ a cut string of length one, and let $k \in \mathbb{N}$ be the fixed order. Then, for every $n \in \mathbb{N}$ there exist compomer sets $\mathcal{C}_n \subseteq \mathcal{C}_+(\Sigma)$ satisfying*

$$|\mathcal{C}_n| = (k+1)(n+1) \quad \text{and} \quad |\Sigma_x| = n+1 \ \text{for}\ \Sigma_x = \{c \in \mathcal{C}_n : c(x) = 0\} \quad (10)$$

such that the corresponding sequencing graph $G_k(\mathcal{C}_n, x)$ *has* $\binom{n+k}{k} = \Theta(n^k)$ *vertices and* $\binom{n+k+1}{k+1} = \Theta(n^{k+1})$ *edges.*

We omit the proofs of this and the following lemma and only note that the sets

$$\mathcal{C}_n := \{\mathrm{A}_m\mathrm{B}_i : m = 0, \ldots, n \text{ and } i = 0, \ldots, k\} \tag{11}$$

satisfy the conditions of the above lemma for $x := \mathrm{B}$.

Lemma 4. *There exist strings* $s \in \Sigma^*$ *over the alphabet* $\Sigma = \{\mathrm{A}, \mathrm{B}\}$ *of length* $O(n^2 + kn)$ *with* $\mathcal{C}_k(s, \mathrm{B}) = \mathcal{C}_n$ *as defined in* (11).

In view of Lemmata 3 and 4, one can suspect that it is impossible to perform de-novo sequencing from compomers. Fortunately, this seems not to be the case neither for random sequences nor for biological sequence data, see Section 5.

4 Algorithm

Let Σ be a constant and finite alphabet where $\mathbf{0}, \mathbf{1} \in \Sigma$ uniquely denote the first and last character of our sample strings. Let $\mathcal{X} = \Sigma^1 \setminus \{\mathbf{0}, \mathbf{1}\}$ be the set of cut strings, and $k \in \mathbb{N}$ the fixed order. We are given sets of compomers $\mathcal{C}_x$ for $x \in \mathcal{X}$ and a set $S \subseteq \Sigma^*$ of strings, and want to solve the Sequencing From Compomers Problem, that is, find all sample strings $s \in S$ satisfying $\mathcal{C}_k(s, x) \subseteq \mathcal{C}_x$ for all $x \in \mathcal{X}$. We further concentrate on the case especially relevant for applications, where

$$S = \{s \in \Sigma^* : l_{\min} \leq |s| \leq l_{\max}, \text{ and } s = \mathbf{0}\, s'\, \mathbf{1} \text{ for some } s' \in (\Sigma \setminus \{\mathbf{0}, \mathbf{1}\})^*\} \tag{12}$$

contains all strings of length in a given interval. This is because we either know the approximate length of the unknown string due to our experimental setup, or we can easily estimate it if necessary.

To solve SFC, we present a depth-first search that backtracks through sequence space, moving along the edges of the sequencing graphs in parallel. In this way, we implicitly build walks in the directed sequencing graphs of order k that are compatible with the constructed sequences. Because of Lemma 2, these sequences are in fact k-compatible with $\mathcal{C}_x$ under x for every $x \in \mathcal{X}$ and, hence, are solutions to SFC. In every recursion step of the algorithm, we attach every character $x \in \mathcal{X}$ to the previously known string s. This forces us to do an edge transition in the sequencing graph $G_k(\mathcal{C}_x, x)$, and we can stop the recursion if this edge transition is not possible. In addition, we do another branch-and-bound check by testing if it will be possible in the future to do edge transitions in all other sequencing graphs.

4.1 Building the Sequencing Graphs

First, we have to build the sequencing graphs $G_x := G_k(\mathcal{C}_x, x)$ for $x \in \mathcal{X}$. For $\Sigma_x := \{c \in \mathcal{C}_x : c(x) = 0\}$ we search for all those vectors $e \in (\Sigma_x)^{k+1}$ that satisfy (8). We make use of the trivial approach here: For every $(k+1)$-tuple $(e_1, \ldots, e_{k+1}) \in (\Sigma_x)^{k+1}$ we test if it satisfies equation (8) with $\mathcal{C} = \mathcal{C}_x$. This can be performed in $O(|\Sigma_x|^{k+1} k^2)$ time using a hash table to check $c \in \mathcal{C}_x$ and, since Σ_x and k are small in applications, this approach is sufficient here. If the condition is satisfied, we add $(e_1, \ldots, e_k)$ and $(e_2, \ldots, e_{k+1})$ to the vertex set of G_x, and we add $(e_1, \ldots, e_{k+1})$ to the edge set of G_x. A faster algorithm for building $G_k(\mathcal{C}_x, x)$ is to iteratively build the graphs $G_\kappa(\mathcal{C}_x, x)$ for $\kappa = 1, \ldots, k$.

For applications, we have to slightly modify the construction of our sequencing graphs G_x: Firstly, $G_k(\mathcal{C}_x, x)$ contains superfluous edges and vertices (cf. Example 6), because we know from (12) that characters $\mathbf{0}, \mathbf{1}$ are uniquely used to denote beginning and end of any string $s \in S$. So, we can limit the above calculations to edges $e = (e_1, \ldots, e_{k+1})$ such that $e_1(\mathbf{0}) \le 1$, $e_{j+1}(\mathbf{0}) = e_j(\mathbf{1}) = 0$ for $j = 1, \ldots, k$, and $e_{k+1}(\mathbf{1}) \le 1$.

More important, there remains one last problem: We do not know where to start in the sequencing graph! To this end, let $* \notin \Sigma_x$ denote a special source character. We add the source vertex $(*, \ldots, *)$ to G_x, but further edges and vertices are necessary to enter the "regular" part of the graph: A *source edge* is an edge $e = (e_1, \ldots, e_{k+1})$ of the de Bruijn graph $B_k(\Sigma_x \cup \{*\})$ such that there exists some $\kappa \in \{2, \ldots, k+1\}$ with:

- $e_j = *$ for $j = 1, \ldots, \kappa - 1$, and $e_j \ne *$ for $j = \kappa, \kappa+1, \ldots, k+1$
- $e_\kappa(\mathbf{0}) = 1$, $e_{j+1}(\mathbf{0}) = e_j(\mathbf{1}) = 0$ for $j = \kappa, \kappa+1, \ldots, k$, and $e_{k+1}(\mathbf{1}) \le 1$
- Equation (8) holds *only for* $\kappa \le i \le j \le k+1$

We add all source edges to the edge set of G_x, and we add all induced vertices to the vertex set of G_x: These induced vertices are of the form $(*, \ldots, *, v_\kappa, \ldots, v_k)$ where $2 \le \kappa \le k$ and $v_\kappa(\mathbf{0}) = 1$. Note that the resulting graph G_x is a subgraph of $B_k(\Sigma_x \cup \{*\})$. Now, we can use the source vertex $(*, \ldots, *)$ of G_x as our start vertex v_x^{start}.

We do not have to explicitly construct a sink, since the recursion below terminates as soon as we can add the end character $\mathbf{1}$. Note again that due to Lim. (vi), the masses of fragments from beginning and end of the sample string in general differ from those of all other fragments in applications, what has to be taken into account when computing the sets $\mathcal{C}_x$.

4.2 The Depth-First Search

Now, we start the recursion with the string $s := \mathbf{0}$. We initialize the current vertices $v_x := v_x^{\text{start}}$ for all $x \in \mathcal{X}$.

In the recursion step, let s be the current sample string, and for all $x \in \mathcal{X}$, let v_x be the current active vertices in the sequencing graph G_x. Let s_x be the unique string satisfying $\text{ord}_x(s_x) = 0$ such that either $x s_x$ is a suffix of s, or $s_x = s$ if $\text{ord}_x(s) = 0$. Set $c_x := \text{comp}(s_x)$.

– If $|s| + 1 \geq l_{\min}$ and we can do an edge transition to an end vertex in all sequencing graphs G_x for $x \in \mathcal{X}$, then **output** $s\mathbf{1}$ as a sequence candidate.
– If $|s| < l_{\max}$, then let $\Sigma_a \subseteq \Sigma$ be the set of admissible characters as defined below. For every admissible character $x \in \Sigma_a$ do a recursion step: Replace s by the concatenation sx; and in the sequencing graph G_x, replace the active vertex $v_x = (v_1, v_2, \ldots, v_k)$ by $(v_2, \ldots, v_k, c_x)$ that is a vertex of G_x.
– Return to the previous level of recursion.

Here we call a character $x \in \mathcal{X}$ *admissible* if the following two conditions hold:

– Let $v_x = (v_1, \ldots, v_k)$ be the active vertex in G_x. Then, the $(k+1)$-tuple $(v_1, \ldots, v_k, c_x)$ must be an edge of the sequencing graph G_x.
– For every $\sigma \in \mathcal{X} \setminus \{x\}$, let $v_\sigma = (v_1, \ldots, v_k)$ be the active vertex in G_σ. Then, there must exist at least one edge $(v_1, \ldots, v_k, c'_\sigma)$ in the sequencing graph G_σ such that $c_\sigma \preceq c'_\sigma$ holds.

We say that we can perform an *edge transition to an end vertex* in a sequencing graph G_x for $x \in \mathcal{X}$ if the following holds: Let $v_x = (v_1, \ldots, v_k)$ be the active vertex in G_x. Set $c'_x := c_x + \mathbf{1}_1$, where $\mathbf{1}_1$ denotes the compomer containing exactly one end character. Then, the $(k+1)$-tuple $(v_1, \ldots, v_k, c'_x)$ must be an edge of the sequencing graph G_x.

Theorem 1. *For fixed order k, $\mathcal{X} := \Sigma \setminus \{\mathbf{0}, \mathbf{1}\}$, and S as defined in (12), the algorithm of this section solves the Sequencing From Compomers Problem by returning all strings $s \in S$ that satisfy $\mathcal{C}_k(s, x) \subseteq \mathcal{C}_x$ for all $x \in \mathcal{X}$.*

Proof. It is clear from the construction that any output string s is compatible with a walk in G_x and, by Lemma 2, k-compatible with $\mathcal{C}_x$ for all $x \in \mathcal{X}$.

It remains to be shown that all strings $s \in S$ that are k-compatible with $\mathcal{C}_x$ for all $x \in \mathcal{X}$, are constructed by the algorithm. To this end, let $s \in S$ be such a string. By Lemma 2, there exist unique walks $p^x = p_0^x p_1^x \cdots p_r^x$ in G_x compatible with s, for every $x \in \mathcal{X}$. We will show by induction that every proper prefix s' of s is an input to the recursion step of the algorithm. This implies that s' with $s'\mathbf{1} = s$ is also an input of the recursion step. Analogously to the reasoning below, one can show that at this point, we can perform edge transitions to an end vertex in every sequencing graph. It follows that $s'\mathbf{1} = s$ is an output of the algorithm.

The induction basis is trivial for $s' = \mathbf{0}$. Assume that $s' = \tilde{s}x$ for some $\tilde{s} \in \Sigma^*$ and $x \in \Sigma$. Let $s_0, \ldots, s_l$ be the x-partitioning of $\tilde{s}$ as defined in (6). The uniqueness of the x-partitioning of s implies $\operatorname{comp}(s_j) = p_j^x$ for $j = 0, \ldots, l$. In addition to the claim above, we claim that in the current recursion step, the active vertex in G_x is set to $(p_{l-k+1}^x, \ldots, p_l^x)$. To simplify matters, we ignore the case $l < k$ that one can solve analogously. By the induction hypothesis, we know that $\tilde{s}$ is an input of the recursion step, and that the active vertex in G_x at this point is still $(p_{l-k}^x, \ldots, p_{l-1}^x)$.

We claim that x is admissible: We know that $(p_{l-k}^x, \ldots, p_l^x)$ is an edge of p^x and, hence, of G_x, so the first condition is satisfied. For $\sigma \in X \setminus \{x\}$, note that the active vertex has not changed since the last time we appended σ. Analogously to above, we can show that $p_0^\sigma, \dots, p_{m-1}^\sigma, c_\sigma$ are the compomers of the σ-partitioning of s'; note again that $\mathrm{ord}_\sigma(s_\sigma) = c_\sigma(\sigma) = 0$. Hence, the active vertex in G_σ is $(p_{m-k}^\sigma, \dots, p_{m-1}^\sigma)$, and $(p_{m-k}^\sigma, \dots, p_m^\sigma)$ is an edge of G_σ. Since s' is a prefix of s, we infer $c_\sigma = \mathrm{comp}(s_\sigma) \preceq p_m^\sigma$, so the second condition is satisfied, too. This implies that x is admissible.

It follows directly from the construction of the algorithm that the new active vertex in G_x is set to $(p_{l-k+1}^x, \dots, p_l^x)$. Consequently, $s' = \tilde{s}x$ is an input of the recursion step, where the new active vertex in G_x is $(p_{l-k+1}^x, \dots, p_l^x)$, as claimed. □

What are time and space requirements of the described algorithm? Clearly, there can be exponentially many solutions to SFC (cf. Example 3), so the worst-case runtime is also exponential in the problem size as well as the maximal length of an output string. In addition, the runtime can still be exponential if there is a unique solution to SFC, or no solution at all. On the contrary, the space requirements are rather moderate: We need $O(m^{k+1})$ memory to store the sequencing graphs, where $m := \max\{|\Sigma_x| : x \in \mathcal{X}\}$. For $n := \max\{|s| : s \in S\}$ we need $O(n)$ memory in the recursion part of the algorithm, because every recursion step uses only constant memory: The reconstructed sequence itself is not stored in the recursion step but only the current character. The critical factor is obviously storing the sequencing graphs, but note that in applications, these graphs are supposedly sparse and a suitable graph representation will allow storing such graphs with much less memory requirements than the worst-case $O(m^{k+1})$ suggests.

The complete process of de-novo sequencing from mass spectrometry data can now be performed as follows: For every cleavage reaction, apply a peak detection algorithm to extract those parts of the measured mass spectrum that most probably correlate to particles in our sample. For every detected peak, calculate all compomers with at most k cleavage bases and with mass sufficiently close to that of the detected peak. Note that we can limit these calculations to compomers containing at most one character $\mathbf{0}, \mathbf{1}$. In this way, we generate compomer sets $\mathcal{C}_x$ for all $x \in \mathcal{X}$. As described in Section 4.1, we build sequencing graphs G_x from these compomer sets. We use the algorithm of Section 4.2 to generate all sequence candidates that are k-compatible with our input compomer sets $\mathcal{C}_x$.

After generating all sequence candidates in this fashion, we want to further evaluate these sequence candidates taking into account the mass spectrometry data available from all cleavage reactions. A simple scoring scheme for doing so was described in [3], where only slight modifications are needed to deal with partial cleavage data. A more advanced scoring scheme could compute likelihood values for the model [reference sequence is s] to calculate a score for every sequence candidate s. Such scoring schemes can use additional information such as peak intensities, overlapping peaks, and peaks corresponding to fragments of order $> k$ to discriminate between sequence candidates. We do not go into the details of this problem here.

Note that for the application of DNA sequencing, we cannot circumvent the complexity of SFC by, say, introducing heuristics with good time complexity: Such heuristics might or might not find (all of) the solutions of SFC. But this is not acceptable in the setting of de-novo sequencing.

5 Results

In the following, we report some preliminary results of our approach; a more detailed evaluation is currently in progress.

In the absence of sufficient data to test the algorithm, we simulated cleavage reactions and mass spectra, and examined the performance of the algorithm from the previous section on this simulated data. We used two data sets to generate the sample DNA: First, we generated random sample DNA sequences proposing that all bases have identical frequency $\frac{1}{4}$ of occurrence. Second, we used the Human LAMB1 gene (`ENSG00000091136`) [16] and chopped it into approximately 400 pieces, using both exons and introns. For a preliminary evaluation of the method, we performed simulations for sequence length $l = 200$ and order $k = 2$, only. Our initial simulations indicate that the presented approach can be used to tackle SFC.

We simulated four cleavage reactions based on real world RNAse cleavage, where we generated only fragments of order at most $k = 2$, supposing that peaks from fragments of order $k + 1$ and higher cannot be detected in the mass spectrum. Then, we calculated masses of all resulting fragments, and addressed Lim. (ii) (calibration and resolution of the mass spectrometer) in the following way: We say that $\delta \geq 0$ is the *accuracy* of the mass spectrometer, where δ is the maximal difference between an expected and the corresponding detected mass. For our initial evaluation we used $\delta = 0.3$ Da corresponding to OTOF mass spectrometry. We perturbed every signal from the expected list of peaks so that its mass differs by at most δ from the expected mass, and for every resulting peak we calculated all compomers of order at most k that might possibly create a peak with mass at most δ off the perturbed signal mass. In this way, we created the sets C_x for $x \in \Sigma$. Note again that we do not take into account the intensities of peaks.

When simulating the mass spectrometry analysis, we simulate neither false positives (additional peaks) nor false negatives (missing peaks) here. The former does not change the results dramatically: Every signal peak can potentially be interpreted as many different compomers due to Lim. (iv). Hence, the compomer lists *do* contain many false positives. The latter, on the contrary, makes it necessary to modify our approach to deal with real world data.

We want to reconstruct our sample DNA from the cleavage reaction data using sequencing graphs of order 2 and the algorithm presented in the previous section. We assumed that the length of the sample sequence is known with a relative error of 10%, so we set $l_{\min} := 180$ and $l_{\max} := 220$. In addition, we assumed that 8 bases at the start and end of the sequence were known in advance, so our sample sequences had a total length of 216 nt. As we learned from these simulations, the most common sequencing error of our approach seems to be the exchange of two bases belonging to a "stutter" repeat (cf. Examples 2 and 6).

# ambiguous bases	random seq.	LAMB1 seq.
0	961 (96.4 %)	341 (90.0 %)
2	30 (3.0 %)	22 (5.8 %)
3	1 (0.1 %)	0 (0 %)
4	5 (0.5 %)	4 (1.1 %)
5	0 (0 %)	1 (0.3 %)
6	0 (0 %)	2 (0.5 %)
8	0 (0 %)	2 (0.5 %)
10+	0 (0 %)	6 (1.8 %)
total	997 (100.0 %)	378 (100.0 %)

Table 1. Results of the simulations for $k = 2$, $l = 200$, and $\delta = 0.3$. For a number m of ambiguous bases, we have listed the absolute and relative number of input sequences where a unique reconstruction of the sequence was possible ($m = 0$) or not possible ($m > 0$).

We present the results of our simulation in Table 1. Here we provide the number of *ambiguous* bases for the given setup: Formally, an ambiguous base is a column in the multiple alignment, taken over all output sequence candidates, where the aligned output sequences differ. One can see that even when the input sequence was not reconstructed uniquely, there are often only a few ambiguities in the output sequences: The average ratio of ambiguous bases was $\frac{0.4}{1000}$ for the random sequences, and $\frac{2.5}{1000}$ for the LAMB1 sequences. As one could have expected, there were no sample sequences with exactly one ambiguous base. By design, the correct sequence was always among the output sequence candidates.

6 Discussion and Improvements

We have introduced the Sequencing From Compomers Problem that stems from the analysis of mass spectrometry data from partial cleavage experiments. Although this problem is computationally difficult in general, we have introduced a computational approach to perform de-novo sequencing from such data. The introduced method uses sub-graphs of the de Bruijn graph to construct all sequences that are compatible with the observed mass spectra. We tested the performance of our approach on simulated mass spectrometry data from random as well as from biological sequences. Surprisingly, our approach is capable of reconstructing the correct sequence in most cases, and ambiguities are limited most of the time to the exchange of two bases in a "stutter repeat." The local information of compomers derived from substrings of the input sequence that contain only a few (here, $k = 2$) cleavage bases, is often sufficient to reconstruct the input string. Our simulations indicate that the presented approach may enable de-novo sequencing with an experimental setup that differs completely from Sanger Sequencing, and still allows for sequencing lengths that are of the same magnitude as those of Sanger Sequencing. Using base-specific cleavage and mass spectrometry for de-novo sequencing has the advantages of high throughput (4 mass spectra can be measured in less than 10 seconds) and potentially increased sensitivity and specificity over classical Sanger Sequencing. Potential additional advantages are the possibility to perform pooling or multiplexing, see below.

As noted in Section 5, the simulation results of that section are only a first step in evaluating the power of our approach. A more thorough simulation analysis is currently in progress.

We mentioned earlier that our setup can easily be extended to cut strings x of arbitrary length. Moreover, it makes sense to replace the cut bases $x \in \Sigma$ by *sets* of cut strings $X \subseteq \Sigma^*$ because there exist enzymes that are specific to, say, both pyrimidines C or T at a certain position. For example, the enzyme *Hin*fI cleaves all sequences of the form GANTC, and here we define $X := \{\mathrm{GAATC}, \mathrm{GACTC}, \mathrm{GAGTC}, \mathrm{GATTC}\}$. The generalization of the presented tools is pretty much straightforward, but note that (a) we can no longer simply calculate the order of a compomer, and (b) we have to cope with cleavage that happens *within* a cut string.

Our approach does at no point depend on the fact that we are sequencing *DNA*. In theory, another application can be de-novo sequencing of proteins, given that we have ways of amino acid (or cut string) specific cleavage of such polypeptides. In reality, this may be difficult because there are 20 characters in the protein alphabet, so a straightforward generalization would call for 20 distinct cleavage reactions. In addition, calculating all compomers for a given mass becomes computationally difficult. However, protein "sequencing" using tryptic digest and a lookup database is a broadly used tool in Proteomics. Note that our approach differs fundamentally from MS/MS peptide sequencing approaches [5, 13]. Furthermore, our approach does not rely on data from MALDI-TOF MS but will work with any method that allows us to determine base compositions of the cleavage reaction products.

Finally, we want to point out that the presented method can easily be adopted for *pooling* as well as *multiplexing*: When pooling sequences, we want to analyze *mixtures* of samples, or heterozygous samples. Our approach will in principal return all sequences found in the mixture, and only a subsequent analysis has to be modified accordingly. For multiplexing, instead of sequencing a single continuous stretch of the sample sequence of length 200 nt, we analyze, say, ten distinct stretches of length 20 nt each in parallel. Here we can iteratively search for 10 appropriate walks in every sequencing graph.

The intensity of a peak in a MS spectrum may indicate the multiplicity of the respective compomer. This motivates the question whether we can uniquely reconstruct the string s from $\mathcal{C}_k(s, x)$ if we define all sets in equations (1–4) to be multisets instead of simple sets.

Clearly, there are more elaborate ways to use sequencing graphs for solving SFC than the depth-first search algorithm we used. In particular, a depth-first search is not fully appropriate: As we have noted in Section 5, the most common ambiguities are stutter repeats where resulting sequences are compatible with walks that are identical except for a short de-tour element.

On the theoretical side of the problem, it would be interesting to characterize transformations of strings that, like stutter repeats, do not change the compomer spectra of the string for some given order k.

Finally, we have mentioned in the previous section that in this paper, we do not take into account the problem of false negative peaks that is common in applications. This will be addressed in a forthcoming paper.

Acknowledgments

I want to thank Zsuzsanna Lipták, Jens Stoye, Matthias Steinrücken, and Dirk van den Boom for proofreading earlier versions of this manuscript, the latter also for many helpful suggestions.

References

1. Jean-Michel Autebert, Jean Berstel, and Luc Boasson. Context-free languages and pushdown automata. In Grzegorz Rozenberg and Arto Salomaa, editors, *Handbook of Formal Languages*, volume 1, pages 111–174. Springer, 1997.
2. William Bains and Geoff C. Smith. A novel method for nucleic acid sequence determination. *J. Theor. Biol.*, 135:303–307, 1988.
3. Sebastian Böcker. SNP and mutation discovery using base-specific cleavage and MALDI-TOF mass spectrometry. *Bioinformatics*, 19:i44–i53, 2003.
4. Michel Cosnard, Jean Duprat, and Afonso G. Ferreira. The complexity of searching in $X+Y$ and other multisets. *Information Processing Letters*, 34:103–109, 1990.
5. Vlado Dančík, Theresa A. Addona, Karl R. Clauser, James E. Vath, and Pavel A. Pevzner. De novo peptide sequencing via tandem mass spectrometry. *J. Comp. Biol.*, 6(3/4):327–342, 1999.
6. Nicolaas G. de Bruijn. A combinatorial problem. In *Indagationes Mathematicae*, volume VIII, pages 461–467. Koninklije Nederlandsche Akademie van Wetenschappen, 1946.
7. Radoje Drmanac, Ivan Labat, Ivan Brukner, and Radomir Crkvenjakov. Sequencing a megabase plus DNA by hybridization: Theory of the method. *Genomics*, 4:114–128, 1989.
8. Ralf Hartmer, Niels Storm, Sebastian Böcker, Charles P. Rodi, Franz Hillenkamp, Christian Jurinke, and Dirk van den Boom. RNAse T1 mediated base-specific cleavage and MALDI-TOF MS for high-throughput comparative sequence analysis. *Nucl. Acids. Res.*, 31(9):e47, 2003.
9. M. Karas and Franz Hillenkamp. Laser desorption ionization of proteins with molecular masses exceeding 10,000 Daltons. *Anal. Chem.*, 60:2299–2301, 1988.
10. Hubert Köster, Kai Tang, Dong-Jing Fu, Andreas Braun, Dirk van den Boom, Cassandra L. Smith, Robert J. Cotter, and Charles R. Cantor. A strategy for rapid and efficient DNA sequencing by mass spectrometry. *Nat. Biotechnol.*, 14(9):1084–1087, 1996.

11. Y. Lysov, V. Floretiev, A. Khorlyn, K. Khrapko, V. Shick, and A. Mirzabekov. DNA sequencing by hybridization with oligonucleotides. *Dokl. Acad. Sci. USSR*, 303:1508–1511, 1988.
12. Allan M. Maxam and Walter Gilbert. A new method for sequencing DNA. *Proc. Nat. Acad. Sci. USA*, 74(2):560–564, 1977.
13. Scott D. Patterson and Ruedi Aebersold. Mass spectrometric approaches for the identification of gel-separated proteins. *Electrophoresis*, 16:1791–1814, 1995.
14. William R. Pearson. Automatic construction of restriction site maps. *Nucleic Acids Res.*, 10:217–227, 1982.
15. Pavel P. Pevzner. *l*-tuple DNA sequencing: Computer analysis. *J. Biomol. Struct. Dyn.*, 7:63–73, 1989.
16. David E. Reich, Michele Cargill, Stacey Bolk, James Ireland, Pardis C. Sabeti, Daniel J. Richter, Thomas Lavery, Rose Kouyoumjian, Shelli F. Farhadian, Ryk Ward, and Eric S. Lander. Linkage disequilibrium in the human genome. *Nature*, 411:199–204, 2001.
17. Charles P. Rodi, Brigitte Darnhofer-Patel, Patrick Stanssens, Marc Zabeau, and Dirk van den Boom. A strategy for the rapid discovery of disease markers using the MassARRAY system. *BioTechniques*, 32:S62–S69, 2002.
18. M. Ronaghi, M. Uhlén, and P. Nyrén. Pyrosequencing: A DNA sequencing method based on real-time pyrophosphate detection. *Science*, 281:363–365, 1998.
19. Frederick Sanger, S. Nicklen, and Alan R. Coulson. DNA sequencing with chain-terminating inhibitors. *Proc. Nat. Acad. Sci. USA*, 74(12):5463–5467, 1977.
20. Mikhail S. Shchepinov, Mikhail Denissenko, Kevin J. Smylie, Ralf J. Wörl, A. Lorieta Leppin, Charles R. Cantor, and Charles P. Rodi. Matrix-induced fragmentation of P3'-N5' phosphoramidate-containing DNA: high-throughput MALDI-TOF analysis of genomic sequence polymorphisms. *Nucleic Acids Res.*, 29(18):3864–3872, 2001.
21. Steven Skiena, Warren D. Smith, and Paul Lemke. Reconstructing sets from interpoint distances. In *Proceedings of Annual symposium Computational geometry*, pages 332–339, 1990.
22. Steven S. Skiena and Gopalakrishnan Sundaram. A partial digest approach to restriction site mapping. *Bulletin of Mathematical Biology*, 56:275–294, 1994.
23. Friedrich von Wintzingerode, Sebastian Böcker, Cord Schlötelburg, Norman H.L. Chiu, Niels Storm, Christian Jurinke, Charles R. Cantor, Ulf B. Göbel, and Dirk van den Boom. Base-specific fragmentation of amplified 16S rRNA genes and mass spectrometry analysis: A novel tool for rapid bacterial identification. *Proc. Natl. Acad. Sci. USA*, 99(10):7039–7044, 2002.
24. Michael S. Waterman. *Introduction to Computational Biology: Maps, sequences and genomes.* Chapman & Hall–CRC Press, 1995.

Bounds for Resequencing by Hybridization

Dekel Tsur

Dept. of Computer Science, Tel Aviv University
dekelts@tau.ac.il

Abstract. We study the problem of finding the sequence of an unknown DNA fragment given the set of its k-long subsequences and a homologous sequence, namely a sequence that is similar to the target sequence. Such a sequence is available in some applications, e.g., when detecting single nucleotide polymorphisms. Pe'er and Shamir studied this problem and presented a heuristic algorithm for it. In this paper, we give an algorithm with provable performance: We show that under some assumptions, the algorithm can reconstruct a random sequence of length $O(4^k)$ with high probability. We also show that no algorithm can reconstruct sequences of length $\Omega(\log k \cdot 4^k)$.

1 Introduction

Sequencing by Hybridization (SBH) [3, 22] is a method for sequencing of long DNA molecules. Using a chip containing all 4^k sequences of length k one can obtain the set of all k-long subsequences of the target sequence: For every sequence in the chip, if its reverse complement appears in the target than the two sequences will hybridize. The set of all k-long subsequences of the target is called the *k-spectrum* (or *spectrum*) of the target. After obtaining the spectrum, the target sequence can be reconstructed in polynomial time [26].

Unfortunately, other sequences can have the same spectrum as the target's. For example, if we assume that the sequence is chosen uniformly from all the sequences of length n, then only sequences of length less than roughly 2^k can be reconstructed reliably [27, 13, 2, 29]. Several methods for overcoming this limitation of SBH were proposed: gapped probes [27, 14, 28, 18, 19, 16, 20], interactive protocols [31, 23, 15, 34], using location information [1, 11, 17, 4, 12, 29], and using restriction enzymes [32, 30].

An additional limitation of SBH is that in practice, there are errors in the hybridization process. Thus, some subsequences of the target do not appear in the experimental spectrum (*false negatives*), and the experimental spectrum contains sequences that do not appear in the target (*false positives*). Several algorithms were given for SBH with errors [26, 21, 9, 7, 5, 8, 6, 10]. The first algorithm with provable performance was given by Halperin et al. [16], and their algorithm was later improved in [33].

In many applications, the target sequence is not completely unknown. For example, the problem of detecting single nucleotide polymorphisms can be considered as finding the sequence of a DNA fragment when most of the sequence

G. Benson and R. Page (Eds.): WABI 2003, LNBI 2812, pp. 498–511, 2003.

(over 99%) is known in advance. Therefore, we study the problem of reconstructing a target sequence given its spectrum and a sequence that is similar to the target sequence (called a *homologous sequence*). This problem is called *resequencing by hybridization (RBH)*. Pe'er and Shamir [25] (see also [24]) gave an algorithm for RBH and showed that the algorithm works well in practice, but did not prove a bound on its performance.

In this work, we give an algorithm for RBH and prove a bound on its performance. We assume that the target sequence is a random sequence, and that the homologous sequence has the following property: Every k-subsequence of the target differs from the corresponding subsequence of the homologous sequence in at most d letter, where $d \leq (\frac{3}{4} - \delta)k$ for an arbitrarily small constant δ. Moreover, we assume that the homologous sequence is generated randomly from the target sequence by selecting positions on the target sequence and then randomly changing the letters in the selected positions. Under these assumptions, our algorithm can reconstruct sequences of length $O(4^k \min(d^{-3/2}, \log k/k))$ with probability close to 1. We also show that that the algorithm can reconstruct sequences of length $O(4^k)$ if the number of different letters between the target and the homologous sequence is $O(n/2^{\epsilon k})$. Moreover, we show that no algorithm can reconstruct sequences of length $\Omega(\log k \cdot 4^k)$ with success probability greater than $\frac{1}{3}$. We also study the RBH problem under the presence of hybridization errors. We give an algorithm for this case whose performance is close to the performance of the algorithm in the errorless case.

Due to lack of space, some proofs are omitted.

2 Preliminaries

For a sequence $A = a_1 \cdots a_n$, let A_i^l denote the l-subsequence $a_i a_{i+1} \cdots a_{i+l-1}$. For two sequences A and B, AB is the concatenation of A and B.

A set $I \subseteq \{1, \ldots, n\}$ is called a *k, d-set* if $|I \cap \{i, \ldots, i+k-1\}| \leq d$ for all i. We denote by $I_{k,d}(n)$ the set of all k, d-sets that are subsets of $\{1, \ldots, n\}$. Two sequences $A = a_1 \cdots a_n$ and $B = b_1 \cdots b_n$ will be called *k, d-equal* if $\{i : a_i \neq b_i\}$ is a k, d-set.

The RBH problem is as follows: Given the k-spectrum of A and a sequence H which is k, d-equal to A, find the sequence A. We study the RBH problem under a probabilistic model. We assume that the target sequence $A = a_1 \cdots a_n$ is chosen at random, where each letter is chosen uniformly from $\Sigma = \{A, C, G, T\}$ and independently of the other letters. The homologous sequence $H = h_1 \cdots h_n$ is built as follows: Some k, d-set I (called the *locations set*) is chosen before the sequence A is chosen. After the sequence A is chosen, for every $i \notin I$ set $h_i = a_i$, and for every $i \in I$, h_i is chosen uniformly from $\Sigma - \{a_i\}$.

We will use $\log x$ to denote the logarithm with base 2 of x.

3 Algorithm for RBH

In this section we give an algorithm for solving RBH, and analyze its performance. We assume that the first and last $k-1$ letters of A are known. Let c be some large integer, and suppose that n is large enough so $ck \leq n/2$. We will use $S = s_1 \cdots s_n$ to denote the sequence that is built by our algorithm. The algorithm is given in Figure 1.

1. Let $s_1, s_2, \ldots, s_{k-1}$ be the first $k-1$ letters of A.
2. Let $s_{n-k+2}, \ldots, s_n$ be the last $k-1$ letters of A.
3. For $t = k, k+1, \ldots, \lceil n/2 \rceil$ do: (forward sequencing)
   (a) Let $\mathcal{B}$ be the set of all sequences of length ck that are k,d-equal to H_t^{ck}.
   (b) Let $\mathcal{B}'$ be the set of all sequences $B \in \mathcal{B}$ such that all the k-subsequences of $S_{t-k+1}^{k-1} B$ appear in A.
   (c) If all the sequences in $\mathcal{B}'$ have a common first letter a, then set $s_t \leftarrow a$. Otherwise, set $s_t \leftarrow h_t$.
4. For $t = n-k+1, n-k, \ldots, \lceil n/2 \rceil + 1$ do: (backward sequencing)
   (a) Let $\mathcal{B}$ be the set of all sequences of length ck that are k,d-equal to H_{t-ck+1}^{ck}.
   (b) Let $\mathcal{B}'$ be the set of all sequences $B \in \mathcal{B}$ such that all the k-subsequences of $B S_{t+1}^{k-1}$ appear in A.
   (c) If all the sequences in $\mathcal{B}'$ have a common last letter a, then set $s_t \leftarrow a$. Otherwise, set $s_t \leftarrow h_t$.
5. Return the sequence S.

Fig. 1. Algorithm A.

As the backward sequencing stage is analogous to the forward sequencing stage, we shall only analyze the latter. A sequence in the set $\mathcal{B}$ in some step t of the algorithm is called a *path (w.r.t. t)*. A path is called *correct* if it is equal to $a_t \cdots a_{t+ck-1}$, and it is called *incorrect* if its first letter is not equal to a_t. An incorrect path in $\mathcal{B}'$ is called a *bad path.* For a path $B \in \mathcal{B}$, a *supporting probe* is a k-subsequence of B which is also a subsequence of the target sequence A. We will use i to denote the supporting probe B_i^k.

Clearly for every t, the correct path is always in $\mathcal{B}'$. Thus, if $a_t = h_t$ then the algorithm will always set s_t to a_t. In other words, the algorithm can fail only at indices t for which $a_t \neq h_t$.

We first give two technical lemmas:

Lemma 1. *For every* $\alpha \geq 0$, *if* $d \leq \left(\frac{3}{4} - \sqrt{\frac{1}{c} + (\frac{1}{2}\log e)\alpha + \frac{\log k}{k}}\right) k$ *then* $k^2 \binom{k}{d} \cdot 3^d e^{\alpha k} / 4^{(1-1/c)k} \leq \frac{1}{2}$.

Proof. Let $a = d/k$. Using Stirling formula we have that

$$\begin{aligned}\binom{k}{d} &\le \frac{1.1\sqrt{2\pi k}\left(\frac{k}{e}\right)^k}{\sqrt{2\pi d}\left(\frac{d}{e}\right)^d \sqrt{2\pi(k-d)}\left(\frac{k-d}{e}\right)^{k-d}} \\ &= \frac{1.1}{\sqrt{2\pi}}\sqrt{\frac{k}{d(k-d)}} \cdot \frac{k^k}{(ak)^{ak}((1-a)k)^{(1-a)k}} \\ &< \frac{1}{2} \cdot \frac{1}{a^{ak}(1-a)^{(1-a)k}} = \frac{1}{2} \cdot 2^{\left(a\log\frac{1}{a}+(1-a)\log\frac{1}{1-a}\right)k}.\end{aligned}$$

Thus, $\log(2\binom{k}{d} \cdot 3^d) \le (a\log\frac{1}{a} + (1-a)\log\frac{1}{1-a} + a\log 3)k$.

Let $f(x) = x\log\frac{1}{x} + (1-x)\log\frac{1}{1-x} + x\log 3$. Using simple calculus, we obtain that $f(\frac{3}{4} - x) \le 2(1-x^2)$. Therefore,

$$\begin{aligned}\log\left(2\binom{k}{d}\cdot 3^d\right) &\le f(a)k \le 2(1-(3/4-a)^2)k \\ &\le 2\left(1 - \frac{1}{c} - \frac{\log e}{2}\alpha - \frac{\log k}{k}\right)k = \log\left(\frac{4^{(1-1/c)k}}{e^{\alpha k}k^2}\right). \quad \square\end{aligned}$$

Lemma 2. *Let* $F = \sum_{j=1}^{k} \frac{(a+j)e^{\alpha j}}{4^j} \sum_{i=0}^{\min(j,d)} \binom{j}{i} 3^i$ *and* $F' = \sum_{j=1}^{k} \frac{e^{\alpha j}}{4^j} \cdot \sum_{i=0}^{\min(j,d)} \binom{j}{i} 3^i$. *If* $a \ge 0$ *and* $0 \le \alpha \le 0.14$ *then* $F \le (2d+6)(a+2d)e^{2\alpha d}$ *and* $F' \le (2d+3)e^{2\alpha d}$.

Proof. Let $f(j) = \sum_{i=0}^{\min(j,d)} \binom{j}{i} 3^i$. For every j,

$$f(j) \le \sum_{i=0}^{j} \binom{j}{i} 3^i = 4^j.$$

If $j > 2d$ then

$$f(j) = \sum_{i=0}^{d} \binom{j}{i} 3^i \le \binom{j}{d} \sum_{i=0}^{d} 3^i \le \frac{3}{2}\binom{j}{d} 3^d.$$

Therefore,

$$F \le \sum_{j=1}^{2d} (a+j)e^{\alpha j} + \sum_{j=2d+1}^{k} (a+j)\left(\frac{e^\alpha}{4}\right)^j \cdot \frac{3}{2}\binom{j}{d} 3^d,$$

where the second sum is empty if $d > (k-1)/2$. We have that

$$\sum_{j=1}^{2d} (a+j)e^{\alpha j} \le \sum_{j=1}^{2d} (a+2d)e^{\alpha\cdot 2d} = 2d(a+2d)e^{2\alpha d},$$

Furthermore, for $j > 2d$,

$$\frac{\binom{j}{d}\left(\frac{e^\alpha}{4}\right)^j}{\binom{j-1}{d}\left(\frac{e^\alpha}{4}\right)^{j-1}} = \frac{j!}{d!(j-d)!} \cdot \frac{d!(j-1-d)!}{j-1!} \cdot \frac{e^\alpha}{4} = \frac{j}{j-d} \cdot \frac{e^\alpha}{4} \le 2 \cdot \frac{e^{0.14}}{4} \le \frac{3}{5},$$

so

$$\binom{j}{d}\left(\frac{e^{\alpha}}{4}\right)^{j} \leq \binom{2d}{d}\left(\frac{e^{\alpha}}{4}\right)^{2d}\left(\frac{3}{5}\right)^{j-2d} \leq 2^{2d}\left(\frac{e^{0.14}}{4}\right)^{2d}\left(\frac{3}{5}\right)^{j-2d}.$$

Thus,

$$\begin{aligned}\sum_{j=2d+1}^{k}(a+j)\left(\frac{e^{\alpha}}{4}\right)^{j}\cdot\frac{3}{2}\binom{j}{d}3^{d} &\leq \frac{3}{2}\cdot 3^{d}\left(\frac{e^{0.14}}{2}\right)^{2d}\sum_{l=1}^{\infty}(a+2d+l)\left(\frac{3}{5}\right)^{l}\\ &= \frac{3}{2}\cdot\left(\frac{e^{0.14}\sqrt{3}}{2}\right)^{2d}\left(\frac{3}{2}(a+2d)+\frac{15}{4}\right)\\ &< \frac{3}{2}\left(\frac{3}{2}(a+2d)+\frac{15}{4}\right) < 6(a+2d),\end{aligned}$$

so the bound on F follows. The bound on F' is proved using similar analysis. □

Theorem 1. *For every $\epsilon > 0$, if $d \leq \left(\frac{3}{4} - \sqrt{1/c + \Omega(\log k/k)}\right)k$ and $n = O(\epsilon 4^k \min(d^{-3/2}, \log k/k))$ then the probability that algorithm A fails is at most ϵ.*

Proof. Let E_t be the event that t is the minimum index for which $s_t \neq a_t$. Assuming that events $E_1, \ldots, E_{t-1}$ do not happen, E_t happens if and only if $a_t \neq h_t$ and there is a bad path w.r.t. t. To bound this probability for some fixed t, we fix a k, d-set $I \subseteq \{1, \ldots, ck\}$ and generate a random path $B' = b'_1 \cdots b'_{ck}$ as follows: If $i \notin I$, then $b'_i = h_{t-1+i}$. If $i > 1$ and $i \in I$ then b'_i is selected uniformly at random from $\Sigma - \{h_{t-1+i}\}$, and if $i = 1$ and $i \in I$ then b'_i is selected uniformly from $\Sigma - \{h_t, a_t\}$. We shall compute the probability that B' is a bad path w.r.t. t, and then we will bound $\mathrm{P}[E_t]$ by roughly $\sum_I 3^{|I|} \cdot \mathrm{P}[B' \text{ is bad}|t, I]$, where the term $3^{|I|}$ bounds the number of possible paths B' given the set I. Note that each letter of B' has a uniform distribution over Σ, and the letters of B' are independent.

Let $B = S^{k-1}_{t-k+1}B'$, and denote $B = b_1 \cdots b_{ck+k-1}$. By definition, B' is a bad path if and only if there are indices $r_1, \ldots, r_{ck}$ such that $B^k_i = A^k_{r_i}$ for $i = 1, \ldots, ck$, so we need to bound the probability that these events happen. We say that supporting probe i is *trivial* if $r_i = t - 1 + i$. Note that probes $1, \ldots, k$ are not trivial as $b'_1 \neq a_t$. We consider two cases: The first case is when there are no trivial supporting probes, and the second case is when there are trivial supporting probes. These cases will be called case I and case II, respectively.

Case I Suppose that there are no trivial supporting probes. The difficulty in bounding the probability that $B^k_i = A^k_{r_i}$ for $i = 1, \ldots, ck$ is that these events are not independent when some of the sequences $A^k_{r_1}, \ldots, A^k_{r_k}$ have common letters, that is, if $|r_i - r_j| < k$ for some pairs (i, j) of indices. We say that two probes r_i and r_j are *strongly adjacent* if $|r_i - r_j| < k$ and $r_j - r_i = j - i$ (in particular, every probe is strongly adjacent to itself). The transitive closure of the strongly adjacency relation will be called the *adjacency* relation. The motivation behind the definitions above is as follows: If r_i and r_j are strongly adjacent probes with $i < j$, then the events $B_i^k = A_{r_i}^k$ and $B_j^k = A_{r_j}^k$ happen if and only if $B_i^{k+j-i} = A_{r_i}^{k+j-i}$. More generally, for each equivalence class of the adjacency relation, there is a corresponding equality event between a subsequence of A and a subsequence of B.

If r_i and $r_{i'}$ are adjacent, then $B_j^k = A_{r_i+j-i}^k$ for every $j = i, \ldots, i'$. Therefore, we can assume w.l.o.g. that $r_j = r_i + j - i$ for $j = i, \ldots, i'$. Thus, each equivalence class of the adjacency relation corresponds to an interval in $\{1, \ldots, ck\}$. More precisely, there are indices $1 = c_1 < c_2 < \cdots < c_x < c_{x+1} = ck + 1$ such that $\{r_{c_i}, r_{c_i} + 1, \ldots, r_{c_{i+1}}\}$ is an equivalence class for $i = 1, \ldots, x$. The sequence B' is a bad path if and only if $B_{c_i}^{k-1+c_{i+1}-c_i} = A_{r_{c_i}}^{k-1+c_{i+1}-c_i}$ for all i. We shall compute the probability that these events happen. Each sequence $A_{r_{c_i}}^{k-1+c_{i+1}-c_i}$ will be called a *block*, and will be denoted by L_i. We denote by $l_i = k - 1 + c_{i+1} - c_i$ the number of letters in the block L_i. To simplify the presentation, we define block L_0 to be the sequence $A_{t-k+1}^{(c+1)k-1}$ ($l_0 = (c+1)k - 1$).

For two blocks L_i and L_j with $0 \leq i < j$ we say that L_j *overlaps with* L_i if the two blocks have common letters, namely if $r_{c_j} \in [r_{c_i} - l_j + 1, r_{c_i} + l_i - 1]$. If the block L_j overlaps with some block L_i, we say that L_j is an *overlapping block.*

We will bound the probability that B' is a bad path in two cases: When there are no overlapping probes, and when there are overlapping probes. In the second case, we will look at the overlapping probe with minimum index, and consider the equality events that correspond to this block and the blocks with smaller indices. However, the analysis of this case can be complicated if the overlapping probe overlaps with two or more blocks. Therefore, we introduce the following definition: A block L_j is called *weakly overlapping* if there is an index $i < j$ such that $|r_{c_j} - r_{c_i}| \leq 2(c+1)k - 4$. In particular, an overlapping block is also a weakly overlapping block. If there are overlapping blocks, define y to be the minimum index such that block L_y is a weakly overlapping block. The definition of a weakly close block ensures that L_y cannot overlap with more than one block.

We consider three cases:

1. There are no overlapping blocks.
2. There are overlapping blocks and $y > 1$.
3. There are overlapping blocks and $y = 1$.

Let $\mathcal{E}_i$ denote the event that there is a bad path that satisfies case i above. We shall bound the probability of each of these events.

Case 1 Suppose that there are no overlapping blocks. In this case, the events $B_{c_1}^{l_1} = A_{r_{c_1}}^{l_1}, \ldots, B_{c_x}^{l_x} = A_{r_{c_x}}^{l_x}$ are independent. Therefore, for fixed t, I, and $r_1, \ldots, r_k$, the probability that event $\mathcal{E}_1$ happens is $\prod_{y=1}^{x} 4^{-l_y} = 4^{-ck-(k-1)x}$. For fixed x, the number of ways to choose $c_1, \ldots, c_x$ is $\binom{ck-1}{x-1}$, and for fixed $c_1, \ldots, c_x$, the number of ways to choose $r_1, \ldots, r_k$ is at most n^x. Thus,

$$\begin{aligned}
\mathrm{P}\left[\mathcal{E}_1|t,I\right] &\leq \sum_{x=1}^{ck}\binom{ck-1}{x-1}\frac{n^x}{4^{ck+(k-1)x}} = \frac{n}{4^{ck+k-1}}\sum_{x=1}^{ck}\binom{ck-1}{x-1}\left(\frac{n}{4^{k-1}}\right)^{x-1} \\
&= \frac{n}{4^{(c+1)k-1}}\left(1+\frac{n}{4^{k-1}}\right)^{ck-1} \leq \frac{n}{4^{(c+1)k-1}}\cdot e^{(n/4^{k-1})\cdot ck}.
\end{aligned}$$

Therefore,

$$\mathrm{P}\left[\mathcal{E}_1|t\right] \leq \sum_{I\in I_{k,d}(ck)} 3^{|I|}\cdot\frac{ne^{(n/4^{k-1})ck}}{4^{(c+1)k-1}}.$$

Let $J_i = \{ik+1,\ldots,ik+k\}$. From the definition of k,d-set we get that

$$\begin{aligned}
\sum_{I\in I_{k,d}(ck)} 3^{|I|} &= \sum_{I\in I_{k,d}(ck)}\prod_{i=1}^{c} 3^{|I\cap J_i|} \\
&= \sum_{j_1,\ldots,j_c=0}^{d} \left|\{I\in I_{k,d}(ck) : |I\cap J_i| = j_i, i=1,\ldots,c\}\right|\cdot\prod_{i=1}^{c}3^{j_i} \\
&\leq \sum_{j_1,\ldots,j_c=0}^{d}\prod_{i=1}^{c}\binom{k}{j_i}\cdot\prod_{i=1}^{c}3^{j_i} = \left(\sum_{j=0}^{d}\binom{k}{j}3^j\right)^c \\
&\leq \left((d+1)\binom{k}{d}\cdot 3^d\right)^c,
\end{aligned}$$

where the last inequality follows from the fact that $d \leq \frac{3}{4}k$.

Since H is k,d-equal to A, it follows that the number of indices i for which $a_i \neq h_i$ is at most $\lceil n/k\rceil d \leq 2dn/k$, so the number of ways to choose t is at most $2dn/k$. Therefore, the probability that event $\mathcal{E}_1$ happens is at most

$$\frac{2dn}{k}\cdot\left(k\binom{k}{d}3^d\right)^c\cdot\frac{ne^{(n/4^{k-1})ck}}{4^{(c+1)k-1}} = \frac{8dn^2}{k\cdot 4^{2k}}\cdot\left(\frac{k\binom{k}{d}3^d\cdot e^{(n/4^{k-1})k}}{4^{(1-1/c)k}}\right)^c < \frac{8n^2}{4^{2k}},$$

where the last inequality follows from Lemma 1.

Case 2 Recall that y is the minimum index such that block L_y is a weakly overlapping block. Let $z = c_y$. Let $\mathcal{E}$ be the event that the equality events corresponding to the blocks $L_1,\ldots,L_{y-1}$ happen, namely $B^{l_i}_{c_i} = A^{l_i}_{r_{c_i}}$ for $i = 1,\ldots,y-1$, and let $\mathcal{E}'$ be the event that $B^k_z = A^k_{r_z}$. As there are no overlapping blocks in $L_1,\ldots,L_{y-1}$, we obtain that

$$\mathrm{P}\left[\mathcal{E}|t,I,z,r_1,\ldots,r_{z-1}\right] = \frac{1}{4^{(k-1)(y-1)+z-1}}.$$

Furthermore, $\mathrm{P}\left[\mathcal{E}'|t,I,z,r_z\right] = 4^{-k}$. This is clear when L_y does not overlap with L_0. To see that this claim is also true when L_y overlaps with L_0, note that event $\mathcal{E}'$ is composed of k equalities $b_{z+i} = a_{r_z+i}$ for $i = 0,\ldots,k-1$.

The probability that such an equality happens given that the previous equalities happen is exactly 1/4 as the letters b_{z+i} and a_{r_z+i} are independent (since probe z is not trivial), and at least one of these two letters is not restricted by the the previous equalities. Therefore, $\mathrm{P}[\mathcal{E}'|t, I, z, r_z] = 4^{-k}$.

Now, we claim that the events $\mathcal{E}$ and $\mathcal{E}'$ are independent. If L_y does not overlap with L_0 then this claim follows from [2, p. 437]. Otherwise, suppose that L_y overlaps with L_0. For each equality $b_{i+i'} = a_{r_{c_i}+i'}$ (where $i = 1, \ldots, y-1$ and $i' = 0, \ldots, l_i - 1$) that is induced by $\mathcal{E}$, the letter $a_{r_{c_i}+i'}$ is not restricted by event $\mathcal{E}'$ (as L_y does not overlap with L_i), and therefore the probability that this equality happen is 1/4. It follows that $\mathcal{E}$ and $\mathcal{E}'$ are independent.

Combining the claims above, we have that

$$\mathrm{P}[\mathcal{E} \wedge \mathcal{E}'|t, I, z, r_1, \ldots, r_z] = \frac{1}{4^{(k-1)(y-1)+z-1+k}}.$$

For fixed y and z, the number of ways to choose $r_1, \ldots, r_{z-1}$ is at most $\binom{(z-1)-1}{(y-1)-1} n^{y-1}$, and the number of ways to choose r_z is at most $(2(2(c+1)k - 4)+1) \cdot (y-1) \leq 4(c+1)kz$. Thus,

$$\begin{aligned}
\mathrm{P}[\mathcal{E} \wedge \mathcal{E}'|t, I, z] &\leq \sum_{y=2}^{z} \binom{(z-1)-1}{(y-1)-1} n^{y-1} \cdot 4(c+1)kz \cdot \frac{1}{4^{(k-1)(y-1)+z-1+k}} \\
&\leq \frac{64(c+1)kzn}{4^{2k+z}} \cdot \left(1 + \frac{n}{4^{k-1}}\right)^{z-2} \leq \frac{64(c+1)kzne^{(n/4^{k-1})z}}{4^{2k+z}}.
\end{aligned}$$

The event $\mathcal{E} \wedge \mathcal{E}'$ depends only on the first z letters of B'. Therefore,

$$\mathrm{P}[\mathcal{E} \wedge \mathcal{E}'] \leq \frac{2dn}{k} \sum_{z=2}^{ck} \sum_{I \in I_{k,d}(z)} 3^{|I|} \cdot \frac{64(c+1)kzne^{(n/4^{k-1})z}}{4^{2k+z}}.$$

Let $z = kz_1 + z_2$ where $1 \leq z_2 \leq k$. Then,

$$\begin{aligned}
\sum_{I \in I_{k,d}(z)} 3^{|I|} &= \sum_{I \in I_{k,d}(z)} \prod_{i=1}^{z_1+1} 3^{|I \cap J_i|} \leq \left(\sum_{i=0}^{d} \binom{k}{i} 3^i\right)^{z_1} \sum_{i=0}^{\min(d,z_2)} \binom{z_2}{i} 3^i \\
&\leq \left(k \binom{k}{d} 3^d\right)^{z_1} \sum_{i=0}^{\min(d,z_2)} \binom{z_2}{i} 3^i.
\end{aligned}$$

Using Lemma 1 and Lemma 2 we obtain that

$$\begin{aligned}
\mathrm{P}[\mathcal{E}_2] &\leq \mathrm{P}[\mathcal{E} \wedge \mathcal{E}'] \\
&\leq \frac{128(c+1)dn^2}{4^{2k}} \sum_{z_1=0}^{c-1} \sum_{z_2=1}^{k} \left(k \binom{k}{d} 3^d\right)^{z_1} \\
&\qquad \cdot \sum_{i=0}^{\min(d,z_2)} \binom{z_2}{i} 3^i \cdot \frac{(kz_1+z_2)e^{(n/4^{k-1})(kz_1+z_2)}}{4^{kz_1+z_2}}
\end{aligned}$$

$$\leq \frac{128(c+1)dn^2}{4^{2k}} \sum_{z_1=0}^{c-1} \left(\frac{k\binom{k}{d} 3^d e^{(n/4^{k-1})k}}{4^k} \right)^{z_1} (2d+6)(kz_1+2d)e^{2(n/4^{k-1})d}$$

$$\leq \frac{128(c+1)d(2d+6)e^{2(n/4^{k-1})d}n^2}{4^{2k}} \sum_{z_1=0}^{c-1} \left(\frac{k^2\binom{k}{d} 3^d e^{(n/4^{k-1})k}}{4^k} \right)^{z_1} (z_1+2d)$$

$$\leq \frac{128(c+1)d(2d+6)e^{2(n/4^{k-1})d}n^2}{4^{2k}} \sum_{z_1=0}^{c-1} \frac{z_1+2d}{2^{z_1}}$$

$$\leq \frac{128(c+1)d(2d+6)e^{2(n/4^{k-1})d}n^2}{4^{2k}} \cdot (2+4d).$$

Case 3 For fixed t, I, and r_1, the probability that $B_1^k = A_{r_1}^k$ is 4^{-k}. We multiply this probability by the number of ways to choose t, the number or ways to choose r_1, and by the number of ways to choose b'_1. Thus,

$$\mathrm{P}\left[\mathcal{E}_3\right] \leq \frac{2dn}{k} \cdot 4(c+1)k \cdot 3 \cdot \frac{1}{4^k} = \frac{24(c+1)dn}{4^k}.$$

Case II We now consider the case when there are trivial supporting probes. Let z be the minimum index such that probe $z+k$ is trivial ($z \geq 1$ as probes $1, \ldots, k$ are not trivial). We will consider only the first $z+k-1$ probes. If there are overlapping blocks among the first $z+k-1$ probes, then event $\mathcal{E}_2$ or $\mathcal{E}_3$ happens, so we only need to consider the case when there are no overlapping blocks among these probes. We denote this event by $\mathcal{E}_4$.

Let x be the number of blocks among probes $1, \ldots, z+k-1$. Then,

$$\mathrm{P}\left[\mathcal{E}_4 | t, I, z\right] \leq \sum_{x=1}^{z+k-1} \binom{(z+k-1)-1}{x-1} \frac{n^x}{4^{(k-1)x+z+k-1}} \leq \frac{16ne^{(n/4^{k-1})(z+k)}}{4^{z+2k}}.$$

Since probe $z+k$ is trivial, we have that $b'_{z+1} = a_{t+z}, \ldots, b'_{z+k} = a_{t+z+k-1}$, so event $\mathcal{E}_4$ depends only on the first z letters of B'. Therefore, by Lemma 2,

$$\mathrm{P}\left[\mathcal{E}_4\right] \leq \frac{2dn}{k} \sum_{z=1}^{(c-1)k} \sum_{I \in I_{k,d}(z)} 3^{|I|} \cdot \frac{16ne^{(n/4^{k-1})(z+k)}}{4^{z+2k}}$$

$$\leq \frac{32dn^2 e^{(n/4^{k-1})k}}{k \cdot 4^{2k}} \sum_{z_1=0}^{c-2} \sum_{z_2=1}^{k} \left(k\binom{k}{d} 3^d \right)^{z_1} \sum_{i=0}^{\min(d,z_2)} \binom{z_2}{i} 3^i \cdot \frac{e^{(n/4^{k-1})(kz_1+z_2)}}{4^{kz_1+z_2}}$$

$$\leq \frac{32dn^2 e^{(n/4^{k-1})k}}{k \cdot 4^{2k}} \sum_{z_1=0}^{c-2} \left(\frac{k\binom{k}{d} 3^d e^{(n/4^{k-1})k}}{4^k} \right)^{z_1} (2d+3)e^{2(n/4^{k-1})d}$$

$$\leq \frac{32d(2d+3)e^{(n/4^{k-1})(k+2d)}n^2}{k \cdot 4^{2k}} \sum_{z_1=0}^{c-2} \frac{1}{2^{z_1}}$$

$$\leq \frac{32d(2d+3)e^{(n/4^{k-1})(k+2d)}n^2}{k \cdot 4^{2k}} \cdot 2.$$

Combining all four cases, we have that the probability that the algorithm fails is $O((d + e^{(n/4^{k-1})k}/k)d^2 e^{2(n/4^{k-1})d} n^2/4^{2k} + dn/4^k)$, so if $n = O(\epsilon 4^k \min(d^{-3/2}, \log k/k))$ then this probability is at most ϵ. □

In some applications, the number of different letters between A and H is much smaller than $\lceil n/k \rceil d$, and in that case, the algorithm performs better:

Theorem 2. *For every $\epsilon > 0$, if $d \leq \left(\frac{3}{4} - \sqrt{1/c + \Omega(\log k/k)}\right) k$, $n = O(\epsilon 4^k)$, and the number of different letters between A and H is $O(n/2^{\epsilon k})$, then the probability that algorithm A fails is at most ϵ.*

4 Upper Bound

In this section, we show an upper bound on the length of the sequences that can be reconstructed from their spectra and homologous sequences.

We use the following lemma:

Lemma 3. *For every sequence P of length k, the probability that P does not appear in the spectrum of a random sequence of length $n \geq 2(k+1)k$ is at most $e^{-\frac{1}{3}n/4^k}$.*

Proof. Let S be a random sequence of length n, and let A_i denote the event that $S_i^k \neq P$. Clearly, the probability that P does not appear in the spectrum of S is $\mathrm{P}\left[\bigwedge_{i=1}^{n-k+1} A_i\right]$. The difficulty in bounding this probability lies in the fact that the events $A_1, \ldots, A_{n-k+1}$ are dependent. Thus, we will split these events into groups such that the events from one group are independent of the events from other groups.

For $i = 1, \ldots, \lfloor n/2k \rfloor$, let $I_i = \{2k(i-1)+1, \ldots, 2k(i-1)+k+1\}$. Let $B_i = \bigwedge_{j \in I_i} A_j$. The events $B_1, \ldots, B_{\lfloor n/2k \rfloor}$ are independent and have equal probabilities, so

$$\mathrm{P}\left[\bigwedge_{i=1}^{n-k+1} A_i\right] \leq \mathrm{P}\left[\bigwedge_{i=1}^{n-k+1} B_i\right] = \prod_{i=1}^{\lfloor n/2k \rfloor} \mathrm{P}[B_i] = \mathrm{P}[B_1]^{\lfloor n/2k \rfloor}.$$

We will now bound $\mathrm{P}[B_1]$. For an index $i \in I_1$, let $C_i = \overline{A_i} \wedge \bigwedge_{j=1}^{i-1} A_j$. The events $\{C_i\}_{i \in I_1}$ are disjoint, so

$$\mathrm{P}[B_1] = 1 - \mathrm{P}\left[\bigvee_{i \in I_1} \overline{A_i}\right] \leq 1 - \mathrm{P}\left[\bigvee_{i \in I_1} C_i\right] = 1 - \sum_{i \in I_1} \mathrm{P}[C_i].$$

For every $i \in I_1$,

$$\begin{aligned}\mathrm{P}[C_i] &= \mathrm{P}\left[\overline{A_i}\right] \cdot \mathrm{P}\left[\bigwedge_{j=1}^{i-1} A_j \,\middle|\, \overline{A_i}\right] = \mathrm{P}\left[\overline{A_i}\right] \cdot \left(1 - \mathrm{P}\left[\bigvee_{j=1}^{i-1} \overline{A_j} \,\middle|\, \overline{A_i}\right]\right) \\ &\geq \mathrm{P}\left[\overline{A_i}\right] \cdot \left(1 - \sum_{j=1}^{i-1} \mathrm{P}\left[\overline{A_j} \mid \overline{A_i}\right]\right).\end{aligned}$$

Clearly, $\mathrm{P}\left[\overline{A_i}\right] = 4^{-k}$. Moreover, for $j < i$, the first $j - i$ letters of S_j^k are independent of the letters of S_i^k, so $\mathrm{P}\left[\overline{A_j} \mid \overline{A_i}\right] \le \frac{1}{4^{i-j}}$. It follows that $\mathrm{P}[C_i] \ge 4^{-k} \cdot (1 - \frac{1}{3})$, hence

$$\mathrm{P}\left[\bigwedge_{i=1}^{n-k+1} A_i\right] \le \left(1 - (k+1) \cdot \frac{2}{3} 4^{-k}\right)^{\lfloor n/2k \rfloor} \le e^{-(k+1)\frac{2}{3}4^{-k} \cdot \lfloor n/2k \rfloor} \le e^{-\frac{1}{3}n/4^k}. \square$$

Theorem 3. *If $n = \Omega(\log k \cdot 4^k)$ then every algorithm for RBH fails with probability of at least $\frac{2}{3}$, even when the homologous sequence differs from the target sequence in one letter (whose position is known).*

Proof. Suppose that $n \ge 3\ln(36k) \cdot 4^k + 2k - 1$. Let $\{k\}$ be the locations set. For a sequence S and an integer i, let $S[i]$ be the set containing S and the 3 sequences that are obtained from S by changing the i-th letter of S. We say that a sequence S of length n is *hard* if for every sequence $T \in S_1^{2k-1}[k]$, all the k-subsequences of T appear in $S_{2k}^{n-(2k-1)}$. By Lemma 3, the probability that a random sequence S is not hard is at most $4k \cdot e^{-\frac{1}{3}(n-2k+1)/4^k} \le \frac{1}{9}$. Therefore, it is suffices to bound the success probability on hard sequences.

Let S be a hard sequence. All the sequences in $S[k]$ have the same spectrum, so their corresponding inputs to the RBH problem are equal. Therefore, every algorithm will fail with probability of at least $\frac{3}{4}$ when given a random sequence from $S[k]$. Since this is true for every hard sequence, it follows that every algorithm fails with probability of at least $\frac{3}{4}$ on a random hard sequence, and therefore any algorithm fails with probability of at least $\frac{8}{9} \cdot \frac{3}{4} = \frac{2}{3}$ on a random sequence of length n. $\square$

5 Hybridization Errors

In this section, we study the RBH problem in a more realistic scenario, in which there are errors in the hybridization data. We assume the following model of errors: Each k-tuple contained in the target appears in the (experimental) spectrum with probability $1-q$, and each k-tuple that is not contained in the target appears in the spectrum with probability p. In other words, the false negative probability is q, and the false positive probability is p. Furthermore, the appearance of a tuple is independent of the other k-tuples.

We say that a sequence S is *simple* if there are no indices $i \ne j$ such that $S_i^k = S_j^k$. Let algorithm B be an algorithm that acts like algorithm A, except that steps 3b and 3c are replaced by the following steps

(3b') Choose a sequences $B \in \mathcal{B}$ such that $S_{t-k+1}^{k-1}B$ is simple, and the number of supporting probes of $S_{t-k+1}^{k-1}B$ is maximal (breaking ties arbitrarily).
(3c') Set set s_t to the first letter of B.

and steps 4b and 4c are replaced by similar steps.

Theorem 4. *For every $\epsilon > 0$ and $c \geq 2$, if $p = O(1/k)$,*

$$d \leq \left(\frac{3}{4} - \sqrt{\frac{1}{c} + \frac{1}{2}\log(1+4q) + \Omega(\log k/k)}\right) k,$$

and $n = O(\epsilon (kd)^{-1} 4^{(1-\frac{c}{4}\log(1+4q))k})$ then the probability that algorithm B fails is at most ϵ.

6 Experimental Results

To complement our theoretical results, we performed simulations with our algorithms. For each value of k and d, we run algorithm A on 1000 random sequences of length n for various values of n, and computed the maximum value of n for which algorithm A returned the correct sequence in at least 90% of the runs. The results are given in Table 1. We also performed simulations with algorithm B, using the parameters $p = q = 0.05$. The results are given in Table 2. We note that further research is needed in order to evaluate the performance of algorithm B on real data. Some modification to the algorithm might be needed as real data do not behave like our probabilistic model.

Table 1. Performance of algorithm A.

k	$d=1$	$d=2$	$d=3$	$d=4$	$d=5$	Classical SBH
7	1900	1340	980	730	470	120
8	7240	5150	3940	3270	2440	240

Table 2. Performance of algorithm B.

k	$d=1$	$d=2$	$d=3$	$d=4$
7	1240	600	290	100
8	4360	2500	1130	390

Acknowledgments

We thank Ron Shamir for helpful discussions.

References

1. L. M. Adleman. Location sensitive sequencing of DNA. Technical report, University of Southern California, 1998.

2. R. Arratia, D. Martin, G. Reinert, and M. S. Waterman. Poisson process approximation for sequence repeats, and sequencing by hybridization. *J. of Computational Biology*, 3(3):425–463, 1996.
3. W. Bains and G. C. Smith. A novel method for nucleic acid sequence determination. *J. Theor. Biology*, 135:303–307, 1988.
4. A. Ben-Dor, I. Pe'er, R. Shamir, and R. Sharan. On the complexity of positional sequencing by hybridization. *J. Theor. Biology*, 8(4):88–100, 2001.
5. J. Błażewicz, P. Formanowicz, F. Glover, M. Kasprzak, and J. Węglarz. An improved tabu search algorithm for DNA sequencing with errors. In *Proc. 3rd Metaheuristics International Conference*, pages 69–75, 1999.
6. J. Błażewicz, P. Formanowicz, F. Guinand, and M. Kasprzak. A heuristic managing errors for DNA sequencing. *Bioinformatics*, 18(5):652–660, 2002.
7. J. Błażewicz, P. Formanowicz, M. Kasprzak, W. T. Markiewicz, and J. Węglarz. DNA sequencing with positive and negative errors. *J. of Computational Biology*, 6(1):113–123, 1999.
8. J. Błażewicz, P. Formanowicz, M. Kasprzak, W. T. Markiewicz, and J. Węglarz. Tabu search for dna sequencing with false negatives and false positives. *European Journal of Operational Research*, 125:257–265, 2000.
9. J. Błażewicz, J. Kaczmarek, M. Kasprzak, W. T. Markiewicz, and J. Węglarz. Sequential and parallel algorithms for DNA sequencing. *CABIOS*, 13:151–158, 1997.
10. J. Błażewicz, M. Kasprzak, and W. Kuroczycki. Hybrid genetic algorithm for DNA sequencing with errors. *J. of Heuristics*, 8:495–502, 2002.
11. S. D. Broude, T. Sano, C. S. Smith, and C. R. Cantor. Enhanced DNA sequencing by hybridization. *Proc. Nat. Acad. Sci. USA*, 91:3072–3076, 1994.
12. R. Drmanac, I. Labat, I. Brukner, and R. Crkvenjakov. Sequencing of megabase plus DNA by hybridization: theory of the method. *Genomics*, 4:114–128, 1989.
13. M. E. Dyer, A. M. Frieze, and S. Suen. The probability of unique solutions of sequencing by hybridization. *J. of Computational Biology*, 1:105–110, 1994.
14. A. Frieze, F. Preparata, , and E. Upfal. Optimal reconstruction of a sequence from its probes. *J. of Computational Biology*, 6:361–368, 1999.
15. A. M. Frieze and B. V. Halldórsson. Optimal sequencing by hybridization in rounds. *J. of Computational Biology*, 9(2):355–369, 2002.
16. E. Halperin, S. Halperin, T. Hartman, and R. Shamir. Handling long targets and errors in sequencing by hybridization. In *Proc. 6th Annual International Conference on Computational Molecular Biology (RECOMB '02)*, pages 176–185, 2002.
17. S. Hannenhalli, P. A. Pevzner, H. Lewis, and S. Skiena. Positional sequencing by hybridization. *Computer Applications in the Biosciences*, 12:19–24, 1996.
18. S. A. Heath and F. P. Preparata. Enhanced sequence reconstruction with DNA microarray application. In *COCOON '01*, pages 64–74, 2001.
19. S. A. Heath, F. P. Preparata, and J. Young. Sequencing by hybridization using direct and reverse cooperating spectra. In *Proc. 6th Annual International Conference on Computational Molecular Biology (RECOMB '02)*, pages 186–193, 2002.
20. H. W. Leong, F. P. Preparata, W. K. Sung, and H. Willy. On the control of hybridization noise in DNA sequencing-by-hybridization. In *Proc. 2nd Workshop on Algorithms in Bioinformatics (WABI '02)*, pages 392–403, 2002.
21. R. J. Lipshutz. Likelihood DNA sequencing by hybridization. *J. Biomolecular Structure and Dynamics*, 11:637–653, 1993.
22. Y. Lysov, V. Floretiev, A. Khorlyn, K. Khrapko, V. Shick, and A. Mirzabekov. DNA sequencing by hybridization with oligonucleotides. *Dokl. Acad. Sci. USSR*, 303:1508–1511, 1988.

23. D. Margaritis and S. Skiena. Reconstructing strings from substrings in rounds. In *Proc. 36th Symposium on Foundation of Computer Science (FOCS 95)*, pages 613–620, 1995.
24. I. Pe'er, N. Arbili, and R. Shamir. A computational method for resequencing long dna targets by universal oligonucleotide arrays. *Proc. National Academy of Science USA*, 99:15497–15500, 2002.
25. I. Pe'er and R. Shamir. Spectrum alignment: Efficient resequencing by hybridization. In *Proc. 8th International Conference on Intelligent Systems in Molecular Biology (ISMB '00)*, pages 260–268, 2000.
26. P. A. Pevzner. l-tuple DNA sequencing: Computer analysis. *J. Biomolecular Structure and Dynamics*, 7:63–73, 1989.
27. P. A. Pevzner, Yu. P. Lysov, K. R. Khrapko, A. V. Belyavsky, V. L. Florentiev, and A. D. Mirzabekov. Improved chips for sequencing by hybridization. *J. Biomolecular Structure and Dynamics*, 9:399–410, 1991.
28. F. Preparata and E. Upfal. Sequencing by hybridization at the information theory bound: an optimal algorithm. In *Proc. 4th Annual International Conference on Computational Molecular Biology (RECOMB '00)*, pages 88–100, 2000.
29. R. Shamir and D. Tsur. Large scale sequencing by hybridization. *J. of Computational Biology*, 9(2):413–428, 2002.
30. S. Skiena and S. Snir. Restricting SBH ambiguity via restriction enzymes. In *Proc. 2nd Workshop on Algorithms in Bioinformatics (WABI '02)*, pages 404–417, 2002.
31. S. Skiena and G. Sundaram. Reconstructing strings from substrings. *J. of Computational Biology*, 2:333–353, 1995.
32. S. Snir, E. Yeger-Lotem, B. Chor, and Z. Yakhini. Using restriction enzymes to improve sequencing by hybridization. Technical Report CS-2002-14, Technion, Haifa, Israel, 2002.
33. D. Tsur. Sequencing by hybridization with errors: Handling longer sequences. Manuscript, 2003.
34. D. Tsur. Sequencing by hybridization in few rounds. In *Proc. WABI '03*, to appear.

Selecting Degenerate Multiplex PCR Primers*

Richard Souvenir[1], Jeremy Buhler[1,2], Gary Stormo[2,1], and Weixiong Zhang[1,2,**]

[1] Department of Computer Science
[2] Department of Genetics
Washington University in St. Louis
St. Louis, MO 63130, USA

Abstract. Single Nucleotide Polymorphism (SNP) Genotyping is an important molecular genetics technique in the early stages of producing results that will be useful in the medical field. One of the proposed methods for performing SNP Genotyping requires amplifying regions of DNA surrounding a large number of SNP loci. In order to automate a portion of this method and make the use of SNP Genotyping more widespread, it is important to select a set of primers for the experiment. Selecting these primers can be formulated as the *Multiple Degenerate Primer Design (MDPD)* problem. An iterative beam-search algorithm, *Multiple, Iterative Primer Selector (MIPS)*, is presented for MDPD. Theoretical and experimental analyses show that this algorithm performs well compared to the limits of degenerate primer design and the number of spurious amplifications should be small. Furthermore, MIPS outperforms an existing algorithm which was designed for a related degenerate primer selection problem.
An implementation of the MIPS algorithm is available for research purposes from the website http://www.cse.wustl.edu/~zhang/software/mips.

1 Introduction

Single Nucleotide Polymorphisms (SNPs) are individual base differences in DNA sequences between individuals. It is estimated that there are roughly three million SNPs in the human genome [12]. Association studies between SNPs and various diseases, as well as differences in how individuals respond to common therapies, promise to revolutionize medical science in the coming years [2]. Recent work suggests there may be only a few hundred thousand "blocks" of SNPs that recombine to provide most of the variability seen in human populations [5]. However, it is still a daunting task to identify the specific genetic variations occurring in specific individuals in order to determine their associations with important phenotypes. Currently, there are many proposed techniques for determining the SNP composition of a given genome. However, in order for these assaying techniques to be effective in large-scale genetic studies of hundreds or thousands of SNPs, they must be scalable, automated, robust, and inexpensive [9].

* We thank Pui Kwok for describing the problem and useful discussions. RS was supported by NIH training grant GM08802 to Washington University Medical School and NSF grant ITR/EIA-0113618, GS was funded by NIH grant HG00249, and WZ was supported by NSF grants IIS-0196057 and ITR/EIA-0113618.

** Corresponding to: zhang@cse.wustl.edu, phone: (314)935-8788.

G. Benson and R. Page (Eds.): WABI 2003, LNBI 2812, pp. 512–526, 2003.

One technique involves the use of multiplex PCR (MP-PCR) to amplify the regions around the SNP. Multiplex PCR is a variation of PCR where multiple DNA fragments are replicated simultaneously. MP-PCR, like all PCR variations, makes use of oligonucleotide primers to define the boundaries of amplification. For each region of DNA that is to be amplified, two primers, generally referred to as the forward and reverse primers, are needed. In MP-PCR, it is necessary to select a forward and reverse primer for each of the regions to be replicated, and for the large-scale amplification required in SNP Genotyping, there can be hundreds, or perhaps thousands, of those regions. The process of selecting such a large set of primers by current methods, including trial-and-error [9], can be time-consuming and difficult.

There are two similar problems in primer selection, the Primer Selection Problem and the Degenerate Primer Design Problem. The Primer Selection Problem [15] involves minimizing the number of primers needed to amplify regions of DNA in a set of sequences. It has been shown that this is an NP-hard problem [6] in reductions from other hard problems, including SET-COVER and GRAPH-COLORING [3]. There have been a number of proposed heuristics to solve this problem, including a branch-and-bound search algorithm [14]. Also, algorithms have been proposed which incorporate biological data about the primers into the search [13, 4].

In an MP-PCR experiment where the number of primers is not optimized, the number of primers needed is equal to twice the number of sequences in the input set. In general, the algorithms mentioned above reduce the number of primers needed to 25-50% of this value, which can still be rather high for the large-scale amplification needed for SNP Genotyping. This leads to the use of degenerate primers. Degenerate primers [10] are primers that make use of degenerate nucleotides. For example, consider this degenerate primer, *ACMCM*, where *M* is a degenerate nucleotide which represents either of the bases, *A* or *C*. This degenerate primer is actually representative of the set of 4 primers {*ACACA, ACACC, ACCCA, ACCCC*}. The number of primers that a degenerate primer represents is referred to as its *degeneracy*. Degenerate primers are as easy to produce as regular primers, and therefore save the molecular biologist time during the primer design phase of the experiment. The use of degenerate primers introduces two new problems. First, the effective concentration of the desired primers is decreased by the presence of undesired primers. Second, the presence of undesired primers can lead to erroneous amplification. Therefore, it is important to use primers of relatively low degeneracy to realize the inherent benefits of degenerate primer design while minimizing the effects of these two problems.

The Degenerate Primer Design Problem (DPD) is the second related problem, however it makes use of degenerate primers. DPD is the decision problem of determining whether or not there exists a single degenerate primer below some given threshold which can amplify regions of DNA for some number of a set of input sequences. There are two variations of DPD. MAXIMUM COVERAGE DPD (MC-DPD) is the related maximization problem where the goal is to find the maximum number of sequences that can be amplified by a degenerate primer whose degeneracy falls below some threshold. MINIMUM DEGENERACY DPD (MD-DPD) is the second variation of DPD whose goal is to find the degenerate primer of minimum degeneracy that amplifies all of the input sequences. Both MC-DPD and MD-DPD have been shown to be NP-Hard problems [11].

In this paper, we describe the Multiple Degenerate Primer Design Problem (MDPD), present an algorithm to solve this problem and describe how the results can be used to select a large set of primers that can be used in MP-PCR for SNP Genotyping. The details of the protocol for applying degenerate primers for genotyping using SNPs can be found in [9]. The basic problem uses a given collection of DNA sequences from genomic regions known to contain SNPs. The regions are chosen such that primers selected from them, one on each side, are not closer to the SNP than a fixed amount and no farther away than a specified distance. We proceed as follows: We first present two variants of MDPD problems. We then describe the *Multiple, Iterative Primer Selector (MIPS)* algorithm and show how MIPS performs relative to another solution in the domain, and the theoretical limits of the problem. We also discuss the issue of erroneous amplification. We specifically study the relationship among the probability of unexpected priming events, the length of degenerate primers and their degeneracy. Finally, we show the results of MIPS on different datasets.

2 Problem Description

Some of the notation from [11] is used to describe the MDPD problem. To maintain consistency, lower-case symbols (i.e. l, b, i) represent numerical values, counting variables, or individual characters in a sequence. Upper-case symbols (i.e. P, S) denote primers, sequences, or subsequences. Finally, calligraphy symbols (i.e. $\mathcal{S}, \mathcal{C}$) represent sets of sequences or primers.

Let $\Sigma = \{A, C, G, T\}$ which is the finite fixed alphabet of DNA. A *degenerate primer* is a string P with several possible characters at each position, i.e., $P = p_1 p_2 \cdots p_l$, where $p_i \subseteq \Sigma, p_i \neq \emptyset$. l is the length of primer P. The *degeneracy* of P is $d(P) = \Pi_{i=1}^{l} |p_i|$. Consider the degenerate primer $P' = \{A\}\{A,C\}\{A,C\}\{C\}$. The length of P' is 4 and $d(P') = 4$. For the sake of clarity, we use the IUPAC symbols for degenerate nucleotides to represent degenerate primers. Therefore, P' can be represented as *AMMC* where M is the degenerate nucleotide which represents $\{A,C\}$. Degenerate primers can be constructed by *primer addition*. For any two primers, P^1 and P^2, their sum, P^3 equals $(p_1^1 \cup p_1^2)(p_2^1 \cup p_2^2) \cdots (p_l^1 \cup p_l^2)$.

For any sequence S_i in an input set $\mathcal{S}$, we say that a degenerate primer P *covers* S_i if there is a substring F of length l in S_i where for each character f_i in F, $f_i \in p_i$.

There are two variants of the MULTIPLE DEGENERATE PRIMER DESIGN problem: PRIMER-THRESHOLD MDPD (PT-MDPD) and TOTAL-THRESHOLD MDPD (TT-MDPD). For both, we are given n sequences, $\mathcal{S} = \{S_1, S_2, \cdots, S_n\}$, and a maximum degeneracy bound α. For PT-MDPD, the goal is to find a set of degenerate primers, $\mathcal{P}$, of minimum size that covers every sequence in $\mathcal{S}$ where, for each degenerate primer, $P_i \in \mathcal{P}$, $d(P) \leq \alpha$. For TT-MDPD, the goal is to find a set of degenerate primers, $\mathcal{P}$, of minimum size that covers every sequence in $\mathcal{S}$ where $\sum d(P_i) \leq \alpha$.

Both PT-MDPD and TT-MDPD are NP-hard [6]. The NP-hardness of PT-MDPD can be shown based on the observation that the Primer Selection Problem (PSP) [15] is a special case of PT-MDPD, where the degeneracy threshold is set to one. The NP-hardness of TT-MDPD can be shown by a reduction from the Weighted Set Covering problem.

3 MIPS: Multiple, Iterative Primer Selector

To overcome the difficulty caused by the NP-hardness of MDPD problems, we propose an iterative beam search algorithm to make a tradeoff between optimality and tractability. In order to solve PT-MDPD and TT-MDPD, MIPS can run in either of two modes, MIPS-PT and MIPS-TT, respectively. This section focuses on MIPS-TT. However, we will highlight how MIPS-PT operates differently.

MIPS progressively constructs a set of primers that covers all the input sequences. Define a *k-primer* to be a degenerate primer that covers k input sequences. The basic algorithm first generates a set of candidate 2-primers, each having some degeneracy value, then iteratively extends all candidate k-primers into $(k+1)$-primers by generalizing them to cover an additional sequence. Generalization stops when no primer can be extended without exceeding the degeneracy threshold α. At this point, all remaining primers cover k_{last} sequences, so we retain the primer of minimum degeneracy, remove the input sequences it covers from consideration, and repeat the algorithm until all sequences are covered.

To guide the search, MIPS uses the degeneracy of a primer as a scoring function. The set of primers that are stored for extension are known as a *beam*. Beam Search [1] differs from greedy or best-first search in that multiple nodes, degenerate primers in this case, are saved for extension instead of just one. This model of progressively adding to a beam of degenerate primers and updating the scoring function is similar to the CONSENSUS motif-finding model [8].

It is important to note that the degeneracy of a given k-primer increases or remains the same, with the addition of additional sequence fragments. This observation permits us to employ a strategy which ignores degenerate primers with high degeneracy, in order to speed up the algorithm. Therefore, the search is restricted only to the primers with the lowest degeneracy. In this algorithm, the number of the candidate primers to restrict the search to at each level can be specified. This constant, b, describes the number of k-primers to save for each level. Increasing b can possibly improve the quality of the solution, but lengthens the running time of the algorithm. In the results section, we examine the effect of this parameter, b, on MIPS.

The pairwise comparison of two fragments is the dominating operation and a rate-limiting step of the algorithm. A majority of these comparisons are between two fragments that share few, if any, nucleotides. To avoid comparisons between dissimilar fragments, the exhaustive pairwise comparison is replaced with a similarity lookup. All of the primer candidates are added to a FASTA-style lookup table. In general, for DNA, a FASTA table fragment length of 6 is recommended [7]. Using the table, each fragment is compared only to the other fragments that are returned.

The constructive search continues until one of two cases occurs. In the first case, all sequences are covered by a single n-primer, where n is the number of sequences in the input set. The algorithm then terminates with that primer as the result. In the second case, no k-primer can be extended to a $(k+1)$-primer without exceeding the degeneracy

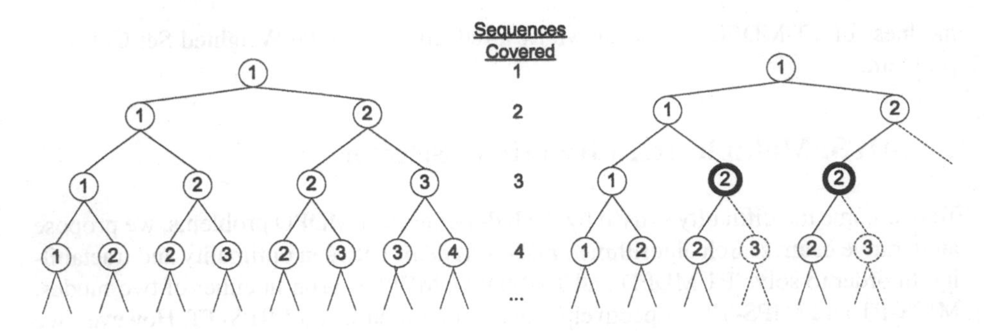

Fig. 1. For these graphs, the depth of a node represents the number of sequences from the input set covered and the number in a node represents the number of degenerate primers that will be used to cover those sequences. Each node can be expanded into two child nodes. The left child represents covering an additional sequence using an existing degenerate primer and the right child represents covering an additional sequence using a new degenerate primer. Graph (a) shows a full search. Graph (b) shows the pruning that takes place in MIPS-TT during the backtracking phrase. Consider the two bold nodes. Both of these cover the same number of sequences with the same number of primers. MIPS-TT will therefore only expand the node whose total score is better. This avoids the exponential expansion seen in (a)

threshold and there exists at least one sequence uncovered. At this point, k_{last} sequences have been covered. The algorithm chooses the best degenerate (k_{last})-primer, P_0, from the set $\mathcal{P}$ of primers sorted by degeneracy value. The problem then reduces to a smaller instance where the input set is the original set of sequences minus those covered by P_0. In MIPS-PT, the degeneracy threshold for this subproblem is equivalent to the original threshold, α. In MIPS-TT, the degeneracy threshold is reduced by the degeneracy of P_0. *The algorithm then restarts on the reduced problem.*

For MIPS-PT, iteratively applying this procedure will eventually return a set of primers to cover the set of input sequences. However, this is not the case for MIPS-TT. After P_0 is discovered and its sequences are removed from consideration, the new threshold may be too low to cover the rest of the sequences. In this case, MIPS-TT *backtracks* to the previous level, $k_{last} - 1$, and selects the next best primer P'_0 for removal. Again, MIPS restarts on the sequences that P'_0 has not covered and with a degeneracy limit that is the original α minus the degeneracy of P'_0. Figure 1 shows, schematically, the execution of MIPS-TT.

A pseudo-code description of MIPS is given in Algorithms 1-3.

4 Analysis

We now examine the theoretical bounds of MIPS and degenerate primer design in general, and investigate the issue of erroneous amplification. In the following, let n be the number of sequences in the input set, m the average length of each sequence, b the beam size, and l the length of the primers.

Algorithm 1 MIPS($\mathcal{S}, \alpha$)

1: *Initialize* **Global variables** (2-D matrices): BEST - candidate fragments; COVERED - sequences covered; ALLOWABLE - remaining degeneracy, $ALLOWABLE(0,0) = \alpha$.
2: **for** $p = 1$ to the number of degenerate primers that will be used **do**
3: Let c = the maximum number of sequences that the (p-1) primers covered
4: **while** $c > 0$ **do**
5: $MIPS\_SEARCH(\mathcal{S} - COVERED(p-1,c), ALLOWABLE(p-1,c), p, c)$
6: if this search covers S, print solution and exit
7: else c=c-1

Algorithm 2 MIPS_SEARCH($\mathcal{S}, \alpha, p, c$)

1: **Input**: Sequence set $\mathcal{S}$, degeneracy bound α, primer number p, sequences covered c.
2: **Output**: total number of sequences covered
3: Initialize priority queue Q of size b;
4: Perform pair-wise comparisons.
5: **for all** sequence $S_i \in \mathcal{S}$ **do**
6: **for all** substring $S_i[j,l]$ **do**
7: Let $C = \{ x | \langle f, x \rangle \in T$ and f is a k-length substring of $S_i[j,l] \}$
8: **for all** fragment $C_k \in C$ **do**
9: $D = S_i[j,L] + C_k$
10: Insert D into queue Q
11: Let $c' = c$
12: **while** queue Q is not empty **do**
13: Let P = the best element of Q
14: **if** degeneracy(P) < degeneracy($BEST(p,c)$) **then**
15: $BEST(p,c') = P$
16: $ALLOWABLE(p,c') = \alpha-$ degeneracy(P)
17: $COVERED(p,c') = COVERED(p-1,c) \cup$ covers($P, \mathcal{S}$)
18: $Q = ONE\_PASS(Q, \mathcal{S}, \alpha)$
19: $c' = c' + 1$
20: return $(c' + 1)$

Algorithm 3 ONE_PASS($Q, \mathcal{S}, \alpha$)

1: **Input**: Priority queue Q, set of sequences $\mathcal{S}$, degeneracy bound α.
2: **Output**: Priority queue Q'
3: **for all** primer $P \in Q$ **do**
4: **for all** sequence $S_i \in \mathcal{S}$ **do**
5: **if** $S_i \notin$ covers(P) **then**
6: **for all** substring $S_i[j,l]$ **do**
7: $D = S_i[j,l] + P$
8: Insert D into queue Q'
9: return Q'

4.1 Algorithm Complexity

Space From the input set, each primer is stored individually which requires space $O(nml)$. In the implementation, there are 4 $n \times n$ matrices that are needed for back-tracking and storing degenerate primers that could eventually become part of the final solution. This adds an additional $O(n^2)$ of storage. Therefore, the total amount of space is $O(n^2 + nml)$.

Time The time complexity is analyzed in a bottom-up fashion. The procedure of comparing the fragments in the beam to the remaining sequences makes $O(bnm)$ primer additions since there are $O(nm)$ total fragments and b fragments in the beam. Each primer addition requires comparing every character in each of the two primer. Therefore, this portion requires $O(bnml)$ time.

The process of generating new beams of k-primers, for increasing k, is called MIPS_SEARCH. MIPS_SEARCH uses the above procedure to build new beams, and could, in the worst case, build n beams. Therefore, the overall time complexity is $O(bn^2ml)$. The number of times MIPS_SEARCH is executed depends on the amount of back-tracking. This is directly related to the number of primers in the final solution. In the best case, if the solution only requires 1 primer, there will be only one call to MIPS_SEARCH. In the worst case, if the solution requires n primers (one primer for each input sequence) there will be $n^2/2$ calls to MIPS_SEARCH. Let p be the number of primers in the final solution. The best approximation to the number of MIPS_SEARCH calls is $O(pn)$. This brings the overall time complexity to $O(bn^3mlp)$. Currently, we are working on a method to reduce the time complexity of comparing the degenerate primers in the beam to the remaining sequences in order to speed up the entire algorithm.

4.2 Limits of Degenerate Primer Design

Multiplex primer design demands that many input sequences share sites complementary to some common (possibly degenerate) primer. The sequences to be co-amplified are not in general homologous, so their complementarity to a common primer is largely a matter of chance. We therefore explored the chance-imposed limits of multiplexing, that is, how many unrelated DNA sequences are likely to be covered by a single PCR primer of a given degeneracy?

Let S be a collection of n DNA sequences of common length m. Call a primer P an *(l, α, k)-primer* for S if it has length l and degeneracy at most α and covers at least k sequences of S. A natural way to quantify the limits of multiplexing is to compute the probability that an (l, α, k)-primer exists for S. However, this probability is difficult to compute, even assuming that S consists of i.i.d. random DNA with equal base frequencies. We instead compute the *expected* number of (l, α, k)-primers for S. If this expectation is much less than one, Markov's inequality implies that S is unlikely to contain any such primer.

We count not the total number of (l, α, k)-primers for S but only the number of *maximal* primers. A primer P of degeneracy at most α is said to be *maximal* if increasing P's degeneracy at any position would cause its total degeneracy to exceed α. The expected number of maximal (l, α, k)-primers for $\mathcal{S}$ is in general less than the total number of (l, α, k)-primers, but a primer of this type exists for $\mathcal{S}$ iff a maximal primer exists. Hence, the former expectation is more useful than the latter for bounding the probability that at least one (l, α, k)-primer exists.

Occurrence Probability for One Fixed Primer Let P be a primer of length l, such that the jth position of P permits $|p_j|$ different bases. Let $\mathcal{S}$ be a collection of n i.i.d. random DNA sequences of common length m with equal base frequencies, and let T be a single l-mer at a fixed position in some sequence $S_i \in \mathcal{S}$. Say that P *matches* T if P would hybridize to T. We have that

$$\Pr[P \text{ matches } T] = \prod_{j=1}^{l} \frac{|p_j|}{4} = \frac{d(P)}{4^l}.$$

The probability that P covers S_i, i.e. that it matches *at least one* l-mer of S_i, depends in a complicated way on P's overlap structure, but if S_i is not too short and $d(P)/4^l \ll 1$ (both of which are typically true), then using Poisson approximation [16],

$$\Pr[P \text{ occurs in } S_i] \approx 1 - e^{-\frac{d(P)}{4^l}(m-l+1)}.$$

Let q be the probability that P matches somewhere in a single sequence of length m, and let $c(P)$ be P's coverage of $\mathcal{S}$, i.e. the number of sequences of $\mathcal{S}$ in which P matches at some position. Because the sequences of $\mathcal{S}$ are independent, the probability that P matches in at least k sequences given by the binomial tail probability

$$\Pr[c(P) \geq k] = 1 - \Pr[B(n, q) < k],$$

where $B(n, q)$ is the sum of n independent Bernoulli random variables, each with probability q of success.

Computing the Expectation Let $\Pi(l, \alpha)$ be the set of all maximal primers of length l and degeneracy at most α. To count the expected number $E_{l,\alpha,k}$ of (l, α, k)-primers for $\mathcal{S}$, observe that

$$E_{l,\alpha,k} = \sum_{P \in \Pi(l,\alpha)} \Pr[c(P) \geq k].$$

Enumerating all $P \in \Pi(l, \alpha)$ to compute this expectation would be computationally expensive, but this enumeration is not needed for i.i.d. sequences with equal base frequencies. Given these assumptions about $\mathcal{S}$'s sequences, the probability that P matches a given l-mer does not change if we rearrange its positions (e.g. "AMC" versus "MCA") or change the precise nucleotides matched (e.g. "RTG" versus "MCA"). Let W be a multiset of l values drawn from $\{1, 2, 3, 4\}$ that lists the degeneracies n_j (in any order) of a primer from $\Pi(l, \alpha)$. Then every primer described by the same W has the same probability of covering at least k sequences in $\mathcal{S}$. Hence, the desired expectation is given by

$$E_{l,\alpha,k} = \sum_W \#(W) \Pr[c(P) \geq k \mid P \text{ described by } W].$$

where the sum ranges over all feasible W for $\Pi(l,\alpha)$ and $\#(W)$ denotes the number of degenerate primers described by W. The probability is computed as described above, so we need only describe how to compute $\#(W)$.

Let W be a multiset with n_1 1's, n_2 2's, n_3 3's, and n_4 4's. If we fix *which* positions in P permit 1, 2, 3, and 4 nucleotides respectively, then there are $4^{n_1} \times 6^{n_2} \times 4^{n_3}$ ways of assigning nucleotide sets to these positions. Hence,

$$\#(W) = \binom{l}{n_1\; n_2\; n_3} 4^{n_1+n_3} 6^{n_2}.$$

Enumerating all feasible W for $\Pi(l,\alpha)$ is straightforward, so the expectation can be computed.

4.3 Mispriming

It is possible that a pair of primers binds to an undesired location and results in an erroneous amplification. *Mispriming* is the occurrence of this event where the unwanted PCR product is indistinguishable, by size, from the targeted products.

Suppose we design a set of degenerate primers with length l, such that the *total degeneracy* of the set is α. We wish to estimate the expected number of mispriming events when our primer set is applied to a genome of length g. For simplicity, we assume that the genome is an i.i.d. random DNA sequence with equal base frequencies, and that a pair of l-mers cause a mispriming event iff they bind to the genome within δ bases of each other, in the appropriate orientations to permit amplification of the sequence between them.

Let i index the positions of the genome on its forward strand. Let the 0-1 random variable x_i indicate the event that an l-mer from our primer set is complementary to the forward strand at position i, and let $\overline{x}_i$ be the event that an l-mer is complementary to the reverse-complement strand at i. For any i, we have that

$$E[x_i] = E[\overline{x}_i] = \frac{\alpha}{4^l}.$$

We say that a mispriming event occurs at i if

$$\overline{x}_i \cap \bigcup_{j=i}^{i+\delta-1} x_j = 1.$$

Denote this event by the 0-1 indicator M_i. The total number of mispriming events M is simply $\sum M_i$, for $i = 1, 2, \cdots, g$.

Note that the two matching events are independent in an i.i.d. random DNA sequence when the two primers do not overlap. To simplify our calculations, we ignore the effect of overlapping primer boundaries. Using Poisson approximation to estimate the probability of the matching event on the forward strand, we have that

$$\begin{aligned} E[M_i] &= E[\overline{x}_i \cap \bigcup_{j=i}^{i+\delta-1} x_j] \\ &= E[\overline{x}_i] E\left[\bigcup_{j=i}^{i+\delta-1} x_j \right] \\ &\approx E[\overline{x}_i] \left(1 - e^{-\sum_{j=i}^{i+\delta-1} E[x_j]} \right). \end{aligned}$$

Finally, setting $\rho = \alpha/4^l$, we derive the expected mispriming rate as

$$\begin{aligned} E[M] &= \sum_{i=1}^{g} E[M_i] \\ &\approx g\rho \left(1 - e^{-\delta\rho}\right). \end{aligned}$$

Using $g = 3 \times 10^9$ for the human genome and $\delta = 500$ bases, we find that a design using 50 degenerate primers of length 20 and average degeneracy 10000 yields about 0.31 expected mispriming events in the genome. The mispriming rate scales linearly with the genome size and roughly quadratically with ρ.

5 Results

MIPS has been applied to both human DNA sequences and randomly generated datasets. We used a dataset containing regions of human DNA sequences surrounding 95 known SNPs. The sequences varied in length from a few hundred nucleotides to well over one thousand. The location of a SNP on a sequence was marked in order to provide a reference for the forward and reverse primers. To ensure effective PCR product analysis, each primer could not be located within 10 bases of the SNP and the entire PCR product length could not exceed 400 bases.

First, we show how MIPS performed relative to the theoretical limits previously discussed. Then, we show how various parameters, such as the beam size and degeneracy threshold, affect the performance, including the number of primers and running time. We then show some results of MIPS on the human dataset. Finally, we compare MIPS to an algorithm designed to solve a similar DPD problem considered in [11].

5.1 Comparison to Theoretical Limits

The theoretical estimates of section 4.2 can be used to evaluate whether a particular primer-design algorithm performs well on the MC-DPD problem, that is, whether it finds degenerate primers with coverage close to the maximum predicted for a given set of input sequences. We evaluated the MIPS algorithm's performance on MC-DPD by comparing the primers it found in random DNA with those expected to exist in theory.

degeneracy α	Avg Coverage	Max Predicted
1000	6.30	7
10000	10.55	12
100000	19.30	26

Table 1. Actual and predicted coverage of 20-mer primers found on sets of 190 random sequences of length 211. Avg Coverage: average coverage of primer found over 20 random trials. Max Predicted: largest coverage m such that $E_{20,\alpha,m} > 1$.

For these experiments, we generated test sets of i.i.d. random DNA sequences with equal base frequencies with $n = 190$, and $m = 211$, so that the number and average length of the test sequences roughly matched those of the human DNA test sequences.

We used MIPS to find a single primer of length $l = 15$ with maximum coverage in each test set, subject to varying degeneracy bounds α. Table 1 compares the average coverage of primers found by MIPS in 20 trials to the largest coverage k such that $E_{l,\alpha,k}$ for test sets of the specified size is > 1. Primers with coverage exceeding this value of k are not expected to occur in the test sets, while primers with slightly smaller coverage may or may not occur frequently.

MIPS proved adept at finding primers close to the maximum predicted coverage for relatively small degeneracies ($\alpha \leq 10000$). We therefore have considerable confidence in its ability to find high-coverage primers if they are present. The gap between the best primers found by MIPS and those predicted to occur in theory grows with the degeneracy bound, but we cannot say with certainty whether this fact represents a limitation of the algorithm or of the theoretical estimates, since primers with expectation greater than one may with significant probability still fail to occur. Moreover, the high degeneracies where MIPS might perform poorly are of less practical interest, since single primers with such high degeneracies are experimentally more difficult to work with.

Overall, MIPS appears to be operating close to the theoretical limit for MC-DPD problems of small degeneracy. Although our analysis does not directly address the MDPD problems, any large gap between the most efficient design and the designs produced by MIPS is unlikely to arise from failure to find single high-coverage primers when they exist.

5.2 Performance

Figure 2 shows the effect of beam size on the solution quality, or number of primers. Figure 2a shows that increasing the beam size linearly increases the running time of the algorithm. These figures show the trade-off between the quality of the solution and the running time of the algorithm. For this particular dataset, there was a decrease of two degenerate primers in the final solution when the beam size was increased from 10 to 100. Moreover, only a slightly better solution was discovered when the beam size was increased to 250. For the average desktop computer, beam sizes larger than a few hundred result in impractical running times. For the input set we used, which contained 95 human DNA sequences, using a beam size of 100 produced a strong answer while keeping the running time reasonable. In general, the beam size should be close to the number of sequences in the input set.

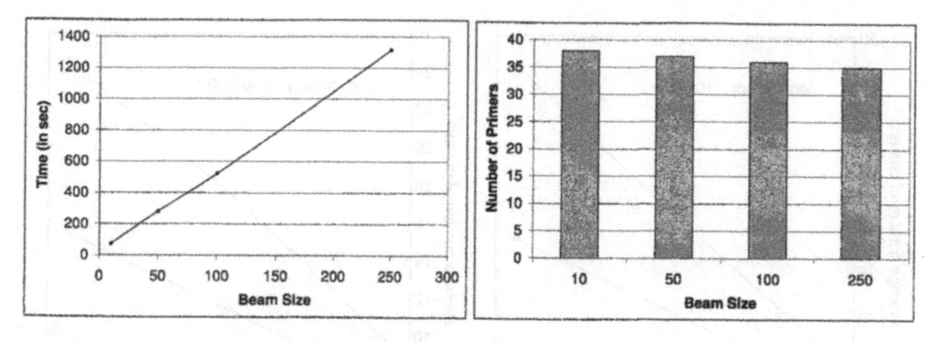

Fig. 2. These graphs shows the effect of beam size on the (a) running time of the algorithm and (b) number of length 20 primers discovered. The algorithm was run on a dataset of 95 sequences which are regions surrounding known SNPs.

PT-MDPD		*TT-MDPD*	
Degeneracy	*# Primers*	*Degeneracy*	*# Primers*
$4^6 \approx 4K$	53	$4^9 \approx 262K$	44
$4^7 \approx 16K$	44	$4^{10} \approx 1M$	37
$4^8 \approx 64K$	36	$4^{11} \approx 4M$	30
$4^9 \approx 262K$	29	$4^{12} \approx 16M$	23

Table 2. Results on a dataset of 95 human SNP regions using primers of length 20 with default settings.

In an unpublished laboratory experiment, a set of degenerate primers of length 20 was manually constructed where each primer was a mixture of 8 specific bases and 12 fully degenerate nucleotides (*e.g. AGTCGGTANNNNNNNNNNNN.*) For this experiment, the total degeneracy would be $\approx 4^{12}$. MIPS was designed to automate this procedure and, possibly, reduce the total degeneracy and/or number of primers used. In practice the desired accuracy in the experiment determines the actual parameter values used for MIPS. Table 2 shows the results a large-scale MP-PCR experiment using primers of length 20. For 95 sequences, 190 primers would be needed in the general case. MIPS-PT decreased the total number of primers to 15% of this unoptimized value for a degeneracy limit of 262,144. Table 2 includes the similar results for PT-MDPD and TT-MDPD.

5.3 Comparison to HYDEN

The HYDEN algorithm [11] is a heuristic designed for finding approximate solutions to the DPD problems. Recall that DPD is a set of problems where the general goal is to find a *single* degenerate primer that either covers the most sequences while having a degeneracy value less than a specified threshold or covers all of the sequences with minimum degeneracy. The DPD problem is the most closely related one to our MDPD problem, and HYDEN is the only published algorithm for DPD that we are aware of.

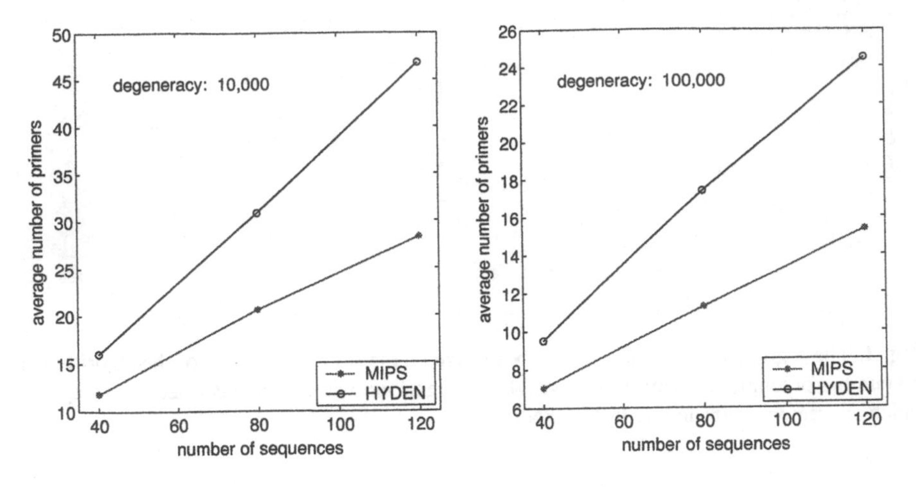

Fig. 3. Both HYDEN and MIPS were tested on 20 randomly-generated datasets in order to solve the PT-MDPD problem. The tests were conducted with different degeneracy thresholds (a) 10,000 and (b) 100,000. The graph shows the number of degenerate primers selected.

HYDEN can solve the PT-MDPD problem indirectly by iteratively solving the MC-DPD problem on smaller and smaller sets. After selecting a pair of degenerate primers under a given bound that covers a certain subset of the sequences in an input set, the algorithm runs again on the remaining sequences. For the reasons described below, iteratively solving MC-DPD is not the most effective way to solve the PT-MDPD problem. However, this was the most reasonable comparison that was possible given the implementation available to us at the time of testing. The graphs in Figure 3 shows the number of primers that each algorithm found from a randomly generated set of sequences of varying lengths with varying degeneracy thresholds. They are uniformly-distributed i.i.d. sequences of equal length. Each program searched for degenerate primers of length 15 without allowing any mismatches at any positions.

In general, HYDEN always produced more primers than MIPS in attempting to solve PT-MDPD. For a primer degeneracy value of 100,000 and over 100 sequences, the difference was as large as 60% more primers. These results can be partially explained by the differing design requirements of the DPD and MDPD problems. Even when applied iteratively, the goal of the DPD problems is to have a result which could be divided into distinct PCR experiments. The goal of the MDPD problems is to have a set of primers for one large-scale PCR experiment. Specifically, to solve the DPD problem, the HYDEN algorithm must ensure that for any given degenerate forward primer that is discovered, exactly one degenerate reverse primer is used to cover the sequences covered by the forward primer. Therefore, a given degenerate forward primer is restricted to which sequences it is reported to cover based on the presence of a suitable degenerate reverse primer, and vice-versa. Moreover, the HYDEN algorithm has an additional restriction in which any given degenerate primer is limited to either covering a set of forward or reverse primers, but not both.

6 Conclusions

We have discussed a problem that arises in large-scale, high-throughput multiplex PCR experiments for SNP Genotyping. We developed an iterative beam-search heuristic, MIPS, for this problem which can be used to select a set of degenerate primers for a given set of sequences. This algorithm compares favorably to an existing algorithm for similar problems. Finally, using both theoretical calculations and experimental analysis, we have shown that MIPS is neither time nor memory intensive and could conceivably be used as a desktop tool for SNP Genotyping. The overall effectiveness of this algorithm will ultimately be determined by the application of the resulting primers in biological experiments, which is our next focus for this research.

References

[1] R. Bisiani. Search, beam. In S. C. Shapiro, editor, *Encyclopedia of Artificial Intelligence*, pages 1467–1468. Wiley-Interscience, New York, NY, 2nd edition, 1992.
[2] F. Collins and V. McKusick. Implications of the human genome project for medical science. *JAMA*, 285:2447–2448, 2001.
[3] K. Doi and H. Imai. A greedy algorithm for minimizing the number of primers in multiple pcr experiments. *Genome Informatics*, pages 73–82, 1999.
[4] K. Doi and H. Imai. Complexity properties of the primer selection problem for pcr experiments. In *Proceedings of the 5th Japan-Korea Joint Workshop on Algorithms and Computation*, pages 152–159, 2000.
[5] S. Gabriel, S. Schaffner, H. Nguyen, J. Moore, J. Roy, B. Blumenstiel, J. Higgins, M. DeFelice, A. Lochner, M. Faggart, S. Liu-Cordero, C. Rotimi, A. Adeyemo, R. Cooper, R. Ward, E. Lander, M. Daly, and D. Altshuler. The structure of haplotype blocks in the human genome. *Science*, 296:2225–2229, 2002.
[6] M. R. Garey and D. S. Johnson. *Computers and Intractability: A Guide to the Theory of NP-Completeness*. Freeman, New York, NY, 1979.
[7] D. Gusfield. *Algorithms on Strings, Trees, and Sequences: Computer Science and Computational Biology*, chapter 15, page 377. Press Syndicate of the University of Cambridge, 1997.
[8] G. Hertz and G. Stormo. Identifying dna and protein patterns with statistically significant alignments of multiple sequences. *Bioinformatics*, 15:563–577, 1999.
[9] P. Kwok. Methods for genotyping single nucleotide polymorphisms. *Annu. Rev. Genomics Human Genetics*, 2:235–58, 2001.
[10] S. Kwok, S. Chang, J. Sninsky, and A. Wang. A guide to the design and use of mismatched and degenerate primers. *PCR Methods and Appl.*, 3:S39–S47, 1994.
[11] C. Linhart and R. Shamir. The degenerate primer design problem. *Bioinformatics*, 18, Suppl. 1:S172–S180, 2002.
[12] G. Marth, R. Yeh, M. Minton, R. Donaldson, Q. Li, S. Duan, R. Davenport, R. Miller, and P. Kwok. Single-nucleotide polymorphisms in the public domain: how useful are they? *Nature Genetics*, 27, 2001.
[13] P. Nicodeme and J.-M. Steyaert. Selecting optimal oligonucleotide primers for multiplex PCR. In *Proceedings of Fifth Conference on Intelligent Systems for Molecular Biology ISMB97*, pages 210–213, 1997.
[14] W. Pearson, G. Robins, D. Wrege, and T. Zhang. A new approach to primer selection problem in polymerase chain reaction experiments. In *Third International Conference on Intelligent Systems for Molecular Biology*, pages 285–291. AAAI Press, 1995.

[15] W. Pearson, G. Robins, D. Wrege, and T. Zhang. On the primer selection problem in polymerase chain reaction experiments. *Discrete Applied Mathematics*, 71, 1996.
[16] M. S. Waterman. *Introduction to Computational Biology*. Chapman & Hall, 1995.

Author Index

Lecture Notes in Computer Science

For information about Vols. 1–2721
please contact your bookseller or Springer-Verlag

Vol. 2722: J.M. Cueva Lovelle, B.M. González Rodríguez, L. Joyanes Aguilar, J.E. Labra Gayo, M. del Puerto Paule Ruiz (Eds.), Web Engineering. Proceedings, 2003. XIX, 554 pages. 2003.

Vol. 2723: E. Cantú-Paz, J.A. Foster, K. Deb, L.D. Davis, R. Roy, U.-M. O'Reilly, H.-G. Beyer, R. Standish, G. Kendall, S. Wilson, M. Harman, J. Wegener, D. Dasgupta, M.A. Potter, A.C. Schultz, K.A. Dowsland, N. Jonoska, J. Miller (Eds.), Genetic and Evolutionary Computation – GECCO 2003. Proceedings, Part I. 2003. XLVII, 1252 pages. 2003.

Vol. 2724: E. Cantú-Paz, J.A. Foster, K. Deb, L.D. Davis, R. Roy, U.-M. O'Reilly, H.-G. Beyer, R. Standish, G. Kendall, S. Wilson, M. Harman, J. Wegener, D. Dasgupta, M.A. Potter, A.C. Schultz, K.A. Dowsland, N. Jonoska, J. Miller (Eds.), Genetic and Evolutionary Computation – GECCO 2003. Proceedings, Part II. 2003. XLVII, 1274 pages. 2003.

Vol. 2725: W.A. Hunt, Jr., F. Somenzi (Eds.), Computer Aided Verification. Proceedings, 2003. XII, 462 pages. 2003.

Vol. 2726: E. Hancock, M. Vento (Eds.), Graph Based Representations in Pattern Recognition. Proceedings, 2003. VIII, 271 pages. 2003.

Vol. 2727: R. Safavi-Naini, J. Seberry (Eds.), Information Security and Privacy. Proceedings, 2003. XII, 534 pages. 2003.

Vol. 2728: E.M. Bakker, T.S. Huang, M.S. Lew, N. Sebe, X.S. Zhou (Eds.), Image and Video Retrieval. Proceedings, 2003. XIII, 512 pages. 2003.

Vol. 2729: D. Boneh (Ed.), Advances in Cryptology – CRYPTO 2003. Proceedings, 2003. XII, 631 pages. 2003.

Vol. 2730: F. Bai, B. Wegner (Eds.), Electronic Information and Communication in Mathematics. Proceedings, 2002. X, 189 pages. 2003.

Vol. 2731: C.S. Calude, M.J. Dinneen, V. Vajnovszki (Eds.), Discrete Mathematics and Theoretical Computer Science. Proceedings, 2003. VIII, 301 pages. 2003.

Vol. 2732: C. Taylor, J.A. Noble (Eds.), Information Processing in Medical Imaging. Proceedings, 2003. XVI, 698 pages. 2003.

Vol. 2733: A. Butz, A. Krüger, P. Olivier (Eds.), Smart Graphics. Proceedings, 2003. XI, 261 pages. 2003.

Vol. 2734: P. Perner, A. Rosenfeld (Eds.), Machine Learning and Data Mining in Pattern Recognition. Proceedings, 2003. XII, 440 pages. 2003. (Subseries LNAI).

Vol. 2735: F. Kaashoek, I. Stoica (Eds.), Peer-to-Peer Systems II. Proceedings, 2003. XI, 316 pages. 2003.

Vol. 2736: V. Mařík, W. Retschitzegger, O.Štěpánková (Eds.), Database and Expert Systems Applications. Proceedings, 2003. XX, 945 pages. 2003.

Vol. 2737: Y. Kambayashi, M. Mohania, W. Wöß (Eds.), Data Warehousing and Knowledge Discovery. Proceedings, 2003. XIV, 432 pages. 2003.

Vol. 2738: K. Bauknecht, A M. Tjoa, G. Quirchmayr (Eds.), E-Commerce and Web Technologies. Proceedings, 2003. XII, 452 pages. 2003.

Vol. 2739: R. Traunmüller (Ed.), Electronic Government. Proceedings, 2003. XVIII, 511 pages. 2003.

Vol. 2740: E. Burke, P. De Causmaecker (Eds.), Practice and Theory of Automated Timetabling IV. Proceedings, 2002. XII, 361 pages. 2003.

Vol. 2741: F. Baader (Ed.), Automated Deduction – CADE-19. Proceedings, 2003. XII, 503 pages. 2003. (Subseries LNAI).

Vol. 2742: R. N. Wright (Ed.), Financial Cryptography. Proceedings, 2003. VIII, 321 pages. 2003.

Vol. 2743: L. Cardelli (Ed.), ECOOP 2003 – Object-Oriented Programming. Proceedings, 2003. X, 501 pages. 2003.

Vol. 2744: V. Mařík, D. McFarlane, P. Valckenaers (Eds.), Holonic and Multi-Agent Systems for Manufacturing. Proceedings, 2003. XI, 322 pages. 2003. (Subseries LNAI).

Vol. 2745: M. Guo, L.T. Yang (Eds.), Parallel and Distributed Processing and Applications. Proceedings, 2003. XII, 450 pages. 2003.

Vol. 2746: A. de Moor, W. Lex, B. Ganter (Eds.), Conceptual Structures for Knowledge Creation and Communication. Proceedings, 2003. XI, 405 pages. 2003. (Subseries LNAI).

Vol. 2747: B. Rovan, P. Vojtáš (Eds.), Mathematical Foundations of Computer Science 2003. Proceedings, 2003. XIII, 692 pages. 2003.

Vol. 2748: F. Dehne, J.-R. Sack, M. Smid (Eds.), Algorithms and Data Structures. Proceedings, 2003. XII, 522 pages. 2003.

Vol. 2749: J. Bigun, T. Gustavsson (Eds.), Image Analysis. Proceedings, 2003. XXII, 1174 pages. 2003.

Vol. 2750: T. Hadzilacos, Y. Manolopoulos, J.F. Roddick, Y. Theodoridis (Eds.), Advances in Spatial and Temporal Databases. Proceedings, 2003. XIII, 525 pages. 2003.

Vol. 2751: A. Lingas, B.J. Nilsson (Eds.), Fundamentals of Computation Theory. Proceedings, 2003. XII, 433 pages. 2003.

Vol. 2752: G.A. Kaminka, P.U. Lima, R. Rojas (Eds.), RoboCup 2002: Robot Soccer World Cup VI. XVI, 498 pages. 2003. (Subseries LNAI).

Vol. 2753: F. Maurer, D. Wells (Eds.), Extreme Programming and Agile Methods – XP/Agile Universe 2003. Proceedings, 2003. XI, 215 pages. 2003.

Vol. 2754: M. Schumacher, Security Engineering with Patterns. XIV, 208 pages. 2003.

Vol. 2756: N. Petkov, M.A. Westenberg (Eds.), Computer Analysis of Images and Patterns. Proceedings, 2003. XVIII, 781 pages. 2003.

Vol. 2758: D. Basin, B. Wolff (Eds.), Theorem Proving in Higher Order Logics. Proceedings, 2003. X, 367 pages. 2003.

Vol. 2759: O.H. Ibarra, Z. Dang (Eds.), Implementation and Application of Automata. Proceedings, 2003. XI, 312 pages. 2003.

Vol. 2761: R. Amadio, D. Lugiez (Eds.), CONCUR 2003 - Concurrency Theory. Proceedings, 2003. XI, 524 pages. 2003.

Vol. 2762: G. Dong, C. Tang, W. Wang (Eds.), Advances in Web-Age Information Management. Proceedings, 2003. XIII, 512 pages. 2003.

Vol. 2763: V. Malyshkin (Ed.), Parallel Computing Technologies. Proceedings, 2003. XIII, 570 pages. 2003.

Vol. 2764: S. Arora, K. Jansen, J.D.P. Rolim, A. Sahai (Eds.), Approximation, Randomization, and Combinatorial Optimization. Proceedings, 2003. IX, 409 pages. 2003.

Vol. 2765: R. Conradi, A.I. Wang (Eds.), Empirical Methods and Studies in Software Engineering. VIII, 279 pages. 2003.

Vol. 2766: S. Behnke, Hierarchical Neural Networks for Image Interpretation. XII, 224 pages. 2003.

Vol. 2768: M.J. Wilson, R.R. Martin (Eds.), Mathematics of Surfaces. Proceedings, 2003. VIII, 393 pages. 2003.

Vol. 2769: T. Koch, I. T. Sølvberg (Eds.), Research and Advanced Technology for Digital Libraries. Proceedings, 2003. XV, 536 pages. 2003.

Vol. 2773: V. Palade, R.J. Howlett, L. Jain (Eds.), Knowledge-Based Intelligent Information and Engineering Systems. Proceedings, Part I, 2003. LI, 1473 pages. 2003. (Subseries LNAI).

Vol. 2774: V. Palade, R.J. Howlett, L. Jain (Eds.), Knowledge-Based Intelligent Information and Engineering Systems. Proceedings, Part II, 2003. LI, 1443 pages. 2003. (Subseries LNAI).

Vol. 2776: V. Gorodetsky, L. Popyack, V. Skormin (Eds.), Computer Network Security. Proceedings, 2003. XIV, 470 pages. 2003.

Vol. 2777: B. Schölkopf, M.K. Warmuth (Eds.), Learning Theory and Kernel Machines. Proceedings, 2003. XIV, 746 pages. 2003. (Subseries LNAI).

Vol. 2778: P.Y.K. Cheung, G.A. Constantinides, J.T. de Sousa (Eds.), Field-Programmable Logic and Applications. Proceedings, 2003. XXVI, 1179 pages. 2003.

Vol. 2779: C.D. Walter, Ç.K. Koç, C. Paar (Eds.), Cryptographic Hardware and Embedded Systems – CHES 2003. Proceedings, 2003. XIII, 441 pages. 2003.

Vol. 2781: B. Michaelis, G. Krell (Eds.), Pattern Recognition. Proceedings, 2003. XVII, 621 pages. 2003.

Vol. 2782: M. Klusch, A. Omicini, S. Ossowski, H. Laamanen (Eds.), Cooperative Information Agents VII. Proceedings, 2003. XI, 345 pages. 2003. (Subseries LNAI).

Vol. 2783: W. Zhou, P. Nicholson, B. Corbitt, J. Fong (Eds.), Advances in Web-Based Learning – ICWL 2003. Proceedings, 2003. XV, 552 pages. 2003.

Vol. 2786: F. Oquendo (Ed.), Software Process Technology. Proceedings, 2003. X, 173 pages. 2003.

Vol. 2787: J. Timmis, P. Bentley, E. Hart (Eds.), Artificial Immune Systems. Proceedings, 2003. XI, 299 pages. 2003.

Vol. 2789: L. Böszörményi, P. Schojer (Eds.), Modular Programming Languages. Proceedings, 2003. XIII, 271 pages. 2003.

Vol. 2790: H. Kosch, L. Böszörményi, H. Hellwagner (Eds.), Euro-Par 2003 Parallel Processing. Proceedings, 2003. XXXV, 1320 pages. 2003.

Vol. 2792: T. Rist, R. Aylett, D. Ballin, J. Rickel (Eds.), Intelligent Virtual Agents. Proceedings, 2003. XV, 364 pages. 2003. (Subseries LNAI).

Vol. 2794: P. Kemper, W. H. Sanders (Eds.), Computer Performance Evaluation. Proceedings, 2003. X, 309 pages. 2003.

Vol. 2795: L. Chittaro (Ed.), Human-Computer Interaction with Mobile Devices and Services. Proceedings, 2003. XV, 494 pages. 2003.

Vol. 2796: M. Cialdea Mayer, F. Pirri (Eds.), Automated Reasoning with Analytic Tableaux and Related Methods. Proceedings, 2003. X, 271 pages. 2003. (Subseries LNAI).

Vol. 2798: L. Kalinichenko, R. Manthey, B. Thalheim, U. Wloka (Eds.), Advances in Databases and Information Systems. Proceedings, 2003. XIII, 431 pages. 2003.

Vol. 2799: J.J. Chico, E. Macii (Eds.), Integrated Circuit and System Design. Proceedings, 2003. XVII, 631 pages. 2003.

Vol. 2801: W. Banzhaf, T. Christaller, P. Dittrich, J.T. Kim, J. Ziegler (Eds.), Advances in Artificial Life. Proceedings, 2003. XVI, 905 pages. 2003. (Subseries LNAI).

Vol. 2803: M. Baaz, J.A. Makowsky (Eds.), Computer Science Logic. Proceedings, 2003. XII, 589 pages. 2003.

Vol. 2805: K. Araki, S. Gnesi, D. Mandrioli (Eds.), FME 2003: Formal Methods. Proceedings, 2003. XVII, 942 pages. 2003.

Vol. 2807: V. Matoušek, P. Mautner (Eds.), Text, Speech and Dialogue. Proceedings, 2003. XIII, 426 pages. 2003. (Subseries LNAI).

Vol. 2810: M.R. Berthold, H.-J. Lenz, E. Bradley, R. Kruse, C. Borgelt (Eds.), Advances in Intelligent Data Analysis V. Proceedings, 2003. XV, 624 pages. 2003.

Vol. 2812: G. Benson, R. Page (Eds.), Algorithms in Bioinformatics. Proceedings, 2003. X, 528 pages. 2003. (Subseries LNBI).

Vol. 2817: D. Konstantas, M. Leonard, Y. Pigneur, S. Patel (Eds.), Object-Oriented Information Systems. Proceedings, 2003. XII, 426 pages. 2003.

Vol. 2818: H. Blanken, T. Grabs, H.-J. Schek, R. Schenkel, G. Weikum (Eds.), Intelligent Search on XML Data. XVII, 319 pages. 2003.

Vol. 2820: G. Vigna, E. Jonsson, C. Kruegel (Eds.), Recent Advances in Intrusion Detection. Proceedings, 2003. X, 239 pages. 2003.

Vol. 2821: A. Günter, R. Kruse, B. Neumann (Eds.), KI 2003: Advances in Artificial Intelligence. Proceedings, 2003. XII, 662 pages. 2003. (Subseries LNAI).

Vol. 2832: G. Di Battista, U. Zwick (Eds.), Algorithms – ESA 2003. Proceedings, 2003. XIV, 790 pages. 2003.

Vol. 2834: X. Zhou, S. Jähnichen, M. Xu, J. Cao (Eds.), Advanced Parallel Processing Technologies. Proceedings, 2003. XIV, 679 pages. 2003.

Vol. 2839: A. Marshall, N. Agoulmine (Eds.), Management of Multimedia Networks and Services. Proceedings, 2003. XIV, 532 pages. 2003.